# Evolutionary Ecology of Weeds

Jack Dekker
New Weed Biology Laboratory
2nd Edition, 2016

Evolutionary Ecology of Weeds is the story of WHAT, WHY and HOW some plant species invade and occupy habitats ripe for exploitation. The nature of weeds is the evolution of adaptive traits for seizing and exploiting locally available opportunity. Weeds are the consequence of human disturbance which creates opportunity spacetime, leaving unused resources eager for invasion by weeds.

The nature of weeds is the story of us, humans. We created highly successful wild-crop-weed complexes that resist control. We created them by channeling natural selection, the driver of biological change. Plants invade by dispersing, colonizing, reproducing and enduring in a locality. Weeds possess mating systems that generate variable genotypes and phenotypes that struggle for existence, the winners take all. Evolution occurs. Adaptation in weed life history is about timing: timing is everything. Adaptation in local plant communities is interference and facilitation animating strategic roles guided by functional traits. Weed community dynamics is community assembly and ecological succession. Complex adaptive weed system formation reveals larger forces of nature: emergent behavior, physical information remembered. Knowledge of weeds is discovered, then represented in several different ways: ecological demography, life history traits. Representation is confounded by the humans that make them, their beliefs, values and models.

Case histories of three weeds explain these concepts: velvetleaf (*Abutilon theophrasti*), triazine resistant rapeseed (*Brassica napus*), and the foxtails (*Setaria* species-group).

[Inside cover leaflet]

# Evolutionary Ecology of Weeds

Evolutionary Ecology of Weeds is the story of WHAT, WHY and HOW some plant species invade and occupy habitats ripe for exploitation. The nature of weeds is the evolution of adaptive traits for seizing and exploiting locally available opportunity. Weeds are the consequence of human disturbance, notably agriculture. Disturbance creates opportunity spacetime, leaving unused resources ripe for invasion by plants with preadapted life history traits expressed at favorable times as the growing season unfolds. Successful weed populations assemble and interact with neighbors. The evolution of successful interactions leads to local adaptation maximizing survival and fitness in the community. The story of the evolutionary ecology of weeds is told in 7 units.

**The nature of weeds in unit 1** is the story of us, humans. We began creating weeds about 10,000 years ago. We created highly successful wild-crop-weed complexes that resist control by hybridization across global metapopulations.

**The evolution of weed populations in unit 2** is the story of natural selection, the driver of evolution, biological change. Plant communities assemble where local opportunity favors a particular plant species at a particular time. To invade, a plant must disperse into a locality, then colonize by completing its life cycle, lastly it must endure. Successful genotypes reproduce, generating variation appropriate for exploitation. Successful phenotypes adapt locally. Evolution occurs.

**Adaptation in weed life history in unit 3** is the timing of trait expression to local opportunity spacetime during embryogenesis, dispersal, recruitment, growth, reproduction. Timing is everything. Mating systems control recombination and minimize mortality. Embryogenesis occurs, seed germinability and dormancy are induced. Heteroblastic seeds disperse to form seed banks, opportunity spacetime is discovered. Plant community assembly is seedling emergence timed to neighbors and disturbances. Phenotypic plasticity is carefully controlled growth in response to local opportunity. Reproduction is success.

**Adaptation in local plant communities in unit 4** is interaction among neighbors. Interference and facilitation animate strategic roles guided by functional traits in crowded communities. Weed biodiversity, population structure and weed species associations predicate community dynamics. Weed community dynamics is community assembly and ecological succession.

**Complex adaptive weed systems in unit 5** provides deeper insights into the nature of weeds. Larger forces of nature give rise to complex adaptive systems. The constructal law guides the formation of complex systems that configure their own architectures that flow more easily with time. Earth's thermodynamic heat-water engine drives seizure of dissolved nutrients and gases caught in evapo-transpiration streams of the xylem. Flow eases movement: a treelike pattern emerges, vascularization. Things in nature respond differently to disturbances: fragile things are vulnerable to uncertainty; antifragile things are fueled by randomness. Complex systems have severe interdependencies: emergent properties are nonlinear, unexpected. Evolution embodies organism-environment information exchange. Information is physical. Biology is physical information with quantifiable complexity: the message is remembered.

**Representation of weed biology in unit 6** explores how knowledge of weeds is discovered, and then modeled. We acquire knowledge with good explanations that are hard to vary: changing the details would ruin the explanation. Knowledge of weeds must consider the uncertainty of complex systems, and the way life history is represented. Plant life history and population dynamics are represented in several different ways: ecological demography and the evolution of life history traits. Experimentally they approach plant life history from very different perspectives. Representation of weed biology is confounded by the human scientists that make them, their beliefs, values and models.

**The experimental case histories of three weeds are presented in unit 7.** They connect concepts in the first six units with experimental examples providing perspective on the evolutionary ecology of weeds. Velvetleaf (*Abutilon theophrasti*) is selfish with the sun, reaching for the sky, putting neighbors in the shade, never ever learning to share. Triazine resistant rapeseed (*Brassica napus*) is a half-wild crop, a chronomutant. It turns water and sun into sugar at the oddest times of day, changes its membranes, keeps leaf stomata open all the time, and has extra-dormant seeds. The foxtails (*Setaria* species-group) has tiny his-her 'nano-thermos bottle' seeds. These seeds have a mysterious oxygen 'sponge' protein, and an 'oxygenator' hull: a Shannon communication system living in dirt.

7.24.16

# Evolutionary Ecology of Weeds

Jack Dekker
New Weed Biology Laboratory
Ames, Iowa  50014  USA
Soiana (PI), 56030  Italy
newweedbiolab@gmail.com

Evolutionary Ecology of Weeds
2nd Edition

Copyright 2016
Jack Dekker

Weeds-R-Us Press
New Weed Biology Laboratory
newweedbiolab@gmail.com
Ames, Iowa  50014  USA
Soiana (PI), 56030  Italy

# SUMMARY

WHAT are weeds?
WHY do we have weeds?
WHY do we have the weed we do? (And not others)
WHY do these weeds look and behave as they do?
HOW did these weeds get to be the way they are?
WHAT is the basis of future weed changes?

Weeds colonize disturbed unoccupied opportunity spacetime. The ecology and evolutionary biology of weeds, colonizing and invasive plants, are the consequence of human activity, notably agriculture. Human and enviromental disturbance (herbicides, atmospheric pollution, frozen winter) creates opportunity spacetime by leaving unused resources in a locality which is seized and exploited by plants with preadapted life history traits expressed at favorable times as the growing season unfolds. Successful weed populations assemble and interact with crop and other weedy neighbors in a locality. The consequences of successful interactions lead to local adaptation maximizing survival and fitness in that plant community.

**Unit 1: The nature of weeds.** The story of weeds is the story of us, humans. We began creating weeds about 10,000 years ago. We continue to create and change them every day. Weeds are us. Humans cannot help classifying and categorizing plants. They are our food and fuel, our medicines and drugs, our clothing and decorations. They can also be our poisons. We survive because we know the differences. We cannot help ourselves, plant love-hate protects our lives, our humanity. But, as humans we resist thinking of weeds as a crucial part of who we are.

**Unit 2: The evolution of weed populations.** Rev. Thomas Robert Malthus proposed that human population growth was limited, expansion prevented by famine and disease. He stimulated Charles Darwin's concept of natural selection as the driver of evolution, biological change. Plant communities assemble when local opportunity favors functional traits in a new species at a particular time. Local opportunity spacetime is the integration of available resources and conditions, neighbor interactions and disturbance. To invade, the plant must first disperse into the habitat, then colonize the site by completing its entire life cycle. Lastly it must endure at that site. Selected individuals reproduce, generating variation appropriate for exploiting a particular locality. This first process of natural selection generates the phenotypic variation needed for the second process resulting in local adaptation: differential survival, reproduction and inheritance. Evolution occurs.

**Unit 3: Adaptation in weed life history.** Weed life history is the expression and timing of traits seizing and exploiting local opportunity spacetime during embryogenesis, dispersal, recruitment, growth, reproduction. Timing is everything. Reproductive life history occurs within the shoot architecture tuned to available opportunity: phenotypic plasticity. Mating systems are carefully timed, controlling recombination, minimizing mortality. Embryogenesis occurs, allocating resources among several seed roles. Germinability and dormancy are induced, modulating signal receptivity from the soil environment. Reproduction ends with abscission, an independent zygote. Heteroblastic seeds are dispersed: seeds preadapted to the unknown future availability of opportunity in diverse habitats. Dispersal is a process of discovery of opportunity spacetime, habitable sites within a heterogeneous landscape. Spatial dispersal determines local population size and structure. Soil seed pools provide temporal dispersal, the source of all future weed infestations. Seed dormancy is dispersal in time.

Seedling emergence timing is driven by variable dormancy, seed heteroblasty. Plant community assembly is determined by seedling emergence timing relative to neighbors and disturbances. Weed seed life history is largely independent of neighbors from flowering through seedling recruitment, after which it interacts with neighbors in the community.

**Unit 4: Adaptation in local plant communities.** Neighbor interactions define the local plant community, a complex and adaptive system. The nature of neighbor interactions is found in interference and facilitation, animating the strategic roles enacted by members in the community in the struggle for existence. Functional traits guide behavioral roles in neighbor interactions affecting development in crowded communities. Weed biodiversity and population structure predicate community dynamics. Genetic connections between individuals, populations and species enable exploitation of local opportunity: weed species associations. Weed behavioral traits form a functional guild, clusters of interacting traits. Weed community dynamics is community assembly and ecological succession. Biodiversity is the pool of populations that invade, seize and exploit local opportunity.

**Unit 5: Complex adaptive systems.** Evolutionary ecology provides an understanding of the nature of weeds. But understanding relies on deeper insights than observations of weed biology. Larger forces need to be understood if evolution is the make sense: forces of nature that give rise to complex adaptive systems. The constructal law guides the formation of complex systems: currents in nature move in configurations that flow more easily over time. Flow systems construct their own architectures and rhythms enabling easy movement, generating designs that evolve to survive (to live). Thermodynamic heat-water flows drive resource seizure by plants in the flow of dissolved nutrients and gases caught in evapo-transpiration streams of the xylem. Flow system self-organization occurs with the construction of flow patterns that ease movement: a treelike pattern emerges throughout nature, vascularization. Things in nature respond differently to disturbances. Fragile things are vulnerable to the disorder, volatility, uncertainty. Antifragile things response with positive sensitivity to randomness: increasing volatility, variability, stress, or uncertainty. Antifragile systems change with time, they evolve. Randomness is information fueling antifragility. Phenomena differ between simple and complex systems. Components of simple systems are not interdependent. Complex systems have severe interdependencies between component parts: emergent properties are nonlinear, the sum different from the parts. In a nonlinear system unexpected structures and events occur; e.g. phenotypic plasticity, skewed plant size hierarchies in crowded communities. Weed-crop communities are complex adaptive systems. A complex adaptive system is a dynamic network made up of a large number of active interacting adaptive agents, acting in parallel, and diverse in both form and capability. The activity of a CAS system of interacting elements results in the emergence of system order, and a nonanticipatory-nonpredictive strategy for adaptation to the environment. Seedling recruitment is a complex adaptive system. The *Setaria* spp.-gp. seed-soil environment communication system times departure from seed dormancy to emerged seedling. Evolution embodies organism-environment information exchange. Information is physical. Biology is physical information with quantifiable complexity. Weedy *Setaria* spp. seeds with heterogeneous dormancy states germinate when they receive specific soil signals: oxy-hydro-thermal time. Information is physical: memory resides in *Setaria* seed functional traits regulating all seed behaviors that encode-decode soil signals. The message is remembered: plants pass these functional traits to their polymorphic progeny.

**Unit 6: Representation of weed biology.** How can knowledge of weeds be discovered, and then represented? The real source of scientific theories is not sensory experience, but conjecture alternating with criticism. Observation enables choice between existing theories, it is not the source of new theory. Creativity is the most important source of variation in explanatory theories. How do we acquire knowledge? It is gained with pre-existing knowledge: what to look at, for; how to look and interpret what one sees. Therefore, theory comes first.

It has to be conjectured, not derived. The substance of scientific theories is explanation. Good explanations are hard to vary, changing the details would ruin the explanation. Explanation attempts to answer the "why" question. The hypothesis is the theoretical tool used to verify explanation. Darwinian evolution is the creation of knowledge through alternating variation and selection. Knowledge embodied in genes is knowledge of how to get replicated at the expense of rivals, often by imparting useful functionality to their organism (e.g. functional traits). Human knowledge and biological adaptations are abstract replicators: once embodied in a physical system they tend to remain so. *Setaria* seed morphology is memory and knowledge. Knowledge and biological adaptations are hard to vary. Knowledge of weeds requires consideration of the randomness, uncertainty and probability of complex systems in the way we represent their life history. The lack of information is uncertainty. Mathematical probability definitions are not defined in terms of the probability of unknown events not previously observed. The normal distribution is the basis of most statistics that evaluate the uncertainty of inductive inferences. Much of biology is nonlinear. There is reason to suspect the additivity of treatment-environmental effects, the independent distribution of experimental errors with a common variance. There exist several different ways of describing, or representing, plant life history and population dynamics: ecological demography and the evolution of selectable life history characteristics of phenotypes in the deme, the local population. Experimentally they approach plant life history from very different perspectives. Representation of the nature of weed biology is confounded by the human scientists that make them, their beliefs, values and models. Human perception structures natural laws. For scientific history there are always conflicting views, uncertain outcomes, unpredicable developments; often due to the role of the emotions, the limits of imagination, the conservatism of institutions.

**Unit 7: Weed case history**. The detailed stories of three weeds are presented. They connect concepts in the first six units with empirical, experimental experiences as examples of a broader perspective on the evolutionary ecology, the nature, of weeds. Velvetleaf (*Abutilon theophrasti*) is selfish with the sun, never letting anyone get in its way. Reaching for the sky if need be, putting all its close neighbors in the shade, stretching up as far as necessary to ensure the first place in the sun. Never ever learning to share. The annual dicot triazine resistant *Brassica napus*, rapeseed, is a half-wild crop, a chronomutant. This weed turns water and sun into sugar at the oddest times of day, changes its membranes, keeps leaf stomata open all the time, has more dormant seeds, on and on. The annual grass *Setaria* species-group his-her seeds are tiny, 'nano-thermos bottles', with a myserious oxygen 'sponge' protein inside an exterior hull in which the surface topography is an oxygenator. All these as parts of a Shannon communication system living in dirt.

An indexed glossary is provided linking important concept definitions with explanatory text sections.

# CONDENSED TABLE OF CONTENTS

| | | |
|---|---|---|
| Foreword | | 31 |
| | | |
| UNIT 1: THE NATURE OF WEEDS | | 37 |
| 1 | The nature of weeds | 41 |
| | Summary | |
| | 1.1 What is a weed? | |
| | 1.2 The definition of a weed | |
| | 1.3 Weeds and human nature | |
| | 1.4 Weedy traits | |
| | 1.5 The origins of weeds | |
| | 1.6 World origins and centers of agriculture, crop domestication and cultivation | |
| | 1.7 Biogeographic regions of crop origination | |
| | 1.8 World crop-weed species-groups | |
| | | |
| UNIT 2: THE EVOLUTION OF WEED POPULATIONS | | 77 |
| 2 | Evolution, natural selection and weedy adaptation | 81 |
| | Summary | |
| | 2.1 Introduction | |
| | 2.2 Evolution | |
| | 2.3 Natural selection and elimination | |
| | 2.4 The process of natural selection | |
| | 2.5 Adaptation | |
| | | |
| 3 | Formation of the local weed population (deme): Precondition to natural selection | 93 |
| | 3.1 Introduction: opportunity and the formation of the local deme | |
| | 3.2 The structure of local weedy opportunity | |
| | 3.3 Habitat heterogeneity and dynamics | |
| | 3.4 Limiting resources and pervasive conditions in local opportunity | |
| | 3.5 The nature of plant invasions of local opportunity | |
| | | |
| 4 | Generation of genotypic and phenotypic variation: First process of natural selection | 125 |
| | 4.1 Genotypes and phenotypes | |
| | 4.2 Generate genetic variation | |
| | 4.3 Generate phenotypic variation | |
| | | |
| 5 | Survival, reproduction and inheritance: Second process of natural selection | 145 |
| | 5.1 Survive, avoid Mortality | |
| | 5.2 Reproduce the fittest, eliminate the others | |
| | 5.3 Inheritance: transmit parental traits to offspring | |
| | 5.4 Mating system and inheritance | |
| | | |
| UNIT 3: ADAPTATION IN WEED LIFE HISTORY | | 161 |
| 6 | Weed life history | 163 |
| | 6.1 Introduction to life history | |

6.2 Plant life history classification systems

| | | |
|---|---|---|
| 7 | Reproductive adaptation | 173 |
| | 7.1 Introduction | |
| | 7.2 Flowering, anthesis, fertilization and birth | |
| | 7.3 Embryo adaptation: embryogenesis and dormancy induction | |
| | 7.4 Propagule adaptation: post-abscission fecundity | |
| 8 | Propagule dispersal in space and time | 195 |
| | 8.1 Introduction | |
| | 8.2 Dispersal in space | |
| | 8.3 Dispersal in time: formation of seed pools in the soil | |
| | 8.4 Propagule germination and recruitment | |

**UNIT 4: ADAPTATION IN LOCAL PLANT COMMUNITIES**    223

| | | |
|---|---|---|
| 9 | Neighbor interactions in the local plant community | 227 |
| | 9.1 Adaptation to neighbors in the community | |
| | 9.2 The nature of neighbor interactions in the community | |
| | 9.3 Strategic roles and traits of interference and facilitation with neighbors | |
| | 9.4 Effects of neighborhood interactions on plant density, growth and form | |
| 10 | Weed community structure, dynamics and biodiversity | 261 |
| | 10.1 Weedy community adaptation | |
| | 10.2 Weed communities | |
| | 10.3 Weed community structure | |
| | 10.4 Weed community dynamics: community assembly and ecological succession | |
| | 10.5 Weed community biodiversity | |

**UNIT 5: COMPLEX ADAPTIVE WEED SYSTEMS**    289

U5.1 The nature of weeds
U5.2 The nature of complexity
U5.3 Formation of complex adaptive systems
U5.4 Complex cdaptive systems
U5.5 Soil-seed communication systems

| | | |
|---|---|---|
| 11 | Complex adaptive systems: formation and nonlinear response strategies | 295 |
| | 11.1 Formation of complex systems: the constructal law | |
| | 11.2 Fragile and antifragile systems | |
| 12 | Weed-Crop Communities as Complex Adaptive Systems | 307 |
| | 12.1 Complexity and complex adaptive systems | |
| | 12.2 System of interacting elements, agents | |
| | 12.3 Emergence of system order | |
| | 12.4 Adaptation in complex adaptive systems: nonanticipatory-nonpredictive strategy | |
| 13 | Weed Seed-Soil Environment Communication Systems | 321 |
| | 13.1 Communication Systems | |
| | 13.2 Biological communication | |
| | 13.3 *Setaria* life history as a complex adaptive soil-seed communication system | |

UNIT 6: REPRESENTATION OF WEED BIOLOGY ... 339
    U6.1 Finding knowledge
    U6.2 Representation, uncertainty and cultural relativity in weed biology

14 Randomness, Probability and Inference in Weed Biology ... 343
    14.1 Randomness, uncertainty and information
    14.2 The probability of events
    14.3 The structure of randomness and probability
    14.4 Conclusions

15 Representation of Weed Life History ... 365
    Summary
    15.1 Representing weed life history
    15.2 The ecological demography of plant population life history dynamics
    15.3 Evolutionary, trait-based, weed life history population dynamics

16 Cultural Relativity of Rationality ... 379
    16.1 Introduction to Cultural Relativity of Rationality
    16.2 The nature of scientific revolutions (Kuhn, 1996)
    16.3 Cultural relativity of rationality in weed science

UNIT 7: WEED CASE HISTORY ... 385

17 Weedy *Setaria* species-group (foxtails) ... 389
    17.1 The general nature of weeds, the specific nature of weedy *Setaria*
    17.2 The nature of weedy *Setaria* seed-seedling life history
    17.3. *Setaria* spatial structure
    17.4. *Setaria* seed-seedling life history behavior

18 Triazine Resistant and Susceptible *Brassica napus* ... 435
    18.1 The nature of s-triazine resistant plants
    18.2 Structure-function change in s-triazine resistant plants
    18.3 Chlorophyll fluorescence in R and S
    18.4 Carbon assimilation in R and S
    18.5 Temperature effects on photosynthetic function in R and S
    18.6 Rubisco activity in R and S
    18.7 Photosynthetic regulation in R and S
    18.8 Evolutionary ecology of s-triazine resistant plants

19 *Abutilon theophrasti* (Velvetleaf) ... 459
    19.1 *Abutilon theophrasti* taxonomy
    19.2 Velvetleaf polymorphism and plasticity
    19.3 Velvetleaf competitive plasticity
    19.4 Velvetleaf photomorphogenic plasticity
    19.5 Velvetleaf seedling emergence

SELECTED READINGS ... 469

REFERENCES CITED 471

INDEXED GLOSSARY 511

# TABLE OF CONTENTS

| | |
|---|---:|
| Foreword | 31 |
| | |
| UNIT 1: THE NATURE OF WEEDS | 37 |
| 1    The nature of weeds | 41 |

    Summary
    1.1 What is a weed?
    1.2 The definition of a weed
    1.3 Weeds and human nature
    1.4 Weedy traits
    1.5 The origins of weeds
        1.5.1 Weeds, planting and crop domestication
        1.5.2 Biogeographic prehistory of agriculture
        1.5.3 Pre-agricultural preadapted wild colonizing species
        1.5.4 Wild-crop-weed complexes
    1.6 World origins and centers of agriculture, crop domestication and cultivation
    1.7 Biogeographic regions of crop origination
        1.7.1 The Americas
            1.7.1.1 South America
            1.7.1.2 Mesoamerica
            1.7.1.3 North America
        1.7.2 Eurasia
            1.7.2.1 Europe
            1.7.2.2 West Asia
            1.7.2.3 South Asia
            1.7.2.4 Central Asia
            1.7.2.5 Southeast Asia
            1.7.2.6 East Asia
        1.7.3 Africa
    1.8 World crop-weed species-groups

| | |
|---|---:|
| UNIT 2: THE EVOLUTION OF WEED POPULATIONS | 77 |
|     Introduction | |
| 2    Evolution, natural selection and weedy adaptation | 81 |

    Summary
    2.1 Introduction
    2.2 Evolution
        2.2.1 Micro- and macroevolution
        2.2.2 Units of evolution and natural selection
    2.3 Natural selection and elimination
        2.3.1 Malthus Postulata
        2.3.2 Natural selection
    2.4 The process of natural selection
        2.4.1 Population formation: Precondition to natural selection
        2.4.2 Representing population growth, the Verhulst-Pearl logistic equation
        2.4.3 Generate variation: Process of natural selection, step 1

    2.4.4 Survival and reproduction: Process of natural selection, step 2
 2.5 Adaptation

3  Formation of the local weed population (deme): Precondition to natural selection  93
Summary
 3.1 Introduction: opportunity and the formation of the local deme
   3.1.1 Population formation examples
   3.1.2 Seizing and exploiting local opportunity
 3.2 The structure of local weedy opportunity
   3.2.1 Weedy habitats
   3.2.2 Niches in the local community
   3.2.3 The niche hypervolume
 3.3 Habitat heterogeneity and dynamics
   3.3.1 Earth's physical geography
     3.3.1.1 Geosphere
     3.3.1.2 Lithosphere
     3.3.1.3 Atmosphere
     3.3.1.4 Hydrosphere
     3.3.1.5 Biosphere
     3.3.1.6 Pedosphere
   3.3.2 Spatial heterogeneity and patchiness
   3.3.3 Temporal division of the environment
   3.3.4 Disturbance
     3.3.4.1 Disturbance possesses dimensionality
     3.3.4.2 Proximity of disturbance
     3.3.4.3 Vulnerability to disturbance
     3.3.4.4 Temporal patterns in disturbance
 3.4 Limiting resources and pervasive conditions in local opportunity
   3.4.1 Thermodynamic Earth
     3.4.1.1 Thermodynamic Earth heat flow
     3.4.1.2 Thermodynamic Earth engines
   3.4.2 Limiting resources
     3.4.2.1 Light
       3.4.2.1.1 Plant phytochromes
       3.4.2.1.2 Photomorphogenesis and phototropism
     3.4.2.2 Water
     3.4.2.3 Mineral nutrients
     3.4.2.4 Gases
   3.4.3 Pervasive conditions in the environment
     3.4.3.1 Heat
     3.4.3.2 Terroir
       3.4.3.2.1 Climate
       3.4.3.2.2 Soil
       3.4.3.2.3 Topography
   3.4.4 Environmental signal spacetime
     3.4.4.1 Signal space dimensions
     3.4.4.2 Photo-thermal modulation of weed seed germination behavior by plant phytochromes

                3.4.4.2.1 The photo-thermal germination window
                3.4.4.2.2 Multiple interacting phytochromes
                3.4.4.2.3 Alternative phytochrome models
    3.5 The nature of plant invasions of local opportunity
        3.5.1 The plant invasion process: seizing, exploiting and occupying opportunity
        3.5.2 Dispersal
        3.5.3 Colonization
        3.5.4 Enduring occupation of a locality
        3.5.5 Extinction
        3.5.6 The perception of plant invasion

4   Generation of genotypic and phenotypic variation: First process of natural selection    125
    Summary
    4.1 Genotypes and phenotypes
        4.1.1 Extended phenotype
        4.1.2 Epigenesis, epigenetics and gene expression
        4.1.3 Genome size, weediness and intra-genomic competition
    4.2 Generate genetic variation
        4.2.1 Sources of genetic diversity
            4.2.1.1 Forces increasing population variability
            4.2.1.2 Forces decreasing population variability
        4.2.2 Speciation
            4.2.2.1 Process of speciation
                4.2.2.1.1 Stage 1: Gene flow is interupted between two populations
                4.2.2.1.2 Stage 2: Completion of genetic isolation
            4.2.2.2 Reproductive isolating mechanisms
            4.2.2.3 Modes of speciation
                4.2.2.3.1 Natural speciation
                4.2.2.3.2 Allopatric speciation
                4.2.2.3.3 Peripatric speciation
                4.2.2.3.4 Parapatric speciation
                4.2.2.3.5 Sympatric speciation
                4.2.2.3.6 Speciation via polyploidization
                4.2.2.3.7 Reinforcement
                4.2.2.3.8 Hybrid speciation
    4.3 Generate phenotypic variation
        4.3.1 Phenotypic plasticity
        4.3.2 Somatic polymorphism
            4.3.2.1 Somatic polymorphism in flowers
            4.3.2.2 Somatic polymorphism in seeds
            4.3.2.3 Somatic polymorphism in leaves
            4.3.2.4 Somatic polymorphism: seasonal dimorphism

5   Survival, reproduction and inheritance: Second process of natural selection    145
    Summary
    5.1 Survive, avoid Mortality
    5.2 Reproduce the fittest, eliminate the others
        5.2.1 Demographic survival and reproduction

    5.2.2 Timing of reproduction
      5.2.2.1 Optimum age of reproduction
      5.2.2.2 Maximizing net reproductive rate
      5.2.2.3 Precocious reproduction
    5.2.3 Plant age and stage structure
    5.2.4 Reproductive value
    5.2.5 Risk of death determines life history
    5.2.6 Influences of plant density on mortality
    5.2.7 Modes of selection and population diversity
  5.3 Inheritance: transmit parental traits to offspring
  5.4 Mating system and inheritance
    5.4.1 Mating system and opportunity
    5.4.2 Evolution of mating systems
      5.4.2.1 Sexual reproduction versus apomixis
      5.4.2.2 Outcrossing versus self-fertilization
      5.4.2.3 Outcrossing: separate sexes versus hermaphrodites
    5.4.3 Sex classification systems
    5.4.4 Types of mating systems
      5.4.4.1 Self-pollenating species
      5.4.4.2 Out-crossing species
      5.4.4.3 Apomictic species
      5.4.4.4 Vegetative clone reproducing species

**UNIT 3: ADAPTATION IN WEED LIFE HISTORY**    161
  Introduction
6  Weed life history    163
  Summary
  6.1 Introduction to life history
    6.1.1 Phenotypic life history traits
      6.1.1.1 Preadaption
      6.1.1.2 Trait basis of the invasion process
    6.1.2 Processes of life history adaptation
  6.2 Plant life history classification systems
    6.2.1 Life span
    6.2.2 Growth and life form
      6.2.2.1 Growth form
      6.2.2.2 Raunkiærian life forms
    6.2.3 Life history strategies
      6.2.3.1 r-selection
      6.2.3.2 K-selection
      6.2.3.3 CSR strategy

7  Reproductive adaptation    173
  Summary
  7.1 Introduction
  7.2 Flowering, anthesis, fertilization and birth
    7.2.1 Parental plant architecture
    7.2.2 Mating systems

7.3 Embryo adaptation: embryogenesis and dormancy induction
    7.3.1 Induction of seed dormancy
    7.3.2 The evolutionary ecology of seed dormancy
    7.3.3 Weed seed dormancy variability and somatic polymorphism
    7.3.4 Evolutionary ecology of seed heteroblasty
    7.3.5 Weed species seed heteroblasty examples
    7.3.6 Observable seed dormancy-germinability regulation life forms
        7.3.6.1 Non-dormant
        7.3.6.2 Vegetative, perinating buds
        7.3.6.3 Environmental seed germination control mechanisms
        7.3.6.4 Hard, gas- and water-impermeable, seed envelope germination inhibition
        7.3.6.5 Light-phytochrome and nitrate stimulated germination
        7.3.6.6 Species with multiple interacting germination control mechanisms
        7.3.6.7 Other seed germination control mechanisms
    7.3.7 Experimental weed seed science
7.4 Propagule adaptation: post-abscission fecundity
    7.4.1 Five roles of seeds
        7.4.1.1 Dispersal and colonization
        7.4.1.2 Persistence
        7.4.1.3 Food reserves for embryo growth
        7.4.1.4 Release of new genetic recombinants into the local deme
        7.4.1.5 Multiplication of the parent plant
    7.4.2 Principle of strategic allocation
    7.4.3 Trade-offs among seed roles
        7.4.3.1 Seed size versus number
        7.4.3.2 Seed size trade-offs
        7.4.3.3 Seed number
        7.4.3.4 Relative weed species seed sizes
        7.4.4.5 Seed size plasticity and stability
            7.4.4.5.1 Variable seed size
            7.4.4.5.2 Small seed size
            7.4.4.5.3 Large seed size
        7.4.4.6 Relationship of seed size to habitat

8   Propagule dispersal in space and time     197
   Summary
   8.1 Introduction
       8.1.1 The evolutionary ecology of dispersal structures
       8.1.2 Seed dispersal trade-offs
       8.1.3 Cost of dispersal
       8.1.4 Space-time dimensions of dispersal
   8.2 Dispersal in space
       8.2.1 Dispersal and post-dispersal processes
       8.2.2 Seed flux at a locality
       8.2.3 Modes of seed and propagule dispersal
           8.2.3.1 Gravity
           8.2.3.2 Wind and air
           8.2.3.3 Water

                8.2.3.4  Animal, non-human
                8.2.3.5  Human
                8.2.3.6  Other modes of dispersal
    8.3  Dispersal in time: formation of seed pools in the soil
        8.3.1  Adaptative roles of soil seed pools
        8.3.2  Population dynamics in the soil seed pool
            8.3.2.1  Life history of a seed
            8.3.2.2  Seed states, fates and seed state transition processes
            8.3.2.3  Seed pool additions, losses and continuity
                8.3.2.3.1  Additions to the seed pool
                8.3.2.3.2  Losses from the seed pool
                8.3.2.3.3  Continuity in the seed pool with time
        8.3.3  Structure of soil seed pools
            8.3.3.1  Spatial distribution in the soil profile
                8.3.3.1.1  Depth in the soil profile
                8.3.3.1.2  Effects of tillage
                8.3.3.1.3  Seed size and depth of burial in the soil
                8.3.3.1.4  Horizontal seed distribution
            8.3.3.2  Floral seed community compostion
            8.3.3.3  Seed pool size
            8.3.3.4  Seed longevity in the soil
    8.4  Propagule germination and recruitment
        8.4.1  Introduction
        8.4.2  Process of recruitment
        8.4.3  Germination micro-sites and safe sites
        8.4.4  Magnitude and duration of seedling emergence
        8.4.5  Patterns of seedling emergence
            8.4.5.1  Single 'flush' period
            8.4.5.2  Bi-modal recruitment
            8.4.5.3  Continuous emergence
            8.4.5.4  Major emergence period with extended, infrequent period
            8.4.5.5  Relative emergence order
        8.4.6  Case studies
            8.4.6.1  *Setaria* seedling emergence in central Iowa
            8.4.6.2  *Chenopodium album* seedling emergence in Europe and North America
        8.4.7  Relationship between seed heteroblasty and recruitment timing

UNIT 4:  ADAPTATION IN LOCAL PLANT COMMUNITIES                                      223
9         Neighbor interactions in the local plant community                          227
          Summary
    9.1  Adaptation to neighbors in the community
        9.1.1  Parental control ends, plant growth resumes
        9.1.2  Recruitment: community assembly
        9.1.3  Interaction with neighbors is harsh
    9.2  The nature of neighbor interactions in the community
        9.2.1  Patterns of neighbor interactions
            9.2.1.1  Causation and complexity
            9.2.1.2  Modal patterns of neighbor interaction in the community

        9.2.1.3 Neutralism
    9.2.2 Interference interactions between neighbors
        9.2.2.1 Competition
        9.2.2.2 Amensalism
        9.2.2.3 Antagonism
            9.2.2.3.1 Parasitism
            9.2.2.3.2 Predation
            9.2.2.3.3 Pathogenic interactions
            9.2.2.3.4 Coevolution in parasitism, predation and pathogenesis
    9.2.3 Facilitative interactions between neighbors
        9.2.3.1 Mutualism
        9.2.3.2 Commensalism
9.3 Strategic roles and traits of interference and facilitation with neighbors
    9.3.1 Strategic roles and traits of interference with neighbors
        9.3.1.1 Genetic foraging for local opportunity
            9.3.1.1.1 Genotypic strategies
            9.3.1.1.2 Phenotypic strategies
        9.3.1.2 Spatial foraging for local opportunity
            9.3.1.2.1 Perennial bud foraging
                9.3.1.2.1.1 Clonal foraging geometry
                9.3.1.2.1.2 Foraging behavior
            9.3.1.2.2 Offensive attack of neighbors
            9.3.1.2.3 Preempt neighbors ability to seize and exploit opportunity
            9.3.1.2.4 Plant movements for spatial foraging
                9.3.1.2.4.1 Plant tropisms
                      9.3.1.2.4.1.1 Light tropisms
                      9.3.1.2.4.1.2 Hydrotropism
                      9.3.1.2.4.1.3 Thermotropism
                      9.3.1.2.4.1.4 Gravitropism
                      9.3.1.2.4.1.5 Thigmotropism
                      9.3.1.2.4.1.6 Chemotropism
                      9.3.1.2.4.1.7 Electrotropism
                      9.3.1.2.4.1.8 Sonotropism
                9.3.1.2.4.2 Nastic movements
                      9.3.1.2.4.2.1 Nastic light movements
                      9.3.1.2.4.2.2 Hydronasty
                      9.3.1.2.4.2.3 Thermonasty
                      9.3.1.2.4.2.4 Nastic growth direction
                      9.3.1.2.4.2.5 Chemonasty
                      9.3.1.2.4.2.6 Thigmonasty
        9.3.1.3 Temporal foraging for local opportunity
    9.3.2 Strategic roles and traits of facilitation with neighbors
        9.3.2.1 Escape and avoid neighbors
            9.3.2.1.1 Refuge from neighbors
                9.3.2.1.1.1 Refuge from physical stress
                9.3.2.1.1.2 Refuge from predation
            9.3.2.1.2 Defense from neighbor stress
            9.3.2.1.3 Segregation from neighbors

                9.3.2.1.3.1 Spatial segregation
                9.3.2.1.3.2 Temporal segregation
        9.3.2.2 Co-exist with neighbors
            9.3.2.2.1 Refuge from competition
            9.3.2.2.2 Ecological combining ability
            9.3.2.2.3 Tolerate neighbors
        9.3.2.3 Co-operate with neighbors
            9.3.2.3.1 Improved resource availability
            9.3.2.3.2 Altruism
9.4 Effects of neighborhood interactions on plant density, growth and form
    9.4.1 Space, neighborhoods and plant density
    9.4.2 Plant density and productivity per unit area.
    9.4.3 Plant density and plant size
        9.4.3.1 Plant size hierarchies in the community
        9.4.3.2 Productivity and plant size
        9.4.3.3 Age- and stage-structure of populations
    9.4.4 Plant density and plant form
        9.4.4.1 Influence of plant density on form and reproduction
        9.4.4.2 Plant form and diversity of a community
    9.4.5 Inference in plant density demography

10   Weed community structure, dynamics and biodiversity                 261
    10.1 Weedy community adaptation
    10.2 Weed communities
    10.3 Weed community structure
        10.3.1 Metapopulation structures
        10.3.2 Plant genetic associations
            10.3.2.1 The origins of weeds: wild-crop-weed plant complexes
            10.3.2.2 Biogeographic population genetic structure
            10.3.2.3 Genotype structuring: species associations for weedy colonization
                10.3.2.3.1 Species-groups
                10.3.2.3.2 Polyploid species clusters
                10.3.2.3.3 Aggregate species
            10.3.2.4 Genetic structuring: pre-adaptive colonizing archetypes
                10.3.2.4.1 Generalist-specialist genotypes
                10.3.2.4.2 Genetic-reproductive colonizing types
        10.3.3 Ecological roles-guilds-trades in weed-crop communities
            10.3.3.1 Guild structure and community organization
            10.3.3.2 Parameters of weed species ecological role and niche
            10.3.3.3 Trait Guild: Relative seedling/bud emergence order
    10.4 Weed community dynamics: community assembly and ecological succession
        10.4.1 Weed community dynamics
        10.4.2 Seizing and exploiting local opportunity
        10.4.3 Propagule recruitment: community assembly
        10.4.4 Plant community ecological succession
        10.4.5 Weed population shifts and ecological succession
            10.4.5.1 Inter-specific population shifts
                10.4.5.1.1 Grassy weed tolerance to 2,4-D

                    10.4.5.1.2 Herbicide-induced life cycle shifts
              10.4.5.2 Intra-specific population shifts
                    10.4.5.2.1 Early flowering
                    10.4.5.2.2 Crop mimicry
                    10.4.5.2.3 Dwarf variants
                    10.4.5.2.4 Biodiversity shifts
              10.4.5.3 Herbicide-induced population shifts
                    10.4.5.3.1 Herbicide resistance genotype shifts: altered herbicide target variants
                    10.4.5.3.2 Herbicide resistance genotype shifts: enhanced herbicide metabolism variants
    10.5 Weed community biodiversity
        10.5.1 Scales of weedy biodiversity
        10.5.2 Biodiversity encountered by interacting neighbors
              10.5.2.1 Perception of neighbor morphology
              10.5.2.2 Perception of neighbor development
              10.5.2.3 Perception of neighbor genotype/phenotype
              10.5.2.4 Perception of microhabitat opportunity
              10.5.2.5 Perception of neighbor diversity
        10.5.3 Weed community biodiversity: complexity, stability and equilibrium
              10.5.3.1 Community complexity
              10.5.3.2 Community stability
              10.5.3.3 Equilibrium in the community
                    10.5.3.3.1 Biodiversity as species richness
                    10.5.3.3.2 Equilibrium theory of island biogeography
                    10.5.3.3.3 Unified neutral theory of biodiversity and biogeography
        10.5.4 Weed communities: conclusions

UNIT 5: COMPLEX ADAPTIVE WEED SYSTEMS                                           289
    U5.1 The nature of weeds
    U5.2 The nature of complexity
    U5.3 Formation of complex adaptive systems
    U5.4 Complex cdaptive systems
    U5.5 Soil-seed communication systems
11  Complex adaptive systems: formation and nonlinear response                  293
    11.1 Formation of complex systems: the constructal law
        11.1.1 First principles: the thermodynamics of flow
        11.1.2 The constructal law
        11.1.3 Flow systems
        11.1.4 Evolution of flow system structure and design
        11.1.5 Flow system structure and movement
        11.1.6 Why do weeds exist?
        11.1.7 How the constructal law predicts the design of weed
        11.1.8 Plant community structure and assembly
        11.1.9 Hierarchy: the treelike structure of flow
        11.1.10 Thermodynamic Earth heat flow
        11.1.11 Thermodynamic Earth engines
    11.2 Fragile and antifragile systems

    11.2.1 The fragile, robust and antifragile
    11.2.2 The random
    11.2.3 The complex
    11.2.4 The nonlinear: concave-convex effects
    11.2.5 The optional

12 Weed-Crop Communities as Complex Adaptive Systems  307
  12.1 Complexity and complex adaptive systems
    12.1.1 Composite systems
    12.1.2 Complex domains
    12.1.3 Characteristics of complex systems
  12.2 System of interacting elements, agents
    12.2.1 Aggregation of agent interactions
    12.2.2 Tagging for aggregation
    12.2.3 Non-linear interactions
    12.2.4 System network flows
    12.2.5 Hierarchical diversity-variability of structures and agents
  12.3 Emergence of system order
    12.3.1 Self-similarity and self-organization
    12.3.2 Scale invariance and self-similarity
    12.3.3 Emergence
  12.4 Adaptation in complex adaptive systems: nonanticipatory-nonpredictive strategy
    12.4.1 Internal models sense environment
    12.4.2 Building blocks of internal models
    12.4.3 Framework for representing adaptive agents
      12.4.3.1 Agents performance system
      12.4.3.2 Adaptation by credit assignment
      12.4.3.3 Adaptation by rule discovery algorithm
    12.4.4 Strategy and environment

13 Weed Seed-Soil Environment Communication Systems  321
  Summary
  13.1 Communication Systems
    13.1.1 Communication and information theory.
    13.1.2 Shannon communication system
    13.1.3 Communication and biological complexity.
      13.1.3.1 Message communication: usefulness and meaning
      13.1.3.2 Complexity of information
  13.2 Biological communication
  13.3 *Setaria* life history as a complex adaptive soil-seed communication system
    13.3.1 Forecasting *Setaria* seed behavior: FoxPatch
      13.3.1.1 Seed state and process model
      13.3.1.2 Soil Signal Controlling Seed Behavior: Oxy-Hydro-Thermal Time
        13.3.1.2.1 The oxy-hydro-time signal: OxSIG
        13.3.1.2.2 The thermal signal: TSIG
        13.3.1.2.3 Oxy-hydro-[thermal]-time: OxSIG & TSIG
      13.3.1.3 Schema of Intrinsic *Setaria* sp. Seed Traits
      13.3.1.4 Germinability-dormancy induction

13.3.1.5 Rules for individual weedy *Setaria* sp. seed behavior
    13.3.1.5.1 The after-ripening rule, and its inverse the dormancy re-induction rule
    13.3.1.5.2 The germination candidate threshold state
    13.3.1.5.3 The seed germination rule
13.3.2 *Setaria* soil-seed communication system
    13.3.2.1 Shannon communication system for weedy *Setaria* seed-seedling
        13.3.2.1.1 Information source (E1): the soil adjacent to the *Setaria* seed
        13.3.2.1.2 Transmitter (E2): water film contact between soil-seed
        13.3.2.1.3 Channel (E3): soil-seed water films
        13.3.2.1.4 Noise source
        13.3.2.1.5 Receiver (E4): the seed interior
        13.3.2.1.6 Destination (E5): the embryo
    13.3.2.2 *Setaria* soil-seed communication transmission algorithm
13.3.3 Seed memory and adaptive evolution

## UNIT 6: REPRESENTATION OF WEED BIOLOGY   339
U6.1 Finding knowledge
U6.2 Representation, uncertainty and cultural relativity in weed biology

**14 Randomness, Probability and Inference in Weed Biology   343**
Summary
14.1 Randomness, uncertainty and information
    14.1.1 Randomness
    14.1.2 Uncertainty and information
    14.1.3 Knowledge and induction
        14.1.3.1 The Problem of induction: The White Swan
        14.1.3.2 Deductive testing of theories
14.2 The probability of events
    14.2.1 Probability
    14.2.2 Probability interpretations
        14.2.2.1 Frequency probability
        14.2.2.2 Subjective probability
    14.2.3 The fundamental problem of the theory of chance
        14.2.3.1 The logical problem of subjective probability
        14.2.3.2 Objective system of probability
        14.2.3.3 Subjective system of probability
    14.2.4 Frequency distributions of events
    14.2.5 Normal probability distributions and inference
        14.2.5.1 Assumptions in the analysis of variance of normal distributions
        14.2.5.2 Inference in normal distributions
    14.2.6 Scalable probability distributions and inference
        14.2.6.1 Exponential frequency distributions
        14.2.6.2 Skewed frequency distributions
        14.2.6.3 Scalable phenomena: fractal randomness
        14.2.6.4 Scalable phenomena: power laws in biology
14.3 The structure of randomness and probability
    14.3.1 Unpredictable, consequential events: The Black Swan

    14.3.2 The domain of unpredictable, consequential events
    14.3.3 Representation of events in the unpredictable, consequential domain
  14.4 Conclusions

15 Represention of Weed Life History               365
  Summary
  15.1 Representing weed life history
  15.2 The ecological demography of plant population life history dynamics
    15.2.1 Weed life history models
    15.2.2 Demographic weed life history population dynamics models
    15.2.3 Representation and information, inference and prediction
      15.2.3.1 Representation and information
      15.2.3.2 Inference
        15.2.3.2.1 The deme
          15.2.3.2.1.1 Population structure
          15.2.3.2.1.2 Individual phenotypic identity
          15.2.3.2.1.3 Local population dynamics
        15.2.3.2.2 Life history development and behavior
          15.2.3.2.2.1 Life history states and processes
          15.2.3.2.2.2 Polymorphism and plasticity
        15.2.3.2.3 Model formalization and measurement metrics
          15.2.3.2.3.1 Hypotheses of local weed population dynamics
          15.2.3.2.3.2 Mathematical, algorithmic, statisitical model formalization and component description
          15.2.3.2.3.3 Random-nonrandom processes
      15.2.3.3 Predicting weed population dynamics
  15.3 Evolutionary, trait-based, weed life history population dynamics

16 Cultural Relativity of Rationality                 379
  16.1 Introduction to Cultural Relativity of Rationality
    16.1.1 History of scientific progress
    16.1.2 Scientific institutions
    16.1.3 Intellectual coherence and truth
    16.1.4 Scientific crisis and change
  16.2 The nature of scientific revolutions (Kuhn, 1996)
    16.2.1 Scientific problems
    16.2.2 The nature of scientific paradigms
    16.2.3 Paradigm: disciplinary matrix
      16.2.3.1 Symbolic generalizations
      16.2.3.2 Commitments to particular models
      16.2.3.3 Shared values
      16.2.3.4 Exemplars of nature
    16.2.4 Scientific paradigm crisis and change
    16.2.5 Knowledge embedded in stimuli-sensation
  16.3 Cultural relativity of rationality in weed science

UNIT 7: WEED CASE HISTORY                   385
17 Weedy *Setaria* species-group (foxtails)             389

17.1 The general nature of weeds, the specific nature of weedy *Setaria*
17.2 The nature of weedy *Setaria* seed-seedling life history
    17.2.1 Seed-seedling life history.
        17.2.1.1 Threshold Events
        17.2.1.2 Germination and seedling emergence
    17.2.2 The nature of *Setaria*: Spatial structure and temporal behavior
    17.2.3 *Setaria* model system of exemplar species
17.3. *Setaria* spatial structure
    17.3.1 Plant morphological structure
        17.3.1.1 Seed morphology
            17.3.1.1.1 Caryopsis
                17.3.1.1.1.1 Embryo
                17.3.1.1.1.2 Endosperm
                17.3.1.1.1.3 Aleurone layer, TACL membrane
                17.3.1.1.1.4 Caryopsis coat
            17.3.1.1.2 Placental pad and pore
            17.3.1.1.3 Hull
            17.3.1.1.4 Glumes
            17.3.1.1.5 Seed
        17.3.1.2 Plant morphology
            17.3.1.2.1 Panicle inflorescence
                17.3.1.2.1.1 Spikelet
                17.3.1.2.1.2 Fascicle
            17.3.1.2.2 Tiller
    17.3.2 Plant genetic structure
        17.3.2.1 Genotype structure
            17.3.2.1.1 Individual plant
            17.3.2.1.2 Intra-specific variation
        17.3.2.2 Population genetic structure
            17.3.2.2.1 Local population (deme)
                17.3.2.2.1.1 *Setaria viridis*
                17.3.2.2.1.2 *Setaria pumila-S. geniculata*
                17.3.2.2.1.3 *Setaria faberi*
                17.3.2.2.1.4 General and specialized genotypes
            17.3.2.2.2 Species associations
            17.3.2.2.3 Metapopulations
17.4. *Setaria* seed-seedling life history behavior
    17.4.1 Life history behaviors: Functions, traits and information
    17.4.2 Seed formation and dormancy induction
        17.4.2.1 Inflorescence and flowering
            17.4.2.1.1 Flowering pattern on panicle main axis
            17.4.2.1.2 Elongation pattern of panicle and culm axis
            17.4.2.1.3 Seasonal flowering pattern
            17.4.2.1.4 Diurnal flowering pattern
        17.4.2.2 Mating and Fertilization
            17.4.2.2.1 Hybridization
            17.4.2.2.2 Asexual reproduction

        17.4.2.2.3 Speciation and reproductive barriers between *Setaria* species
    17.4.2.3 Seed formation and embryogenesis
        17.4.2.3.1 Hull development
        17.4.2.3.2 Placental pore and pad development
        17.4.2.3.3 Caryopsis development
        17.4.2.3.4 Embryo development
        17.4.2.3.5 Dormancy induction
            17.4.2.3.5.1 Seed germinability
            17.4.2.3.5.2 Caryopsis germinability
            17.4.2.3.5.3 Embryo germinability
            17.4.2.3.5.4 Multiple germinability phenotypes
            17.4.2.3.5.5 Light and dormancy induction
            17.4.2.3.5.6 Morpho-physiological dormancy traits
            17.4.2.3.5.7 Compartmentalization of seed germinability
        17.4.2.3.6 Seed abscission
    17.4.2.4 Seed fecundity and plasticity
17.4.3 Seed rain dispersal
    17.4.3.1 Seed germinability-dormancy heteroblasty: Dispersal in time
    17.4.3.2 Soil seed pool formation: Dispersal in space
    17.4.3.3 Dispersal and local adaptation: Evolutionary history of *Setaria* invasion
17.4.4 Seed behavior in the soil
    17.4.4.1 Control of seed germinability
        17.4.4.1.1 Regulation of weedy *Setaria* seed behavior
        17.4.4.1.2 Movement of water-gas between soil and seed
    17.4.4.2 Seed behavior
        17.4.4.2.1 Annual seed-dormancy-germinability cycling in the soil
        17.4.4.2.2 Seed longevity in the soil
17.4.5 Seedling recruitment
    17.4.5.1 Seedling emergence and community assembly
        17.4.5.1.1 Seedling emergence
        17.4.5.1.2 Seed germination depth from the soil
    17.4.5.2 Patterns of seedling emergence
        17.4.5.2.1 General seedling recruitment pattern
        17.4.5.2.2 Early season recruitment
        17.4.5.2.3 Parental and environmental influences on recruitment
            17.4.5.2.3.1 Parental influences
            17.4.5.2.3.2 Year
            17.4.5.2.3.3 Seed age in the soil
    17.4.5.3 Seed germinability-dormancy heteroblasty blueprints seed behavior in the soil.
        17.4.5.3.1 Relationship of seed heteroblasty and seedling emergence
        17.4.5.3.2 Seedling emergence hedge-betting for assembly in agro-communities

18    Triazine Resistant and Susceptible *Brassica napus*     435
Summary
18.1 The nature of s-triazine resistant plants

18.2 Structure-function change in s-triazine resistant plants
    18.2.1 Structural change in s-triazine resistant plants
    18.2.2 Functional change in s-triazine resistant plants
        18.2.2.1 Resistance to s-triazine herbicides
        18.2.2.2 Differences in carbon assimilation efficiency
        18.2.2.3 Complex *psbA* mutant phenotype
        18.2.2.4 *psbA* mutant pleitropic reorganization
18.3 Chlorophyll fluorescence in R and S
18.4 Carbon assimilation in R and S
18.5 Temperature effects on photosynthetic function in R and S
    18.5.1 Diurnal temperature effects
    18.5.2 Controlled leaf temperature effects
18.6 Rubisco activity in R and S
18.7 Photosynthetic regulation in R and S
18.8 Evolutionary ecology of s-triazine resistant plants
    18.8.1 Pleiotropy in R
    18.8.2 R adaptation to the environment and regulation of carbon assimilation

**19 *Abutilon theophrasti* (Velvetleaf)     459**
Summary
19.1 *Abutilon theophrasti* taxonomy
19.2 Velvetleaf polymorphism and plasticity
19.3 Velvetleaf competitive plasticity
    19.3.1 Branching and flowering plasticity predicate dry weight and seed yield plasticity
        19.3.1.1 Introduction
        19.3.1.2 Experimental methods
    19.3.2 Branching
    19.3.3 Flowering
    19.3.4 Dry matter accumulation
    19.3.5 Seed weight at harvest
19.4 Velvetleaf photomorphogenic plasticity
    19.4.1 Phototropic and photomorphogenic plasticity of velvetleaf leaf architecture predicates soybean yield losses
    19.4.2 Introduction
19.5 Velvetleaf seedling emergence
    19.5.1 Seedling emergence plasticity in space and time
        19.5.1.1 Introduction
        19.5.1.2 Methods
    19.5.2 Seedling emergence depth
    19.5.3 Seedling emergence numbers and pattern
    19.5.4 Seedling growth and soil pH

**SELECTED READINGS     469**

**INDEXED GLOSSARY     471**

**REFERENCES CITED     511**

# FOREWORD

**A biological explanation**

This book is an explanation of the ecology and evolutionary biology of weeds and other colonizing and invasive plants. Weed biology is the ecology and evolution of plants in localities influenced by human activity, notably agriculture. The focus is on these big WHY, HOW and WHAT questions of weed biology:

What are weeds?
Why do we have weeds?
Why do we have the weed species that we do? (And not others)
Why do these weeds look and behave as they do?
How did the weeds we have get to be the way they are?
What is the basis of future changes in weeds?

The goal of this book is to provide comprehensive explanations of factual information about weed biology in an evolutionary context as the basis for understanding and management of local weed communities of the future. The goal is also to provide the reader with a dynamic framework to guide understanding of new observations in the future: a mental 'toolkit' to focus observations of new weed phenomena, a way to understand the fundamental forces in nature that cause weediness.

Nothing in biology makes sense unless seen in the light of evolution (Dobzhansky, 1973). Weed and crop management is the management of selection and elimination leading inexorably to the weed adaptations that plague our fields and interfere with our crops. To understand what we observe in agriculture and want to manage more wisely and efficiently, we need to understand how the evolutionary process works in weed communities.

*Weeds are plants too.* The principles of weedy invasion and colonization are same as for all plants regardless of the time they appear in a locality during ecological succession. Weeds colonize disturbed unoccupied opportunity spacetime, while later successional species colonize opportunity spacetime created by earlier-appearing species in those same localities. The same underlying processes and locality pertain, only the traits and opportunity change.

*Seizing and exploiting opportunity.* The thesis of this book is that human disturbance (e.g. tillage, herbicides, atmospheric pollution, frozen winter) creates opportunity spacetime by leaving unused resources (e.g. nitrogen, water, light) in a local field with few or no plant neighbors. Opportunity spacetime is seized and exploited by heterogeneous plant phenotypes with preadapted life history traits expressed at favorable times as the growing season unfolds. Successful weed populations assemble and interact with crop and other weedy neighbors in their particular locality. The consequences of successful interactions lead to local adaptation maximizing survival and fitness in that plant community.

**How I came to write this book**

The story of this book is the story of the evolution of my understanding of weed biology and how I came to the ecological evolutionary perspective presented. This book is also the story of the evolution of weed control since the mid-20th century: from a wide diversity of herbicides and numerous agricultural chemical companies to consolidation with the seed industry and the dominance of proprietary transgenic crops in North America today.

Understanding weed biology began with my observations of herbicide activity, which inexorably drew me to the complexities of weed biology. A summer internship with Elanco testing trifluralin on wheat in North Dakota in the mid-1970's as a Univ. of Minnesota undergraduate began my experience with weeds. It was the golden era of weed science. Ph.D. level jobs abounded in both the growing agricultural chemical industry and an expanding commitment by public universities to weed science. After graduating from Michigan State Univ. in 1980 I took a job at the Univ. of Guelph, Ontario Agriculture College in Ontario, Canada. I developed a research program evaluating herbicides and studying weed biology. It was a fertile environment in which to learn about both plant physiology and evolutionary ecology. Evaluating the numerous herbicides available in the 1980's was a wonderful opportunity to learn about plant physiology. Why did the crop live and weed die? Why did the crop injury appear in one situation and not another? Why did herbicides fail? Herbicide evaluation was also a wonderful opportunity to learn about plant evolutionary ecology. Why did herbicide resistant weeds appear? Why did weed populations in a particular crop 'shift' over years when a new herbicide was introduced?

The crucial role of functional traits to plant growth and development was revealed in the process of evaluating herbicides and their physiology, and documenting the appearance of resistance weeds. Herbicides are physiological probes with considerable specificity and activity in plants. Atrazine taught me about photosynthesis and the pleiotropic consequences of a single chloroplast gene mutation. Glyphosate, the aryl-oxy-phenoxy and cyclohexenones taught me about phloem translocation and apical dominance in perennial weeds. I learned about microtubule cell ultrastructure scaffolding from trifluralin, about free radical quenching in plant tissues from paraquat, about very rapid microbial degradation of thiocarbamates in 'history' soils, about leaf lipids from alachlor, and about auxins from 2,4-D and 2,4,5-T. I learned about glutathione-s-transferase and oxidative degradation of several herbicides, including resistant weed variants.

With time new insights from herbicides into these important functional traits decreased. I found myself becoming more interested in their consequences in the local crop-weed community. The behavior of my untreated experimental controls provided more insight than those with herbicides.

Predicting soybean (*Glycine max*) yield losses due to velvetleaf (*Abutilon theophrasti*) infestation was my first tentative step into weed ecology. I became fascinated with C.T. de Wit's 'On Competition' and the replacement series experimental design to characterize plant competition. I read of 'density-yield' functions and other experimental designs of the pioneering Japanese researchers (e.g. Hozumi, Kira, Shinozaki, Yoda). Utilizing these approaches taught me the perils of observation. I measured growth and yield as aggregated quantities to which I applied statistics: I concluded velvetleaf inhibited soybeans by means of allelopathy. I was dead wrong. It was easy to measure plant growth in the field when and how I wanted, lump these quantities into means, and finally to digest them with statistics. Much easier than to observe with my eyes what the weeds and crops were really doing. Subsequent studies by William Akey clearly showed that phototropic stem internode elongation in response to canopy light resulted in velvetleaf growing taller than soybeans. Yield losses were due to shading effects. The crucial functional trait that makes the velvetleaf phenotype so successful is its ability to grow however tall it has to in order to shade its neighbor and seize and exploit light opportunity spacetime. Photomorphogenesis was so much less scientifically sexy than alleopathy. My suspicion of the convenience of statistics over observing was just beginning.

Observing s-triazine resistance in weeds provided my first big understanding of weed evolutionary ecology. Ecology became apparent with human tools interacting with specific plant systems. First, revealing the mechanisms that provided tolerance to atrazine. Then came deeper insights of how these R mutants conducted their life histories in the absence of herbicides. Interactions of weeds with humans were expressions of functional traits of successful individual phenotypes. Heterogeneous collections of different phenotypes in local populations provided the excess individuals amongst which the best were able survive and exploit local opportunity. Evolution by means of natural selection and elimination was the only plausible explanation of the nature I observed.

In 1989 I began to focus my research on the weedy foxtails, Iowa's number one weed, a group I came to know as the *Setaria* species-group. I observed certain crucial functional traits expressed by the *Setaria* seed phenotype revealed the same biology as herbicide resistant weeds: the best variants of the population with those traits seized, exploited and reproduced to their neighbors' disadvantage.

When herbicides were first commercially available beginning in the late 1940's their efficacy was often evaluated on their effects on 'weeds' and 'grasses'. My generation improved on this, but we still lumped 'foxtails' and 'pigweeds' into the same category. With a closer look I realized there was more variation in weed populations that I first thought, the specific components of these species-groups were not all quite the same. There exists a tradition in other pest disciplines, Plant Pathology and Entomology, for a researcher to devote effort to a single species or group. What is lost in narrowed focus is paid back in depth of understanding of the pest organism and its life history. No such tradition exists in Weed Science. Generic solutions to common weed problems have a strong appeal, solutions that can applied across a wide range of production systems. These generic solutions can be further enabled with deeper understandings of the wide variety of functional traits responsible for weedy success in different weeds: different weeds that exploit different opportunities in different fields in different cropping production systems at different times of the year in different years. We eventually outgrow 'one-size-fits-all' clothing.

I learned that humans create opportunity in space and time in many habitats around the world. Weeds are those plant species that seize and exploit those opportunities by means of rapid evolution of keystone functional traits finely tuned to each unique locality in invades. The traits I came to know best were those involved with seed dormancy and seedling emergence: the soil-weed seed communication system. I learned these weedy plant species most frequently arise from the human process of plant domestication and the formation of robust wild-crop-weed complexes that exchange genes and form world metapopulations resistant to all human control efforts.

**Learning from students**

The contributions of all past students in Agronomy 517, Weed Biology, from 1992 to 2014, have been a crucial component in the development of this book. They came locally and at a distance: Iowa and the U.S. corn belt; California, Texas and Georgia; Sweden, Czech Republic, Iran and Iceland. Student projects, often focused on a single weed species, as well as student discussions, questions, insights (especially those from their own experience) and examination responses have strongly influenced this book.

**By definition**

Scientific jargon is informative, extensive and can be very confusing. Scientific terminology often has different meanings in different disciplines. Terms are sometimes used promiscuously, causing misunderstanding and often leading to incorrect mental models of how systems work (e.g. invasive species and biodiversity). For this reason, definitions of most of the important concepts are provided in the indexed glossary, with alternative meanings provided to highlight where confusion and misunderstanding within the sciences arises (e.g. functional trait, Violee et al., 2007). Discussion of these differing usages provided much insight in the classroom. Understanding the variety of student perspectives on definitions is gained by this comparative etymology.

**On the shoulders of giants**

Harper's 'Population Biology of Plants' (1977) provided a broad view of plant biology, especially weed biology. It is now out of print. This textbook was the original source I used in developing and teaching weed biology. There is no replacement that provides the scope and detail this classic reference provided. In the intervening years I borrowed much of Harper's concepts in this book. I also relied on Jonathan Silvertown's two demographic-centric textbooks (Silvertown and Doust, 1993; Silvertown and Charlesworth, 2001) to fill out the scope of the course. The structural organization of this book has been guided by evolutionary principles clearly elucidated by Ernst Mayr in "What Evolution Is" (2001), especially his clear presentation of the component processes and conditions by which natural selection operates in biological systems. I dedicate this book to Charles Darwin, John L. Harper, Ernst Mayr and Ivanovitch Vavilov.

**Knowledge and representation**

How can knowledge of weed biology be discovered and then represented? Understanding the nature of weeds relies on deeper insights than those provided by observations of weed biology. There exist larger forces in nature that need to be understood if evolution is the make sense. We can characterize these larger forces of nature as those that give rise to complex adaptive systems. This book utilizes case histories of three weed species to illuminate the nature of complex weedy adaptive systems, providing examples of representation based on keystone functional traits.

Lastly, this book's ambition is to move towards a theory of weed science, a natural philosophy of weeds (Deutsch, 2011; Snyder, 2011); Uglow, 2002).

So, we begin. The first task is to define what weeds are.

"To Adam he said, "Because you listened to your wife and ate from the tree about which I commanded you, 'You must not eat it.'

"Cursed is the ground because of you; through painful toil you will eat of it all the days of your life. It will produce thorns and thistles for you, and you will eat the plants of the field. By the sweat of your brow you will eat your food until you return to the ground, since from it you are taken; for dust you are and to dust you will return." "
Genesis 3:17-19, Anonymous, 1984.


"The Parable of the Weeds

Jesus told them another parable:

The kingdom of heaven is like a man who sowed good seed in his field. But while everyone was sleeping, his enemy came and sowed weeds [tares, darnel, *Lolium temulentum*] among the wheat [*Triticum* sp.], and went away. When the wheat sprouted and formed heads, then the weeds also appeared.

The owner's servants came to him and said, "Sir, didn't you sow good seed in your field? Where then did the weeds come from?"

'An enemy did this,' he replied.

The servants asked him, 'Do you want us to go and pull them up?'

'No,' he answered, 'because while you are pulling the weeds, you may root up the wheat with them. Let both grow together until harvest. At that time I will tell the harvesters: First collect the weeds and tie them in bundles to be burned; then gather the wheat and bring it into my barn.'"
Matthew 24-30, Anonymous, 1984.

"The Parable of the Weeds Explained

Then he left the crowd and went into the house. His disciples came to him and said, "Explain to us the parable of the weeds in the field."

He answered, "The field is the world, and the good seed stands for the sons of the kingdom. The weeds are the sons of the evil one, and the enemy who sows them is the devil. The harvest is the end of the age, and the harvesters are angels.

As the weeds are pulled up and burned in the fire, so it will be at the end of the age. The Son of Man will send out his angels, and they will weed out of his kingdom everything that causes sin and all who do evil. They will throw them into the fiery furnace, where there will be weeping and gnashing of teeth. Then the righteous will shine like the sun in the kingdom of their Father. He who has ears, let him hear."
Matthew 13:36-43, Anonymous, 1984.

# UNIT 1: THE NATURE OF WEEDS

*The stocky woman leaned and shook the foxtail seedheads gently into her hemp apron. The patter and spatter of seeds dimpled the cloth as well as her cheeks. The patch was here again, where it had been for many years, its first autumnal seeds now ready for her. The seeds fell quickly, loosened from the panicles as they matured. The woman gathered carefully, ensuring continuous weekly harvests from the patch. This early harvest was the biggest, dedicated to making beer. Her husband would be pleased. There would be plenty of seeds for him to brew foxtail beer for the upcoming harvest festival and rituals. Soon the clan families would gather at the traditional camp here along the Yellow (Huang He) River in northeast China. At autumn's end she would carefully let the last seeds fall to the ground to reseed the next year's patch. It was her patch, everyone recognized her right, and responsibility to it. So she protected it, nurturing the foxtail seedheads as they gave up their bounty, just as she nurtured and protected her young children.*

**The weeds' story**. I am foxtail.

In the beginning I was wild. Seizing and exploiting opportunity in those infrequent patches where nature conveniently tore up the landscape, leaving it barren for me to conquer. Then, in several places around the north temperate Earth I found a helper in my struggle, you, the human. One of those early favorable places was near your camps and settlements bordering the Yellow (Huang He) River in northeast China where you interacted with hunter–gatherer humans from the arid north. 9000 years ago you began to use my seeds to make beer, one of your favorite intoxicants. I was flattered you wanted me so much. Around 7500 years ago you began to cultivate me as a domesticated crop along with your dogs and pigs. You picked the ones in my family that hung tightly to the seedheads until you could gather me in. You picked those of my family who germinated right away, not those that lingered for days and years in the dirt, waiting to grow. You grew me during your human 'Early Neolithic' period and I was your principal crop for at least four millennia. As time passed you became more systematic in your cultivation of the crop brother-sisters. And my wild and weedy family tagged along for the free ride you provided us in your crop fields.

Then came Houji, 3500-4000 years ago during your Xia dynasty, Lord of Millet. You actually named your most powerful and important leader of the time after me. I was thrilled! With the coming of these ancient human leaders of civilization also came the ancient Chinese tradition of beer making. It was important to you for ancestral worship, funeral and other rituals of your Xia, Shang and Zhou dynasties. In one of your oracle bone scripts you called it Lao Li (醪醴). Rice was first seen here in the north about 2500 years ago, but it never was as popular as it became in the hot humid south of China. I was your favorite 'eater' until wheat came and slowly replaced me. Our family is still here in northern China, my wild and weedy and crop family are all together, but we are no longer your favorites. That's OK, we conquered Eurasia together.

All this has passed now, but it remains written in my genes. I waited for you. You came and gathered me to eat. You saved your favorite seeds and planted them in soil free of neighbors. Now my world family is diverse: wild relatives from before, 'eater' crops cultivated in your fields, and our weedy members who promiscuously mate with everyone in the family.

We evolved and conquered the Earth together, always in your image. I could never have done it without you.

**Summary**. The story of weeds is the story of us, we humans. We began creating them about 10,000 years ago. We continue to create and change them every day. They were here before we came, they will be here long after we are gone. This book is their story, our story. Weeds are us.

Why do we humans look upon weeds as something separate from ourselves? Aliens. Foreigners. Bad. "… everything that causes sin and all who do evil …" As humans we are endowed at birth with inherent intuitions about the nature that surrounds us and that we observe. We cannot help but classify and categorize plants we see. They are our food, our medicines, our intoxicants, our decorations. They can also be our poisons. We humans now alive on this planet are the ones who survived because we knew the difference between these different roles plants play. We eat transgenic maize, not hemlock. We use foxglove for heart rate arrhythmia, not nightshade berries. We smoke marijuana, not jimsonweed. Our dining room table has fragrant flowers, not foxtail panicles. The survivors are the winners able to successfully classify neighboring plants. We continue to classify weeds: invasive plants, aliens. We cannot help ourselves. It is a successful ecological evolutionary adaptation that protects our lives, our humanity.

As humans we have resisted thinking of weeds as a crucial part of who we are. To embrace weeds as who we are puts us in a bad light, makes us look like animals, unable to control the nature around us. Are not humans something special? Something more than mere animals? That is our belief. This belief can get in the way of our understanding of nature. The combination of inherent plant classifying intuition and belief has prevented us from seeing the real nature of our weeds: as integral, inseparable part of an inter-fertile genetic metapopulation, the wild-weed-crop complex. Isn't it time to fully embrace Darwin's evolution?

Ivanovich Vavilov was killed by starvation in 1943 in a prison in the old USSR because of his view of nature, of plant genotypes as consequences of natural selection of genetically heritable traits. It is a severe irony for an agronomist to starve to death. Josef Stalin was threatened by what this science revealed about the future of Soviet agriculture. Or more importantly, by what it didn't reveal. Vavilov was probably the most famous crop geneticist in the world during the brief flowering of Bolshevik science and society during 1920's and early 1930's. He and his institute traveled the world collecting crop plants in search of the 'hearths' of crop domestication. His world crop germplasm collection was second to none. He thought of crop improvement as the natural (which includes human) selection of desirable genetic traits, acquisition of genotypes with desirable heritable traits. No one today would be sent to prison for such an enlightened view. Evolutionary understanding was the key.

**Picture U1**. Ivanovich Vavilov (left); Josef Stalin (middle); Trofim Denisovich Lysenko (right).

In the 1930's a jealous and ambitious Trofim Denisovich Lysenko thought otherwise. He had little formal scientific training. But he knew how to get the attention of Stalin. He argued that crops could be improved by exposing them to environmental stress. The plastic, phenotypic effects of stress on a parent crop plant would thereby be heritable in its' seeds. The USSR possesses vast land areas of marginally arable land. Dry, salty, cold. Why not expose crops to these stresses and use their now-improved seed to grow more grain in the following year? Stalin was completing the collectivization of his empire's land, notably in the Ukraine. He dispossessed individual landowners and farmers from their land and organized crop production in massive, centralized state-controlled farms. Controlled by Moscow bureaucrats. Millions of people died of starvation. Stalin and his Bolshevik regime were desperate. He had to provide more food or lose power. Lysenko's approach seemed like the answer to a prayer. Or at least a propaganda answer to a political prayer. Darwinian approaches to crop improvement were now officially wrong. Lamarckian evolution of acquired, plastic, phenotypic traits was officially right. There was nothing wrong about Lysenko's ecology, but his genetics was terrible. Vavilov had to go, along with his institute. And go they did.

All this seems a bit historical for our modern (post-modern) enlightened age. Or is it? Why is this the first book in the world on the evolutionary ecology of weeds? Many fine examples of evolutionary thinking dominate such areas as herbicide resistance and management of weeds. But considerable resistance to evolutionary thinking exists in the world of today, notably here in the USA. This has had consequences. For example, where is the evolutionary thinking in areas like demographic models of weed growth and development? How can you count plants and make inferences about the future of a local population when every generation is potentially a new set of genotypes (e.g. dioecious plant species)? Evolutionary understanding of weed ecology is the key. This book is the story of that understanding. It is written in the hope that Ivanovich Vavilov didn't die in vain in that cold Soviet prison.

# Chapter 1: The Nature of Weeds

**The weeds' story.** I am foxtail (*Setaria* species-group).

I am a closely-knit family of wild, crop and weedy relatives. Our parents are green foxtail (*Setaria viridis*) which came out of Africa and conquered Eurasia without your human help. After the ice had melted you humans picked out your favorites and cultivated foxtail millet (*Setaria viridis*, subspecies *italica*). With time our specialized cousins yellow foxtail (*Setaria pumila*), knotroot foxtail (*Setaria geniculata*), bristly foxtail (*Setaria verticillata*) and giant foxtail (*Setaria faberi*) arose and invaded. We followed you closely for 10,000 years, until we conquered the earth. I couldn't have done it without you. You gave me opportunity, a home. You killed me, you ate me, you made me strong. You carefully created me in your image.

How did you change me, make me so successful? What tricks do you humans possess that makes me so adaptable to the places you disturb and dominate?

**Summary.** The concept of a weed plant is inherently human in two different ways. In an immediate sense, they are the plants we define as weedy. In a historical sense, they are plants that arose as a consequence of agricultural crop domestication, an inevitable result of the singular human act of planting a seed (or propagule). Understanding the intimate relationship of humans and plants in both of these ways can provide a more comprehensive understanding of the nature of these plant species.

Weeds are defined as a plant out of place, thriving in habitats disturbed by humans, possessing competitive behavior, and capable of mass movement from one area to another. Human values related to disturbed and agricultural habitats, appearance, utility and biological traits dominate how we define a plant as weedy.

What is the relationship between human nature and the nature of weeds? The nature of weeds is an inevitable evolutionary consequence of agricultural natural selection under the influence of human nature. Human nature includes inherent intuitions about the natural world: taxonomy and the classification of plants we observe, eat and utilize. Human cognition is finely tuned to discern both good and bad qualities about the specific plants with which we interact. We humans are very sensitive to plant behaviors whose 'form and powers' we do not appreciate, especially in our managed landscapes. The evolutionary consequences of these human traits are the major crop-weed groups of contemporary world agriculture. Unfortunately for us, separating weedy species from desirable species is often genetically impossible.

The nature of weeds can be understood at a deeper level than definitions and human nature by observing the biological and adaptative characteristics that lead to their success. Defining what a weed is, and appreciating the traits they possess, is a good start to understanding the nature of these plants. Evolution acts on individuals in a population, so understanding the nature of the weeds we have today requires an understanding of where and when particular plants species became weeds.

Most of the common and widespread weed species we now have came as a consequence of crop domestication, planting and cultivation. These agricultural processes began about 12,000 years ago. They occurred on different continents and involved different native species available for selection as crops. Since those early origins both crops and their weeds have spread throughout the world. These crop-weed groups are the most successful invasive species in human history. The processes of plant domestication, planting and cultivation created new plant communities featuring the crop genotypes they desired. These domesticated species typically came from preexisting wild relatives selected for their crop qualities. The wild relatives interbred with their new crop derivatives and new variants joined these heterogeneous communities, forming metapopulations extending across the landscape. Some of these new weeds were in turn again domesticated, others not. Over 12 millennium this promiscuous inter-fertility and gene flow led to the world crop-weed groups we now have. The most common pattern for the origins of agricultural plants is the inter-fertile wild-crop-weed (w-c-w) plant complex in which both crop and weed were derived from the same wild progenitor species.

What plant species evolved under such close human scrutiny and management? The nature of weeds is ultimately revealed in the particular weed species that have survived to plague human-managed ecosystems to the present day. It is the properties and stability of these successful weeds that define the nature of weeds most precisely.

What exactly are these crop-weed groups? A comparison of the origins of specific crop species with the current weed flora infesting contemporary agriculture reveals the close genetic relationship of most of our major weed species and crops: crop-weed groups. It is to these crop-weed groups that we shall look throughout this book to understand the nature of weediness. The major crop-weed species groups are described here: local native cultivated and/or domesticated plant species reported in seminal publications of the original "hearths" of agriculture are compared to contemporary world weed species of the same genus. The crops and their centers of origin, diversity, cultivation and domestication as reported in older classical sources. The weeds from older classical weed flora taxonomy and identification sources.

This chapter concludes with a list of these highly adapted crop-weed groups, many of which will be examined more closely throughout this book to understand the nature of weediness: weed population evolution, adaptation in weed life history, and adaptation of weeds in local plant communities.

## 1.1 What is a weed?

The concept of a weed plant is inherently human in two different ways. In an immediate sense, they are the plants we define as weedy. In a historical sense, they are plants that arose as a consequence of agricultural crop domestication, an inevitable result of the singular human act of planting a seed (or propagule). Predictable, systematic disturbances in agricultural and other human-managed habitats create opportunity for colonizing plant species over vast areas of the earth. Understanding the intimate relationship of humans and plants in both of these ways can provide a more comprehensive understanding of the nature of these plant species.

What is the relationship between human nature and the nature of weeds? What plant species evolved under such close human scrutiny and management? To begin to answer these questions we need to look closely at the most successful invasive plant species of the last 12,000 years: crops, and their genetic fellow-traveler weeds. The roots of the nature of weeds can be found in human sociobiology, plant biogeography, archaeology, anthropology and ethnography. These scientific disciplines can provide clues to the impact of human nature on plant communities. A comparison of the origins of specific crop species with the current weed flora infesting contemporary agriculture reveals the close genetic relationship of most of our major weed species and crops: crop-weed groups. It is to these crop-weed groups that we shall look throughout this book to understand the nature of weediness.

## 1.2 The definition of a weed

Human desires, values, and most importantly economic needs are what drive a plant being defined as a weed. The qualities by which humans define plants as weeds include disturbance, aesthetics, utility or biological characteristics. All of these definitions are the consequence of interactions with humans. Many of these definitions are anthropogenic, plant qualities as perceived by humans. As such they reveal the plants' relationship to us, and tell us much of how we view nature. The nature of weeds and weediness begins by understanding the basis on which these types of plants are defined.

Based on human values:

**weed:**
1: any plant that is objectionable or interferes with the activities or welfare of man (Anonymous, 1994)
2: a plant out of place, or growing where it is not wanted (Blatchley, 1912)
3: a plant growing where it is not desired (Buchholtz, 1967)
4: a very unsightly plant of wild growth, often found in land that has been cultivated (Thomas, 1956)
5: useless, unwanted, undesirable (Bailey and Bailey, 1941)
6: a herbaceous plant, not valued for use or beauty, growing wild and rank, and regarded as cumbering the ground or hindering growth of superior vegetation (Murray et al., 1961)
7: a plant whose virtues have not yet been discovered (Emerson, 1878)

Based on human disturbance:

**weed:**
8: a generally unwanted organism that thrives in habitats disturbed by man (Harlan and deWet, 1965)
9: opportunistic species that follow human disturbance of the habitat (Pritchard, 1960)
10: a plant that grows spontaneously in a habitat greatly modified by human action (Harper, 1944)
11: a plant is a weed if, in any specified geographical area, its populations grow entirely or predominantly in situations markedly disturbed my man (without, of course, being deliberately cultivated plants) (Baker, 1965, p. 147)

Based on plant behavior:

**weed:**
12: pioneers of secondary succession of which the weedy arable field is a special case (Bunting, 1960)
13: competitive and aggressive behavior (Brenchley, 1920)
14: appearing without being sown or cultivated (Brenchley, 1920)
15: persistence and resistance to control (Gray, 1879)

The nature of weeds is also found in synonymous terminology:

**ruderal:** a plant inhabiting a disturbed site (Lincoln et al., 1998)

**agrestal:** growing on arable land (Lincoln et al., 1998)

**feral plants:** a plant that has reverted to the wild from a state of cultivation or domestication; wild, not cultivated or domesticated (Lincoln et al., 1998)

**colonizing species:** a plant, typically 'r'-selected, which invades and colonizes a new habitat or territory (Lincoln et al., 1998)

**invasive species:**

1: organism undergoing a mass movement or encroachment from one area to another (Lincoln et al., 1998)
2: an alien species whose introduction does or is likely to cause economic or environmental harm or harm to human health (Anonymous, 1999)
3: a species that is non-native (or alien) to the ecosystem under consideration and whose introduction causes or is likely to cause economic or environmental harm or harm human health (Anonymous, 2004)

It is the competitive relationship between humans and plants that define it as a weed.

## 1.3 Weeds and human nature

> "Weeds have been constant and intimate companions of man throughout his history and could tell us a lot more about man, where he has been and what he has done, if only we knew more about them."
> J. Harlan, 1992.

The dominant theme that emerges from these definitions of weediness is human: plant behavior as a consequence of human values and behavior. The sociobiology of human nature can provide insight into the nature of weeds. Brown (1991; in Wilson and Keil, 1999) has compiled a list of "human universals", universal behaviors exhibited by all humans and human cultures noted in ethnographic studies. Two may be relevant to the nature of weeds: 'classification of flora' and 'taxonomy'. These universal human traits include behaviors to classify and organize the plants with which we observe, eat, utilize and interact. These traits arise from the way the human brain conceptualizes the world.

Saying a plant is out of place is an essential cognitive function, or faculty, of the brain. Cognitive functions of the human brain are our core intuitions, our reasoning faculties, the way we conceptualize our world. The brain consists of multiple reasoning faculties, each based on a core intuition that is suitable for analyzing the world in which we evolved. Examples include a spatial sense, a number sense, language and a sense of probability (Pinker 1994, 1999, 2002). Among these core intuitions we are born with is:

> "An intuitive version of biology or natural history, which we use to understand the living world. Its core intuition is that living things house a hidden essence that gives them their form and powers and drives their growth and bodily functions." S. Pinker, 2002.

We humans are very sensitive to plant behaviors whose 'form and powers' we do not appreciate, especially in our managed landscapes. Human perception of plants changed profoundly with changes in land use from hunter-gatherer to agricultural ways of life. Widely dispersed, migratory or nomadic hunter-gatherers viewed land as held in common by many people. Land attitudes were tolerant to many uses: live and let live, harvest what you need. Agriculture brought changes. It meant sedentary existence on a particular piece of land, land ownership and its use by particular individuals. This brought changes in land use perceptions with exclusive ownership: what species do I like? Which do I not like?

Unfortunately for us, separating weedy species from desirable species is often genetically impossible. After defining particular plant species as weeds, the next step in defining the nature of weeds is discerning their biological and adaptative characteristics. What functional traits they possess makes them so successful?

## 1.4 Weedy traits

The nature of weeds can be understood at a deeper level than definitions and human nature by observing the biological and adaptative characteristics that lead to their success. These weedy traits are expressed at different times in the plants development, during its life history (table 1.1).

**Table 1.1** Biological and adaptative traits of weeds; B, Baker's 'ideal weed concept' of the world's worst weeds (1965, 1974); W, de Wet's characteristics common to all weeds; P, Patterson's (1985) adaptive characteristics of agronomic weeds.

| Life History Phase | Functional Trait (Reference) |
|---|---|
| Reproduction | 1) Self-compatibility but not complete autogamy or apomixy (B) |
| | 2) Cross-pollination by unspecialized visitors or wind (B) |
| | 3) Pollination by wind or generalized insect visitors (P) |
| | 4) Breeding systems that provide some outcrossing but also allow self-fertilization (P) |
| | 5) Continuous seed production for as long as growing conditions permit (B) |
| | 7) All weeds form part of a wild colonizer or crop complex (W) |
| | 8) Very high seed output in favorable environmental circumstances (B) |
| | 9) Production of some seed in wide range of environmental conditions; tolerance and plasticity (B) |
| | 10) Copious seed production under favorable conditions with some seed production over a range of favorable and stressful conditions (P) |
| | 11) Timing of seed maturity to coincide with crop harvest (P) |
| Propagule dispersal in space | 1) Adaptations for short-distance and long-distance dispersal (B) |
| Propagule dispersal in time | 1) Seed dormancy, longevity in soil; discontinuous germination over long periods of time (P) |
| Propagule germination-recruitment | 1) Germination requirements fulfilled in many environments (B) |
| | 2) Discontinuous germination (internally controlled) and great longevity of seed (B) |
| | 3) High relative growth rates in seedling stage (P) |
| Neighbor interactions | 1) Rapid growth through vegetative phase to flowering (B) |
| | 2) High rates of photosynthesis (P) |
| | 3) Rapid development of exploitative root systems (P) |
| | 4) Rapid partitioning of photosynthate into new leafs (P) |
| | 5) Rapid vegetative growth to reproductive phase (P) |
| | 6) Freedom from environmental constraints ("general purpose genotype"); high capacity for acclimation to changing environment (P) |
| | 7) Wide environmental tolerance (W) |
| | 8) Adaptation to man-made, permanently disturbed habitats (W) |
| | 9) Phenotypic flexibility (W) |
| | 10) Perennial: vigorous vegetative reproduction or regeneration from fragments (B) |
| | 11) Perennial: brittlenenss so as not to be drawn from ground easily (B) |
| | 12) Resistance to mechanical control; regeneration from rhizomes or other vegetative propagules (P) |

13) Ability to compete interspecifically by special means (e.g. rosette, choking growth, allelochemicals) (B)
14) Special "weapons" for interference (P)
15) Morphological and physiological similarity to crop (P)
16) Resistance or tolerance to chemical herbicides (P)

## 1.5 The origins of weeds

"The reader should be warned that as yet there is no one to whom he can turn for an orderly history of weeds; by a strange paradox these commonest of plants are comparatively unknown ... The vegetation of many a remote mountain range is better understood than the common flowers and weeds in your garden."
E. Anderson, 1954


"Weeds are adapted to habitats disturbed by man. They may be useful in some respects and harmful in others. They may be useful to some people and hated and despised by others. There are weed races of most of our field crops and these interact genetically with cultivated races as well as truly wild races. This interaction probably results ultimately in better crops and more persistent weeds. Although some weeds have evolved elegant adaptations under the influence of man, many had weedy tendencies before man existed. Weeds are products of organic evolution; they exist in intermediate states and conditions. They are also genetically labile and phenotypically plastic. Weeds have been constant and intimate companions of man throughout his history and could tell us a lot more about man, where he has been and what he has done, if only we knew more about them."
J.R. Harlan, 1992


Defining what a weed is, and appreciating the traits they possess, is a good start to understanding the nature of these plants. Evolution acts on individuals in a population, so understanding the nature of the weeds we have today requires an understanding of where and when particular plants species became weeds.

Most of the most common and widespread weed species we now have came as a consequence of crop domestication, planting and cultivation. These agricultural processes began about 12,000 years ago. They occurred on different continents and involved different native species available for selection as crops. Since those early origins both crops and their weeds have spread throughout the world. These crop-weed groups are the most successful invasive species in human history.

**1.5.1 Weeds, planting and crop domestication**. Weeds are members of plant communities, colonizing species in disturbed habitats. Disturbed habitats include any human-managed plant community: crop and horticulture fields, pastures, forests, rangeland, aquatic habitats, lawn and turfgrass. They also include other managed habits such as urban landscapes, right-of-ways and industrial space. Human activity created opportunities for these types of plants in the vast new disturbed habitats they created with agriculture and permanent settlement. The processes of plant domestication, planting and cultivation created new plant communities featuring the crop genotypes they desired. These domesticated species typically came from preexisting wild relatives selected for their crop qualities. The wild relatives interbred with their new crop derivatives and new variants joined these heterogeneous communities, forming metapopulations extending across the landscape. Some of these new weeds were in turn again domesticated, others not. Over 12 millennium this promiscuous inter-fertility and gene flow led to the world crop-weed groups we now have.

**1.5.2 Biogeographic prehistory of agriculture.** Why did agriculture start 10-12,000 years ago and not earlier of later. The reason is that there was no opportunity for agriculture before that time. The first photosynthetic organisms arose around 3.5 billion years ago (table 1.2). Over time they polluted the young earth's atmosphere with its waste products, most importantly oxygen. Oxygen is toxic to biology in its unprotected state. The biology of today protects itself from oxygen in many ways, but also rely on it for respiration. The history of the earth has also been the history of glaciation inhibiting biology. 650 million years ago the earth was almost entirely frozen, under layers of ice. With global warming around 500-550 million years ago an explosion of biological diversity occurred, the Cambrian period. Soon after that the first terrestrial plants appeared, an around 60-140 million years ago the first modern flowering plants appeared. For the last 400,000 years climatic fluctuations have regularly occurred, cycles of ice ages and global warming.

**Table 1.2** Plant pre-history in years before present (B.P.), period or epoch (Graham, 1993).

| Years B.P. | Eon/Era/Period/Epoch | Plants |
|---|---|---|
| 3,500 million | Archean Eon | First oxygenic photosynthesis |
| 2,400 million | Proterozoic Eon | Oxygenation of earth's atmosphere |
| 650 million | Neoproterozoic Era | Frozen "snowball" earth surface |
| 500-550 million | Paleozoic Era | Cambrian period life explosion |
| 458 million | Ordovician Period | First land plants |
| 428 million | Silurian Period | First vascular plants (e.g. *Cooksonia*) |
| 60-140 million | Cretaceous Period | First flowering plants, angiosperms |
| 10,000 | Recent Epoch | First weeds; human agriculture |

The peak of the last ice age occurred around 20,000 years ago (Fagan, 2004; Mithin, 2003). The human population was very small and struggled with a hostile climate in isolated refugia. Massive ice sheets covering much of North America and northern Eurasia were caused by subtle changes in the earth's orbit around the sun. The earth was a very dry and cold place to live. Sea levels had fallen to expose vast barren coastal plains. By 15-10,000 years ago (before present, B.P.) global warming began, but with many fluctuations. The course of human history was changed profoundly by this global warming. By 7000 B.P. many people lived by farming with domesticated plants and animals. They inhabited permanent settlements and supported specialized craft, political and religious occupations in addition to the cultivators of the land. These changes in human existence arose in many regions of the earth at roughly similar times. Civilized life on earth as we now know it had begun. With civilization came weeds.

**1.5.3 Pre-agricultural preadapted wild colonizing plants.** Wild colonizing plants existed before agriculture, able to seize opportunity when natural disturbances destroyed or altered exisiting plant communities. Ever since the origin of land plants glaciations, erosion, landslides and fire have created disturbed habitats for exploitation by wild colonizing species (de Wet, 1966). The natural ranges of distribution of the most successful colonizers were the same areas of the earliest known centers of agriculture. It was these vigorous colonizing species that humans gathered and selected among for the very earliest crops.

Both weeds and crops often begin with a common progenitor (Harlan, 1992). Some pre-human wild colonizing species possessed pre-adaptations that made them ideal weeds with the introduction of agriculture. Some other wild colonizing species were not as good as weed colonizers with the advent of agriculture as before. Pre-agriculture wild colonizing plants can be broken into two groups: wild plants that thrived in human selected disturbed habitats (agriculture) and formed wild-crop-weed complexes; wild species whose colonizing pre-adaptatations were not selected for, they were not as fit, and therefore did not thrive in agricultural habitats.

### 1.5.4 Wild-crop-weed complexes.

"Then they tried the bizarre argument that herbicide-reisistant crops might cross-breed with wild plants and result in a 'super' weed that is impossible to kill – with a herbicide." (Ridley, 2011)

The most common pattern for the origins of agricultural plants is the inter-fertile wild-crop-weed (w-c-w) plant complex in which both crop and weed were derived from the same wild progenitor species (Harlan, 1992). There exists extensive genetic interaction between wild, weed, and cultivated races, each acting as a genetic reservoir for the other. Continuous introgression between weeds and crops in some complexes has completely masked the original wild forms, and only weed-crop complexes remain. In others the cultivar, weed and wild colonizer races are very distinct (Harlan, 1966). The weed phenotype seems to originate by means of extensive hybridization and selection for genetic flexibility (de Wet, 1966). In some cases the wild and weed types are easily distinguished from the crop, in others it is very difficult. The species lines are fuzzy within a particular w-c-w group due to gene flow and introgression, and this fuzziness varies between different groups and different mating systems. For example, Harlan (1992) described "... four clear-cut and morphologically recognizable categories for *Sorghum*: (i) truly wild races that can tolerate considerable disturbance of the habitat and are mildly weedy, (ii) shattercanes that are derived from wild X cultivated crossing and are serious pests, (iii) shattercanes that are derived secondarily from cultivated races and are also serious pests, and (iv) semidomesticated to fully domesticated races that are grown under cultivation. The range of variation is not continuous, however. The shattercanes resemble domesticated races more closely than wild races."

To better understand the relationship between crops and weeds we need to look to the origins of crop domestication cultivation in more detail. It is from these early agricultural roots that the genetically stable wild-crop-weed groups can be seen as the sources of our current weedy problems.

### 1.6 World origins and centers of agriculture, crop domestication and cultivation

"Geographic patterns of variation have historically been used to trace the origin and evolution of cultivated plants." "... Vavilov (1992) thought that areas of maximum genetic diversity represented centers of origin and that the origin of a crop could be identified by the simple procedure of analyzing variation patterns and plotting regions where diversity was concentrated. It turned out that centers of diversity are not the same as centers of origin, yet many crops do exhibit centers of diversity." Harlan, 1992

Several theories have been proposed about crop origins, none of which completely explains their beginnings. The first comprehensive theory was proposed by Nickolay Ivanovich Vavilov (1992), who identified eight independent, primary "hearths" of crop domestication: China, India, Near East, highland Guatemala, Andes, Sudan-Abyssinia and the South American tropics. He hypothesized that the areas with the greatest concentration of domesticated plant varieties of a particular species were the centers of that crop's origin. The story of Vavilov is one of the most interesting in agronomy (Popovsky, 1984). Supported by the new Bolshevik government in Rusia during the 1920's, he led a group of agricultural research centers and germplasm collection-preservation facilities far beyond any in the world at that time. This lofty scientific achievement was followed by arrest and imprisonment by Josef Stalin in 1940, and death by starvation in a Soviet prison in 1943. It is a tragic tale of human nature, the scientific rivalry with Trofim Denisovich Lysenko over the heritability of traits for crop improvement. The scientific debate was settled in the short term by Josef Stalin, but Vavilov's legacy endures.

Since Vavilov's time many improvements on the centers-of-origin, and centers-of-diversity approach to the roots of agricultural crop domestication have been proposed (Harlan, 1992; MacNeish, 1992; table 1.3). There is no single model, or theory, of the origins of agriculture. Food crops arose from weeds, and some weeds arose from crops. One interesting theory proposed that desirable crop genotypes arose in middens or dump-heaps near human settlements (Anderson, 1952, 1956). Anderson's conclusions were based on his observations and those of N.I. Vavilov. Archaeological data suggested that this genotypic diversification occurred in the kitchen middens of sedentary people in open habitat regions. He proposed that most domesticated plants, usually seed plants, are "open-habitat" plants the readily diversify when grown in disturbed soils, providing heterogeneous types among which to choose. Humans collected the desirable variants from these locations and planted them, the so-called "dump heap theory" of agricultural beginnings. Plants were domesticated for ritual-ceremonial-religious purposes, others for drugs-intoxicants, some for both purposes (e.g. *Datura* spp.). None of these sources explain all of the cultivated plants we now grow. The origins and geographic centers of crop cultivation and diversity are impossible to trace in reality. Many, probably most, had multiple centers of origin (e.g. foxtail millet, *Setaria* sp.; Kawase and Sakamoto, 1984, 1987).

Many theories have been proposed for the origins of agriculture, but all hold in common the singular human behavior of 'planting seeds' (or other types of propagules) with the intention of harvesting them at maturity. The essential evolutionary consequences of the act of planting arose from the qualities of the seeds choosen, the traits that proliferated with agriculture. Whether intended or not, weeds inevitably arose from this singular event.

Jack Harlan (1992) probably was correct when he proposed the "no-model model" of agricultural origins, the centers of diversity, origin, cultivation and domestication. Every model proposed has evidence against it. Some crops were derived from weeds, and some weeds from crops, but the most common pattern is crop-weed complexes in which both are derived from a progenitor species (the wild-crop-weed complex). Untangling the genotype history of these groups is problematic. An answer is unnecessary for our purposes in understanding from where weeds come. It is necessary though to see what the consequences of all this agricultural history is, what we have inherited today from the activities of our ancestors. What has evolved over the last 12,000 years to form today's weed communities: the crop-weed groups of today.

**1.7 Biogeographic regions of crop origination.** Crops originated in many continents of the world, a sample of which is included in tables 1.3 and 1.4. A partial sample of the earliest domestication and cultivation events is described below.

**Table 1.3 World regions of crop origination.** Geographic regions of the world in which crops were first domesticated and cultivated (MacNeish, 1992; Vavilov, 1992)

|  | Table 1.4 abbreviations |
|---|---|
| American continents | |
|     South America (MacNeish, 1992; Vavilov, 1992) | SA |
|     Mesoamerica (MacNeish, 1992; Vavilov, 1992) | MA |
|     North America, Southwest U.S. (MacNeish, 1992) | NA-SW |
|     North America, Eastern U.S. (MacNeish, 1992) | NA-EAST |
| | |
| Eurasian continent | EA |
|     Europe (MacNeish, 1992) | EA-EUR |
|     Mediterranean (Vavilov, 1992) | EA-MED |

| | |
|---|---|
| Near East, west Asia (MacNeish, 1992; Vavilov, 1992) | EA-WEST |
| Central Asia (Vavilov, 1992) | EA-CENT |
| Far East, east Asia (MacNeish, 1992; Vavilov, 1992) | EA-EAST |
| Southeast Asia (MacNeish, 1992; Vavilov, 1992) | EA-SE |
| South Asia (Vavilov, 1992) | EA-SOUTH |
| African Continent | AFRICA |
|    Abyssinia (Vavilov, 1992) | |
|    East Africa (MacNeish, 1992) | |
|    North savanna (MacNeish, 1992) | |
|    Tropical central Africa (MacNeish, 1992) | |
|    Tropical west Africa and Congo basin (MacNeish, 1992) | |

### 1.7.1 The Americas.

"The Azteca did not settle on that island in the lakeside swamp [Tenochtitlan in Lake Texcoco] because their god gave them any sign, and they did not go there joyfully. They went because there was nowhere else to go, and because no one else had cared to claim that pimple of land surrounded by marshes." "So for a long time your ancestors existed – just barely existed – by eating revolting things like worms and water insects, and the slimey eggs of those creatures, and the only edible plant that grew in that miserable swamp. It was mexixin, the common cress or peppergrass, a scraggly and bitter-tasting weed [*Lepidium* sp.]. But if your forebearers had nothing else ... they had a mordant sense of humor. The began to call themselves, with wry irony, the Mexica." "But they never forgot that humble weed which had sustained them in the beginning, the mexixin, and they never afterward abandoned the name they had adopted from it. Mexica is a name now known and respected and feared throughout our world, but it means only ... The Weed People!" "... for weeds may be unsightly and unwanted, but they are fiercely strong and almost impossible to eradicate." Jennings (1980)

Crop domestication occurred on all American continents: south, meso and north. A partial sample of the earliest events is presented below. The first cereal crops were *Chenopodium*, *Amaranthus* and *Setaria*. Maize, *Curcurbita* and *Phaseolus* replaced them in time.

**1.7.1.1 South America.** Plant domestication in Andes, 10,500-5000 B.C. began first with quinoa (*Chenopodium quinoa*), canihua (*Chenopodium pallidicaule*), potatoes (*Solanum tubersosum*), oca (*Oxalis tuberosa*), maca (*Lepidium meyenii*) and amaranth (*Amaranthus caudatus*) (MacNeish, 1992; Mithin, 2003)

**1.7.1.2 Mesoamerica.** Harvesting native grasses and nuts in central America began as early as 6,000 B.P. (Fagan, 2004). The first domesticated cereal in Mesoamerica was foxtail (*Setaria geniculata*, Callen, 1967; *Setaria parviflora*, Austin, 2006). The beginnings of cultivating and domesticating plants is ancient in the highlands of Mesoamerica: pumpkin (*Curcurbita pepo*), 8750-7840 B.C.; maize (*Zea mays*), 7200-6800 B.P.; amaranth (*Amaranthus* sp.) and squash (*Curcurbita* sp.), 6700-6570 B.P.; common beans (*Phaseolus* sp.), 6217-6025 B.P.; sunflower (*Helianthus* sp.), 4500-3800 B.P. (MacNeish, 1992). In the Oaxaca valley, Mexico, domestication of maize (*Zea mays*), squash (*Curcurbita* sp.) and beans (*Phaseolus* sp.) dates to 12,500-7000 B.C. (Mithin, 2003). Farming in Ocampo, Mexico, dates to 6000-4300 B.P.: opuntia (*Opuntia* sp.), agave (*Agave* sp.), wild squash and pumpkins (*Curcurbita* sp.), runner beans (*Phaseolus* sp.), and setaria millet (*Setaria* sp.). During this period the size of setaria millet and squash seeds increased indicating crop domestication (MacNeish, 1992). Evidence of farming in Guerra, Mexico (3800-3400 B.P.) shows the increasing use of teosinte

(*Zea* sp.), and a hybrid of teosinte and maize (*Tripsacoid chapalote*), while at the same time large seeded setaria millet disappeared (MacNeish, 1992). Utilization of wild setaria seeds (*Setaria* sp.) and opuntia pods (*Opuntia* sp.) from Ajuereado, Tehuacan valley, Mexico dates from between 14,000-9000 B.P. by pre-settlement peoples (MacNeish, 1992).

**1.7.1.3 North America.** In southern Illinois around 3000 B.P. farming villages appeared with newly domesticated plants. At first they utilized descendants of marsh elder (*Iva* sp.), sunflowers (*Helianthus* sp.), and goosefoot (*Chenopodium* sp.); much later adapting maize (*Zea mays*) and squash (*Curcurbita* sp.) from central America (Mithin, 2003). The origins of cultivated and domesticated plants in the eastern U.S. include: gourds (*Curcurbita* sp.), 7300 B.P.; pumpkin (*Curcurbita pepo*), 7000-5100 B.P.; sunflower (*Helianthus* sp.), 4200-4000 B.P.; lambsquarters, goosefoot (*Chenopodium* sp.), 4000-2500 B.P.; swampweed-marshelder (*Iva* sp.) and knotweed (*Polygonum* sp.), 2800-2400 B.P.; maize (*Zea mays*), 2400 B.C.-2000 B.P.; amaranth (*Amaranthus* sp.) and beans (*Phaseolus* sp.), 1900-1700 B.P.; and ragweed (*Ambrosia* sp.), 1900-1500 B.P. (MacNeish, 1992).

**1.7.2 Eurasia.** Crop domestication occurred in many regions of Eurasia. Many of these domesticated arose in several regions of the continent. For example, barley-wheat-goat (*Capra* sp.) agriculture spread rapidly from its origins in west asia to Europe, Central and South Asia.

**1.7.2.1 Europe.** Farming beginnings in central Europe occurred during 8000-6400 B.P., spreading from the fringes of the Hungarian plain around 7500 B.P., growing wheat (*Triticum* sp.), barley (*Hordeum* sp.), lentils (*Lens* sp.) and peas (*Pisum* sp.). (Mithin, 2003). The first weeds (e.g. *Plantago* sp.) and stone age agriculture appear around 4000-5000 B.P. in Europe (Fagan, 2004). Proso millet (*Panicum miliaceum*) was a major crop in the Ukraine from 5800-4900 B.P. (Harlan, 1992).

**1.7.2.2 West Asia.** Groups in southwestern Asia were growing cultivated cereals 12,000 B.P. as means of coping with drought (Fagan, 2004). The earliest evidences of farming in west Asia was possibly that in the foothills of the Zagros mountains in Iraq where presumably wild plant remains of cereals, peas (*Pisum* sp.), lentils (*Lens* sp.) and vetch (*Vicia* sp.) were found around 11400 B.P., with these same species fully domesticated in the same villages around 10,000 B.P. (Mithin, 2003)

**1.7.2.3 South Asia.** Early farming arrived on the Indus plain in south Asia about 9500-7000 B.P. with barley (*Hordeum* sp.) and wheat (*Triticum* sp.) from peoples migrating from west Asia (Mithin, 2003). Foxtail millet (*Setaria* sp.) was domesticated in India around 8000 B.P., a similar time as that in China. It has been almost completely replaced as a cereal grain in south and east Asia by rice.

**1.7.2.4 Central Asia.** Early farming arrived in Afghanistan and Turkmenistan, central Asia around 8250 B.P. with barley (*Hordeum* sp.) and wheat (*Triticum* sp.) (Mithin, 2003).

**1.7.2.5 Southeast Asia.** Cultivation of taro (*Colocasia esculenta*; Araceae) occurred in the highlands of New Guinea, S.E. Asia, around 7500 B.P. (Mithin, 2003).

**1.7.2.6 East Asia.** The roots of cultivating and domesticating cereals is ancient in China: foxtail millet (*Setaria* sp.) and rice (*Oryza* sp.), date to around 8000-7000 B.P. (MacNeish, 1992). Rice (*Oryza* sp.) farming originated along the Yangtze River, southern China, between 13,500-8,500 B.P. with the cultivation of wild rice (*Oryza rufipogon*) found in swampy areas (Mithin, 2003). Foxtail millet was the first cereal of China, slowly replaced over time by rice.

**1.7.3 Africa.** Cereal agriculture arrived in the Nile valley and north Africa around 7500-6000 B.P. with barley (*Hordeum* sp.) and wheat (*Triticum* sp.) (Mithin, 2003).

## 1.8 World Crop-Weed Species-Groups

The nature of weeds is ultimately revealed in the particular weed species that have survived to plague human-managed ecosystems to the present day. It is the properties and stability of these successful weeds that define the nature of weeds most precisely.

What exactly are these crop-weed groups? The major crop-weed species-groups are described using authoritative weed sources (table 1.4). Local native cultivated and/or domesticated plant species reported in seminal publications of the original "hearths" of agriculture (Ford, 1985; Harlan, 1971, 1982, 1992; MacNeish, 1992; Mangledorf et al., 1964; Renfrew, 1973; Vavilov, 1992) are compared to contemporary world weed species of the same genus as reported in classical weed floras (Alex and Switzer, 1976; Anonymous, 1970; 1981; Behrendt and Hanf, 1979; Hanf, 1974, 1983; Holm et al., 1997; Stubbendieck et al., 1995; Stucky et al., 1980; Whitson et al., 1991; Vavilov, 1992) (table 1.4). The crops and their centers of origin, diversity, cultivation and domestication as reported in older classical sources. The weeds from older classical weed flora taxonomy and identification sources.

How many weed species exist in the world? The most informed estimates are that fewer than 8000 species behave as agricultural weeds, but only about 250 of these are important for world agriculture (Holm, et al., 1979).

Included here is an important caveat about taxonomic nomenclature. In general, the taxonomic binomial used is that reported in the original publication. The original authors species identification was used unless it was obviously changed (e.g. Vavilov (1992) nomenclature). In some cases the older taxonomic description is used to preserve a close relationship when the genus has been reclassified subsequent to publication (e.g. *Agropyron-Elymus*; *Solanum lycopersicum-Lycopersicon* sp.). Alternatively, not reported are many crop-weed groups that are inter-fertile, or near-interfertile, with weed and crop species of closely related genuses (e.g. *Zea-Tripsicum*; *Triticum-Aegilops*). Inclusion of these closely related genuses would only broaden the genetic base of these crop-weed groups.

**Table 1.4 World crop-weed species-groups.** Original, local, native, cultivated and/or domesticated plant species (Ford, 1985; Harlan, 1971, 1982, 1992; MacNeish, 1992; Mangledorf et al., 1964; Renfrew, 1973; Vavilov, 1992). Contemporary world weed species of the same genus (Alex and Switzer, 1976; Anonymous, 1970; 1981; Behrendt and Hanf, 1979; Hanf, 1974, 1983; Holm et al., 1977, 1979, 1997; Stubbendieck et al., 1995; Stucky et al., 1980; Whitson et al., 1991; Vavilov, 1992). Weed species in **bold** type listed as world's worst weeds in Holm, et al. (1977).
**Crop world locations**.
<u>Americas</u>: South America, SA; Mesoamerica, MA; North America-Southwest U.S., NA-SW; North America-Eastern U.S., NA-EAST.
<u>Eurasia</u>: Europe, EA-EUR; Mediterranean , EA-MED; Near East, west Asia, EA-WEST; Central Asia, EA-CENT; Far East, east Asia, EA-EAST; Southeast Asia, EA-SE; South Asia, EA-SOUTH.
<u>Africa</u>.

---

| ORIGINAL CROP SPECIES | | CONTEMPORARY WEED SPECIES |
|---|---|---|
| Plant Family<br><u>Crop species</u> (<u>common name</u>) | Crop<br><u>Location</u> | <u>Weed species</u> (<u>common name</u>) |
| **Agavaceae** | | |
| *Yucca elephantipes* (vegetable yucca) | MA | *Yucca elata*<br>*Yucca glauca* (Great Plains yucca)<br>*Yucca torreyi* |
| *Sagittaria sagittifolia* (elephant ear) | EA-EAST | *Sagittaria aginashi* |
| *Sagittaria sinensis* (arrowhead root) | EA-EAST | *Sagittaria calycina*<br>*Sagittaria chilensis*<br>*Sagittaria falcata*<br>*Sagittaria guyanensis* (swamp potato)<br>*Sagittaria latifolia* (common arrowhead)<br>*Sagittaria montevidensis*<br>*Sagittaria platyphylla* |

*Sagittaria pygmaea* (pygmy arrowhead)
*Sagittaria sagittifolia* (Chinese arrowhead)
*Sagittaria subulata*
*Sagittaria trifolia*

**Amaranthaceae**

| | | |
|---|---|---|
| *Amaranthus anardana* (amaranth) | EA-SOUTH | *Amaranthus albus* (tumble pigweed) |
| *Amaranthus blitum* (strawberry blite) | EA-SOUTH | *Amaranthus angustifolius* |
| *Amaranthus caudatus* (amaranth) | SA/MA | *Amaranthus australis* |
| *Amaranthus cruentus* (quelite) | MA | *Amaranthus blitoides* (prostrate pigweed) |
| *Amaranthus frumentaceus* (amaranth) | EA-SOUTH | *Amaranthus blitum* (purple amaranth) |
| *Amaranthus gangeticus* (Indian amaranth) | EA-SOUTH | *Amaranthus caudatus* |
| *Amaranthus hypochondriacus* (pigweed) | NA-SW/EAST | *Amaranthus chlorostachys* |
| *Amaranthus leucocarpus* (amaranth) | MA | *Amaranthus crassipes* |
| *Amaranthus mangostanus* (amaranth) | EA-SE | *Amaranthus crispus* (crispleaf amaranth) |
| *Amaranthus paniculatus* (purple amaranth) | MA | *Amaranthus cruentus* |
| *Amaranthus speciousus* (amaranth) | EA-SOUTH | *Amaranthus deflexus* (largefruit amaranth) |
| *Amaranthus tricolor* (Indian amaranth) | EA-SOUTH | *Amaranthus dubius* |
| *Amaranthus sp.* (amaranth) | NA-SW | *Amaranthus gangeticus* |

*Amaranthus giganticus*
*Amaranthus graecizans* (prostrate amaranth)
**Amaranthus hybridus (smooth pigweed)**
*Amaranthus hypochondriacus*
*Amaranthus lividus* (wild amaranth)
*Amaranthus macrocarpus*
*Amaranthus mangostanus*
*Amaranthus mitchelli*
*Amaranthus muricatus* (African amaranth)
*Amaranthus oleraceus*
*Amaranthus palmeri* (Palmer amaranth)
*Amaranthus paniculatus*
*Amaranthus patulus*
*Amaranthus polygamus*
*Amaranthus powellii* (green pigweed)
*Amaranthus quitensis*
*Amaranthus retroflexus* (redroot pigweed)
*Amaranthus rudis* (common waterhemp)
**Amaranthus spinosus (spiny amaranth)**
*Amaranthus standleyanus*
*Amaranthus thunbergii*
*Amaranthus tricolor*
*Amaranthus tristis*
*Amaranthus tuberculatus* (tall waterhemp)
*Amaranthus viridis* (slender amaranth)

**Araceae**

| | | |
|---|---|---|
| *Colocasia esculenta* (taro) | EA-SE | *Colocasia esculenta* |
| | | *Colocasia gigantea* |

**Asparagaceae**

| | | |
|---|---|---|
| *Asparagus lucidus* | EA-EAST | *Asparagus lucidus* |
| *Asparagus officinalis* | EA-MED | *Asparagus officinalis* |

**Asteraceae**

| | | |
|---|---|---|
| *Ambrosia trifida* (giant ragweed) | NA-EAST | *Ambrosia acanthicarpa* (annual borage) |
| | | *Ambrosia artemisiifolia* (common ragweed) |
| | | *Ambrosia bidentata* (lanceleaf ragweed) |
| | | *Ambrosia cumanensis* |
| | | *Ambrosia grayi* (woollyleaf bursage) |

|  |  |  |
|---|---|---|
|  |  | *Ambrosia maritime* |
|  |  | *Ambrosia paniculata* |
|  |  | *Ambrosia peruviana* |
|  |  | *Ambrosia polystachya* |
|  |  | *Ambrosia psilostachya* (western ragweed) |
|  |  | *Ambrosia tarapacana* |
|  |  | *Ambrosia tenuifolia* |
|  |  | *Ambrosia tomentosa* (skeletonleaf bursage) |
|  |  | *Ambrosia trifida* (giant ragweed) |
| *Chrysanthemum coronarium* (vegetable chrysanthemum) | EA-EAST | *Chrysanthemum coronarium* (crown daisy) |
|  |  | *Chrysanthemum indicum* |
|  |  | *Chrysanthemum leucanthemum* (oxeye daisy) |
|  |  | *Chrysanthemum myconis* |
|  |  | *Chrysanthemum nauseosus* (gray rabbitbush) |
|  |  | *Chrysanthemum parthenium* |
|  |  | *Chrysanthemum segetum* (corn marigold) |
|  |  | *Chrysanthemum suaveolens* |
|  |  | *Chrysanthemum tanacetum* |
|  |  | *Chrysanthemum viscidiflorus* (Douglas rabbitbush) |
|  |  | *Chrysanthemum vulgare* |
| *Helianthus annuus* (sunflower) | NA-SW/EAST | *Helianthus annuus* (common sunflower) |
| *Helianthus tuberosus* (Jerusalem artichoke) | NA-EAST | *Helianthus californicus* |
|  |  | *Helianthus ciliaris* (Texas blueweed) |
|  |  | *Helianthus grosseserratus* (sawtooth sunflower) |
|  |  | *Helianthus maximiliani* (Maximilian sunflower) |
|  |  | *Helianthus petiolaris* (prairie sunflower) |
|  |  | *Helianthus tuberosus* (Jerusalem artichoke) |
| *Iva annua* (swampweed) | NA-EAST | *Iva axillaris* (poverty sumpweed) |
| *Iva macrocarpa* (marsh elder) | NA-EAST | *Iva ciliata* |
|  |  | *Iva xanthifolia* (marshelder) |
| *Lactuca denticulata* (Chinese vegetable leaf) | EA-EAST | *Lactuca biennis* (tall blue lettuce) |
| *Lactuca indica* (Indian lettuce) | EA-SOUTH | *Lactuca canadensis* (tall lettuce) |
| *Lactuca sativa* (lettuce) | EA-EUR/MED/WEST | *Lactuca capensis* |
| *Lactuca* sp. (asparagus lettuce) | EA-EAST | *Lactuca dissecta* |
|  |  | *Lactuca formosana* |
|  |  | *Lactuca indica* |
|  |  | *Lactuca intybacea* |
|  |  | *Lactuca muralis* (wall lettuce) |
|  |  | *Lactuca oblongata* (blue lettuce) |
|  |  | *Lactuca orientalis* |
|  |  | *Lactuca perennis* (perennial lettuce) |
|  |  | *Lactuca pulchella* (blue lettuce) |
|  |  | *Lactuca runcinata* |
|  |  | *Lactuca saligna* |
|  |  | *Lactuca sativa* (lettuce) |
|  |  | *Lactuca scariola* (prickly lettuce) |
|  |  | *Lactuca serriola* (prickly lettuce) |
|  |  | *Lactuca taraxacifolia* |
|  |  | *Lactuca tatarica* |
|  |  | *Lactuca versicolor* |
|  |  | *Lactuca viminea* |
| *Tagetes erecta* (marigold; medicine) | MA | *Tagetes minuta* (wild marigold) |
| *Tagetes patula* (marigold; medicine) | MA | *Tagetes patula* |
|  |  | *Tagetes tenuifolia* |

| | | |
|---|---|---|
| *Xanthium strumarium* (vegetable cocklebur) | EA-EAST | *Xanthium ambrosioides* |
| | | *Xanthium argentums* |
| | | *Xanthium brasilicum* |
| | | *Xanthium californicum* |
| | | *Xanthium cavanillesii* |
| | | *Xanthium chinense* |
| | | *Xanthium echinatum* |
| | | *Xanthium italicum* |
| | | *Xanthium macroparpum* |
| | | *Xanthium occidentale* |
| | | *Xanthium orientale* (cocklebur) |
| | | *Xanthium pensylvanicum* (common cocklebur) |
| | | *Xanthium pungens* |
| | | *Xanthium saccharatum* |
| | | *Xanthium speciosum* |
| | | ***Xanthium spinosum* (spiny cocklebur)** |
| | | ***Xanthium strumarium* (common cocklebur)** |

**Brassicaceae**

| | | |
|---|---|---|
| *Brassica aloglabra* (Chinese kale) | EA-EAST | *Brassica adpressa* |
| *Brassica campestris* (rapeseed) | EA-EUR/WEST/EAST | *Brassica alba* |
| *Brassica carinata* (Abyssinian mustard) | AFRICA | *Brassica armoraciodides* |
| *Brassica cernua* (leafy vegetable) | EA-EAST | *Brassica campestris* (wild turnip) |
| *Brassica chinesis* (oil cabbage; bok choi) | EA-EAST | *Brassica hirta* (white mustard) |
| *Brassica glauca* (Indian colza) | EA-SOUTH | *Brassica incana* |
| *Brassica japonica* (water mustard) | EA-EAST | *Brassica juncea* (Indian mustard) |
| *Brassica juncea* (Indian mustard, sarson) | EA-WEST/EAST/SE/SOUTH | *Brassica kaber* (wild mustard) |
| *Brassica napiformis* (napa cabbage) | EA-EAST | *Brassica nigra* (black mustard) |
| *Brassica napus* (rape seed; swedes) | EA-MED/EAST | *Brassica napus* |
| *Brassica narinosa* (broadbeaked mustard) | EA-EAST | *Brassica oleracea* (cabbage) |
| *Brassica nigra* (black mustard) | EA-MED/WEST/SOUTH | *Brassica rapa* (birdsrape mustard) |
| *Brassica nipposinica* (mizuna) | EA-EAST | *Brassica rugosa* |
| *Brassica oleracea* (cabbage) | EA-MED/WEST/EAST | *Brassica sinapistrum* |
| *Brassica pekinensis* (celery cabbage) | EA-EAST | *Brassica tournefortii* |
| *Brassica rapa* (turnip) | EA-EAST | |
| | | |
| *Lepidium meyenii* (maca) | SA | *Lepidium africanum* |
| *Lepidium sativa* (garden cress) | EA-EUR/MED/WEST; AFRICA | *Lepidium auriculatum* |
| | | *Lepidium bipinnatifidum* |
| | | *Lepidium bonariense* |
| | | *Lepidium chalepense* |
| | | *Lepidium campestre* (field pepperweed) |
| | | *Lepidium densiflorum* (greenflower pepperweed) |
| | | *Lepidium graminifolium* (grassleaf pepperweed) |
| | | *Lepidium hyssopifolium* |
| | | *Lepidium latifolium* (perennial pepperweed) |
| | | *Lepidium nitidum* |
| | | *Lepidium perfoliatum* (claspingleaved peppergrass) |
| | | *Lepidium repens* |
| | | *Lepidium ruderale* (narrow-leaved pepperwort) |
| | | *Lepidium sagittulatum* |
| | | *Lepidium sativa/um* (hoary cress) |
| | | *Lepidium schinzii* |
| | | *Lepidium virginicum* (Virginia pepperweed) |
| | | |
| *Raphanus caudatus* (oil radish) | EA-SOUTH | *Raphanus microcarpus* |
| *Raphanus sativus* (radish) | EA-WEST/EAST | *Raphanus raphanistrum* (wild radish) |
| | | *Raphanus sativus* (wild radish) |

|  |  |  |
|---|---|---|
|  |  | *Raphanus sylvestris* |
| *Sinapis alba* (white mustard) | EA-MED | *Sinapis alba* (white mustard) |
|  |  | *Sinapis arvensis* (charlock) |
|  |  | *Sinapis juncea* |

**Cactaceae**

| | | |
|---|---|---|
| *Opuntia exaltata* (cactus) | SA | *Opuntia aurantiaca* |
| *Opuntia ficus-indica* (prickly pear) | MA | *Opuntia borinquensis* |
| *Opuntia megacantha* (prickly pear) | MA | *Opuntia dillenii* |
| *Opuntia streptacantha* (prickly pear) | MA | *Opuntia elatior* |
| | | *Opuntia engelmanni* |
| | | *Opuntia ficus-indica* |
| | | *Opuntia fragilis* (brittle cactus) |
| | | *Opuntia fulgida* |
| | | *Opuntia humifusa* |
| | | *Opuntia imbricate* |
| | | *Opuntia inermis* |
| | | *Opuntia leptocaulis* |
| | | *Opuntia lindheimeri* |
| | | *Opuntia macrorhiza* (bigroot pricklypear) |
| | | *Opuntia megacantha* |
| | | *Opuntia polyacantha* (plains pricklypear) |
| | | *Opuntia schumannii* |
| | | *Opuntia spinosior* |
| | | *Opuntia spinulifera* |
| | | *Opuntia streptacantha* |
| | | *Opuntia stricta* |
| | | *Opuntia tardospina* |
| | | *Opuntia tomentosa* |
| | | *Opuntia tuna* |
| | | *Opuntia versicolor* |
| | | *Opuntia vulgaris* |
| | | *Opuntia vulpina* |

**Cannabaceae**

| | | |
|---|---|---|
| *Cannabis sativa* (hemp) | EA-WEST/EAST | *Cannabis ruderalis* |
| *Cannabis indica* (hashish, hemp, marijuana) | EA-CENT | *Cannabis sativa* (marijuana; wild hemp) |

**Chenopodiaceae**

| | | |
|---|---|---|
| *Beta vulgaris* (beet) | EA-EUR/MED/WEST | *Beta vulgaris* (beet) |
| *Chenopodium album* (grass seed) | NA-EAST | *Chenopodium acuminatum* |
| *Chenopodium bushianum* (goosefoot) | NA-EAST | **_Chenopodium album_ (common lambsquarters)** |
| *Chenopodium nuttaliae* (apazote grain) | MA | *Chenopodium ambrosioides* (Mexican tea) |
| *Chenopodium pallidicaule* (canahua) | SA | *Chenopodium anthelminticum* |
| *Chenopodium quinoa* (quinoa) | SA | *Chenopodium aristatum* |
| *Chenopodium* sp. (chenopod) | NA-SW | *Chenopodium berlandieri* (net-seeded lamb's-quarters) |
| | | *Chenopodium bonei* |
| | | *Chenopodium bonus-henricus* (good King Henry) |
| | | *Chenopodium botrys* (Jerusalem-oak) |
| | | *Chenopodium capitatum* (strawberry blite) |
| | | *Chenopodium carinatum* |
| | | *Chenopodium cordobense* |
| | | *Chenopodium cristatum* |
| | | *Chenopodium fasciculosum* |
| | | *Chenopodium ficifolium* (fig-leaved goosefoot) |
| | | *Chenopodium foetidum* |
| | | *Chenopodium foliosum* |

|  |  |  |
|---|---|---|
|  |  | *Chenopodium giganto-spermum* (maple-leaved goosefoot) |
|  |  | *Chenopodium glaucum* (oak-leaved goosefoot) |
|  |  | *Chenopodium hircinum* |
|  |  | *Chenopodium hybridum* (mapleleaf goosefoot) |
|  |  | *Chenopodium leptophyllum* |
|  |  | *Chenopodium macrospermum* |
|  |  | *Chenopodium multifidum* |
|  |  | *Chenopodium murale* (nettleleaf goosefoot) |
|  |  | *Chenopodium nuttalliae* (Mexican tea) |
|  |  | *Chenopodium oahuense* |
|  |  | *Chenopodium opulifolium* (grey goosefoot) |
|  |  | *Chenopodium paganum* (pigweed) |
|  |  | *Chenopodium paniculatum* |
|  |  | *Chenopodium polyspermum* (all-seed) |
|  |  | *Chenopodium pratericola* |
|  |  | *Chenopodium procerum* |
|  |  | *Chenopodium rubrum* (red goosefoot) |
|  |  | *Chenopodium simplex* (mapleleaf goosefoot) |
|  |  | *Chenopodium strictum* |
|  |  | *Chenopodium triangulare* |
|  |  | *Chenopodium urbicum* (upright goosefoot) |
|  |  | *Chenopodium vulvaria* (stinking goosefoot) |
|  |  | *Chenopodium zobelii* |

**Cleomaceae**

| | | |
|---|---|---|
| *Cleome serrulata* (Rocky Mountain bee weed) | NA-SW | *Cleome aculeate* |
|  |  | *Cleome aspera* |
|  |  | *Cleome brachycarpa* |
|  |  | *Cleome burmanni* |
|  |  | *Cleome chelidonii* |
|  |  | *Cleome diffusa* |
|  |  | *Cleome gynandra* (spider flower) |
|  |  | *Cleome hirta* |
|  |  | *Cleome icosandra* |
|  |  | *Cleome integrifolia* |
|  |  | *Cleome lutea* (yellow spiderflower) |
|  |  | *Cleome monophylla* |
|  |  | *Cleome moritziana* |
|  |  | *Cleome rudidosperma* |
|  |  | *Cleome serrata* |
|  |  | *Cleome serrulata* (Rocky Mountain beeplant) |
|  |  | *Cleome spinosa* |
|  |  | *Cleome stenophylla* |
|  |  | *Cleome viscose* |
|  |  | *Cleome welwitschii* |

**Convolvulaceae**

| | | |
|---|---|---|
| *Ipomoea aquatica* (water spinach) | EA-EAST | *Ipomoea acuminate* |
| *Ipomoea batatas* (sweet potato) | SA | *Ipomoea alba* |
|  |  | *Ipomoea amoena* |
|  |  | *Ipomoea angustifolia* |
|  |  | *Ipomoea aquatica* (swamp morningglory) |
|  |  | *Ipomoea asperifolia* |
|  |  | *Ipomoea barbigera* |
|  |  | *Ipomoea batatas* |
|  |  | *Ipomoea blepharosepala* |
|  |  | *Ipomoea cairica* |
|  |  | *Ipomoea calobra* |

|  |  |  |
|---|---|---|
|  |  | *Ipomoea caloneura* |
|  |  | *Ipomoea cardiosepala* |
|  |  | *Ipomoea chryeides* |
|  |  | *Ipomoea coccinea* (red morningglory) |
|  |  | *Ipomoea congesta* |
|  |  | *Ipomoea coptica* |
|  |  | *Ipomoea cordofana* |
|  |  | *Ipomoea coscinosperma* |
|  |  | *Ipomoea crassifolia* |
|  |  | *Ipomoea cymosa* |
|  |  | *Ipomoea cynanchifolia* |
|  |  | *Ipomoea digitata* |
|  |  | *Ipomoea eriocarpa* |
|  |  | *Ipomoea fistulosa* |
|  |  | *Ipomoea gossypioides* |
|  |  | *Ipomoea gracilis* |
|  |  | *Ipomoea hardwikii* |
|  |  | *Ipomoea hederacea* (ivyleaf morningglory) |
|  |  | *Ipomoea hederifolia* |
|  |  | *Ipomoea hirsutula* (woolly morningglory) |
|  |  | *Ipomoea indivisa* |
|  |  | *Ipomoea involucrata* |
|  |  | *Ipomoea lacunosa* (pitted morningglory) |
|  |  | *Ipomoea leari* |
|  |  | *Ipomoea maxima* |
|  |  | *Ipomoea muelleri* |
|  |  | *Ipomoea muricata* |
|  |  | *Ipomoea nil* |
|  |  | *Ipomoea obscura* |
|  |  | *Ipomoea pandurata* (bigroot morningglory) |
|  |  | *Ipomoea pes-caprae* |
|  |  | *Ipomoea pes-tigridis* |
|  |  | *Ipomoea plebeian* |
|  |  | *Ipomoea polyantha* |
|  |  | *Ipomoea purpurea* (tall morningglory) |
|  |  | *Ipomoea quamoclit* (cypressvine morningglory) |
|  |  | *Ipomoea setifera* |
|  |  | *Ipomoea stolonifera* |
|  |  | *Ipomoea tiliacea* |
|  |  | *Ipomoea trichocarpa* |
|  |  | *Ipomoea trifida* |
|  |  | *Ipomoea triloba* (three-lobe morningglory) |
|  |  | *Ipomoea tuba* |
|  |  | *Ipomoea tuboides* |

**Curcucbitaceae**

| | | |
|---|---|---|
| *Curcurbita ficifolia* (figleaved gourd) | MA | *Curcurbita andreana* |
| *Curcurbita maxima* (pumpkin) | SA | *Curcurbita digitata* |
| *Curcurbita mixta* (pumpkin) | MA | *Curcurbita foetidissima* (buffalo gourd) |
| *Curcurbita moschata* (squash, melon) | MA; EA-EAST | *Curcurbita pepo* |
| *Curcurbita pepo* (gourd) | EA-WEST | |

**Cyperaceae**

| | | |
|---|---|---|
| *Cyperus esculentus* (earth-almonds) | EA-MED | *Cyperus alopecuroides* |
|  |  | *Cyperus alternifolius* (umbrella plant) |
|  |  | *Cyperus amabilis* (souchet gracieux) |
|  |  | *Cyperus amuricus* |
|  |  | *Cyperus aristatus* |
|  |  | *Cyperus articulates* (jointed flatsedge) |

*Cyperus atrovirens* (club rush)
*Cyperus babakan*
*Cyperus brevifolius* (kyllinga)
*Cyperus bulbosus*
*Cyperus castaneus*
*Cyperus compactus*
*Cyperus compressus* (flathead sedge)
*Cyperus corymbosus*
*Cyperus cuspidatus* (souchet a pointes aigues)
*Cyperus cyperinus*
*Cyperus cyperoides*
*Cyperus diandrus* (galingale)
**Cyperus difformis (smallflower umbrellaplant)**
*Cyperus diffusus*
*Cyperus digitatus* (digitate cyperus)
*Cyperus distans* (slender cyperus)
*Cyperus eragrostis*
*Cyperus erythrorhizos*
**Cyperus esculentus (yellow nutsedge)**
*Cyperus exaltatus*
*Cyperus flavidus*
*Cyperus fulvus*
*Cyperus fuscus*
*Cyperus giganteus*
*Cyperus globosus*
*Cyperus gracilinux*
*Cyperus gracillis*
*Cyperus hakonensis*
*Cyperus haspan* (sheathed cyperus)
*Cyperus hermaphroditus*
*Cyperus imbricatus* (souchet imbrique)
**Cyperus iria (umbrella sedge)**
*Cyperus javanicus*
*Cyperus kyllingia*
*Cyperus laetus*
*Cyperus laevigatus*
*Cyperus longus* (sweet cyperus)
*Cyperus luzulae*
*Cyperus malaccensis* (shichito matgrass)
*Cyperus maranguensis*
*Cyperus maritimus*
*Cyperus microiria*
*Cyperus monti*
*Cyperus nipponicus*
*Cyperus novae-hollandiae*
*Cyperus odoratus*
*Cyperus pangorei*
*Cyperus papyrus*
*Cyperus pilosus*
*Cyperus polystachyos*
*Cyperus procerus*
*Cyperus pulcherrimus*
*Cyperus pumilus*
*Cyperus pygmaeus*
*Cyperus radians*
*Cyperus reduncus*
*Cyperus retzii*
*Cyperus rigidifolius*
**Cyperus rotundus (purple nutsedge)**

|  |  | *Cyperus sanguinolentus* |
|  |  | *Cyperus schweinfurthianus* |
|  |  | *Cyperus seemannianus* |
|  |  | *Cyperus sesquiflorus* |
|  |  | *Cyperus sphacelatus* (roadside flatsedge) |
|  |  | *Cyperus strigosus* (sedge) |
|  |  | *Cyperus tegetiformis* |
|  |  | *Cyperus tegetum* |
|  |  | *Cyperus tenuiculmis* |
|  |  | *Cyperus tenuispica* |
|  |  | *Cyperus trialatus* |
|  |  | *Cyperus truncates* |
|  |  | *Cyperus uncinatus* |
|  |  | *Cyperus zollingeri* |
| *Eleocharis tuberosa* (water chestnut) | EA-EAST/SE | *Eleocharis acicularis* (slender spikerush) |
|  |  | *Eleocharis acuta* |
|  |  | *Eleocharis acutangula* (scirpe a angles aigus) |
|  |  | *Eleocharis afflata* |
|  |  | *Eleocharis atropurpurea* (junco purpurea) |
|  |  | *Eleocharis attenuate* |
|  |  | *Eleocharis cellulose* |
|  |  | *Eleocharis chaetaria* |
|  |  | *Eleocharis congesta* |
|  |  | *Eleocharis dulcis* (Chinese waterchestnut) |
|  |  | *Eleocharis elegans* (elegant spikerush) |
|  |  | *Eleocharis equisetina* |
|  |  | *Eleocharis erecta* |
|  |  | *Eleocharis filiculmis* (scirpe a tige filiforme) |
|  |  | *Eleocharis geniculata* (spikerush) |
|  |  | *Eleocharis interstineta* |
|  |  | *Eleocharis kuroguwai* |
|  |  | *Eleocharis mamillata* |
|  |  | *Eleocharis multicaulis* |
|  |  | *Eleocharis mutate* |
|  |  | *Eleocharis nodulosa* |
|  |  | *Eleocharis ochreata* |
|  |  | *Eleocharis ovata* |
|  |  | *Eleocharis palustris* (creeping spikerush) |
|  |  | *Eleocharis parvula* |
|  |  | *Eleocharis pellucida* (scirpe pellucid) |
|  |  | *Eleocharis plantaginea* (Chinese waterchestnut) |
|  |  | *Eleocharis quadrangulata* |
|  |  | *Eleocharis retroflexa* |
|  |  | *Eleocharis sphacelata* |
|  |  | *Eleocharis subtilis* |
|  |  | *Eleocharis tetraquetra* |
|  |  | *Eleocharis tuberosa* (Chinese waterchestnut) |
|  |  | *Eleocharis variegate* |
|  |  | *Eleocharis wolfii* |
| *Momordica charantia* (balsam pear) | EA-SE | *Momordica balsamina* |
|  |  | *Momordica charantia* (balsam apple) |
|  |  | *Momordica tuberosa* |

**Gramineae**

| *Agropyron* sp. (wheatgrass) | EA-EUR/WEST | *Agropyron caninum* (bearded couch grass) |
|  |  | *Agropyron cristatum* (crested wheatgrass) |
|  |  | *Agropyron intermedium* (intermediate wheatgrass) |

| | | |
|---|---|---|
| | | *Agropyron kamoji* |
| | | *Agropyron ramosum* |
| | | **Agronpyron repens (quackgrass)** |
| | | *Agropyron semicostatum* |
| | | *Agropyron smithii* (western wheat-grass) |
| | | *Agropyron tsukushiense* |
| | | |
| *Avena abyssinica* (Ethiopian oats) | EA-WEST; AFRICA | *Avena alba* |
| *Avena brevis* (sand oats) | EA-MED/WEST | *Avena barbata* (slender oat) |
| *Avena byzantina* (Mediterranean, red oats) | EA-MED/WEST | *Avena byzantine* |
| *Avena nubibrevis* (naked oats) | EA-WEST | *Avena cultiformis* |
| *Avena nuda* (naked-seeded oats) | EA-EAST | **Avena fatua (wild oat)** |
| *Avena sativa* (common white oats) | EA-EUR/WEST | *Avena ludoviciana* (winter wild oat) |
| *Avena strigosa* (fodder, sand oats) | EA-EUR/MED/WEST | *Avena sativa* |
| *Avena weistii* (desert oats) | EA-WEST | *Avena sterilis* (sterile oat) |
| | | *Avena strigosa* (bristle oat) |
| | | |
| *Brachiaria deflexa* (guinea millet) | AFRICA | *Brachiaria brizantha* |
| | | *Brachiaria ciliatissima* |
| | | *Brachiaria deflexa* (guinea millet) |
| | | *Brachiaria distachya* (armgrass millet) |
| | | *Brachiaria eruciformis* (sweet signalgrass) |
| | | *Brachiaria extensa* |
| | | *Brachiaria fasciculata* (browntop millet) |
| | | *Brachiaria lata* (fula fulfulde) |
| | | *Brachiaria milliiformis* |
| | | **Brachiaria mutica (buffalograss)** |
| | | *Brachiaria paspaloides* (common signalgrass) |
| | | *Brachiaria piligera* |
| | | *Brachiaria plantaginea* (alexandergrass) |
| | | *Brachiaria platyphylla* (broadleaf signal grass) |
| | | *Brachiaria ramosa* (browntop millet) |
| | | *Brachiaria reptans* |
| | | *Brachiaria repens* (runninggrass) |
| | | *Brachiaria subquadripara* |
| | | *Brachiaria texana* (Texas millet) |
| | | |
| *Cynodon aethiopicum* (stargrass) | AFRICA | **Cynodon dactylon (Bermudagrass)** |
| *Cynodon dactylon* (Bermuda grass) | AFRICA | *Cynodon hirsutus* |
| | | *Cynodon plectostachyum* |
| | | |
| *Digitaria cruciata* (millet) | EA-SE | *Digitaria abyssinica* |
| *Digitaria decumbens* (pangolagrass) | AFRICA | *Digitaria adscendens* |
| *Digitaria exilis* (hungry rice) | AFRICA | *Digitaria biformis* |
| *Digitaria iburua* (black hungry rice) | AFRICA | *Digitaria ciliaris* (tropical crabgrass) |
| | | *Digitaria debilis* |
| | | *Digitaria decumbens* |
| | | *Digitaria didactyla* |
| | | *Digitaria fauriei* |
| | | *Digitaria filiformis* (slender crab-grass) |
| | | *Digitaria fuscenscens* (yellow crabgrass) |
| | | *Digitaria hayatae* |
| | | *Digitaria horizontalis* (Jamaican crabgrass) |
| | | *Digitaria insularis* |
| | | *Digitaria ischaemum* (smooth crabgrass) |
| | | *Digitaria longiflora* (lesser crabgrass) |
| | | *Digitaria macractenia* |
| | | *Digitaria magna* |
| | | *Digitaria microbachne* |

| | | |
|---|---|---|
| | | *Digitaria pentzii* |
| | | *Digitaria pruriens* |
| | | *Digitaria pseudo-ischaemum* |
| | | *Digitaria radicosa* (trailing crabgrass) |
| | | ***Digitaria sanguinalis* (large crabgrass)** |
| | | ***Digitaria scalarum* (blue couch)** |
| | | *Digitaria sericea* |
| | | *Digitaria setigera* (East Indian crabgrass) |
| | | *Digitaria ternata* (black-seed fingergrass) |
| | | *Digitaria timorensis* |
| | | *Digitaria velutina* (velvet finger grass) |
| | | *Digitaria violascens* (violet crabgrass) |
| *Echinochloa crus-galli* (barnyard millet) | EA-EUR | ***Echinochloa colona* (junglerice)** |
| *Echinochloa frumentacea* (Japanese millet) | EA-EAST/SE | ***Echinochloa crus-galli* (barnyardgrass)** |
| | | *Echinochloa crus-pavonis* (gulf cockspur) |
| | | *Echinochloa helodes* |
| | | *Echinochloa holubii* |
| | | *Echinochloa macrocarpa* |
| | | *Echinochloa macrocorvi* |
| | | *Echinochloa orizicola* |
| | | *Echinochloa orizoides* |
| | | *Echinochloa phyllopogon* |
| | | *Echinochloa pungens* (barnyard-grass) |
| | | *Echinochloa pyramidalis* (antelopegrass) |
| | | *Echinochloa stagnina* |
| | | *Echinochloa walteri* |
| *Eleusine coracana* (finger millet) | EA-SOUTH; AFRICA | *Eleusine aegyptiaca* |
| *Eleusine indica* (goosegrass) | EA-SOUTH | *Eleusine africana* |
| | | *Eleusine compressa* |
| | | *Eleusine coracana* (Indian millet) |
| | | ***Eleusine indica* (goosegrass)** |
| | | *Eleusine tristachya* |
| *Hordeum distichion* (naked 2-row barley) | EA-WEST | *Hordeum berteroanum* |
| *Hordeum nodosum* (wild barley) | AFRICA | *Hordeum brachyantherum* |
| *Hordeum pusillum* (desert barley) | NA-SW/EAST | *Hordeum bulbosum* |
| *Hordeum sativum* (barley) | EA-MED; AFRICA | *Hordeum glaucum* |
| *Hordeum vulgare* (naked 6-row barley) | EA-WEST | *Hordeum hystrix* |
| | | *Hordeum jubatum* (foxtail barley) |
| | | *Hordeum leporinum* (hare barley) |
| | | *Hordeum murinum* (barley grass) |
| | | *Hordeum pusillum* (little barley) |
| | | *Hordeum marinum* |
| | | *Hordeum secalinum* |
| | | *Hordeum spontanum* |
| | | *Hordeum vulgare* |
| *Lolium* sp. (rye grass) | EA-EUR | *Lolium multiflorum* (Italian ryegrass) |
| | | *Lolium perenne* (perennial ryegrass) |
| | | *Lolium persicum* (Persian darnel) |
| | | *Lolium remotum* (hardy rye-grass) |
| | | *Lolium rigidum* (Wymmera rye-grass) |
| | | ***Lolium temulentum* (darnel)** |
| *Oryza glaberrima* (African rice) | AFRICA | *Oryza alta* |
| *Oryza sativa* (rice) | EA-EAST/SE/SOUTH | *Oryza barthii* (wild rice) |
| | | *Oryza minuta* |

| | | |
|---|---|---|
| | | *Oryza perennis* (wild rice) |
| | | *Oryza punctata* (wild rice) |
| | | *Oryza rufipogon* (wild red rice) |
| | | *Oryza sativa* (red rice) |
| *Panicum frumentaceum* (Japanese millet) | EA-EAST | *Panicum adspersum* |
| *Panicum maximum* (guinea grass) | AFRICA | *Panicum agrostoides* |
| *Panicum mileaceum* (broomcorn millet) | EA-EUR/EAST | *Panicum ambigium* |
| *Panicum miliare* (slender millet) | EA-SE | *Panicum amplexicaule* |
| *Panicum sonorum* (sauwi) | MA/NA-SW | *Panicum antidotale* (giant panic) |
| | | *Panicum arizonicum* |
| | | *Panicum attenuatum* |
| | | *Panicum auritum* |
| | | *Panicum austroasiaticum* |
| | | *Panicum bergii* |
| | | *Panicum bisulcatum* (chaff panic) |
| | | *Panicum brevifolium* (short-leaved panicum) |
| | | *Panicum capillare* (witchgrass) |
| | | *Panicum clandestinum* (deertongue dichnthelium) |
| | | *Panicum caudiglume* |
| | | *Panicum chloroticum* |
| | | *Panicum ciliatum* |
| | | *Panicum dichotomiflorum* (fall panicum) |
| | | *Panicum effusum* |
| | | *Panicum elephantipes* |
| | | *Panicum erectum* |
| | | *Panicum fasciculatum* (browntop signalgrass) |
| | | *Panicum gattingeri* (Gattinger witch-grass) |
| | | *Panicum glabrescens* |
| | | *Panicum gouini* |
| | | *Panicum hallii* |
| | | *Panicum hemitomun* |
| | | *Panicum hygrocharis* |
| | | *Panicum laevifolium* (sweet grass) |
| | | *Panicum kerstingii* |
| | | *Panicum laevifolium* |
| | | *Panicum lancearum* |
| | | *Panicum laxum* |
| | | *Panicum luzonense* |
| | | **Panicum maximum (guinea grass)** |
| | | *Panicum meyerianum* |
| | | *Panicum miliaceum* (wild proso millet) |
| | | *Panicum millegrana* |
| | | *Panicum monostachyum* |
| | | *Panicum montanum* |
| | | *Panicum obtusum* (vine mesquite) |
| | | *Panicum paludosum* |
| | | *Panicum phyllopogon* |
| | | *Panicum pilipes* |
| | | *Panicum polygonatum* |
| | | *Panicum prionitis* |
| | | *Panicum psilopodium* |
| | | *Panicum queenlandicum* |
| | | **Panicum repens (torpedograss)** |
| | | *Panicum sabulorum* |
| | | *Panicum sarmentosum* (scrambling panicgrass) |
| | | *Panicum spectabile* |
| | | *Panicum texanum* (Texas signalgrass) |
| | | *Panicum torridum* |

| | | |
|---|---|---|
| | | *Panicum trichocladum* |
| | | *Panicum trichoides* (ticklegrass) |
| | | *Panicum trypheron* |
| | | *Panicum turgidum* (desert grass) |
| | | *Panicum umbellatum* |
| | | *Panicum urvilleanum* |
| | | *Panicum virgatum* (switch grass) |
| *Paspalum dilatatum* (dallisgrass) | SA | *Paspalum ciliatifolium* (fringeleaf paspalum) |
| *Paspalum scrobicalatum* (slender millet) | EA-SE/SOUTH | *Paspalum commersonii* |
| | | **Paspalum conjugatum (sourgrass)** |
| | | *Paspalum dasypleurum* |
| | | **Paspalum dilatatum (Dallis grass)** |
| | | *Paspalum distachyon* |
| | | *Paspalum distichum* (knotgrass) |
| | | *Paspalum fasciculatum* |
| | | *Paspalum fimbriatum* (Panama paspalum) |
| | | *Paspalum fluitans* (water paspalum) |
| | | *Paspalum geminatum* |
| | | *Paspalum haenkeanum* |
| | | *Paspalum laeve* (field paspalum) |
| | | *Paspalum lividum* (longtom) |
| | | *Paspalum longifolium* (long-leaved paspalum) |
| | | *Paspalum maculosum* |
| | | *Paspalum maritimum* |
| | | *Paspalum melanospernum* |
| | | *Paspalum millegrana* |
| | | *Paspalum nicorae* |
| | | *Paspalum notatum* (bahia grass) |
| | | *Paspalum nutans* |
| | | *Paspalum orbiculare* |
| | | *Paspalum paniculatum* |
| | | *Paspalum paspaloides* (knotgrass) |
| | | *Paspalum plicatulum* (brownseed paspalum) |
| | | *Paspalum polystachyum* |
| | | *Paspalum proliferum* |
| | | *Paspalum pumilum* |
| | | *Paspalum quadrifarium* |
| | | *Paspalum racemosa* |
| | | *Paspalum scrobiculatum* (kado millet) |
| | | *Paspalum thunbergii* (Japanese paspalum) |
| | | *Paspalum urvillei* (vasegrass) |
| | | *Paspalum vaginatum* (saltwater couch) |
| | | *Paspalum virgatum* (talquezal) |
| *Pennisetum americanum* (pearl millet) | AFRICA | *Pennisetum alopecuroides* (Chinese pennisetum) |
| *Pennisetum clandestinum* (kikuyu grass) | AFRICA | *Pennisetum americanum* (pearl millet) |
| *Pennisetum purpureum* (elephant grass) | AFRICA | *Pennisetum chilense* |
| *Pennisetum spicatum* (pearl millet) | AFRICA | **Pennisetum clandestinum (kikuyu grass)** |
| | | *Pennisetum fructescens* |
| | | *Pennisetum hordeodes* |
| | | *Pennisetum japonicum* |
| | | *Pennisetum macrourum* |
| | | **Pennisetum pedicellatum (kayasuwa grass)** |
| | | **Pennisetum polystachyon (West Indian pennisetum)** |
| | | **Pennisetum purpureum (napiergrass)** |
| | | *Pennisetum ruppellii* |
| | | *Pennisetum seaceum* (fountaingrass) |
| | | *Pennisetum setosum* |

|  |  |  |
|---|---|---|
|  |  | *Pennisetum villosum* (feathertop) |
| *Phalaris arundinaceae* (reed canary grass) | EA-EUR/WEST | *Phalaris angusta* |
| *Phalaris caroliniana* (maygrass) | EA-MED | *Phalaris aquatica* (tuberous canarygrass) |
| *Phalaris tuberosa* (hardinggrass) | EA-MED | *Phalaris arundinaceae* (reed canarygrass) |
|  |  | *Phalaris brachystachys* (short-spiked canarygrass) |
|  |  | *Phalaris canariensis* (canarygrass) |
|  |  | *Phalaris minor* (little seeded canarygrass) |
|  |  | *Phalaris paradoxa* (bristle-spiked canarygrass) |
|  |  | *Phalaris tuberosa* |
| *Saccharum officinarum* (sugar cane) | EA-SE | *Saccharum arundinaceum* |
|  |  | *Saccharum benghalense* |
|  |  | *Saccharum narenga* |
|  |  | *Saccharum officinarum* |
|  |  | *Saccharum spontaneum* (serio grass) |
| *Secale cereale* (rye) | EA-EUR/WEST/CENT | *Secale cereale* |
|  |  | *Secale montanum* (wild rye) |
| *Setaria geniculata* (knotroot foxtail) | MA | *Setaria acromelaena* |
| *Setaria pumila* (korali ) | EA-SOUTH | *Setaria aequalis* |
| *Setaria viridis*, subsp. *italica* (foxtail millet) | EA-EAST | *Setaria aurea* |
|  |  | *Setaria conspersum* |
|  |  | *Setaria barbata* (East-Indies foxtailgrass) |
|  |  | *Setaria faberi* (giant foxtail) |
|  |  | *Setaria geniculata* (knotroot foxtail) |
|  |  | *Setaria gigantean* |
|  |  | *Setaria grisebachii* |
|  |  | *Setaria homonyma* |
|  |  | *Setaria incrassate* |
|  |  | *Setaria intermedia* |
|  |  | *Setaria italica* |
|  |  | *Setaria longiseta* |
|  |  | *Setaria pallide-fusca* (pigeon grass) |
|  |  | *Setaria palmifolia* (broadleaved bristlegrass) |
|  |  | *Setaria plicata* |
|  |  | *Setaria poiretiana* |
|  |  | *Setaria pumila* (yellow foxtail) |
|  |  | *Setaria sphacelata* (golden timothy) |
|  |  | **Setaria verticillata (bristly foxtail)** |
|  |  | *Setaria verticilliformis* (ambiguous foxtailgrass) |
|  |  | **Setaria viridis, subsp. viridis (green foxtail)** |
|  |  | *Setaria welwitschii* |
| *Sorghum bicolor* (sorghum) | AFRICA | *Sorghum affine* |
| *Sorghum halepense* (Johnsongrass) | EA-EUR | *Sorghum almum* (almum grass) |
|  |  | *Sorghum arundinaceum* |
|  |  | *Sorghum bicolor* (shattercane) |
|  |  | **Sorghum halepense (Johnsongrass)** |
|  |  | *Sorghum miliaceum* |
|  |  | *Sorghum pumosum* |
|  |  | *Sorghum propinquum* |
|  |  | *Sorghum sudanese* |
|  |  | *Sorghum verticilliflorum* (wild Sudan grass) |
|  |  | *Sorghum virgatum* |
|  |  | *Sorghum vulgare* |

| | | | |
|---|---|---|---|
| *Tripsacum andersonii* (forage) | MA | *Tripsacum laxum* (Guatemala grass) | |
| | | | |
| *Triticum aestivum* (bread wheat) | EA-WEST | *Triticum aestivum* | |
| *Triticum compactum* (club wheat) | EA-CENT | *Triticum cylindricum* (einkorn) | |
| *Triticum dicoccum* (emmer) | EA-MED; AFRICA | *Triticum monococcum* (jointed goatgrass) | |
| *Triticum durum* (hard wheat) | EA-MED/WEST; AFRICA | *Triticum ramosum* | |
| *Triticum macha* (masha wheat) | EA-WEST | | |
| *Triticum monococcum* (einkorn) | EA-WEST | | |
| *Triticum orientale* (Khorassan wheat) | EA-WEST | | |
| *Triticum persicum* (Persian wheat) | EA-WEST | | |
| *Triticum polonicum* (Polish wheat) | EA-MED; AFRICA | | |
| *Triticum spelta* (spelt) | EA-MED | | |
| *Triticum sphaerococcum* (shot wheat) | EA-CENT | | |
| *Triticum timopheevi* (Sanduri wheat) | EA-WEST | | |
| *Triticum turgidum* (poulard wheat) | EA-WEST; AFRICA | | |
| *Triticum vulgare* (soft wheat) | EA-WEST/CENT | | |
| | | | |
| *Zea mays* (soft corn) | MA | *Zea mays* (volunteer corn) | |

**Leguminoceae**

| | | | |
|---|---|---|---|
| *Arachis hypogaea* (peanuts) | SA | *Arachis hypogaea* | |
| | | | |
| *Cassia angustifolia* (senna) | EA-SE | *Cassia absus* | |
| | | *Cassia auriculata* | |
| | | *Cassia bauhinioides* | |
| | | *Cassia bicapsularis* | |
| | | *Cassia brewsterii* | |
| | | *Cassia chamaecrista* (showy partridgepea) | |
| | | *Cassia eremophilia* | |
| | | *Cassia fasciculata* (partridge pea) | |
| | | *Cassia fistula* | |
| | | *Cassia floribunda* | |
| | | *Cassia hirsute* | |
| | | *Cassia leiophlla* | |
| | | *Cassia letocarpa* | |
| | | *Cassia lechenaultiana* | |
| | | *Cassia ligustrina* | |
| | | *Cassia marilandica* (wild senna) | |
| | | *Cassia mimosoides* | |
| | | *Cassia newtonii* | |
| | | *Cassia nictitans* (partridge pea) | |
| | | *Cassia obtusa* | |
| | | *Cassia occidentalis* (coffee senna) | |
| | | *Cassia pleurocarpa* | |
| | | *Cassia senna* | |
| | | *Cassia sophera* | |
| | | *Cassia surattensis* | |
| | | *Cassia tomentosa* | |
| | | *Cassia tora* (sicklepod) | |
| | | | |
| *Glycine max* (soybeans) | EA-EAST/SE | *Glycine hedysaroides* | |
| | | *Glycine soja* | |
| | | | |
| *Lens esculenta* (large-seeded lentils) | EA-MED/WEST; AFRICA | *Lens lenticula* | |
| | | *Lens kotschyana* | |
| | | *Lens nigricans* | |
| | | *Lens orientalis* | |
| | | | |
| *Medicago sativa* (alfalfa) | EA-WEST | *Medicago agrestis* | |

|  |  |  |
|---|---|---|
|  |  | *Medicago arabica* (spotted medick) |
|  |  | *Medicago ciliaris* |
|  |  | *Medicago dentatus* |
|  |  | *Medicago denticulate* |
|  |  | *Medicago disciformis* |
|  |  | *Medicago falcate* |
|  |  | *Medicago granatensis* |
|  |  | *Medicago hispida* |
|  |  | *Medicago intertexa* |
|  |  | *Medicago lupulina* (black medic) |
|  |  | *Medicago maculata* |
|  |  | *Medicago minima* (bur medick) |
|  |  | *Medicago obscura* |
|  |  | *Medicago orbicularis* (blackdisk medick) |
|  |  | *Medicago polymorphus* (toothed medick) |
|  |  | *Medicago sativa* (lucerne) |
|  |  | *Medicago scutellata* (snail medick) |
|  |  | *Medicago tribuloides* |
|  |  | *Medicago turbinata* |
| *Melilotus officinalis* (yellow sweet clover) | EA-EUR/WEST | *Melilotus alba* (white sweet clover) |
| *Melilotus* spp. (sweet clovers) | EA-EUR/WEST | *Melilotus altissima* (tall yellow sweet clover) |
|  |  | *Melilotus indica* (Indian sweetclover) |
|  |  | *Melilotus messanensis* |
|  |  | *Melilotus officinalis* (yellow sweet clover) |
|  |  | *Melilotus segetalis* (corn melilot) |
|  |  | *Melilotus sicula* |
|  |  | *Melilotus sulcata* (Mediterranean sweetclover) |
| *Phaseolus aconitifolius* (mat beans) | EA-SOUTH | *Phaseolus aconitifolius* (mat beans) |
| *Phaseolus acutifolius* (tepary beans) | MA | *Phaseolus adenanthus* |
| *Phaseolus angularis* (Adzuki beans) | EA-EAST | *Phaseolus atropurpurens* |
| *Phaseolus aureus* (mung bean) | EA-SOUTH | *Phaseolus aureus* |
| *Phaseolus calcaratus* (rice bean) | EA-SOUTH | *Phaseolus lathyroides* |
| *Phaseolus cracalla* (bertoni bean) | SA | *Phaseolus linearis* |
| *Phaseolus lunatus* (lima beans) | SA/MA | *Phaseolus lunatus* |
| *Phaseolus multiflorus* (runner beans) | MA | *Phaseolus tribolus* |
| *Phaseolus mungo* (gram) | EA-SOUTH |  |
| *Phaseolus vulgaris* (common beans) | SA/MA; EA-EAST |  |
| *Pisum sativum* (garden pea) | EA-MED/WEST; AFRICA | *Pisum elatius* (wild pea) |
|  |  | *Pisum sativum* (garden pea) |
| *Trifolium alexandrinum* (berseem clover) | EA-MED | *Trifolium amabile* |
| *Trifolium incarnatum* (crimson clover) | EA-MED | *Trifolium agrarium* (yellow clover) |
| *Trifolium repens* (white clover) | EA-MED | *Trifolium angustifolium* (narrowleaf crimson clover) |
| *Trifolium resupinatum* (Persian clover) | EA-WEST | *Trifolium arvense* (hare's-foot) |
| *Trifolium* sp. (true clover) | EA-EUR | *Trifolium campestre* (low hop clover) |
|  |  | *Trifolium cherleri* (cupped clover) |
|  |  | *Trifolium dubium* (little hop clover) |
|  |  | *Trifolium elegans* |
|  |  | *Trifolium filiforme* |
|  |  | *Trifolium fimbriatum* (coast clover) |
|  |  | *Trifolium fragiferum* (strawberry clover) |
|  |  | *Trifolium glomeratum* |
|  |  | *Trifolium hybridum* |
|  |  | *Trifolium incarnatum* (crimson clover) |
|  |  | *Trifolium lappaceum* (burdock clover) |
|  |  | *Trifolium macraei* |

|  |  |  |
|---|---|---|
|  |  | *Trifolium medium* |
|  |  | *Trifolium montanum* |
|  |  | *Trifolium nigrescens* |
|  |  | *Trifolium pratense* |
|  |  | *Trifolium procumbens* (low hop clover) |
|  |  | *Trifolium purpureum* |
|  |  | *Trifolium repens* (white clover) |
|  |  | *Trifolium resupinatum* |
|  |  | *Trifolium scabrum* (rough clover) |
|  |  | *Trifolium stellatum* (starry clover) |
|  |  | *Trifolium strepens* |
|  |  | *Trifolium striatum* (soft trefoil) |
|  |  | *Trifolium subterraneum* (subterranean clover) |
|  |  | *Trifolium tomentosum* |
| *Vicia ervilia* (French erse, lentils) | EA-MED/WEST | *Vicia atropurpurea* |
| *Vicia faba* (faba bean) | EA-MED | *Vicia angustifolia* (narrowleaf vetch) |
| *Vicia pannonica* (Hungarian vetch) | EA-WEST | *Vicia articulata* (one-flower tare) |
| *Vicia sativa* (vetch) | EA-MED/WEST | *Vicia benghalensis* (reddish tufted vetch) |
| *Vicia villosa* (hairy vetch) | EA-WEST | *Vicia bithynica* (Bithynian vetch) |
|  |  | *Vicia calcarata* |
|  |  | *Vicia cracca* (tufted vetch) |
|  |  | *Vicia disperma* (European vetch) |
|  |  | *Vicia enuissima* (slender hare) |
|  |  | *Vicia ervilia* (blister vetch) |
|  |  | *Vicia faba* |
|  |  | *Vicia gracilis* |
|  |  | *Vicia graminea* |
|  |  | *Vicia grandifolia* (large yellow vetch) |
|  |  | *Vicia hirsute* (hairy tare) |
|  |  | *Vicia hybrid* |
|  |  | *Vicia lathyroides* |
|  |  | *Vicia lutea* (yellow vetch) |
|  |  | *Vicia monantha* |
|  |  | *Vicia monanthos* |
|  |  | *Vicia narbonensis* (French vetch) |
|  |  | *Vicia onobrychioides* |
|  |  | *Vicia peregrina* (wandering vetch) |
|  |  | *Vicia pannonica* (Hungarian vetch) |
|  |  | *Vicia sativa* (vetch) |
|  |  | *Vicia sepium* (bush vetch) |
|  |  | *Vicia sibthorpii* |
|  |  | *Vicia tenuifolia* |
|  |  | *Vicia tetrasperma* (smooth tare) |
|  |  | *Vicia villosa* (hairy vetch) |

**Liliaceae**

| | | |
|---|---|---|
| *Allium bakeri* (Chinese shallot) | EA-EAST | *Allium angulosum* |
| *Allium cepa* (onion) | EA-EUR/MED/WEST | *Allium ampeloprasum* (roundheaded garlic) |
| *Allium chinense* (Chinese onion) | EA-EAST | *Allium canadense* (wild onion) |
| *Allium fistulosum* (Japanese leek) | EA-EAST | *Allium cepa* |
| *Allium kurrat* (Egyptian kurrat) | EA-MED | *Allium flavum* |
| *Allium macrostemon* (syao-suan) | EA-EAST | *Allium macrostemon* |
| *Allium pekinense* | EA-EAST | *Allium neapolitanum* (ail de Naples) |
| *Allium porrum* (leek) | EA-EUR/MED/WEST | *Allium nigrum* (black onion) |
| *Allium ramosum* (Chinese leek) | EA-EAST | *Allium oleraceum* (field garlic) |
| *Allium sativum* (garlic) | EA-EUR/MED | *Allium roseum* |
| *Allium* sp. (shallots) | AFRICA | *Allium rotundum* (ail arrondi) |
|  |  | *Allium rubellum* |

|  |  |  |
|---|---|---|
|  |  | *Allium sativum* |
|  |  | *Allium sphaerophalon* (round-headed leek) |
|  |  | *Allium subhirsutum* |
|  |  | *Allium textile* (wild onion) |
|  |  | *Allium tricoccum* (wild leek) |
|  |  | *Allium triquetrum* |
|  |  | *Allium vineale* (crow garlic) |

**Malvaceae**

| | | |
|---|---|---|
| *Abutilon avicennae* (Chinese jute) | EA-EAST | *Abutilon asiaticum* |
| | | *Abutilon fruticosum* |
| | | *Abutilon glaucum* |
| | | *Abutilon graveolens* |
| | | *Abutilon indicum* |
| | | *Abutilon molle* |
| | | *Abutilon oxycarpum* |
| | | *Abutilon striatum* |
| | | *Abutilon theophrasti* (velvetleaf) |
| | | *Abutilon trisulcatum* |
| | | |
| *Gossypium arboretum* (cotton) | EA-SE | *Gossypium tomentosum* |
| *Gossypium barbadense* (Egyptian cotton) | SA | |
| *Gossypium herbaceum* (African cotton) | AFRICA | |
| *Gossypium hirsutum* (cotton) | MA | |
| *Gossypium mustelinum* (cotton) | SA | |
| *Gossypium tomentosum* (cotton) | Hawaii | |
| | | |
| *Hibiscus cannabinus* (kenaf fiber) | EA-SE/SOUTH; AFRICA | *Hibiscus abelmoschus* |
| *Hibiscus esculentus* (bamia; okra) | AFRICA | *Hibiscus articulates* |
| *Hibiscus sabdariffa* (roselle) | AFRICA | *Hibiscus cannibinus* |
| | | *Hibiscus elatus* |
| | | *Hibiscus esculentus* (bamia) |
| | | *Hibiscus trionum* (venice mallow) |
| | | *Hibiscus ficulneus* |
| | | *Hibiscus lasiocarpus* |
| | | *Hibiscus mastersianus* |
| | | *Hibiscus micranthus* |
| | | *Hibiscus obtusilobus* |
| | | *Hibiscus palustris* |
| | | *Hibiscus panduraeformis* |
| | | *Hibiscus rosa-sinensis* |
| | | *Hibiscus rugosus* |
| | | *Hibiscus sabdariffa* |
| | | *Hibiscus tetraphyllus* |
| | | *Hibiscus tiliaceus* |
| | | *Hibiscus vitifolius* |
| | | |
| *Malva sylvestris* (high mallow) | EA-EAST | *Malva crispa* |
| *Malva verticillata* (vegetable mallow) | EA-EAST | *Malva hispanica* |
| | | *Malva moschata* (musk mallow) |
| | | *Malva neglecta* (common mallow) |
| | | *Malva nicaeensis* |
| | | *Malva parviflora* |
| | | *Malva pusilla* |
| | | *Malva rotundifolia* |
| | | *Malva sylvestris* |
| | | *Malva verticillata* |

**Oxalidaceae**

| | | |
|---|---|---|
| *Oxalis tuberosa* (oca) | SA | *Oxalis acetosella* |
| | | *Oxalis anthelmintica* |
| | | *Oxalis articulate* |
| | | *Oxalis bahiensis* |
| | | *Oxalis barrelieri* |
| | | *Oxalis bowiei* |
| | | *Oxalis cernua* |
| | | *Oxalis compressa* |
| | | *Oxalis cordobensis* |
| | | ***Oxalis corniculata* (creeping woodsorrel)** |
| | | *Oxalis corymbosa* (pink woodsorrel) |
| | | *Oxalis debilis* |
| | | *Oxalis dilenii* (gray-green woodsorrel) |
| | | *Oxalis dombeyi* |
| | | *Oxalis europeaea* (yellow wood sorrel) |
| | | *Oxalis flava* |
| | | *Oxalis florida* (yellow wood sorrel) |
| | | *Oxalis latifolia* (trebol) |
| | | *Oxalis mallobolba* |
| | | *Oxalis martiana* |
| | | *Oxalis micrantha* |
| | | *Oxalis neaei* |
| | | *Oxalis obliquifolia* |
| | | *Oxalis pes-caprae* (buttercup oxalis) |
| | | *Oxalis purpurata* |
| | | *Oxalis purpurea* |
| | | *Oxalis repens* |
| | | *Oxalis rubra* |
| | | *Oxalis semiloba* |
| | | *Oxalis sepium* |
| | | *Oxalis stricta* (upright yellow wood sorrel) |
| | | *Oxalis violacea* |
| **Papaveraceae** | | |
| *Papaver somniferum* (opium poppy) | EA-EUR/WEST/EAST | *Papaver aculeatum* |
| | | *Papaver argemone* (long prickly-headed poppy) |
| | | *Papaver dubium* (long-headed poppy) |
| | | *Papaver hybridum* (round prickly-headed poppy) |
| | | *Papaver lecoqii* (yellow-juiced poppy) |
| | | *Papaver pinnatifidum* (Mediterannean poppy) |
| | | *Papaver rhoeas* (corn poppy) |
| | | *Papaver somniferum* |
| **Plantaginaceae** | | |
| *Plantago psyllium* (medicine psyllium) | EA-EUR | *Plantago afra* (glandular plantain) |
| | | *Plantago arenaria* (whorled plantain) |
| | | *Plantago aristata* (bracted plantain) |
| | | *Plantago asiatica* |
| | | *Plantago bicallosa* |
| | | *Plantago camtschatica* |
| | | *Plantago coronopus* (buck's horn plantain) |
| | | *Plantago hirtella* |
| | | *Plantago indica* |
| | | *Plantago lagopus* (Mediterranean plantain) |
| | | ***Plantago lanceolata* (buckhorn plantain)** |
| | | ***Plantago major* (broadleaf plantain)** |
| | | *Plantago media* (hoary plantain) |
| | | *Plantago patagonica* (wooly plantain) |
| | | *Plantago psyllium* |

*Plantago purshii* (woolly plantain)
*Plantago pusilla*
*Plantago rugelii* (blackseed plantain)
*Plantago serraria* (toothed plantain)
*Plantago tomentosa*
*Plantago varia*
*Plantago virginica* (hoary plantain)

**Polygonaceae**

| | | |
|---|---|---|
| *Fagopyrum esculentum* (buckwheat) | EA-EAST | *Fagopyrum esculentum* (buckwheat) |
| *Fagopyrum tataricum* (tatar buckwheat) | EA-EAST | *Fagopyrum tataricum* (tatar buckwheat) |
| *Polygonum erectum* (knotweed) | NA-EAST | *Polygonum acre* |
| *Polygonum hydropiper* (vegetable knotweed) | EA-EAST | *Polygonum acuminatum* |
| *Polygonum tinctorium* (Chinese indigo) | EA-EAST | *Polygonum alatum* |

*Polygonum achoreum* (striate knotweed)
*Polygonum amphibium* (water smartweed)
*Polygonum aquisetiferme*
*Polygonum arenostrum* (common knotweed)
*Polygonum argyrocoleon* (silversheath knotweed)
*Polygonum aviculare* (prostrate knotweed)
*Polygonum barbatum*
*Polygonum bellardia*
*Polygonum bistorta*
*Polygonum bistortoides*
*Polygonum Chinese*
*Polygonum cilinode*
*Polygonum coccineum* (swamp smartweed)
*Polygonum conspicuum*
**Polygonum convolvulus (wild buckwheat)**
*Polygonum cuspidatum* (Japanese knotweed)
*Polygonum densiflorum*
*Polygonum dumetorum*
*Polygonum erecto-minus*
*Polygonum erectum* (erect knotweed)
*Polygonum flaccidum*
*Polygonum fugax*
*Polygonum glabrum*
*Polygonum hastato-sagittatum*
*Polygonum higegaweri*
*Polygonum hydropiper* (marshpepper smartweed)
*Polygonum hydropiperoides* (mild water pepper)
*Polygonum japonicum*
*Polygonum lapathifolium* (pale smartweed)
*Polygonum linicola*
*Polygonum longisetum*
*Polygonum maackianum*
*Polygonum maritime*
*Polygonum minus*
*Polygonum mite*
*Polygonum neglectum*
*Polygonum nepalense*
*Polygonum nipponense*
*Polygonum orientale* (princes-feather)
*Polygonum pensylvanicum* (Pennsylvania smartweed)
*Polygonum perfoliatum*
*Polygonum persicaria* (ladysthumb smartweed)
*Polygonum plebeium*
*Polygonum pubescens*

|  |  |  |
|---|---|---|
|  |  | *Polygonum punctatum* |
|  |  | *Polygonum ramosissimum* (bushy knotweed) |
|  |  | *Polygonum sachalinense* |
|  |  | *Polygonum salicifolium* |
|  |  | *Polygonum scabrum* |
|  |  | *Polygonum scandens* (hedge bindweed) |
|  |  | *Polygonum segetum* |
|  |  | *Polygonum senegalense* |
|  |  | *Polygonum senticosum* |
|  |  | *Polygonum serrulatum* |
|  |  | *Polygonum setaceum* (bog smartweed) |
|  |  | *Polygonum sieboldii* |
|  |  | *Polygonum taquettii* |
|  |  | *Polygonum thunbergii* |
|  |  | *Polygonum viscosum* |

**Portulacaceae**

| | | |
|---|---|---|
| *Portulaca oleracea* (purslane) | EA-EUR/MED/WEST | *Portulaca cryptopetala* |
|  |  | *Portulaca formosana* |
|  |  | *Portulaca grandifolia* |
|  |  | *Portulaca lanceolata* |
|  |  | **Portulaca oleracea (common purslane)** |
|  |  | *Portulaca pilosa* |
|  |  | *Portulaca quadrifida* |

**Rosaceae**

| | | |
|---|---|---|
| *Crataegus pubescens* (tejocote) | MA | *Crataegus* sp. (hawthorne) |
| *Crataegus* sp. (hawthorne) | EA-EUR | *Crataegus crus-galli* |
|  |  | *Crataegus douglasii* |
|  |  | *Crataegus marshallii* |
|  |  | *Crataegus oxyacantha* |
|  |  | *Crataegus rivularis* |
|  |  | *Crataegus saligna* |
|  |  | *Crataegus succulenta* |
| *Rosa canina* (rosehip) | EA-EUR | *Rosa arkansana* (prairie wildrose) |
| *Rosa centifolia* (cabbage rose) | EA-WEST | *Rosa bracteata* |
| *Rosa damascene* (Damascene rose) | EA-MED | *Rosa californica* |
|  |  | *Rosa canina* |
|  |  | *Rosa eglanteria* |
|  |  | *Rosa gymnocarpa* |
|  |  | *Rosa laevigata* |
|  |  | *Rosa multiflora* (multiflora rose) |
|  |  | *Rosa nutcana* |
|  |  | *Rosa rubiginosa* |
|  |  | *Rosa sulphurea* |
|  |  | *Rosa tomentosa* |
|  |  | *Rosa woodsii* (western wildrose) |

**Solanaceae**

| | | |
|---|---|---|
| *Capsicum annuum* (chile pepper) | MA/SA | *Capsicum annuum* |
| *Capsicum baccatum* (pepper) | SA | *Capsicum baccatum* |
| *Capsicum chinense* (pepper) | SA | *Capsicum frutescens* |
| *Capsicum frutescens* (chile pepper) | MA/SA |  |
| *Capsicum pubescens* (pepper) | SA |  |
| *Datura meteloides* (jimsonweed) | NA-SW | *Datura arborea* |
| *Datura stramonium* (jimson/loco weed) | MA/SA | *Datura discolor* |
|  |  | *Datura ferox* (Chinese thorn-apple) |

|  |  |  |
|---|---|---|
|  |  | *Datura innoxia* (sacred datura) |
|  |  | *Datura leichhardtii* |
|  |  | *Datura metel* |
|  |  | *Datura meteloides* |
|  |  | *Datura stramonium* (jimsonweed) |
| *Lycopersicum cerasiforme* (cherry tomato) | MA | *Lycopersicum hirsutum* (wild tomato) |
| *Lycopersicum esculentum* (bush tomatoes) | SA |  |
| *Nicotiana attenuate* (tobacco) | NA-SW | *Nicotiana alata* |
| *Nicotiana bigelovii* (tobacco) | NA-SW | *Nicotiana glauca* |
| *Nicotiana otophora* (tobacco) | SA | *Nicotiana glutinosa* |
| *Nicotiana paniculata* (tobacco) | SA | *Nicotiana longiflora* |
| *Nicotiana rustica* (Aztec tobacco) | SA/MA/NA-EAST | *Nicotiana paniculata* |
| *Nicotiana sylvestris* (tobacco) | SA | *Nicotiana suaveolens* |
| *Nicotiana tabaccum* (tobacco) | SA/MA | *Nicotiana tabacum* |
| *Nicotiana tomentosiformis* (tobacco) | SA | *Nicotiana trigonophylla* |
| *Nicotiana trigonophylla* (tobacco) | NA-SW |  |
| *Nicotiana undulata* (tobacco) | SA |  |
| *Physalis aequata* (tomatillo) | MA | *Physalis alkekengi* (Chinese lantern plant) |
| *Physalis ixocarpa* (husk tomato) | MA | *Physalis angulata* (cutleaf groundcherry) |
| *Physalis peruvianum* (uchuba berry) | SA | *Physalis divaricata* |
|  |  | *Physalis heterophylla* (clammy groundcherry) |
|  |  | *Physalis ixocarpa* |
|  |  | *Physalis lagascae* |
|  |  | *Physalis lanceolata* |
|  |  | *Physalis lobata* |
|  |  | *Physalis longifolia* (longleaf groundcherry) |
|  |  | *Physalis macrophysa* |
|  |  | *Physalis mendocina* |
|  |  | *Physalis micrantha* |
|  |  | *Physalis minima* (pigmy groundcherry) |
|  |  | *Physalis nicandroides* |
|  |  | *Physalis peruviana* (Peruvian groundcherry) |
|  |  | *Physalis pubescens* (husk tomato) |
|  |  | *Physalis solanacae* |
|  |  | *Physalis subglabrata* (smooth groundcherry) |
|  |  | *Physalis turbinata* |
|  |  | *Physalis virginiana* (Virginia groundcherry) |
|  |  | *Physalis viscosa* (starhair groundcherry) |
|  |  | *Physalis wrightii* (Wright groundcherry) |
| *Solanum andigenum* (Andean potatoes) | SA | *Solanum aculeastrum* |
| *Solanum aethiopicum* (African tomato) | AFRICA | *Solanum alatum* |
| *Solanum incanum* (bitter tomato) | AFRICA | *Solanum americanum* (American nightshade) |
| *Solanum macrocarpon* (nightshade) | AFRICA | *Solanum atropurpureum* |
| *Solanum melongena* (eggplant) | EA-EAST/SOUTH | *Solanum auriculatum* |
| *Solanum muricatum* (melon pear) | SA | *Solanum balbisii* |
| *Solanum phureja* (potatoes) | SA | *Solanum biflorum* |
| *Solanum quitoense* (naranjilla) | SA | *Solanum bonariense* |
| *Solanum topiro* (cocona) | SA | *Solanum capsicastrum* |
| *Solanum tuberosum* (common potatoes) | SA | *Solanum caribaeum* |
|  |  | *Solanum carolinense* (horsenettle) |
|  |  | *Solanum chacoense* |
|  |  | *Solanum chenopodioides* |
|  |  | *Solanum ciliatum* |
|  |  | *Solanum cinereum* |
|  |  | *Solanum commersonii* |

*Solanum diflorum*
*Solanum dubium*
*Solanum dulcamara* (bitter nightshade)
*Solanum elaeagnifolium* (silver nightshade)
*Solanum ellipticum*
*Solanum esuriale*
*Solanum flagellare*
*Solanum glaucum*
*Solanum gracile*
*Solanum grossedentatum*
*Solanum hamulosum*
*Solanum hirtum*
*Solanum hispidum*
*Solanum hystrix*
*Solanum incanum*
*Solanum indicum*
*Solanum jamaicense*
*Solanum laciniatum*
*Solanum lepidotum*
*Solanum mammosum*
*Solanum marginatum*
*Solanum mauritianum*
*Solanum melongena*
*Solanum miniatum*
**Solanum nigrum (black nightshade)**
*Solanum nodiflorum*
*Solanum orthocarpum*
*Solanum panduriforme*
*Solanum paniculatum*
*Solanum photeinocarpum*
*Solanum poeppigianum*
*Solanum psuedocapsicum*
*Solanum ptycanthum* (eastern black nightshade)
*Solanum pygmaeum*
*Solanum retroflexum*
*Solanum rostratum* (buffalobur)
*Solanum seaforthianum*
*Solanum semiarmatum*
*Solanum sisymbrifolium*
*Solanum sodomenum*
*Solanum stramoniifolium*
*Solanum sturtianum*
*Solanum subinerme*
*Solanum surattense*
*Solanum sarachoides* (cupped nightshade)
*Solanum torvum* (turkeyberry)
*Solanum triflorum* (cutleaf nightshade)
*Solanum tuberosum*
*Solanum verbascifolium*
*Solanum villosum* (hairy nightshade)
*Solanum welwitschii*
*Solanum xanthocarpum*

**Tiliaceae**

| | | |
|---|---|---|
| *Corchorus capsularis* (jute) | EA-SE; AFRICA | *Corchorus aestuans* |
| *Corchorus olitorius* (tossa jute) | AFRICA | *Corchorus anolensis* |
| | | *Corchorus antichorus* |
| | | *Corchorus argutus* |
| | | *Corchorus capsularis* |

|  |  | *Corchorus depressus* |
|  |  | *Corchorus fascicularia* |
|  |  | *Corchorus hirsutus* |
|  |  | *Corchorus hirtus* |
|  |  | *Corchorus olitorius* (tossa jute) |
|  |  | *Corchorus orinocensis* |
|  |  | *Corchorus siliquosus* |
|  |  | *Corchorus tridens* |
|  |  | Corchorus trilocularis |

**Umbelliferae**

| Crop species | Location | Weed species |
|---|---|---|
| *Daucus carota* (carrot) | EA-EUR/WEST | *Daucus aureus* (golden carrot) |
|  |  | *Daucus carota* (Queen Anne's lace; wild carrot) |
|  |  | *Daucus crinitus* (long haired daucus) |
|  |  | *Daucus glochidiatus* |
|  |  | *Daucus guttatus* |
|  |  | *Daucus maximus* |
|  |  | *Daucus montanus* |
|  |  | *Daucus montevidensis* |
|  |  | *Daucus muricatus* |
|  |  | *Daucus pussillus* (rattlesnake weed) |
|  |  | *Daucus setulosus* |

**Vitaceae**

| Crop species | Location | Weed species |
|---|---|---|
| *Vitis amurensis* (black Amur grape) | EA-EAST/CENT | *Vitis aestivalis* |
| *Vitis labrusca* (fox grape) | NA-EAST | *Vitis candicans* |
| *Vitis vinifera* (grape) | EA-MED | *Vitis hastate* |
|  |  | *Vitis rotundifolia* |
|  |  | *Vitis rupestris* |
|  |  | *Vitis tiliaefolia* |
|  |  | *Vitis trifolia* |
|  |  | *Vitis vulpina* |

| Plant Family | Crop |  |
|---|---|---|
| Crop species (common name) | Location | Weed species (common name) |
| **ORIGINAL CROP SPECIES** |  | **CONTEMPORARY WEED SPECIES** |

Many, if not most, of the important weed genuses of the world are also among the top domesticated crop species (table 1.5). Of the 30 major crops listed in Harlan (1992), 24 are included in table 1.5. The plant genus is listed only once in table below, but many have crop members in several different categories (e.g. *Brassica, Cannabis, Solanum, Vicia*).

Some crop species have been abandoned since their early cultivation by humans. Notable is *Setaria* sp. which was the major cereal crop in Mesoamerica before the use of maize, and before the widespread adoption of rice in East Asia (Austin, 2006). Domestication and introgression with wild-weed relatives seems to have made it ideally suited as a global invasive weed species since that abandonment.

**Table 1.5** Plant crop-weed species-groups genuses (table 1.4) classified within several domesticated crop categories (Harlan, 1992). Genus in **BOLD**: world's 30 leading food crops in terms of estimated edible dry matter (Harlan, 1992).

| **Cereals** | **Drugs** | **Fiber** |
|---|---|---|
| *Avena* (oat) | *Cannabis* (marijuana) | *Abutilon* (Chinese jute) |
| *Fagopyron* (buckwheat) | *Cassia* (senna) | *Corchorus* (jute) |
| *Hordeum* (barley) | *Datura* (jimsonweed) | *Gossypium* (cotton) |
| *Oryza* (rice) | *Nicotiana* (tobacco) | *Hibiscus* (kenaf; okra) |
| *Secale* (rye) | *Papaver* (opium) | |
| *Sorghum* | *Plantago* (medicine psyllium) | |
| *Triticum* (wheat) | *Rosa* (rosehip) | |
| *Zea* (maize) | | |
| **Forages** | **Fruits & Nuts** | **Millets** |
| *Agropyron* (wheatgrass) | *Arachis* (peanut) | *Brachiaria* (guinea millet) |
| *Cynodon* (Bermuda grass) | *Capsicum* (chile pepper) | *Digitaria* (hungary rice millet) |
| *Lolium* (ryegrass) | *Crataegus* (tejocote) | *Echinochloa* (Japanese millet) |
| *Medicago* (alfalfa) | *Curcurbita* (squash) | *Eleusine* (finger millet) |
| *Melilotus* (sweet clover) | *Lycopersicum* (tomato) | *Panicum* (broomcorn millet) |
| *Phalaris* (reed canary grass) | *Mormordica* (bitter gourd) | *Papsalum* (slender millet) |
| *Trifolium* (clover) | *Opuntia* (prickly pear) | *Pennisetum* (pearl millet) |
| *Tripsacum* (forage) | *Physalis* (tomatillo) | *Setaria* (foxtail millet) |
| | *Vitis* (grape) | **Millets** (as a group) |
| **Oil** | **Psuedocereals** | **Pulses** |
| *Brassica* (oilseed rape) | *Ambrosia* (ragweed) | *Glycine* (soybean) |
| *Helianthus* (sunflower) | *Amaranthus* (amaranth) | *Lens* (lentils) |
| | *Chenopodium* (quinoa) | *Phaseolus* (beans) |
| | *Iva* (swampweed) | *Pisum* (peas) |
| | *Polygonum* (knotweed) | *Vicia* (vetch) |
| **Root & Tuber** | **Sugar** | **Vegetables & Spices** |
| *Beta* (beet) | *Saccharum* (sugar cane) | *Allium* (onion) |
| *Colocasia* (taro) | | *Asparagus* |
| *Cyperus* (earth-almonds) | | *Chrysanthemum* (vegetable) |
| *Daucus* (carrot) | | *Cleome* (Rocky Mt. beeweed) |
| *Eleocharis* (water chestnut) | | *Lactuca* (lettuce) |
| *Ipomoea* (sweet potato) | | *Lepidium* (gardencress) |
| *Oxalis* (oca) | | *Malva* (vegetable malva) |
| *Raphanus* (radish) | | *Portulaca* (purslane) |
| *Sagittaria* (elephant ear) | | *Sinapis* (mustard) |
| *Solanum* (potato) | | *Tagetes* (marigold) |
| | | *Xanthium* (vegetable cocklebur) |
| | | *Yucca* (vegetable yucca) |

The goal of the remainder of the book is to explain how and why these weed species became what they are now, to us. The answer, as for all of life, is evolution: natural selection and adaptation.

Let's take a look.

## UNIT 2: THE EVOLUTION OF WEED POPULATIONS

"Adaptation is a word too loosely used in ecological writing. Often to say that a feature of an organism's life or form is adaptive is to say no more than that the feature appears to be a good thing, judged on the basis of an anthropomorphic attitude to the problems that the organism is seen to face. More accurately, adaptations are those features of an organism that in the past improved the fitness of its ancestors and so were transmitted to descendants. Adaptation is always retrospective. Fitness itself is relative - it is defined by the numbers of descendants left by an individual relative to its fellows. An organism will be more fit if its activities reduce the number of descendants left by neighbors, even if the activities do nothing to the number of descendants that it itself leaves. The point is easily made by considering the evolution of height in plants. Within a population of plants growing densely and absorbing the larger part of incident light, success depends on placing leaves high in the canopy and shading and suppressing neighbours. There is no intrinsic advantage to the individual from being high (there are some real disadvantages in the amount of non-reproductive tissue to be supported), only an advantage from being higher than neighbours. It is being higher, not just high, that pays. Similarly a genetic change that gave a plant a larger and earlier root system might bring no advantages to the possessor other than the relative advantage over the neighbors that it is able to deprive. If an activity of an organism brings no direct benefit but hinders the chance the neighbors will leave descendants, the activity will increase fitness - it will be "adaptive".

This argument may be important in understanding evolutionary processes. Often the process is seen as in some way optimizing the behavior of descendants - in some way making them "better" or "adjusted to the environment". There is in fact nothing innate in a process that maximizes evolutionary fitness, that necessarily "optimizes" physiological function. Indeed a genetic change that resulted in an organism immobilizing mineral nutrients in old tissue until it died instead of returning them to the cycle within the ecosystem would almost certainly confer fitness provided that potentially competing neighbors were deprived of needed nutrients by this activity.

A theory of natural selection that is based on the fitness of individuals leaves little room for the evolution of populations or species towards some optimum, such as better use of environmental resources, higher productivity per area of land, more stable ecosystems, or even for the view that plants in some way become more efficient than their ancestors. Instead, both the study of evolutionary processes and of the natural behaviour of populations suggest that the principles of "beggar my neighbor" and "I'm all right Jack" dominate all and every aspect of evolution. Nowhere does this conclusion have more force than when man takes populations that have evolved in nature under criteria of individual fitness, grows them in culture as populations and then applies quite different criteria of performance - productivity per unit area of land. Natural selection is about individuals and it would be surprising if the behavior that favoured one individual against another was also the behavior that maximized the performance of the population as whole. For this to happen, selection would have to act on groups. It is an interesting thought that group selection which is believed to be extremely rare or absent in nature (Maynard Smith, 1964) may be the most proper type of selection from improving the productivity of crop and forest plants. Plant breeding would then be concerned to undo the results of selection for selfish qualities of individual fitness and focus on the performance of populations." J.L. Harper, 1977

**The weeds' story**. I am foxtail.

We conquered the earth together. You gathered my seeds, the ones you eat, you left the rest. In the spring you killed and tilled and made me a home in the soil. You planted the ones of me you liked, you left the rest of me buried in the ground. My seedlings emerged. You could not tell us apart. We grew, we had sex with ourselves, then you took our children. You gathered and ate the ones you liked: the biggest, the easy to germinate, the ones clinging to my seedheads. You tried to kill the rest: the ones you could see. You rid me of the weak. I inherited the strong: those hiding in the ground, the last to emerge, those readily falling to earth. You gave me opportunity. Despite yourself, you created me in your image. You eat me, kill me, you make me strong. I couldn't have done it without you.

**Summary**. Rev. Thomas Robert Malthus (1798-1826) proposed that human population growth was limited, expansion prevented by famine and disease. Pierre François Verhulst published a logistic function of human population growth. Malthus as a primary stimulus to Darwin and Wallace conceptualizing natural selection and elimination as the driver of evolution. Verhulst providing a mathematical model of population growth. Both were founded on human population biology. Humans, animals, differ from plant populations, mitigating against the use of the logistic function to describe population growth. By means of phenotypic plasticity and somatic polymorphism traits, plant architecture is capable of generating a very wide range of sizes of the individual. Population structure also differs from that of animals: plants don't move, animals can move to exploit opportunity elsewhere. These two crucial aspects of plants compromise the utility of the logistic equation to represent plant and weed populations.

Plant communities assemble when opportunity in a locality favors functional traits in a new species at a particular time. Local opportunity spacetime is the integration of available resources and conditions, neighbor interactions and disturbance in that local niche. The population of this species must first be able to disperse into the habitat, then colonize the site by completing its life cycle from recruitment to reproduction. Lastly it must endure at that site, with individuals selected or eliminated. Without performing all of these processes invasion is not successful. Without all these processes evolution of the population will not occur. For weeds, disturbance in the broadest sense is what creates this original opportunity. All local populations eventually go extinct.

Weed communities assemble in a new location seizing and exploiting local opportunity spacetime. The invading members of a weed species have preadapted traits that allow them to colonize the site. Once established they mate, recombining their genes in meiosis, dispersing their variable progeny into the habitat. The new genetic offspring interact with the opportunity provided by the local environment and produce new phenotypes which live or die. The generation of new weed genotype-phenotype genetic variation is required for natural selection and elimination to drive local adaptation of the population. Generation of genetic variation appropriate for exploiting a particular locality allow an individual weed species to seize available opportunity spacetime in an efficient manner. The mating system, or mode of fertilization, of a weed species is crucial to the generation of appropriate amounts and types of genotypic variants. The mating system of a plant is the mechanism creating genetic diversity among its progeny. Parent plants generate variable genotypes producing variable phenotypes depending on environmental conditions. Phenotypic plasticity is the capacity of an organism to vary as a result of environmental opportunity. Phenotypic variation is also expressed in somatic polymorphisms. Weed species possess the ability to express phenotypic variation from a single genotype. These multiple interactions of genes are called epistasis. This first process of natural selection generates the phenotypic variation needed for the second process resulting in local adaptation: differential survival, reproduction and inheritance.

Heterogeneous weed populations invade, colonize and endure in a locality by surviving their developmental life history, culminating in reproduction of offspring inheriting their adaptive traits. Survival and reproduction can be represented demographically with the net reproductive rate, the summation of survival times fecundity at each age of the individual's life history. The optimum age for individual reproduction depends on how reproduction at a particular age would affect later survival and reproduction. Mortality drives the time of reproduction. Opportunity spacetime drives mortality. Evolution shapes the timing of reproduction by a careful compromise of fecundity optimization and mortality risk minimization. The last step in natural selection is heredity, the transmission of specific characters or traits, genes and genetic information, from the ancestral parent to the offspring: the basis of adaptive evolution. New or altered traits in individuals can be produced by mutations, the transfer of genes between populations or between species, or by genetic recombination in sexual reproduction. The transmission of genes to the next generation is accomplished in weeds by their mating system. The mating system of a successful weed species generates appropriate amounts and types of genotypic variants among its progeny. Mating systems indirectly control species diversity, population genetic structure at all spatial levels, community diversity and assembly with the dispersal of these offspring. The search for opportunity spacetime is accomplished with genetic variability expressed as phenotypic heterogeneity within populations of a species. A local population relies on its mating system to adjust genetic recombination to the fabric of the exploited disturbed space. Mating systems for colonizing species varies from obligate outcrossing (e.g. monoecy, dioecy) to self-pollinating and apomictic species.

## Chapter 2: Evolution, Natural Selection and Weedy Adaptation

**The weeds' story**. I am foxtail.

We conquered the Earth together. We evolved together. At first you wanted all of me, unconditionally. Then you picked your favorites, the quick to germinate, the easiest to gather, the tastiest. You gathered my seeds, stripped the ground of everything and planted these favorites. The rest of us un-favorites lay quietly hidden beneath you and waited. Some of us were impatient, we died. Some of us waited, and lived.

You grew fat, prosperous on our bounty. You became greedy, planting more land. You killed each other, greedy for our bounty. You killed us, ensuring your bounty. In return I gave you civilization in China, in India, in Mesopotamia, in the Balkans, in America. We all grew stronger, fatter, richer, greedier as we endured.

Nature enriched our opportunities in some places, removing it in others. We grew, had sex, changed, moved, died. Those you ate became fatter, docile to your needs, clinging tightly to our seedhead nursery. Those of us you tried to kill became smaller, leaner, able to hide, growing in your fields when you weren't looking. We had better sex. Our children inherited these new ways of living, escaping your relentless killing. We became better at spreading, better at enduring. We conquered the earth together.

**Summary**. Rev. Thomas Robert Malthus, a Church of England curate, published 'An Essay on the Principle of Population' (1798-1826) in which he proposed that human population growth was limited, expansion prevented by famine and disease. He proposed this gloomy outlook in opposition to the popular 18$^{th}$ century European view that society was improving and in principle perfectible.

Pierre François Verhulst was a Belgian mathematician of number theory at the University of Ghent who published a logistic function or logistic curve, a common special case of the more general sigmoid function, and its relationship to human population growth. In the initial stage its growth is approximately exponential; then, as saturation begins, the growth slows, and at maturity, growth stops. This logistic function presented the important concepts of intrinsic growth rate and the carrying capacity (numbers of individuals) of a local environment.

The ideas of these two men had large impacts on future science: Malthus as a primary stimulus to Darwin and Wallace conceptualizing natural selection and elimination as the driver of evolution; and Verhulst providing a mathematical model of population growth. Both were revolutionary ideas. Both created confusion as their ideas were subsequently adopted by evolutionary biology. For our present story of weeds confusion was generated by the fact that both these ideas were founded on human population biology. Humans, animals, differ from plant populations in several important ways that mitigate against the use of the logistic function to describe population growth. First, the plastic architecture of a plant is capable of generating a very wide range of sizes of the individual by means of phenotypic plasticity and somatic polymorphism traits. Population structure also differs from that of animals: plants don't move, animals can move to exploit opportunity elsewhere. These two crucial aspects of plants compromise the utility of the logistic equation to represent plant and weed populations. This confusion fully explored in subsequent chapters (e.g. Unit 6). This confusion does not lessen the importance of these ideas, but it obligates us to look upon demographic representations of weed biology with caution.

## 2.1 Introduction

"All the world's a stage,
And all the men and women merely players.
They have their exits and their entrances,
And one man in his time plays many parts …"
Shakespeare, W. (1623)

The nature of weeds is fully revealed in the story of the evolutionary processes that determine their adaptive form and behavior. The primary character in this story is human, the antagonist. Human activity affects directly or indirectly much of the arable land of the earth. Agricultural fields, forests, residential spaces, urban landscapes, industrial factory tracts, land-air-water transportation systems, golf courses, parks, school yards and amusement parks are all managed habitats disturbed directly and indirectly by human activity, whether these actions are intended or not. The atmosphere is filled with the gases humans generate and oceans, rivers and lakes are home to the waste products humans flush into them. The community structure dominated by herbivores and predators that once balanced the abundant plant life of continental ecosystems are gone. The balance now is the balance of the human omnivore.

In all these changes the human antagonist has become the administrator of natural selection in those habitats. Managed habitats fail to utilize all the resources made available by human disturbance and thereby create opportunities for other organisms to exploit at particular times (opportunity spacetime).

The protagonists of this story are the weedy organisms humans create by their actions to control and manage nature: weeds are us. The weedy organisms that seize and exploit these opportunities do so because their phenotypes possess traits that are well suited to those localities and conditions. Natural selection in, and adaptation to, these opportunity spacetimes are the fundamental means by which weedy plants evolve and exploit the changing conditions they are confronted with in disturbed habitats. This chapter explores the fundamental forces and processes that form weed communities and drive the appearance and changes in the weeds we have created.

The story of weed evolution unfolds in a local setting, the activities of the population lead to adaptive resolutions in future generations. The setting of weed evolution, the stage upon which diversifying evolution takes place, is a local population of variable phenotypes of a weed species in a particular locality. The nature of a particular locality is defined by the opportunity spacetime it affords the weed population to survive and reproduce. Opportunity spacetime of a local habitat for a population is defined by its available resources and conditions, interactions with neighbors, and disturbances: the local niche.

The action and characterization of the weed evolution story is divided into two linked processes. First, the chance production of various heritable traits in combinations within diverse unique individual phenotypes. The second process is the evolutionary plot, the necessary consequences of natural selection and elimination among the excess individual phenotypes of the population in every generation. The plot of the story is the resolution of the central problem, survival and reproduction by a small fraction of the local population in the next generation. The resolution is adaptation and fitness of those favored by natural selection and elimination.

## 2.2 Evolution

In biology, evolution is the process of change in the inherited traits of a population of organisms from one generation to the next:

**evolution**:
1: any gradual directional change

2: change in the properties of populations of organisms over time (Mayr, 2001)
3: any cumulative change in characteristics of organisms or populations from generation to generation; descent or development with modification
4: the opportunistic process of change in the characteristics (traits) of individual organisms/phenotypes and their local populations (demes) from generation to generation by the process of natural selection
5: change in the frequency of genes in a population
6: the genetic turnover of the individuals of every population from generation to generation (Mayr, 2001)
7: the gradual process by which the living world has been developing following the origin of life (Mayr, 2001)

**variational evolution**: a population or species changes through continuous production of new genetic variation and through elimination of most members of each generation because they are less successful either in the process of nonrandom elimination of individuals or in the process of sexual selection (i.e. they have less reproductive success) (Mayr, 2001)

**co-evolution**:
1: reciprocal evolution as a consequence of two (or more) kinds of organisms interacting with each other such that each exerts a selection pressure on the other; much of the process of evolution occurs through coevolution (Mayr, 2001)
2: parallel evolution of two kinds of organisms that are interdependent, like flowers and their pollinators, or where at least one depends on the other, like predators on prey or parasites and their hosts, and where any change in one will result in an adaptive response in the other (Mayr, 2001)
3: change of a biological object triggered by the change of a related object (Wikepedia, 5.09)

**2.2.1 Micro- and macroevolution.** Evolutionary phenomena are sometimes categorized as microevolution or macroevolution.

**microevolution**:
1: minor evolutionary events usually viewed over a short period of time, consisting of changes in gene frequencies, chromosome structure or number within a population over a few generations (Lincoln et al., 1998)
2: the occurrence of small-scale changes in allele frequencies in a population, over a few generations, also known as change at or below the species level (Wikipedia, 5.08)
3: evolution at or below the species level (Mayr, 2001)

**macroevolution**:
1: major evolutionary events or trends usually viewed through the perspective of geological time; the origin of higher taxonomic categories; transspecific evolution; macrophylogenesis; megaevolution (Lincoln et al., 1998)
2: a scale of analysis of evolution in separated gene pools; change that occurs at or above the level of species; the occurrence of large-scale changes in gene frequencies in a population over a geological time period (Wikipedia, 5.08)
3: evolution above the species level; the evolution of higher taxa and the production of evolutionary novelties, such as new structures

**2.2.2 Units of evolution and natural selection.** The lowest level of living organization to evolve is the population. Every sexually reproducing species is composed of numerous local populations, within which every individual is uniquely different from every other individual.

**population**:

1: all individuals of one or more species within a prescribed area;
2: a group of organisms of one species, occupying a defined area and usually isolated to some degree from other similar groups

**deme**: a local population of potentially interbreeding individuals of a species at a given locality

**local community**: all the interacting demes in a locality

The variable population of phenotypes in a particular local opportunity spacetime defines the fundamental unit of evolution, the setting of evolution, the stage upon which diversifying evolution occurs.
     The individual phenotype member of a local population, including its extended phenotype, is the unit of natural selection and elimination, the object of selection, the target of elimination.

**phenotype**:
1: the sum total of observable structural and functional properties of an organism; the product of the interaction between the genotype and the environment.
2: the total of all observable features of a developing or developed individual (including its anatomical, physiological, biochemical, and behavioral characteristics). The phenotype is the result of interaction between genotype and the environment (Mayr, 2001)
3: the characters of an organism, whether due to the genotype or environment
4: the manifested attributes of an organism, the joint product of its genes and their environment during ontogeny; the conventional phenotype is the special case in which the effects are regarded as being confined to the individual body in which the gene sits (Dawkins,1999, p.299).
5: phenotypic function consists always of two universes, the physical (quantitative, formal structure; physiological) and the phenomenal (quantities that constitute a 'world') (Sacks, 1998, p.129)

Each individual phenotype in a sexually reproducing population is unique. This same unique individual changes continuously throughout its lifetime, and when placed into a different environments. Among these heterogeneous members of the population differential survival and reproduction result in adaptation to that local opportunity spacetime.
     Richard Dawkin (1976) argues cogently that the allele (within the context of the individual organism's entire genome) may be the unit of natural selection-elimination: the concept of 'selfish genes' or 'selfish DNA'.

## 2.3 Natural Selection and Elimination

### 2.3.1 Malthus postulata

"It is an obvious truth ... that population must always be kept down to the level of the means of subsistence ..."

"I think I can fairly make two postulata.

First, That food is necessary to the existence of man.

Secondly, That the passion between the sexes is necessary, and will remain nearly in its present state."

"Assuming, then, my postulata as granted, I say, that the power of population is indefinitely greater than the power in the earth to produce subsistence for man.

Population, when unchecked, increases in a geometrical ratio. Subsistence increases only in an arithmetical ratio. A slight acquaintace with numbers will show the immensity of the first power in comparison of the second.

By that law of our nature which makes food necessary to the life of man, the effects of these two unequal powers must be kept equal.

This implies a strong and constantly operating check on population from the difficulty of subsistence. This difficulty must fall some where, and must necessarily be severely felt by a large portion of mankind.

Through the animal and vegetable kindgoms, nature has scattered the seeds of life abroad with the most profuse and liberal hand. She has been comparatively sparing in the room and the nourishment necessary to rear them. The germs of existence contained in this spot of earth, with ample food and ample room to expand in, would fill millions of worlds in the course of a few thousand years. Necessity, that imperious all pervading law of nature, restrains them within the prescribed bounds. The race of plants and the race of animals shrink under this great restrictive law. And the race of man cannot, by any efforts of reason, escape from it. Among plants and animals its effects are waste of seed, sickness, and premature death." Malthus, 1798.

### 2.3.2 Natural selection

"Can it, then, be thought improbable, seeing that variations useful to man have undoubtedly occurred, that other variations useful in some way to each being in the great and complex battle of life, should occur in the course of many successive generations. If such do occur, can we doubt (remembering that many more individuals are born than can possibly survive) that individuals having any advantage, however slight, over others, would have the best chance of surviving and of procreating their kind? On the other hand, we may feel sure that any variation in the least degree injurious would be rigidly destroyed. This preservation of favourable individual differences and variations, and the destruction of those which are injurious, I have called Natural Selection, or the Survival of the Fittest. Variations neither useful nor injurious would not be affected by natural selection, and would be left either a fluctuating element, as perhaps we see in certain polymorphic species, or would ultimately become fixed, owing to the nature of the organism and the nature of the conditions." Darwin, 1859

Malthus was primarily interested in the social consequences of human overpopulation, but his work inspired both Darwin and Wallace: excess individuals in innately variable populations provided the pool in which natural selection and elimination could act. Overpopulation provides a powerful driving force, the natural selection and elimination among individual neighbors for seizing and exploiting limiting local opportunity spacetime.

Natural selection is the increase in frequency of the fittest individual phenotypes of a diverse local population relative to less well-adapted individuals. Natural selection is the process of elimination of those less well-adapted phenotypes. The consequences of natural selection is the adaption of certain individual weeds to the local opportunity spacetime their populations exist in over generations.

**natural selection:**
1: process by which forms of organisms in a population that are best adapted to the environment increase in frequency relative to less well-adapted forms over a number of generations

2: the process by which phenotypes of individual organisms in a local population that are best adapted to the local opportunity spacetime increase in frequency relative to less well-adapted phenotype neighbors over a number of generations

3: the non-random and differential reproduction of different genotypes acting to preserve favorable variants and to eliminate less favorable variants; viewed as the creative force that directs the course of evolution by preserving those variants or traits best adapted in the face of natural competition

4: essence of theory of evolution by natural selection is that genotypes with higher fitness leave a proportionately greater number of offspring, and consequently their genes will be present in a higher frequency in the next generation

**artificial selection**: natural selection by humans; domestication; selective breeding

## 2.4 The Process of Natural Selection and Elimination

Adaptative evolution results from the virtually simultaneous actions of two seemingly opposed causations, chance (variation) and necessity (elimination). Evolution is characterized by both chance (contingency) in the generation of population variability, and necessity (adaptation) in survival and reproduction of the fittest progeny. All evolutionary phenomena can be assigned to one or the other of two major evolutionary processes: the origin and role of organic diversity, and the acquisition and maintenance of adaptedness.

Natural selection/elimination is a two-step process that requires five conditions to occur, which results in an adapted local population of weed phenotypes (table 2.1).

**Precondition 1**: Excess local phenotypes compete for limited opportunity spacetime

**Process step 1**: Produce phenotypic variation
    Condition 1: generate variation in individual traits
    Condition 2: generate variation in individual fitness

**Process step 2**: Survive and reproduce the fittest phenotypes
    Condition 3: survive to reproduce the fittest offspring, eliminate others
    Condition 4: reproductive transmission of parental traits to offspring (inheritance)

**Adaptation** arises in the variable local population of fit phenotypes

**Table 2.1** The process of natural selection of the fittest phenotypes resulting in an adapted local population of weeds.

| The Process of Natural Selection for Local Adaptation | | | |
|---|---|---|---|
| PROCESS | | CONDITION | |
| | | Precondition 1 | Excess local phenotypes compete for limited opportunity spacetime |
| Process step 1 | Produce phenotypic variation | Condition 1 | Generate variation: individual traits |
| | | Condition 2 | Generate variation: individual fitness |

| Process step 2 | Survive and reproduce the fittest phenotypes | Condition 3 | Survive to reproduce the fittest offspring, eliminate others |
| --- | --- | --- | --- |
| | | Condition 4 | Reproductive transmission of parental traits to offspring (inheritance) |
| Adaptation | Adaptation arises in the variable local population of fit phenotypes | | |

**2.4.1 Precondition to natural selection.** A necessary precondition to the process of natural selection is the formation of a local population, a deme.

**precondition 1**: excess individuals in a local deme competing for limited local opportunity spacetime

A weed species invades a susceptible opportunity spacetime with excess numbers of members than the local habitat can support, setting the stage for evolution to occur through the process of natural selection and elimination. These weeds produce many more seed than will survive. Many more seeds germinate and form seedlings than will mature to produce their own seeds. Only the successful competitors will reproduce, mortality is very high.

**2.4.2 Representing population growth, the Verhulst-Pearl logistic equation.** Verhulst (1839) and Pearl and Reed (1920) provided a formal mathematical representation of population growth in a limited environment: the Verhulst-Pearl standard logistic sigmoid function equation:

$$dN/dt = rN (K - N/K)$$

$$dNumbers/dtime = (\text{intrinsic rate of natural increase}) (\text{degree N falls short of maximum K})$$

This form for a set of specified conditions indicates that the rate of population growth (change [d] in numbers [$N$] over time [$t$]) is equal to the potential rate of increase of the population per unit time times the degree to which the population ($N$) falls short of its maximum attainable value ($K$). Two potentially measurable parameters together define ideal population behavior. '$r$' is the intrinsic rate of natural increase, a measure of the potential rate a population may increase provided unimpeded access to opportunity. The $r$-phase population growth is often used to characterize that phase after a disturbance or new colonization into an unexploited environment (e.g weeds). '$K$' defines the upper limit which a population may develop, the level it may not exceed for the amount of opportunity available in that environment. The $K$-phase of population growth is often used to characterize when it is prevented from further expansion as a consequence of its size, populations under stress of neighbors (e.g a mature forest).

**Figure 2.1.** Standard logistic sigmoid function ($f(x)$) for values of $x$ in the range of real numbers from $-\infty$ to $+\infty$; a logistic function or logistic curve is a common special case of the more general sigmoid function (Wikipedia, June, 2014).

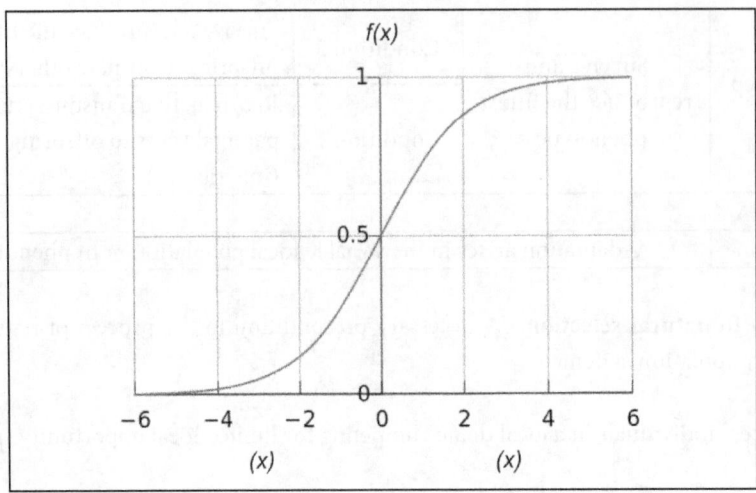

"There are few plant populations that are at all suitable for studying a process of continuous population growth, not only because most plants have life cycles and therefore discete jumps of population size, but because the fixed rooted habit of most plants prevents them from moving about in search of the limited resources and escaping from their neighbours. The result of clumped distributions is that some members of a population are under density stress even though unexploited resources may be present nearby. The choice of a suitable plant as a model for continuous population growth is therefore limited to free-living algae ... or free-floating aquatic plants ... or water ferns ... These are all odd and exceptional plants." Harper, 1977; p.3

The intent of Malthus, Verhulst, Pearl and Reed was the representation of population growth in humans and animals. The qualitative features of animal growth superficially resembles that of plant populations. Much confusion has resulted from the application of these types of growth models to plants, as indicated in the Harper quote above. Differences in individual plant architecture, population structure and functional behaviors have failed to prevent the widespread misapplication of these growth models.

A fundamental flaw arises in models that ignore assumptions central to normal frequency distributions (see ch. 13.2.5.1, assumptions in the analysis of variance in normal distributions), violations of the central limit theorem (ch. 13.2.5, normal probability distributions and inference). Conditions producing normal distributions of observations include many factors which occur independently of each other and are independent in effect (their effects are additive) and these factors contribute equally to the variance (population variances are homogeneous). Plant and weed behaviors are typically complex, and arise from factors very much dependent on interacting biological and environmental phenomena (see ch. 13.2.6, scalable probability distributions and inference). Several concepts in future sections reveal this confusion and neglect: demographic representation of survival and reproduction (e.g. ch. 5.2.1, net reproductive rate; ch. 14.2.2, demographic weed life history population dynamics models), as well as plant density and productivity per unit area (e.g. ch. 9.4.2; law of constant final yield). These confusions are systematically and fully explored in Unit 6, Representation of Weed Biology.

**2.4.3 Generate variation: Process of natural selection, step 1**. The first step in the process of natural selection is the production of phenotypic variation that provides the material for the selection and elimination processes in step two.

**step 1:**   production of phenotypic variation in the local population

New variation is produced in the first step, which consists of all the processes leading to the production of a new zygote, including meiosis, gamete formation and fertilization. Variation arises from several sources in the weed genotype/phenotype: mutation of the zygote (from fertilization to death) caused by errors of replication during cell division; recombination via crossing-over during the first division of meiosis; random movement of homologous chromosomes during the second division of meiosis; and, to random aspects of mate choice and fertilization. The production of variation that provides the material for the selection process is dominated by stochastic processes (chance, contingency, accident) (Mayr, 2001).

The first condition necessary for evolution to occur is that there must be genetic variation in heritable characters (traits) among the excess offspring (phenotypes) produced.

**condition 1**:     generation of variation in individual characters/traits among phenotypes of the local population

**trait:**
1: a character: any detectable phenotypic property of an organism
2: any character or property of an organism
3: a characteristic feature or quality distinguishing a particular person or thing
4: predictors (proxies) of organismal performance (Darwin, 1859)

**functional trait**: morpho-physio-phenological traits which impact fitness indirectly via their effects on growth, reproduction or survival, the three components of individual performance (Violle et al., 2007)

The second condition in step one of the natural selection process is variation in fitness of the excess phenotypes of the population.

**condition 2**:     variation in fitness of individual phenotypes of the local population

**fitness:**
1: the average number of offspring produced by individuals with a certain genotype, relative to the numbers produced by individuals with other genotypes
2: the relative competitive ability of a given genotype conferred by adaptive morphological, physiological or behavioral characters, expressed and usually quantified as the average number of surviving progeny of one genotype compared with the average number of surviving progeny of competing genotypes; a measure of the contribution of a given genotype to the subsequent generation relative to that of other genotypes (Lincoln, et al., 1998)
3: the relative ability of an organism to survive and transmit its genes to the next generation (or some defined number of future generations)

**Darwinian fitness:**
1: the relative probability of survival and reproduction for a genotype
2: a measure of the relative contribution of an individual to the gene pool of the next generation
3: the relative reproductive success of a genotype as measured by survival; fecundity or other life history parameters

Productivity and fecundity are often confused with fitness (e.g. Silvertown and Charlesworth, 2001). This confusion is understandable in crop science where crop yield is directly related to crop fitness.

**productivity:**
1: the potential rate of incorporation or generation of energy or organic matter by an individual, population or trophic unit per unit time per unit area or volume; rate of carbon fixation;
2: often used loosely for the organic fertility or capacity of a given area or habitat

**fecundity:**
1: the potential reproductive capacity of an organism or population, measured by the number of gametes or asexual propagules (Lincoln et al., 1998)
2: potential fertility or the capability of repeated fertilization. Specifically the term refers to the quantity of gametes, generally eggs, produced per individual over a defined period of time.

**2.4.4 Survival and reproduction: Process of natural selection, step 2.** The second step in the process of natural selection is the differential survival and reproduction of offspring. The "survival of the fittest" from among the excess, unique and variable phenotypes of the local population is to a large extent determined by genetically based characteristics.

**step 2**: non-random aspects of survival and reproduction. Specific sets of parental characters/traits are transmitted to the offspring who resemble them by the process of reproduction; and the fittest progeny survive to reproduce themselves.

At the second step of natural selection the quality of the new individual weed phenotype is constantly tested for its entire life history, from embryo to vegetative plant to reproductive plant. Those individuals most capable of coping with the environment, competing with other members of the local community (same species, other species) for opportunity spacetime, will have the best chances to survive to reproduce progeny (Mayr, 2001). During this selection and elimination process certain phenotypes are clearly superior to others: the fittest to survive. This second step is primarily deterministic, but there are also many chance factors of elimination including natural catastrophes (floods, hurricanes and tornados, volcanic eruptions, lightning, blizzards and violent storms) and the loss of superior genes in small populations due to unpredictable disturbance events. The "survival of the fittest" is to a large extent determined by genetically based characteristics.
The third condition necessary for evolution to occur is for the parent plant to survive to reproduction, and then to produce more offspring than can normally survive.

**condition 3:** reproduction: the act or process of producing offspring/progeny

**reproduction:** the act or process of producing offspring

The net (average) result of reproduction is that a parent plant leaves one descendant that reproduces, yet many more die than are produced.
The fourth condition necessary for evolution to occur is the transmission of parental traits of the "fittest" phenotypes that survive to the successful progeny.

**condition 4:** heredity, the transmission of specific characters/traits/genetic information from the ancestral parent to the descendants/offspring/progeny

The offspring must tend to resemble their parents.

**heredity**: the mechanism of transmission of specific characters or traits from parent to offspring.

**inheritance**: the transmission of genetic information from ancestors or parents to descendants or offspring.

## 2.5 Adaptation

Adaptation arises as a consequence of natural selection and elimination among excess phenotypes in a local population as the seize and occupy opportunity spacetime.

**adaptation:**
1: the process of adjustment of an individual organism to environmental stress; adaptability
2: process of evolutionary modification which results in improved survival and reproductive efficiency
3: any morphological, physiological, developmental or behavioral character that enhances survival and reproductive success of an organism
4: a positive characteristic of an organism that has been favored by natural selection and increases the fitness of its possessor (Wikipedia, 5.08)
5: any property of an organism believed to add to its fitness (Mayr, 2001)

Adaptation by elimination of the less fit individuals is dominated by the processes of survival and sexual selection favoring those most successful in seizing and exploiting local opportunity spacetime: superior success of certain phenotypes throughout their life history (survival selection); nonrandom mate choice, and everything that enhances reproductive success of certain phenotypes (sexual selection). A considerable amount of random elimination occurs simultaneously with these non-random processes.

Adaptation is not an active process. Adaptation is a passive consequence of the process of natural selection and elimination. It is a property of the local population of phenotypes favored by non-elimination (figure 1.1).

**Figure 1.1** Schematic diagram of the formation of a locally adapted population (deme) by the processes of natural selection and elimination; relative population size, 0-1, shaded area; local opportunity, niche for particular phenotypes/traits. Process of natural selection with time:

| | |
|---|---|
| Time 0: | no population |
| Time 1: | seed (propagule) dispersal into locality |
| Time 2: | death of less fit individual phenotypes |
| Time 3: | locally adapted phenotypes |
| Time 4/5/6: | shift of population to new local adapted phenotypes |

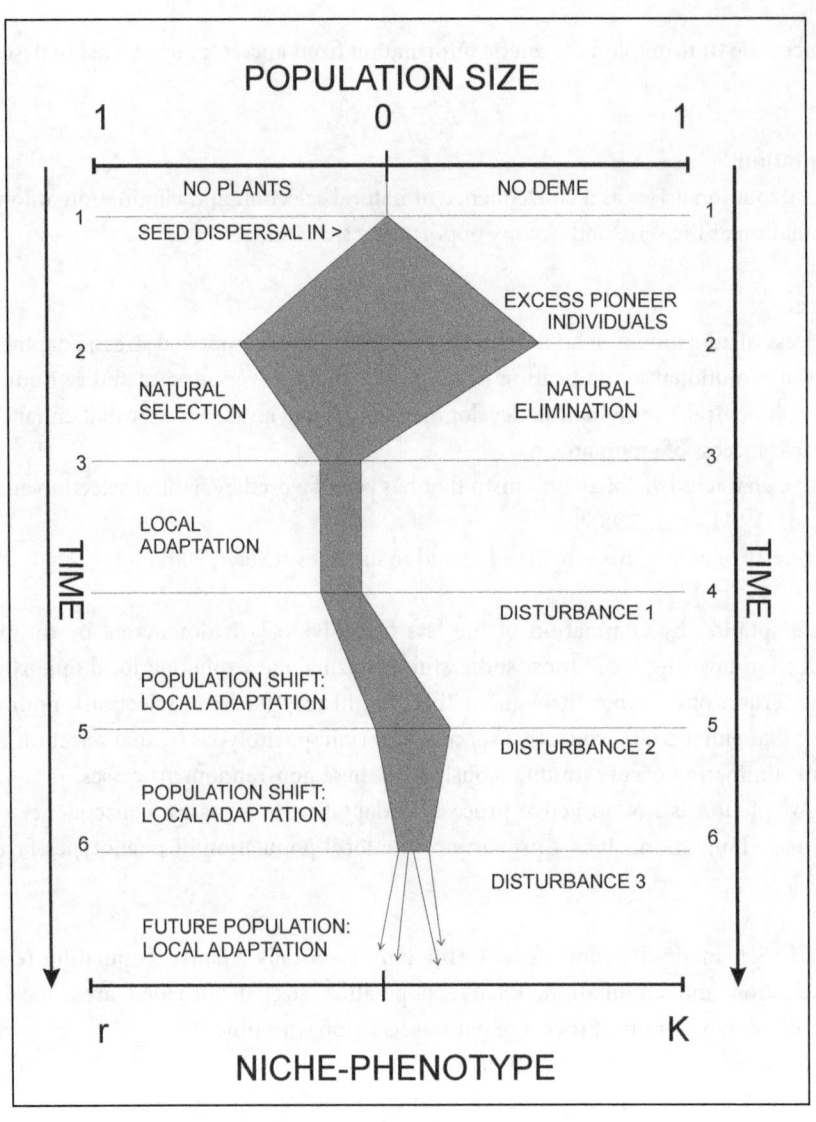

## Chapter 3: Formation of the Local Population (Deme): Precondition to Natural Selection

**The weeds' story**. I am foxtail.

In the beginning there was wild me in tropical Africa, waiting on the glaciers for opportunity. I looked like green foxtail back then, before you. I dispersed to Eurasia when the ice receded. Life was harder then, without your helpful disturbing ways. I settled in local patches, where there was no one else.

We first met in Eurasia. Right away you picked your favorites from among me, foxtail millet. I followed, wild and weedy. Then we invaded the Earth together, but it took time and effort. First you sought out locations to grow. You were greedy for food, so you found sunny moist fields with ample fertility and heat. You killed all the other plants, dug up the soil and planted your favorite foxtail millet. And our weedy family members followed you right into the field. You found us a pest with those traits we were developing: ready exploitation of unused essentials, competitive with neighbors and crop cousins. You killed the weedy of us, but left behind the most clever: the ones who hid, or escaped, or looked just like your favorite crops. Then the survivors mated and reproduced children that inherited all those clever new traits. In most of those fields we endured, in some few we went extinct.

We invaded China 6000 years ago, Europe 3500 years ago. A sudden change in my genes, a doubling, expanded my family to include cousins yellow foxtail and bristly foxtail. My yellow foxtail was wild at first. Then in India you picked your favorites, and again we invaded and endured. Then we met in the Americas, twice actually. The first time is still unclear, my wild antecedent knotroot foxtail coming from Asia about 10-12,000 years ago. Your prehistoric migration from central Asia brought me across the Bering Straits land bridge to North America. This new native American invader made me the oldest cultivated cereal in the new world 8000 years ago. We came more deliberately the second time, 500 years ago, following your European invader Columbus. Our green foxtail emigrating with you from Europe, our yellow foxtail from Asia. Our weeds still dominating much of the Americas today.

Conquering the Earth with you wasn't all easy. You became fickle, greedy for more attractive plants to eat. In Asia you chose rice, abandoning us to the dry and salty fringes. In Europe it was wheat and barley and rye you chose. In the Americas maize was your new favorite. Our foxtail millet went extinct in so many fields around the earth then. But our weedy family endured and became even stronger. We invaded these new crops too, and we found a home. Your success in global trade and travel has accelerated our success beyond our wildest dreams. My gene doubling gave birth to our cousin giant foxtail (*Setaria faberi*) allowing us dominate the richest fields of North America. Your invention of herbicides neatly uncovered our shy, preadapted, resistant siblings. They are our proudest new members, brought from obscurity by your dedicated efforts. We couldn't have done any of it without you.

**Summary**. Plant communities assemble when opportunity in a locality favors functional traits in a new species at a particular time. Local opportunity spacetime is the integration of available resources and conditions, neighbor interactions and disturbance in that local niche. The population of this species must first be able to disperse into the habitat, then colonize the site by completing its life cycle from recruitment to reproduction. Lastly it must endure at that site, with individuals selected or eliminated. Without performing all of these processes invasion is not successful. Without all these processes evolution of the population will not occur. For weeds, disturbance in the broadest sense is what creates this original opportunity. All local populations eventually go extinct.

Every interbreeding local population of a species has an origin, a time and place it began. Every population also goes extinct at some time. This invasion process begins with dispersal of propagules (seeds, annual or perennial vegetative meristems) into the new space. If opportunity exists for the weed species, and if it possesses functional traits appropriate to exploit the new opportunity, the population can become established in that locality. If these early colonizing plants successfully reproduce, or disperse in sufficient numbers of propagules, then the species can begin an enduring occupation of that habitat. When an excess number of individuals of the first colonizing weed species in a local habitat begin competing with each other the preconditions for natural selection-elimination are met and a local deme is formed. A weed species invades a susceptible opportunity spacetime with excess numbers of members than the local habitat can support, setting the stage for evolution to occur through the process of natural selection and elimination. These weeds produce many more seeds than will survive. Many more seeds germinate and form seedlings than will mature to produce their own seeds. Only the successful competitors will reproduce, mortality is very high. Local adaptation has begun.

## 3.1 Introduction: Opportunity and the Formation of the Local Deme

"… the capacity to choose in each fleeting moment of a situation that which is … most opportune."
Count Otto von Bismarck (Pflanze, 1990)

"… guiding power of that conventional trinity of crime investigation: motive, means and opportunity."
Reginald Hill, 2001

"vedere moneta, toccare camello"
"see the money, touch the camel"
Italian folk wisdom

**3.1.1 Population formation examples**. When do new populations form in a locality? There are many scenarios by which new populations seize and exploit new habitats, new opportunity. A good example is that of virgin land which occurred in vast areas of North America and Eurasia around 15-10,000 B.P. (see ch. 1.5.2, Biogeographic prehistory of agriculture). These new lands provided an opportunity for both agriculture to develop as the earth warmed, and the opportunity for weed populations to develop. Another more recent example would be the dispersal of Eurasian species to the New World over the last 10,000 years, and especially in the post-Columbian period (after 1492 A.D.) of European, and then Asian, human invasion. With the newly cultivated crop fields of New England, Georgia, Brazil and Argentina came their fellow traveler weeds. With the new seeds came the new weeds.

Another example of the formation of new demes is the change in conditions associated with changes in cropping practices. Shifts from winter annual crops to summer annual crops, extended or shortened crop species rotations, or the introduction of new types of crops as herbicide resistant crops (e.g. glyphosate resistant maize, soybeans in Iowa) result in a change in opportunity. Past weed species favored by a particular crop or cropping system may no longer be so favored under the new regime. When the first selective herbicide 2,4-D was introduced in the late 1940's cropping systems changed: formerly dominant dicotyledenous weed species (e.g. *Chenopodium album*) were selectively eliminated. New opportunity was thus created in those fields for the more herbicide resistant monocotyledonous grassy species (e.g. weedy *Setaria*). In these fields excess numbers of new seeds were produced, the raw material from which natural selection could pick winners and losers. Local adaptation to the new herbicidal opportunity had begun. And it continues around the world to the present.

Another example is that of range expansion of an existing population across the landscape. New fields adjacent to existing demes can be colonized by neighbors. Given enough time and success these populations are connected to each other across the landscape, region and globe to form a continuous metapopulation of local populations sharing successful traits and alleles, enriching their ability to adapt to a wider range of opportunity, habitats.

Local plant community assembly is the process indicated in all these examples. The first step in community assembly begins with colonization (see ch. 3.5, the nature of plant invasions of local opportunity) and ends with the terminal phases of ecological succession in the fully mature community. Community assembly begins with propagule recruitment (see ch. 9.1.2, recruitment: community assembly). With seedling emergence the local weed-crop community begins to be assembled. With community assembly begins interaction with neighbor plants and other organisms in the local community. These neighbor interactions frame, determine community assembly. During plant community ecological succession earlier successional species facilitate the invasion of later successional species by the opportunity spacetime they create in the habitat (see ch. 10.4.4). The driving forces of Earth thermodynamics ultimately determines community structure and assembly (see ch. 11.1.8, plant community structure and assembly).

The multi-step process by which these colonizing events form new demes is developed fully at the end of this chapter (section 3.5, The nature of plant invasions of local opportunity). First we need to look at what these new weed species seize and exploit: opportunity.

**3.1.2 Seizing and exploiting local opportunity.** Why do some populations succeed and others fail? Why do some species never successfully invade a particular habitat and others fail before they successfully occupy a habitat? What is needed for successful occupation of a locality? The answer is that the habitat must provide something that can be seized and utilized to allow a complete life cycle of establishment, growth and reproduction. Next, the weed species must possess functional traits expressed at appropriate times over the growing season to exploit that which is available in the habitat. The newly invading species must possess these traits before the invasion process begins in order to take advantage of local opportunity:

**preadaptation**:
1: the possession by an organism of characters or traits that would favor its survival in a new or changed environment (Lincoln et al., 1998)
2: a situation where a species evolves to use a preexisting structure inherited from an ancestor for a potentially unrelated function (Wikipedia, 11.10)
3: a new character or function arising from a mutation in an organism which becomes a useful adaptation after an environmental change, often occurring many generations later (King and Stansfield, 1985)

**exaptation**: shifts in the function of a trait during evolution; for example, a trait can evolve because it served one particular function, but subsequently it may come to serve another (Wikipedia, 11.10)

A good weed example of preadaptation or exaptation can be found in s-triazine resistance in plants (see ch. 6.1.1.1, preadaptation). Pre-existing traits in *Brassica* spp. conferring altered diurnal photosynthetic activity and shade-adapted leaf morphology (*psbA* chloroplast mutants) allowed more efficient productivity in stressful and shaded micro-habitats (e.g. Dekker, 1993; Dekker and Sharkey, 1992; see ch. 18, triazine resistant and susceptible *Brassica napus*). These preadapted traits later conferred resistance to s-triazine herbicides utilized in crop production fields in the 1960's (e.g. Holt and LeBaron, 1990).

The setting of weed evolution, the stage upon which diversifying evolution takes place, is a local population of variable phenotypes of a weed species in a particular locality. The nature of a particular locality is defined by the opportunity spacetime it affords the weed population to survive and reproduce. Opportunity spacetime of a local habitat for a population is defined by its available resources and conditions, interactions with neighbors, and disturbances: the local niche.

**opportunity spacetime**: locally habitable space for an organism at a particular time which includes its available resources (e.g. light, water, nutrients, gases), pervasive conditions (e.g. heat, terroir), disturbance history (e.g. tillage, herbicides, frozen winter soil), and neighboring organisms (e.g. crops, other weed species, diseases) with which it interacts

Therefore, weedy plant life history behavior in a deme and local community is a consequence of natural selection and reproductive success among excess variable phenotypes (and their functional traits) in response to the structure, quality and timing of locally available opportunity spacetime. Plants will fill any available and habitable growing space. The structure of available and habitable space to an invading plant is also opportunity space at a particular time. Opportunity space-time is not general, it is always particular to a locality and time. Opportunity space-time is synomymous with habitat and niche. The more difficult task is to characterize fully the nature of opportunity spacetime.

The successful weed phenotype is comprised of functional traits allowing it to seize and exploit local opportunity spacetime, mitigate/attenuate/tolerate stress, avoid mortality, reproduce and transmit those traits to progeny. Little discussed in evolutionary theory is what animates an individual organism to live. What drives the will to live in an organism? What is the physical basis of greed in nature? The search for the answer to this fundamental question can take many forms, most not conducive to science. The will to live can be seen as a matter of belief, of philosophy, or an emergent property of a complex adaptive system (see unit 5, complex adaptive weed systems). All three are empirically intractable.

"The world is my representation".
"… one can thus no longer separate the perceiver from the perception …"
A. Schopenhauer, 1844.

The philosophy of Arthur Schopenhauer provides some insight into a connection between nature and an organism's will to live. It holds that all nature, including man, is the expression of an insatiable *will to life*: "desire," "striving," "wanting," "effort," and "urging." To Schopenhauer, the 'Will' is a malignant, metaphysical existence which controls not only the actions of individual, intelligent agents, but ultimately all observable phenomena. The desire for more is what causes an organism to seize and exploit available opportunity. The functional traits that a weed possesses provides the means to actualize that will to live. Nature is perceived by a weed (its representation of the world) by the stimulation of opportunity. Nature is perceived by means of those functional traits which allow opportunity to be exploited. Those responses to how opportunity is sensed, then exploited, determine how that organism perceives nature as an object: the world is its representation. The less difficult task is to characterize the structure of local weed opportunity spacetime.

The nature of opportunity spacetime, or the niche (or more comprehensively, the niche hypervolume) can be characterized by a description of a local habitat in terms of its spatial and temporal heterogeneity, disturbance, the differential use of resources and conditions, and the local community of species with which it interacts. How local opportunity spacetime is seized and exploited is revealed in the multi-step process of plant invasion.

## 3.2 The Structure of Local Weed Oppportunity

Plants perceive local opportunity space as a unified signal over time. Opportunity is an integrated signal of all its components: qualities of the locality, resources, conditions, disturbance, and neighbors. All components are sensed by a plant as one, at once, during its life history. All act together during development allowing the individual in a population to seize and exploit opportunity.

**3.2.1   Weedy habitats.**   Local opportunity space-time can be represented as a weedy habitat.

**locality**: the geographic position of an individual population or collection

**habitat:**
1: the locality, site and particular type of local environment occupied by an organism
2: local environment
3: the physical conditions that surround a species, or species population, or assemblage of species, or community (Clements and Shelford, 1939).
4: an ecological or environmental area that is inhabited by a particular species; the natural environment in which an organism lives, or the physical environment that surrounds (influences and is utilized by) a population (Wikipedia, 5.08).

**microhabitat:**
1: a physical location that is home to very small organisms (e.g. a seed in the soil); microenvironment is the immediate surroundings and other physical factors of an individual plant or animal within its habitat
2: a very localized habitat (e.g. on the size scale of an individual seed in the soil seed bank)

**microenvironment**: the immediate surroundings and other physical factors of an individual plant or animal within its habitat; analogous to a microhabitat

**microsite**: analogous to a microhabitat; e.g. the local spatial environment immediately perceived by a seed in the seed bank, or a seedling in a field.

Habitats are constructed within the space-time matrix. A habitat is the place, the site, the locality and particular type of local environment occupied by weeds. As an agronomist I think of habitats mostly as crop production fields, like a field of corn early in the growing season. This book discusses weeds in crop fields as well as other disturbed habitats (a residential lawn, a waste field near a factory, a highway roadside, etc.). Weeds are found everywhere, not just in corn fields. Any place weeds thrive is any important habitat, with its own unique set of environmental opportunities and restrictions. Habitats are also defined by land use factors, the type and level of management of the habitat by humans. These land use factors are closely related to disturbance, which is discussed in detail in chapter 3.3.4. One of the most important characteristics of a habitat is the biological community of the locality. In agroecosystems this is often dominated by the crop and other weeds.

**3.2.2 Niches in the local community.** Local opportunity space-time can be represented as a niche.

**niche:**
1: the ecological role of a species in a community; conceptualized as the multidimensional space, of which the coordinates are the various parameters representing the condition of existence of the species, to which it is restricted by the presence of competitor species;
2: loosely as an equivalent of microhabitat

3: the relational position of a species or population in its ecosystem; how an organism makes a living; how an organism or population responds to the distribution of resources and neighbors and how it in turn alters those same factors (Wikipedia, 5.08)

Different weed species can occupy similar niches in different locations and the same species may occupy different niches in different locations. It is this plastic response to opportunity that makes the definition of its structure so complex. The different dimensions, or plot axes, of a niche represent different biotic and abiotic variables. These factors may include descriptions of the organism's life history, habitat, trophic position (place in the food chain), and geographic range. According to the competitive exclusion principle, no two species can occupy the same niche in the same environment for a long time (see ch. 9.3.2.2.2, ecological combining ability).

**competitive exclusion principle**:
1: two species competing for the same resources cannot coexist if other ecological factors are constant (Hardin, 1960); complete competitors cannot coexist
2: a theory which states that two species competing for the same resources cannot stably coexist, if the ecological factors are constant; complete competitors cannot coexist. Either of the two competitors will always take over the other which leads to either the extinction of one of the competitors or its evolutionary or behavioral shift towards a different ecological niche (Wikipedia, 5.08)

Ecological factors in this sense are never constant. It is for this reason that opportunity spacetime can be imagined as infinite. The full range of environmental conditions (biological and physical) under which an organism can exist describes its fundamental niche:

**fundamental niche:** the entire multidimensional space that represents the total range of conditions within which an organism can function and which it could occupy in the absence of competitors or other interacting (neighbor) species

As a result of pressure from, and interactions with, other organisms (e.g. superior competitors), species are usually forced to occupy a niche that is narrower than this, and to which they are mostly highly adapted. This is termed the realized niche:

**realized niche:** that part of the fundamental niche q.v. actually occupied by a species in the presence of competitive or interactive (neighbor) species

The niche concept is fundamental to understanding the evolutionary ecology of weeds. There exists certain connotations implicit in the term that prevent full understanding of the concept underlying niche. Niche is a passive term, it connotes receptivity, a state, the yin of yin-yang if you will. And of course, it is, after all, French. The concept of opportunity spacetime is conceptually analogous to niche. The word concept 'opportunity' contains something not implicit in niche. Opportunity connotes exploitation, the potential for action or a process physically seizing something in the habitat, the yang in yin-yang. And of course, it is, after all, very American.

**3.2.3 The niche hypervolume**. The structure of local opportunity can be defined by a space-time matrix: the niche hypervolume. Niche differentiation can be formalized as the dimensions of time and space in which resources and conditions are used in a locality. Opportunity space-time can therefore be expressed as an n-dimensional niche hypervolume (Hutchison, 1957):

**n-dimensional niche hypervolume:**
1: the multi-dimensional space of resources and conditions available to, and specifically used by, organisms in a locality
2: the phenotype is described by the niche hypervolume; phenotype = Genotype x Environment = realized niche; the selection pressure consequence of the G x E interaction
3: the limits or borders within which a species has adapted
4: experimentally defined by the testable parameters (dimensions) one can evaluate; the parameters determining the form of existence of a plant

What is the structure of opportunity space-time that weeds seize and exploit? How do we represent it? Harper (1977) wrote that resource heterogeneity drives niche differentiation. He discussed elements contributing to the diversity of plant populations and to community structure:

"Specializations that we most commonly see within communities of plants are with respect to the dimensions of time and space in which resources are used."

"Most of the niche differentiation that has occurred has been interpreted in relation to heterogenous distribution of resources in space and time."

Harper concluded that plant community diversity is dependent on four dimensions of the niche hypervolume: the lateral heterogeneity of environments, the vertical heterogeneity of environments, the temporal division of the environment, and differential use of resources. His spatial dimensions (patchiness, lateral, vertical) can be conveniently combined, and a fourth, disturbance, included. The dimensionality or structure of opportunity space-time, the niche hypervolume, can be represented by four aspects of opportunity operating in habitats:

1: spatial heterogeneity
2: temporal use of environment
3: role of local habitat disturbance
4: differential use of resources and conditions

Natural selective forces guide a weed population in its search for, and exploitation of, opportunity space-time (disturbance, resources-conditions, neighbors). It is to these that weeds respond and adapt. The cumulative effect of weed populations interacting within these aspects of opportunity result in the observed local community structure over time. These themes will be fully developed in ch. 4.3.1, reaction norm; ch. 9.3, strategic roles and traits of interference and facilitation with neighbors; and ch. 10.3.3, ecological roles-guilds-trades in weed-crop communities.

## 3.3 Habitat Heterogeneity and Dynamics

The local plant community habitat is in the first instance determined by Earth's physical geography and the forces of nature therein that drive plant evolution. From these first fundamental influences the spatial and temporal heterogeneity of the local environment and the role of disturbance in a broad sense are considered.

**3.3.1 Earth's physical geography.** The physical geography of the Earth is the natural science of processes and patterns in the natural environment as opposed to the cultural or built environment, the domain of human geography (Wikipedia, 1.16). The physical geography of Earth is often split into several spheres or environments: the geosphere, lithosphere, atmosphere, hydrosphere and cryosphere, biosphere and pedesphere.

**3.3.1.1 Geosphere.** The geosphere is the collective name for the lithosphere, hydrosphere (including the cryosphere, water in solid form), atmosphere and biosphere.

**3.3.1.2 Lithosphere.** Earth's lithosphere includes the crust and the uppermost mantle, which constitute the hard and rigid outer layer of the Earth. The lithosphere is subdivided into tectonic plates. The uppermost part of the lithosphere that chemically reacts to the atmosphere, hydrosphere and biosphere through the soil forming process is called the pedosphere

**3.3.1.3 Atmosphere.** The atmosphere of Earth is the layer of gases surrounding the planet Earth that is retained by Earth's gravity. The atmosphere protects life on Earth by absorbing ultraviolet solar radiation, warming the surface through heat retention (greenhouse effect), and reducing temperature extremes between day and night (the diurnal temperature variation). Atmospheric circulation is the large-scale movement of air through the troposphere, and the means (with ocean circulation) by which heat is distributed around and on the surface of the Earth. The large-scale structure of the atmospheric circulation varies from year to year, but the basic structure remains fairly constant because it is determined by Earth's rotation rate and the difference in solar radiation between the equator and poles.

**3.3.1.4 Hydrosphere.** The hydrosphere describes the combined mass of water found on, under, and over the surface of a planet, the Earth. It includes water in liquid, gaseous and frozen forms in groundwater, glaciers, oceans, lakes and streams, and the atmosphere.

The hydrosphere as a closed system. The hydrosphere is intricate, complex, interdependent, all-pervading and stable and appears built for regulating life on Earth. Water is a basic necessity of life. The total amount of water has probably not changed since geological times. There is no evidence that water vapor escapes into space. The turnover of water on Earth involves that which evaporates from the oceanic surface and from land. The same amount of water falls as atmospheric precipitation on the ocean and on land. The difference between precipitation and evaporation from the land surface represents the total runoff of the Earth's rivers and direct groundwater runoff to the ocean. These are the principal sources of fresh water to support life necessities and man's economic activities.

Hydrosphere plays an important role in the existence of the atmosphere in its present form. When the Earth was formed it had only a very thin atmosphere rich in hydrogen and helium. Later the gases hydrogen and helium were expelled from the atmosphere. The gases and water vapor released as the Earth cooled became our present atmosphere. Other gases and water vapor released by volcanoes also entered the atmosphere. As the Earth cooled the water vapor in the atmosphere condensed and fell as rain. The atmosphere cooled further as atmospheric carbon dioxide dissolved in to rain water. In turn this further caused the water vapor to condense and fall as rain. This rain water filled the depressions on the Earth's surface and formed the oceans. The first life forms began in the oceans. Later, the photosynthetic process of conversion of carbon dioxide into food, and the splitting of water and release of oxygen, began. Our atmosphere is a fundamental requirement for life on Earth.

The water cycle, also known as the hydrological cycle or the $H_2O$ cycle, describes the continuous movement of water on, above and below the surface of the Earth. The mass of water on Earth remains fairly constant over time. The transfer of water from one state to another partitions the water into the major reservoirs of ice, fresh water (rivers, lakes, groundwater, subterranean aquifers, polar icecaps and saturated soil), saline water (oceans) and atmospheric water (snow, rain and clouds). Solar energy, in the form of heat and light, and gravity cause the transfer from one state to another over periods from hours to thousands of years. The water moves from one reservoir to another by the physical processes of evaporation, condensation, precipitation, infiltration, runoff, and subsurface flow. The water cycle involves the exchange of energy, which leads to temperature changes. These heat exchanges influence climate. The flow of liquid water and ice transports minerals across the globe. The water cycle is also essential for the maintenance of most life and ecosystems on the planet. The sun, which drives the water cycle, heats water in oceans and seas. Water evaporates as water vapor into the air.

Evapotranspiration is water transpired from plants and evaporated from the soil. Most evaporation comes from the oceans and is returned to the earth as snow or rain. Sublimation refers to evaporation from snow and ice. Transpiration refers to the expiration of water through the minute pores or stomata of trees. Evapotranspiration includes the three processes together, transpiration, sublimation and evaporation.

Transpiration is the process of water movement through a plant and its evaporation from aerial parts, such as leaves, stems and flowers. Water is passively transported into the roots and then into the xylem. The forces of cohesion and adhesion cause the water molecules to form a column in the xylem. Water moves from the xylem into the mesophyll cells, evaporates from their surfaces and leaves the plant by diffusion through the stomata. Water is necessary for plants but only a small amount of water taken up by the roots is used for growth and metabolism. The remaining 99-99.5% is lost by transpiration. Leaf surfaces are dotted with pores called stomata. Transpiration occurs through the stomatal apertures, the opening of the stomata allows the diffusion of carbon dioxide gas from the air for photosynthesis. Transpiration also cools plants, changes osmotic pressure of cells, and enables mass flow of mineral nutrients and water from roots to shoots. Mass flow of liquid water from the roots to the leaves is driven in part by capillary action, and any dissolved mineral nutrients travel with it through the xylem. Plants regulate the rate of transpiration by controlling the size of the stomatal apertures. The rate of transpiration is also influenced by the evaporative demand of the atmosphere surrounding the leaf such as boundary layer conductance, humidity, temperature, wind and incident sunlight. Soil water supply and soil temperature can influence stomatal opening, and thus transpiration rate. The amount of water lost by a plant also depends on its size and the amount of water absorbed at the roots. Transpiration accounts for most of the water loss by a plant by the leaves and young stems. Transpiration serves to evaporatively cool plants.

**3.3.1.5 Biosphere.** The biosphere is the zone of life on Earth, the global sum of all ecosystems. It is a closed system (apart from solar and cosmic radiator and heat from the interior of the Earth), and largely self-regulating. The biophysiological definition of the biosphere is the global ecological system integrating all living beings and their relationships, including their interaction with the elements of the lithosphere, geosphere, hydrosphere, and atmosphere.

**3.3.1.6 Pedosphere.** The pedosphere is the outermost layer of the Earth that is composed of soil and subject to soil formation processes. It exists at the interface of the lithosphere, atmosphere, hydrosphere and biosphere. The sum total of all the organisms, soils, water and air is termed as the "pedosphere". The pedosphere is the skin of the Earth and only develops when there is a dynamic interaction between the atmosphere (air in and above the soil), biosphere (living organisms), lithosphere (unconsolidated regolith and consolidated bedrock) and the hydrosphere (water in, on and below the soil). The pedosphere is the foundation of terrestrial life on this planet. The pedosphere is a dynamic interface of all terrestrial ecosystems. The pedosphere acts as the mediator of chemical and biogeochemical flux into and out of these respective systems and is made up of gaseous, mineralic, fluid and biologic components. As part of the larger global system, any particular environment in which soil forms is influenced solely by its geographic position on the globe as climatic, geologic, biologic and anthropogenic changes occur with changes in longitude and latitude. The pedosphere lies below the vegetative cover of the biosphere and above the hydrosphere and lithosphere. The soil forming process (pedogenesis) can begin without the aid of biology but is significantly quickened in the presence of biologic reactions. Soil formation begins with the chemical and/or physical breakdown of minerals to form the initial material that overlies the bedrock substrate. Biology quickens this by secreting acidic compounds (dominantly fulvic acids) that help break rock apart. Many other inorganic reactions take place that diversify the chemical makeup of the early soil layer. Once weathering and decomposition products accumulate, a coherent soil body allows the migration of fluids both vertically and laterally through the soil profile causing ion exchange between solid, fluid and gaseous phases. As time progresses, the bulk geochemistry of the soil layer will deviate away from the initial composition of the bedrock and will evolve to a chemistry that reflects the type of reactions that take place in the soil.

The roles played by the Earth's physical geography in complex adaptive weed systems is developed in both ch. 3.4.1 (thermodynamic Earth) and ch. 11.1 (formation of complex systems: the constructal law).

**3.3.2 Spatial heterogeneity and patchiness.** The niche hypervolume is defined and confined by the spatial dimensionality of the local habitat weeds exploit. The diversity of local community structure is dependent on spatial heterogeneity and patchiness. The diversity of weed populations and plant community structure is a consequence of natural selection and adaptation to heterogeneous local opportunity.

There are many lateral and vertical factors that comprise the spatial diversity of habitats that weeds exploit. The horizontal spatial variability over the landscape is a mosaic. It consists of heterogeneous soil gradients, topography factors and global position, to name but a few. Vertical spatial variability includes the soil below and the atmosphere above. Some few examples are presented in table 3.1.

**Table 3.1** Habitat-disturbance-neighbor-resource/condition factors responsible for setting the scale of local spatial heterogeneity/patchiness to which weed populations adapt and evolve.

| FACTOR | SUB-FACTOR | HETEROGENEITY |
|---|---|---|
| HABITAT | Soil | type, structure, tilth, aggregation, fertility, moisture (% moisture/drainage/drought risk) organic matter %, soil pH, depth, sub-soil, geology, age |
| | | seed pool distribution; ground cover |
| | Topography | slope, gradient, aspect, elevation, landscape-watershed position |
| | Atmosphere | gases, moisture, pollution |
| | Location | latitude, longitude, solar exposure |
| | | microsite |
| DISTURBANCE | Farmer-Land Manager Practices | crop production disturbances such as tillage and planting; combines and harvesting; herbicides and cultivation |
| | Non-human | cycles, crashes, catastrophes |
| NEIGHBORS | Seeds | seed heteroblasty; seed dispersal pattern |
| | Community | competition and interactions with neighbors |
| | | population phenotype diversity |
| RESOURCES & CONDITIONS | Resources | nutrients: nitrogen; phosphorus; potassium; micronutrients |
| | | light: quality, quantity, plant canopy |
| | | water: quantity, quality (salinity, toxins) |
| | | gases: oxygen, carbon dioxide |
| | Conditions | temperature, soil matrix |

An essential part of understanding habitats is the scale at which habitats exist. For example, to a farmer trying to manage a field, there are very big differences between habitats at the microsite scale and the field scale. The farmer is most interested in scale from a practical perspective: tractor scale and field scale. This viewpoint is the essence of site-specific crop and field management. Habitats exist in an ordered spatial hierarchy at many scales: global > continental > regional or State > landscape > farm > field > locality or site > microsite.

Weed populations in local fields and habitats usually exhibit spatial patchiness. Natural selection on patchy and spatially heterogenous weed populations acts in many niches, providing opportunity space-time for a complex community structure. Many factors set the scale of natural selection in a plant population's spatial diversity (e.g. table 3.1).

Seed dispersal mechanisms, as well as perennial weed ramet foraging, set the scale to a plant populations spatial heterogeneity. They are the means by which the plant "reaches out" to contact neighbors. They determine the selection pressures the plant will meet.

**3.3.3 Temporal division of the environment.** The structure of opportunity spacetime, the niche hypervolume, is apparent in the temporal use of the environment made by weed populations. Resources and conditions in a locality are unevenly distributed over time, the basis of differential life histories of weed species and the temporally differential expression of traits during life history development of the individual weed plant. Some few examples are presented in table 3.2.

**Table 3.2** Factors responsible for setting the scale of local temporal heterogeneity to which weed populations adapt and evolve.

| TIME | FACTOR | PLANT BEHAVIOR |
|---|---|---|
| SHORT-TERM | Phenology: timing of development and growth | seedling emergence times; leaf, branch and tiller timing |
| | Phenology: reproductive timing | flowering timing; seed production periodicity; time to seed maturity; pollenation timing |
| | Disturbance | tolerance to farming practices: e.g. tillage, herbicides; winter freezing |
| SEASONAL | Risk of mortality: temporal avoidance of neighbors, disturbance, stress | life cycle time and duration: perennials, biennials, summer annuals, winter annuals |
| LONG-TERM | Plant ecological succession | annual colonizing species succeeded by herbaceous perennials, by woody perennials |
| | Environmental adaptation | long-lived species (e.g. perennials) experience full range of yearly climate; annual species experience, and adapt to, seasonal periods of active growth |

**3.3.4 Disturbance.** The dimensionality of opportunity spacetime, the niche hypervolume, includes the dominant role of local habitat disturbance. The creation of new, and the destruction of old, opportunity spacetime for plant invasion can be a consequence of disturbance, a change in the local spacetime matrix, a natural selection pressure fostering invading and colonizing populations.

Changes in plant community structure, and the process of plant invasion, require an appreciation of the broad role disturbance plays in creating and destroying opportunity for new individuals to exploit. Competitive exclusion by extant individuals within a plant community puts invading species at a disadvantage in establishment. The role of disturbance in creating and destroying opportunity is crucial to community structure. It is often not fully considered in the ecology of apparently undisturbed habitats. This is reflected in the definitions provided for disturbance:

**disturbance**
1: the act of disturbing or the state of being disturbed (Anonymous, 1979, 2001)
2: an interruption or intrusion (Anonymous, 1979, 2001)
3: destruction of biomass by any natural or human agency (Silvertown and Charlesworth, 2001)

Disturbance cannot be avoided in studying agroecosystems, and the profound changes in plant community it causes. Herein a more inclusive definition of disturbance is provided:

**disturbance**
4: an interruption or intrusion with direct and indirect spatial, temporal, biological or abiological effects that alters or destroys a biological individual or community

Perturbation is closely related to disturbance:

**perturb**:
1: to disturb the composure of (Hanks et al., 1979)
2: to throw into disorder (Hanks et al., 1979)

**perturbation**:
1: the act of perturbing or the state of being perturbed (Hanks et al., 1979)
2: a cause of disturbance or upset (Hanks et al., 1979)

Disturbance is more than the direct cause of damage or mortality to a plant. It also includes the indirect effects of the abiotic environment, and the biological community (neighbors), with which the individual phenotype interacts. Disturbance of plant communities can be human-mediated or not. Disturbance can be divided into those due to apparently random causes (stochastic disturbance; e.g. fires, flooding, windstorms, and insect population explosions) and those caused by humans (e.g. crop tillage, herbicides, de-forestation). The boundary between these is not clear, limited by our ability to detect causation.

**3.3.4.1 Disturbance possesses dimensionality**. Disturbance can be understood by considering the biological community structure, and the abiotic environment, influencing the community at a locality (the population) and microsite (the individual) (Table 3.3).

**Table 3.3**. Dimensions of disturbance (spatial, temporal, biological community, abiotic environment), disturbance factors within each dimension, and examples of factors.

| Disturbance Dimension | Disturbance Factor | Examples |
|---|---|---|
| SPATIAL | -proximity of effect: direct or indirect<br>-localized or widespread<br>-heterogeneity and fragmentation | -direct, localized: lightning strike spot in field<br>-indirect, widespread: highway corridor effects on adjacent forests<br>-variable erosion and drainage effects with landscape elevation |
| TEMPORAL | -severity: quantity, frequency and duration<br>-regularity and predictability of patterns | -cycles: annual winter soil freezing<br>-crashes: yearly tillage of crop field<br>-catastrophes: removal of tropical rain forests |
| BIOLOGICAL COMMUNITY | -competitive neighbor interactions<br>-specificity and vulnerability: sensitivity and resistance<br>-change in biodiversity | -competitive exclusion by earliest emerging seedling in field<br>-response to predators, parasites and diseases<br>-increase in prairie fires with loss of large herbivores |
| ABIOTIC ENVIRONMENT | -resource availability<br>-inhibitors and stress<br>-climate and weather | -drought<br>-herbicides<br>-winter freezing of soil |

**3.3.4.2 Proximity of disturbance.** Disturbance can have proximate and distal effects in the creation and elimination of opportunity. Disturbance is more than the direct cause of damage or mortality to a plant. It also includes the indirect effects of the abiotic environment, and the biological community (neighbors), with which the individual phenotype interacts and is affected. Disturbance of plant communities can be human-mediated or not. The scale of the disturbance is also important: local versus widespread.

**3.3.4.3 Vulnerability to disturbance.** The susceptibility and sensitivity of a locality or microsite to invasion varies with the robustness and resistance of a local community to the traits possessed by an invading species. The vulnerability of habitats to invasion is often a function of the extent of direct and indirect disturbance by humans. Ironically, many agro-ecosystems have stable weed communities that resist invasion by new species. Weed populations often are stable due to the high, consistent level of disturbance management of these controlled systems. Population shifts are most likely to occur in these agriculture fields when crop management tactics change, e.g. introduction of new herbicides or herbicide resistant crops (Dekker and Comstock, 1992). The disturbances accompanying cropping systems creates very large selection pressures which open and close large amounts of opportunity spacetime depending on the life history traits of crops, control tactics and the temporal sequence of the systems components. The impact of these strong forces of selection are enhanced in annual cropping systems which eliminate above ground vegetation every year, leaving vast open opportunity spacetime available to weed infestation and invasion. Less disturbed habitats are often more vulnerable to invasion due to the fact that direct and indirect disturbance can dramatically change the ecological balance within these unmanaged biological communities, creating new opportunities (e.g. plant community changes due to the loss of large herbivores with human colonization of North America).

**3.3.4.4 Temporal patterns in disturbance.** Harper defined a fifth selection force driving variability in weed population, the evolutionary consequences of disturbance, which included differentiating crashes, cycles, catastrophes (Harper, 1977, pg. 769-774).

Catastrophes are rare, once in a lifetime events:

**catastrophes:**
1: an event subverting the order or system of things; significant population decrease, possible local extinction
2: disaster, a horrible event (Wikipedia, 5.08)

Catastrophe examples include the October blizzard of 1947, and the Armistice Day Freeze when 90% of Iowa apple (*Malus* sp.) trees died. Catastrophes are not cyclical, they occur on a time scale such that an organism can't adapt to their unpredictability. They can produce genetic changes that are themselves not adaptive as populations struggle to become reestablished (e.g. genetic bottleneck). Catastrophes are outside of the "memory" of organisms; events for which no member of the soil seed pool or population has adapted to (no pre-adaptations to prevent local extinction).

Of tangential interest, raising interesting speculations about disturbance and community structure predicability, is:

**catastrophe theory:** a field of mathematics that studies how the behavior of dynamic systems can change drastically with small variations in specific parameters (Wikipedia, 5.08)

In mathematics, catastrophe theory is a branch of bifurcation theory in the study of dynamical systems; it is also a particular special case of more general singularity theory in geometry. Bifurcation theory studies and classifies phenomena characterized by sudden shifts in behavior arising from small changes in circumstances, analysing how the qualitative nature of equation solutions depends on the parameters that appear in the equation. This may lead to sudden and dramatic changes, for example the unpredictable timing and magnitude of a landslide (e.g Bak, 1996). Small changes in certain parameters of a nonlinear system can cause equilibria to appear or disappear, or to change from attracting to repelling and vice versa, leading to large and sudden changes of the behavior of the system. However, examined in a larger parameter space, catastrophe theory reveals that such bifurcation points tend to occur as part of well-defined qualitative geometrical structures (Wikipedia, 5.08). See related topics in unit 5, complex adaptive weed systems; ch. 12.3.3, emergence; and ch. 14.3, the structure of randomness and probability.

Another consequence of disturbance is a crash:

**crash**:
1: an event of collapse or sudden failure
2: a decrease in a population, within the time scale [life history] of an organism

Potentially an organism is pre-adapted to crashes. Crashes are within an organism's evolutionary experience; potentially some variants with pre-adaptations will allow continuation in a locality. Examples include tillage, herbicide application, grazing, harvesting; longer or shorter seed rain period, dispersal; drought, flooding, freezing, fire, tornado, lightening; alien landing sites (e.g. crop circles); pathogen attack; diseases (epidemics).

Cycles occur on many time scales: diurnal: daily temperature cycles; seasonal: tillage, planting, cultivating, harvesting; and cycles of predator-prey population changes.

**cycle**:
1: happening at regular intervals
2: an interval of space or time in which one set of events or phenomena is completed
3: a complete rotation of anything
4: a process that returns to its beginning and then repeats itself in the same sequence
(Wikipedia, 5.08)

Cycles are predictable (predictable crashes), therefore weeds and other organisms are well adapted to them. The cycle crucial to weed populations is their life history, developed in detail in unit 3, adaptation in weed life history.

These three types of temporal patterns (catastrophes, crashes, cycles) can overlap. Also, these temporal patterns can be divided into those 'in' and 'out' of a local population's experience. Are there members of the local seed pool that have already experienced the disturbance type; is it in the seed bank's 'memory'? Are past adapted variants in the seed pool or is this experience completely new? All of these have important evolutionary consequences.

### 3.4 Limiting Resources and Pervasive Conditions in Local Opportunity

"Plants interfere with the distribution of neighbors by depleting limited resources. An effect will occur only when the depletion zone created by one plant includes the zone available to another. Resources (or supply factors) are exhaustible and contrast with conditions (quality factors) such as temperature that are not exhaustible. The nature of a resource (light, water, nutrients, $O_2$, $CO_2$) determines how a plant may affect a neighbors' growth; the diffusion of light, gases and nutrients are at very different rates and determine how far away from an individual depletion effects are sensed. Light, water and nitrates are probably the three resources most commonly involved but the interaction between resource factors makes it unrealistic to isolate any one resource as that for which competition occurs: the extent of interference below ground is not well understood." Harper (1977; summary, p. xvii; Ch. 10: pp. 305-345)

**3.4.1 Thermodynamic Earth.** From first principles, the most important global consideration in plant, weed, structure and function is thermodynamic Earth heat flow driving thermodynamic Earth engines. Once plant structure-function evolves in concert with Earth heat flow engines the next considerations are resources and pervasive conditions in locally available opportunity spacetime. The final consideration is the summation of these apparently separate influences into integrated environmental signals perceived by local plants in space and time.

**3.4.1.1 Thermodynamic Earth heat flow.** Plant shape and structure can be predicted from the universal tendency to facilitate flow access (Bejan and Zane, 2012; ch. 11.1.3). It's hard to recognize terrestrial plants as flow systems: water is flowing. Terrestrial plants happen because that is where the water is and must flow (upward), not because 'plants like water'. Plants do not exist in service to themselves but in service to the global thermodynamic flow that facilitates cyclical flow of water in nature: the Earth water cycle (see ch. 3.3.1, Earth's physical geography). Because of the second law of thermodynamics, water is governed by the natural tendency to equilibrate all the moisture in the environment. The terrestrial plant is a design for moving water. Only a small fraction is consumed by the plant; 99-99.5% is lost by transpiration. The bulk of it is pumped back into the atmosphere. Trees and forests, plants and their communities, are pumping stations operating all the time to move water from the ground to the air. Beginning with the roots that pull water from the surrounding area, to the stem that conveys water to the branches, to the leaves that release it when they open their pores (stomates) to capture carbon dioxide and sunlight for photosynthesis, the design of the tree is geared toward performing this work efficiently. The second law of thermodynamics proclaims that nature should manifest the tendency to move water from wet to dry both locally and globally.

Flow designs are global engines that have arisen to enable currents that flow through them to move easily across the landscape (Bejan and Zane, 2012). The source of almost all movement, all life, on Earth is the sun. For all the diversity we find in nature, the history of our planet is the unfolding story of the interaction between solar energy and the mass it sets in motion.

The sun shoots streams of energy in all directions. Some of these streams are intercepted and absorbed by the Earth. Altogether they represent one current of energy that flows out of the sun and into the Earth. This current flows from sun to Earth because the sun's temperature is higher than the Earth's. Similarly, a current of energy flows from the Earth to the sky, because the Earth is warmer than the sky. The plant is an intermediate stop for the train of energy from the hot sun to the cold sky. The total heat current that is rejected into the sky is the same as the heat current received from the sun (figure 3.1). Because solar energy heats the Earth unevenly, the heat on Earth flows in accordance with the second law of thermodynamics (from hot to cold; ch. 11.1.1, first principles: the thermodynamics of flow) and the constructal law (with evolving design; ch. 11.1.2). One word for this constantly morphing design is climate. The constructal law accounts for the fact that all live systems on Earth intercept and use this energy from the sun. As a result, the entire Earth is flowing, especially in its spherical shells that house the designs that we observe and interest us: the hydrosphere, atmosphere, lithosphere, and biosphere. They all flow by acquiring configurations that evolve in time.

**3.4.1.2 Thermodynamic Earth engines.** Anything that moves on Earth does so because it is driven (Bejan and Zane, 2012). The driving is done by very subtle engines, one engine for each flow. These engines include atmospheric circulation that brings the snow and the rains on the mountains and plains so that water will flow in the rivers. Solar heat falls on the warm zones and drives the ocean currents. No matter how numerous and diverse, all these engines are driven by fuel that comes from the sun in the form of the heat current intercepted by the Earth. All of these engines convert fuel into heat to perform work. The solar heat current hits the Earth and ultimately sinks completely into the cold universe. The Earth temperature settles at a steady level between the sun temperature and the sky temperature.

**Figure 3.1.** Schematic diagram of the Earth thermodynamic engine: one-way flows from high (hot, wet) to low (cold, dry).

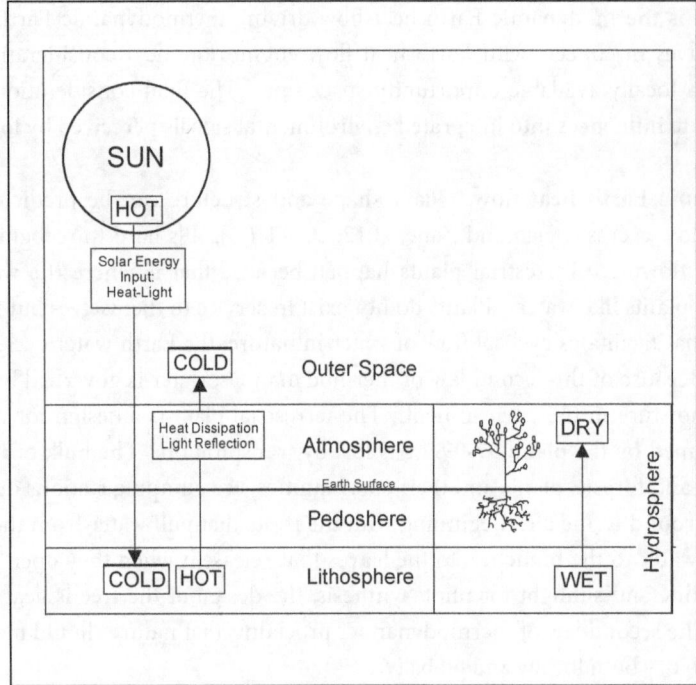

When heat flows from high temperature to low temperature, and momentum flows from fast fluid to slow fluid, the result is not just the movement of mass but the mixing and churning of hot and cold, fast and slow. This flow moves in one direction, from the entity that possesses it (whether it's momentum, warmth, chemical species, knowledge, food, culture, etc.) to that which does not. Our planet is an enormous river basin driven by the sun with a hierarchical distribution of evolving multiscale animate and inanimate channels, whose size and distribution are balanced to facilitate flow. The constructal law (see ch. 11.1.2) reveals that natural convection and turbulent flow are designs, the first designs, that emerged to facilitate the movement and mixing of currents (mass).

Life on earth is a tapestry of engines (which drive every flowing current) and brakes (all the resistances and losses that the currents encounter). All these designs, the engines and their brakes, evolve hand in glove and are governed by the laws of physics that include the constructal law. The whole earth is an engine-and-brake system, containing innumerably smaller engine-and-brake systems (winds, ocean currents, animals, and human-and-machine species; plants). All the engines are part of the single engine we call global design, or nature, whose flows get better and better in time. We do not see "engines" in what moves around us. The winds and the rivers seem to move by themselves. They are thought of as free and natural because we do not "pay" for moving them.

The design of nature is an engine-and-brake system for using and destroying the useful energy (exergy) streaming to Earth from the sun, and by the partial destruction of this flow in the animate and inanimate engines, followed by the complete destruction of the remaining useful energy stream in the interactions with the environment (the brakes). The second part of the design of nature is the movement that occurs against resistances that constantly try to stop it. Without resistances, the objects driven by the work would accelerate forever and spin out of control. All driven things dissipate all the driving work in the brakes that form between the moving objects and their immediate surroundings; the environment of any engine is a brake.

All the flow systems on Earth function as converters of useful energy (fuel and food) into mass moved. The engine-and-brake design of nature has evolved with time by generating a flow architecture that distributes imperfections through the flow space and endow it with configuration. The engine part evolves in time toward generating more power (or less dissipation), and as a consequence, the brake part evolves toward more dissipation. All flow systems improve over time, so that we find the evolution of better-flowing engines (lower dissipation) and more effective brakes (higher dissipation and higher rates or irreversibility or entropy generation).

The thermodynamic flows of water and heat through plants are the drivers, engines, through which weeds exploit opportunity spacetime. The Earth water cycle drives resource seizure by plants in the flow of dissolved nutrients and gases caught in evapo-transpiration streams of the xylem. The flow of heat in Earth systems both drives the water cycle and plant metabolism. Plants, weeds, channel these Earth water and heat flows to seize and exploit opportunity spacetime resources and conditions. Resource acquisition is driven by water flow through the plant xylem: nutrients and gases dissolved in soil water. Local conditions provide heat driving plant metabolism: the growth and development of the plant. Light resource is captured by the plant shoot architecture consequential to this growth and development.

The structure of opportunity spacetime, the habitat, the niche hypervolume, can be represented by the differential use of resources and conditions in local habitats. The diversity of community structure is dependent on their differential use. Plants interfere with neighbors by depleting limited resources available to another. Resources are exhaustible (supply factors) and contrast with pervasive conditions (quality factors) such as temperature (heat) that are not exhaustible. The presence of a plant changes the resource-condition environment of its neighbor, affecting the growth and form of both. Neighbor interactions can take many modal forms, including competition, interference, coexistence, cohabitation and synergism, topics developed in ch. 9, neighbor interactions in the local plant community.

**3.4.2 Limiting resources in the environment**. Seedling plant growth depends in the beginning on internal resources supplied by parents: stored seed food reserves (e.g. cotyledons, endosperm). Independent autotrophic growth by the seedling and mature plant depends on its ability to extract external resources from the environment: consumable supply factors such as light, nutrients, water and gases.

These resources are limited by neighbor effects, competition. It is very difficult to untangle what mechanism is involved when neighbor plants interfere with the growth and development of each other. Relief of density stress by the addition of a resource may not be sufficient evidence that the resource was the limiting factor in the stress. For example, the addition of water may increase nitrate availability and relieve shortages of both.

**3.4.2.1 Light as resource.** Supply of light is the most reliable of the environmental resources because of the regularity of diurnal and annual photocycles. Although light may be available above the canopy of leaves of a community, it can be limited below by shading from neighbor plants. Light does not accumulate. Light varies in intensity, duration, quality, direction and angle of incidence both in daily and annual cycles.

Light intercepted by plants can be reflected. It can be absorbed and converted into photosynthetic product or heat. The light profile of the leaf canopy changes as it passes down through the canopy to the soil. The canopy leaves can transmit and filter the light so that the lower leaves receive less quanta, and the light quality or spectra altered (e.g., enriched far red). Canopy shading can occur as self-shading of upper and lower leaves on the same plant, and mutual shading by neighboring plants. The intensity of light under plant vegetation commonly is below the compensation point (photosynthesis is balanced with respiration).

Light passing down throught the canopy is not a continuous gradient but a moving, dappled pattern of direct light added to a background of diffuse light. Leaf morphology varies between species, and on the same plant, in terms of the angle at which they are borne and consequently in the time of day at which they cast the greatest shadow. Leaf movements affect the amount of light that is intercepted. For example there exist diurnal leaf and shoot solar tracking/avoidance movements in *Helianthus* spp. and *Abutilon theophrasti*.

Photosynthesis in the canopy of a population depends on the affects of neighbors. In an isolated, individual leaf that is free of neighbor effects, the rate of photosynthesis increases linearly with increasing light intensity until a maximum value is reached. Beyond this maxima no further photosynthetic productivity increases occur with increasing light. The whole canopy may respond differently. There may occur increased photosynthesis in a population with increasing light intensity over a much wider range of light intensities. In a canopy the upper leaves become saturated more quickly as light intensity increases. The lower leaves in the shade may still respond to increasing light intensity that penetrates through the light-saturated upper canopy. Canopies are a population of leaves, not individuals. The population acquires a holistic physiology within which the individual plant is subordinated to the physiology of the whole. Plant populations adjust their structure and growth rate to the resources available. Perfect adjustment is impossible because the environment changes. Canopies are compromises, balancing respiration and photosynthesis.

Interactions among species, plants and parts of plants occurs through the structure of the plant body, plant architecture. Height above ground, lateral extension of branches or prostrate stems close to the ground, tillering and branching, all define what light is captured by what leaves. The effects of neighbors in light competition occur between individual leaves, not individual plants, permitting individual leaves to act as discrete units often interfering directly with other parts of the same plant. For example, different species may use light quality differently: shade and sun-loving competitors can coexist. For example, *Rosa* sp. (multi-flora rose) can form clumps in a grassy pasture. The rose shades out the grass, and the community structure appears clumpy with both present.

Light is a resource to fuel photosynthetic carbon fixation needed for growth and development. It also plays a crucial role as an environmental signal stimulating or inhibiting many plant developmental processes including photomorphogenesis and seed germination. Light as a signal is mediated by means of phytochromes.

**3.4.2.1.1 Plant phytochromes.** Light signal transduction is accomplished by means of plant phytochromes. Phytochrome is a photoreceptor, a pigment that plants use to detect light (Wikipedia, 11.10, 4.14). It is sensitive to light in the red and far-red region of the visible spectrum. Many flowering plants use it to regulate the time of flowering based on the length of day and night (photoperiodism) and to set circadian rhythms. It also regulates other responses including the germination of seeds (see ch. 7.3.6.5, light-phytochrome and nitrate regulated seed germination), elongation of seedlings, the size, shape and number of leaves, the synthesis of chlorophyll, and the straightening of the epicotyl or hypocotyl hook of dicot seedlings. It is found in the leaves of most plants.

Phytochromes are characterised by a red/far-red photochromicity. Photochromic pigments change their "color" (spectral absorbance properties) upon light absorption. Phytochromes are proteins with a light absorbing pigment attached (chromophore). The chromophore is a linear tetrapyrrole called phytochromobilin. The phytochrome apoprotein is synthesized in the *Pr* form. Upon binding the chromophore, the holoprotein becomes sensitive to light. If it absorbs red light it will change conformation to the biologically active *Pfr* form. The *Pfr* form can absorb far red light and switch back to the *Pr* form. In the case of phytochrome the ground state is $P_r$, the $_r$ indicating that it absorbs red light particularly strongly. The absorbance maximum is a sharp peak 650–670 nm, so concentrated phytochrome solutions look turquoise-blue to the human eye. But once a red photon has been absorbed, the pigment undergoes a rapid conformational change to form the $P_{fr}$ state. Here '$_{fr}$' indicates that now not red but far-red (also called "near infra-red"; 705–740 nm) is preferentially absorbed. This shift in absorbance is apparent to the human eye as a slightly more greenish colour. When $P_{fr}$ absorbs far-red light it is converted back to $P_r$. Hence, red light makes $P_{fr}$, far-red light makes $P_r$. In plants at least $P_{fr}$ is the physiologically active or "signalling" state. Phytochromes, photochromic pigments, change color, spectral absorbance properties upon light absorption. $P_r$ and $P_{fr}$ are unusually stable ($P_{fr}$ half-life hours to days).

$P_r$, the ground state, strongly absorbs red light (650-670 nm absorbance maximum):

$P_r$ → absorbs red, 650-670 nm → $P_{fr}$     rapid conformational change to active form

This conversion typically occurs in daylight. Red promotes seed germinaton and flowering; far-red reverses these actions.

$P_{fr}$, the physiologically active or "signaling" or excited state, referentially absorbs far-red (near infra-red; 705-740 nm absorbance maximum):

$P_{fr}$ → absorbs far-red, 705-740 nm → $P_r$    conformational change to inactive form

This conversion typically occurs slowly at night. This conversion occurs in shade (under plant canopies) where leaves absorb red, and reflect/transmit far-red.

In the red/far-red system, plants contain multiple blue light photoreceptors which have different functions. Higher plants contain at least four, and probably five, different blue light photoreceptors. Cryptochromes were the first blue light receptors to be isolated and characterized from any organism. Two cryptochromes have been identified in plants. Cryptochromes control stem elongation, leaf expansion, circadian rhythms and flowering time. Recent experiments indicate that a fourth blue light receptor exists that uses a carotenoid as a chromophore. This new photoreceptor controls blue light induction of leaf stomatal opening. However, the gene and protein have not yet been found.

Cryptochromes also perceive long wavelength UV irradiation (UV-A) in addition to blue light photoreceptors. Phototropin is the blue light photoreceptor that controls phototropism. Only one phototropin has been identified so far (NPH1). Phototropin also perceives long wavelength UV irradiation (UV-A) in addition to blue light (Wikipedia, 4.14).

Phytochromes are a family of photo-receptors (photo-thermal receptors). More than one type of phytochrome exist. Most plants have multiple phytochromes encoded by different genes. The different forms of phytochrome control different responses but there is also a lot of redundancy so that in the absence of one phytochrome, another may take on the missing functions. *Arabidopsis* spp. has five phytochrome genes (PHY A-E; possibly representing diploid dicots; Clack and Sharrock, 1994). Rice has only three (PHY A-C; possibly representing diploid monocots). Polyploidy can change this, increasing the alleles for, and numbers of, phytochromes in a species (Wikipedia, 2009).

An alternative way of viewing the photo-thermal responses is presented in De Petter et al. (1988) of work on the species *Kalanchoë blossfeldiana* Poelln. Three or four different seed germination responses were elicited by variable light fluence conditions, closely related to phytochrome responses, and revealing the intimate relationship between light and temperature signals (table 3.7).

**Table 3.7.** Variable light fluence conditions and signals for *Kalanchoë blossfeldiana* seed germination (De Petter, 1988).

| LIGHT FLUENCE | SIGNALS |
| --- | --- |
| VLFR (very low fluence response) | single light pulse; heat shock, temperature treatment |
| LFR (low fluence response) | imbibed, several daily light pulses; phytochrome |
| HFR (high fluence response) | prolonged irradiation (> 1h); insensitive to short irradiation |
| HIR (high irradiance response) | |

**3.4.2.1.2 Photomorphogenesis and phototropism.**

**photomorphogenesis**: light-mediated plant development (Wikipedia, 4.14)

**phototropism**: the growth or movement of an organism or plant part in response to light

Phototropism is the growth or movement of a plant part in response to a source of light (see ch. 9.3.1.2.4.1.1 light tropisms). It is most often observed in plants, but can also occur in other organisms such as fungi. The cells on the plant that are farthest from the light have a chemical called auxin that reacts when phototropism occurs. This causes the plant to have elongated cells on the farthest side from the light. Phototropism is one of the many plant tropisms or movements which respond to external stimuli. Growth towards a light source is called positive phototropism, while growth away from light is called negative phototropism. Most plant shoots exhibit positive phototropism, and rearrange their chloroplasts in the leaves to maximize photosynthetic energy and promote growth. Roots usually exhibit negative phototropism, although gravitropism may play a larger role in root behavior and growth. Some vine shoot tips exhibit negative phototropism, which allows them to grow towards dark, solid objects and climb them. The combination of phototropism and gravitropism allow plants to grow in the correct direction (see ch. 9.3.1.2, spatial foraging for local opportunity) (Wikipedia, 4.14). *Abutilon theophrasti* is highly responsive to shading by neighboring plants. Internodal stem length increases overtop those neighbors, a phototropic and photomorphogenic trait to preferentially capture light (see ch. 16.4 velvetleaf photomorphogenic plasticity).

**3.4.2.2 Water as resource.** The supply of water in a locality is often the least reliable of all the plant growth resources. It can be in excess or in limited supply. Plants draw water primarily from stored moisture, the soil acting as a buffer against the uncertainty of its availability. Plants act as wicks, drawing stored water reserves from the soil into the atmosphere via the xylem and leaf stomates (transpiration). With time and plant growth (leaf area, root system size), the size of this wick increases, increasing water loss rates. Bare soil dries quickly and impedes the continued loss of water from the soil. The dry surface no longer acts as a continuous wick, becoming to some extent self-sealing. Light and water are intimately related. Heat from solar radiation drives transpiration while light quanta drive photosynthesis. Stomatal leaf function links both together: it is impossible to separate the roles of light and water (and thus also soil nutrients) in limitations to plant growth. There exists an intimate role of $CO_2$ uptake with these phenomena: both pass into, out of, the plant through stomates.

The amount of water available to a plant is determined by the extensiveness of the root system. Dense plant populations suffer water shortages earlier in life than sparsely dispersed populations. Sparse populations cover the surface later with leaves, and water is conserved in the soil longer. Sparse populations suffer less neighbor stress by shading and nutrient depletion. These widely spaced plants are more vigorous and develop more extensive root systems than densely packed populations.

Variable plant shoot development above ground is reflected in below ground root system growth. Therefore water stress occurs differentially by these different individuals. Greater interference occurs between root systems of different individuals in a population than between parts of one root system, in contrast to the potentially greater interference between leaves on the same plant above ground. Different species develop different root systems at different times. Earlier germinating species (e.g. *Chenopodium album*) can use up water and nutrient resources sooner, depriving later emerging species. Different species may exploit different zones of the soil profile and avoid interference for water.

Water use affects community structure. For example, *Carnegiea gigantea* (sagaro cactus) and *Artimesia tridentata* (sagebrush) have different strategies in storing and seeking out water which determine their community structure. Saguaro holds water in its body and has a shallow root system. Sagebrush doesn't store water but has a deep and extensive root system to find water. Another example is found in wetlands. *Lythrum salicaria* (purple loosestrife) thrives in wetlands in which the water level is lower and dry at certain times of the year. *Typha latifolia* (cattail) can't tolerate periods of no water. Communities that contain both species indicate a response to selection pressure for a dry period tolerant cattail (ch. 9.3.2.2.2, ecological combining ability).

**3.4.2.3 Mineral nutrients as resources.** There are 17 essential plant nutrients (Wikipedia, 2010). Carbon and oxygen are absorbed from the air, while other nutrients including water are obtained from the soil. Plants obtain the other mineral nutrients from the soil (Barker and Pilbeam, 2007). They are the three primary macronutrients, three secondary macronutrients, and several micronutrients or trace minerals (table 3.4). The macronutrients are consumed in larger quantities and are present in plant tissue in quantities from 0.2% to 4.0% (on a dry matter weight basis). Micronutrients are present in plant tissue in quantities measured in parts per million, ranging from 5 to 200 ppm, or less than 0.02% dry weight. Functions of plant nutrients are described in table 3.4. Each of these nutrients is used in a different place for a different essential function.

**Table 3.4** Essential plant nutrients (Taiz and Zeiger, 2002).

**Macronutrients**

| | |
|---|---|
| **carbon** | forms the backbone of many plants biomolecules, including starches and cellulose. Carbon is fixed through photosynthesis from the carbon dioxide in the air and is a part of the carbohydrates that store energy in the plant. |
| **hydrogen** | necessary for building sugars and building the plant. It is obtained almost entirely from water. |
| **oxygen** | necessary for cellular respiration. Cellular respiration is the process of generating energy-rich adenosine triphosphate (ATP) via the consumption of sugars made in photosynthesis. Plants produce oxygen gas during photosynthesis to produce glucose but then require oxygen to undergo aerobic cellular respiration and break down this glucose and produce ATP. |
| **phosphorus** | important in plant bioenergetics. As a component of ATP, phosphorus is needed for the conversion of light energy to chemical energy (ATP) during photosynthesis. Phosphorus can also be used to modify the activity of various enzymes by phosphorylation, and can be used for cell signalling. Since ATP can be used for the biosynthesis of many plant biomolecules, phosphorus is important for plant growth and flower/seed formation. |
| **potassium** | regulates the opening and closing of the stomata by a potassium ion pump. Since stomata are important in water regulation, potassium reduces water loss from the leaves and increases drought tolerance. Potassium deficiency may cause necrosis or interveinal chlorosis. |
| **nitrogen** | an essential component of all proteins. Nitrogen deficiency most often results in stunted growth. |
| **sulphur** | a structural component of some amino acids and vitamins, and is essential in the manufacturing of chloroplasts. |
| **calcium** | regulates transport of other nutrients into the plant and is also involved in the activation of certain plant enzymes. Calcium deficiency results in stunting. |
| **magnesium** | an important part of chlorophyll, a critical plant pigment important in photosynthesis. It is important in the production of ATP through its role as an enzyme cofactor. There are many |

| | |
|---|---|
| | other biological roles for magnesium. Magnesium deficiency can result in interveinal chlorosis. |
| **silicon** | deposited in cell walls and contributes to its mechanical properties including rigidity and elasticity. |

**Micronutrients**

| | |
|---|---|
| **iron** | necessary for photosynthesis and is present as an enzyme cofactor in plants. Iron deficiency can result in interveinal chlorosis and necrosis. |
| **molybdenum** | a cofactor to enzymes important in building amino acids. |
| **boron** | important for binding of pectins in the RGII region of the primary cell wall, secondary roles may be in sugar transport, cell division, and synthesizing certain enzymes. Boron deficiency causes necrosis in young leaves and stunting. |
| **copper** | important for photosynthesis. Symptoms for copper deficiency include chlorosis. Involved in many enzyme processes. Necessary for proper photosythesis. Involved in the manufacture of lignin (cell walls). Involved in grain production. |
| **manganese** | necessary for building the chloroplasts. Manganese deficiency may result in coloration abnormalities, such as discolored spots on the foliage. |
| **sodium** | involved in the regeneration of phosphoenolpyruvate in CAM and C4 plants. It can also substitute for potassium in some circumstances, it can replace potassium's regulation of stomatal opening and closing |
| **zinc** | required in a large number of enzymes and plays an essential role in DNA transcription. A typical symptom of zinc deficiency is the stunted growth of leaves, commonly known as "little leaf" and is caused by the oxidative degradation of the growth hormone auxin. |
| **nickel** | in higher plants, essential for activation of urease, an enzyme involved with nitrogen metabolism that is required to process urea. Without Nickel, toxic levels of urea accumulate, leading to the formation of necrotic lesions. In lower plants, Nickel activates several enzymes involved in a variety of processes, and can substitute for Zinc and Iron as a cofactor in some enzymes. |
| **chlorine** | necessary for osmosis and ionic balance; it also plays a role in photosynthesis. |

**Other Nutrients**

| | |
|---|---|
| **cobalt** | beneficial to at least some plants, but is essential in others, such as legumes where it is required for nitrogen fixation. |
| **vanadium** | may be required by some plants, but at very low concentrations. It may also be substituting for molybdenum. |

Plant nutrition is a difficult subject to understand completely, partially because of the variation between different plants and even between different species or individuals of a given clone. An element present at a low level may cause deficiency symptoms, while the same element at a higher level may cause toxicity. Further, deficiency of one element may present as symptoms of toxicity from another element. An abundance of one nutrient may cause a deficiency of another nutrient.

Plants take up essential elements from the soil through their roots (dissolved in water) and from the air (mainly consisting of carbon and oxygen) through their leaves (Wikipedia, 2010). Nutrient uptake in the soil is achieved by cation exchange, where in root hairs pump hydrogen ions ($H^+$) into the soil through proton pumps. These hydrogen ions displace cations attached to negatively charged soil particles so that the cations are available for uptake by the root. In the leaves, stomata open to take in carbon dioxide and expel oxygen. Green plants obtain their carbohydrate supply from the carbon dioxide in the air by the process of photosynthesis. Though nitrogen is plentiful in the Earth's atmosphere, relatively few plants engage in nitrogen fixation (conversion of atmospheric nitrogen to a biologically useful form). Most plants therefore require nitrogen compounds to be present in the soil in which they grow. These can either be supplied by decaying organic matter, nitrogen fixing (soil-born, plant root nodules) bacteria, animal waste (manure), or through the agricultural application of purpose-made fertilizers. Nutrients are moved inside a plant to where they are most needed. For example, a plant will try to supply more nutrients to its younger leaves than its older ones. So when nutrients are mobile, the lack of nutrients is first visible on older leaves. However, not all nutrients are equally mobile. When a less mobile nutrient is lacking, the younger leaves suffer because the nutrient does not move up to them but stays lower in the older leaves. Nitrogen, phosphorus, and potassium are mobile nutrients, while the others have varying degrees of mobility. This phenomenon is helpful in determining what nutrients a plant may be lacking.

Plants obtain mineral resources primarily from the soil (exceptions include atmospheric nitrogen-fixing Leguminoseae family), and many of the conditions that govern their availability are similar to those affecting water. Soil minerals are held in the soil by physical and chemical linkages with insoluable soil components and are in a rapid, dynamic, equilibrium with ions in the soil water solution. When nutrients are removed by a root, there is a local lowering of its concentration, a diffusion gradient is created, nutrients diffuse along this gradient. Nitrate ions are an exception, not being held to soil colloids and being wholly mobile in the soil solution.

The transpiration stream in plants from the roots absorbing water from the soil, through the vascular system (xylem) to the leaves, creates a mass flow of soil solution towards the absorbing roots. Mass flow and nutrient diffusion in the soil maximize nutrient flow towards plants with the greatest growth: they tap the largest soil volume and they have the greatest transpiration rates. Some plants have "luxury" consumption of nutrients, the excess over immediate needs being absorbed when the plant is young and subsequently redistributed in the plant as it grows. This can occur when a plant takes up larger quantities of nutrients early in the competitive interaction with other species, preventing timely uptake by neighbors. *Chenopodium album* (common lambsquarters) is believed to be such a species.

Mycorrhizal associations with plant roots can affect nutrient availability, increasing uptake. A mycorrhiza is a symbiotic (generally mutualistic, but occasionally weakly pathogenic) association between a fungus and the roots of a vascular plant (Wikipedia, 4.14). In a mycorrhizal association, the fungus colonizes the host plant's roots, either intracellularly as in arbuscular mycorrhizal fungi (AMF or AM), or extracellularly as in ectomycorrhizal fungi. They are an important component of soil life and soil chemistry. Mycorrhizae form a mutualistic relationship with the roots of most plant species. While only a small proportion of all species has been examined, 95% of those plant families are predominantly mycorrhizal. They are named after their presence in the plant's rhizosphere (root system). This mutualistic association provides the fungus with relatively constant and direct access to carbohydrates, such as glucose and sucrose. The carbohydrates are translocated from their source (usually leaves) to root tissue and on to the plant's fungal partners. In return, the plant gains the benefits of the mycelium's higher absorptive capacity for water and mineral nutrients due to the comparatively large surface area of mycelium: root ratio, thus improving the plant's mineral absorption capabilities. Plant roots alone may be incapable of taking up phosphate ions that are demineralized in soils with a basic pH. The mycelium of the mycorrhizal fungus can, however, access these phosphorus sources, and make them available to the plants they colonize. In some cases, the transport of water, carbon, and nutrients could be done directly from plant to plant through mycorrhizal networks that are underground hyphal networks created by mycorrhizal fungi that connect individual plants together. The mechanisms of increased absorption are both physical and chemical. Mycorrhizal mycelia are much smaller in diameter than the smallest root, and thus can explore a greater volume of soil, providing a larger surface area for absorption. Also, the cell membrane chemistry of fungi is different from that of plants (including organic acid excretion which aids in ion displacement). Mycorrhizas are especially beneficial for the plant partner in nutrient-poor soils. Mycorrhizas are present in 92% of plant families studied (80% of species), with arbuscular mycorrhizas being the ancestral and predominant form, and indeed the most prevalent symbiotic association found in the plant kingdom. The structure of arbuscular mycorrhizas has been highly conserved since their first appearance in the fossil record, with both the development of ectomycorrhizas, and the loss of mycorrhizas, evolving convergently on multiple occasions

It is very difficult to identify the effects of nutrients as a limiting resource in plant populations. There exists an intimate relationship between water and nutrient availability. Supplying nutrients to plants may just speed up the time that light becomes limiting to growth. Enhanced fertility may result in increased root system size speeding up the time water availability becomes limiting to growth.

Nutrient resource use can affect community structure in many ways. For example, in a legume-grass pasture, leguminous clover fixs atmospheric nitrogen. Excess nitrogen in the soil solution is used by the grassy species. A stable, heterogeneous community structure arises from this differential supply and use of soil nitrogen.

**3.4.2.4 Gases as resources**. Nitrogen, water, carbon dioxide and oxygen.

*Nitrogen*. Over 75% of the atmosphere on earth is nitrogen, a neutral gaseous matrix.

*Water*. $H_2O$ in the gas phase plays many of the same roles as in the liquid phase. Humidity in the atmosphere and soil matrix play essential roles.

*Carbon dioxide*. The amount of carbon dioxide supplied to a leaf can control the rate of photosynthesis in that leaf, therefore it can be a limiting resource to plant growth. Determining whether $CO_2$ can be a limiting resource in the plant leaf canopies depends on two factors: evidence that levels of $CO_2$ fall in photosynthesizing canopies; and evidence that rate of photosynthesis falls in response to these decreased $CO_2$ levels. These situations are not typical of most above ground terrestial plant communities.

Movement and flux of $CO_2$ within plant canopies to leaf surfaces occurs by gaseous diffusion and by turbulent transfer (e.g. wind). There can sometimes occur diurnal cycles of measurable zones of $CO_2$ depletion and enrichment within canopies. Depletion zones are the greatest, and extend deep into canopy. Evening $CO_2$ levels can rise as respiration exceeds photosynthesis, enriching the local atmosphere. Wind can increase plant assimilation rates in canopies by resupplying depleted $CO_2$.

Differences in carbon metabolism occur in C3 and C4 plant species. In these species, photosynthetic rates can differ due to the role of the $CO_2$ compensation point. The compensation point is the $CO_2$ level below which photosynthesis does not occur. In C3 plants the compensation point is about 30-70 PPM $CO_2$. In C4 plants it is 5 PPM $CO_2$ or lower. C3/C4 plants utilize different biochemical pathways to assimilate $CO_2$. C4 plants can maintain very low $CO_2$ levels in the intercellular spaces of the leaf, creating a steeper diffusion gradient to the atmosphere which speeds up the movement of carbon to the sites of photosynthesis. The enhanced $CO_2$ uptake in C4 plants results in a narrower stomatal opening for less time. Smaller stomatal aperture results in more efficient water usage, C4 transpiring less water per unit dry weight gain more than C3 plants. C4 plants continue to increase photosynthesis with increasing light beyond the point C3 plants reach a plateau. C4 plants have a higher optimal temperature for photosynthesis than C3 plants. The net effect in plant populations is greater efficiency of C4 plant. This efficiency is reflected not in producing larger plants, but in a more effective water-conserving plant with a greater reproductive efficiency.

Overall, atmospheric $CO_2$ level is of lesser importance than understanding the factors that influence intercellular $CO_2$ levels in plants. Photosynthesis proceeds on the basis of carbon availability in the chloroplast in many instances. Intercellular $CO_2$ is a function of stomatal aperature, which in turn is a function of the plant metabolic system, water availability, transpiration, carbon utilization in the plant, and many more factors. Atmospheric $CO_2$ probably is relatively unimportant in terms of a limiting resource in most instances. Intercellular $CO_2$ is a very important limiting resource in plant growth and development; its availability is another excellent example of the dynamic relationship between all the resources in limiting productivity (water, light, nutrients)

*Oxygen as resource.* It is difficult to imagine oxygen being limiting to above ground plant parts. Oxygen may become limiting to plant growth below ground, in the soil. Diffusion of $O_2$ in water is 10,000 times slower than in air. The presence of water films in soil and on plant roots can slow or stop diffusion pathways in soil, and can hinder $O_2$ movement in soil. These factors are balanced by the extremely high affinity of plant terminal oxidase systems for oxygen. The net effect is that aerobic respiration in roots unlikely to be hindered unless the $O_2$ concentration approaches zero near the root. Local zones of $O_2$ depletion may arise in water-saturated soils. Short (1 day) periods of anaerobiosis can seriously damage roots. Neighbor roots may exaggerate depletion of $O_2$. Oxygen can be limiting in the soil in anaerobic conditions of flooding, soil compaction, unusually high rates of respiration by soil organisms. Oxygen solubility in water is inversely related to temperature: cool, wet spring soils are oxygen-rich; warm, dry summer soils are relatively oxygen-poor. Oxygen dissolved in soil water films plays a crucial role in seed germination (see ch.13, weed seed-soil environment communication systems).

**3.4.3 Pervasive Conditions in the Environment.** Heat and terroir are two pervasive conditions weeds experience in the environment.

**3.4.3.1 Heat.** The most common condition in the environment is temperature, or the presence of heat to support or hinder growth and development. Heat penetrates soil, plant and air so it is ubiquitous. Heat in the environment can affect the availability of oxygen dissolved in water, and the amount of water in the soil and atmosphere. Heat can interact with light, and in some cases synergize or substitute for light as an environmental signal.

**3.4.3.2 Terroir.** Terroir is the condition the individual plant finds itself in (Wikipedia, 2010). It is a helpful concept because it integrates many environmental conditions into a single perspective of what forms the local phenotype. The concept originated in French wine production, but herein we use it in the same manner for the conditions that influence the development of the individuals unique phenotypic growth, development, mortality and reproduction in a local deme.

**terroir**: the special characteristics that geography and geology bestow upon particular plants; "a sense of place," which is embodied in certain characteristic qualities; the sum of the effects that the local environment has had on the plant (Wikipedia, 2010)

Agricultural sites in the same region share similar soil, weather conditions, and farming techniques, which all contribute to the unique qualities of the crop. Particular consideration is given to the natural elements that are generally considered beyond the control of humans. Components often described as aspects of *terroir* include climate, soil type and topography.

**3.4.3.2.1 Climate.** The climate of a location is affected by its latitude, terrain, and altitude, as well as nearby water bodies and their currents. Climates can be classified according to the average and the typical ranges of different variables, most commonly temperature and precipitation.

**weather**: the state of the atmosphere, to the degree that it is hot or cold, wet or dry, calm or stormy, clear or cloudy; the present condition of climate and their variations over periods up to two weeks (Wikipedia, 2010)

**climate**: the statistics of temperature, humidity, atmospheric pressure, wind, rainfall, atmospheric particle count and other meteorological elements in a given region over long periods of time (Wikipedia, 2010)

The interaction of climate and *terroir* is generally separated into the macroclimate of a larger area, and the mesoclimate of a smaller subsection of that region, and even to the individual microclimate of a particular site.

**macroclimate**: the regional climate of a broad area, it can include an area on the scale of tens to hundreds of kilometers (Wikipedia, 2010)

**mesoclimate**: the climate of a particular site, generally restricted to a space of a tens or hundreds of meters (Wikipedia, 2010)

**microclimate**: the specific environment in a small restricted space, such as a crop row or the environment around an individual plant (Wikipedia, 2010)

**3.4.3.2.2 Soil**. Terroir includes the soil matrix, both the composition and the intrinsic nature of the soil (e.g. fertility, drainage, ability to retain heat). Edaphic factors include the soil type (sand, silt, clay, organic matter composition), soil pH, soil nutrient and moisture status, salinity, and soil aeration. Soil structure compaction (bulk density) and tilth determine the soil matrix structure of solids and the intervening gaseous atmosphere influencing plant growth.

**3.4.3.2.3 Topography**. Terroir includes the local affects of slope, aspect, latitude, longitude, elevation or altitude. Topography refers to the natural landscape features like mountains, valleys and bodies of water, which affect how the climate interacts with the region, and includes elements of aspect and altitude of the location

**3.4.4 Environmental signal spacetime**. The thermodynamic flows of water and heat through plants are the drivers, engines, through which weeds exploit opportunity spacetime. The Earth water cycle drives resource seizure by plants in the flow of dissolved nutrients and gases caught in evapo-transpiration streams of the xylem (hydro-nutrient-gas spacetime). The flow of heat in Earth systems both drives the water cycle and plant metabolism (hydro-thermal spacetime). Plants, weeds, channel these Earth water and heat flows to seize and exploit opportunity spacetime resources and conditions. Resource acquisition is driven by water flow through the plant xylem: nutrients and gases dissolved in soil water. Local conditions provide heat driving plant metabolism: the growth and development of the plant. Light resource is captured by the plant shoot architecture consequential to this growth and development (photo spacetime).

Resources and conditions in the local environment play many roles. Many are intimately related to each other and many can act in concert. They can act as supply factors for growth and development at relatively larger concentrations and availability, and they can act as signals stimulating biological responses at low concentrations; or vice-versa. Resources and conditions are sensed, perceived, as integrated signals by receptive plants. Light, water, heat and mineral nutrients are often impossible to separate (e.g. nitrates; photo-hydro-thermal-nitrate-time signal) in competitive interactions between neighbor plants. Light, water and heat (photo-hydro-thermal-time signal) affect leaf stomatal aperture opening and therefore photosynthetic carbon fixation. Conductance of moisture via soil-water films around soil particles regulates the amount of water-heat-oxygen-dissolved nutrient (oxy-hydro-thermal-nutrient-time signal) uptake from the soil by the roots (see ch. 17.4, *Setaria* seed-seedling life history behavior). Plants respond to particular integrated signals, but their sensitivity changes at different times over the course of their life history. Effective environmental signals stimulate responses in different ways at different times. The same integrated signal can have a stimulatory, and inhibitory, or no effect on the plant depending on its physiological and morphological sensitivity at that particular time in its life cycle. Environmental signals affecting plant responses occur at the scale of the microsite, at the specific cellular or tissue location of reception and action. Resources and conditions can transduce signals from the surrounding environment that stimulate or inhibit plant behavior. The nature of weeds is a atmosphere-soil-plant communication system (e.g. ch. 13, weed seed-soil environment communication systems; Dekker, 2014).

**signal transduction**: a mechanism that converts a mechanical/chemical stimulus to a cell into a specific cellular response. Signal transduction starts with a signal to a receptor, and ends with a change in cell function (Wikipedia, 11.10)

The dependability of a particular integrated signal to stimulate plant life history behaviors is one of the hallmarks of adaptation by the phenotype and species. The evolution of weedy adaptation is the evolution of plant traits specifically sensitive to the spacetime of dependable signals. For example, increasing temperature (thermal-time signal) in the spring is a grossly dependable signal in temperate regions of the world. Flucuations in this general trend vary considerably so immediate responses can be harmful to a plant if cooler conditions follow (e.g. flowering followed by a frost event). Other immediate signals can be beneficial to the plant. For example, high concentrations of nitrate in the soil solution (hydro-nitrate-time signal) can have an immediate stimulatory effect on *Chenopodium album* seed germination when coupled with adequate light, heat and moisture (photo-hydro-thermal-nitrate-time signal) (Altenhofen, 2009; Altenhofen and Dekker, 2013).

**3.4.4.1 Signal space dimensions**. What are the dimensions of environmental signal space? Trewavas (2000) had suggested there are several distinguishable environmental signals to which individual plants are sensitive: 1) water; 2-4) heat; 5-6) primary minerals: N, P, K, etc.; 8) light; 9) gravity; 10) soil structure; 11) neighbor competition; 12) herbivory; 13) disease; 14) allelopathy; 15) wind; 16) gases; oxygen and carbon dioxide. This incomplete list is an alternative way of expressing the dimensions of a niche hypervolume. Is there a better way of expressing this seemingly infinite range of parameters?

Weed seeds and underground propagules require three things to germinate and resume growth and development: water, heat and oxygen (oxy-hydro-thermal-time signal). Some weed seeds respond strongly to light, especially in the presence of dissolved nitrate in the soil solution (photo-hydro-nitrate-time signal). Emerged, autotrophic plants require gaseous carbon dioxide for photosynthesis, and light (photo-$CO_2$-time signal) for the induction of developmental processes such as flowering and physiological changes to the reproductive mode. Oxygen and mineral nutrients in the soil are only absorbed into living plant tissue when dissolved in water (oxy-hydro-nitrate-time signal), an integrated signal drawn into the plant or seed via continuous soil-water films. Carbon dioxide is absorbed directly in the gaseous phase, but must be dissolved in liquid solutions once inside the leaf to be utilized for photosynthesis.

The signal space for plant behavior can be defined specifically (table 3.5). The weed and crop growth environmental signal for shoot growth is photo-hydro-thermal-$CO_2$-time; and for underground root growth oxy-hydro-thermal-nutrient-time. Examples of integrated environmental signals for seed germination are found in table 3.6.

**Table 3.5**. Environmental signals affecting plant growth and development.

| Resource/Condition | Signal | Description |
|---|---|---|
| WATER | $H_2O_{QUANTITY}$ | water quantity |
| | $H_2O_{NUTRIENT}$ | soluable soil nutrients; e.g. nitrate ($NO_3$) dissolved in water |
| | $H_2O_{GAS}$ | soluable gases; e.g. oxygen dissolved in water |
| HEAT | $HEAT_{GROW}$ | heat for growth, development and |

|  |  | protective mechanisms |
|---|---|---|
|  | HEAT$_{GAS-SOLUBILTY}$ | gas solubility changes with temperature; e.g. gaseous oxygen in water |
|  | HEAT$_{SOLID-SOLUBILTY}$ | solid mineral solubility changes with temperature; e.g. nitrate (NO$_3$) dissolved in water |
|  | HEAT$_{LIQUID-SOLUBILTY}$ | liquid solubility changes with temperature; e.g. water in solid fraction of plant tissues |
| LIGHT | LIGHT$_{QUANTITY}$ | quantity, photon flux density/time |
|  | LIGHT $_{DURATION}$ | duration, photoperiod |
|  | LIGHT$_{QUALITY}$ | quality, wavelength |
| CARBON DIOXIDE | CO$_2$ | gaseous carbon |
| OXYGEN | H$_2$O$_{OXYGEN}$ | oxygen dissolved in water |

Table 3.6 Integrated environmental signals stimulating seed germination examples.

| Species | Signal | Description |
|---|---|---|
| *Zea mays* (maize) | hydro-thermal-time | porous seed envelopes allow unrestricted oxygen entry; rapid germination with adequate heat and water in the dark |
| *Setaria faberi* (giant foxtail) | oxy-hydro-thermal-time | gas- and water-tight seed envelopes restrict entry to small seed pore opening |
| *Chenopodium album* (common lambsquarters) | photo-oxy-hydro-thermal-nitrate-time | phytochrome light regulation in the presence of nitrate coupled with some seed envelope water entry restrictions |

**3.4.4.2 Photo-thermal modulation of weed seed germination behavior by plant phytochromes.** Germination (dormancy, recruitment timing, seedling establishment) is a fundamental life history trait. Seedling establishment time predicates realized fitness (Heschel et al., 2007). Seeds sense environmental conditions to ensure germination at the appropriate time of the year in their life history (Dekker, 2004b). The time of emergence is a major determinant of lifetime fitness in plants, explaining the majority (72%) of variance in fitness among an experimental population of *Arabidopis thaliana* genotypes (Donohue et al., 2007). Light signal transduction is accomplished by means of plant phytochromes. Phytochrome is a photoreceptor, a pigment that plants use to detect light and in some species to regulate germination of seeds (see ch. 3.4.2.1, light as resource; ch. 7.3.6.5, light-phytochrome and nitrate regulated seed germination; and ch. 13, weed seed-soil environment communication systems).

Phytochrome-mediated germination pathways simultaneously respond to light and temperature signals, cues, that affect germination (Wikipedia, 2008). This system is located in the living cell cytosol, and P$_{fr}$ may act directly with transcription factors controlling gene expression (Wikipedia, 2008). Phytochrome-mediated germination pathways simultaneously respond to light and temperature signals, cues, that affect germination. In some instances heat can substitute for light in seed germination (e.g. *Chenopodium album*; Altenhofen, 2009). Little is known about the integration of light and temperature information to elicit germination (Heschel et al., 2007). Phytochrome A, phyA, plays an important role promoting germination in warmer temperatures (Heschel et al., 2007, *Arabidopsis thaliana*). Phytochrome A photo-irreversibly triggers the photoinduction of seed germination after irradiation with extremely low fluence light in a wide range of wavelengths (UV-A to far-red) (Shinomura, 1997, *Arabidopsis thaliania*). Phytochrome B, phyB, important to germination across a range of temperatures (Heschel et al., 2007, *Arabidopsis thaliana*). Phytochrome B mediates the well-characterized photoreversible reaction, responding to red and far-red light of fluencies four orders of magnitude higher than those PhyA responds to (Shinomura, 1997, *Arabidopsis thaliania*). Far-red absorbing $P_{fr}$ form is involved in light-stimulated seed germination (Whitelam and Devlin, 1977, *Arabidopsis thaliana*). Phytochrome E, phyE, important to germination at colder temperatures (Heschel et al., 2007, *Arabidopsis thaliana*). It is not found in monocots (Matthews and Sharrock, 1997).

### 3.5 The Nature of Plant Invasions of Local Opportunity

The nature of opportunity spacetime provides the foundation for understanding how it is seized and exploited: the invasion process. The invasion process is the process of local plant community assembly. The first step in community assembly begins with colonization and ends with the terminal phases of ecological succession in the fully mature community. Community assembly begins with propagule recruitment (see ch. 9.1.2, recruitment: community assembly). With seedling emergence the local weed-crop community begins to be assembled. With community assembly begins interaction with neighbor plants and other organisms in the local community. These neighbor interactions frame, determine community assembly. During plant community ecological succession earlier successional species facilitate the invasion of later successional species by the opportunity spacetime they create in the habitat (see ch. 10.4.4). The driving forces of Earth thermodynamics ultimately determines community structure and assembly (see ch. 11.1.8, plant community structure and assembly). The multi-step process by which these colonizing events form new demes is developed fully herein.

The concept of 'invasive species' has broader social, economic and political implications, emphasizing the differences in how humans perceive weedy and colonizing species (Dekker, 2005). The most significant historical biological invasion was the 'Columbian Exchange' occurring after Christopher Columbus's four voyages to the Caribbean, and subsequent Spanish colonization and enduring occupation of the Americas. The exchange of diseases, domestic animals, crops and weeds was both a gift and catastrophe that irreversibly connected the New and Old Worlds (Crosby, 1972). As examples, the introduction and utilization of potatoes (*Solanum tuberosum*) exploited the hunger of the Russians and Irish, while smallpox exploited the vulnerability of native Americans.

**3.5.1 The plant invasion process: seizing, exploiting and occupying opportunity spacetime**. The plant invasion process is the mass movement or encroachment of an organism from one area to another (table 3.8). Successful plant invasion is the consequence of the presence of a particular species possessing life history traits suitable to exploit an opportunity space in a particular locality. Given these conditions, an individual invading species must successfully survive three processes: dispersal of that species into that locality, followed by colonization and enduring occupation of that habitat. Lastly, the species succeeding in occupying a locality must be perceived by humans as being problematic. Without the occurrence of all three processes, a plant species is not labeled invasive.

**Table 3.8** Plant invasion in a nutshell: opportunity, selection, invasion and extinction.

| OPPORTUNITY CREATION | Individual phenotypes respond to opportunities created by disturbances affecting locally available resource-conditions, to neighbors (or lack thereof), and to mortality | |
|---|---|---|
| NATURAL SELECTION FAVORS INDIVIDUALS | Natural selection favors individual phenotypes able to preferentially take advantage of these opportunities at the expense of their neighbors | |
| PREADAPTED LIFE HISTORY PHENOTYPES INVADE | Plants seize local opportunity by timing their life history to optimize the invasion process | **dispersal** of propagules into an opportunity |
| | | **colonization**: recruitment and establishment |

| | | **enduring occupation** of that locality for some time |
| --- | --- | --- |
| EXTINCTION | | Local populations become extinction ultimately from any locality |

Plant invasions are events in the ecology of community assembly and succession, as well as in the evolution of niche differentiation by speciation. There is not meaningful difference between the invasion process and these processes except the scale of attention humans bring to their observations. In all these processes disturbance is a prime motivator of change. Habitat disturbance as a direct or indirect consequence of human activity is of central importance. The scale of habitats in time and space is continuous; and all communities are inter-related.

Plant invasion can be succinctly described by four processes: dispersal, colonization, enduring occupation and extinction (table 3.9). The invasion process specifically consists of three sub-processes followed by extinction. Given a plant species with certain life history traits and a vulnerable local opportunity space, the invasion process is successful only when all of these are accomplished. Most invading species probably fail to complete all three steps, and there is little experimental information estimating the failure rate. All local plant populations become extinct eventually.

**Table 3.9** Invasion processes, life history activities and examples of incomplete invasion (invasion termination before completion).

| INVASION PROCESS | LIFE HISTORY ACTIVITY | Example |
| --- | --- | --- |
| INVASION | Dispersal | Propagule (e.g. seed, vegetative bud, spore, pollen) movement from one continent (or locality) to another and fails to reproduce |
| COLONIZATION | All events must occur:<br>- recruitment<br>- establishment<br>- reproduction | Volunteer maize (*Zea mays* L.) lives for only one generation ($F_2$) in a field, failing to colonize due to lack of dormancy |
| ENDURING OCCUPATION | Several modes possible:<br>- enduring presence for more than one generation<br>- range expansion<br>- formation of soil propagule (e.g. seed) pool | Successful, long-term, agricultural weeds; e.g. North America: *Amarathus* spp.-gp.; *Setaria* spp.-gp. |
| EXTINCTION | Mortality | Population shift from susceptible to resistant weed species biotypes with the widespread use a herbicide |

**3.5.2 Dispersal**. The first activity in invasion is successfully introducing propagules (seeds, vegetative buds, etc.) into a candidate opportunity space. Definition:

**dispersal**
1: outward spreading of organisms or propagules from their point of origin or release; one-way movement of organisms from one home site to another (Lincoln et al., 1998)
2: the act of scattering, spreading, separating in different directions (Anonymous, 2001)
3: the spread of animals, plants, or seeds to new areas (Anonymous, 1979)
4: the outward extension of a species' range, typically by a chance event; accidental migration (Lincoln et,

1998)

5: the search by plant propagules (e.g. seeds, buds) for opportunity in space and time

**3.5.3 Colonization.** The process of colonization includes three activities: recruitment, establishment and reproduction at the new locality. Definitions:

**colonization:**
1: the successful invasion of a new habitat by a species (Lincoln et al., 1998)
2: the occupation of bare soil by seedlings or sporelings (Lincoln et al., 1998)
3: (of plants and animals) to become established in (a new environment) (Anonymous, 1979)
4: all events must occur: recruitment, establishment, reproduction

**recruitment**
1: seedling and bud shoot emergence
2: the influx of new members into a population by reproduction or immigration [dispersal in] (Lincoln et al., 1998)

**establishment**: growing and reproducing successfully in a given area (Lincoln et al., 1998)

**3.5.4 Enduring occupation of a locality.** Several modes of long-term presence at a locality are possible. An invading species can have an enduring presence for more than one generation in the same locality. This long-term presence is often facilitated by plant traits that allow the formation of soil propagule (e.g. seed) pools. A species present in one locality can also expand its range into new localities.

Local selection and adapted phenotypes. Once a species successfully occupies a local site of some time period, the action of selection pressures result in local adaptation in favor of particular genotypes and phenotypes. The selection pressures these populations experience in the invasion and occupation phases derives from both biological, abiotic and human selection pressures. This local selection also acts on the variable phenotypes of that invading species and selects adapted biotypes that occupy that space into the future. Some of the consequences of this local evolution and adaptation include increases in locally-adapted phenotypes, range expansion beyond the locality, and population shifts in the local community as a consequence of altered neighbor interactions.

**3.5.5 Extinction.** All plant populations go extinct at some time. Definition:

**extinction**
1: the process of elimination, as of less fit genotypes
2: the disappearance of a species or taxon from a given habitat or biota, not precluding later recolonization from elsewhere

All local populations become extinct. The important considerations for an individual species are on what spatial and time scale these extinction events occur. Many of our most common crop field weeds (e.g. *Setaria*) have been around for thousands of years. But every local population goes extinct at some point. For example, many herbicide susceptible phenotypes have disappeared from local fields and were replaced by either herbicide resistant populations of the same species, or by other species in that locality. Within any field an individual weed species is spatially located in patches. These patches can change from year to year. On this local spatial scale extinction occurs continuously with re-invasion of adjacent areas. Most weed species accomplish this process of patch movement continuously on many spatial and temporal scales. Plant community succession is a series of invasions and extinctions. As the colonizers become established they create opportunity space for later successional plant species. On and on it goes.

### 3.5.6 The perception of plant invasion.

"We must make no mistake. We are seeing one of the great historical convulsions in the world's fauna and flora." Elton, 1958.

"It may seem that how we talk about nature is irrelevant to how we deal with it, a mere sematic gloss. But where alien species are concerned semantics is everything." Burdick, 2005

"I want to examine the concept of "native plants" within this framework [the crucial differences between the origin and later use of a biological feature, and the origin and later use of an idea], for this notion encompasses a remarkable mixture of sound biology, invalid ideas, false extensions, ethical implications, and political usages both intended and unanticipated. Clearly, Nazi ideologues provided the most chilling uses. In advocating native plants along the *Reichsautobahnen*, Nazi architects of the Reich's motor highways explictly compared their proposed restriction to Aryan purification of the people. By this procedure, Reinhold Tuxen hoped to "cleanse the German landscape of unharmonious foreign substance." In 1942 a team of German botanists made the analogy explicit in calling for the extirpation of *Impatiens parviflora*, a supposed interloper: "as with the fight against Bolshevism, our entire Occidental culture is at stake, so with the fight against this Mongolian invader; an essential element of this culture, namely the beauty of our home forest, is at stake." Gould, 1998

"The garden that I created myself shall ... be in harmony with their landscape environment and the racial characteristics of its inhabitants. They shall express the spirit of America and therefore shall be free of foreign character as far as possible. The Latin and Oriental crept and creeps more and more over our land, coming from the South, which is settled by Latin people, and also from other centers of mixed masses of immigrants. The Germanic character of our cities and settlements was overgrown ... Latin spirit has spoiled a lot and still spoils things every day." (in a statement by Jens Jensen published in a German magazine in 1937; quoted in Gould, 1998 and Wolschke-Bulmahn, 1995)

The biology of the invasion process as presented in the section above is rational and experimentally tractable. What is less apparent is the human component of the selection process that creates opportunity spaces into which invasive species disperse (e.g. Cockburn, 2015; Dekker, 2005; Gould, 1998). Of critical importance is the role human perception plays in selection and creation of opportunity space for invasive species. There exists a perception that invasive species are increasing of late due to increased global movement of people, trade, and transport of biological and agricultural commodities and novel plant materials. The terminology used by those interested in invasion biology is often defined somewhat differently by these respective groups. The perception of a plant species as invasive by humans is a complex, often highly subjective process. Plant invasions are events in the ecology of community assembly and succession, as well as in the evolution of niche differentiation by speciation. There is no meaningful difference between the invasion process and these processes except the scale of attention humans bring to their observations. In all these processes disturbance is a prime motivator of change.

# Chapter 4: Generation of Genotypic and Phenotypic Variation: First Process of Natural Selection

**The weeds' story**. I am foxtail.

You created opportunity for me as you civilized the Earth with your new food plants. You selected your favorites and killed most of the others. This was your gift to us, letting the best escape, hide, resist. Only a few of us follow you to new lands, new fields. These new explorer invaders are less variable than all of us. Each new field you create has unique opportunities, so we likewise must become unique to invade, exploit and endure.

Our secret as foxtails is to mate carefully with ourselves. This gift enables us to become locally unique by generating new siblings with new qualities. We begin by mixing our own genes and recombining them in each sibling. Then we turn them loose in the fields you created, and see what opportunity brings. Each new recombinate interacts with its unique environment. You kill most of the new plants that develop, thank you. You miss some of the new recombinates and they endure, thank you. Sometimes, rarely, we use our genes in a special way: we create a new species for our group by doubling our genes. We did it to make our specialist cousins: yellow, bristly, knotroot and giant foxtail.

Our other secret as foxtails is to grow precisely within the opportunity you and nature offer. Before we met you, each of us had preadapted growth traits. We sense local opportunity, then we grow just the right amount of leaves, stems and flowers for that provided. We also can form parts of our body in different ways, like our seeds, to maximize individual opportunity.

We use our secret traits and create a precise amount of newness in our different siblings. We test the newness out on you, to see which you kill and which survive. The clever new ones make us stronger. We owe it all to you.

**Summary.** Weed communities assemble in a new location seizing and exploiting local opportunity spacetime. The invading members of a weed species have preadapted traits that allow them to colonize the site. Once established they mate, recombining their genes in meiosis, dispersing their variable progeny into the habitat. The new genetic offspring interact with the opportunity provided by the local environment and produce new phenotypes which live or die.

The generation of new weed genotype-phenotype genetic variation is required for natural selection and elimination to drive local adaptation of the population. Generation of genetic variation appropriate for exploiting a particular locality allow an individual weed species to seize available opportunity spacetime in an efficient manner. The mating system, or mode of fertilization, of a weed species is crucial to the generation of appropriate amounts and types of genotypic variants. The mating system of a plant is the mechanism creating genetic diversity among its progeny. Additionally, several forces operate in nature to increase and decrease population genetic diversity.

Environmental changes external to the individual, as well as internal allelic forces, bring about changes in gene frequencies in a population. The primary internal allelic forces increasing variability are recombination of parental genes during meiosis, and mutation. Gene flow and introgression, as well as segregation distortion, also drive variation. Random sources of decreasing genetic diversity in a population include genetic drift, bottleneck and founder effects. Over time genetic diversity can lead to the formation of new species arising from older ones. There are several processes or modes by which a phylogenic lineage can diverge, all relying ultimately one of several types of reproductive isolating mechanisms.

Parent plants generate variable genotypes which confront the environment in their life history producing variable phenotypes. The expression of phenotypes from genotypes during development often has much flexibility. In weed species these phenotypes are often very different under varying environmental conditions. Phenotypic plasticity is the capacity of an organism to vary morphologically, physiologically or behaviorally as a result of environmental opportunity (e.g. extent of branching in an individual plant). Phenotypic variability can be expressed as discrete classes (canalization) or continuous variation of a trait. Phenotypic variation is also expressed in somatic polymorphisms (e.g. sexual dimorphism of male and female plants in dioecious *Amarathus* spp.-gp.).

Genotypes produce phenotypes in their interactions with the environment, and variation in phenotypes is usually associated with concomitant variation in genotype. But this is not always the case, especially in weed species which possess the ability to express phenotypic variation from a single genotype. How does a single genotype produce a wide array of phenotypes? Each gene does not act independently, numerous interactions among genes produce the phenotype. Many genes may simultaneously affect several aspects of the phenotype. In other instances a particular aspect of the phenotype my be affected by several different genes. These multiple interactions of genes are called epistasis.

Epigenetics is the study of heritable changes in gene activity that are not caused by changes in the DNA sequence. There exists a distinction between architecture, blueprint, and self-assembly during embryogenesis. Embryology works by self-assembly during development. Order, organization, and structure all emerge as by-products of rules which are obeyed locally and many times over, not globally. It is all done by local rules, at various levels but especially the level of the single cell.

The first process of natural selection generates the phenotypic variation needed for the second process resulting in local adaptation: differential survival, reproduction and inheritance.

## 4.1 Genotypes and Phenotypes

Individual organisms are the units of natural selection. The diversity found among individual weed plants in populations and agricultural floral communities arises from both genetic, somatic and behavioral variation. This diversity of individual plants and plant characters is revealed in the weed genotype and phenotype.

**genotype:**
1: the hereditary or genetic constitution of an individual; all the genetic material of a cell, usually referring only to the nuclear material
2: all the individuals sharing the same genetic constitution; biotype
3: the set of genes of an individual (Mayr, 2001)
4: the specimen on which a genus-group taxon is based; the primary type of the type species

**phenotype:**
1: the sum total of observable structural and functional properties of an organism; the product of the interaction between the genotype and the environment.
2: the total of all observable features of a developing or developed individual (including its anatomical, physiological, biochemical, and behavioral characteristics). The phenotype is the result of interaction between genotype and the environment (Mayr, 2001)
3: the characters of an organism, whether due to the genotype or environment

4: the manifested attributes of an organism, the joint product of its genes and their environment during ontogeny; the conventional phenotype is the special case in which the effects are regarded as being confined to the individual body in which the gene sits (Dawkins,1999)

5: phenotypic function consists always of two universes, the physical (quantitative, formal structure; physiological) and the phenomenal (quantities that constitute a 'world') (Sacks, 1998, p.129)

The relationship between the genotype (G), the phenotype (P) and the environment (E) is expressed as:

$$P = G \times E$$

Genes provide a 'blueprint' for developmental expression as phenotypes. The phenotype that arises from an individual weed genotype is highly variable, or plastic, in many traits. This variable and plastic gene expression is one of the key biological features of weeds. They possess the ability to respond to their local conditions in a very precise and fine-scale manner to maximize their productivity for the opportunity available in the environment.

**4.1.1 Extended phenotype.** A much broader view of the phenotype has been expressed by Richard Dawkins (1999), a concept he calls the 'extended phenotype'.

**extended phenotype**: all effects of a gene upon the world; 'effect' of a gene is understood as meaning in comparison with its alleles; the concept of phenotype is extended to include functionally important consequences of gene differences, outside the bodies in which the genes sit; in practice it is convenient to limit 'extended phenotype' to cases where the effects influence the survival chances of the gene, positively or negatively (Dawkins, 1999).

The extended phenotype is a biological concept introduced by Richard Dawkins in a 1982 book with the same title (Dawkins, 1999). The main idea is that phenotype should not be limited to biological processes such as protein biosynthesis or tissue growth, but extended to other aspects such as cognitive activities and more generally to the behavior of the animal in its environment. In the main portion of the book, Dawkins argues that the only thing that genes control directly is the synthesis of proteins. He points to the arbitrariness of restricting the idea of the phenotype to apply only to the phenotypic expression of an organism's genes in its own body. Dawkins develops this idea by pointing to the effect that a gene may have on an organism's environment through that organism's behavior, citing as examples caddis insect houses and beaver dams. He then goes further to point to first animal morphology and ultimately animal behavior, which can seem advantageous not to the animal itself, but rather to a parasite which afflicts it. Dawkins summarizes these ideas in what he terms the Central Theorem of the Extended Phenotype (Wikipedia, 11.10).

**central theorem of the extended phenotype**: an animal's behavior tends to maximize the survival of the genes "for" that behavior, whether or not those genes happen to be in the body of the particular animal performing it

The Beaver's Tale (and a bird's nest) in Dawkins (2004) provides a good example of the extended phenotype concept:

> "But beaver genes have special phenotypes quite unlike those of tigers, camels or carrots. Beavers have lake phenotypes, caused by dam phenotypes. A lake is an extended phenotype. The extended phenotype is a special kind of phenotype."

"Anatomical structures have no special status over behavioural ones, where 'direct' effects of genes are concerned. Genes are 'really' or 'directly' responsible only for proteins or other immediate biochemical effects. All other effects, whether on anatomical or behavioural phenotypes, are indirect."

"Any consequence of a change in alleles, anywhere in the world, however indirect and however long the chain of causation, is fair game for natural selection, so long as it impinges on the survival of the responsible allele, relative to its rivals."

"… only beavers have brain clockwork for dam-building … It only evolved because the lakes produced by dams are useful. … a lake provides a beaver with a safe place to build its lodge, out of reach from most predators, and a safe conduit for transporting food." "Those genetic variants that resulted in improved dams were more likely to survive in beaver gene pools."

"… behaviour … this too is a perfectly respectable phenotype. Differences in building behaviour are without doubt manifestations of differences in genes. And, by the same token, the consequences of that behaviour are also entirely allowable as phenotypes of genes. What consequences? Dams, of course. And lakes, for these are consequences of dams." "We may say that the characteristics of a dam, or of a lake, are true phenotypic effects of genes, using exactly the logic we use to say that the characteristics of a tail are phenotypic effects of genes."

"Conventionally, biologists see the phenotypic effects of a gene as confined within the skin of the individual bearing that gene. The Beaver's Tale shows that this is unnecessary. The phenotype of a gene, in the true sense of the word, may extend outside the skin of the individual. Birds' nests are extended phenoptypes. Their shape and size, their complicated funnels and tubes where these esixt, all are Darwinian adaptations, and so must have evolved by the differential survival of alternative genes."

See ch. 9.2.2.3.4, coevolution in parasitism, predation and pathogenesis, for an example of weed-human extended phenotypes overlapping in response to herbicide innovation.

**4.1.2 Epigenesis, epigenetics and gene expression.** Genotypes produce phenotypes in their interactions with the environment, and variation in phenotypes is usually associated with concomitant variation in genotype. But this is not always the case, especially in weed species which possess the ability to express phenotypic variation from a single genotype. How does a single genotype produce a wide array of phenotypes? Each gene does not act independently of other genes, there exist numerous interactions among genes to produce the phenotype. Many genes may simultaneously affect several aspects of the phenotype. In other instances a particular aspect of the phenotype my be affected by several different genes. These multiple interactions of genes are called epistasis.

**epistasis:**
1: a class of interactions between pairs of genes in their phenotypic effects; technically the interactions are non-additive which means, roughly, that the combined effect of the two genes is not the same as the sum of their separate effects; for instance, one gene might mask the effects of the other. The word is mostly used of genes at different loci, but some authors use it to include interactions between genes at the same locus, in which case dominance/recessiveness is a special case (Dawkins,1999)
2: the interaction of non-allelic genes in which one gene (epistatic gene) masks the expression of another at a different locus (Lincoln, et al.)

3: the nonreciprocal interaction of nonallelic genes; the situation in which one gene masks the expression of another
4: interactions between two or more genes (Mayr, 2001)
5: the interaction between genes; the effects of one gene are modified by one or several other genes (modifier genes) (Wikepedia, 5.09)

**polygenic inheritance (polygeny):**
1: inheritance of a trait govered by several genes (polygenes or multiple factors); their effect is cumulative (Mayr, 2001)
2: quantitative inheritance; multifactorial inheritance; inheritance of a phenotypic characteristic (trait) that is attributable to two or more genes and their interaction with the environment; polygenic traits do not follow patterns of Mendelian inheritance (qualitative traits). Instead their phenotypes typically vary along a continuous gradient depicted by a bell curve (Gaussian frequency distribution) (Wikepedia, 5.09)

A concept related to epistasis is pleiotropy. See ch. 18, triazine resistant and susceptible *Brassica napus* for a detailed case study of the many pleiotropic effects of a single mutation of the chloroplast *psbA* gene and the dynamic reorganization of structure and function in those plants.

**pleiotropy:**
1: pertaining to how a gene may affect several aspects of the phenotype (Mayr, 2001)
2: a single gene influences multiple phenotypic traits; a mutation in a gene may affect some all or all the traits simultaneously; selection on one trait may favors one allele while selection on another trait favors another allele (Wikepedia, 5.09)

Epistatic interactions of genes also implies a stable, dynamic reorganization among the many properties of the affected phenotype. All living organisms depend on maintaining a complex set of interacting metabolic chemical reactions. From the simplest unicellular organisms to the most complex plants and animals, internal processes operate to keep the conditions within tight limits to allow these reactions to proceed. Homeostatic processes act at the level of the cell, the tissue, and the organ, as well as for the organism as a whole (Wikepedia, 4.14).

**homeostasis:**
1: the property of a system (typically a living organism), either open or closed, that regulates its internal environment and tends to maintain a stable, constant condition; multiple dynamic equilibrium adjustment and regulation mechanisms make homeostasis possible (Wikepedia, 12.10)
2: the maintenance of metabolic equilibrium within an animal by a tendency to compensate for disrupting changes (Collins English Dictionary, 1979; Collins, Glasgow, UK)
3: the tendency for the internal environment of an organism to be maintained constant (Chambers Sci Dictionary)

Epigenesis occurs as an organism develops, guided by genes interacting with their local environment.

**epigenesis:**
1: 'in addition to' genetic information encoded in DNA sequence
2: heritable changes in gene function without DNA change
3: a theory of the development of an organism by progressive differentiation of an initially undifferentiated whole

4: the unfolding development in an organism, and in particular the development of a plant, fungus or animal from a seed, spore or egg through a sequence of steps in which cells differentiate and organs form (Wikipedia, 4.14)
5: each embryo or organism is gradually produced from an undifferentiated mass by a series of steps and stages during which new parts are added

There exists a distinction between architecture, blueprint, and self-assembly during embryogenesis. Embryology works by self-assembly. Order, organization, and structure all emerge as by-products of rules which are obeyed locally and many times over, not globally. It is all done by local rules, at various levels but especially the level of the single cell (see ch. 12.2, system of interacting elements, agents; and ch. 12.3, emergence of system order).

The originator of the theory of epigenesis was Aristotle in his book *On the Generation of Animals*. The theory was not given much credence in former times because of the dominance for many centuries of Creationist theories of life's origins. In the history of biology, preformationism (or preformism) is the idea that organisms develop from miniature versions of themselves. Instead of assembly from parts, preformationists believe that the form of living things exist, in real terms, prior to their development. It suggests that all organisms were created at the same time, and that succeeding generations grow from homunculi, or animalcules, that have existed since the beginning of creation. Epigenesis is the notion that "each embryo or organism is gradually produced from an undifferentiated mass by a series of steps and stages during which new parts are added." Epigenesis is still used, on the other hand, in a more modern sense, to refer to those aspects of the generation of form during ontogeny that are not strictly genetic, or, in other words, epigenetic (Wikipedia, 4.14): 'in addition to' genetic information encoded in DNA sequence, heritable changes in gene function without DNA change (Dawkins, R. 2009).

**epigenetics:**
1: the study of the mechanism that produces phenotypic effects from gene activity, processes involved in the unfolding development of an organism, during differentiation and development, or heritable changes in gene expression that do not involve changes in gene sequence
2: the study of how environmental factors affecting a parent can result in changes in the way genes are expressed in the offspring, heritable changes in gene function without DNA change
3: the study of reversible heritable changes in gene function that occur without a change in the sequence of nuclear DNA: how gene-regulatory information that is not expressed in DNA sequences is transmitted from one generation (of cells or organisms) to the next.
4: changes in gene expression due to mechanisms other than changes in DNA sequence
5: the study of the mechanisms of temporal and spatial control of gene activity during the development of complex organisms; thus epigenetic can be used to describe anything other than DNA sequence that influences the development of an organism (Holliday, 1990)

Epigenetics is the study of heritable changes in gene activity that are not caused by changes in the DNA sequence; it also can be used to describe the study of stable, long-term alterations in the transcriptional potential of a cell that are not necessarily heritable. Unlike simple genetics based on changes to the DNA sequence (the genotype), the changes in gene expression or cellular phenotype of epigenetics have other causes. The term also refers to the changes themselves: functionally relevant changes to the genome that do not involve a change in the nucleotide sequence. Examples of mechanisms that produce such changes are DNA methylation and histone modification, each of which alters how genes are expressed without altering the underlying DNA sequence. These epigenetic changes may last through cell divisions for the duration of the cell's life, and may also last for multiple generations even though they do not involve changes in the underlying DNA sequence of the organism. Iinstead, non-genetic factors cause the organism's genes to behave (or "express themselves") differently (Wikipedia, 4.14).

Epigenetics (as in "epigenetic landscape") was coined by C. H. Waddington in 1942 as a portmanteau of the words epigenesis and genetics. Epigenesis is an old word that has more recently been used to describe the differentiation of cells from their initial totipotent state in embryonic development. When Waddington coined the term the physical nature of genes and their role in heredity was not known; he used it as a conceptual model of how genes might interact with their surroundings to produce a phenotype; he used the phrase "epigenetic landscape" as a metaphor for biological development. Waddington held that cell fates were established in development much like a marble rolls down to the point of lowest local elevation. Waddington suggested visualising increasing irreversibility of cell type differentiation as ridges rising between the valleys where the marbles (cells) are travelling (Wikipedia, 4.14).

The epigenome includees transcription factor binding events and epigenetic phenomenona (Watters, 2006):

"Our DNA ... is regarded as the instruction book for the human body. But genes themselves need instructions for what to do, and where and when to do it." "Those instructions are found not in the letters of the DNA itself but on it, in an array of chemical markers and switches, know collectively as the epigenome, that lie along the length of the double helix. These epigenetic switches and markers in turn help switch on or off the expression of particular genes."

"... the epigenome is sensitive to cues from the environment." "... epigenetic signals from the environment can be passed on from one generation to the next, sometimes for several generations, without changing a single gene sequence." "... the epigenetic changes wrought by one's diet, behavior, and surroundings can work their way into the germ line and echo far into the future." "... the epigenome can change in response to the environment throughout an indivdual's lifetime."

"One form of epigenetic change physically blocks access to the genes by altering what is called the histone code." "Alterations to this packaging cause certain genes to be more or less available to the cell's chemical machinery and so determine whether those genes are expressed or silenced. A second, well-understood form of epigenetic signaling, called DNA methylation, involves the addition of a methyl group ... to particular bases in the DNA sequence. This interferes with the chemical signal that would put the gene into action and thus effectively silences the gene."

**4.1.3 Genome size, weediness and intra-genomic competition.** Dawkins (1999) has hypothesized, based on the work of several others, that weediness is associated with small genome size due to the heightened intra-genomic competition in these aggressive species. The significance of genome size in phylogeny is not understood, the so-called 'C-value paradox' (Orgel and Crick, 1980). Cavalier-Smith (1978) asserts there is a good correlation between low C-values and strong r-selection, selection for weedy qualities. Dawkins (1999) suggests that intragenomic selection pressure may lead to a decrease or elimination of "junk" DNA (untranslated introns) and therefore smaller genome size in colonizing, invasive species like foxtails.

The selfish gene theory postulates that natural selection will increase the frequency of those genes whose phenotypic effects ensure their successful replication. Generally, a gene achieves this goal by building, in cooperation with other genes, an organism capable of transmitting the gene to descendants.

"Each gene promotes its own selfish welfare, by cooperating with the other genes in the sexually stirred gene pool which is that gene's environment, to build shared bodies" Dawkins, 2004.

"The selfish DNA hypothesis is based on an inversion of this assumption: phenotypic characters are there because they help DNA to replicate itself, and if DNA can find quicker and easier ways to replicate itself, perhaps bypassing conventional phenotypic expression, it will be selected to do so." Dawkins, 1999.

Intragenomic conflict arises when genes inside a genome are not transmitted by the same rules, or when a gene causes its own transmission to the detriment of the rest of the genome. This last kind of gene is usually called selfish genetic element, or ultra-selfish gene or parasitic DNA (Wikipedia, 5.08).

"Mutation in the larger sense, however, includes more radical changes in the genetic system, minor ones such as inversions, and major ones such as changes in chromosome number or ploidy, and changes from sexuality to asexuality or vice versa. These larger mutations 'change the rules of the game' but they are still, in various senses, subject to natural selection. Intragenomic selection for selfish DNA belongs in the list of unconventional types of selection, not involving choice among alleles at a discrete distance." Dawkins, 1999.

Compare Trivers (1971) earlier work on selfish genes and altruism with Dawkins (1976, 1999); see ch. 9.3.2.3.2, altruism.

## 4.2 Generate Genetic Variation

The generation of new weed genotype-phenotype genetic variation is required for natural selection and elimination driving evolution. The mating system, or mode of fertilization, of a weed species is crucial to the generation of appropriate amounts and types of genotypic variants (see ch. 5.4, mating systems and inheritance). Sexual reproduction involves three steps: meiosis to form haploid cells; formation of gametes; and the fusion of gametes to form a new zygote, during which recombination occurs. Half the gametes produced contain a copy from the maternal, half from the paternal, dividing individual. Thus an individual produces a great range of gamete genotypes, from which a huge variety of diploid genotypes can arise. This is true for progeny derived from genetically different parents, and also for progeny of self-fertilization of genotypes in which some of the genes are heterozygous (Silvertown and Charlesworth, 2001). Thus is new novelty created in a species in a deme, the raw material for natural selection.

The mating system of a plant is the mechanism creating its genetic diversity, and several genetic forces result in increases and decreases in population diversity. Amounts of appropriate generation of genetic variation allow an individual weed species to seize available opportunity spacetime in an efficient manner.

Appropriate generation of genotypic variation by an individual plant, or a population, can be an effective hedge-betting strategy for enhanced fitness of progeny. Evolutionarily, hedge-betting is a strategy of spreading risks to reduce the variance in fitness, even though this reduces intrinsic mean fitness. Hedge-betting is favored in unpredictable environments where the risk of death is high because it allows a species to survive despite recurring, fatal, disturbances. Risks can be spread in time or space by either behavior or physiology. Risk spreading can be conservative (risk avoidance by a single phenotype) or diversified (phenotypic variation within a single genotype) (Jovaag, 2006; Cooper and Kaplan, 1982; Philippi and Segar, 1989; Seger and Brockman, 1987).

Maintenance of genetic variability in populations may be accounted for in two ways. Variation is selectively neutral: diversity and polymorphism observed are the result of slow random changes in the frequencies of alleles (Kimura, 1983). Or, variability is maintained within populations by natural selection. Generation, and conservation, of genetic inheritance is controlled in the individual plant by its mating system.

**4.2.1 Sources of Genetic Diversity.** Their exist genomic trade-offs between fidelity and mutability in the generation and loss of genetic diversity in a population or species. The forces in nature that drive genetic diversity can be either external or internal to the plant. To understand these changes in genetic diversity in a locality the following important concepts are defined:

**Hardy-Weinberg equilibrium**: the maintenance of more or less constant allele frequencies in a population through successive generations; genetic equilibrium

**Hardy-Weinberg law**: that allele frequencies will tend to remain constant from generation to generation and that genotypes will reach an equilibrium frequency in one generation of random mating and will remain at that frequency thereafter; demonstrating that meiosis and recombination do not alter gene frequencies

*External.* Much of the variation in populations and species is retained due to changes in the environment (adaptive traits are only good in some environments, and not in others). Hardy-Weinberg Law above indicates that the original variability in a population will be maintained in the absence of forces that tend to decrease or increase this variability: changes in gene frequencies are brought about by outside forces in the plants' environment.

*Internal.* The ultimate sources of variation, the source of new heritable characteristics in populations and species, are due to mutation and recombination in chromosomes, genes, DNA. The frequency of a gene, or allele, in a population is due to number of forces: forces increasing variability, and forces decreasing variability.

**4.2.1.1 Forces increasing population variability.** Four important allelic forces drive population heterogeneity: mutation, recombination, gene flow and segregation distortion. Community variation is also increased with speciation; in plants polyploidization plays an important role in generating new species, often founding polyploid species clusters.

**mutation:**
1: a sudden heritable change in the genetic material, most often an alteration of a single gene by duplication, replacement, insertion or deletion of a number of DNA base pairs;
2: an individual that has undergone such a mutational change; mutant

Many, most, mutations are deleterious; many, most, mutants are unfit. The most important force increasing population variation by far is recombination.

**recombination:**
1: any process that gives rise to a new combination of hereditary determinants, such as the reassortment of parental genes during meiosis through crossing over; mixing in the offspring of the genes and chromosomes of their parents.
2: event, occurring by crossing over of chromosomes during meiosis, in which DNA is exchanged between a pair of chromosomes of a pair. Thus, two genes that were previously unlinked, being on different chromosomes, can become linked because of recombination, and linked genes may become unlinked.

Recombination doesn't change gene frequency, but it does lead to combinations of different genes that could be better than others. Also, the number of genetic recombinations is infinitely larger than the possible number of mutations. Most new types in populations arise from recombination.

**gene flow:**
1: the exchange of genetic factors within and between populations by interbreeding or migration; incorporation of characteristics into a population from another population
2: in population genetics, gene flow (also known as gene migration) is the transfer of alleles of genes from one population to another (Wikipedia, 5.08).

**introgression, introgressive hybridization:**

1: the spread of genes of one species into the gene pool of another by hybridization and backcrossing (Lincoln et al., 1998)
2: the incorporation of genes of one species into the gene pool of another. If the ranges of two species overlap and fertile hybrids are produced, they tend to backcross with the more abundant species. This process results in a population of individuals most of whom resemble the more abundant parents but which possess also some characters of the other parent species (King and Stansfield, 1985)

Migration of genes into or out of a population may be responsible for a marked change in allele frequencies (the proportion of members carrying a particular variant of a gene). Immigration may also result in the addition of new genetic variants to the established gene pool of a particular species or population. There are a number of factors that affect the rate of gene flow between different populations. One of the most significant factors is mobility, as greater mobility of an individual tends to give it greater migratory potential. Animals tend to be more mobile than plants, although pollen and seeds may be carried great distances by animals or wind. Maintained gene flow between two populations can also lead to a combination of the two gene pools, reducing the genetic variation between the two groups. It is for this reason that gene flow strongly acts against speciation, by recombining the gene pools of the groups, and thus, repairing the developing differences in genetic variation that would have led to full speciation and creation of daughter species. Example, if a field of genetically modified corn is grown alongside a field of non-genetically modified corn, pollen from the former is likely to fertilize the latter (Wikipedia, 5.08).

**segregation distortion**: the unequal segregation of genes in a heterozygote due to:
1: an aberrant meiotic mechanism; e.g. meiotic drive: any mechanism operating differentially during meiosis in a heterozygote to produce the two kinds of gametes with unequal frequencies;
2: other phenomena that result in altered gametic transmission ratios; e.g. in pollen competition where one allele results in a more slowly growing pollen tube than an alternate allele. Gametes bearing this allele will therefore show up in zygotes at a frequency less than 50%, as will all genes linked to the slow growing pollen tube allele (Wendel, pers. comm., 1998).

*Meiotic drive.* All nuclear genes in a given diploid genome cooperate because each allele has an equal probability of being present in a gamete. This fairness is guaranteed by meiosis. However, there is one type of gene, called a segregation distorter, that "cheats" during meiosis or gametogenesis and thus is present in more than half of the functional gametes. True meiotic drive is found in other systems that do not involve gamete destruction, but rather use the asymmetry of meiosis in females: the driving allele ends up in the ovocyte instead of in the polar bodies with a probability greater than one half. This is termed true meiotic drive, as it does not rely on a post-meiotic mechanism. The best-studied examples include the neocentromeres (knobs) of maize, as well several chromosomal rearrangements in mammals (Wikipedia, 5.08).

**4.2.1.2 Forces decreasing population variability.** Natural selection and elimination is the central non-random source of decreasing genetic diversity. Random sources of decreasing genetic diversity in a population include genetic drift, bottleneck and founder effects.

Genetic drift or allelic drift is the change in the frequency of a gene variant (allele) in a population due to random sampling.

**genetic drift**:
1: the occurrence of random changes in the gene frequencies of small isolated populations, not due to selection, mutation or immigration; drift; Sewall Wright effect; equivalent to static noise in system; adaptive alleles can be lost in process, especially in small populations

2: in population genetics, genetic drift (or more precisely allelic drift) is the evolutionary process of change in the allele frequencies (or gene frequencies) of a population from one generation to the next due to the phenomena of probability in which purely chance events determine which alleles (variants of a gene) within a reproductive population will be carried forward while others disappear (Wikipedia, 5.08).

The alleles in the offspring are a sample of those in the parents, and chance has a role in determining whether a given individual survives and reproduces. A population's allele frequency is the fraction of the copies of one gene that share a particular form. Genetic drift may cause gene variants to disappear completely and thereby reduce genetic variation. When there are few copies of an allele, the effect of genetic drift is larger, and when there are many copies the effect is smaller. Vigorous debates occurred over the relative importance of natural selection versus neutral processes, including genetic drift. Ronald Fisher held the view that genetic drift plays at the most a minor role in evolution, and this remained the dominant view for several decades. Motoo Kimura (1968, 1983) rekindled the debate with his neutral theory of molecular evolution, which claims that most instances where a genetic change spreads across a population (although not necessarily changes in phenotypes) are caused by genetic drift (Wikipedia, 5.14).

Genetic drift is especially relevant in the case of small populations. The statistical effect of sampling error during random sampling of certain alleles from the overall population may result in an allele, and the biological traits that it confers, to become more common or rare over successive generations, and result in evolutionary change over time. The concept was first introduced by Sewall Wright in the 1920s, and is now held to be one of the primary mechanisms of biological evolution. It is distinct from natural selection, a non-random evolutionary selection process in which the tendency of alleles to become more or less widespread in a population over time is due to the alleles' effects on adaptive and reproductive success (Wikipedia, 5.08).

A population bottleneck is when a population contracts to a significantly smaller size over a short period of time due to some random environmental event resulting in genetic drift. Bottlenecks can occur at the foundation of the deme or later (e.g. disturbance).

**bottleneck effect**: fluctuations in gene frequencies occurring when a large population becomes very small, passes through a contracted stage and then expands again with an altered gene pool (usually one with reduced variability) as a consequence of genetic drift

In a true population bottleneck, the odds for survival of any member of the population are purely random, and are not improved by any particular inherent genetic advantage. The bottleneck can result in radical changes in allele frequencies, completely independent of selection. The impact of a population bottleneck can be sustained, even when the bottleneck is caused by a one-time event such as a natural catastrophe. After a bottleneck, inbreeding increases. This increases the damage done by recessive deleterious mutations, in a process known as inbreeding depression (Wikipedia, 5.14).

The founder effect is a special case of a population bottleneck, occurring when a small group in a population splinters off from the original population and forms a new one.

**founder effect**:
1. changes in allele frequencies caused by sampling at the foundation of new demes (Silvertown and Charlesworth, 2001)
2: that only a small fraction of the genetic variation of a parent population or species is present in a small number of founder members of a new colony or population (Lincoln et al., 1998)
3: the principle that when a small sample of a larger population establishes itself in a newly isolated entity, its gene pool carries only a fraction of the genetic diversity represented in the parental population. The evolutionary fates of the parental and derived populations are thus likely to be set along different pathways

because the different evolutionary pressures in the different areas occupied by the two populations will be operating on different gene pools (King and Stansfield, 1985)

The random sample of alleles in the newly formed new colony is expected to grossly misrepresent the original population in at least some respects. It is even possible that the number of alleles for some genes in the original population is larger than the number of gene copies in the founders, making complete representation impossible. When a newly formed colony is small, its founders can strongly affect the population's genetic make-up far into the future (Wikipedia, 5.14).

**4.2.2 Speciation**. No single weed species dominates a crop production field or an agroecosystem. Usually several weed species coexist in a field to exploit the diverse opportunity unused by crop plants (inter-specific diversity). Within a single weed species, a diverse population of genotypes and phenotypes interfere with crop production (intra-specific diversity). Given sufficient time and other factors, new species can arise from within current weed populations. Unused niche space left by homogeneous crop populations, diverse and fit weed populations, as well as crop management practices, provide ample opportunity for new weeds and the process of speciation.

**speciation**
1: the formation of new species
2: the splitting of a phylogenetic lineage
3: acquistion of reproductive isolating mechanisms producing discontinuities between populations
4: process by which a species splits into two or more species
5: the evolutionary process by which new biological species arise (Wikipedia, 5.08)

**species**
1: a group of organisms, minerals or other entities formally recognized as distinct from other groups
2: a taxon of the rank of species; in the hieracrchy of biological classification the category below genus; the basic unit of biological classification; the lowest principal category of zoological classification
3: a group of morphologically similar organisms of common ancestry that under natural conditions are potentially capable of interbreeding
4: a group of interbreeding natural populations that are reproductively isolated from other such groups (Lincoln et al., 1998)
5: the basic units of biological classification and a taxonomic rank; a group of organisms capable of interbreeding and producing fertile offspring (Wikipedia, 5.08)

The species concept becomes ambiguous in many detailed views of plant taxonomy (e.g. *Amaranthus* species-group; see also ch. 4.2.2.3.6, speciation via polyploidization). The biological species concept applies to biparental organisms exclusively. Uniparental organisms have arrays of clones and are not included in this concept (Grant, 1971).

**4.2.2.1 Process of speciation**. The process of speciation is a two stage process in which reproductive isolating mechanisms (RIM's) arise between groups of populations (table 4.1). Reproductive isolation is the condition in which interbreeding between two or more populations is prevented by factors intrinsic to their situation.

**4.2.2.1.1 Stage 1: Gene flow is interupted between two populations**. The absence of gene flow allows two populations to become genetically differentiated as a consequence of their adaptation to different local conditions (genetic drift also can act here too). As the populations differentiate, RIMs appear because different gene pools are not mutually coadapted. Reproductive isolation appears primarily in the form of postzygotic RIMs: hybrid failure. These early RIMs are a byproduct of genetic differentiation, and are not yet directly promoted by natural selection.

**4.2.2.1.2 Stage 2: Completion of genetic isolation.** Reproductive isolation develops mostly in the forms of prezygotic RIMs. The development of prezygotic RIMs is directly promoted by natural selection: alleles favoring intraspecific fertility will be increased over time at the expense of interspecific fertilization alleles.

**4.2.2.2 Reproductive isolating mechanisms.**

**reproductive isolating mechanism:** a cytological, anatomical, physiological, behavioral, or ecological difference, or a geographic barrier which prevents successful mating between two or more related groups of organisms.

**reproductive isolation**
1: the absence of interbreeding between members of different species
2: the condition in which interbreeding between two or more populations is prevented by intrinsic factors

Two types of RIMs facilitate speciation: prezygotic and postzygotic. Natural selection favors development of RIMs, especially prezygotic RIMs. Less favored by natural selection are postzygotic RIMs, which waste more energy.

**Table 4.1** Reproductive isolating mechanisms.

| **Prezygotic RIMs** prevent the formation of hybrid zygotes | Ecological isolation | populations occupy the same territory but live in different habitats, and thus do not meet |
| --- | --- | --- |
| | Temporal isolation | mating or flowering occur at different times, whether in different seasons, time of the year, or different times of the day |
| | Mechanical isolation | pollen transfer is forestalled by the different size, shape or structure of flowers |
| | Gametic isolation | female and male gametes fail to attract each other, or the pollen are inviable in the stigmas of flowers |
| **Postzygotic RIMs** reduce the viability or fertility of hybrids | Hybrid inviability | hybrid zygotes fail to develop or at least to reach sexual maturity |
| | Hybrid sterility | hybrids fail to produce functional gametes |
| | Hybrid breakdown | the progenies of hybrids (F2 or backcross generations) have reduced viability or fertility |

**4.2.2.3 Modes of speciation.** There are four modes of natural speciation, based on the extent to which speciating populations are geographically isolated from one another: allopatric, peripatric, parapatric, and sympatric. Speciation may also be induced artificially, through animal husbandry or laboratory experiments. (Wikipedia, 5.08).

**4.2.2.3.1 Natural speciation.** All forms of natural speciation have taken place over the course of evolution, though it still remains a subject of debate as to the relative importance of each mechanism in driving biodiversity. There is debate as to the rate at which speciation events occur over geologic time. While some evolutionary biologists claim that speciation events have remained relatively constant over time, some palaeontologists such as Niles Eldredge and Stephen Jay Gould have argued that species usually remain unchanged over long stretches of time, and that speciation occurs only over relatively brief intervals, a view known as 'punctuated equilibrium'.

**4.2.2.3.2 Allopatric speciation.** During allopatric speciation, a population splits into two geographically isolated allopatric populations (for example, by habitat fragmentation due to geographical change such as mountain building or social change such as emigration). The isolated populations then undergo genotypic and/or phenotypic divergence as they become subjected to dissimilar selective pressures or they independently undergo genetic drift. When the populations come back into contact, they have evolved such that they are reproductively isolated and are no longer capable of exchanging genes. Observed instances include island genetics, the tendency of small, isolated genetic pools to produce unusual traits. This has been observed in many circumstances, including insular dwarfism and the radical changes among certain famous island chains, like Komodo and Galápagos, the latter having given rise to the modern expression of evolutionary theory, after being observed by Charles Darwin. Perhaps the most famous example of allopatric speciation is Darwin's Galápagos Finches.

**4.2.2.3.3 Peripatric speciation.** In peripatric speciation, new species are formed in isolated, small peripheral populations which are prevented from exchanging genes with the main population. It is related to the concept of a founder effect, since small populations often undergo bottlenecks. Genetic drift is often proposed to play a significant role in peripatric speciation. Observed instances include the London Underground mosquito. This is a variant of the mosquito *Culex pipiens* which entered in the London Underground in the nineteenth century. Evidence for its speciation include genetic divergence, behavioral differences, and difficulty in mating.

**4.2.2.3.4 Parapatric speciation.** In parapatric speciation, the zones of two diverging populations are separate but do overlap. There is only partial separation afforded by geography, so individuals of each species may come in contact or cross the barrier from time to time, but reduced fitness of the heterozygote leads to selection for behaviours or mechanisms which prevent breeding between the two species. Ecologists refer to parapatric and peripatric speciation in terms of ecological niches. A niche must be available in order for a new species to be successful. Observed instances include the grass *Anthoxanthum* has been known to undergo parapatric speciation in such cases as mine contamination of an area.

**4.2.2.3.5 Sympatric speciation.** In sympatric speciation, species diverge while inhabiting the same place. Often cited examples of sympatric speciation are found in insects which become dependent on different host plants in the same area. However, the existence of sympatric speciation as a mechanism of speciation is still hotly contested. People have argued that the evidences of sympatric speciation are in fact examples of micro-allopatric, or heteropatric speciation. The most widely accepted example of sympatric speciation is that of the cichlids of Lake Nabugabo in East Africa, which is thought to be due to sexual selection. Sympatric speciation refers to the formation of two or more descendant species from a single ancestral species all occupying the same geographic location. Until recently, there has a been a dearth of hard evidence that supports this form of speciation, with a general feeling that interbreeding would soon eliminate any genetic differences that might appear. Sympatric speciation driven by ecological factors may also account for the extraordinary diversity of crustaceans living in the depths of Siberia's Lake Baikal.

**4.2.2.3.6 Speciation via polyploidization.** A new species can arise by means of a polyploidization event. Polyploidy refers to a special arithmetic relationship between the chromosome numbers of related organisms which possess different numbers. The polyploidy organism, in the simplest case, has a chromosome number which is twice that of some related form (Grant, 1971).

**ploidy:** the number of sets of chromosomes present (e.g. haploid, diploid, tetraploid)

**polyploidy:** multiple sets of homologous chromosomes in an organism (e.g. tetraploid, octaploid)

**polyploid complex, polyploid series:**
1: a genus, section, or species group containing (e.g.) diploids and tetraploids (Grant, 1971)
2: a group of interrelated and interbreeding plants that also have differing levels of ploidy that can allow genetic exchanges between unrelated species

A diploid cell undergoes failed meiosis, producing diploid gametes, which self-fertilize to produce a tetraploid zygote. Polyploidy is a mechanism often attributed to causing some speciation events in sympatry. Not all polyploids are reproductively isolated from their parental plants, so an increase in chromosome number may not result in the complete cessation of gene flow between the incipient polyploids and their parental diploids. Polyploidy is observed in many species of both plants and animals. In fact, it has been proposed that all of the existing plants and most of the animals are polyploids or have undergone an event of polyploidization in their evolutionary history.

The polyploid complex was first described by Babcock and Stebbins (1938). In *Crepis* and some other herbaceous perennial species, a polyploid complex may arise where there are at least two genetically isolated diploid populations, in addition to auto- and allopolyploid derivatives that coexist and interbreed (hybridize). Thus a complex network of interrelated forms may exist where the polyploid forms allow for genetic exchange between the diploid species that are otherwise unable to breed (Stebbins, 1940). A polyploid complex has also been well described in *Glycine* (Doyle, 1999). Classical examples include *Gossypium* and the *Nicotiana tabacum* group. This complex situation does not fit well within the biological species concept of Ernst Mayr which defines a species as "groups of actually or potentially interbreeding natural populations which are reproductively isolated from other such groups" (Lincoln et al., 1998; Wikipedia, 11.10). The polyploid complex and weed community structure is developed in more detail in ch. 10.3.2.3.2.

A weed example is the instantaneous formation of giant foxtail, *Setaria faberi*, when two diploid genomes (probably the ancestral *Setaria viridis* and an unknown *Setaria* species) hybridized in a polyploidization event (probably in Southern China) to form the new, fertile and viable, weed species that subsequently found niche not fully exploited by green foxtail (Darmency and Dekker, 2011). It now is a major weedy pest plaguing the US Corn Belt.

**4.2.2.3.7 Reinforcement.** Reinforcement is the process by which natural selection increases reproductive isolation. It may occur after two populations of the same species are separated and then come back into contact. If their reproductive isolation was complete, then they will have already developed into two separate incompatible species. If their reproductive isolation is incomplete, then further mating between the populations will produce hybrids, which may or may not be fertile. If the hybrids are infertile, or fertile but less fit than their ancestors, then there will be no further reproductive isolation and speciation has essentially occurred (e.g., as in horses and donkeys.) The reasoning behind this is that if the parents of the hybrid offspring each have naturally selected traits for their own certain environments, the hybrid offspring will bear traits from both, therefore would not fit either ecological niche as well as the parents did. The low fitness of the hybrids would cause selection to favor assortative mating, which would control hybridization. If the hybrid offspring are more fit than their ancestors, then the populations will merge back into the same species within the area they are in contact. Reinforcement is required for both parapatric and sympatric speciation. Without reinforcement, the geographic area of contact between different forms of the same species, called their "hybrid zone," will not develop into a boundary between the different species. Hybrid zones are regions where diverged populations meet and interbreed. Hybrid offspring are very common in these regions, which are usually created by diverged species coming into secondary contact. Without reinforcement the two species would have uncontrollable inbreeding. Reinforcement may be induced in artificial selection experiments as described below.

**4.2.2.3.8 Hybrid speciation.** Hybridization between two different species sometimes leads to a distinct phenotype. This phenotype can also be fitter than the parental lineage and as such natural selection may then favor these individuals. Eventually, if reproductive isolation is achieved, it may lead to a separate species. However, reproductive isolation between hybrids and their parents is particularly difficult to achieve and thus hybrid speciation is considered an extremely rare event. Hybridization without change in chromosome number is called homoploid hybrid speciation. It is considered very rare but has been shown in sunflowers (*Helianthus* spp.). Polyploid speciation, which involves changes in chromosome number, is a more common phenomena, especially in plant species.

### 4.3 Generate Phenotypic Variation

Phenotypic variation (due to underlying heritable genetic variation) is a fundamental prerequisite for evolution by natural selection (Wikipedia, 9.11). It is the living organism as a whole that contributes (or not) to the next generation, so natural selection affects the genetic structure of a population indirectly via the contribution of phenotypes. Without phenotypic variation, there would be no evolution by natural selection. Natural selection and elimination acts in the first, immediate, instance on phenotypes, not genotypes. The interaction between genotype and phenotype has often been conceptualized by the following relationship:

$$\text{genotype (G)} + \text{environment (E)} \rightarrow \text{phenotype (P)}$$

A more nuanced version of the relationship is:

$$\text{genotype (G)} + \text{environment (E)} + \text{genotype \& environment interactions (GE)} \rightarrow \text{phenotype (P)}$$

Genotypes often have much flexibility in the modification and expression of phenotypes; in many organisms these phenotypes are very different under varying environmental conditions (i.e. phenotypic plasticity).

Discontinuous phenotypic differences are variation that falls into discrete phenotype classes, canalization. Such traits show less variability within a genotype (caused by environment) than between different genotypes.

**canalization**: a measure of the ability of a population to produce the same phenotype regardless of variability of its environment or genotype

Canalization is the relative ability of a population to produce the same phenotype regardless of environmental variability or genotype: robustness. The term canalisation was coined by C. H. Waddington (1942), who used the word to capture the fact that "developmental reactions, as they occur in organisms submitted to natural selection...are adjusted so as to bring about one definite end-result regardless of minor variations in conditions during the course of the reaction". He used this word rather than robustness to take into account that biological systems are not robust in quite the same way as, for example, engineered systems. Biological robustness or canalisation comes about when developmental pathways are shaped by evolution. Waddington introduced the epigenetic landscape, in which the state of an organism rolls "downhill" during development. In this metaphor, a canalised trait is illustrated as a valley enclosed by high ridges, safely guiding the phenotype to its "fate". Waddington claimed that canals form in the epigenetic landscape during evolution, and that this heuristic is useful for understanding the unique qualities of biological robustness (Wikipedia, 5.14).

Continuous phenotypic differences are those without distinctly demarcated phenotypic classes. These quantitative characters are often subject to strong environmental effects (Silvertown and Charlesworth, 2001).

Individual weed plants possess a high degree of variability based on both their response to local environment and on their genetic constitution. Two important sources of phenotypic variation in weed plants are phenotypic plasticity and somatic polymorphism. Instances and examples of these fundamental weed traits are fully developed in units 3 and 4, adaptation in weed life history and local plant communities.

**4.3.1 Phenotypic plasticity.** Phenotypic plasticity is the capacity for marked variation in the individual plant phenotype as a result of environmental influences on the genotype during development. An individual plant can grow larger of smaller depending on the resources and conditions available to it in its habitat. Phenotypic plasticity can be expressed through epigenetic mechanisms in weed plants.

**phenotypic plasticity:**
1: the capacity of an organism to vary morphologically, physiologically or behaviorally as a result of environmental flucuations; reaction type
2: the capacity for an individual for marked variation in the phenotype as a result of environmental influences on the genotype during development, during the plant's life history

The ability of an organism with a given genotype to change its phenotype in response to changes in the environment is called phenotypic plasticity. Such plasticity in some cases expresses as several highly morphologically distinct results (canalization); in other cases, a continuous norm of reaction describes the functional interrelationship of a range of environments to a range of phenotypes. The term was originally conceived in the context of development, but is now more broadly applied to include changes that occur during the adult life of an organism, such as behavior.

Organisms of fixed genotype may differ in the amount of phenotypic plasticity they display when exposed to the same environmental change. Hence phenotypic plasticity can evolve and be adaptive if fitness is increased by changing phenotype. Immobile (sessile) organisms such as plants have well developed phenotypic plasticity, giving a clue to the adaptive significance of plasticity.

The range of possible phenotypes that a single genotype produces is often expressed experimentally in terms of a 'reaction norm' or 'norms of reaction':

**reaction norm:**
1: set of phenotypes expressed by a singe genotype, when a trait changes continuously under different environmental and developmental conditions
2: phenotype space; opportunity space; hedge-bet structure
3: a norm of reaction describes the pattern of phenotypic expression of a single genotype across a range of environments (Wikipedia, 5.08)

In ecology and genetics, a norm of reaction describes the pattern of phenotypic expression of a single genotype across a range of environments. One use of norms of reaction is in describing how different species—especially related species—respond to varying environments. But differing genotypes within a single species will also often show differing norms of reaction relative to a particular phenotypic trait and environment variable. For every genotype, phenotypic trait, and environmental variable, a different norm of reaction can exist; in other words, an enormous complexity can exist in the interrelationships between genetic and environmental factors in determining traits (see ch. 3.2.3, the niche hypervolume).

Scientifically analyzing norms of reaction in natural populations can be very difficult, simply because natural populations of sexually reproductive organisms usually do not have cleanly separated or superficially identifiable genetic distinctions. However, seed crops produced by humans are often engineered to contain specific genes, and in some cases seed stocks consist of clones. Accordingly, distinct seed lines present ideal examples of differentiated norms of reaction. In fact, agricultural companies market seeds for use in particular environments based on exactly this.

Suppose the seed line A contains an allele b, and a seed line B of the same crop species contains an allele B, for the same gene. With these controlled genetic groups, we might cultivate each variety (genotype) in a range of environments. This range might be either natural or controlled variations in environment. For example, an individual plant might receive either more or less water during its growth cycle, or the average temperature the plants are exposed to might vary across a range.

A simplification of the norm of reaction might state that seed line A is good for "high water conditions" while a seed line B is good for "low water conditions". But the full complexity of the norm of reaction is a function, for each genotype, relating environmental factor to phenotypic trait. By controlling for or measuring actual environments across which monoclonal seeds are cultivated, one can concretely observe norms of reaction. Normal distributions, for example, are common. Of course, the distributions need not be bell-curves.

A simplification of the norm of reaction might state that seed line A is good for "high water conditions" while a seed line B is good for "low water conditions". But the full complexity of the norm of reaction is a function, for each genotype, relating environmental factor to phenotypic trait. By controlling for or measuring actual environments across which monoclonal seeds are cultivated, one can concretely observe norms of reaction (Wikipedia, 5.08).

**4.3.2 Somatic Polymorphism.** Somatic variation pertains to diversity in the plant body or any non-germinal cell, tissue, structure or process. Somatic polymorphism is the genetically controlled expression of diverse plant parts and processes independent of environment: e.g. cotyledons and true leaves in plants are both leaf tissue.

**somatic polymorphism**
1: production of different plant parts (morph), or different plant behaviors, within the same individual plant; the expression of somatic polymorphism traits is not much altered by the environmental conditions it encounters (as opposed to phenotypic plasticity)
2: the occurrence of several different forms of a structure-organ of a plant body; distinctively different forms adapted to different conditions.

Somatic (body) polymorphism (many forms) is a type of phenotypic biodiversity. It is the production of different plant parts, or different plant behaviors, within the same individual plant.

Polymorphism also refers to the occurrence of structurally and functionally more than two different types of individuals. Morphs must occupy the same habitat at the same time and belong to a panmictic population (one with random mating). In polyphenism, an individual's genetic make-up allows for different morphs, and the switch mechanism that determines which morph is shown is environmental. In genetic polymorphism, the genetic make-up determines the morph. A polyphenic trait is a trait for which multiple, discrete phenotypes can arise from a single genotype as a result of differing environmental conditions. It is therefore a special case of phenotypic plasticity,

Somatic diversity is ultimately genetic, heritable, it is the product of differential genetic expression due to differential penetrance and expressivity of plant genes. Unlike phenotypic plasticity, somatic polymorphism is always expressed in plants that possess those traits, and is not altered much by the environment conditions it encounters. It usually functions to retain variety of form in a population living in a varied environment.

Somatic polymorphisms are important traits for weeds to possess. Somatic diversity can be found in leaves on the same plant, seed from the same plant, and the form of the plant at different times of its life cycle. The most common example is sexual dimorphism (phenotypic difference between males and females of the same species), which occurs in many important weed species (e.g. *Amaranthus, Cannibis*).

**4.3.2.1 Somatic polymorphism in flowers.** *Daucus carota* (wild carrot) has two types of flowers. The predominate flowers are small and dull white, clustered in flat, dense umbels (Wikipedia, 1.15). There may be a second type of reddish flower in the centre of the umbel. *D. carota* was introduced and naturalised in North America, where it is often known as "Queen Anne's lace". Both Anne, Queen of Great Britain, and her great grandmother Anne of Denmark are taken to be the Queen Anne for which the plant is named. It is so called because the flower resembles lace; the red flower in the center is thought to represent a blood droplet where

Queen Anne pricked herself with a needle when she was making the lace. The function of the tiny red flower, coloured by anthocyanin, is to attract insects.

The domestic sunflower (*Helianthus annuus*) possesses a single large inflorescence atop an unbranched stem. This "flower head" or pseudanthium consists of numerous small individual five-petaled flowers ("florets") of two different types, ray and disk. The outer flowers, which resemble petals, are called ray flowers. Each "petal" consists of a ligule composed of fused petals of an asymmetrical ray flower. They are sexually sterile and may be yellow, red, orange, or other colors. The flowers in the center of the head are called disk flowers. These mature into fruits (sunflower "seeds"). The disk flowers are arranged spirally.

**4.3.2.2 Somatic polymorphism in seeds.** Foxtail plants shed seed with different germination requirements, an important form of seed somatic polymorphism (see ch. 7.3, embryo adaptation: embryogenesis and dormancy induction; see ch. 7.3.3, seed seed dormancy variability and somatic polymorphism). Most *Setaria faberi* (giant foxtail) seed shed in the fall is dormant, but some can germinate immediately (see ch. 7.3, embryo adaptation: embryogenesis and dormancy induction; ch. 8.4.6.1, *Setaria* seedling emergence in central Iowa; ch. 17.4.5.3, seed germinability-dormancy heteroblasty blueprints seed behavior in the soil). Some try to germinate even earlier, seed vivipary. Maize (*Zea mays* premature germination), and not just weeds, possess this type of somatic diversity, seed vivipary. In corn, plant breeders have purposefully bred this trait out to the best of their ability. Still, every once in a while some corn will germinate before it should. Common cocklebur (*Xanthium strumarium*) seed capsules contain two seeds. One can germinate in the first year following dispersal, the other is more dormant and germinates later. This is an important form of seed dormancy somatic polymorphism: different seed dormancy in seeds from the same plant. *Avena fatua* (wild oat) also shed seed with different germination requirements, dormancy. Seed from different parts of the panicle possess different levels of dormancy. *Chenopodium album* (common lambsquarters) shed both black and brown seed. The darker seed is usually more dormant.

**4.3.2.3 Somatic polymorphism in leaves.** *Glycine max* (soybeans) provide a common example of leaf somatic diversity. The first set of leaves are the seed leaves (cotyledons), then the unifoliates emerge, then a continuing series of trifoliates. Soybeans possess different leaf forms, for different functions, at different times of the life cycle.

*Setaria* species-group (weedy foxtails) possess the ability to tiller, produce additional shoots after the emergence of the primary parental shoot. These stem branches allow it to take advantage of opportunities. Each subsequent tiller has a different shape and size. Stem length, panicle size, all seem to get smaller as tillering proceeds.

**4.3.2.4 Somatic polymorphism: seasonal dimorphism.** Another form of somatic diversity is seasonal dimorphism: two distinct phases of growth of an individual plant, each adapted to a specific season (seasonally dimorphic phenotype). Two growth, or plant, forms occur in the *Arctium minus* (common burdock) during its biennial life history. In the first year the plant grows leaves, stems, and roots (vegetative structures), then it enters a period of dormancy over the colder months. Usually the stem remains very short and the leaves are low to the ground, forming a rosette. During the next spring or summer, the stem elongates greatly, or "bolts". The plant then flowers, producing fruits and seeds before it finally dies. Dandelion (*Taraxacum officinale*) can exist as a summer (emerging in the spring) or winter (emerging in the autumn) annual. Many weed species in the mustard (*Brassiceae*) family can exist as summer and winter annuals as well as biennials.

# Chapter 5: Survival, Reproduction and Inheritance: Second Process of Natural Selection

**The weeds' story**. I am foxtail.

I cover the civilized Earth you provide for me. I thrive globally, I eat locally. I mate with myself to ensure I gather all the opportunity you provide for me in the field of a season. My self-mating generates new and different children. We have sex by recombining carefully, at the appropriate time. The many bodies of my children's growth is plastic, appropriate to the opportunity you provide, and no more. They avoid your fatal hand, escape your inconstant efforts, resist your tools, fool you in any way they can. You still manage to kill most of my children, but the rest mate. They form new seeds which inherit all of our surviving traits. We survive, mate and inherit in your image. We conquered the civilized Earth together.

I foxtail have endured. How did you make me so successful? For over 10,000 years I have invaded the civilizing opportunity you provided and formed local populations. In each you disturbed and provided what I needed to grow. I mated and gave birth to unique children, each with special gifts. You killed most of them, thereby making my survivors uniquely adapted to your local field, to your local farmer.

I foxtail evolve. What remains to be revealed is the nature of these special gifts we have coevolved together over all these millennia. What special traits do I have? When in my life do I express them? How do my local crop and weed neighbors react when confronted by my actions? How does my complexity ensure my endurance, my anti-fragility? How should you humans portray me to understand me? These are my mysteries yet to be revealed.

**Summary**. Heterogeneous weed populations invade, colonize and endure in a locality by surviving their developmental life history, culminating in reproduction of offspring inheriting their adaptive traits. Survival and reproduction can be represented demographically with the net reproductive rate, the summation of survival times fecundity at each age of the individual's life history. There exist several life history features of reproduction and survival that determine the timing of reproduction. The optimum age for individual reproduction depends on how reproduction at a particular age would affect later survival and reproduction. The optimum age of reproduction is reached when no further increase in the net reproductive rate can be obtained by further delay. Plants reproductive value changes with age and it depends on the plant's life span (annual, biennial, perennial). Mortality drives the time of reproduction. Opportunity spacetime drives mortality. Evolution shapes the timing of reproduction by a careful compromise of fecundity optimization and mortality risk minimization. Individuals reproducing too early are replaced by those that maximize season length opportunity. Individuals reproducing too late die or are less fecund. Whether to reproduce early in season or wait depends on the particular ecological conditions of a locality: unused late season opportunity will eventually be seized by some species, and the fuller a species uses available opportunity in the face of death the more it will be able to exploit.

The last step in natural selection is heredity, the transmission of specific characters or traits, genes and genetic information, from the ancestral parent to the offspring: the basis of adaptive evolution. New or altered traits in individuals can be produced by mutations, the transfer of genes between populations or between species, or by genetic recombination in sexual reproduction. Evolution occurs when these heritable differences become more common or rare in a population. The transmission of genes to the next generation is accomplished in weeds by their mating system, their breeding system. The mating system, or mode of fertilization, of a successful weed species generates appropriate amounts and types of genotypic variants among its progeny. The mating system directly controls inheritance of traits by reproduction of fit progeny. Mating systems indirectly control species diversity, population genetic structure at all spatial levels, community diversity and assembly with the dispersal of these offspring. Mating system direct control of sexual reproduction indirectly regulates population variability, either to conserve local adaptation and fitness, or to generate novelty in the face of change. The search for opportunity spacetime is accomplished with genetic variability expressed as phenotypic heterogeneity within populations of a species.

A local population relies on its mating system to adjust genetic recombination to the fabric of the exploited disturbed space. A plant species' particular mating system controls recombination to generate an appropriate amount of variation to the opportunity spacetime being exploited locally. Mating systems for colonizing species varies from obligate outcrossing (e.g. monoecy, dioecy) to self-pollinating and apomictic species

## 5.1 Survive, Avoid Mortality

In the second process of natural selection some individuals in a local population survive and reproduce the fittest phenotypes, while the others die and are eliminated. This non-random evolutionary process is conditional: individuals must survive (ch. 5.1) to reproduce (ch. 5.2) and then reproduce to inherit (ch. 5.3). Populations adapt to local opportunity when these successful individuals survive, reproduce and their offspring inherit their genes and traits. Successful completion of this survival-reproduction process results in weedy life history adapatation (unit 3) and adaptation in local plant communities (unit 4).

Evolution shapes the timing of reproduction by a careful compromise of fecundity optimization and mortality risk minimization. Mortality, and the risk of mortality, are the primary shapers of weed life history. Mortality determines a plant's life duration and the timing of its developmental events: e.g. annual (winter, summer), biennial, perennial (ch. 6.2, plant life history classification systems). Mortality and it's risk predicate a plant's reproduction, flowering time and duration of mating (ch. 5.2).

## 5.2 Reproduce the Fittest, Eliminate the Others

The life history of a particular plant is correlated with two major components of fitness: survival and reproduction. Survival and reproduction have been represented in several different ways. Population growth in a limited environment can be represented by the Verhulst-Pearl logistic equation (see ch. 2.4.1, precondition to natural selection). In this section survival and reproduction are represented with demography by the net reproductive rate ($R_o$), the summation of reproduction and survival/mortality (see also ch. 15.2.2, demographic weed life history population dynamics models). Population growth and size in a local space obey the law of constant final yield (biomass, seed weight) in which the total plant yield of a high density population in a specified area tends to approach a constant weight (see ch. 9.4.2, plant density and productivity per unit area). Much confusion has resulted from the application of these types of growth models to plants (Harper, 1977; Dekker, 2011a). These confusions are systematically explored in unit 6, representation of weed biology.

**5.2.1 Demographic survival and reproduction**. Demography is one of several ways of viewing survival and reproduction. It is widely used in demographic ecology studies (e.g. Silvertown & Doust, 1993, Ch. 9).

The net reproductive rate ($R_o$) is the summation of reproduction and survival/mortality. Fitness includes reproduction and survival which are the consequences of trade-offs of conflicting goals/ends for a plant's life history. There is much more to fitness than net reproductive rate discussed here. Using $R_o$ as equaling net reproductive rate, an optimal life history under particular ecological conditions can be defined as one which maximizes:

$$\Sigma l_x m_x$$

Components of $\Sigma l_x m_x$:
$\Sigma l_x m_x = R_o$ = net reproductive rate
$l_x$ = proportion of individuals surviving to age x
$m_x$ = fecundity of an individual of age x

The net reproductive rate is equal to the summation of survival times fecundity at each age of the individual's life history.

**5.2.2 Timing of reproduction.** There exist several life history features of reproduction and survival that stimulate a plant to precocious reproduction, or cause it to wait to reproduce, in a particular set of ecological conditions

**5.2.2.1 Optimum age of reproduction.** The optimum age for individual reproduction depends on how reproduction at a particular age would affect later survival and reproduction. The optimum age of reproduction is reached when no further increase in the net reproductive rate ($l_x m_x$) can be obtained by further delay.

**5.2.2.2 Maximizing net reproductive rate.** $\Sigma l_x m_x$ may be maximized by delaying reproduction until plant reaches a size to be able to survive and complete the first phases of reproduction. Survival expectation mitigates otherwise precocious reproduction. Delaying reproduction incurs a demographic penalty when the annual rate of population size increase is greater than 1. The demographic penalty is the inability to exploit available opportunity space. When the annual rate of population size increases (> 1) the penalty for delay increases: plants with precocious reproduction are better able to exploit local opportunity. When the annual rate of population size decreases (< 1) there is an advantage to delaying reproduction.

**5.2.2.3 Precocious reproduction.** If reproduction incurs no costs, and a population is in a phase of increase (> 1), the earlier reproduction occurs the better for fitness. But reproduction does incur costs. The cost of reproduction itself may slow its vegetative growth. Fecundity and yield are closely correlated to biomass and plant size at the time of flowering (Dekker, 2004b). Maximizing the duration of vegetative growth to exploit local opportunity may be compromised by premature reproduction. Precocious flowering may also increase the risk of death and slow growth as smaller plants become more vulnerable to neighbor plant interference.

**5.2.3 Plant age and stage structure.** Not all individuals make an equal contribution to the annual, finite, rate of population increase because birth and death rates vary with individual age, size and stage.

**age structure:**
1: the number or percentage of individuals in each age class of a population; age distribution; age composition
2: structure in populations where recruitment is a frequent event compared to the lifespan of adults (e.g. forest trees) (Silvertown and Charlesworth, 2001)

**stage structure:** individuals of the same age but with different local environments may be at different stages of their life life cycles as a consequence of highly plastic plant rates of growth and development (Silvertown and Charlesworth, 2001)

The forces of natural selection and eliminaton on survival varies with age: fitness is much less sensitive to changes in survival late in life than it is to changes that occur earlier. Differences in seedling emergence timing leads to different age-stage population structure in a local community. These differences determine a plant's competitive advantage, and forms the basis of size-frequency distribution hierarchy asymmetry observed later in the growing season (see ch. 9.4.3.1, plant size hierarchies in the community). Some plants age but don't develop in the normal time due to other aspects of the environment in which it finds itself:

**oskar:**
1: stunted juveniles that linger around the photosynthetic compensation point waiting for favorable growing conditions (Silvertown and Charlesworth, 2001)
2: juveniles that age but do not grow (Grass, 2010)

How long does an annual plant live? How old is it when it dies? Annual plants actually can live to very old age, older than mature trees. Seedbanks are elongated waiting periods for seeds, providing potentially very old plants as seedlings. For example, velvetleaf (*Abutilon theophrasti*) can live for 20, 50, 100 years in the seed bank before it germinates and completes its life cycle as vegetative plant.

What is the oldest plant in nature? Candidates include: bristlecone pine trees (3 *Pinus* spp.), 5000+ years); olive trees (*Olea europaea*, 2000+ years); oak trees (*Quercus* sp., 1000+ years); redwood trees (*Sequoia* sp., *Sequoiadendron* sp., 1000's of years); lotus seed (*Nelumbo nucifera*) viable after 1300+ years; and individual aspen trees (*Populus* sp.) live 40-150 years, will their long-lived clonal root systems can live over 80,000+ years. The oldest weed from an evolutionary perspective might be the primitive *Equisetum* with spores, rhizomes, primitive photosynthetic leaves, scales. Perennial weeds like quackgrass (*Elymus repens*), johnsongrass (*Sorghum halpense*), hemp dogbane (*Apocynum cannabinum*), milkweed (*Asclepias syriaca*) and Canada thistle (*Cirsium arvense*) could be older than mature trees in the later phases of succession: these vegetative clones may never die. They may be decades, hundreds, or thousands of years old since originally arising from seed. If they are very old, somatic mutations may affect different parts of plant: the plant may be composed of different genotypes in different tissues resulting in the evolution of different genotypes in same individual plant.

### 5.2.4 Reproductive value.

"… the relative effects on fitness of changes in [the proportion of individuals in a cohort surviving to a particular age] or [the fecundity of an individual at that particular age] at different ages … provides the ground rules for the evolution of life histories." Silvertown and Charlesworth, 2001

An organisms reproductive value changes with age. Seed from a plant produced at different times in its life cycle differ in their contributions to the future growth of the population.

**scenescence:** a decline in physiological state with age, manifested by an increase in mortality with age (Silvertown and Charlesworth, 2001)

This is true for successional species (perennial woody plants, trees) and for annual weeds. Plants reproductive value changes with age and it depends on the plant's life span (annuals: summer, winter; biennials; perennials: herbaceous, woody). The time of reproduction for an individual species in its life history and its effect on reproductive value has been developed in demographic models:

$$V_x = \text{reproductive value}$$

Reproductive value is the contribution an average individual aged x will make to the next generation before it dies (Fisher, 1958). Reproductive value ($V_x$) is the sum of the average number of offspring produced in the current age interval ($m_x$) plus the sum of the average number produced in later age intervals ($m_{x+i}$), allowing for the probability that an individual now of age x will survive to each of those intervals ($l_{x+i}/l_x$) (Silvertown and Doust, 1993).

Reproductive value ($V_x$) has two components:

($m_x$) = current fecundity

which is the average number of offspring produced in the current age interval.

($V_x - m_x$) = residual reproductive value

which is the potential reproductive contribution an individual might yet make; equivalent to the chances that remain for it to produce further offspring in following seasons.

The species that uses all the opportunity spacetime available to it over the entire season is the one that will predominate. Precocious growth and quitting early lets those that continue to ultimately win because there is unused opportunity, even with the greater hazards of mortality by continuing.

Think of it from the point of view of two identical plants "deciding" whether to reproduce early or wait until the end of the season. The one that reproduces early shifts its energy to flowers, its vegetative phase neighbor shades it out right away, so the loss to the early reproducer happens right away. Larger, vegetatively growing, individuals that survive have potentially much greater seed yield from their greater biomass than earlier reproducers.

Because $l_x$, the proportion of individuals in a cohort surviving to age x, decreases with each passing time interval, small changes in fecundity will have larger effects on fitness the earlier in life they occur. For example *Setaria faberi*, giant foxtail, produces seeds from August through October or November in Iowa, USA conditions. Most of that seed is produced early, with declining fecundity as the freezing conditions of late autumn approach (Haar and Dekker, 2011).

Other things being equal, this should favor the evolution of the earliest possible age of reproductive maturity and concentration of reproduction at this age, ultimately causing semelparity. To understand why plants delay reproduction and why semelparity is the exception rather than the rule we need to consider how the costs of reproduction and mortality from various sources shape life history.

**semelparity**:
1: organisms that have only one brood during the lifetime; big-bang reproduction (Lincoln et al., 1998)
2: reproduction that occurs only once in the life of and individual (King and Stansfield, 1985)

**iteroparity**:
1: organisms that have repeated reproductive cycles (Lincoln et al., 1998)
2: repeated periods of reproduction during the life of an individual (King and Stansfield, 1985)

Tradeoffs between reproduction and growth/survival lead to tradeoffs between 'reproduction now' or 'reproduction later'. The age at first reproduction varies greatly among species and tends to be positively correlated with lifespan suggesting they co-evolve because early reproduction negatively affects subsequent survival. The optimum age at which an individual should begin reproduction depends upon how reproduction at a particular age would affect later survival and reproduction. The cost of reproduction at a young age may slow growth and increase the chance of death because small plants are more vulnerable. Conversely, delaying maturity may sometimes expose a plant to increased mortality risk that prematurely curtails reproduction, causing selection for an earlier age of first reproduction

**5.2.5 Risk of death determines life history**. Mortality drives the time of reproduction. Opportunity spacetime drives mortality. Evolution shapes the timing of reproduction by a careful compromise of fecundity optimization and mortality risk minimization. Individuals reproducing too early are replaced by those that maximize season length opportunity. Individuals reproducing too late die or are less fecund. Whether to reproduce early in season or wait depends on the particular ecological conditions of a locality: unused late season opportunity will eventually be seized by some species, and the fuller a species uses available opportunity in the face of death the more it will be able to exploit.

The greatest hazard to summer annual plants in the north temperate regions of the world is frozen winter soil, temperatures unable to support growth (e.g. < 5°C). Opportunity lies within the remainer of the year. Risk is great early in the cropping season when seedbed preparation, soil tillage and herbicide treatment occurs. Plants surviving these early disturbances are at a distinct advantage (escape, avoidance, resistance). For example, 'layby' in Iowa maize fields occurs in June or early July and is the time after which farmers can't get equipment in the field because the crop is too large: they "layby" their equipment. Individual weed plants surviving past, or emerging after, layby have a much reduced risk of farmer mortality until the time of crop harvest. The crucial value of survival after layby was demonstrated with common lambsquarters (*Chenopodium album*) seedling emergence in multiple European and North American crop fields (Andujar et al., 2013).

**5.2.6 Influences of plant density on mortality.** Mortality is influenced by the stresses induced by crowding of plants in a local deme. Plant density (numbers of plants per unit area) itself is a proximate, not the primary, cause of death for an individual. Mortality in crowded populations is directly correlated with the productivity of that area. Crowding leads to mortality of some plants, freeing space allowing increased growth potential for the survivors or replacements. Plant growth in crowded habitats is developed in Ch. 9.4, effects of neighbor interactions on plant density, growth and form. Definitions:

**mortality**: death rate as a proportion of the population expressed as a percentage or as a fraction; mortality rate; often used in a general sense as equivalent to death

There exist two categories of mortality: density-independent and density-dependent.

**density-dependent mortality**:
1: the increasing risk of death associated with increasing population density
2: a decrease in population density (numbers per unit area) due to the effects of population density (self-thinning)

**density-independent mortality**:
1: the increasing risk of death not associated with population density change
2: a decrease in population density due to any factor which is independent of population density

**density dependence**: a change in the influence of an environmental factor (a density dependent factor) that affects population growth as a population density changes, tending to retard population growth (by increasing mortality or decreasing fecundity) as density increases or to enhance population growth (by decreasing mortality or increasing fecundity) as density increases

**density independent factor**: any factor affecting population density, the influence of which is independent of population density

Some mortality factors fall neatly within these definitions, others do not, still other may overlap. Predation of plants in a locality may be increased by high density abundance perceived by a grazing herbivore or insect: density-dependent mortality. Prairie fires can kill plants in an area of variable density: overlapping of density-dependent and independent factors. Lightning striking a tree at the top of a hill, herbicide spray application, and being hit by a hailstone are most probably density-independent mortality. The utility of these definitions lay in there ability to focus our attention on ultimate rather than proximate causes of death. The should be used with caution when defining strict experimental parameters.

The risk of death in a high-density population often assumes a slope of -1.5 (the "3/2 power law"). This rule says that while the number of individuals is decreasing, the weight of the population as a whole is increasing. The rate of growth of individuals more than compensates for the decrease in numbers. The risk of mortality does not change with age. There is a constant risk of death for both individual plants and their component parts (e.g. leaves).

Pre-emption of space (resources) by developing seedlings of the same species in a population, density-dependent mortality, is also called "self-thinning" (Ross, 1968). Self-thinning mortality is greater in high fertility environments. Survival is greater in high light regimes than in low light regimes. The mechanism of self-thinning is not understood. Populations derived from large seed suffer more rapid mortality than those derived from small seed. The faster-growing, larger, more vigorous seedlings produce a more intense density stress among themselves than in populations of the same density but from smaller, slower growing, seedlings. "Alien-thinning" is density-dependent mortality in one species that can be ascribed to the stress of density of an neighbor species.

**5.2.7 Modes of selection and population diversity.** Natural selection alters the frequency of phenotypes in a population and species, including those of human disturbances in agro-ecosystems. Several modes of selection can be discerned in population dynamics.

**directional selection**: selection for an optimum phenotype resulting in a directional shift in gene frequencies of the character concerned and leading to a state of adaptation in a progressively changing environment; dynamic selection; progressive selection

In population genetics, directional selection occurs when natural selection favors a single phenotype and therefore allele frequency continuously shifts in one direction. Selection for resistant traits in a weed population subjected to continuous use of one herbicide is a good example of directed selection towards a tolerant variant. Under directional selection, the advantageous allele will increase in frequency independently of its dominance relative to other alleles (i.e. even if the advantageous allele is recessive, it will eventually become fixed). Directional selection stands in contrast to balancing selection where selection may favor multiple alleles, and is the same as purifying selection which removes deleterious mutations from a population, in other words it is directional selection in favor of the advantageous heterozygote. Directional selection is a particular mode or mechanism of natural selection (Wikipedia, 5.08).

**disruptive selection**: selection for phenotypic extremes in a polymorphic population, which preserves and accentuates discontinuity; centrifugal selection; diversifying selection.

Disruptive selection, also called diversifying selection, is a descriptive term used to describe changes in population genetics that simultaneously favor individuals at both extremes of the distribution. When disruptive selection operates, individuals at the extremes contribute more offspring than those in the center, producing two peaks in the distribution of a particular trait (Wikipedia, 5.08). For example, common cocklebur (*Xanthium strumarium*) often disperses seeds with different amounts of inherent dormancy in a single capsule, one germinates readily while the other remains dormant in the soil.

**stabilizing selection**: selecting for the mean, mode or intermediate phenotype with the consequent elimination of peripheral variants, maintaining an existing state of adaptation in a stable environment; centripetal selection; normalizing selection.

Stabilizing selection, also referred to as purifying selection or ambidirectional selection, is a type of natural selection in which genetic diversity decreases as the population stabilizes on a particular trait value. Put another way, extreme values of the character are selected against. This is probably the most common mechanism of action for natural selection. Stabilizing selection operates most of the time in most populations. This type of selection acts to prevent divergence of form and function. In this way, the anatomy of some organisms, such as sharks and ferns, has remained largely unchanged for millions of years (Wikipedia, 5.08). A primary goal in crop breeding and domestication is the production of seeds from a single cultivar in which all germinate at the same time: stabilizing selection.

**5.3 Inheritance: Transmit Parental Traits to Offspring**. The last step in natural selection is heredity, the transmission of specific characters or traits, genes and genetic information, from the ancestral parent to the offspring.

The genes that are passed on to an organism's offspring produce the inherited traits that are the basis of adaptive evolution. Mutations in genes can produce new or altered traits in individuals, resulting in the appearance of heritable differences between organisms, but new traits also come from the transfer of genes between populations, as in dipersal, or between species, in horizontal gene transfer. In species that reproduce sexually, new combinations of genes are produced by genetic recombination, which can increase the variation in traits between organisms. Evolution occurs when these heritable differences become more common or rare in a population.

Over many generations, adaptations occur through a combination of successive, small, random changes in traits, and natural selection of those variants best-suited for their environment. In contrast, genetic drift produces random changes in the frequency of traits in a population. Genetic drift results from the role chance plays in whether a given individual will survive and reproduce. Though the changes produced in any one generation by drift and selection are small, differences accumulate with each subsequent generation and can, over time, cause substantial changes in the organisms. (Wikipedia, 5.08)

**heritability**: an attribute of a quantitative trait in a population that expresses how much of the total phenotypic variation is due to genetic variation.
    a: in the **broad sense**, heritability is the degree to which a trait is genetically determined, and it is expressed as the ratio of the total genetic variance to the phenotypic variance.
    b: in the **narrow sense**, heribability is the degree to which a trait is transmitted from parents to offspring, and is expressed as the ratio of the additive genetic variance to the total phenotypic variance. The concept of additive genetic variance makes no assumption concerning the mode of gene action (expression) involved (King and Stansfield, 1985); narrow-sense heritability can be estimated from the resemblance between relatives (Silvertown and Charlesworth, 2001)

**5.4 Mating systems and inheritance**. The transmission of genes to the next generation is accomplished in weeds by their mating system, their breeding system. The mating system, or mode of fertilization, of a successful weed species generates appropriate amounts and types of genotypic variants among its progeny (see ch. 4.2, generate genetic variation). The mating system directly controls inheritance of traits by reproduction of fit progeny. Mating systems indirectly control species diversity, population genetic structure at all spatial levels, community diversity and assembly with the dispersal of these offspring. Mating system direct control of sexual reproduction indirectly regulates population variability, either to conserve local adaptation and fitness, or to generate novelty in the face of change. Sexual reproduction produces offspring with new genotypes, and novel combinations of genes.

**5.4.1 Mating system and opportunity**. The search for opportunity spacetime is accomplished with genetic variability expressed as phenotypic heterogeneity within populations of a species. Variation in heritable characteristics in populations and species arise from selection, mutation of genetic material, recombination of genes and chromosomes, gene flow by hybridization and migration, segregation distortion of alleles, and genetic drift in small populations.

A plant species' particular mating system controls recombination to generate an appropriate amount of variation to the opportunity spacetime being exploited locally. Mating systems for colonizing species varies from obligate outcrossing (e.g. monoecy, dioecy) to self-pollinating and apomictic species (Harper, 1977). The spatial structure of opportunity being explored and exploited favors certain mating systems and rates of hybridization. The purpose of a plant species mating system is to generate variation appropriate to a locally available habitable locality. A local population relies on its mating system to adjust genetic recombination to the fabric of the exploited disturbed space:

"We would predict on the basis of such work that different genetic strategies of colonization will be evolved depending upon the statistical pattern of the environment. Populations with low genetic variability for a character will more often be successful in environments requiring frequent radical alteration of phenotype, whereas populations with high genetic variability will more often leave successful colonies in environments which, although radically different from the original species range, are in themselves rather stable." (Lewontin, 1965; p. 91).

A genetically plastic mating system has to generate variation sized exactly to the space being searched and seized. If the mating system generates intra-specific variants unsuited to the breadth and grain of the environment explored they are at a disadvantage in colonizing that space. Even self-pollinated species, and apomictic races within a species, produce a "carefully considered" amount of variation needed for the space searched, seized and occupied. For these reasons, the responses of a plant's mating system to an opportunity spacetime creates the population genetic structure of a locally adapted population.

Mating system control of population genetic structure is a consequence of the exploration for opportunity space, and the degree of variation or no variation generated by a particular species mating system is a strong inferential clue to the space being explored. The spatial scale of hybridizing populations, demes and metapopulations spans the globe, continents, regions, landscapes, localities and microsites. Highly successful colonizing plant species also exploit the fragmentary spatial and temporal structures they confront, be it island geography or a landscape of farmlands amidst roads, residences and woodlands:

"Colonization can be an effective strategy under the following conditions. If a species is able to populate many semi-isolated regions, such as a system of oceanic islands, its populations will be tested by a variety of physical and biotic environments. Over a period of time, the greater the number of regions colonized, the higher the probability that the species will survive somewhere, so the endemic foci persist out of which colonization can proceed again." (E.O. Wilson, 1965; p. 14)

This experience preadapts a species to colonize neighboring regions. It also reveals a potential answer to the question posed by Lewontin (1965; p.88) concerning:

"... what the optimal genetic structure is for a colonizer to maintain a positive rate of increase in a new environment, while at the same time maintaining itself in the original territory of the species. The same question can be asked over time rather than in space. Given a species which is reduced sharply in numbers at more or less regular intervals by unfavorable conditions, what is the optimal genotypic structure of the population which will both guarantee its survival in low numbers during unfavorable times and allow it to maintain very large populations at peak periods?"

Allard has argued (1965, p. 49) that:

"There is, however, one feature which the great majority of these notably successful colonizers share: a mating system involving predominant self-fertilization."

Mating, or breeding, systems in weed species range from obligate clones (e.g. apomixis in *Taraxicum officinale*, dandelion) to obligate out-crossing species (e.g. dioecy in *Amaranthus tuberculatus*, tall waterhemp). Self pollination is the most common mating system in weed species. This certainly has been observed amongst weeds of North American agroecosystems (e.g. *Setaria* species-group; Dekker, 2003, 2004a). Since the 1960's, with the introduction of numerous selective herbicides, a mating system population shift has occurred. Outbreeding species such as *Lolium* and *Amaranthus* have been favored. They are highly effective in generating variability in search of rare, yet highly advantageous, biotypes resistant to those chemicals.

A plant's mating system is an important mechanism by which is responds to changes in its locality, and the way it searches for opportunity space in the habitats it lives in. When conditions are within historical experience of the population little new variation is needed for enduring occupation. When changes occur beyond the pre-adapted variation of a population, new genotypes and phenotypes may be better. So, this new variation may increase its fitness in that locality.

A plant's mating system affects its ability to handle changes, crashes, in the environment. Self-pollination, protects against quick changes in progeny. Some types of mating system buffers against longer term changes: colonizers are more likely to be self-pollenated. Selfers protect themselves against effects of large genetic display, against big swings in population composition. Hermaphrodites avoid tracking transient adaptive optima in a season; they keep to a longer term strategy. Polyploids are more likely to be later successional species, while diploid species are more likely colonizers. Why is this so? Phenotypic plasticity enables a single genotype to form different phenotypes depending on conditions.

Weed species generate a typical amount of variation in each generation. Producing too many new unsuccessful genotypes/phenotypes can waste precious resources of the parent plant. Too little new variation can lead to competitive disadvantage with neighbors who can seize these new opportunities. As such, almost all weed species generate some degree of variation each generation. Mating systems are a compromise between conflicting ends: the means to control the balance between conservative repetition (selfing) and novel new forms (genotype experiments).

Survival of a local population involves variation in individual genotype-phenotypes as hedge-bets against an uncertain future. There is no such thing as a "perfect" weed phenotype. Fitness changes with changing local opportunity spacetime: resource-conditions, disturbance, community neighbors.

All mating systems end up in the same place: be diverse or die. Mating systems maintain genetic diversity within populations by different means: co-adapted inbred lines, apomictic races; co-adapted gene complexes (partly inbreeding species), or ecological differentiation between sexes (outbreeding species). Gene flow is not critical to maintain or generate diversity, it is readily found in apomicts and inbreeders. Diversity is maintained within populations in many different ways, ways that compensate for apparent differences between breeding systems. Natural populations have genetic diversity, even within breeding systems not considered as diverse. Natural selection favors diversity in the long run because so many slightly different niches are available to a species at a locality. The net effect of natural selection is to create more variability: survival depends on many different hedge-bet "roulette chip" genotype-phenotypes to confront risk, despite the fact that the immediate effect of selection 'seeks' the "perfect" type for the immediate opportunity afforded to that population.

Several patterns of genotypic novelty generation can be discerned in examples from the mating systems of typical weed species.

**5.4.2 Evolution of mating systems**. There exist three issues in mating system evolution: sex versus apomixis?; for sexual reproduction: why outcross instead of self-fertilization?; and for outcrossing, why two sexes instead of hermaphroditism? (Silvertown and Charlesworth, 2001). Hermaphroditism is the predominant sexual state for angiosperms. Evolution of sex depends strongly on the process of transmitting genes to progeny. The evolution of selfing rates can be understood in terms of two major genetic factors: transmission of genes and the relative survival of inbred versus outbred progeny. The evolution of separate sexes involves third genetic factor: allocation of resources between male and female functions. The breeding system does not evolve in order to control levels of genetic diversity, although diversity patterns are strongly affected by breeding systems. Individual advantages and disadvantages seem to drive these evolutionary changes (Maynard Smith, 1978; Holsinger, 2000). Major factors influencing the evolution of plant breeding systems include: transmission of genes; relative survival of inbred versus outbred progeny; allocation of resources; and mate reproductive availability.

**5.4.2.1 Sexual reproduction versus apomixis.** There exists a need to explain why sexual reproduction is not lost, why it occurs. What prevents females from producing all-female families, which would give them the advantage of transmitting more genes to the progeny generation? The answer is the 'two-fold cost' of refraining from asexual reproduction. What prevents a hermaphroditic plant from producing apomictic mutants that make ovules instead of ovules and pollen? Such new apomict types should rapidly invade populations. The answer is the two-fold cost of males: recombinate variety and loss of undesirable mutations. The advantages of sexual reproduction include the ability to produce genetically varied offspring (recombination) to confront variable and changing opportunity. Sexual reproduction may also be critical to the ability to evolve rapidly to survive attacks of fast-evolving pathogens; to rapidly incorporate advantageous mutations; to genetically 'forage' for desirable traits. The advantages also include the greater ability to rid themselves of disadvantageous mutations.

**5.4.2.2 Outcrossing versus self-fertilization.** Self-fertilization has many ecological advantages: it spares organisms the need to seek mates; they do not need to expend resources attracting and courting mates, or pollinators. This may be important in low-density populations where selfing offers reproductive assurance. Assurance is provided under conditions of low pollinator service or low/absent density of potential mates. This is especially relevant following long distance dispersal (e.g. genetic bottlenecks), wherein outcrossing can lead to pollination failure. Selfing allows locally adapted genotypes to persist, while outcrossing between too distant populations sometimes leads to low progeny fitness, outbreeding depression. The cost of outcrossing include the lower transmission costs of selfing. A selfing plant provides pollen for creation of its progeny as well as provide pollen to other plants. Outcrossers can only send pollen to another plant, or receive pollen from another plant. Given equal total reproductive success (in terms of seeds matured, and seed sired on other plants), this results in a considerable (50%) transmission advantage to selfers (Fisher, 1941).

The disadvantages of selfing includes inbreeding depression (low survival or fertility of inbred progeny). The evolution of separate sexes may allow the optimization of resources by specializing in one sex function. The reduced pollen export, pollen discounting, of selfing reduces its advantage compared to plants able to fertilize other plants more readily.

Intermediate selfing rates can evolve if the disadvantage of selfing increases with successive generations of inbreeding. Local adaptation can also lead to a stable balance between selection for selfing and outcrossing. When populations are patchily distributed spatially, biparental inbreeding between different plants involving close relatives will occur. When it does the transmission advantage of selfing relative to outcrossing is reduced, leading to the maintenance of intermediate selfing levels. Different selfing rates also imply other differences to help maintain the different phenotypes in a population.

**5.4.2.3 Outcrossing: separate sexes versus hermaphrodites.** With the evolution of separate sexes, self-fertilization is prevented by having separate male and female plants, dioecy. Unisex plants avoid inbreeding depression. Unisexuals may differ from hermaphrodites in the amount of resources expended on reproduction: females save the cost of producing pollen. Dioecy is often found in arid or other harsh conditions (Barrett, 1992) in which resources are limited and plants can't maintain both sex functions.

**5.4.3 Sex classification system.** Mating by weed species can be sexual or asexual. The breeding system of a plant relies on three important criteria: sexual reproduction or asexual; if sexual reproduction, the sexes of the flowers; and, floral features that effect whether individual plants will self-fertilize or outcross (table 5.1).

**Table 5.1** Plant sex system classification (Silvertown and Charlesworth, 2001).

| | | |
|---|---|---|
| **asexual** | apomictic | *Taraxacum officinale* (dandelion) |
| | | *Poa pratensis* (Kentucky bluegrass) |
| **sexual monomorphism** | hermaphrodite | *Setaria viridis* (green foxtail) |
| (individuals alike) | monoecious | *Zea mays* (maize) |
| | | *Manihot esculenta* (cassava) |
| | | *Alnus* sp. (alder) |
| | gynomonoecious | *Solidago* spp. (goldenrod) |
| | andromonoecious | *Daucua carota* (wild carrot) |
| **sexually polymorphic** | dioecious | *Cannabis sativa* (wild hemp) |
| (different individuals) | | *Bouteloua dactyloides* (buffalograss) |
| | | *Asparagus officinalis* (asparagus) |
| | | *Pistachia vera* (pistachio) |
| | gynodioecious | *Glenchoma* spp. (ground ivy) |
| | androdioecious | (rare) *Mercurialis annua* |

Some weedy plant genuses contain both inbreeding and outcrossing species: *Lolium* spp., *Plantago* spp. (Silvertown and Charlesworth, 2001).

**5.4.4 Types of mating systems.** Sexual breeding systems include self-pollinating and out-crossing species. Asexual mating systems include apomictic species and vegetative clone reproducing species.

**5.4.4.1 Self-pollinating species.** Self-pollinating mating systems with sexual monomorphism:

**hermaphrodite**:
1: flowers have both male and female functions
2: plant that has only bisexual reproductive units (flowers, conifer cones, or functionally equivalent structures); i.e. perfect flowers

In angiosperm terminology a synonym is monoclinous from the Greek "one bed".

Many annual weedy plant species reproduce almost exclusively by inbreeding, self-pollination. Their anthers rarely emerge, and pollen is shed directly on its own stigmas for self-fertilization. Genetic diversity arises in a local population of a species despite this selfing. There are as many potential adaptive optima (opportunity microsites) within a habitat as there are individuals. Therefore there is no perfect phenotype, therefore population variety is maintained. Different optima in the same locality drives different variants needed to exploit those opportunities.

Genetic selection is more complex than selection for individuals. Selection acts to structure the genetic resources of a local population into highly interactive allelic complexes: diversity at the level of co-adapted gene complexes (the 'selfish' trait) rather than strictly at the level of inbred genetic lineages. For example, *Fescue microstachys* is a grass species with almost no outbreeding. Natural populations are as genetically variable as other species. This genetic heterogeneity is explained by supposing nearly as many adaptive optima within a habitat as there are individuals.

*Setaria* spp.-gp. (foxtails) and *Elymus repens* (quackgrass) self-fertilization mating system affects its population genetic structure. Self-fertilization provides high local population genetic homogeneity; high landscape scale genetic heterogeneity provides rare outcrossing opportunity for inter-locality variation when confronted with environmental change.

Baker (see ch. 1.4, weedy traits) has argued many of the most successful weed species in the world are self-compatible but not completely autogamous (sexual self-fertilization) or apomictic (asexual reproduction through seeds) (Baker, 1965, 1974). When cross-pollinatws, out-crossing, it is accomplished by wind or unspecialized visitors. Other examples include the early flowers of *Viola* sp. (violet) which are cleistogomous: completely enclosed, sealed and protected from foreign pollen.

**5.4.4.2 Out-crossing species.** Outcrossing weed species are prevented from self-fertilization by physical or temporal separation.

**outcrossing mating system**: breeding system with self-incompatibility, spatial separation of male and female reproductive organs, or temporal separation of male and female flower activity

Out-crossing species with sexual monomorphism:

**monoecious**:
1: separate sex flowers on the same individual plant
2: having separate male and female reproductive units (flowers, conifer cones, or functionally equivalent structures) on the same plant; from Greek for "one household"

Individuals bearing separate flowers of both sexes at the same time are called simultaneously or synchronously monoecious. Individuals that bear flowers of one sex at one time are called consecutively monoecious; plants may first have single sexed flowers and then later have flowers of the other sex.

**andromonoecious**: both hermaphrodite and male (female sterile) flower structures occur on the same individuals

**gynomonoecious**: both hermaphrodite and female (male-sterile) flower structures occur on the same individuals

**trimonoecious (polygamous)**: male, female, and hermaphrodite structures all appear on the same plant

Out-crossing species with sexual polymorphism:

**dioecious**:
1: separate male and female sex individuals
2: having unisexual reproductive units with male and female plants (flowers, conifer cones, or functionally equivalent structures) occurring on different individuals; from Greek for "two households". Individual plants are not called dioecious: they are either gynoecious (female plants) or androecious (male plants)

**androdioecious**: individuals either male or hermaphrodite

**gynodioecious**: individuals either female or hermaphrodite

**subdioecious**: a tendency in some dioecious species to produce monoecious plants. The population produces normally male or female plants but some are hermaphroditic, with female plants producing some male or hermaphroditic flowers or vise versa. The condition is thought to represent a transition between hermaphroditism and dioecy.

**androecious**: plants producing male flowers only, produce pollen but no seeds, the male plants of a dioecious species.

**subandroecious**: plant has mostly male flowers, with a few female or hermaphrodite flowers.

**gynoecious**: plants producing female flowers only, produces seeds but no pollen, the female of a dioecious species. In some plant species or populations all individuals are gynoecious with non sexual reproduction used to produce the next generation.

**subgynoecious**: plant has mostly female flowers, with a few male or hermaphrodite flowers.

Out-crossing species with sexual polymorphism in which sex can change:

**protoandrous**: describes individuals that function first as males and then change to females

**protogynous**: describes individuals that function first as females and then change to males.

Genetic variation is expected with outbreeding species. Outbreeding provides genetic novelty that explores for subtle differentiation of phenotypes to fill different niches. Obligate outcrossing provides continuous genetic novelty from local genotype pool. The cross-pollenating mating system provides maximum variation to searches for highly adapted traits (e.g. herbicide resistance alleles).

In general long-lived plants tend to be outcrossers, and annuals tend to be inbreeders. Out-crossers can have perfect flowers (hermaphrodites), or single sex flowers (dioecy, monoecy). In outcrossing species with perfect flowers some percent of them experience pollen transfer from another individual. The rate of hybridization with pollen from another plant varies from low-to-100%: 1-10% in *Avena fatua* (wild oat), to 100% in obligate outcrossers *Lolium multiflorum* (or *L. perenne*) or *Kochia scoparia*.

Other mating system mechanisms (reproductive isolating mechanisms) exist to ensure outcrossing. Some obligate out-crossers control hybridization with physical or chemical mechanisms to prevent self-fertilization. Obligate outbreeders also include species with separate male and female flowers and/or plants. Monecious species have their flowers separated on same plant. Dioecious species have separate sex plants (Wikipedia, 5.08):

Dioecy hinders local adaptation, hybridization aways generates new variants. This mode of reproduction demands very high selection pressure to maintain local adaptation. It allows different sex plants to take different ecological roles: different seasonal niches, differential competition, different times of scenescence, and it allows differential exploitation of the light enviroment.

Species examples include *Rumex asetosella*. Like *cannabis sativa/indica* (hemp, marijuana) different sex plants exploit different seasonal and temporal niches. The males arise earlier, die and scenesce earlier, thereby allowing greater light penetration in the canopy as they complete their earlier life cycle. The male plants of *Asparagus officinalis* are utilized for their edible shoots, but they die early and don't compete with females (season niche differentiation). *Mercurialis perennis* is a perennial with male and female plants in different parts of woodlands: those in shaded and open areas exploit different light environments. Other important weedy dioecious species include *Amaranthus tuberculatus* and *A. rudis* (tall and common waterhemp).

**5.4.4.3 Apomictic species.** Asexual breeding systems include those that perform apomixis. Apomictic species do not utilize sexual reproduction, but populations contain considerable heterogeneity for exploiting opportunity space.

**apomixis:**
1: seeds have the same genotype as their mother; in angiosperms pollination and fertilization of the endosperm (psuedogamy) is common
2: apomixis (also called apogamy) is asexual reproduction, without fertilization. In plants with independent gametophytes (notably ferns), apomixis refers to the formation of sporophytes by parthenogenesis of gametophyte cells. Apomixis also occurs in flowering plants, where it is also called agamospermy. Apomixis in flowering plants mainly occurs in two forms: agamogenesis, adventitious embryony

**agamogenesis:** (also called gametophytic apomixis), the embryo arises from an unfertilized egg that was produced without meiosis.

**adventitious embryony:** a nucellar embryo is formed from the surrounding nucellus tissue.

**microspecies:**
1: genotype that is perpetuated by apomixis; small population with limited genetic variability (Wiktionary, 5.14)
2: apomictic plants genetically identical over generations in which each generic lineage has some characters of a true species; distinctly different apomictic lineages having intra-generic differences much smaller than normal (Wikipedia, 5.14)

**aggregrate species:**
1: a group of species that are so closely related that they are regarded as a single species.
2: a named species that represents a range of very closely related organisms (Wiktionary, 5.14)
3: a named species consisting of a large group of identified and named microspecies (Wikipedia, 5.14)

Apomictically produced seeds are genetically identical to the parent plant. As apomictic plants are genetically identical from one generation to the next, each has the characters of a true species, maintaining distinctions from other congeneric apomicts, while having much smaller differences than is normal between species of most genera. They are therefore often called microspecies. In some genera, it is possible to identify and name hundreds or even thousands of microspecies, which may be grouped together as aggregate species, typically listed in floras with the convention "Genus species agg." (e.g., the bramble, *Rubus fruticosus* agg.). Examples of apomixis can be found in the genera *Crataegus* (hawthorns), *Amelanchier* (shadbush), *Sorbus* (rowans and whitebeams), *Rubus* (brambles or blackberries), *Hieracium* (hawkweeds) and *Taraxacum* (dandelions). Although the evolutionary advantages of sexual reproduction are lost, apomixis does pass along traits fortuitous for individual evolutionary fitness (Wikipedia, 5.08). See also ch. 10.3.2.3.3 aggregate species.

The best example of a weed species that utilizes apomixis for its mating system is *Taraxacum officinale* (dandelion), an aggregate species with many apomictic microspecies with strictly maternal inheritance. One would expect genetically uniform population structure from this species, but this is not so. Natural populations contain an assortment of apomictic races varying from site to site. Races are differentiated on the basis of precocity of flowering, seed output, and longevity. The high apparent genetic homogeneity provides a low-cost (minus the costs of sexual reproduction) generation of successful genotypes.

Hawkweed refers to any species in the very large genus *Hieracium* and its segregate genus *Pilosella*, in the sunflower family (Asteraceae). They are common perennials, occurring worldwide . They are usually small and weedy. Only a few are ornamental plants. Most are considered to be troublesome weeds. They grow to 5-

100 cm tall, and feature clusters of yellow, orange or red flower heads, similar to dandelions, atop a long, fuzzy stalk. Few genera are more complex and have given botanists such a headache due to the great number of apomictic species. Through speciation by rapid evolution, polyploidy, and possibly also hybridisation, this variable genus has given rise to thousands of small variations and more than 10,000 microspecies, each with their own taxonomic name, have been described (Wikipedia, 5.08). Another example of a plant with this type of reproductive mode is *Rubus fruticosus*, another aggregate species composed of apomictic microspecies. Small populations contain a mixture of several distinct microspecies.

**5.4.4.4 Vegetative clone reproducing species**. Vegetative clonal growth is an asexual reproductive mode in many important weed species. The branches of a clonal herbaceous plant are ramets capable of independent existence. Several terms are useful in discussing clonal reproduction:

**ramet**:
1: an individual in a plant genet
2: a member or modular unit of a clone, that may follow an independent existence if separated from the parent organism

**genet**:
1: a unit or group derived by asexual reproduction from a single original zygote, such as a seedling or a clone
2: a clonal colony, a group of genetically identical individuals that have grown in a given location, all originating vegetatively (not sexually) from a single ancestor

**ortet**: the original organism from which a clone was derived

Some of the advantages of clonal growth are that it permits rapid increase in size of genets; mobility; capture of opportunity spacetime; storage; vegetative reproduction without the costs of sex; and vegetative reproduction without the costs of recruitment. Clonal growth may postpone rather than replace sexual reproduction, ultimately increasing lifetime fecundity in a species. Clonal growth should be expected to be absent where selection favors early reproduction, for instance in highly disturbed habitats where the risk of adult mortality is high.

*Asexual clonal and sexual reproductive mode combinations.* Many perennial weed species rely on combinations of both vegetative meristem, and sexual seed, reproductive modes (e.g. *Elymus repens*).

*Clonal reproduction and spatial foraging.* Vegetative clonal reproduction is at the same time a means by which perennial species seize and exploit opportunity spacetime by means of several types of vegetative foraging traits. This topic is developed more fully in ch. 9.3.1.2, spatial foraging for local opportunity.

# UNIT 3: ADAPTATION IN WEED LIFE HISTORY

"The life cycle is the fundamental unit of description of the organism." Caswell, 2001

**The weeds' story**. I am foxtail.

I foxtail evolve. I have told you how I change with the opportunity you humans have provided me for thousands of years. How I invade your fields and conquer the earth with you. What I have not told you is how I do it, how I behave during my life, during a season. I have not revealed the special abilities we have coevolved together over all these millennia. What special traits do I have? When in my life do I express them? How do my local neighbors in a field react when confronted by my actions? How does my complexity ensure my endurance, my infragility? These are my mysteries I have yet to reveal.

**Summary**. The nature of weeds is revealed in the traits expressed during their life history allowing them to seize and exploit local opportunity. Successful weed phenotypes are fit at every step of their life history, there are no weak points. Weed life history is defined by life processes (birth, dispersal, recruitment, growth, reproduction) performed by morphological structures during development. Phenotype fitness is due to the timing of functional traits expression. Timing is everything.

Reproductive life history begins with floral primordia initiation in the developing parent plant shoot architecture. Plant size and structure are sensitively tuned to available local opportunity by means of phenotypic plasticity. Mating systems are carefully timed, controlling pollen-egg recombination, minimizing the risks of mortality. The embryogenic process allocates resources among five seed roles: dispersal, persistence, food reserves for early growth, genetic novelty, multiplication. Embryogenesis occurs. Germinability and dormancy is induced. Seed morphogenesis modulates signal receptivity from the soil environment. Reproduction ends with seed abscission: the zygote is independent. Heteroblastic seeds are dispersed: seeds preadapted to the unknown future availability of opportunity in diverse habitats.

Propagule dispersal begins with abscission and ends with seedling-bud emergence. Dispersal traits provide advantages to the seed: independence from parents, escape from neighbors and unfavorable locations or times, movement to favorable times and sites for exploitation. Dispersal is a process of discovery of opportunity spacetime. Spatial dispersal determines local population size and structure. Dispersal mechanisms reveal how habitable sites within a heterogeneous landscape are sought. Successful local colonization is determined by available habitable sites, seed dispersibility, and the speed of discovery. Seeds are dispersed by gravity, wind and air, water, animals and humans. Temporal dispersal is provided by pools of seeds in the soil, the source of all future weed infestations. Seed dormancy is dispersal in time: escape until conditions are favorable. Weed seeds are often dispersed at different times within a season. Seedling emergence begins with germination in the soil. The single most important determinant of plant assembly into communities is the timing of weed seedling emergence relative to neighbors and disturbances. Seedling emergence timing is a direct consequence of variable dormancy (seed heteroblasty).

The life history of an individual weed seed-propagule from flowering and reproduction through life in the soil seed-propagule pool and seedling recruitment are in many ways independent of neighbors. Therefore unit 3 presents an overview of weed life history, reproductive adaptation, and propagule dispersal in space and time. In unit 4, the life history of weed plants from seedling recruitment through reproduction are revealed in the process of community interactions with neighbors, adaptations to weed community structure, dynamics and biodiversity.

# Chapter 6: Weed Life History

**The weeds' story**. I am foxtail.

I do everything right my whole life: mating with myself, spreading my seed with your help and wind, growing fat on the richness of your fields, and birthing many new children for next year. Many many of us die, leaving only those with the best characters to dominate the opportunity you provide us. We grow every spring after emerging from hiding in the winter soil. We mate with ourselves and send a new generation of children out to hide again, waiting for the right moment to resume. We do all these things at exactly the right time, or else we die. If we live we pass everything on to our children once again. We live fast, die young and spread lots of children across the landscape to do it all again, over and over.

**Summary**. The nature of weeds is revealed in the several traits allowing them to seize and exploit local opportunity spacetime over the course of their lives, their life history. Successful weed phenotypes must be fit at every step of their life history, there are no weak points. Natural selection and elimination would act on such vulnerabilities leading to adaptation or extinction. Several simultaneous, overlapping aspects define the individual phenotype life history: life processes (birth, dispersal, recruitment, vegetative growth and seed reproductive growth); the morphological structures accomplishing these processes; and the developmental phases and behavioral activities each performs. Phenotype fitness is accomplished with functional traits, and by the time they are expressed. Timing is everything, it defines an individual plant's life history. Plant life histories can be classified into several general systems based on life span (e.g. annual, biennial, perennial), growth (e.g. grasses, forbs, woody plants) and life form, and life history strategy. The life history strategy is the schedule and duration of key events in an organism's lifetime shaped by natural selection to produce the largest possible number of surviving offspring. One life history strategy is the $r$- and $K$-strategy theory. This scheme proposes that species vary along a continuum from those with short generation time, high fecundity and good dispersal ($r$ strategists) that are adapted to colonize habitats where mortality is density-independent. There exist species with the opposite traits which are adapted to thrive in habitats where mortality is density-dependent ($K$-strategists). This generalized, simplistic, dichotomy proposes tradeoffs between dispersal and competitive ability.

## 6.1 Introduction to Life History

The nature of weeds is revealed in the evolutionary processes that determine the fittest phenotypes in a local population, deme. These processes select those individuals with the combination of traits that allow them to seize and exploit locally available opportunity spacetime over the entire course of their lives, their life history. Nature selects only those individuals capable of seizing and exploiting opportunity spacetime at every instant of their life cycle. Nature eliminates the others.

**life history:**
1: the significant features of the life cycle through which an organism passes, with particular reference to strategies influencing survival and reproduction

2: how long a plant typically lives, how long it usually takes to reach reproductive size, how often it reproduces and a number of other attributes that have consequences for demography and fitness (Silvertown and Charlesworth, 2001)

**life cycle**:
1: the sequence of events from the origin as a zygote, to the death of an individual
2: those stages through which an organism passes between the production of gametes by one generation and the production of gametes by the next

**ontogeny**: the origination and development of an organism, usually from the time of fertilization of the egg to the organism's mature form; also the study of the entirety of an organism's lifespan (ontogenesis or morphogenesis) (Wikipedia, 5.14)

**phenology**: the study of periodic plant and animal life cycle events and how these are influenced by seasonal and interannual variations in climate and habitat factors; principally concerned with the dates of first occurrence of biological events in their annual cycle (e.g. date of emergence of leaves and flowers); in ecology, the term is used more generally to indicate the time frame for any seasonal biological phenomena (Wikipedia, 5.14)

**development**: the biological study of the process by which organs grow and develop; closely related to ontogeny; genetic control of cell growth, differentiation and morphogenesis, which is the process that gives rise to tissues, organs and anatomy; developmental genetics studies the temporal and spatial control of gene expression, the effect that genes have in a phenotype, given normal or abnormal epigenetic parameters (Wikipedia, 5.14)

To be fit means that every step of the life history is successful. There are no weak points in the course of a life history that can be exploited. If there were such weaknesses, or vulnerabilities, natural selection and elimination would act on them leading to adaptation or extinction: 'what kills me makes me stronger'.

The goal of this chapter is to introduce several overarching concepts of the weed life history to provide a foundation for the following three chapters on the specific life history traits during reproduction (ch. 7), dispersal in space and time (ch. 8), and neighbor interactions in the community (ch. 9). These life history concepts include life history traits, processes of life history adaptation, and classification systems used in organizing plant life histories.

The life history of a weed species can be viewed from the point of view of four simultaneous, overlapping aspects that define the individual phenotype: the life processes that each successful weed undergoes, the morphological structures by which these processes are accomplished, the developmental phases and activities each performs and the behavior that defines the steps of the life cycle. Table 6.1 below summarizes the plant morphological structures, developmental (physiological, morphogenic) processes and whole plant phenotypic activities that occur during the plant life history processes of birth, dispersal, recruitment, vegetative growth and seed reproductive growth.

**Table 6.1** Plant life history processes, morphological plant structures, developmental changes and whole plant behaviors.

| Process | Morphology | Development | Behavior |
|---|---|---|---|
| BIRTH | flower/meristem; seed/vegetative bud | ▪fertilization<br>▪zygote formation | ▪seed and bud formation |

| | | ▪embryogenesis<br>▪bud morphogenesis<br>▪dormancy induction | |
|---|---|---|---|
| DISPERSAL | seed/vegetative bud (independent ramet; parental ortet) | ▪dormancy maintenance | ▪spatial dispersal<br>▪spatial foraging (ortet)<br>▪seed or bud pool formation (dispersal in time) |
| RECRUITMENT | seedling/bud shoot (juvenile) | ▪germination or bud growth<br>▪emergence from soil<br>▪first leaf greening | ▪establishment |
| VEGETATIVE GROWTH | vegetative plant (adult) | ▪growth<br>▪meristem morphogenesis<br>▪senescence of some tissues | ▪interactions with neighbors |
| REPRODUCTIVE GROWTH | flowering plant (adult) | ▪flower formation<br>▪senescence<br>▪meristem morphogenesis | ▪pollen dispersal |

**6.1.1 Phenotypic life history traits.** The expression of genes, alleles, results in the phenotype. The phenotype demonstrates its fitness by means of critical traits, characteristics. The adaptative advantage of these traits is determined by the time they are expressed relative to that of their neighbors behavior and development. Timing is everything. Timing of trait expression defines an individual plants life cycle, its life history.

Phenotypes and traits act inevitably to fill opportunity spacetime in disturbed localities. Selection favors individual phenotypes and traits that preferentially take advantage of these opportunities at the expense of their neighbors. Selected phenotypes dominate their neighbors because the timing of their life history optimizes their relative fitness and minimizes mortality. The character of these opportunity spacetimes can be deduced by observing the new phenotypes adapted to these new opportunies, and what traits they possess allowing such ready invasion.

Given an opportunity to invade in a locality, the second condition necessary for plant invasion is the presence of propagules of a particular species possessing life history traits suitable to exploit that spacetime. A life history perspective provides some advantages in understanding how invasion occurs in a community. Plants experience the same general life history processes (birth, dispersal, recruitment, vegetative and seed reproductive growth). This life cycle can be described by the underlying plant morphological structures, developmental processes and whole plant activities that occur during each of these phases (table 6.1). A plant's life cycle is a Markov chain or process in which the state of the plant at any one time is a direct consequence of its state in the previous time period (Dekker, 2004b). Failure at any time in the life history ends the invasion process.

**Markov chain or process**: a sequence of events, usually called states, the probability of which is dependent only on the event immediately preceeding it (Borowski and Borwein, 1991)

The life history of an annual weedy plant consists of several discrete threshold events, events that can provide strong experimental inferences in life history studies: anthesis, fertilization and zygote formation, abscission from the parent plant, seedling emergence time and death (table 6.2). The consistent temporal occurrence of these threshold events make them ideal for experimental comparisons between and among individuals and populations during their life history.

**Table 6.2** Processes of life history, natural selection (see table 2.1), and invasion biology underlying functional trait expression in weeds; life history **threshold events in bold**.

| | PROCESSES | |
|---|---|---|
| LIFE HISTORY | NATURAL SELECTION | INVASION BIOLOGY |
| PROPAGULE REPRODUCTION<br>Parental Architecture<br>Floral induction | Condition 3: survive to produce the fittest offspring | Colonization & Extinction |
| **Anthesis**<br>**Fertilization** | Condition 1: generate variation in individual traits<br>Condition 4: reproductive transmission of parental traits to offspring | |
| Embryogenesis | | Enduring occupation |
| **Abscission** | Condition 2: generate variation in individual fitness | |
| PROPAGULE DISPERSAL<br>Spatial dispersal | Condition 3: survive to produce the fittest offspring | Colonization & Extinction |
| Temporal dispersal | Pre-Condition 1: excess local phenotypes compete for limited opportunity | Invasion<br><br>Enduring occupation |
| RECRUITMENT<br>Seed germination<br>**Seedling emergence** | Pre-Condition 1: excess local phenotypes compete for limited opportunity<br>Condition 3: survive to produce the fittest offspring | Invasion, Colonization & Extinction |
| VEGETATIVE GROWTH<br>Neighbor Interactions:<br>Growth-Development<br>Stress Responses | Pre-Condition 1: excess local phenotypes compete for limited opportunity<br>Condition 3: survive to produce the fittest offspring | Colonization & Extinction |

**6.1.1.1 Preadaption**. Many preexisting traits in weed species can assume a new role when crop management practices, disturbances or the environment change. A good example would be oxidative-degradative metabolic systems which play a role in plant environmental stress tolerance. They sometimes also provide the ability to degrade herbicides prior to any exposure to these chemicals. This is a good example of preadaptation (see ch. 3.1.2, seizing and exploiting local opportunity). What phenotypic traits confer weedy success? All traits depend on the context (genotype) within which they are found. All traits in a weed interact with each other, and the norm is that there are trade-offs among them in reaching a balance within an individual plant (homeostasis). Many traits that seem non-weedy can lead to weed success if they are mixed with certain other traits in the same individual.

**6.1.1.2 Trait basis of the invasion process**. Why study the invasion process in terms of traits? They are identifiable as phenotypes, functions and structures. They occur as predictable life history threshold events.

They are experimentally tractable. They are selectable and heritable. Life history provides some important advantages to organizing morphological traits. "Timing is Everything": when a plant performs important developmental processes and activities, relative to its neighbors, is a key to its success in the invasion process. If a particular invading plant is at the right place at the right time, it is the traits that it expresses at those times that make it a winner or a loser over its neighbor. Trade-offs among these traits that compete within the individual phenotype are apparent when we organize them into similar times in their life history.

**6.1.2 Processes of life history adaptation.** Adaptation in weed life history consists of three, simultaneous, overlapping processes: the plant's life cycle, natural selection, and the biological processes of weed invasion and exploitation of local opportunity spacetime (table 6.2).

**6.1.3 Life history trade-offs.** Adaptive characteristics of plants during evolution have been shaped by trade-offs among the several requirements at each phase of its life cycle (see ch. 7.4.2, principle of strategic allocation). Each individual weed species has a unique life history which is the evolutionary consequence of reproductive and survival trade-offs. The allocation of resources for vegetative versus reproductive growth within the developing plant can conflict (see ch. 5.2.5, risk of death determines life history): clonal propagules or seed production in herbaceous perennial species. Strategic allocation of limited resources reveals conflicting goals in reproduction: when to reproduce, how often to reproduce, seed size versus numbers (see ch. 7.4.3, weed seed role tradeoffs), vegetative versus propagule growth. Plant development reveals conflicting goals in life history timing: seed dormancy or immediate germination; seed dormancy in the soil or variable seasonal emergence timing; continuous flowering or rapid completion of reproduction.

## 6.2 Plant Life History Classification Systems

Plant life histories can be classified into several general systems, including those based on life span, growth and life form and life history strategy.

**6.2.1 Life span.** The life span of a weed is its longevity, the maximum or mean duration of life of an individual or group (Lincoln et al., 1998).

**annual**:
1: having a yearly periodicity; living for one year (Lincoln et al., 1998)
2: plants which germinate, bloom and produce seeds, and die in one growing season (Alex and Switzer, 1976)

**winter annual**:
1: usually germinate in the autumn and overwinter as seedlings before renewing growth in the spring (Stubbendieck et al., 1995)
2: plants which germinate and produce a leafy rosette in the autumn and then bloom, seed, and die the following summer (Alex and Switzer, 1976)

**summer annual**: plants which germinate and grow in the spring or summer, bloom, produce seed and die in the autumn

**biennial**: lasting for two years; occurring every two years; requiring two years to complete the life cycle (Lincoln et al., 1998)

**perennial**:
1: used of plants that persist for several years with a period of growth each year;
2: occurring throughout the year (Lincoln et al., 1998)

**herbaceous perennial**: plants which live a number of years, develop each year from underground stems, or roots, or crowns, usually flowering each year but dying back to the ground each year (Alex and Switzer, 1976)

**woody perennial**: trees, shrubs and woody vines which live for many years, producing new growth each year from their aboveground stems, branches and twigs, and in some cases from underground stems, or roots, or crowns (Alex and Switzer, 1976)

Annuals are commonest in arid areas and habitats disturbed by human activity (Silvertown and Charlesworth, 2001). Most annuals have evolved from iteroparous perennial ancestors. The advantage of the annual life history depends on the ratio of adult to juvenile survival. If adult and juvenile survival are equivalent, this means that an annual can match the fitness of a perennial by producing just one more seed than the perennial does (Cole, 1954). The one extra seed is the annual parent plant investment, perennials have preexisting parent plant when the season begins. This is the key to the success of annuals and explains why they typically allocate more resources to reproduction than do perennial relatives (Silvertown and Dodd, 1997). Iteroparous perennials will be favored in environments where seedling mortality is high, typically mesic habitats with a closed cover of vegetation (light interference). Annuals are favored in habitats that have low vegetation cover as a consequence of aridity or disturbance where seedlings can attain high rates of survival if germination is opportunely timed.

**6.2.2 Growth and life form**. Life-form and growth-form are essentially synonymous concepts, despite attempts to restrict the meaning of growth-form to types differing in shoot architecture. Most life form schemes are concerned with vascular plants only. Weed life form is the characteristic structural features and method of perennation of a plant species. It is the result of the interaction of all life processes, both genetic and environmental (Lincoln et al., 1998). Plant life-form schemes constitute a way of classifying plants as an alternative to the species-genus-family scientific nomenclature classification system. In colloquial speech, plants may be classified as trees, shrubs, herbs (forbs and graminoids), etc. The scientific use of life-form schemes emphasizes plant function in the ecosystem and that the same function or "adaptedness" to the environment may be achieved in a number of ways, i.e. plant species that are closely related phylogenetically may have widely different life-form. Conversely, unrelated species may share a life-form through convergent evolution. The most widely applied life-form scheme is the Raunkiær system (Wikipedia, 11.10). See also ch. 10.3.3, ecological roles-guilds-trades in weed-crop plant communities.

**6.2.2.1 Growth form**. Weed life history takes a growth form, the characteristic appearance of a plant under a particular set of environmental conditions (Lincoln et al., 1998; Stubbendieck et al., 1995).

**grasses**: hollow or occasionally solid culms or stems with nodes; leaves are two-ranked and have parallel veins; flowers are small and inconspicuous

**grasslike plants**: resembling grasses, but generally have solid stems without nodes; the leaves are 2- or 3-ranked and have parallel veins; flowers small and inconspicuous

**forbs**: herbaceous plants other than grasses and grasslike plants; herbaceous plants die after one season of growth (annuals) or die back near the soil surface (perennials or biennials); generally forbs have solid or pithy stems and broad leaves that are usually net-veined; flowers may be small or large, colored, and showy

**succulents**: fleshy plants that have thick, water-retaining stems that may resemble pads; large amounts of water may be stored in the stems or pads and utilized by the plants during periods of insufficient soil moisture; flowers are often showy, and pads are frequently armed with sharp spines; examples include cacti

**woody plants**: trees and shrubs; solid stems (with growth rings) and secondary growth from aerial stems which live throughout the year; the aerial stems may be dormant part of the year; leaves of broadleaf trees are generally net-veined, while evergreen trees usually have needle-like leaves; flowers may be either inconspicuous or showy

**6.2.2.2 Raunkiærian life forms**. Christen C. Raunkiær's classification (1904) recognized life-forms (first called "biological types") on the basis of plant adaptation to survive the unfavorable season, be it cold or dry, that is the position of buds with respect to the soil surface (figure 6.3). In subsequent works, he showed the correspondence between gross climate and the relative abundance of his life-forms.

The subdivisions of the Raunkiær system are based on the location of the plant's growth-point (bud) during seasons with adverse conditions (cold seasons, dry seasons):

**phanerophytes**: projecting into the air on stems – normally woody perennials - with resting buds more than 25 cms above soil level, e.g. trees and shrubs, but also epiphytes, which Raunkiær separated out as a special group in later versions of the system. These may be further subdivided according to plant height in megaphanerophytes, mesophanerophytes and nanophanerophytes and other characters, such as duration of leaves (evergreen or deciduous), presence of covering bracts on buds, succulence and epiphytism

**chamaephytes**: buds on persistent shoots near the ground – woody plants with perennating buds borne close to the ground, no more than 25 cms above soil surface, (e.g. bilberry and periwinkle)

**Figure 6.3**. Raunkiær's life forms (1904): 1, phanerophyte; 2-3, chamaephytes; 4, hemicryptophyte; 5-9, cryptophytes: 5-6, geophytes; 7, helophyte; 8-9, hydrophytes. Therophyte, aerophyte and epiphyte not shown (Wikepdia, 11.10)

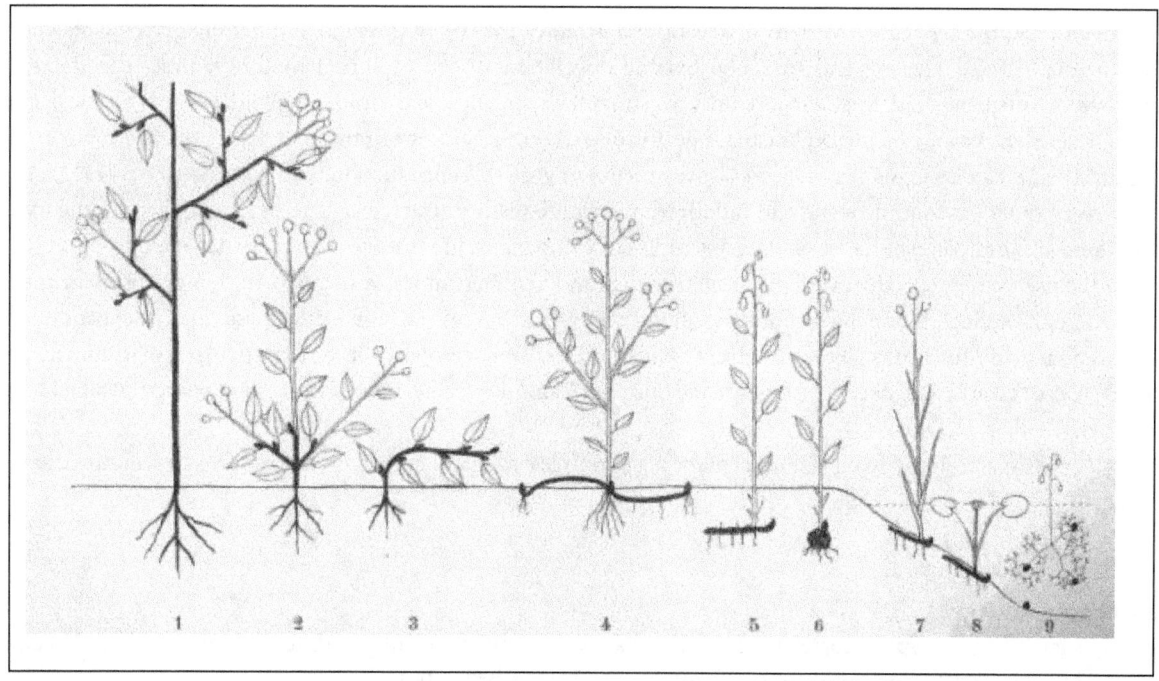

**hemicryptophytes**: buds at or near the soil surface , e.g. daisy, dandelion. These may be further subdivided int protohemicryptophytes (only stem leaves), partial rosette plants (both stem and basal rosette leaves), rosette plants (only basal rosette leaves)

**cryptophytes**: below ground or under water - with resting buds lying either beneath the surface of the ground as a rhizome, bulb, corm, etc., or a resting bud submerged under water

Cryptophytes are divided into 3 groups:

    **geophytes**: resting in dry ground, e.g. crocus, tulip. May be further subdivided into rhizome, stem-tuber, root-tuber, bulb and root geophytes.

    **helophytes**: resting in marshy ground, e.g. reedmace, marsh-marigold.

    **hydrophytes**: resting by being submerged under water, e.g. water-lily, frogbit.

**therophytes**: annual plants which survive the unfavorable season in the form of seeds and complete their life-cycle during favorable seasons. Annual species are therophytes. Many desert plants are by necessity therophytes.

**aerophytes**: new addition to the Raunkiaer lifeform classification. Plant that obtains moisture (though not through haustoria) and nutrients from the air and rain; usually grows on other plants but not parasitic on them. This includes some *Tillandsia* species, as well as staghorn ferns.

**epiphytes**: originally placed in Phanerophytes (above) but then separated because of irrelevance of soil position

    G.E. Du Rietz (1931) reviewed the previous life-form schemes in 1931 and strongly criticized the attempt to include functional characters. He tabulated six parallel ways of life-form classification: general plant physiognomy; growth-forms/shoot architecture; periodicity life-form-seasonal physiognomic variation; bud height life-form (Raunkiær's scheme); bud structure life-form; and, leaf life-form (form, size, duration, texture of leaves) (Wikipedia, 11.10).

**6.2.3 Life history strategies**. Life history theory posits that the schedule and duration of key events in an organism's lifetime are shaped by natural selection to produce the largest possible number of surviving offspring (Wikipedia, 5.14). These events, notably juvenile development, age of sexual maturity, first reproduction, number of offspring and level of parental investment, senescence and death, depend on the physical and ecological environment of the organism. Life history characteristics are traits that affect the life table of an organism, and can be imagined as various investments in growth, reproduction, and survivorship. The goal of life history theory is to understand the variation in such life history strategies. The key to life history theory is that there are limited resources available. Examples of some major life history characteristics include age at first reproductive event; reproductive lifespan and aging; and number and size of offspring. Variations in these characteristics reflect different allocations of an individual's resources (i.e., time, effort, and energy expenditure) to competing life functions. For any given individual, available resources in any particular environment are finite. Time, effort, and energy used for one purpose diminishes the time, effort, and energy available for another.

    See related topic in ch. 12.4, adaptation in complex adaptive systems: non-anticipatory-nonpredictive strategy.

**life history strategy**:
1: a specialized phenotype of correlated traits that has evolved independently in different populations or species exposed to similar selection pressures (Silvertown and Charlesworth, 2001)
2: the schedule and duration of key events in an organism's lifetime shaped by natural selection to produce the largest possible number of surviving offspring

The best developed life history strategy concept is the *r*- and *K*-strategy theory (MacArthur and Wilson, 1967; Pianka, 1970). This scheme proposes that species vary along a continuum from those with short generation time, high fecundity and good dispersal (*r* strategists) that are adapted to colonize habitats where mortality is density-independent. Contrasted to this there exist species with the opposite traits which are adapted to thrive in habitats where mortality is density-dependent (*K*-strategists). This generalized, simplistic, dichotomy proposes the existence of an expected tradeoff between dispersal and competitive ability.

This scheme was developed based on symbols used in demographic population biology models (Silvertown and Charlesworth, 2001) (see the Verhulst-Pearl logistic equation in ch. 2.4.1, population formation: precondition to natural selection):

**r**: in population biology, the intrinsic rate of increase

**$K_i$**: the maximum sustainable density of a species in a mixture; subscripts denote different species.

**6.2.3.1 r-selection**. In this life history strategy, selection occurs for the qualities needed to succeed in unstable, unpredictable environments, where ability to reproduce rapidly and opportunistically is at a premium, and where there is little value in adaptations to succeed in competition. A variety of qualities are thought to be favoured by *r*-selection, including high fecundity, small size, and adaptations for long-distance dispersal. Weeds, and their animal equivalents, are examples. Contrasted to this is *K*-selection. It is customary to emphasize that *r*-selection and *K*-selection are the extremes of a continuum, most real cases lying somewhere between. Ecologists enjoy a curious love/hate relationship with the *r/K* concept, often pretending to disapprove of it while finding it indispensable (Dawkins, 1999). *r* selection favors such traits as fecundity, precocity, allocating resources mostly to seed: colonizers arising after a crash or new invasion. Examples include *Arabidopsis*, "live fast, reproduce quick, die young". Winter annuals are considered *r*-strategists where the main struggle is with winter survival. Autumn emergence is relatively easy with no competition. In the spring they bolt and set seed before competition starts. Environmental stress is the biggest challenge. Another weed species example is *Kochia scoparia* which is often the first species to emerge in those communities in its range (e.g. drier areas of Midwestern U.S.). It is adapted to dry areas wherein environment is the main challenge. It's a weak competitor with subsequently emerging weeds. It's an exception to competitive exclusion where the first seedlings up capture resources preferentially.

**6.2.3.2 K-selection**. In this strategy, selection occurs for the qualities needed to succeed in stable, predictable environments where there is likely to be heavy competition for limited resources between individuals well-equipped to compete, at population sizes close to the maximum that the habitat can bear. A variety of qualities are thought to be favored by *K*-slection, including large size, long life, and small numbers of intensively cared-for offspring. The '*K*' and '*r*' are symbols in the conventional algebra of population biologists (Dawkins, 1999). *K* selection is "live slow, reproduce slow, die old" (e.g. redwoods). They put effort into a competitive vegetative body. Perennials utilizing a combination of vegetative reproduction plus seed reproduction is a *K* strategy in which biological, density-dependent, competition is more important than environment. They invest in a competitive body first, and seed investment is second. They favor competition over precocity and fecundity.

**6.2.3.3 CSR strategy**. Grime (1977, 1979) devised the CSR scheme, a three-strategy scheme intended as a refinement of *r/K* theory which classifies plants in the established phase of their life history as competitors (similar to *K*-strategists), stress-tolerators that grow slowly but are successful in nutrient-poor habitats, and ruderals (similar to *r*-strategists).

## Chapter 7: Reproductive Adaptation

**The weeds' story**. I am foxtail.

I am born anew, every year, across Eurasia, across North America. Anywhere and everywhere in the temperate fields of Earth you prepare for me with your many disturbances. After the solstice, in the long days of summer, my mom-dad sex organs ripen and we mate with myself in the closed confines of my flowers, protected. Every so often the man next door manages to slip in and commit adultery with my mother. It all works out in the end: my dad has been known to slip quietly over to the next door neighbors too. Those pollenacious men seem to like windy days for these exploits.

Each of my children is special. I raise each and every one of them carefully, making sure to instill in them a nice protective coat. The ones you eat, those destined to be grown by you helpful humans, I send out naked to ensure they start their new lives right away in the spring. That's the way you like them, isn't it? Things get a bit more complicated with my weedy children. For them I provide a nice strong coat to resist all those scratchy dirt clods in their new home. Under their winter coats I make a raincoat, I wrap them up to their neck tightly to make sure nothing gets inside too easily. In the spring the children want to undress right away and go crazy in the early spring fields. They get fooled by the easy warmth and exhilarating oxygen-rich soil water. But as we all know, parental control is necessary to keep our children's wild impulses in check. So their coats have no buttons. They must wait until I say they can grow again, to become parents themselves. But only when I say it is OK. And I make sure that each of my children grows again, at the right time in the spring, and summer, and even a few in the autumn. I like to spread them around, it works out better that way. You humans just wait every year for my children to come out and play. And what do you do to them? You kill them, as many as you can. That's OK, what kills me makes me stronger. The children you miss, the ones you leave, are in your image.

I spend my hot summer days growing my children. It's not easy, and it goes on and on as the summer days lengthen into the cool autumn nights after the equinox. I don't mind. Most of my kids are born and leave home early, but those late children are sometimes my favorite, lingering on in my old age until the cold northern winds chill me, and then kill me. But my children keep raining down on the ground, seeking the safety and protection of the deep dark soil.

I try to have as many children as I can. I give them each what they need to grow, to survive, to succeed, but not a bit more than is necessary. There are so many good home sites available in your neighborhoods, the ones you create for me. I just can't stop myself, I have to fill all those houses up, every year, every field. So my children rain down on your rich warm neighborhoods and go to work. Some grow this year, but I make sure some wait for next year, and the following year. I can't be too careful with you murderous humans after all. It's all OK though, it only takes one seed child in the right place at the right time to have thousands and thousands of new children.

We endure. Thank you. I couldn't have done it without you.

**Summary**. The reproductive life history of a weed begins with the initiation of floral primordia in the developing parent plant shoot architecture. The size and structure of the plant body are sensitively tuned to available local opportunity by means of phenotypic plasticity. First, shoot branching-tillering is encouraging or aborted. Then flowers are aborted or left to develop as mature seeds. Flowerheads emerge on branches or tillers. Morphogenesis of pollen and egg/ovules occurs. Flowering and seedhead formation precede mating and are carefully timed for survival.

The operation of the mating system is carefully timed to minimize the risks of mortality. The mating system controls recombination between pollen and egg, a balance of genetic foraging with genetic conservation appropriate to the opportunity being exploited. The new zygote is formed soon after anthesis and the fertilization of receptive ovules. The embryogenic process is adapted to the allocation of resources among five seed roles. Trade-offs during the species evolution balance propagule dispersal to favorable localities, persistence, embryonic food reserves for early growth, release of genetic novelty, and the multiplication of successful parental genotypes.

Embryogenesis occurs. The embryo acquires germinability competency and dormancy is induced in many weed species. The morphogenesis of germinability-dormancy structures occurs: either none; or protective envelopes modulating signal reception from the soil environment; or development of light sensitive photo-receptors; or complex combinations of all of these. The qualitative result in many weed species is the formation of heteroblastic seeds: variability in germinability capacity among individual siblings on a single seedhead.

Developmental processes during embryogenesis include contributions from two potentially different individual genotypes: the parental genome forming tissues surrounding the new zygotic genome. Seed traits that maximize parental reproductive success are quite often different than those that might maximize the success of an individual progeny seed. This apparent conflict is only resolved when the seeds themselves become parents.

The evolution of seed size, seed dispersal and seed dormancy are all traits under the control of the parental plant, traits which determine the success of parent plants in future generations. Embryo quantity and seed fecundity are preadapted to the unknown future availability of opportunity in diverse habitats. Adaptation in seed size of a particular species anticipates habitat opportunity. Smaller, more numerous seeds allows exploitation of many unoccupied local sites. While fewer large seeds allow seizure of sites with more competing neighbors.

Reproductive processes end with the threshold life history event of seed abscission, physiological separation from the parent plant. The zygote has now developed into an independent organism. Abscission from the parent plant leads to dispersal of heterogeneous seeds from the seedhead in both space and time to locations with favorable opportunity spacetime.

## 7.1 Introduction

A plant is born, begining its life history, when the egg is fertilized by pollen. Events of critical importance prior to the formation of the zygote include development of the flower primordial inside the plant during the vegetative phase. Flowering and seed head formation preceed mating: anthesis and fertilization. With the new zygote embryogenesis occurs: acquisition of germinability competency and dormancy induction. All these developmental processes include contributions from two potentially different individual genotypes: the parental genome forming tissues surrounding the new zygotic genome. These reproductive processes end with the threshold life history event of seed abscission. Abscission from the parent plant leads to dispersal of heterogeneous seeds from the seedhead in both space and time (ch. 8).

Adaptation in propagule reproduction is the consequence of the three overlapping processes of weed life history, natural selection and invasion biology (table 7.1).

**Table 7.1** Overlapping processes during propagule (seeds, buds) reproduction: life history development, natural selection-elimination (see table 2.1), and invasion biology underlying functional traits of weeds.

| LIFE HISTORY | NATURAL SELECTION | INVASION BIOLOGY |
|---|---|---|
| Parental Architecture<br>Floral induction | Condition 3: survive to produce the fittest offspring | Colonization & Extinction |
| Anthesis<br>Fertilization (threshold event) | Condition 1: generate variation in individual traits<br>Condition 4: reproductive transmission of parental traits to offspring | |
| Embryogenesis | | Enduring occupation |
| Abscission (threshold event) | Condition 2: generate variation in individual fitness | |

Traits important to reproductive adaptation are those that fulfill roles that generate heritable genotypic, phenotypic and speciation variation in fitness appropriate to the opportunity spacetime being exploited by the weed species (table 7.2). This includes morphogenesis of traits occuring during reproduction whose expression is crucial in subsequent times of life history.

Table 7.2 Weedy life history roles, developmental processes and functional traits for adaptation in reproduction.

| PLANT ROLE | TRAIT PROCESS | TRAITS |
|---|---|---|
| **Mating system**: control recombination: balance genetic foraging with genetic conservation appropriate to the opportunity being exploited | Genetic foraging | -outcrossing; monoecy; dioecy<br>-speciation and reproductive isolating mechanisms<br>-external pollination vectors morphology |
| | Genetic conservation | apomixis; self-pollenation; vegetative reproduction |
| **Floral quality**: control flower production, plant architecture | Floral morphogenesis | -seedhead architecture and morphogenesis<br>-floral organ expression and anthesis |
| | Perennial fecundity | balance production of vegetative buds and sexual seeds |
| **Fertilization**: zygote formation, fecundity, inheritance | (threshold event) | |
| **Timing of reproduction** | Seasonal timing | -floral induction and morphogenesis<br>-fertilization<br>-plant reproductive period duration and times of abscission |
| | Periodicity | -embryogenic duration<br>-time to embryo germination competence |
| **Embryo-propagule quality** | Propagule morphogenesis | -seed dormancy and heteroblasty induction<br>-seed size<br>-protective structures enveloping embryo<br>-nutritive tissues supporting embryo (e.g. cotyledons, endosperm)<br>-dispersal structures |
| **Propagule fecundity**: | Propagule number | -plastic production of flowers, flowering |

| | | |
|---|---|---|
| maximizing/optimizing seizing of local opportunity spacetime | | branches and seed number<br>maximizing/optimizing plant size<br>-reallocation of plant food-nutrients to propagules<br>-propagule productivity and duration maximizes seasonal conditions-resources<br>-propagule number appropriate to future safe sites available in locality |
| **Abscission**: separation of propagule from parent plant | (threshold event) | -dispersal mechanisms |

## 7.2 Flowering, Anthesis, Fertilization and Birth

Parental plant architecture and mating systems provide the form for fertilization and zygote formation functions.

**7.2.1 Parent plant floral architecture.** The body of the plant allows the individual to seize and exploit local opportunity spacetime. Its size is a function of its phenotypic plasticity exploiting available resources and conditions, including light capture and exclusion of neighbors, and positioning and display of flowers. Weed architecture is about tillering, branching and canopy development, floral primordial development, flowering and the development of the seedhead. All these processes in the individual phenotype are strongly influenced by relationships to neighbors at the time they occur. The expression of these functional traits during weed life history determine the above and below ground form of the plant when it begins reproduction, and is a significant determinate of its subsequent reproductivity.

**7.2.2 Mating systems.** See ch. 5.4 (mating system and inheritance) for a complete development of the roles the weed breeding system plays in reproductive adaptation.

## 7.3 Embryo Adaptation: Embryogenesis and Dormancy Induction

Once a weed seed is fertilized it begins the process of embryogenesis, the formation, growth and development of the zygotic seed or bud, the propagule. During this critical process the new plant is formed. In many, if not most, weed species this is also the time of seed dormancy induction, a process inseparable from development of the rest of the embryo and seed. Because seed dormancy-germinability commences in embryogenesis, these traits are discussed here. The seeds' most important roles are expressed following the threshold life history event of abscission from the parent plant. The subsequent life history of seed dormancy-germinability traits are presented in ch. 8.3 (dispersal in time: formation of seed pools in the soil), and ch. 8.4 (propagule germination and recruitment)

**7.3.1 Induction of seed dormancy.** Seed dormancy-germinability regulation is induced in weed seeds during embryogenesis. This induction is the beginning of dispersal in time which continues with the formation of soil seed pools, and ends with seedling recruitment. Dormancy mechanisms are the drivers of, the reasons why, seed pools form in the soil (or not). Much of the material on dormancy mechanisms overlaps with seed pools and recruitment. Soil seed pools are the inevitable consequence of dormancy mechanisms. Recruitment, seed germination and seedling emergence/establishment are also inevitable consequences of dormancy mechanisms.

Seed dormancy-germinability literature is filled with imperfect and confusing descriptions (e.g. Bradford and Nonogaki, 2007) and uninformative and confusing terminology (e.g. Baskin and Baskin, 1998, 2004). Herein a clear, simple use of terminology and concepts is utilized to expose the underlying functional traits (morphology, behavior, physiology) important to the control of seeds in the soil. In order to bring understanding to these complex phenomena, emphasis in terminology will be on the biology and activity of weed seeds and buds (i.e. germinability) rather than on what is not occurring (i.e. dormancy). Some basic seed dormancy-germinability definitions include:

**seed dormancy** (and bud dormancy)
1: a state in which viable seeds (or buds; spores) fail to germinate under conditions of moisture, temperature and oxygen favorable for vegetative growth (Amen, 1968);
2: a state of relative metabolic quiescence; interruption in growth sequences, life cycle (usually in embryonic stages)
3: dispersal in time

The category of primary dormancy (unable to germinate at abscission) includes those phenomena that arise within the parent plant during embryogenesis. Definitions include:

**innate dormancy**
1: unable to germinate under any normal set of environmental conditions
2: the condition of seeds (born with) as they leave the parent plant, and is a viable state but prevented from germinating when exposed to warm, moist aerated conditions by some property of the embryo or the associated endosperm or parental structures (e.g. envelopes) (implies a time lag) (after-ripening)
2: characteristic of genotype, species; dormancy when leaving the parent
3: innate dormancy is caused by: incomplete development; biochemical 'trigger'; removal of an inhibitor; physical restriction of water and/or gas access

The category of secondary dormancy (acquisition of dormancy after abscission) includes those phenomena that arise from environmental conditions affecting seeds after abscission. Definitions include:

**induced dormancy**
1: an acquired condition of inability to germinate caused by some experience after abscission
2: acquired dormancy after abscission due to some environmental condition(s)

**enforced dormancy**
1: an inability to germinate due to an environmental restraint (e.g. shortage of water, low temperature, poor aeration)
2: unfavorable conditions prevent germination of a seed that would otherwise be able to germinate
3: conditional dormancy; germination only in a narrow range of conditions (both in the loss of dormancy and reacquisition of dormancy in dormancy cycling through the year)

The biological definitions of inactive seeds provides little or no insight into what is, will, or can occur in nature. As such they are experimentally intractable (see ch. 7.3.7, experimental weed seed science). Behavior is observed in the processes of seed development.

**seed germinability:** the capacity of an seed, bud or spore to germinate under some set of conditions

**after-ripening**
1: changes in seeds, spores and buds that cannot germinate (even under favorable conditions) in the period after dispersal which allow it to germinate
2: the poorly understood chemical and/or physical changes which must occur inside the dry seeds of some plants after shedding or harvesting, if germination is to take place after the seeds are moistened (Chambers scientific and technology dictionary, 1988)

3: the period after a seed has been dispersed when it cannot germinate, even if conditions are favorable, and during which physiological changes occur so that it can germinate (Henderson's dictionary of biological terms, 1979).

**vivipary:** germinating while still attached to the parent plant; in crops its called pre-harvest sprouting, or precocious germination

**viviparous:** producing offspring from within the body of the parent

**7.3.2 The evolutionary ecology of seed dormancy.** Plants have seed (or bud) dormancy because it is a successful strategy for dealing with periods of increased risk of mortality (e.g. frozen winter soil). Seed dormancy is dispersal in time to avoid mortality (winter, farming practices), allowing the seed to wait for the right conditions, or until the favorable habitat, reappears. The existence and duration of seed dormancy evolved by adopting one of three high-cost options, or alternative strategies, to survival during unfavorable conditions. They are somewhat similar and overlap: some species have adopted all 3 ecological strategies for survival.
*Option 1: seeds or bud dormancy.* Annuals produce dormant seed, herbaceous perennial species produce dormant underground rhizome, rootstock, etc., to avoid unfavorable conditions.
*Option 2: seasonally dimorphic phenotypes.* Perennial and biennial plant species that adopt different vegetative bodies at different times of the year (somatic polymorphism over time). Herbaceous perennials display a dormant underground rhizome, rootstock, etc. in the winter; and shoot and underground organs in summer. Biennial plants display an overwintering rosette; in the spring the shoot elongates with reproduction.
*Option 3: homeostatic growth form that is tolerant of the entire range of environmental conditions in the plant's life cycle.* Woody perennials (trees) display a permanent shoot year-round, but in the winter have dormant leaf shoot buds. An exception is found with evergreen, conifer, leaves that don't drop in the autumn (but even their leaves are dormant in cold, dry conditions).

Seed and bud dormancy afford a lowered struggle for existence, a lower cost, but at the expense of fitness by the delay. Dormancy allows plants to survive during, escape from, harsh winters. Seed dormancy is the least stressful form for a plant during harsh times, and is also the smallest. The dormant plant isn't using, demanding, resources when they are unavailable, or at their lowest level. Seed dormancy delays the start in producing progeny to more favorable times. The dormant soil seed pool is an adaptation that provides a genetic buffer, long-term continuity, for the weed species.

Seed and bud dormancy differ from each other in some ways. The differences are revealed in the roles seed play (see ch. 7.4.1, five roles of seeds). Buds have no hulls, few protective structures, less tissue differentiation; seeds do. Longevity in seeds is usually greater than that in buds. There exists less spatial movement, dispersal in buds. In terms of genetic novelty of new generation, buds have no genetic novelty (they are clones). Seeds are usually produced with sexual mating and recombination, and are therefore potentially new genotypes.

Seeds and vegetative buds are similar to each other in other ways. The similarities are revealed in the five roles propagules play. Both seeds and buds are for multiplication or individuals, and for storage of energy for growth. Both are profoundly influenced by their parents until germination. Buds are physiologically connected to the parental plant system, by which they are regulated and dependent. Seeds are usually enclosed by dormancy-inducing parental tissues regulating germination.

**7.3.3 Weed seed dormancy variability and somatic polymorphism.** The importance of seed dormancy heterogeneity (seed heteroblasty; Silvertown, 1984; see ch. 4.3.2.2, somatic polymorphism in seeds) as a functional trait in weed life history is revealed in several insights by Harper (1964):

"The discovery of clines in seed polymorphism ... suggests that seed polymorphism may be a most sensitive indicator of evolution in weedy species."

"There is increasing evidence that specialized properties of the seed, its shape, its size and form, and the nature of its surface determine the type of contact the seed makes with the soil and on this depends the ability of the seed to germinate successfully."

"Present knowledge of the behavior of invading species, both successful and unsuccessful, suggests that it is in phases of germination and seedling establishment that their success or failure is most critically determined and it would seem reasonable therefore, to look particularly closely at seeds and seedlings of invading species. The considerations set out in this paper lead one to expect that it is in properties such as seed number, seed size, seed polymorphisms, and precise germination requirements that the most sensitive reactions of a species to an alien environment are likely to occur."

"Seed polymorphisms seem particularly likely to be sensitive indicators of evolutionary change in alien invaders."

Definitions:

**somatic polymorphism of seed dormancy**: the occurrence of several different forms (amounts) of seed dormancy produced by an individual plant; distinctively different dormancy forms (amounts) adapted to different conditions

**seed heteroblasty**: variability in dormancy-germinability capacity among individual seeds at abscission shed by a single parent plant; seed germination heteromorphism

**seed dormancy capacity**: the dormancy-germinability state of an individual seed at abscission, as well as the amount of environmental signals (water, heat, light, oxygen, etc.) required to stimulate a change in state (after-ripening)

Most of the differences in dormancy-germinability among seeds are not apparent externally (e.g. seed size, color, shape).

**7.3.4 Evolutionary ecology of seed heteroblasty.** Why is seed dormancy variable? What is the purpose, or selective advantage, of a plant shedding seed with variable dormancy? Soil seed pools filled with seeds with slightly different dormancy-germinabilty capacities provide seedling available to seize and exploit all recruitment opportunity spacetime microsites (safe sites) of the local habitat. Individuals resume growth and development at times appropriate to fully seize and exploit local opportunity (see ch. 8.4.7 relationship between seed heteroblasty and recruitment timing).

Seed dormancy polymorphisms can be seen as a hedge betting strategy for an uncertain future. This strategy has a roulette wheel analogy, if you place a bet or chip on every number (time of emergence) you can't lose. If you have a means of remembering (dormancy-germinabilty capacity), over time selection occurs for the best number to bet on, or the best combination of numbers. Each winning seed once again produces seeds with same ratio; but over time any mutation/variation ratio can shift to more favorable ratio for local conditions. Selection for dormancy somatic polymorphism, instead of genotype polymorphism, for dormancy allows each different weed genotype to produce an array of dormancy state individuals. The mechanisms by which they are produced can be epigenetic (e.g. weedy *Setaria*). More variety and flexibility can be produced with different phenotypes than with different genotypes. A species only needs one genotype at a location to colonize it fully, exploit all good emergence times. Different genotypes would require all of them being there to spread their pollen for full exploitation of that locality.

How variable is seed dormancy in a species, a parent plant, a local population? This is poorly understood, yet the answer for any individual is critical to understanding and predicting its life history and fitness. Is variation greater between plants or between habitats? Weed germination variation differences are greater between plants than between habitats (Cavers and Harper, 1966). Why? Wide amounts of variation in seed dormancy is necessary at any site it is capable of growing in, in order to take advantage of the variable opportunities, micro-sites in the field. Habitat to habitat variability is less than intra-field variability (Harper (1977, Ch. 3).

Germination heteromorphism (over more than one season) often is correlated with variation in seed size (Hendrix, 1984; Silvertown, 1984). It is favored when there is variation in offspring success between seasons, in order to spread the risk of mortality, to hedge your bets (Cohen, 1966; Venable and Brown, 1988; Rees, 1994). There exists a negative correlation between seed dormancy and adult longevity, and between seed size and seed dispersal (Rees, 1993).

**7.3.5 Weed species seed heteroblasty examples**. Occurance of seed heterblasty in weed species is common, if not poorly documented (Harper, 1977; Baker, 1974; Silvertown, and Charlesworth. 2001. p.144).

Common cocklebur (*Xanthium strumarium*, Harper, 1977, p.69) has a two-seeded capsule, each with different dormancy: the first germinates in the first, the second in the second, year. One seed is larger than the other. When the first seed germinates the capsule partially splits and pulls the capsule out of soil with emerging cotyledons, which re-disperses the capsule remnant and the second, dormant seed. A mutant from Mississippi with five seeds per capsule has been observed. Which seed might be the one with more dormancy? The first non-dormant seed has more food reserve at the start; extra energy to break out of the capsule. The second, smaller seed is more dormant and emerges later. It has less investment and lives longer. Alternatively, there may be no order to their dormancy, and differential stimulation depends on the microsite conditions.

Common lambsquarters (*Chenopodium album*) has been reported to have four different dormancy-germinability types: two color morphs and two surface morphs (Harper, 1977). The two seed color types have the most variation in dormancy. The brown represent 3% of the seed produced which are less dormant and have a larger size. They are easily germinated, cool temperatures and a thin seed coat wall controls germinability. The brown seed have a shorter longevity in soil. Altenhofen (2009, 2014) has reported these as immature seeds. Black seed represent the majority of seed produced. They are more dormant and they germinate in cooler temperatures. They are stimulated to germinate in light in the presence of nitrate. Altenhofen has reported these as mature seeds. Two types of seed surface have been observed: reticulate and smooth seed coats. The reticulate coat sometimes is pealed away by mechanical action exposing the shiny black surface underneath.

Curled dock (*Rumex crispus*) and salt bush (spreading orache; *Atriplex* spp.) in dry parts of the USA shows differences in dormancy among its seeds (Harper, 1977). Smaller black seeds are the earliest produced, while larger brown seed are produced later. Environmental conditions during seed formation, embryogenesis, determine the ratio of brown:black produced on parent plant and their relative dormancy-germinability.

Common poppy (*Papaver somniferum*) can produce dormant and non-dormant genotypes (Harper, 1977). In each type, the dormant minority is about 3%. Selection for domesticated poppy genotypes still results in the expression of the dormancy minority fraction, they never lose all their dormancy. Poppy has two types of dormancy: seed dormancy; and summer and winter annual life cycle dimorphism in which the non-dormant seed is shed in the fall. Winter mortality is high depending on weather, but survivors in the spring produce 10 times as much seed as the other type.

Weed species in the *Cruciferae* family exhibit differences in seed dehiscence which result in dispersal in time differences (Harper, 1977). The indehiscent, closed silique produces one seed per silique. The dehiscent type opens and sheds many seed per silique, resulting in quick soil germination. Other dormancy differences also occur.

Wild oats (*Avena fatua, A. ludoviciana*) has seed heteroblasty as a result of the seeds position on the spikelet (Harper, 1977). The seed on the end of spikelet lacks dormancy, while those below the spikelet end are deeply dormant.

*Dimorphotheca* sp. is like other composite family sunflowers wherein the disk and ray flowers on capitulum are different, with different levels of germinability (Harper, 1977). *Gymnarrhena* sp. (field marigold) is an extreme case. It has two types of flowers and patterns of flowering. Small, open flowers flower when more rain is available. Closed, cleistogamous flowering occurs inside the flower (not open to outside pollen) below ground when rainfall is low. Velvetleaf, (*Abutilon theophrasti*) possesses differences in seed germination based on subtle differences in weight, presumably in seed coat thickness (Nurse and DiTomasso, 2005). Variable dormancy in the smartweeds (*Polygonum* spp.) is revealed by an after-ripening treatment (Harper, 1977).

Weedy foxtails (*Setaria* spp.-gp.) possess seed heteroblasty. A range of dormancy-germinability types have been found on a single individual plant from viviparous to very dormant (Atchison, 2001; Dekker, 2014; Haar, 1998; Haar et al., 2012, 2014; Haar and Dekker, 2012, 2014; Jovaag et al., 2012a). For example, among 17 different locally adapted *S. faberi* populations in central Iowa, USA in 1998 a range of responses to after-ripening was observed (figure 7.1, top left). Analysis grouped these populations into three different general responses to after-ripening (figure 7.1, top right). These responses can be expressed as a phenotypic reaction norm, germination hedge-bet structure capable of searching and exploiting local opportunity spacetime (figure 7.1, bottom). These observations are developed in ch. 17, weedy *Setaria* species-group case history. For another example, different dormancy phenotypes are shed during the growing season by different tillers which follow the "early-early" rule: the earliest seed formed on and individual panicle of any age is the most dormant; and, the earliest panicles (primary; in Iowa seed formed in August) are the most dormant while later panicles (secondary, tertiary; in Iowa seed formed in September-November) are increasingly less dormant (figure 7.2). Seed shed within different years also differ. For example, seeds shed in 1998 (figure 7.2, top row) were more germinable than those of 1999 (figure 7.2, bottom row) at comparable harvest times.

This somatic polymorphism of seeds is caused by three different, interacting, germination control mechanisms in *Setaria*. First, the surrounding seed envelope layers restricts water-gas entry into the seed to the narrow placental pore opening (TACL; Rost, 1971b, 1973; Rost and Lersten, 1970). Small amounts of water freely diffuse in and out of the opening, but oxygen uptake is limited by its solubility in the water taken up. The second somatic polyolymorphism is that of a heme-containing peroxidase scavenging oxygen in the seed and preventing after-ripening and germination (Sareni, 2002). Seed germinability somatic polyolymorphism is also expressed in hull rugosity (surface wrinkles) and shape which affect soil-seed contact, water accumulation on the seed, and transport of water surface films across the seed to the placental pore and uptake into the seed (Dekker, 2014; Donnelly et al., 2012, 2014).

**7.3.6 Observable seed dormancy-germinability regulation life forms**. Weed germination occurs when an individual seed embryo receives the appropriate signal (quality, quantity) from its immediate environment to stimulate the resumption of growth (Dekker, 2013; see ch. 3.4.4, environmental signal spacetime). Dormant seeds require an after-ripening period before actual germination can occur. Categories of germinability-dormancy regulation mechanisms in weed and other plant propagules can be distinguished based on functional, observable, endogenous qualities that inhibit, modulate or regulate these requisite environmental signals. The environmental signals that stimulate viable weed seed/bud germination are heat, water, light, nitrate and oxygen. No propagules can control ambient heat, light or nitrate in the soil so they are not amenable to control. This leaves only water and oxygen that seed traits can adapt to regulate endogenous behavior. Oxygen can only enter living cells in the seed/bud dissolved in water. Heat plays two roles in germination: controlling oxygen solubility in water (low temperature, high solubility; high temperature, low solubility); and, heat for growth. Some species have specialized physiological and morphological systems that are stimulated by light by means of phytochrome (see ch. 3.4.4.2, photo-thermal modulation of weed seed germination behavior by plant phytochromes). In some weed species these more complex phytochrome-containing seeds require, or are stimulated to greater germination by, nitrate. Based on this, endogenous weed seed dormancy is regulated by the inhibition of two (or 3 or 4) requisite factors. It is to these soil environment signals that traits have evolved and adapted to control propagule germination (table 7.4).

**Figure 7.1.** Germination versus after-ripening (AR) duration of seventeen locally adapted central Iowa, USA *Setaria faberi* populations assayed in 1998 (Atchison, 2001; Jovaag et al., 2012a).
**Top left**: Cumulative germination (%, least square mean, smoothed curves) versus after-ripening duration for seventeen individual 1998 populations; 1, individual population with unusual pattern, dashed line (W98-32).
**Top right**: Cumulative seed germination (%) versus after-ripening (AR) duration for the seventeen 1998 populations within 3 groups with similar after-ripening durations (inter-AR groups, numbers 1-3).
**Bottom**: Reaction norm schematic diagram of cumulative seed germination versus after-ripening (AR) duration: seed heteroblasty hedge-bet structure response regions. A: High initial germinability, rapid germination response to after-ripening. B1: Low initial germinability, increasing germinability with increasing AR duration until a plateau of high germinability is reached. B2: Increasing germinability with increasing AR, but less overall germination than in B1. C: Little or no early germinability, but some increase with long AR durations. D: "Perfect" crop response: all seeds immediately germinable.

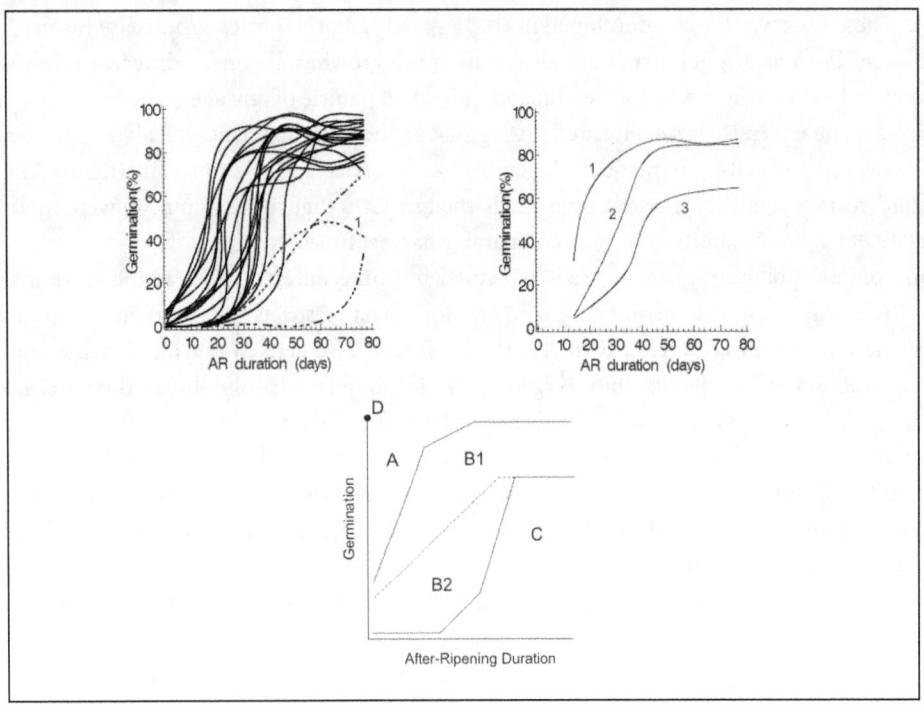

**Figure 7.2.** Germination versus after-ripening (AR) duration of several locally adapted *Setaria faberi* populations from Ames, Iowa, USA harvested in 1998 and 1999 (Atchison, 2001).
**Top row**: Cumulative seed germination (%) versus after-ripening duration for three *Setaria faberi* populations harvested in August (C98-32), September (C98-36) and October (C98-40) 1998.
**Bottom row**: Cumulative seed germination (%) versus after-ripening duration for three *Setaria faberi* populations harvested in August (C99-35), September (C99-38) and October (C99-42) 1999.

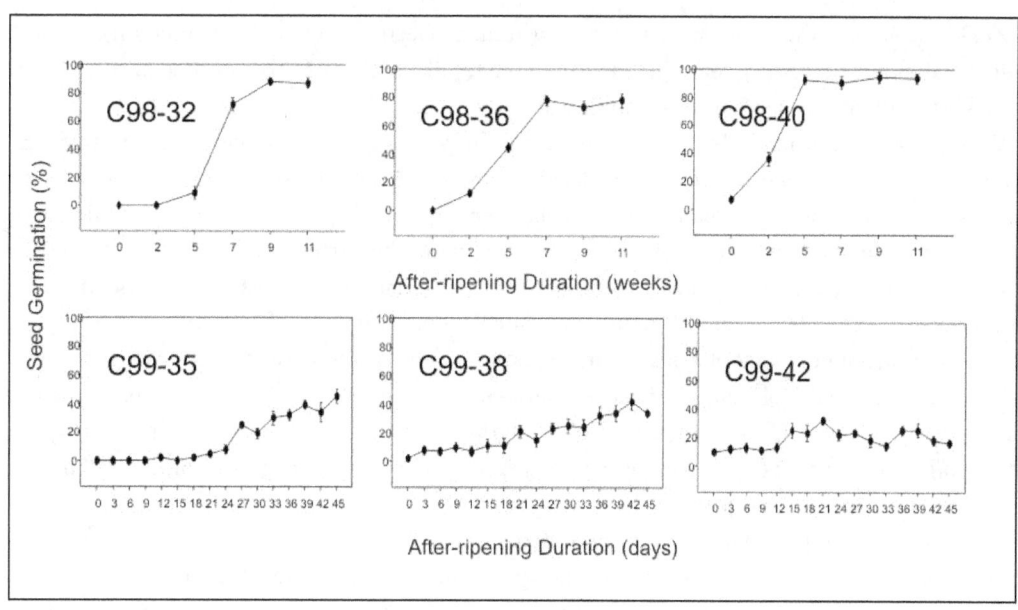

Table 7.4 Observable propagule life forms for dormancy-germinability regulation and some typical weed species examples; endogenously regulated, REG; unregulated, UNREG; and unregulatable, UNREGABLE.

| LIFE FORM | REG | UNREG | UNREGABLE | EXAMPLE |
|---|---|---|---|---|
| Non-dormant | | $H_2O$ $O_2$ | heat | *Kochia scoparia* commercial maize seed |
| Vegetative buds | | $H_2O$ $O_2$ | heat | *Circium arvense* *Taraxacum officinale* |
| Environmental | chemicals fire soil microbes | | heat | |
| Hard, gas- & water-tight, envelopes | $H_2O$ $O_2$ | | heat | *Abutilon theophrasti* *Setaria faberi* |
| Phytochrome seeds | light $NO_3$ | | heat | *Solanum nigrum* |
| Multiple mechanisms | $H_2O$ $O_2$ light $NO_3$ | | heat | *Chenopodium album* |

These categories are not separate, and several can pertain to a particular weed species. For example, germination in most of the heteroblastic seed of *Setaria faberii* is controlled by inhibition of water-oxygen signals from the soil. But, some fraction of the heterogeneous seed is readily germinable at abscission. In other atypical situations, light plays a role in germination. As such it can be considered in different instances to be either non-dormant, water-oxygen inhibited, light inhibited, or possessing multiple control traits.

**7.3.6.1 Non-dormant.** Weed species whose propagule germination is unregulated by heat, water and oxygen inhibition. Non-dormant seed types, like crop seeds, do not inhibit heat, water or oxygen entry into the seed, and therefore germinate readily when these signals are present for some period of time. Weed examples include kochia (*Kochia scoparia*), *Galinsoga* spp., and crops (human selection for rapid germination). Dust-like seed with no internal food reserves (e.g. parasitic plants; epiphytes like orchids; ferns) must acquire an external food source quickly (or attain independent autotrophic growth) or die.

**7.3.6.2 Vegetative, perinating buds.** Species whose vegetative bud or meristem germination is unregulated by heat, water and oxygen inhibition, but is regulated by internal plant physiological control (e.g. intra-plant hormonal apical dominance). Perennial herbaceous weed species begin growth when sufficient heat and moisture is present in the immediate vicinity of the perennating bud (rhizome, rootstock, etc.). Most of the buds of these species are exposed to the soil so water-oxygen signals are not restricted. Individual buds in an intact plant are also regulated by apical dominance along a hormone (auxin) gradient from the parent plant shoot to the terminal shoots. Shoot buds include stolons. Some of the perennating propagule types include rootstocks (dicot perennials; e.g. Canada thistle (*Circium arvense*), milkweeds (*Asclepias* spp.)); rhizomes (grass perennials; e.g. quackgrass (*Elymus repens*), Johnsongrass (*Sorghum halepense*)); taproots (e.g. dandelion (*Taraxacum officinale*); tubers (e.g. nutsedge (*Cyperus* spp.)); corms, bulbs (e.g. wild onion (*Allium* spp.)) and bulblets (e.g. wild garlic (*Allium* spp.)).

Three types of bud dormancy can be discerned.

*Apical dominance in the intact plant system.* Dormancy is induced by auxin translocation from actively growing shoots (photosynthetic source) to the actively foraging terminal buds of the rhizome (photosynthetic sink). Auxin transport inhibits buds closest to the dominant shoot: the basal or crown region. Germination is more likely the farther the bud is from these basal shoots, and most likely towards the terminal buds, farthest from the shoots. Auxins suppress basal buds while new shoots and terminal buds grow. Perennial buds can be dormant while others on same plant at same time are actively growing. Herbicide inhibition is modulated by apical dominance. Less herbicide accumulates in dormant buds which are therefore less inhibited. Herbicides tend to translocate to the actively growing terminal buds of the rhizome system (Dekker and Chandler, 1985; Harker and Dekker, 1988a, b).

*Apical dominance in perennial bud fragments.* Perennial rhizomes systems are frequently broken up with tillage. There exists differences in lingering parental apical dominance in these independent fragments: buds can range from readily germinable (terminal) to dormant (basal).

*Winter dormancy.* Rhizome buds (fragments or intact system) experience enforced dormancy in those habitats that are cold or frozen soil during the winter.

**7.3.6.3 Environmental seed germination control.** Factors external to the seed/bud can stimulate or inhibit germination such as disturbance, chemicals, fire, soil microflora and macrofauna. These exogenous factors can interact with other control mechanisms and traits controlling germination.

*Disturbance.* Soil disturbance can stimulate seed and bud germination in several ways. Tillage can move light-sensitive seeds to the soil surface, or bury them. Tillage can fragment perennial bud systems releasing many from apical dominance and encouraging germination.

*Chemicals.* Substances in the immediate vicinity of the propagule can stimulate germination. Endogenous chemical inhibitors in weed seeds have been proposed as dormancy-germinability regulators, but there existence is questionable: no clear experimental evidence for them exists. External inhibitors in the soil would not exert dynamic control of germination, all seeds would remain dormant or die. Germinability would not be reversible or adaptable to changing conditions with time. Allelopathic chemicals from the soil environment, or other species, could influence germination (e.g. Gressel and Holm, 1964).

*Fire.* Disturbance by fire can affect seed germination in many ways. Seed directly in the fire are destroyed, elevated temperature near the fire can either stimulate germination (cracking hard seed coats) or kill it. Fire increases soil temperature that can favor germination. Fire increases light penetration to the soil surface by reducing the plant canopy and exposing light-sensitive seeds to stimulation. Fire can physically change the seed position, and also change the microsite around seed. Pine cones are released from dormancy with fire when the seed is freed from the cone structure. Chemicals affecting germination can be released with fire. They can both fertilize the emerged plants with residual combustion products, and smoke stimulates the germination of some species seed (e.g. Flematti et al., 2004). For example, karrikins are a group of plant growth regulators of the butenolide class found in the smoke of burning plant material from forest fires had the ability to stimulate the germination of seeds (Wikipedia, 5.14).

*Soil microflora*. Soil microflora could interact with weed seeds/buds and affect germination. Endophytic fungi inside some weed species may also play a regulatory role.

*Macrofauna*. Animal consumption and digestion of seeds can increases germination (e.g. birds, herbivores). The digestive chemical action, plus stomach-bowel-crop/gizzard scarification of surrounding propagule envelopes can increase or decrease seed germination, or kill seed viability.

**7.3.6.4 Hard, gas- and water-impermeable, seed envelope germination inhibition**. Many of the most common weed species that form soil seed pools utilize impermeable, or semi-impermeable, propagule envelopes to inhibit the uptake of water and oxygen. Species that utilize these control mechanisms depend on hard seed coats (e.g. *Abutilon theophrasti*, *Trifolium* spp.). Gas- and water-tight inhibition is accomplished by specialization of seed enveloping tissues that act by restricting water (therefore water-oxygen) entry to different types of openings specific to the species. For example, in *Setaria* spp.-gp. entry is through the placental pore and transfer aleurone cell layer membrane. In dicots like *Abutilon theophrasti* entry is through the chalazal slit. Natural (e.g. animal digestion) and experimental scarification often increases the germination of this type of seed. Three experimental phenomena point to the crucial role water-oxygen inhibition plays in regulating germination in species with impermeable seed envelopes: scarification (physically damaging seed envelopes), stratification (cold, moist, oxygen-rich water treatment) and alternating temperatures.

Examples of species that rely on water-oxygen inhibition by seed envelopes include: morningglorys: field (*Convolvulus arvensis*) and hedge bindweeds (*Calystegia sepium*); sunflower (*Helianthus* spp.; commercial seed companies perform scarification to ensure cultivar germination); velvetleaf (*Abutilon theophrasti*) seed coats can be cracked with heat (50°C) or sulfuric acid (e.g. Dekker and Meggitt, 1983a, b); smartweeds (*Polygonum* spp.) with hard seed coat scarification followed by cold temperatures); tomato, (*Lycopersicum* spp.) the endosperm forms a hard seed coat with a mucilage plug in the entry hole which inhibits germination and can be opened by soaking-fermenting); *Solanum* spp., (nightshade berry, buffalobur), hard seed coat with mucilage stickyness around seeds; clover (*Trifolium* spp.), scarification physically abraids the thin hard seed coat increasing germination. In the weedy *Setaria* species-group specialized seed envelopes restrict oxygen and water signals stimulating after-ripening and germination (Dekker, 2014). Dormancy induction, maintenance, and re-induction are due to restriction of oxygen (dissolved in water) availability in the seed interior. Germination occurs when sufficient oxygen-water reaches the interior.

Velvetleaf (*Abutilon theophrasti*) germination was associated with extended periods during which individual rainfall events do not exceed 125 mm (see ch. 19.5, velvetleaf seedling emergence; Dekker and Meggitt, 1986). Seedling emergence depth (and possibly number) may be stimulated by seasonal periods of shallow soil drying. Shallow soil drying in some manner may open the chalazal slit, the hilum-like opening of the intact (hard-coated) seed to which water entry into the interior embryo is restricted. Opening of the hard coat may stimulate velvetleaf seed germination.

**7.3.6.5 Light-phytochrome and nitrate regulated seed germination**. The seeds of many weed species are stimulated to germinate in the presence of light (photoblasty) and soil water nitrate, especially those that lay on the soil surface or are shallowly buried under soil or crop residues (e.g. Ballare and Casal, 2000). Seed phytochromes regulate these germination responses. Phytochrome can stimulate or inhibit seed germination depending on light duration, quantity and quality. The complex molecular nature and behavior of plant phytochromes was presented more completely in ch. 3.4.2.1, light as resource; and ch. 3.4.4.2, photo-thermal modulation of weed seed germination behavior by plant phytochromes.

Weed species with light-phytochrome regulationed germination include pigweeds (*Amaranthus* spp.-gp.), common lambsquarters (*Chenopodium album*; Altenhofen, 2009, 2014), many horticultural crops (marigolds (genus *Tagetes*), poppies (*Papaver* sp.), lettuce (*Lactuca sativa*) cv. *Grand Rapids*), and curled dock (*Rumex crispus*). Light and soil water nitrate ($NO_3^-$) often increases germinaton in these species, especially in N poor soils. Pigweed spp. (*Amaranthus*) germination decreased with an ammonia N form ($NH_3$) but nitrogen N alone has no stimulation effect. Night tillage, tractor headlights in night tillage, stimulate germination. Nightshade (*Solanum*) late season emergence occurs in canopy holes when light strikes the soil surface in mid-season.

**7.3.6.6 Species with multiple interacting germination control mechanisms.** Many of the successful weed species of the world possess several different, interacting, mechanisms controlling their seed behavior. Why should a species have more than one mechanism to control dormancy and germination? The more interacting systems a species has the more able it is to respond precisely to new, changing, conditions and avoid false signals that stimulate unfit behaviors. Weed seed germinability-dormancy control in weed species is poorly understood.

Common lambsquarters (*Chenopodium album*) disperses heteroblastic seed (black, brown, reticulate, smooth), each potentially with different environmental signals required to stimulate germination: black dormant seed requiring cold temperatures (high oxygen solubility in water), nitrate, light; brown, less dormant, thinned walled, seed only requiring cold temperatures. The advantages for *C. album* are considerable, allowing them to seize and exploit over a wide range of opportunities on several continents (e.g. Andujar et al., 2013; see also ch. 8.4.5, patterns of seedling emergence). The weedy foxtails (*Setaria* spp.-gp.) rely on complex oxygen-water-thermal signals for correspondingly complex emergence patterns finely adapted to local opportunity in fields (Dekker, 2014; Jovaag et al., 2011, 2012a, b).

**7.3.6.7 Other seed germination control mechanisms.** Other means of regulating seed and propagule dormancy exist (incomplete embryo development at abscission), and others have not been yet discovered.

**7.3.7 Experimental weed seed science.** An alternate title to this section might be "Beware of the weed seed dormancy literature" or "Why is seed dormancy science such a confusing mess?" Long experience with seed dormancy literature has left me with a skeptical approach to past reported research. There are many reasons why this is so, but the over-riding consideration is that most of dormancy and germination for our major weed species is unknown. The phenomena of dormancy is very complex and difficult to study experimentally. The literature is filled with articles that make claims, inferences and conclusions that are not entirely justified because they are riddled with experimental artifacts and omissions.

One of the most compromising types of artifacts in weed seed experiments is a lack of seed characterization and description (allowing repeatability) both before and after testing for treatment-parameter effects. What was the history of the experimental seed: collection date, habitat and location; in what conditions, for what duration were they subsequently stored? What was the experimental genotype, and where was it collected? Is the species description correct? Was germination assessed at harvest, subsequently during storage, and immediately prior to experimentation allowing inferences to be made? Were proper genotype and environmental parameter controls used for comparative inferences (e.g. lack of light control in Leon and Owen, 2003)? These are only a partial list of questions that may be asked in a critical analysis of inferences made in weed seed germination studies.

To study dormancy there is a need to collect and document seed immediately after abscission to avoid confounding effects due to subsequent after-ripening or dormancy induction in storage. For example, *Setaria faberii* was assessed for germination after harvest in an Iowa study in which unusually high germination was observed. The authors let seed absciss and accumulate in pollination bags tied around the panicles for the entire growing season. Seeds from the same panicle were collected together in these bags for the several months of the seed rain period. Some seed experienced extended periods of dry, hot conditions in the bags (earliest maturing) while others were only recently matured. Germination data was reported not as a function of storage age after dispersal (panicle bags in the field) but as a function of imposed parameters in subsequent experimentation (Kegode and Pearce, 1998). A study by the same group inferred genotypic and seed germination differences between *S. faberii* accessions when confounding differences in the year of collection and storage mitigated these conclusions (Kegode et al., 1998). Seed storage conditions experienced by parents are not traits inherited by their progeny (i.e. Lamarckian evolution; see unit 2). These mitigating factors were not clearly reported in the journal, leading readers to infere the uninferable (e.g. Leon and Owen, 2003).

Other essential considerations include whether the seed is heteroblastic, and if so, what is the variability in germination of the seeds used? Can this phenotypic variation be separated from treatment effects, do variable seed respond to imposed experimental parameters in the same manner? Is the mean germination behavior used to make inferences? Mean behaviors can easily mask what is going on in germinability reaction norms. Is this variability normally distributed or not? If not, have the appropriate statistics been used? For this reason, studies of populations require large numbers of individuals. Germination variability is masked by data presented as bulk means (i.e. many plants, seedheads) without component variances (not to mention individual seed responses). Often, researchers refuse to embrace variability of weed seed because weeds behave so differently than crop seeds. Often heard is "… if my c.v. (coefficient of variation) is high then I must not be doing things right …" (i.e. if my experimental seed is highly variable I must be doing something wrong). Conversely, expressing germination as mean behavior induces a reassuring sense of normality in the experimentalist. Embracing, explaining, this heteroblasty will reveal what weeds do best, anticipate anything we throw at them, the things they have adapted to so well. Population phenomena is confused with individual behavior, means mask the real action.

Conducting studies with a model or hypothesis of what the dormancy mechanism of the species can often guide the experimenter to clues of its regulatory nature, therefore guiding the choice of appropriate parameters to elucidate germination reaction norms. On the other hand, germination assays with no idea of causation fill the libraries. What can others learn from them? There exists no single, good model of weed seed/bud dormancy-germinability because every weed species has potentially different means of regulating its seed life history. Until we discover those mechanisms we will not understand how weeds assemble in agricultural communities. Seed dormancy is highly variable within an individual plant, within a seed rain season, within an individual soil seed pool. It is this heteroblasty that makes them so successful, so highly adaptive, so supremely fit.

Dormancy is a confusing term ("the biology of what isn't happening"), it confuses and obscures the mechanisms underlying behavior. Many germinability control mechanisms exist in weeds, and many weeds have several interacting in combination for more sensitive control of behavior. Many, if not most, experimentation begins with an unjustified assumption of what controls germination. For example, hydro-thermal (or just thermal, e.g. Murdoch, 2010; Schutte et al., 2014) are frequently used as models of germination behavior because of their simplicity and apparent tractability. Seed science is dominated by research on crop species with uniform germination (ideally). Are the same assumptions viable for weed seeds with heteroblastic seed?

What of experimental hypotheses explicitly studying interactions between heteroblastic weed seed and the multiple environmental parameters controlling the complex regulation of its germination? For example, several of these limitations in weed seed research were discussed in reference to germinability studies on *Chenopodium album* (common lambsquarters; Altenhofen, 2009; Altenhofen and Dekker, 2014):

"The history of seed prior to experimentation may have an effect on the germinability of seed, but this information is often incomplete or missing in published research. A complete seed history should include the following: a) the date of harvest, b) the ecological description of the population habitat, c) seed harvest and storage preparation description, d) storage condition and e) duration of storage pre-experiment. Of these necessary pieces of information, only 4% of the 24 common lambsquarters germination studies reviewed herein included all five details, 29% included four, 25% included three, 25% included two, and 17% included only one or no details of the seed history. This lack of seed history information seriously compromises the repeatability, comparability and interpretation of these published studies.

Another gap in our understanding of *C. album* germination comes from the limited number of parameters (e.g. light, temperature, after-ripening, nitrate, and water) which have been examined within any one study. Of the 24 studies focused on *C. album* germination reviewed here, only 12% included five or more parameters, 21% included four, 29% included three, and 38% included only one or two. Additionally, only 4 of the studies compared multiple populations of common lambsquarters."

What of oxygen? Oxygen restriction may be the single most important means by which weed seeds time their reentry into the above-ground plant community. Yet their exists almost no weed seed literature on oxygen (the oxy-hydro signal) and germination. What of light-heat-nitrate regulation by phytochrome controlled seed behavior? Modern molecular analyses indicate this is a complex subject, with multiple phytochromes responding to different light-heat quality/quantity signals in different species at different times and situations (see section 3.4.4.2, photo-thermal modulation of weed seed germination behavior by plant phytochromes). Yet these new insights have not penetrated the weed seed literature. What of experimentation embracing in its hypotheses the complete collection of factors and mechanisms controlling germination in complex regulatory interactions of heteroblastic *Chenopodium album* weed seed (e.g. Altenhofen, 2009; Altenhofen and Dekker, 2014)?

Most of weed seed germinability control remains unknown. This is surely one of the richest opportunities in weed science if appropriate models, seed characterization, and variation are fully embraced experimentally. It is also a science replete with poorly designed experiments leading to overly ambitious inferences (e.g. Schulte et al., 2014; see ch. 8.4.6.2).

**7.4 Propagule Adaptation: Post-abscission Fecundity**

One of the most crucial threshold events in weed seed life history is abscission from the parent plant. Prior to this time the traits expressed by the parental genome guide the new offspring's embryogenic development. The direct physiological relationship ends with abscission. After this threshold life history event the seed is an independent organism and important functional traits concerning propagule reproductive fecundity are displayed. The evolution of seed size, seed dispersal and seed dormancy are all traits under the control of the parental plant, traits which determine the success of parent plants in future generations. Seed traits that maximize parental reproductive success are quite often different than those that maximize the fitness of individual progeny seeds. This apparent conflict is only resolved when the seeds themselves become parents. The seed is a unique plant form in the life history of the weed, and it performs unique roles as it progresses from parental abscission to independent autotrophic growth as a seedling. The seed in flowering plants consists of sporophytic (the entire body except the embryo sac tissues) and gametophytic tissues (embryo sac which is often differentiated into seed envelope tissues). Hence parental plants nuture seeds from both paternal and maternal tissues.

**7.4.1 Five roles of seeds**. Their exist five roles that a seed has to accomplish to be fit. These five seed roles are intimately related to the five roles played by soil seed pools (see ch. 8.3.1, adaptive roles of soil seed pools).

1: Dispersal and colonization to the "right" place
2: Persistence to avoid mortality and predation for longevity
3: Food reserves for embryo growth
4: Release of novel genetic recombinants at the seed stage in the life cycle
5: Multiplication of parent to continue the successful genotype

These roles compete and conflict with each other for their parent plant's limited resources, resulting in complex adaptive trade-offs shaped by natural selection (table 7.5).

**Table 7.5** Five roles played by weed seeds and associated evolutionary process roles and seed traits.

| Evolutionary Role | Seed Role | Seed Traits |
|---|---|---|
| Exploit opportunity spacetime: survive & reproduce until growth & development resume | Dispersal and colonization to the "right" place | -specialized dispersal structures<br>-palatability to encourage consumption, movement |
| | Persistence:<br>  -longevity in soil seed pools<br>  -avoid mortality, predation | -seed envelopes for protection and dormancy<br>-phytochrome |
| | Food reserves for embryo growth | -energy quantity & quality for growth |
| Generate genotypic & phenotypic variation | New genetic recombinants released into deme:<br>  -novelty for change<br>  -conservation of local adaptation | -seed heteroblasty<br>-parental mating system |
| Transmit fitness to progeny with inheritance | Seed number: multiplication of successful genotype parent | -seed/bud numbers<br>-seasonal time, duration of fecundity |

**7.4.1.1 Dispersal and colonization.** The seed must be able, on its own, to move spatially to a location in which there exists opportunity spacetime to seize and exploit. This can be accomplished in many ways (see ch. 8.2, dispersal in space) including remaining in the same location (e.g. gravity) or movement by means of specialized seed structures (wings, pappus; floatation). Seed palatability by herbivores encourages consumption and movement by the dispersal vector if it is excreted intact and viable (e.g. indigestable seed coat).

**7.4.1.2 Persistence.** Persistence is accomplished by enduring occupation of a locality when seeds accumulate in the soil and form pools (see ch. 8.3, dispersal in time: formation of seed pools in the soil). Many important traits associated with seed dormancy and longevity provide this persistence, including hard, gas- and water-impermeable seed envelopes; and seed heteroblasty.

**7.4.1.3 Food reserves for embryo growth.** Energy stored in the seed provides food for the developing embryo during germination and recruitment. The amount of energy stored in cotyledons (dicots) and the endosperm (grasses) is a large determinant of seed size (along with specialized seed envelopes for dispersal and dormancy control). Parasitic weeds, and very small seeds, have no stored food reserves. They require energy from their host or environment at the time of germination. Weeds in general possess a relatively small seed size and small reserves compared to woody perennials. Herbaceous perennial species initially obtain food for growth from the perinating tissue (e.g. rhizomes, rootstocks).

**7.4.1.4 Release of new genetic recombinants into the local deme.** The seed is the time in the life history of a plant when new genetic recombinants are released (see ch. 4, generation of genotypic and phenotypic variation). The mating system provides the means by which seeds can provide novelty for changing conditions, or conserve successful past genotypes, or both (heteroblasty). Sexual reproduction introduces novelty. Mating systems all introduce different levels of novelty.

**7.4.1.5 Multiplication of the parent plant.** Seed fecundity, numbers, provide the means to continue successful genotypes into the future. The numbers of seeds/buds produced by a parent plant range from many to few. The timing, duration and frequency of the seed/bud germination and seedling/shoot emergence varies considerably between species and individuals. The seed number produced is that which is appropriate to the opportunity spacetime potentially available, safe germination sites.

How many offspring does the average plant leave in the course of its life? The answer is one: if it was less than 1 it would go extinct; if it was more than 1 it would take over world. There exists a very large range in the numbers of seeds produced by individual plants. This fantastic variation in the range of reproductive capacities of plants extends from 1 to $10^{10}$ (infinity for vegetative clone propagule production). Most weed species produce relatively large numbers of seeds, but fecundity is highly plastic and dependent on local opportunity. Estimates of seed number per plant per year from some weed species are listed in table 7.6 below.

**Table 7.6** Estimates of seed numbers per plant produced by various weed species.

| SPECIES | NUMBER PER PLANT |
| --- | --- |
| *Amaranthus tuberculatus* (tall water hemp) | 1,800,000 (per season) |
| *Circium arvese* (Canada thistle) | 680/plant (Canada thistle) but with experimental collection losses; 4 inflorescences per stalk; 47 seeds per inflorescence; = 1955 achenes/seeds per plant |
| *Elymus repens* (quackgrass) | 15-400 (seed); 206 (rhizome buds) |
| *Glechoma hederacea* (ground ivy) | perennial, stolons; fruits per ramet, 4 seeds per fruit; 500-900 seeds per year per ramet per plant; gynodioecious (both perfect, bisexual, flowers as well as female only flowers) |
| *Kochia scoparia* (kochia) | 12,000-14,000 |
| *Lythrum salicaria* (purple loosestrife) | ave. 2.5 million seeds |
| *Panicum dichotomiflorum* (fall panicum) | 2,000 |
| *Polygonum persicaria* (ladysthumb smartweed) | 880-4010; 2898 (average) |
| *Portulaca oleracea* (purslane) | +200,000/plant (maximum, optimum); 175,000/plant (average); 100 seed per flower |
| *Rosa multiflora* (multi-flora rose) | 200,000 (per season) |
| *Sequoia* sp. (redwood) | 109-110 |
| *Solanum* sp. (nightshade) | 178,000 (800,000 maximum) |
| *Sorghum halepense* (Johnson grass) | 1 panicle = 400 seed; 2 yrs: 28,000 seeds per plant |

**7.4.2 Principle of strategic allocation.** Different weed species partition their limited resources among the five seed roles in different ways. Individual plants can't do everything. Within an individual plant three things conflict in seed allocation: limited time, resources and energy. These conflicts result in an adaptive compromise:

**principle of strategic allocation**: organisms under natural selection optimize partitioning of limited resources and time in a way that maximizes fitness (by means of tradeoffs between partitions over time).

Strategic allocation in a plant is the when resources are allocated among alternative demands. These conflicting demands lead to trade-offs between different activities. Natural selection acts to optimize the form of the compromises in such a way to maximize individual fitness.

**7.4.3 Trade-off among seed roles.** The compromises between the five roles of seeds of a species involves trade-offs. Differences in fecundity, seed/bud numbers, is only one component acted upon by natural selection. Emphasis on one role is likely to involve compromises in others. Table 7.7 provides a combinatoric schemata of the trade-offs a species faces in seed adaptation. The most relevant to evolutionary adaptation is the 5-way interaction. Some insight can be gained by decomposing this very complex interaction into component 2-4 role interactions to aid comprehension.

**Table 7.7** Trade-off combinations between five roles (N°, 2-5 role number interactions) played by weed seeds; D, dispersal and colonization; P, persistence and soil seed pools; S, seed food reserve size; V, genetic recombinants released to the deme; N, multiplication of parent seed numbers.

| N° | Disperse | Persist | Seed Food | Vary | Seed Number |
|---|---|---|---|---|---|
| 5 | D | P | S | V | N |
| 4 | D | P | S | V |   |
|   | D | P | S |   | N |
|   | D | P |   | V | N |
|   | D |   | S | V | N |
|   |   | P | S | V | N |
| 3 | D | P | S |   |   |
|   | D | P |   |   | N |
|   | D |   |   | V | N |
|   |   |   | S | V | N |
|   | D | P |   | V |   |
|   | D |   | S | V |   |
|   |   | P | S | V |   |
|   | D |   | S |   | N |
|   |   | P | S |   | N |
|   |   | P |   | V | N |
| 2 | D | P |   |   |   |
|   | D |   |   |   | N |
|   |   |   |   | V | N |
|   | D |   | S |   |   |
|   |   | P | S |   |   |
|   | D |   |   | V |   |
|   |   |   | S | V |   |
|   |   | P |   | V |   |
|   |   | P |   |   | N |
|   |   |   | S |   | N |

**7.4.3.1 Seed size versus number.** The principle of strategic allocation dictates that a plant can package its reproductive effort into a few large seeds or many small ones given a fixed amount of energy, obeying the size-number tradeoff (Silvertown and Charlesworth, 2001). The trade-offs between seed number and seed size are represented by these five interactions in tables 7.7 and 7.8. Seed size is determined by trade-offs among dispersal, persistence and food reserve commitments (table 7.9). Fecundity trade-offs are determined by the other two seed roles: multiplication of parent numbers and the display of genetic novelty.

**Table 7.8** Examples of combinatoric interactions (N°) of trade-offs between five roles played by weed seeds; D, dispersal and colonization; P, persistence and soil seed pools; S, seed food reserve size; V, genetic recombinants released to the deme; N, multiplication of parent seed numbers.

| Nº | D | P | S | V | N | Trade-off Examples |
|----|---|---|---|---|---|---------------------|
| 5  | D | P | S | V | N | seed size vs. number: D/P envelopes-food reserves vs. genetic variation-number |
| 4  | D | P | S | V |   | seed size vs. genetic variation: D/P envelopes-food vs. recombinant novelty |
|    | D | P | S |   | N | seed size vs. multiplication of parent: |
|    | D | P |   | V | N | dispersal vs. persistence vs. number: D/P envelopes vs. recombinants-multiply |
|    | D |   | S | V | N | dispersal vs. food reserves vs. number: envelopes vs. food vs. number |
|    |   | P | S | V | N | persistence vs. food reserves vs. genetic variation display vs. multiplication |
| 3  | D | P | S |   |   | seed size: dispersal structure-persistence envelopes-food reserve |
|    | D | P |   |   | N | dispersal vs. persistence structures vs. multiplication of parent numbers |
|    | D |   |   | V | N | dispersal structures vs. genetic novelty display vs. multiplication of parents |
|    |   |   | S | V | N | embryo food reserves vs. genetic novelty display vs. multiplication of parents |
|    | D | P |   | V |   | dispersal vs. persistence structures vs. genetic novelty display |
|    | D |   | S | V |   | dispersal structures vs. embryo food storage vs. genetic novelty display |
|    |   | P | S | V |   | persistence vs. food reserves vs. genetic novelty display |
|    | D |   | S |   | N | dispersal structures vs. embryo food storage vs. parent plant multiplication |
|    |   | P | S |   | N | persistence vs. food reserves vs. multiplication of parent plants |
|    |   | P |   | V | N | longevity protective structures vs. genetic display vs. multiplication of parents |
| 2  | D | P |   |   |   | dispersal structures vs. persistence envelopes |
|    | D |   |   |   | N | dispersal structures vs. multiplication |
|    |   |   |   | V | N | seed number: genetic variability display and multiplication of parent |
|    | D |   | S |   |   | dispersal vs. food reserve: light/float-food weight |
|    |   | P | S |   |   | persistence vs. food reserve: protective envelopes vs. food reserves |
|    | D |   |   | V |   | dispersal structure vs. display of genetic novelty |
|    |   |   | S | V |   | food reserves vs. display of genetic novelty |
|    |   | P |   | V |   | persistence envelopes vs. display of genetic novelty |
|    |   | P |   |   | N | Abscission dispersal timing vs. number: early/less-late/more seed rain |
|    |   |   | S |   | N | food reserves vs. multiplication |

**7.4.3.2 Seed size trade-offs.** Seed size and number are often examined for insights into how a weed species apportions its limited resources into fecundity. Seed size is itself a trade-off among three of the five seed roles: dispersal, persistence and embryonic food reserve. Food for the developing embryo when it germinates may be the largest commitment, but specialized structures for spatial dispersal and other modifications to seed envelope tissues for protection and enduring occupation in soil seed pools can also demand considerable investment. So seed size trade-offs can be represented (table 7.8) using relevant portions of table 7.7. Examples of seed role trade-offs are presented in table 7.8

**Table 7.9** Seed size trade-offs involving combinations between three roles (Nº) played by weed seeds; D, dispersal and colonization; P, persistence and soil seed pools; S, seed food reserve size.

| Nº | Dispersal | Persistence | Seed Food |
|----|-----------|-------------|-----------|
| 3  | D         | P           | S         |
| 2  | D         | P           |           |

|   | D |   | S |
|---|---|---|---|
|   |   | P | S |

Small investments in food reserves, no specialized spatial dispersal structures, and moderate envelope modifications for protection are typical of many common weed species allowing relatively small seed sizes. This three-way role interaction is the foundation for seed size-number trade-off evaluations.

**7.4.3.3 Seed number.** There exists and inherent conflict between the parent plant and its progeny. It may be to the future advantage of the individual embryo in a seed to grow as large as possible, but the parent controls the flow of resources to the seeds, such that fitness costs to the parent place an upper limit on the advantage of large seeds (Casper, 1990). The optimum size of a seed is unaffected by the resources available to a plant. When size varies, plants respond by changing seed number rather than seed size (Smith and Fretwell, 1974).

Four variables determine number of seeds produced by an individual plant:

1: weight of the plant
2: proportion of the plant allocated to seeds
3: number of seeds per unit plant weight
4: energetic chemical composition of the seed weight (e.g. protein, carbohydrate, lipid, other biochemicals)

**7.4.3.4 Relative weed species seed sizes.** Table 7.10 provides some examples of weed seed sizes. In general, seeds (and buds) of perennial species are relatively larger than those of annual weed species. Annual weed species seeds are relatively smaller, which facilitates dispersal into a greater number of opportune microsites covering the field.

**Table 7.10** Estimates of weed seed size produced by various weed species.

| SPECIES | SEED SIZE |
|---|---|
| Thistles | achene with one seed + pappus, except plumeless thistle |
| *Circium* arvense (Canada thistle) | 2.5-4 mm L,1-1.5 mm W, 288,254 seeds/lb. |
| *Circium vulgare* (bull thistle) | 3-4 mm L, 1.3-1.6 mm W; tall thistle, 4.5-6 mm L,1.5-2 mm W; |
| Circium flodmanii (Flodman's thistle) | 3-4 mm L , 1.5-2 mm W; mean, 3.25-4.5mm L, 1.3-1.8 mm W |
| *Carduus* nutans (musk thistle) | 3-4 mm L; plumeless, 2.5-3 L; mean, 2.75-3.5mm L |
| *Galinsoga* sp. (galinsoga) | 1.5 mm |
| *Glechoma hederacea* (ground ivy) | 1.5 x 1 mm |
| *Setaria* spp.-gp. (weedy foxtails) | green, 1.5 mm length; giant, 3 mm length |
| *Sorghum halepense* (Johnson grass) | 3 mm |

**7.4.4.5 Seed size plasticity and stability.** Seed size is often conserved within a weed species because it is the consequence of trade-offs among the five seed roles optimized by natural selection in many different localities. Weed seed size is one of the least plastic plant life forms during its life history. There is a strong stability to seed size, which is not the case in seed numbers which are highly plastic. For example, species with a relatively constant, stable, non-plastic seed size include *Abutilon theophrasti* (velvetleaf), *Galinsoga* sp. (galinsoga), *Glechoma hederacea* (ground ivy), *Gossypium* sp. (cotton), *Lythrum persicaria* (purple loosestrife), *Setaria* spp. (weedy foxtail), *Sorghum bicolor* (shattercane), and *Sorghum halepense* (Johnson grass). Species with relatively different seed sizes include *Ambrosia* sp. (ragweed), *Bidens pilosa* (black jack), *Glycine max* (soybean),

*Helianthus* sp. (sunflower), *Rosa multiflorum* (multiflora rose), *Rumex crispus* (curled dock), *Xanthium strumarium* (cocklebur), and *Zea mays* (maize). Keep in mind though, that seed size stability in a species is scale relative. Some seeds are plastic but within very narrow range of sizes.

      Seed size stability is a consequence of success in the habitat to which the weed species is adapted (see below). Changes in seed size are the last, the consequence, of other tradeoffs that occur previously in development: seed number is highly plastic to allow the individual plant to grow to a size appropriate to available resources. Seed size is a matter of scale. Small changes in seed size may not be apparent to us, but to the seeds of a species small differences can have big implications. For example, in *Abutilon theophrasti* (velvetleaf), very small individual weight differences translate into more seed hard seed coat, enhancing dormancy (Nurse and Di Tomasso, 2005). Experimentally this fine-scale size heterogeneity are missed when we characterize seed by the average weight of 100 seeds (a common metric in seed science). Seed size may be much more important evolutionarily than seed number. Plants initiate more seeds than they produce, abortion during development is plastic response to stresses of competition. There exists a strong stabilizing selection for one constant size: quality is not compromised by quantity.

      Size is related to the habitats and time in ecological succession that a species thrives in: colonizer versus stable environment (see ch. 6.2.3, life history strategies). Seed size is a heritable trait. There exists an optimal size for a species. The rules about seed size change with cultivated crops. Human selection can lead to different seed sizes not found in the wild progentor species. For instance, cultivated sunflower (*Helianthus* sp.) has a 6-fold difference in ray and disk seeds (Harper, 1977; p.669). Plasticity of parent plant ovary tissue (capsule, silique, etc.) size compared to the size of the seed within reveals that *Abutilon theophrasti* (velvetleaf) has plastic capsule size but little seed size plasticity. Is this plasticity difference related to developmental sequences: capsule first, flower primordial within second, seed fill and or abortion third?

**7.4.4.5.1 Variable seed size**. Variation can occur in seed size in a single species. For example, *Trifolium* sp. (clover) can have 17-fold differences in size in seeds from same plant (Black, 1959). When seed size does change it is often an adaptive polymorphism (e.g. very small differences in *Abutilon theophrasti* (velvetleaf) seed weight is correlated with dormancy amount). It indicates a different allocation between large (high investment category) and small (low investment category) seeds. Variable seed size implies two distinct optima (somatic polymorphism), the operation of disruptive selection that allows it to exploit two different niches in a habitat. For example, it is common in Compositae weeds with a capitum containing large disk flowers (for distance dispersal) as well as small ray flowers and seeds that stay with seedhead and do not disperse (also observed in *Senecio jacobaea* (tansy ragwort), *Centaurea* sp. (knapweed)). Another example is the coffee tree (*Coffea* sp.) with two beans per fruit. These can be two half rounds (unable to roll down steep hills for dispersal), or they can be one round that can roll and disperse; or they can be $^1/_3$ and $^2/_3$; they are never uniform. Seed size is not critical at low population density and competition. But, small and large seed competing together at high population density probably will favor the large seed size in struggle due to greater embryonic capital. Seed weight differences between flowering plant species can be as high as ten orders of magnitude: $10^{-6}$ gram (*Goodyera repens* (orchid); *Orobanche* sp. (broomrape)) to $10^4$ grams (*Lodoicea maldivica* (double coconut palm). North American weeds can have three orders of magnitude ($10^{-5}$ to $10^{-2}$ gram) difference in weight (Harper et al., 1979).

**7.4.4.5.2 Small seed size**. Why do some species have relatively small seed sizes? There are many reasons, such as "because it works". Orchids, parasites like *Orobanche* sp., fern species (Pteridophyta) have seeds that "hunt" for places to germinate. They have no need of stored energy. They all have anomalous nutrition, food arrives exactly at where they germinate. For example: early mycorrhizal associations, parasitic plants, and saprophytes. Many of these are induced to germinate when assured of food supply by germination stimulants. There small size allows them to be compared to viruses: a collection of selfish genes, a bag of DNA not making a complete individual that lets the host do the work. An invalid reason to explain small seed size is that they possess a need to produce large numbers to overcome difficulty in establishing themselves with low food reserves: there is no survival value in death.

**7.4.4.5.3 Large seed size**. Relatively larger seeds confer benefits upon their seedlings: increased seedling survival in the shade, more embryo food for height growth; better mineral nutrition; less nutrional dependence on mycorrhizal symbionts; and, greater tolerance to drought and herbivory. These benefits accrue at the expense of fewer numbers of plants in a locality to exploit unused opportunity, opportunity that will be exploited by a species with sufficient numbers to seize all available safe germination sites.

**7.4.4.6 Relationship of seed size to habitat.** The size of a seed is closely correlated with its seedling's success in competitive situations and when germinating in the shade. There exists a seed size for a particular habitat allowing a plant the best chance for success. Seed size is correlated to the availability of opportunity spacetime in the habitat being seized and exploited. Large-seeded plant species are at a numerical disadvantage to those that produce many small seeds in agricultural fields with large numbers of safe sites. Because small-seeded species produce greater numbers they can seize a larger number of opportune microsites. For early successional habitats small seed size is best in open colonizable habitats. These species must depend on their own independent photosynthesis from an early age. They are often more widely dispersed, and the large seed number allows rapid colonization of many available sites. Species that thrive in intermediate successional communities have relatively larger seed size. Later successional species, like those in woodlands and forests have the largest seeds. The large embryo food reserves allow emergence from greater depth, survive longer, grow to more aggressive size in environments low on resources (especially light) with much fewer available safe germination sites. They can devote more resources to maximize individual survival rather than exploiting many safe microsites. For example, *Trifolium*, clover, has different seed sizes depending on where in succession the plant is establishing itself.

## Chapter 8: Propagule Dispersal in Space and Time

**The weeds' story**. I am foxtail seed.

Those of us you humans grow to eat are round, fat and shiny. We are known as the 'eaters'. Our life is very safe, waiting patiently under your roof all winter long. Until you plant us together in neat rows in your nicely groomed and fertilized fields. Once we are snugly in the warm moist spring soil, just before we germinate together, we notice our weedy brother-sisters buried beneath us. How curious. They're smaller, darker, skinny and wrinkled. We are so eager to grow, so impetuous, we can't wait, so we don't. The weedy seeds, the ones forced to live their entire life underground, they must wait. Why not join us? Come play in the sun with us!

Those of us weedy seeds see you pampered family members being placed carefully in the field. How can you stand to live with those giant humans, those murderers? We know when we have landed in just the right place, human fields have all we could wish for. We watch you 'eater' cousins, your impetuous germination and emergence. But it's too early for us, we wait. The human farmer tends to you, but its death for us if they catch us unawares. So we wait, for the right moment. We come up early in large numbers, trying to overwhelm you, hoping you will miss just a few of us 'early-birds'. The risk is great, but if we escape the rewards are fantastic: thousands upon thousands of children to overrun the earth. Best to have a very large family than none at all, right? Some of us wait till the summer solstice. We know from past experience that you get overconfident then, put your tools away till harvest and go on vacation. So we arise just before then, knowing that if we can find a spot in the field not dominated by our 'eater' cousins, or some other weedy riff-raff neighbors, we have a clear shot till harvest. Some others of us emerge just before harvest. Its not the best time, but if we stay short and avoid the blades of harvest we can grow and bear children in the autumn. Better to have a small family than none at all, right? Then there are the really late ones. They come up in the autumn, hidden in the drying crop and weed residues. The sun comes out for fewer and fewer hours every day late in the season. For some strange reason this makes us mate with ourselves very very early. Sometimes when we have only a few leaves, we are only centimeters high. We late-comers have one or two seeds sometimes, a very tiny family. Better to have a tiny family than none at all, right?

We weedy seeds are born leaner and longer and smaller. We are wrinkly and rough, short antennae to sense the best opportunity of the season or year. We are born with coats to protect us in the soil, and to tell us when to germinate and regrow. Our parents can't wait to see us gone once we are grown up. They push us out of the house right away, all season long. Most of us fall to the ground at the feet of our parents. Out into the world, alone at first. But as we begin to sense our surroundings we notice we are not really alone. Our family of seeds are everywhere here in the dirt. Some are old and gnarly and have experienced many seasons underground. We are gathered in great pools of family, each with our own destiny. It is reassuring knowing we will endure, together, despite our differences. Because of our differences. Some of us are restless, hungry for travel, adventure, new worlds to conquer. We hitchhike around the world on the fur of dogs, in the vortices of tornados, in the whirlpools of rivers, but mostly as hidden stowaways in the vehicles of humans. We are the elite invaders of the family. We conquer the Earth by our wits alone.

We weedy and 'eater' seeds are different, we have different jobs. We 'eater' seeds like to put our effort into fat seeds, lots of endosperm inside that the humans love so much. And not very much effort at all into coats to protect us. We don't need it. Almost all of us are carefully collected by the attentive human farmers who host our lives. We reciprocate the favor by dutifully remaining on our mother-father seedheads together till the humans come for us in the autumn. No life in the nasty brutish soil for us. We are the choosen favorites or our human masters. We conquer the Earth on the coat-tails of our human gods.

**Summary**. Propagule dispersal begins with abscission and physiological independence from the parent plant (a threshold event) and ends with seedling or bud shoot emergence (a threshold event). The advantages of dispersal in a variable environment include a means of escaping the negative consequences of neighbors in the local community, escape from unfavorable locations or seedling emergence times until more favorable opportunity spacetime exists, or directing propagules to favorable times and sites for germination and recruitment. Functional dispersal traits include those providing propagule independence from parent plants, exploitation of locally available establishment sites, structures and mechanisms for spatial and temporal dispersal in the soil to escape and exploit opportunity spacetime. Dispersal of seeds is a process of discovery of habitable sites with time. Discovery depends on the spatial and temporal (seasonal) distribution of habitable areas, and on the dispersibility of seeds.

Spatial dispersal is critical to population size and structure of a locality. Dispersal mechanisms of a plant species seed indicate how it seeks habitable sites over a landscape with heterogeous patchs of opportunity spacetime. The numbers of an individual species able to colonize in a locality are determined by the number and spatial distribution of habitable sites, the dispersibility of seeds, and the speed with which they are discovered and colonized over time. There are six modes, or ways, seeds and propagules are dispersed: gravity, wind and air, water, animal (non-human), human, and other types.

Dispersal in time is accomplished by the formation of pools of propagules in the soil. The source of all future weed infestations in a locality are soil seed pools (and dispersal in from other seed pools). Seed/bud pools, or banks, are propagules with a long- or short-term occupancy in the soil awaiting either seedling emergence or death. Seed dormancy is dispersal in time. It provides an escape until conditions are more favorable to continue growth and development. Seeds, especially in weed species, are dispersed within a season at differential times of shattering (abscission and physical separation from the parent plant).

Seedling emergence is called seedling recruitment: enlisting seeds from the soil to resume autotrophic growth leading to reproduction. It is the life history of a weed seed from germination to seedling emergence from the soil and recommencement of growth as a seedling. The single most important determinant of agricultural weed community assembly, and subsequent community structure, is the timing of weed seedling emergence relative to that of crop emergence and related crop management activities (e.g. planting, tillage, herbicide use). Seedling emergence and plant establishment are a direct consequence of the inherent heterogeneous dormancy of individual seeds (heteroblasty) and the environmental conditions that modulate the behavior of those dormant seeds. Dispersal ends with seed/bud germination and recruitment, the first moments of independent autotrophic growth.

## 8.1 Introduction

Propagule dispersal can increase fitness within a weed population that has invaded a heterogenous habitat by several means (Silvertown and Charlesworth, 2001). Dispersal is advantageous in a variable environment by providing a means of escaping the negative consequences of crowded inter- and intra-specific population conditions in the local community. Dispersal allows weeds to escape unfavorable locations or times of seedling emergence until more favorable opportunity exists. Dispersal can direct offspring to favorable times of, or sites for, germination and recruitment.

**dispersal**
1: outward spreading of organisms or propagules from their point of origin or release; one-way movement of

organisms from one home site to another (Lincoln et al., 1998)
2: the act of scattering, spreading, separating in different directions (Anonymous, 2001)
3: the spread of animals, plants, or seeds to new areas (Anonymous, 1979)
4: the outward extension of a species' range, typically by a chance event; accidental migration (Lincoln et al., 1998)
5: the search by plant propagules (e.g. seeds, buds) for opportunity in space and time

Propagule dispersal in space and time is the consequence of the three overlapping processes of weed life history, natural selection and invasion biology (table 8.1).

Table 8.1 Propagule dispersal: overlapping processes of life history, natural selection and invasion biology underlying functional traits of weed seed/bud dispersal.

| | PROCESSES | |
|---|---|---|
| LIFE HISTORY | NATURAL SELECTION | INVASION BIOLOGY |
| **Progagule dispersal** | Condition 3: survive to produce the fittest offspring | Colonization & Extinction |
| Spatial dispersal | Pre-Condition 1: excess local phenotypes compete for limited opportunity | Invasion |
| Temporal dispersal | | Enduring occupation |

Traits important to propagule dispersal in space and time are those that fulfill roles of propagule independence from parent plants and exploitation of available establishment sites. Functional dispersal traits also include formation of structures and mechanisms for spatial dispersal and temporal dispersal in the soil to escape and exploit appropriate opportunity spacetime. These dispersal traits differ among specific weed species (table 8.2) (Harper, 1977: ch. 2; summary p. XIV).

Table 8.2 Weedy life history functional roles and traits for propagule dispersal in space and time.

| TRAIT ROLES | TRAITS |
|---|---|
| **Propagule independence** | **Seed shattering:** commence dispersal soon after abscission |
| | **Perennial ramet formation:** ramet bud independence from parent plant; ramet-ortet ratio balance |
| **Seize local opportunity spacetime** | **Optimize propagule size and number** appropriate to amount of safe soil microsites available for exploitation |
| **Spatial propagule dispersal:** spatial foraging appropriate to opportunity spacetime being exploited | **Structures and mechanisms:** move propagules in space via: 1. Gravity: no structures to ensure local placement 2. Wind-air: pappus; wings 3. Water: flotation 4. Animal: attachment burs; attraction 5. Human: mimic crop seed |

|  | **Perennial ramet foraging**: underground bud dispersal via parent vegetative tissue spatial foraging |
| --- | --- |
| **Temporal propagule dispersal**: form enduring seed pools in the soil of opportunity spacetime being exploited | **Propagule dormancy-heteroblasty**: responsiveness to environmental signals stimulating propagule behavior in the soil: germinability-dormancy state seasonal cycling; germination |
|  | **Escape to survive**: soil life: longevity; soil depth tolerance-preference; self-burial |
|  | **Escape to exploit**: genetic resource reservoir:<br>1. propagules for future local and distant invasion<br>2. genetic foraging novelty for future out-crossing, recombination<br>3. genetic buffering against short-term local environment changes<br>4. genetic memory of successful past phenotypes |

**8.1.1 The evolutionary ecology of dispersal structures.** Evolution favors development of dispersal structures and mechanisms when there is a greater chance of colonizing a site more favorable than the one that is presently inhabited.

The evolution of cocklebur (*Xanthium strumarium*) dispersal provides a good example. This important weeds species evolved along river banks with two dispersal mechanisms: barbed burrs (the inspiration for Velcro™) to snag a macrofaunal vector, and the ability to float and move down river to a favorable establishment site. In recent times cocklebur has appeared in crop fields where floating (or barb) dispersal may be less important, or not at all. If barbs are not important to dispersal, over time we could predict that crop field biotypes of cocklebur would adapt to the new (non-aquatic) conditions and invest less energy in expensive dispersal mechanisms. Barbs would be reduced, as well as morphology for floating in water. It is interesting to speculate that this might have already occurred to some extent. Typically *X. strumarium* capsules contain two seeds, although biotypes in Mississippi have been observed with five seeds per capsule. This could be evidence of adaptation to crop fields wherein more numbers of seeds (versus size) are favored because of the many more microsites available in crop fields.

**8.1.2 Seed dispersal trade-offs.** One of the most important trade-offs an individual species makes is the allocation of resources during reproduction between seed number and size: few large or many small? (see ch. 7.4.3, trade-offs among seed roles). The most important forces of natural selection acting on that individual species are the number of habitable sites and the speed with which they are discovered and colonized. Weeds tend to favor relatively more numbers of relatively smaller sizes because the number of habitable sites in a crop field are very large.

**8.1.3 Cost of dispersal.** Dispersibility has a cost to the plant, but not always. Energy invested in specialized structures cannot be used for more numerous seed production, or for embryo or food reserves in that same seed. The cost of these structures is a measure of the fitness advantage they confer to their ancestors, ancestors who have gained by placing descendants at a distance from, rather than close by, a parent. Weed species usually have no specialized structures for dispersal because distant dispersal is not the primary force of selection they face in agricultural and disturbed localities. They have already arrived at opportune sites. Later succession plants do invest in distant dispersal mechanisms/structures because they are often in competition with parents at the local site where they are born. For example: trees; shrubs; weeds like *Asclepius* sp. (milkweed), *Circium* and *Carduus* sp. (thistles); and obligate long distance dispersers like *Verbascum* sp. (mullein).

**8.1.4 Space-time dimensions of dispersal.** Dispersal of seeds is a process of discovery of habitable sites with time. Discovery depends on the spatial and temporal (seasonal) distribution of habitable areas, and on the dispersibility of seeds. There are several ways to look at dispersal. The first is dispersal that expands the range a species colonizes and occupies. The second is dispersal that leads to increasing population size of an invading species in an area. They are both part of the process by which an established and stabilized population maintains itself. These two are parts of the same intertwined, inseparable, whole.

## 8.2 Dispersal in Space

Seed are dispersed in four spatial dimensions. Two are horizontal on a field surface (length, width). Two are vertical in a locality (soil depth, height above the soil surface in the air). Spatial dispersal is critical to population size and structure of a locality. Dispersal mechanisms of a plant species seed indicate how it seeks habitable sites over a landscape with heterogeous patchs of opportunity spacetime. The numbers of an individual species able to colonize in a locality are determined by the number and spatial distribution of habitable sites, the dispersibility of seeds, and the speed with which they are discovered and colonized over time.

Most weed species disperse their seed at the base of the parent plant (gravity), with decreasing numbers with distance from the parent ('what's good for the parents, is good for the offspring'; 'we're already where we want to be'). Isolated, widely dispersed, plant populations have a different spatial dispersal structure, their seed is dispersed widely because the spatial structure of favorable sites is widely dispersed. Those species need to send their seed across the landscape to search for those favorable sites.

**8.2.1 Dipersal and post-dispersal processes.** The processes of dispersal and post-dispersal are often differentiated, although there is little inherent difference between these two overlapping concepts. Dispersal clearly begins with physiological independence from the parent plant (abscission), and just as clearly is over when a seed germinates or becomes established and capable of autotrophic growth. Both are discrete threshold events in a weed's life history. The shortest dispersal time imaginable would be vivipary, when the seed germinates while still attached to the parent plant). Seed that are initially dispersed to a location can experience post-dispersal movement, such as that caused by tillage, cropping practices. Dispersed seed can be subsequently consumed by a wild or domesticated animal, passed through their digestive system, and then enter the soil. This animal post-dispersal process can compromise the weed seed coat (e.g. scarification) providing enhanced germinability in the soil seed pool; or it can kill the seed.

Purple loosestrife (*Lythrum salicaria*) has a complex dispersal process, complicating the difference between dispersal and post-dispersal processes. In the beginning it disperses from the parent plant and falls to the bottom of open water where the parent lives (dispersal event 1). When it germinates it floats to the establishment site (dispersal event 2). *Galinsoga* sp. begins germination on the parent plant (pre-abscission, vivipary; dispersal-establishment event 1), then falls to ground with germination already begun (a head start; dispersal-establishment event 2).

**8.2.2 Seed flux at a locality.** The flux of seed into, and out of, a locality determines the potential population size and species composition of the plant community at that site. The flux at a site is a consequence of the number already there, the number that disperse in, the number that leave or die, and the replacement and rearrangement by dispersal of seed within that site. The net flux out for gravity weeds may not be quantitatively significant. The flux out for long distance dispersal mechanism possessing plants may be quantitatively important.

**8.2.3 Modes of seed and propagule dispersal.** There are six modes, or ways, seeds and propagules are dispersed. These dispersal vectors are gravity, wind and air, water, animal (non-human), human, and other miscellaneous types.

**8.2.3.1 Gravity.** Most seed of our common crop field weed species have no specialized structures and mechanisms (e.g velvetleaf (*Abutilon theophrasti*), weedy foxtails (*Setaria* sp.), *Galinsoga* sp., ground ivy (*Glechoma hederacea*), purple loosestrife (*Lythrum salicaria*). As such, seed falls to the ground by gravity, but other modes may also act (e.g. wind, animal, fall into soil cracks and self-plant). Most gravity seed dispersal acts to leave seed at the base of the parent plant, with decreasing numbers of seeds with distance. Seeds move as a horizon, a front away from the source. Gravity dispersal is found with weeds that favor their present site, no dispersal structures, mechanisms needed.

**8.2.3.2 Wind and air**. There are four different morphological adaptations, seed structures, for wind dispersal. Three involve an energy cost to the plant, one is for for free.

**Dust**. Dispersal of dust-like seed is for free, there are no trade-offs in adopting dust as a dispersal mode. Dust seed is so light it can stay aloft even in still air. Example species include poppies (*Papaver* sp.), fungal spores, ferns, parasitic plants, orchids. Dust seed can dissolve in water, a rain drop, and disperse that way.

**Plumes**. Plumes are like a feather, or feather-like structure. Examples include thistles (*Circium, Carduus* spp.).

**Pappus**. A pappus is a circle or tuft of bristles, hairs, or feathery processes (could be plume-like) in place of a calyx. They are typical of the Compositae family (e.g. dandelion (*Taraxacum officinale*), milkweed (*Asclepias* sp.), sowthistles (*Sonchus* sp.)

**Wings**. Wings with concentrated central mass are adapted to still air, lift and distance. They have stable flight and glide. Examples include the lianes (woody vines; e.g. *Metrosideros* sp.) and tropical forest trees. Winged seeds and fruits which rotate when they fall can have a symmetrical and asymmetrical flight which affects where fall (ex. maple (*Acer* sp.) seeds).

Different adaptations in weeds can produce the same wind dispersal effects as these structures. Selective evolutionary pathways producing the same wind dispersal effect include a decrease in weight, and increase in the ratio of pappus to achene, improvement of the drag efficiency of the pappus, and release of the seed from a higher place. Soil surface roughness affects how far the wind can blow a seed. Wind dispersed seed tend to accumulate along fence rows, along furrows: traps. Seed can also move still attached to the parent plant (e.g. tumbleweed or Russian thistle (*Salsola kali*), tumble pigweed (*Amaranthus albus*), *Kochia scoparia*. I observed velvetleaf (*Abutilon theophrasti*) capsules with seeds still attached to branches blowing over snow drifts in a Michigan crop field into the neighbor's yard. A tornado can move any seed. Specialized wind dispersal mechanisms, structures, don't colonize and move as a front or horizon, but as isolated individuals over a greater distance.

**8.2.3.3 Water**. Seed can be dispersed by water in four different ways.

**Float**. Floating seeds are ones that can have a low specific gravity seed (milkweed, *Asclepias* sp.), or a flattened seed shape (e.g. the corky seed wings of Rumex crispus (curled dock). Floating seed tend to concentrate on edges of water, possibly an ideal establishment site (e.g. cocklebur (*Xanthium strumarium*), waterhemps (*Amaranthus* sp.). Coconut (*Cocos nucifera*) is a species that can float in the ocean for long distance movement, it is a heavy seed but with low specific gravity.

**Surface water**. Movement of low specific gravity seeds occurs on the surface waters of irrigation, rivers, and lakes. Relatively higher specific gravity seed (denser) can also be transported, but with faster moving water.

**Flooding** can have a similar effect and has a big influence on long distance movement.

**Movement over soil surface with surface water (erosion)**. This can stir and mix seed in the soil for germination, which affects formation of seed banks and can shift the distribution of seed in the soil. Slippery seed move easily over soil surfaces. An example is found with mangrove (e.g. *Rhizophoraceae* family) at the mouth of a river. The mangrove propagule is a stick that floats upright. It is held at the right height to germinate when it comes against the land's edge. The seed is at the bottom of the propagule stick at just the right location to germinate at the shore. Mangrove roots breathe air.

**8.2.3.4 Animal, non-human**. Non-human animals disperse seed if several ways.

**Eating, digestability and viability** by means of animal-mediated chemical and physical actions. A seed eaten by an animal can experience different fates. It can be eaten and disperse as a viable seed, or be eaten and destroyed by digestion. Eating can also remove a dormancy factor (e.g. physical scarification by gut action, acids, abrasion, etc. of hard seed coat dormancy type species). After being eaten a seed can be dispersed with animal feces, the feces can then provide nutrients and/or a favorable microsite for germination and establishment. Some seeds are attractive to animal eating, specialized attraction structures that come at an energy cost; e.g. for pollenation. These species specific feeding patterns affect seed dispersal. Migratory birds (ducks, geese) can vector long distance dispersal. Seeds can accumulate at bird roosting sites (e.g. red cedar (e.g. *Juniperus virginianum*) trees in Oklahoma). Birds sit on fences and defecate out seeds where they germinate in these protected sites. Animals can break open fruit and expose seeds for dispersal without eating the seeds. Plants can possess attractants, fruits and flowers, that draw animals to them that in turn disperse seed (e.g. *Trilium* sp. has eliasomes, which ants eat with the oil body structure and attached seed after collecting).

**Animal seed storage.** Animals collect and store seeds as food which can affect dispersal distance, the spatial concentration of seed, and its location. Ants cache seed in their nests. When these germinate they move them out of the nest into a more favorable zone of germination. Mice and other rodents cache seed. A manure pile can be a favorable germination site.

**Specialized movement structures** of seeds can facilitate their movement by animals. These structures include burrs (*Xanthium strumarium* (cocklebur), *Cenchrus* sp. (sandbur)), spines, barbs, hooks (*Hordeum jubatum* (foxtail barley), *Bidens* sp. (beggarticks)). Seed can become stuck to animals and fur with mud.

These three modes can lead to special spatial patterns of dispersal. These spatial distribution patterns are related to how animals move seed (e.g. clumping at a storage or defecation site such as under a fence).

**8.2.3.5 Human**. All the dispersal modes of non-human animals above applies to humans. Humans and their technology, trade and restless global movement provide many many other ways by which seed are dispersed. Humans are arguably the best weed seed vectors on earth. Just a few examples include local dispersal with farm equipment (cultivators, tillage, combines, planters). Seed are carried along this way with mud, grain or feed on equipment too. Weeds movement as contaminants with irrigation or attached to humans by burrs, barbs, etc.

Long distance dispersal occurs in things moved around the world like in ship ballast, soil and seed, farm equipment, crop seed, grain sales, campers in wilderness areas. Waffle-tread hiking boots pick seeds up easily. Imported crops that turn to weeds once they are introduced include *Opuntia* cactus in Australia as border plant, aquatic plants, morningglory (*Convolvulus arvensis*, *Calystegia sepium*) flowers, *Pueraria* sp. (kudzu) as forage, *Sorghum halepense* (johnsongrass) as forage, hemp (*Cannibis sativa*) and velvetleaf (*Abutilon theophrasti*) as fiber crops, multiflora rose (*Rosa multiflorum*) as border plant or hedge, *Plantana cantaria* is a hedge introduced to Kenya that is now a weed, ground ivy (*Glechoma hederacea*) was a cover crop that was introduced and is now a weed. Other species introduced as crops or ornamental plants include butter & eggs (*Linaria* spp.), crown vetch (*Securigera varia*), Japanese honeysuckle (*Lonicera japonica*), honeysuckle (*Caprifoliaceae* family) escaped in woods as weed, purple loosestrife (*Lythrum salicaria*) was once a horticultural crop, and water hyacinth (*Eichhornia crassipes*).

Crops with weed seed in it can be dispersed with poor harvest combine separation, poor sieving (e.g. soybeans (*Glycine max*) with nightshade (*Solanum* sp.) seed stuck to it). Weed seed in hay is dispersed as it is moved. Commercial seed can contain weed seed contamination. Weed seed can mimic crop seeds (crop seed mimicry). It looks like crop seed and is moved with crop seed (e.g. *Echinochloa crus-galli* (barnyardgrass) in rice (*Oryza sativa*)) Weed seed can contaminate when it has the same size and shape as crop seed (e.g *Solanum nigrum* (black nightshade) berries in dry beans (*Phaseolus vulgaris*) in Michigan).

Maybe the best way a weed can disperse itself is to become a crop, or a plant that humans like (e.g. medicinal plants, ornamentals, drugs). Plants domesticated by humans as crops are clearly the most successful invasive species in history. Look at the success of maize (*Zea mays*), rice (*Oryza sativa*) and wheat (*Triticum aestivum*), all weeds in their day.

**Dispersal anecdote: capitalism as dispersal vector.** In 1992 I visited the Czech Republic along the Elbe River collecting weed seed germplasm. This river was a major route of grain traffic from USA to the USSR eastern bloc in the Cold War years. The banks of the river and adjacent areas were covered in *Amaranthus* and *Setaria* plants that looked very similar to the biotypes I had seen in midwestern USA corn and soybean fields. My Czech host complained bitterly about these "invasive weed species" and how some people felt their introduction was a capitalist-imperialist plot to disperse weed seeds in Communist countries. The irony was missed by me:

*Setaria* was introduced to the Americas from the Eurasia, an Old World invader re-invading the Old World: "What goes around, comes around" or "If you buy your grain from your neighbors you get their weeds."

#### 8.2.3.6 Other modes of dispersal.

**Seed ejaculation** disperses mistletoe (*Viscum album*) and other parasitic weeds. Some rely on water potential differences in the seed and plant parts. They disperse by explosion (propulsion, click). Milkweed (*Asclepias* sp.) capsules crack and the seeds with pappus are partially ejected because the seeds are highly compressed inside the capsule. For example: fungal puffballs (*Morganella* sp.).

**Sticky seed** stick to bird feet on a plant branch. Nightshade (*Solanum* sp.) berries can break in the harvest combine and stick to soybeans (*Glycine max*). The nightshade's sticky juice dries and sticks its seed to the crop seed. Rugose, rough seed adheres to mud easier than smooth seed. Other interactions of seed surfaces and soil roughness create seed aggregation traps.

**Hygroscopic awns** are a mechanism for a seed to self-plant itself by the twisting of awns (*Avena fatua*; *Erodium* spp.) with areas of differential water and humidity absorption.

**Self-burial**. The pappus or plume holds a seed in a soil crack and keeps it near the soil surface: a proper, shallow planting depth. In these species the specialized dispersal structures serve a dual role with wind dispersal. Chaff of the prairie grass bluestem (e.g. *Andropogon gerardii*) allow the seed glumes around the seed to hold it in cracks, openings that are favorable positions for germination.

**Fire**. Fire stimulates germination (see ch. 7.3.6.7, other seed germinaton control mechanisms ) and propagule dispersal.

### 8.3 Dispersal in Time: Formation of Seed Pools in the Soil

> "Seed populations in soil are usually described as a seed bank, a term which gives the erroneous impression that this is a place for safekeeping for plant genotypes. In fact, the seed bank is constantly plundered by predators and is highly prone to attack by fungi. Even when they germinate, few seeds make a successful escape. Plants which invest their progeny in the seed bank retrieve only a tiny fraction of them, and each of these is worth (in terms of contribution of genes to future generations) much less than other seeds which have already successfully produced progeny of their own rather than remaining dormant in the soil. In effect, the soil is like a bank that offers little protection from thieves or kidnappers and which guarantees that the value of what is left of a deposit when it is withdrawn will be severely eroded by inflation." (Silvertown and Charlesworth, 2001)

Seed/bud pools, or banks, are propagules with a long- or short-term occupancy in the soil awaiting either seedling emergence or death. Despite the potentially large numerical investment losses, the potential paybacks of successful seed pool survivors can be huge. Most weed species can attain large individual plant sizes given ample opportunity. Large plant biomass is directly correlated with fecundity (Dekker, 2005). In the end result, a small numerical investment in a risky soil seed bank can yield exponential returns for the deme. Those rich investment profits are realized by the fittest individuals, enriching the soil with locally adapted phenotypes, while the less fit are naturally eliminated. Dispersal in time in the soil is not a bank, but a lottery pool controlled by natural selection and performed with stochasticity.

Seed are dispersed in time in several ways. Seed dormancy is dispersal in time, temporal dispersal. It provides an escape until conditions are more favorable to continue growth and development. Seeds, especially in weed species, are dispersed within a season at differential times of shattering (abscission and physical separation from the parent plant). Perennial species can disperse seed each year, over several years. For example, some perennial species have 'mast' years, a year in which seed production is exceptionally high, usually followed by a several years in which production is low. There are several advantages to producing different quantities of seeds in different years. Seed production in a locality varies with plant community succession, variable dispersal by time and species.

The source of all future weed infestations in a locality are soil seed pools (and dispersal in from other seed pools). Endogenous dormancy-germination traits of individual species are the primary mechanisms responsible for the formation of local soil seed pools (or not). These internal germination timing traits are induced during embryogeneis (see ch. 7.3, embryo adaptation: embryogenesis and dormancy induction) and are retained in seeds for their entire life: dormancy is not 'overcome', or 'broken'. Seed pools are the inevitable consequence of the dormancy-germinability capacities of individual seeds accumulating in the soil of a local deme. Recruitment (seed germination and seedling emergence/establishment) is also the inevitable consequence of these seed traits (or death).

**8.3.1 Adaptative roles of soil seed pools**. Soil seed pools play five adaptive roles in the evolution of a local weed population.

1: continuity of a species in a locality
2: a reservoir of seed for future reproduction
3: a refuge for survival during periods unfavorable for growth (e.g. frozen winter)
4: a buffer of the species genotype composition at that site against seasonal changes in the local environment (against extinction caused by selection and drift (del Castillo, 1994)) which results in:
    4a: maintenance of successful, locally adapted genotypes and genetic variation when challenged by the vagaries of year-to-year population shifts
    4b: storage of novel genotypes from unusual years in anticipation of the recurrence of those conditions (pre-adaptation)
    4c: a memory of past successful genotypes/phenotypes in the local community; prevention of genetic "drift"
5: the source of variable genotypes for outcrossing and genotypic novelty, and as a source of seeds for invasion of other localities in the range expansion of that species

The five roles played by soil seed pools are intimately related to the five roles played by seeds (see ch. 7.4.1, five roles of seeds).

**8.3.2 Population dynamics in the soil seed pool.**
**8.3.2.1 Life history of a seed.** Soil seed banks formed by individuals with diverse germinability-dormancy states and capacities (seed heteroblasty) results in several observed behaviors (figures 8.1, 8.2). Seedling emergence occurs during the growing season: summer annuals in the spring and early summer, with many weed species emerging late in the growing season. Seed become after-ripened by the environment they experience in the soil; typically only a fraction germinate and emerge every year. Seeds can become ready for germination (a germination "candidate") in after-ripening conditions (e.g. cool, moist), but fail to germinate without a change to somewhat different environmental conditions specifically favorable to germination (warm, moist). Higher soil temperatures of summer reverse the earlier after-ripening effects and secondary dormancy can be induced. Taken together these environmental conditions result in an annual cycling between dormant and highly germinable states within the heterogeneous seeds remaining in the soil seed bank (see Baskin and Baskin, 1985).

**Figure 8.1** Schematic diagram of life history changes in individual weed seed germination-dormancy states with time from embryogenesis to germination or death; dashed (---) vertical line represents abscission, an irreversible threshold life history event; axes not to scale.

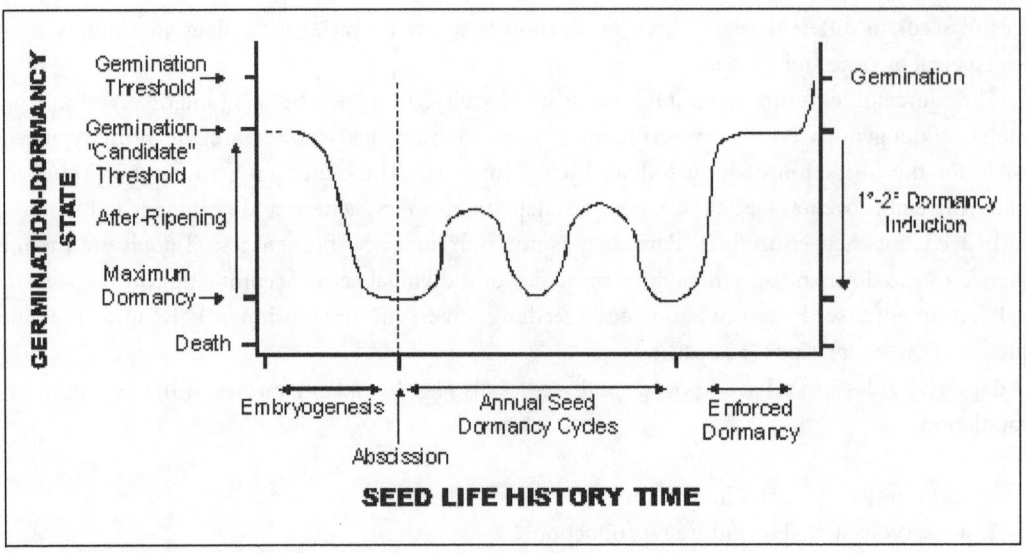

**Figure 8.2** The annual life cycle of weed seed in the soil seed pool.

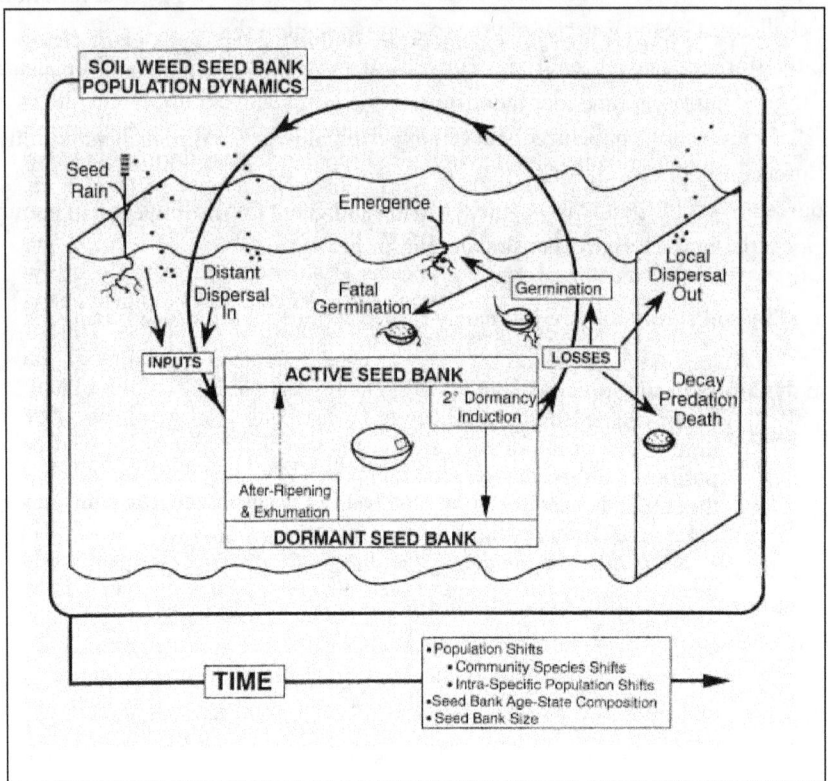

**Figure 8.3** The annual life cycle of weedy *Setaria* species-group (foxtails) seed pool in the soil: processes causing changes in seed states and fates with time.

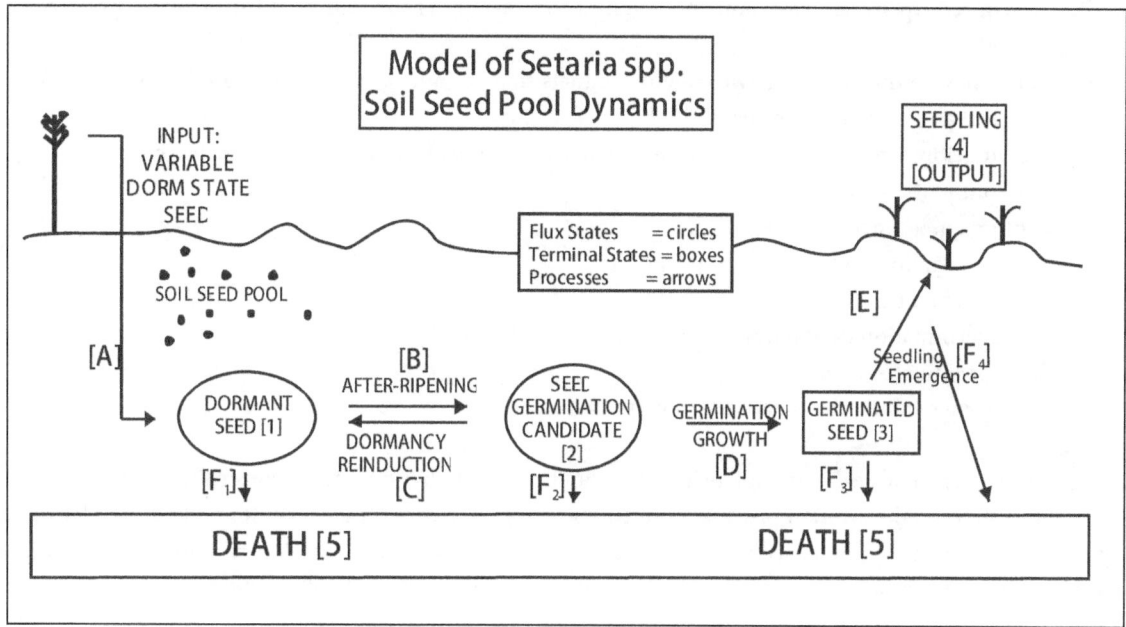

**8.3.2.2 Seed states, fates and seed state transition processes.** Heteroblastic seed entering the soil seed pool experience several fates which change under the influence of environmental signals. For example that of weedy foxtails (*Setaria* spp.; figure 8.3).

States of weed seeds in soil pools:

1: **dormant seed**: seed with dormancy-germinability capacity induced on the parent plant; can be primary (post-abscission) or secondary (environmentally re-induced) dormancy state
2: **seed germination candidate**: fully after-ripened seed ready to germinate given minimum favorable environmental signals (enforced dormancy)
3: **germinated seed**: in the process of germination and emergence from the soil
4: **emerged seedling**: germinated seed that has emerged from the soil; recruitment ends when the emerged seedling begins autotrophic growth (independent photosynthesis)
5: **dead seed**: mortality can occur at any time, including fatal germination (death during the germination process)

Seed states in the soil can be in flux during annual dormancy cycling (figure 8.3): [1] dormant; [2] germination candidate, when states are capable of changing in response to seasonal soil signals increasing germinability (after-ripening; [B]), or inducing secondary dormancy [C]). Terminal, irreversible, seed states are those found in germinated seed ([3]), emerged seedlings ([4]) and dead seeds ([5]).

After induction of primary dormancy during embryogenesis and abscission, heteroblastic seed in the soil seed pool experience changes in state, processes stimulated by the influence of environmental signals (figure 8.3).

Seed state transition processes in the soil seed pool are:

A: **dispersal of seed** into the soil, either from local or distant parent plants

B: **seed after-ripening**: environmental signals increasing germinability and decreasing primary dormancy

C: **dormancy re-induction**: environmental signals that re-induce (secondary) dormancy in after-ripened seed and decrease germinability

D: **germination**: emergence of the seed axes (shoot and root) from the seed stimulated by favorable soil environmental signals

E: **seedling emergence** from the soil

F: **death**: mortality can happen at any time:
- $F_1$, to dormant seeds
- $F_2$, to germination candidates
- $F_3$, to germinated seeds
- $F_4$, during seedling emergence ('fatal germination')

Additionally, dispersal of seed in the soil in any state can be moved physically to a new location or deme.

Seed states and transition processes can be taken together to represent the life cycle of weed seed in a locality. Seed states, fates and processes with time in the soil for several *Setaria faberii* populations in central Iowa, USA have been reported (Jovaag et al., 2011). It is experimentally difficult to observe seed life history in the soil, even with the careful spatial control of seeds of this study. The size of state pools, and the signals stimulating state transitions can both be calculated (Dekker et al. 2003, 2012b). Genetic analysis (e.g. molecular fingerprinting) of seed input to, and fecundity of successfully reproducing seedlings emerging from, the soil can provide an estimate of evolutionary change and local adaptation by a species.

An example of how soil matrix signals can drive seed state transitions has been reported previously (Dekker, 2013, 2014; Dekker et al. 2003, 2012b) and more fully developed in ch. 13.3, *Setaria* life history as a complex adaptive soil-seed communication system.

**8.3.2.3 Seed pool additions, losses and continuity**. The consequences of these state transition processes results in dynamic changes in the soil seed pool and local population: losses, additions, and continuity as well as evolutionary changes in the genetic composition of the seed pool with time.

**8.3.2.3.1 Additions to the seed pool**. Pool additions include seed dispersal into a local seed pool and the seasonal seed rain of occupant plants of the community.

**8.3.2.3.2 Losses from the seed pool**.

**Germination and seedling emergence**. Once a seed germinates it is committed to resuming growth and development. It can proceed and emerge as a seedling (recruitment) or die (fatal germination). Fatal germination can occur for a number of reasons, including insufficient seed energy reserves to grow and emerge from deep in the soil, predation and herbicides (typically a layer incorporated into, or directly on, the soil surface).

**Mortality**. Death can occur at any time. In addition to fatal germination and herbicidal death, seeds can die from predation, decay (rotting increases with time, moisture, temperature, slow and/or stressed germination), scenscence (buried alive until death, "old age"; mutation leading to death), other chemicals (allelopathy, soil fumigants such as methyl bromide, and cropping system disturbances (e.g. tillage bring seed to surface, then eaten; physical destruction by tillage equipment, other cropping disturbances).

**Dispersal out of the local seed pool**. Many dispersal mechanisms can remove weed seed from a local pool, including movement with farming operations, and soil or water or wind movement.

The speed of seed lost from the seed pool varies. Seed losses from the seed bank occur at an exponential rate in most instances. This means that when the seed pool is relatively large, large numbers are lost quickly. It also means that when seed pools are relatively small, the losses are small and the last seeds may never disappear due to the low loss rate. The consequence of this is that it is very difficult to entirely eliminate the seed pool. All it takes is one patch in a field to escape until harvest and the exponential increase in seed production that weeds possess can quickly fill the seed pool. The germination and recruitment losses every year are a function of the species (recruitment rate is a heritable trait in weeds). *Kochia scoparia* seed are non-dormant and all seeds either germinate, die or are prevented from germination by an unfavorable environment. *Setaria* seed range from 5-50% germination and recruitment per year depending on environmental conditions during dormancy, heteroblasty, induction and the growing season.

**8.3.2.3.3 Continuity in the seed pool with time.** Continuity in the seed pool is provided by the dormant seed already in the pool from past seed rains. There exists an age, and dormancy state, structure to this diversity.

**8.3.3 Structure of soil seed pools.**

**8.3.3.1 Spatial distribution in the soil profile.**

**8.3.3.1.1 Depth in the soil profile.** Weed seed are located, distributed, in soils horizontally and vertically. Most weed seeds fall on, are dispersed to, the soil surface (e.g. top 10 cm). Predation losses are enormous on the seed surface from macrobiotic predators as mice, beetles (e.g. carabid), earthworms, etc. Some small minority of seed become incorporated into the soil with time. These numbers are greatest when tillage incorporates them, and least in no-till production systems. The density of weed seeds declines exponentially with depth, as does microbial (seed degrading) biomass.

**8.3.3.1.2 Effects of tillage.** The distribution of seed in the soil profile changes when tillage changes from moldboard plowing to no-till. The vertical distribution changes to a shallower seed location. Phytochrome-regulated light/nitrate requiring surface germinators are favored as they remain susceptible to light. Shallower soil depths also favor oxygen stimulated species in zones of moist surface layer soils. Changes in tillage systems from no-till to moldboard-plowed production systems result in deeper seed placement, seed distributed more evenly at many levels in the soil, and longer seed longeveity (possibly due to less oxidative stress at depth). It does not favor surface germinators like light-phytochrome-nitrate types (no light signal). It does favor large-seeded dicots able to emerge from depth (more food reserve for deeper emergence), and seed protected for soil degradation by hard coats.

**8.3.3.1.3 Seed size and depth of burial in the soil.** Small seed are more likely to be found at depth. They are more likely to fall into soil cracks and holes. Movement of smaller seeds by worms and animals increase their distribution. These movements are a function of tillage, soil surface crop litter, animal vectors and soil type.

**8.3.3.1.4 Horizontal seed distribution.** The horizontal distribution of seed in the soil is highly variable. *Papaver* sp. (poppy) seed are very small and are distributed fairly evenly throughout Spanish crop fields where they can occur. Most other weed species, with larger seeds, are distributed in a highly patchy and variable distribution. The horizontal distribution of weed seed in the soil may be experimentally intractable, despite numerous reports relying on sub-sampling estimates.

**8.3.3.2 Floral seed community composition.** The floral plant species composition of a seed bank is determined by several inherent biases intrinsic to the locality (see unit 4, adaptation in local plant communities). It is biased by the history of past vegetation. It is biased by the weed species differences in dispersability, reproductive output (number per plant, numbers of plants), seed longevity and predation. It is biased by the weed species differences in dormancy and dormancy heterogeneity (seed heteroblasty). It is biased by the heterogeneity in environmental-physical qualities of the soil niche and safe microsites of the locality.

**8.3.3.3 Seed pool size.** The size of a soil seed pool at a locality is a product of its qualities, environment and the species that occupy it: local opportunity spacetime. Seed pool size is an indirect consequence of several heritable traits of a weed species, including seed heteroblasty. Colonizing species, especially herbaceous annual weeds, thrive in disturbed localities and tend to have relatively large seed pools because there exist so many microsite opportunities to exploit and fill. Grazing animals tend to increase soil pool size size due to the open habitat, manure and seed movement, and the effects of animal trampling of soil. Forests tend to have small seed pools due to lack of safe establishment sites. Prairies also tend to have relatively smaller seed pools because herbaceous perennials are able to establish in the spring more rapidly with underground vegetative propagules (e.g. rhizomes, rootstocks). Prairie plants have low seed production investment relative to that invested in vegetative propagules. The common feature in these later two system types is that both are relatively long-lived, later successional stage, communities.

Table 8.3 Qualities of relatively large and small soil seed pools.

| Large Seed Pools | Small Seed Pools |
| --- | --- |
| annuals | perennials |
| colonizers | later successional species: forests, prairies |
| small seeded species | large seeded species |
| tilled fields | (versus) prairies, forests |
| tilled fields | (versus) pastures and hay fields |
| pastures (herbivory, grazing) | (versus) hay fields (biomass harvested) |
| acidic, poorly drained, water-logged soils | neutral, well drained soils |
| resource rich | (versus) resource poor |

Soil seed pool sizes vary in different habitats (Silvertown and Charlesworth, 2001; table p. 137). Decay contributes to the loss of seeds from soil seed pools (Silvertown and Charlesworth, 2001; p.138 top; Roberts and Feast. 1973). Herbaceous species in persistant British and European soil seed banks tend to have smaller, rounder seeds than species with persistant seed banks (Thompson et al., 1993; Silvertown and Charlesworth, 2001, figure p.140).

**8.3.3.4 Seed longevity in the soil.** The longevity of seeds in the soil is highly variable. *Kochia scoparia* is non-dormant and lives typically only one year. Velvetleaf (*Abutilon theophrasti*) can live 50 or more years in the soil. Some examples include: *Galinsoga* sp., 1 year; *Chenopodium album* (common lambsquarters), 1600 years in archaeological digs; *Portulaca oleracea* (purslane) up to 40 years; *Solanum* sp. (nightshade) up to 30 years, with a mean of 5-10 years; and *Polygonum* sp., 5-10 years, usually less than 5 years. Viable seed has been found in the Egyptian pyramids in jars by mummies. *Lotus* sp. have remained viable for 1,000 years. Seed longevity increases when interred in a peaty, acidic, anaerobic bog. Biennials species tend to last longer than annuals. Mutagenesis (radon, chemical mutagens, microflora toxins) can harm very old seed in the soil. The Beal experiment at Michigan State University has been measuring seed viability of some common weed seeds buried in glass jars since the 19th century, some species are still viable although the storage method is atypical of crop fields (Kivillan and Bandurski, 1973, 1981).

Experimentally estimating the longevity of the seed of a species in the soil is difficult, and there are several considerations that arise because of the heterogeneous nature of the seed of a species (including heteroblasty). Although mortality is, in general, exponential with time, three different parameters are needed to characterize longevity of species in a seed pool. The longest time is that taken for the last seed of a cohort to die, this may be an atypical number when compared to the rest of the seeds. The second is the mean half-life, the time when 50% of the seed are gone. The third is how many seeds are left after year 1, year 2, etc.

Some species' soil seed banks are longer-lived than others. Natural selection acts on the longevity of a species seed in unpredictable environment.

Table 8.4 Qualities of relatively long- and short-lived seeds in the soil.

| Long-lived | Short-lived |
| --- | --- |
| seed that thrive in tilled, disturbed habitats | seed that thrive in undisturbed habitats |
| lower soil oxygen conditions; anaerobic | higher soil oxygen conditions; aerated soil |
| deeper tillage and seed burial | shallower tillage and seed burial |
| trapped in soil aggregates | brought up from lower depths |

| | |
|---|---|
| compacted soil | loose, lower specific gravity, soil structure |
| acidic soil | neutral to basic reaction soil |
| cold soil | warm soil |
| dry soil (less oxidation, microbial activity) | wet but aerated soil |
| | soils whose conditions change significantly |

Relatively smaller seeds tend to be longer-lived than large in soil seed pools. Most small seeded species are colonizers of disturbed habitats. There is relatively less predation of small seed by macrobiota because it is harder for birds and big animals to get gather. Larger seed size provides more stored food reserve to eat. Small seeds are easier to bury and fall down cracks to soil depth. Small seeds are more likely to be incorporated into soil aggregates, a protective condition.

## 8.4 Propagule Germination and Recruitment

Seedling emergence is called seedling recruitment: enlisting seeds from the soil to resume autotrophic growth leading to reproduction. It is the life history of a weed seed from germination to seedling emergence from the soil and recommencement of growth as a seedling.

**8.4.1 Introduction.** The single most important determinant of agricultural weed community assembly, and subsequent community structure, is the timing of weed seedling emergence relative to that of the crop emergence and related crop management activities (e.g. planting, tillage, herbicide use) (see ch. 9, neighbor interactions in the local plant community). Seedling emergence and plant establishment are a direct consequence of the inherent heterogeneous dormancy of individual seeds (heteroblasty) and the environmental conditions that modulate the behavior of those dormant seeds.

Propagule germination and seedling recruitment is the consequence of the three overlapping processes of weed life history, natural selection and invasion biology (table 8.5).

**Table 8.5** Propagule recruitment: processes of life history, natural selection and invasion biology underlying functional traits of weeds (see table 8.2).

| | PROCESSES | |
|---|---|---|
| LIFE HISTORY | NATURAL SELECTION | INVASION BIOLOGY |
| **Recruitment** | Pre-Condition 1: excess local phenotypes compete for limited opportunity | Invasion, Colonization & Extinction |
| Seed germination | | |
| Seedling emergence (threshold event) | Condition 3: survive to produce the fittest offspring | |

Traits important to propagule germination and recruitment are those that fulfill roles of timing of emergence and assembly in local agricultural communities (table 8.6).

**Table 8.6** Weedy life history functional roles and traits for propagule germination and recruitment.

| TRAIT ROLES | TRAITS |
|---|---|
| **Propagule germination timing:** in the soil | **Propagule germinability-dormancy**: responsiveness to seasonal environmental signals stimulating germination in the soil:<br>1. non-dormant<br>2. perennial species bud dormancy |

| | |
|---|---|
| | 3. hard seed coat dormancy<br>4. light-nitrate stimulation<br>5. oxygen-water restriction<br>6. other and multiple control |
| **Seedling emergence & community assembly**: number, timing and pattern to maximize subsequent local opportunity spacetime exploitation | **Seed heteroblasty**: fine-scale timing of emergence appropriate to locality<br>**Emergence ability**:<br>1. horizontal: seed number-size appropriate to exploit all local safe microsites<br>2. vertical: emergence from depth; soil surface |
| THRESHOLD EVENT: SEEDLING EMERGENCE ||

### 8.4.2 Process of recruitment.

Seedling recruitment and establishment is characterized by four events:

1: *Loss of dormancy* is the widening of the germinability 'window' of individual seeds of a species at particular times of the year, and among years by means of after-ripening.

2: *Seed germination*, once a seed is after-ripened, is determined by conditions, resources and stimuli in immediate environment. The specific environmental signals that stimulate germination in largely unknown for most common weed species, but many clues exist. The minimum requirements for all seeds are moisture, heat and oxygen. Other dormancy regulating mechanisms, signals stimulating germination, may include light-photoperiod-phytochrome, physical factors affecting seed envelopes, carbon dioxide and other soil gases, and microbial activity, depending on the species.

3: *Seedling emergence* from the soil.

4: *Photosynthetic independence* of the seed from parental seed food reserves for growth and development, autotrophic growth of its leaves.

This characterization of the seedling recruitment can be enhanced by viewing it as an evolutionary process in which disturbance in a locality creates both risk-mortality as well as opportunity space-time. Weed seed traits provide the mechanisms and means by which they can exploit opportunity space-time created by disturbance. As an example, weedy *Setaria* species are stimulated to germinate by oxy-hydro-thermal units over time. The amount of oxygen dissolved in water imbibed into the seed embryo over time determines its behavior. In the spring, cool oxygen-rich water diffuses into the seed and oxygenation of the embryo is at its yearly maxima. These oxygen-rich seeds await warming temperatures that stimulate germination in the most available seed, the numbers precisely determined by local microsite moisture and temperature conditions. As the soil warms, oxygen solubility decreases, resulting in a lessening of germination due to less available oxygen despite the favorably warm soil temperatures. As summer approaches oxygen levels are at their seasonal minima, and secondary dormancy is induced in the previously germinable seed.

**8.4.3 Germination microsites and safe sites.** Seed germination and seedling emergence occurs on the scale of the size of the seed. Seed-scale micro-sites in the soil that are favorable are termed 'safe-sites' for germination. Safe sites are those microsites with the specific conditions that allow the seeds of a particular species to emerge successfully from the soil (Harper et al., 1965). Safe site and niche concept both share the quality that neither can be reliably identified until after they have been successfully occupied. In theory, germination safe sites might explain the macrodistribution of species, as well as their microdistribution: opportunity spacetime is scalar. Safe micro-sites in the soil are very heterogeneous across a seed pool or field. They are characterized by conditions, resources and stimuli in immediate environment conducive to germination. They are habitable sites that avoid or reduce the risk of predation and decay. The qualities that make a site safe vary by the requirements of each individual species and seed. The qualities safe for a surface germinating species differ from those of deeper germinating species. For example, surface germinators place more reliance on seed shape and seed structures for favorable soil-water contact.

**8.4.4 Magnitude and duration of seedling emergence.** The seedling emergence of a weed species at a locality can be described in terms of the magnitude (numbers per unit time, period or cohort), duration (time duration of emergence) and pattern (structure of recruitment periodicity: cohorts with similar seasonal times of emergence). Seedling emergence is a consequence of the interaction of phenotypic traits controlling seed germinability-dormancy characteristic of a weed species and the immediate soil opportunity spacetime of individual seeds. For example, the influence of seed heteroblasty on recruitment behavior was apparent in that *Setaria faberi* populations with higher dormancy at the time of dispersal had lower emergence numbers the following spring, and in many instances occurred later, compared to those less dormant (Jovaag et al., 2012b). Heteroblasty was thus the first determinant of behavior, most apparent in recruitment number, less so in pattern. Environment modulated seedling numbers, but more strongly influenced pattern. The resulting pattern of emergence revealed the actual "hedge-bet" structure for *Setaria* seedling recruitment investment, its realized niche, an adaptation to the predictable mortality risks caused by agricultural production and interactions with neighbors. These complex patterns in seedling recruitment behavior support the conjecture that the inherent dormancy capacities of *S. faberi* seeds provides a germinability 'memory' of successful historical exploitation of local opportunity, the inherent starting condition that interacts in both a deterministic and plastic manner with environmental signals to define the consequential heterogeneous life history trajectories.

**8.4.5 Patterns of seedling emergence.** The pattern of seedling recruitment is a heritable, phenotypic, trait arising from the interaction of seed heteroblasty with the soil. Several modal patterns of seasonal seedling emergence can be observed among common weed species. The timing of emergence may be over an extended, or in a relatively short, period.

**8.4.5.1 Single 'flush' period.** Perennials often have new shoots that appear early in spring. Winter annuals emerge in the fall before they overwinter. Precocious species germinate immediately after leaving parent plant. Seeds without dormancy all germinate soon after shed. They may also be delayed by winter conditions (enforced dormancy), but germinate in a flush when conditions are favorable (e.g. *Kochia scoparia*). Opportunistic emergence occurs when seeds are regulated by enforced dormancy, only awaiting the right conditions to germinate. Some species require a specific event or specific environmental signal for germination to occur. These simple signals may include a fire event, a rainfall event (e.g. desert plants), disturbance, or light (e.g. nightshade germination stimulated by light through an opening in the surrounding, mid-season, leaf canopy). In all these instances of a single germination event, the range of emergence times might be only apparently narrow.

**8.4.5.2 Bi-modal recruitment.** Dandelion (*Taraxicum officinale*) has two times of seedling emergence, spring and fall flushes.

**8.4.5.3 Continuous emergence.** *Bidens pilosa* germinates anytime in the tropics. Conditions in some areas are relatively constant all year and may be conducive to continuous emergence. Crabgrass (*Digitaria* sp.) emerges continuous all year in turfgrass and home lawn habitats.

**8.4.5.4 Major emergence period with extended, infrequent period.** This may be the most typical pattern of seedling emergence in many temperate region agro-ecosystems. In Iowa, USA as well as most major grain producing areas of North America, a large number of weed seedlings are recruited in the spring, followed by much smaller numbers of seedlings for the remainder of the season until the soil freezes again. This is the typical emergence pattern for the *Setaria* spp. foxtails), *Chenopodium album* (lambsquarters) (see section 8.4.6, case studies) and *Sorghum bicolor* (shattercane).

**8.4.5.5 Relative emergence order.** In any particular agricultural area there are certain weed species that appear consistently over time. Their presence, relative abundance, dominance or minority status, in a field can change but they compose the weed floral community of that area. There exists a consistent relative order of emergence in the growing season over time, and this emergence sequence is a consequence of the timing of expression of the traits these weed species possess. For example, in Iowa the relative seedling emergence sequence in table 8.7 has been proposed (Hartzler et al., 1997). These particular species appear when they do for many reasons, some inherent in the species seed in the seed pool (e.g. seed dormancy heteroblasty), others a consequence of the micro-site, agronomic and environmental qualities of the locality.

Much could be revealed about the process of weed community assembly (see ch. 9.1.2, recruitment: community assembly) if this list was analyzed in terms of traits that affected both the relative emergence timing and pattern, but also the traits that made the species at an advantage in emerging when it does relative to the neighbors that had emerged prior, at the same time, and later in the season. For instance, *Kochia scoparia*, kochia, has no inherent seed dormancy and emerges very early in the season (group 1). It is apparently cold-tolerant of those early conditions, but does not appear to be display vigorous growth and photosynthesis as species that emerge later. Kochia's growth traits are a trade-off between early emergence stress tolerance and low growth/photosynthetic growth rates.

*Seasonal recruitment pattern.* Table 8.7 reveals the time of first seedling emergence of a species. Each of these species emerges for longer periods of time in the season. The variation and duration in seasonal timing also differs by weed species. For example, *Kochia scoparia* emergence is probably completed early in the season, while *Setaria faberii* continues into the autumn. Of crucial importance to the farmer are weeds emerging after layby, the time when farming operations cease and the main obstacle to seizing opportunity spacetime by weeds is competitive interactions with neighbors. To reveal these later emergence events and their importance for the evolving local population a more detailed list of recruitment times is required (see table 8.8, *Setaria faberi*, Iowa; and table 8.9, *Chenopodium album*, Finland)

**Table 8.7** Relative seedling emergence time and order for the common weeds of summer annual crops in Iowa and adjacent areas; associated calendar dates and farm operation timing; proposed ecological roles played by recruitment timing cohorts (groups 0-7).

| GROUP 0 | GROUP 1 | GROUP 2 | GROUP 3 | GROUP 4 | GROUP 5 | GROUP 6 | GROUP 7 |
|---|---|---|---|---|---|---|---|
| > | | | TIME OF SEASON | | > | | |
| AUTUMN | EARLY SPRING | | SPRING | | LATE SPRING | | EARLY SUMMER |
| October | April | | May | | June | | July |
| Julian Week: | | | JW | | | | |
| 40-45 | 14-15 | 16-17 | 18-19 | 20-21 | 22-23 | 24-25 | 26-27 |
| | | | FARM OPERATION DISTURBANCES | | | | |
| Post-Harvest | Prior to Planting | Maize Planting | Soybean Planting | Planting Time | Post-Planting | Layby | Post-Harvest |
| GROUP 0 | GROUP 1 | GROUP 2 | GROUP 3 | GROUP 4 | GROUP 5 | GROUP 6 | GROUP 7 |

| GROUP 0 | GROUP 1 | GROUP 2 | GROUP 3 | GROUP 4 | GROUP 5 | GROUP 6 | GROUP 7 |
|---|---|---|---|---|---|---|---|
| horseweed, marestail | foxtail barley | quack-grass | smooth brome | Canada thistle | green foxtail | black nightshade | fall panicum |
| downy brome | kochia | orchard-grass | common ragweed | giant foxtail | common milkweed | shattercane | crabgrass |
| field pennycress | prostrate knotweed | giant ragweed | woolly cupgrass | common cocklebur | hemp dogbane | common sunflower | morning-glories |
| shepard's purse | wild mustard | Pennsylvania smartweed | velvetleaf | yellow nutsedge | wirestem muhly | Venice mallow | jimsonweed |
| biennial thistles | dandelion | ladys-thumb smartweed | wild buckwheat | redroot pigweed | barnyard-grass | waterhemp | witchgrass |
| wild carrot | Russian thistle | common lambsquarters | | | yellow foxtail | smooth ground- | |

| | | | | | | cherry | |
|---|---|---|---|---|---|---|---|
| dandelion (from seed) | white cockle | wild oats | | | wild proso millet | Jerusalem artichoke | |
| | | hairy night-shade | | | | | |
| **GROUP 0** | **GROUP 1** | **GROUP 2** | **GROUP 3** | **GROUP 4** | **GROUP 5** | **GROUP 6** | **GROUP 7** |

| ECOLOGICAL ROLE | | | | | | | |
|---|---|---|---|---|---|---|---|
| •Escape neighbors by autumn emergence •survive winter | •Dominate neighbors by early emergence: -light competition -hasten reproduction | •Dominate neighbors by early emergence: -light competition -luxury nutrient consumption) | •Dominate neighbors by: -light competition -high growth rates | •Dominate neighbors by: -light competition -high growth rates | •Dominate neighbors by: -light competition -high growth rates | •Dominate neighbors with high growth rates & canopy (light) closure •Exploit later season light opportunities | •Dominate neighbors with canopy (light) closure •Exploit later season light opportunities |
| **GROUP 0** | **GROUP 1** | **GROUP 2** | **GROUP 3** | **GROUP 4** | **GROUP 5** | **GROUP 6** | **GROUP 7** |

| GROUP | ANNUALS | BIENNIALS | PERENNIALS |
|---|---|---|---|
| 0 | **WINTER ANNUALS:** **Poaceae:** •*Bromus tectorum* (downy brome) **Cruciferae:** •*Capsella bursa-pastoris* (Shepard's purse) •*Thlaspi arvense* (field pennycress) | **Composite:** Biennial thistles: •*Carduus* sp. •*Cirsium* sp. **Umbelliferae:** •*Daucus carota* (wild carrot) | **Compositae:** •*Conyza canadensis* (horseweed; marestail) •*Taraxacum officinale* (dandelion); from seed |
| 1 | **Polygonaceae:** •*Polygonum aviculare* (prostrate knotweed) **Chenopodiaceae:** •*Kochia scoparia* (kochia) •*Salsola kali* (Russian thistle) **Cruciferae:** •*Brassica kaber* (wild mustard) | **Caryophyllaceae:** •*Lychnis alba* (white cockle) | **Poaceae:** •*Hordeum jubatum* (foxtail barley) **Compositae:** •*Taraxacum officinale* (dandelion) |
| 2 | **Poaceae:** •*Avena fatua* (wild oat) **Compostitae:** •*Ambrosia trifida* (giant ragweed) **Polygonaceae:** •*Polygonum pensylvanicum* (Pennsylvania smartweed) •*Polygonum persicaria* (ladysthumb) **Solanaceae:** •*Solanum physalisfolium* (hairy nightshade) **Chenopodiaceae:** •*Chenopodium album* (common lambsquarters) | | **Poaceae:** •*Elymus repens* (quackgrass) •*Dactylis glomerata* (orchardgrass) |
| 3 | **Poaceae:** •*Eriochloa villosa* (woolly cupgrass) **Compositae:** •*Ambrosia artemisiifolia* (common ragweed) **Polygonaceae:** •*Polygonum convolvulus* (wild buckwheat) **Malvaceeae:** •*Abutilon theophrasti* (velvetleaf) | | **Poaceae:** •*Bromus inermis* (smooth brome) |
| 4 | **Poaceae:** •*Setaria faberii* (giant foxtail) **Compositae:** •*Xanthium strumarium* (common cocklebur) **Amaranthaceae:** •*Amaranthus retroflexus* (redroot pigweed) | | **Compositae:** •*Circium arvense* (Canada thistle) **Cyperaceae:** •*Cyperus esculentus* (yellow nutsedge) |
| 5 | **Poaceae:** •*Echinochloa crus-galli* (barnyardgrass) •*Panicum miliaceum* (wild proso millet) •*Setaria pumila* (yellow foxtail) •*Setaria viridis* (green foxtail) | | **Poaceae:** •*Muhlenbergia frondosa* (wirestem muhly) **Apocynaceae:** •*Apocynum cannabinum* (hemp dogbane) **Asclepiadaceae:** •*Asclepias syriaca* (common milkweed) |
| 6 | **Poaceae:** | | **Compositae:** |

|   |   |   |   |
|---|---|---|---|
|   | •*Sorghum bicolor* (shattercane)<br>**Compositae:**<br>•*Helianthus annuus* (common sunflower)<br>**Solanaceae:**<br>•*Solanum ptycanthum* (eastern black nightshade)<br>**Amaranthaceae:**<br>•*Amaranthus rudis* (common waterhemp)<br>**Malvaceae:**<br>•*Hibiscus trionum* (Venice mallow) |   | •*Helianthus tuberosus* (Jerusalem artichoke)<br>**Solanaceae:**<br>•*Solanum subglabrata* (smooth groundcherry) |
| 7 | **Poaceae:**<br>•*Digitaria sanguinalis*, (large crabgrass)<br>•*Digitaria ischaemum* (smooth crabgrass)<br>•*Panicum capillare* (witchgrass)<br>•*Panicum dichotomiflorum* (fall panicum)<br>**Solanaceae:**<br>•*Datura stramonium* (jimsonweed)<br>**Convolvulaceae:**<br>•*Ipomoea hederacea* (ivyleaf morningglory)<br>•*Ipomoea purpurea* (tall morningglory) |   |   |

**8.4.6 Case studies.** To reveal the importance of the entire emergence periods of weed species, and their significance to the evolving local population, a more detailed list of recruitment times and associated local cropping practices is required (see table 8.8, *Setaria faberi*, Iowa, USA; and table 8.9, *Chenopodium album*, Finland). Several observations about seedling recruitment pattern in these two different species, cropping systems and continental locations can be made. First, emergence patterns seem to extend over the entire growing season, taking advantage of all opportunities the particular cropping system has to offer (unused, exploitable opportunity). The majority of the seedlings emerge early, when the potential reproductive payback at the end of the season is greatest for those that survive the several farm operations. In both locations a significant seedling investment occurs just prior to, and just after, layby. Layby is the seasonal time when farming operations cease because crop canopy closure and height restrict the entry of farm equipment into the field. From layby to harvest opportunity spacetime is available if interactions with crop and weed neighbors can be overcome or survived. A much smaller recruitment cohort occurs late in the season, around harvest time, when small seedlings can emerge and escape harvest equipment for a short period of low-output reproduction and fecundity contributing to the local soil seed pool.

It is important to observe that local adaptation by each of these weed species occurs by natural selection favoring those that emerge at opportune times and locations. Although the individual species responds directly to the environment of the particular field and cropping system, these conditions are a signal guiding them to the right timing. Environment is a signal the local plants use to change seed responsiveness to ensure proper emergence timing. Environment guides but does not rule behavior. Natural selection rules as it exploits environmental signals to change and guide internal dormancy capacity. For this reason a fixed environmental model (e.g. hydro-thermal time; or worse, a simple thermal model (e.g. Schutte et al., 2014)) to predict emergence will never reveal fine scale seasonal timing, nor will it reveal seasonal emergence pattern cohorts. Weed seed heteroblasty is constantly evolving in a locality to exploit those particular times of the cropping system that provide opportunity spacetime: the locality and time of recruitment in which subsequent reproduction can succeed. The population evolves, phenotypes are constantly changing as they replenish the adapted and adaptable soil seed pool, the source of all future weed infestations.

**8.4.6.1 *Setaria* seedling emergence in central Iowa.** Insights about weedy hedge-betting strategies of *Setaria faberi* for invasion, colonization and enduring occupation of central Iowa, USA agricultural fields have been reported (Jovaag et al., 2012b). Fitness in *S. faberi* is conferred by strategic diversification of seedling recruitment when confronted by risks from agricultural disturbance and heterogeneous, changeable habitats. Evolutionarily, hedge-betting is a strategy of spreading risks to reduce the variance in fitness, even though this reduces intrinsic mean fitness. Hedge-betting is favored in unpredictable environments where the risk of death is high because it allows a species to survive despite recurring, fatal, disturbances. Risks can be spread in time or space by either behavior or physiology. Risk spreading can be conservative (risk avoidance by a single phenotype) or diversified (phenotypic variation within a single genotype) (Cooper & Kaplan, 1982; Philippi & Seger, 1989; Seger & Brockman, 1987).

Hedge-betting of seedling recruitment time and pattern reduces temporal variance in fitness of *S. faberi* in agroecosystems characterized by annual cycles of disturbance and unpredictable climatic conditions. Earlier emergence allows for greater biomass and potentially explosive seed production at season's end (Dekker 2004), but also has a high risk of death from agricultural practices. Later emergence has a lower potential for seed production, but also a lower risk of death. Thus variation in germination time results in both a decrease in the maximum and an increase in the minimum potential seed production, reducing fitness variability and ensuring enduring occupation of the locality. Variation in germination time has the additional benefit of reducing sibling competition, which enhances individual fitness (Cheplick 1996).

Hedge-betting by *S. faberi* begins with the birth of seeds and the induction of heteroblastic differences that are subsequently echoed in seedling emergence behavior. The complex pattern of emergence observed in Jovaag et al. (2012b) reveals the actual hedge-bet structure for central Iowa, USA *S. faberi* seedling recruitment, its realized niche. The relative seedling investment made by *S. faberi* at different times during the season is a consequence of the changing balance between the risk of mortality and the potential fecundity during those periods.

There exists some predictability to the primary sources of mortality and recruitment opportunities for *S. faberi* in Iowa agroecosystems over the course of their annual life histories (table 8.8; see ch. 17.4.5, seedling recruitment). The majority of seedlings are recruited in the spring when the risk of mortality is very high from crop establishment practices (seedbed preparation, planting, weed control) and the rewards are the greatest. Weed seedlings emerging early have the greatest time available for biomass accumulation and competitive exclusion of later emerging neighbors. Subsequent fitness devolves on those individual *S. faberi* plants that escape these disturbances.

Seedling recruitment in the late spring and early summer is affected by the cessation of all agricultural disturbances and weed control tactics, a time termed 'layby' in the Midwestern US. Seeds emerging after layby avoid mortality from cropping disturbances, and the risk of death shifts to that from interactions with neighbours. Freed from cropping disturbance risk, seedlings emerging post-layby face increasing competition from crop and weed neighbours and decreasing time for biomass production. Subsequent fitness devolves on those plants freed from human intervention that compete effectively with neighbors until harvest, an advantageous opportunity space-time apparent from the seedling investment observed.

Very small *S. faberi* seedling investments are made late in the season when the risks of mortality are much lower. Seedlings recruited late experience a senescing crop light canopy. Those emerging during or after harvesting experience mortality risks only from the approaching winter climate. Floral induction in *S. faberi* is increasingly stimulated by decreasing photoperiod, allowing a seedling with as few as 2-3 leaves to produce a few seeds very late in the season. This justifies a small hedge-bet investment yielding low numerical returns but ensuring some potential increase in soil seed pool size for future infestations.

**Table 8.8.** Calendar of historical, seasonal times (Julian week, month) of agricultural field disturbances (seedbed preparation; planting; weed control, including tillage and herbicides; time after which all cropping operations cease, layby; harvest and autumn tillage), and seedling emergence timing for central and southeastern Iowa, USA, *Setaria faberi* population cohorts (all *S. faberi* combined; Jovaag et al., 2012b)

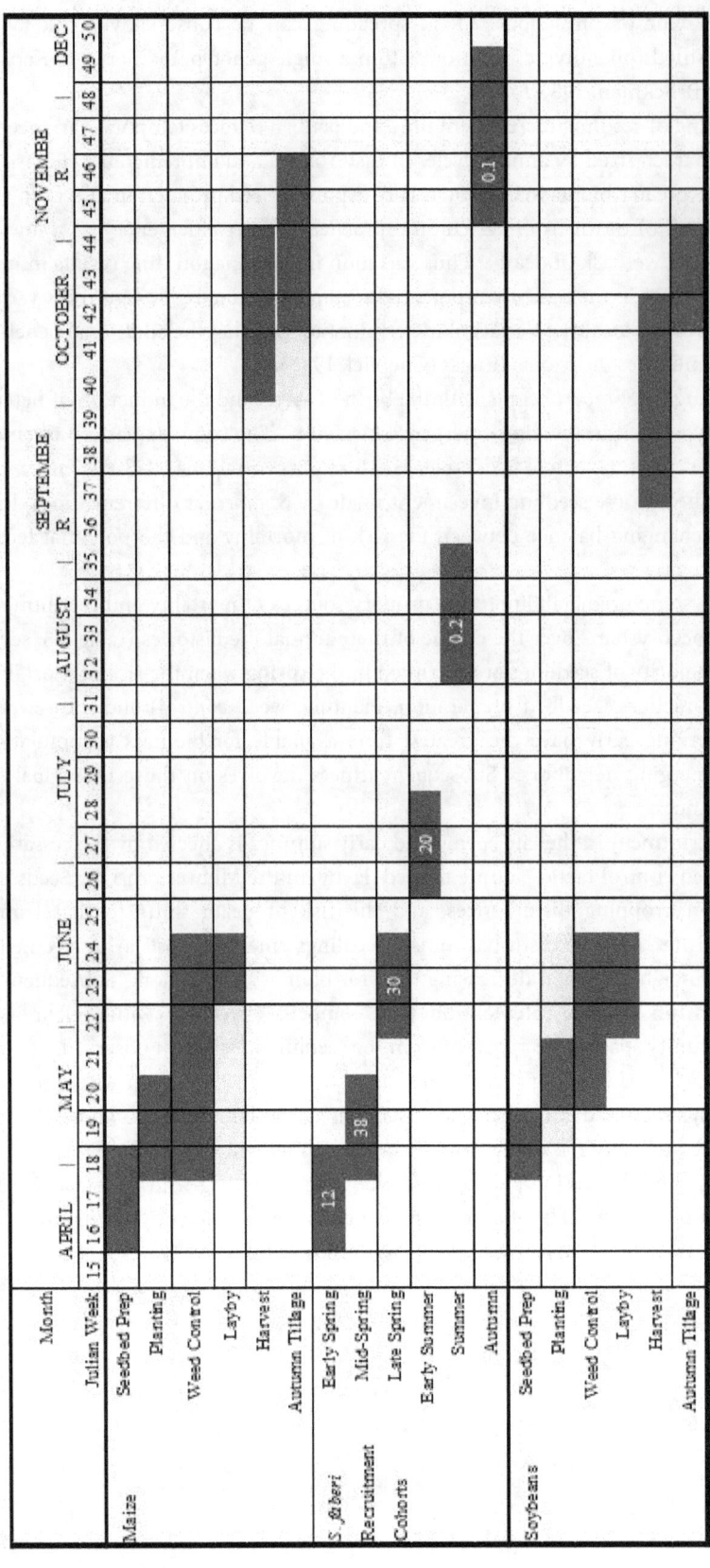

**8.4.6.2 *Chenopodium album* seedling emergence in Europe and North America.** *Chenopodium album* seedling emergence studies were conducted at nine European and two North American locations comparing local populations with a common population from Denmark (DEN-COM) (Andujar et al., 2013). Recruitment from Finland appears in table 8.9 as an example of these studies.

**Mortality risk.** Weedy plant life history conforms to the time of mortality risk from either the environment (e.g. frozen soil in the winter) or cropping systems practices (e.g. herbicide use). *C. album* life histories are constrained by the environmental and cropping system practices of a locality. It was hypothesized that *C. album* seedling recruitment timing and magnitude have adapted to these local conditions and are expressed in emergence behavior. Recruitment periodicity and quantity were found to be closely related to predictable cropping system disturbances, experimentally induced disturbances, and seasonally-available plant resources and conditions. Recruitment patterns were the consequence of seasonal limitations in the habitat (filter 1), predictable cropping system disturbances (filter 2) and experimentally induced disturbances to simulate early season tillage. As a consequence of these filters *C. album* seized locally available seedling emergence opportunity.

**Filter 1: Habitat.** Limitations in the habitat (filter 1) are reflected in local *C. album* population recruitment season length: Julian weeks (JW) 2-37, duration 12-37 weeks. Generally, the duration of seedling recruitment (undisturbed, experimentally disturbed) of both populations (local; DEN-COM) increased with decreasing latitude, north-to-south. In general, compared to the local population, DEN-COM recruitment at locations north of Denmark was longer and south of Denmark was shorter, and ended sooner. The early season onset of recruitment was more consistent than that in the late season, which had smaller numbers and greater variability over longer times. Autumnal recruitment in the burial year (2005) occurred at a few locations.

**Filter 2: Cropping disturbance.** Predictable, historical, seasonal times of local cropping system disturbances (CSD) provided the second filter of seedling recruitment opportunity. Generally, the CSD period increased with decreasing latitude, with some exceptions. The total duration of the CSD period was over twice as long in the south as that in the north. After layby there occurred a mid-season period of CSD inactivity at all locations, the post-layby (PL) period, when cropping activity ceases. This post-layby period was from JW 21-34, with durations of 2-13 weeks, depending on location. No CSD's occurred at any location during JW 26-29, excepting the UK.

**Seedling emergence structure: time, number.** Recruitment at each locality possessed seasonal structure (time, number) consisting of 2-4 discrete seasonal cohorts, a consequence of habitat and disturbance limitations. The spring cohort (JW 12-24) was the largest recruitment investment, with consistent local times of seasonal onset. A short gap in recruitment occurred between the spring and summer cohorts at all locations, for both populations, and coincided with layby and summer solstice (LSS), the PL period, crop canopy closure, and a period of increasing heat and potentially decreasing moisture. This LSS gap was characterized by the consistent absence of seedling emergence during JW 25-26 in undisturbed, and in a small number of experimentally disturbed, instances. The gap was unexpected at this favorable recruitment time of the season. The specific timing of the recruitment hiatus varied with location (latitude), population and experimental disturbance. The short cessation of seedling recruitment may be a consequence of a unique seasonal change in light, the most important environmental signal regulating *C. album* seed germination. Photoperiod length and intensity increases before, and then decreases after, summer solstice, a threshold event, a solar switch. The coincidence of all these anthrogenic, astronomical and environmental factors at the LSS may be an important basis by which *C. album* seedling emergence is 'fine-tuned' consistent with local opportunity. Late recruitment consisted of the second largest cohort in the summer (JW 25-37). This cohort was smaller than, longer and more variable than, and consistently separated from, the spring cohort. Recruitment during the autumn cohort (burial year, following year) occurred at only a few locations, in low numbers. Winter recruitment only occurred in warm, irrigated, multi-cropped Portugal.

**Emergence cohorts.** Local recruitment opportunity was seized by *C. album* structured in seasonal cohorts whose timing was consistent with habitat and cropping disturbance periodicity. Distinct recruitment cohorts arose around the temporal interfaces separating CSD operations. This may be an adaptive means by which *C. album* searches for, and exploits, recruitment opportunity just prior to, and after, predictable disturbances. These patterns coincide with intervening periods of no disturbance, opportunity: spring and summer cohorts are separated by the LSS gap; the early summer cohort at the beginning of the post-layby (PL) period of mid-season opportunity; and at harvest. Recruitment periodicity differed by population at each local common nursery. Experimental disturbance affected the recruitment cohort structure in several contradictory ways. In some locations it was extended, others shortened, in others stimulated the appearance of new late season cohorts

**Seed control of emergence.** The control of *C. album* seedling emergence is contained in the heteroblastic traits of its locally adapted seeds, and is stimulated by a complex interaction of light (photoperiod duration, intensity and quality), heat (growth; oxygen-water and nitrate-water solubility), water, nitrate and oxygen signals inherent in the local environment. Observations of complex recruitment patterns occurring at critical cropping times is strong evidence that *C. album* possesses a flexible and sensitive germination regulation system adaptable to opportunity in many different Eurasian and North American agricultural habitats.

**Summer solstice light signal.** A potential role for summer solstice as a rapid astronomical signal, a threshold switch, regulating photoperiod length, as well as the direction of change in diurnal length (longer, shorter) was shown for the first time. This solstice photoperiod switch may provide a rapid light signal stimulating or inhibiting recruitment, a dependable seasonal signal ensuring the separation of seedling cohorts before and after layby and canopy closure.

**A linear model for understanding superficial soil disturbance-induced *C. album* seedling emergence.** An alternative linear, demographic, representation of *Chenopodium* album seedling emergence was subsequently reported in 2014 (Schulte et al.) by substantially the same authors from the seedling germination working group of the European Weed Research Society as the previous study (Andujar et al., 2013). Several concerns are raised about the appropriateness of this type of representation of seedling recruitment. Seedling recruitment as reported in 2013 clearly revealed a complex, non-linear reaction of *C. album* to habitat and predictable annual local cropping disturbances (in addition to systematic, experimental, superficial soil disturbance treatments). As such the choice of linear models (i.e. GLMM, LMM) inherently violate assumptions underlying the use of analysis of variance statistical techniques (see ch. 14.2.5.1): that observations are drawn from normally distributed populations and errors are normally distributed within each treatment population and are independent of each other; and that observations are random samples from the populations and independence of error effects; and, the variances of the populations are homogeneous. The conversion and transformation of time of emergence to time-heat (day-degrees) of emergence was both confusing and obfuscating: *C. album* germination is controlled by several seed traits that respond to heat, water, soil nitrogen and oxygen, and light signals (Altenhofen, 2009; Altenhofen and Dekker, 2014). No presentation of both undisturbed and disturbed emergence with time (as well as no presentation of emergence with time-heat at nine of ten sites) made interpretation of the study more difficult: the observational foundation upon which the effects of superficial soil disturbance might be inferred. Homogenization of seedling emergence data into mean behaviors averaged over site (latitude, continent), population (local, Denmark common), time (year, cohort), light (e.g. latitudinal photoperiod), and soil water-nitrogen-oxygen results in a loss of clarity, confounding inference.

There exist few appropriate statistical techniques to represent complex phenomena like seedling emergence in habitats with predictable local production practices which vary from farm to farm, country to country, latitude to latitude, continent to continent. Demographic representations of weed behavior present their own set of artifacts (see unit 6 and ch. 15.2). Adaptive behavior is non-linear, the emergent product of complex interactions among local opportunity space-time factors, habitat, cropping system, and local *C. album* phenotype. This was apparent in the weekly seedling emergence versus time calendars presented in the 2013 study. As such, sometimes historical narratives are the only, best, current way to explain complex phenomena: generalization is powerful, but exacts a heavy price, destroying the concreteness of history (Auyang, 1998).

**Table 8.9.** Calendar of historical, seasonal times (Julian week, month) of agricultural field disturbances (seedbed preparation; planting; weed control, including tillage and herbicides; time after which all cropping operations cease, layby; harvest and autumn tillage), and seedling emergence timing for Finland, *Chenopodium album* population cohorts (studies conducted by J. Salonen, reported in Andujar et al., 2013)

**8.4.7 Relationship between seed heteroblasty and recruitment timing.** The relationshipe between *Setaria faberii* seed heteroblasty and recruitment timing has been reported (Atchison, 2001; Jovaag et al., 2011, 2012a, b) In brief, these studies demonstrate weedy *Setaria* seed dormancy capacity heterogeneity at abscission (seed heteroblasty) provides a "blueprint" for those subsequent behaviors. The majority of the 39 populations studied were differentiated from each other; this seed heteroblasty variation indicated a fine scale adaptation to different local environments. Heteroblastic germination responses were used to rank populations based on their dormancy level to facilitate later comparisons with emergence behavior. Fine scale differences in seed dormancy capacities (heteroblasty), dormancy cycling in the soil, and seedling emergence behavior were observed among *Setaria* populations. The differences were influenced by both genotypic (time of embryogenesis, species) and environmental (year, nursery, seed age in the soil) parameters. *S. faberii* seed dormancy heterogeneity strongly influenced the relationship between heteroblasty and emergence in terms of the seedling numbers: more dormant populations had lower emergence numbers during the first year after burial than less dormant populations. Evidence of this relationship between heteroblasty and emergence number was also provided by the positive Spearman correlation between dormancy capacity at abscission and the cumulative number of seeds emerged during the first year after burial for both the 1998 and 1999 *S. faberii* populations. Individual *S. faberii* seed populations retained heteroblastic qualities throughout their annual germinability cycle (high spring germinability, summer dormancy re-induction, increased fall germinability). Retention of heteroblasty was evident in the variable proportions of candidate seeds within a single population at individual seasonal times. The influence of heteroblasty was also apparent in the heterogeneous decline in germinability from early spring to early summer. Further, the germinability pattern (high in spring, low in summer, somewhat increased in fall) was consistent with the emergence pattern (high in spring and early summer, low in mid summer, none in late summer-early fall, low in mid-late fall). The life history trajectory of the individual *Setaria* seed, from anthesis and embryogenesis to germination and seedling emergence is determined in the first instance by the inherent, parentally-induced, dormancy capacity of the seed, but is sensitive to the environmental conditions confronted on different time scales. At the widest temporal scale studied, heteroblastic qualities largely determine the numbers of seeds that are recruited in a single year and locality. On a finer temporal scale (weekly or daily), the distribution and timing (pattern) of seedling emergence is strongly influenced by the unfolding seasonal environment. The life history trajectory is therefore a pathway whose course is first set by inherent dormancy qualities. The subsequent pathway is continually adjusted in response to fine scale temporal conditions encountered. Sensitive, short-term responses to these signals accumulate, and the dormancy states of seeds change on a weekly and daily basis. The consequent life history trajectory is apparent in dormancy-germinability cycling. The life history trajectory is therefore both deterministic and plastic, conditional on starting conditions (parentally induced dormancy capacity) and modulated by conditions confronted on shorter time scales. The resulting recruitment is an emergent property of the complex component pathways of individual temporal commitments.

# UNIT 4: ADAPTATION IN LOCAL PLANT COMMUNITIES

"In short, it was all beautiful, as neither nature nor art could contrive, but as only happens when they united together, when nature's chisel puts its final touch to the often unintelligently heaped up labor of man, relieves the heavy masses, destroys the all too crudely palpable symmetry and the clumsily conceived gaps through which the unconcealed plan reveals itself so nakedly, and imparts a wonderful warmth to everything that has been created by the cold and carefully measured neatness and accuracy of human reason." N.V. Gogol, 1842.

**The weeds' story**. I am foxtail.

I pop up out of the ground every year. All the time. In those wonderful fields you so carefully prepare for me. I can't help myself. But I am careful, very careful when I emerge. You are so predictable.

All winter you sit in your warm farmhouse and look out at the frozen fields. The sun turned over, hiding its face. You farmers are dreaming, impatiently waiting. Drinking weak coffee, listening to the crop reports and herbicide ads on the radio. Then comes the spring equinox, the sun peaking out of one eye, glowing anew. You get on your tractor and ride! Plow, till, harrow, sweep, spray, chisel. You kill all the neighbors who managed to survive the winter. Clean fields you call them. I call them opportunity to eat, grow and have lots and lots of babies. Some of us very brave and impatient ones emerge just after you prepare your seedbed. Chances are we will die soon, but some of you miss. These are the big winners. We early-bird guy-girls who escape till harvest have gigantic, huge families. Thousands of babies if we hit this big early lotto jackpot. Thank you for your laziness, you lack of attention, your inconstancy.

You plant your 'eaters' at the same time every year. You are so predictable. We wait. Some of us emerge after planting, but this is the danger zone for us. You get out those tall lanky loose floppy sprayers and hose the ground with herbicides and god knows what other horrible expensive chemicals you buy. You shouldn't believe everything you hear on the radio and TV. No problem. You kill all these post-planting girl-guys with your chemicals and feel proud. What you don't notice, at least at first, is some of us are very different indeed. Some of us rare unusual girl-guys have a trick. Resistance! We block you from entry, right at our most sensitive molecules. Or we eat your chemicals in our guts before you can hurt us. Whatever, resistance! Now you are really doomed. Maybe not this year so much, but keep listening to all those herbicide ads on TV all winter and see where it gets you.

The sun rolls over at summer solstice, face to the sky, open and happy. It's hot. The 'eaters' are getting big and there are no weeds in sight. At least none are apparent to you as you drive down the road looking up your clean crop rows. Time for vacation! We love it! This is one our most favorite times of the season. You park your big monster tractors in the barn, load up the car and off you go on vacation with the kids. All is well with the world as you go to Disney World for some fun. But we are still back on the farm, listening very carefully. No sound from the farmhouse. Nothing going on in the barn. It must be 'layby', layby your equipment in the barn and go on vacation. Some handsome young seeds appear soon in the field under the 'eaters', under the early-bird 'lotto winners', under the 'resistors', under all those other riff-raff weeds that got the jump on us 'laybys'. But, if we work hard, pick our homes carefully, its a clean pure homerun to harvest for us mid-season emergers. Why? You are on vacation! By the time you get home you won't really notice us under the late season canopy. Hurray!

The grinding shout of machinery echos across the field. The sun has turned sideways and the autumnal equinox is looking down and winking at all us foxtails waving in the soft breezes of an Indian summer. It's harvest time. OK, OK, we know what you are going to say. The combine kills lots of us guy-girls as you sweep up your 'eaters' into your harvesters. The foxtails are gone now for sure you say. Ha ha ha (LOL)! We fooled you again. The big guys have shed most of their seed in pre-harvest August. More of our seeds still rain down on your fields but we did the job. And, the rest of us get caught up in your harvester, some manage to get carried along with the 'eaters'. Directly into your grain bins. Directly into your food chain distribution network. Directly around the town, around the county, around the state, around the country, around the world! Ha ha ha (LOL)!

We see the farmer standing by his grain bin packed full of zillions and bazillions of 'eater' seeds. He is content. Life is good. Another bumper harvest. Time to pay back the bank. The season is done. Time for that coffee while listening to those trans-genic seed ads on the radio. But wait. What's happening? Some of our little timid girl-guy seeds begin to poke their shy heads out of the soil and emerge. The crafty ones that come out just a few days before the harvester chops and whacks and sucks seeds out of the sky. The crafty ones are up and alive, but so short that the big clean oily dusty combine blades shave inches over our newly emerged heads. Fwing! You missed us! Ha ha ha (LOL)! So we grow and watch you standing content by the bulging grain bins. You never notice us late season arrivistas. We stay small. The short days stimulate us to flower right away. Little seeds heads with just a few seeds. But a one-seed bet late in the season is a sure thing. Payback? It's a 1-for-3, or 1-for-5, or even a 1-for-10 bet. It's all good. Every single seed put back in the soil seed bank is an investment for the future. Better returns than Wall Street.

So now it's time for us to sleep and rest for the winter. The sun is flopped over and starting to snore. Our grain bins, the soil of your fields, are bulging with early-risers, resistors, laybys and post-harvest midgets. Life is good. We owe it all to you. Your eyes that are only able to see fields and crop rows, almost never down on your hands and knees looking were we hide from you. We owe it your laziness, the gigantic size of your farms, your inability to get off your backside and out of your air-conditioned tractors, your deep abiding belief in herbicide and trans-genic seed ads on the winter TV, your goverment subsidies and cheap grain prices. Inconstant, unattentive, greedy, spoiled and self-satisfied. Please, don't ever change. We love you, and accept you, just as you are. You're special.

Thank you humans. We couldn't have done it without you.


**Summary**. Life history from embryo fertilization to seedling emergence was seen in chapters 7 and 8 from the perspective of the individual seed-plant, outside the influence of the local community. With the emergence of the seedling autotrophic growth begins and individual life history is increasingly affected by neighbors and community interactions. Unit 4 on weedy adaptation in the local community continues appropriately in the context of neighbors and community. Adaptation to life with neighbors in a local plant community, and expression of life history traits by the individual phenotype, is highly variable and complex.

The goal of chapter 9 is to explain neighbor interactions that define the local plant community as a complex and adaptive system. Biological interactions occur between species in a local community. The nature of neighbor interactions is found in patterns of interference and synergy. Interference and facilitation animate the strategic roles enacted by individual plants and species in the community in the struggle for existence. Neighbor interactions are based on functional traits, behavioral roles of individual weed species. Interfering with neighbors involves foraging, 'reaching out' to seize and exploit local opportunity spacetime. Interactions in crowded communities has effects on plant growth and plant form. Demographic descriptions of growth, form and mortality can provide a quantitative basis for understanding the nature of weeds in local communities.

In chapter 10, the nature of weeds is seen as the consequence of adaptive evolution resulting in enduring biodiversity, community structure and behavioral dynamics caused by seizing and exploiting local opportunity. Weed biodiversity and population structure predicate community dynamics. Weed population biodiversity has structure, changing with time in response to the adaptive culminations in the community. Genetic connections, gene flow, between individuals, populations and species exist that provide advantages in the struggle to exploit local opportunity: weed species associations. Weed community population structure reveal larger scale phenomena emerging from smaller scale community structure: behavioral traits assume a role, forming a functional guild based on clusters of interacting traits that allow them to dominate. Weed community dynamics is the formation of a local plant community, including the processes of community assembly and ecological successsion. Biodiversity is the source of all future local weed populations. Weed biodiversity is the pool of potential candidate populations that might invade, seize and exploit local agricultural opportunity. Weed biodiversity in the community has been linked with the concepts of stability and equilibrium.

## Chapter 9: Neighbor Interactions in the Local Plant Community

**The weeds' story**. I am foxtail.

When I appear in your fields every spring I am very happy. No longer confined to the soil, no longer oppressed by my parents. Free to grow in the warm spring sun. Of course this doesn't last very long. As if you humans and your big hands and machines weren't problems enough. I have to put up with all the riff-raff neighbor crops and weeds too. They are up early every morning at dawn making a commotion. All day long trying to eat my food and steal my place in the world. Up late planning the next day's trouble no doubt. But I put up with it. I am not just sitting around taking the abuse without a fight. Some of us have learned to steal the good stuff better than those incorrigible neighbors. Some of us have even learned to get along together with them. It is especially nice when our cousins come and hang out with us for the summer. The big guy-girl giant foxtail especially loves extra human-made nitrogen. The yellow foxtail girl-guy does a bit better resting a long time in the soil, and when it gets dry. The green foxtail guy-girl is a plant for all seasons. I love it when the neighbor's dog gets snagged by the reverse barbs on the girl-guy bristly foxtail seed in the autumn. We tend to gang up on the other weeds and crops in the fields. They call us 'bully'. Well, bully for us!

Then along come you human scientists with your science and grad students. You think that counting us will reveal all our trade secrets. Ha! Don't spend too long counting any one time, you'll miss all the action. Which you do. We could care less. All it takes is one of us to make it to the finish line in the autumn. Then that single little plant you so carefully counted starts producing thousands of babies, drenching the open harvested fields with seeds. And ruining your additive hypotheses with our non-linear plasticity. We are ready for another winter knowing you haven't a clue of how we do it. Year after year, millenia after millenia. We owe it all to you, your pride, your confused science, your heavy-handed farmers.

Thanks for the ride!

**Summary**. The assembly and formation of the local weed-crop community is a consequence of seedling/bud emergence and the subsequent life history interactions among neighboring organisms. The goal of this chapter is to explain several crucial aspects of neighbor interactions that define the local plant community as a complex and adaptive system; no single manner of representation can explain the dynamic consequences.

Biological interactions are the relationships between species in a local community. They can be categorized based either on the effects or on the mechanism of the interaction. The nature of neighbor interactions is found in patterns of interference (competition, amensalism, antagonism) and synergy (mutualism, commensalism).

Neighbor interactions are based on the functional traits, and behavioral roles of individual weed species in the community. Interfering with neighbors involves the strategic roles of foraging (genetic-phenotypic; spatial), the 'reaching out' of a plant to seize and exploit local opportunity spacetime. Strategic roles and traits provide a qualitative basis to understand the evolving adaption of changing populations in the community.

Interactions in crowded communities has effects on plant growth and plant form. A way of explaining the nature of community interactions is a demographic description of the effects of neighbors on plant density, growth, form, reproduction and mortality. Demographic descriptions of growth, form and mortality can provide a quantitative basis for understanding the nature of weeds in local communities, but inferences derived from demography are confounded.

## 9.1 Adaptation to neighbors in the community

The assembly and formation of the local weed-crop community is a consequence of seedling/bud emergence and the subsequent interactions among neighboring organisms that determine the life history of a weed from seedling to flowering plant. The goal of this chapter is to explain several crucial aspects of neighbor interactions that define the local plant community as a complex and adaptive system. Neighbor interactions in the weed-crop community are complex, and no single manner of representation can explain the dynamic consequences. This chapter is divided into four parts to emphasize the different ways of understanding the vegetative weed plant and its interactions with neighbors. The first explores the nature of neighbor interactions as these roles are performed, the patterns of inference and synergism among community members. The second part describes the roles individual weed species play in the community, and the traits that underlie these behavioral roles. The third section concerns spatial and temporal foraging for local opportunity in a habitat. In the last part, the effects of interactions in crowded communities on plant growth and plant form are presented.

**9.1.1 Parental control ends, plant growth resumes.** Beginning with fertilization and the formation of the zygote, development was controlled by the parent plant nuturing the developing embryo. With abscission and dispersal the offspring began life as an separate organism in which little apparent growth occurred due to restrictions placed on germination by the accompanying parental tissues (e.g. seed envelopes). The seed or bud spent most of its time maintaining itself while the soil environment induced changes in its metabolic state in preparation for its roles of survival and resumption of growth when conditions were favorable to subsequent reproduction. Seedling emergence is the resumption of growth and development of the individual weed plant leading to reproduction.

**9.1.2 Recruitment: community assembly.** Community assembly is colonization of a local opporunty spacetime (see ch. 3.5, the nature of plant invasions of local opportunity). The local weed-crop community begins assembly with seed/bud recruitment (see ch. 8.4, propagule germination and recruitment). Assembly continues with the interaction with neighbor plants and other organisms in the local community. With time ecological succession proceeds, with earlier successional species facilitating invasion of later successional species by the opportunity spacetime they create in the habitat (see ch. 10.4, weed community dynamics: community assembly and ecological succession). Community assembly ends with the terminal phases of ecological succession in the fully mature community, or extinction. The driving forces of Earth thermodynamics ultimately determine community opportunity spacetime, structure and assembly (see ch. 11.1.8, plant community structure and assembly)

With seedling emergence the local weed-crop community begins to be assembled. With community assembly begins interaction with neighbor plants and other organisms in the local community.

**neighborhood**:
1: used in population genetics for the area around an individual within which the gametes that produced the individual can be considered to have been drawn at random (Lincoln et al., 1998)
2: the plants in an area adjacent to an individual target plant that exert a competitive influence (Silvertown and Charlesworth, 2001)
3: the plants in an area adjacent to an individual target plant that exert a facilitating, neutral or competitive influence

**community**
1: any group of organisms belonging to a number of different species that co-occur in the same habitat or area and interact through trophic and spatial relationships; typically characterized by reference to one or more dominant species (Lincoln)
2: in ecology, an assemblage of populations of different species, interacting with one another; sometimes limited to specific places, times, or subsets of organisms; at other times based on evolutionary taxonomy and biogeography; other times based on function and behavior regardless of genetic relationships (Wikipedia, 6.08)
3: in ecology, assemblages or associations of populations of two or more different species occupying the same geographical area and in a particular time; in its simplest form, groups of organisms in a specific place or time (Wikipedia, 6.14)

**priority effect**: in ecology, the impact that a particular species can have on community development due to prior arrival at a site:
> **inhibitory priority effects** occur when a species that arrives first at a site negatively impacts a species that arrives later by reducing the availability of space or resources
> **facilitative priority effects** occur when a species that arrives first at a site alters abiotic or biotic conditions in ways that positively impact a species arriving later

The most typical community that weeds invade is the agricultural community. The hallmark of agricultural production fields is the elimination of all plants that might compete with the annual crop being established. As such weeds find enormous opportunity (priority effect) to seize and exploit in these managed, fertile, habitats. Neighbors begin with the crop species, and include other competing weed species as the season progresses (see ch. 8.4.5.5, relative emergence order). Crop fields are the beginning, the starting point, of plant community ecological succession, see ch. 10.4.2.2.

Assembly, emergence, of weed species in agro-communities is based on the plant species composition of a seed pools in the soil underlying the local field (see ch. 8.3.3.2, floral seed community composition).

Floral composition of soil seed pools in agro- and other eco-systems is typically a community composed of a majority subset of dominant species, and subsets of minority species. This results in a skewed distribution of species, a hierarchy of large numbers of a few dominant species and smaller numbers of other minority species. This observation of skewed distributions consisting of numbers of dominant and minority species is a common theme in the mating systems of weeds, mostly observed in herbaceous annuals, but also in a few perennial species. This theme of skewed distribution in species composition is reflected also in skewed competititive interactions in plant communities that results in skewed distributions of individual plant size within a species in a locality: a few big individuals, and a large number of small individuals.

The species that are dominant or in a minority in any particular soil seed pool varies as function of the qualities and environment of that locality. What is most typical is that this skewed hierarchy of dominant sized and numbers exists across such a wide range of agricultural and disturbed localities. For example, in Iowa, USA corn fields the dominant grassy weeds include foxtails (*Setaria* spp.-gp.), bromegrass (*Bromus* sp.), woolly cupgrass (*Eriochloa villosa*), and yellow nutsedge (*Cyperus esculentus*; sandy soils, river bottom soils). Dominant dicot weeds include common lambsquarters (*Chenopodium album*), velvetleaf (*Abutilon theophrasti*), common cocklebur (*Xanthium strumarium*), smartweeds (*Polygonum* spp.; seasonal, depends on year), pigweeds (*Amaranthus* spp.-gp.; including waterhemp, common north of I-80; tall waterhemp south of I-80), giant ragweed (*Ambrosia trifida*), and common ragweed (*Ambrosia artemisiifolia*). Minority species in Iowa are found either low numbers in lots of fields, or dominants in a few number of fields. They include the grassy weeds wild proso millet (*Panicum mileaceum*) and other *Panicum* sp., barnyardgrass (*Echinochloa crus-galli*), shattercane (*Sorghum bicolor*). Minority dicot weeds in Iowa include prickly sida (*Sida* sp.; sandy soils), shepard's purse (*Capsella bursa-pastoris*), mustards (e.g. *Brassica* spp.), sunflowers (*Helianthus* spp.), milkweeds

(*Asclepias* spp.), hemp dogbane (*Apocynum cannabinum*), curled dock (*Rumex crispus*), nightshades (*Solanum* spp.). In Iowa volunteer corn (*Zea mays*) has been a minority species until recently. In the past proper harvest combine setting has prevented the return of seed to the field. With the advent of herbicide resistant crops volunteer corn has become a major weed infesting herbicide resistant soybeans (*Glycine max*; Dekker and Comstock, 1992). Species composition changes with a changes in tillage from moldboard plowing to no-till favor surface germinating species (light-requiring, light/nitrate requiring, oxygen stimulated), winter annuals (e.g. shepards purse; *Capsella bursa-pastoris*), perennials, common lambsquarters (*Chenopodium album*; an early emerger) and the pigweeds (*Amaranthus* spp.-gp.).

**9.1.3 Interaction with neighbors is harsh**. Once the weed seeding has emerged from the soil and commenced photosynthesis, autotrophic growth, its life becomes markedly different. In the beginning relatively few interactions with neighbors occurs among uncrowded seedlings. With time they begin to interact with nearby neighbor plants: siblings of the same species, other weed species and crops. Once established, interactions with neighbors will dominate the life history of the weed plant until it dies or reproduces. The struggle to survive, grow, and reproduce increases as they compete with neighbors for limited opportunity spacetime. Crowding in the locality affects plant growth, development and plant form. The presence of neighbors acts as a feedback influence on itself as well as on subsequent seedling recruitment (e.g. shading). Life in the community is harsh.

## 9.2 The nature of neighbor interactions in the community

"Plants may modify the environment of their neighbors by changing conditions, reducing the level of available resources or by adding toxins to the environment. All these effects can be shown quite clearly to operate in experiments with artificial populations, but there are probably no examples of plant interactions in the field in which the mechanism has been clearly and unambiguously demonstrated. Interactions dominate most situations that have been analysed in the field and the search for unique factors of competiton may not be very sensible. Death or a reduced growth rate are often attributable to competive interactions, but interpretation of competition in the field may depend on the recognition of much more specific symptoms." (Harper, 1977: summary, p.xvii-xviii; Ch. 11: pp. 347-381).

Biological interactions are the relationships between species in an ecosystem (Wikipedia, 12.10). These relationships can be categorized into many different classes of interactions based either on the effects or on the mechanism of the interaction. The interactions between two species vary greatly in these aspects as well as in duration and strength. Species may meet once in a generation (e.g. pollination) or live completely within another (e.g. endosymbiosis). Effects may range from one species eating the other (predation), to mutual benefit (mutualism). The interactions between two species need not be through direct contact. Due to the connected nature of ecosystems, species may affect each other through intermediaries such as shared resources or common enemies.

### 9.2.1 Patterns of neighbor interactions

**9.2.1.1 Causation and complexity**. Interactions dominate most field situations, and the search for unique factors of competiton (beyond death or reduced growth) may not be possible. Plants modify the environment of their neighbors by changing conditions, reducing the level of available resources or by adding toxins to the environment. These effects can be demonstrated in artificial populations, but demonstrating them in the field has not been successful. Most of these interactions occur through some aspect of an intermediating factor (e.g. resources, conditions, microorganisms, pollinators, herbivores). These interactions can be dissected into the effect each of the species has on the abundance of the intermediary, and its response to an increase in the abundance of the intermediary (Goldberg, 1990).

Cause and effect is nearly impossible to correlate in neighbor interactions. Attempts to understand the nature of weed-crop interactions is difficult, if not impossible. Experimentally we attempt to ascribe the success of one weed over another, or over a crop, to a particular morphological feature, a particular pattern of life cycle, or a simple physiological trait. Comparisons of particular factors influencing the interaction of species is very difficult, and will usually not reveal the nature how they affect each other. Correlation between individual factors is the best we often do experimentally. Differences between plants and species include small and large contributions, the size of those factors is often hard to determine. The summation of the contributing factors to a plant-plant interaction is often greater than the individual components: synergistic phenomena emerge as a non-linear consequence of their mutual interaction. The methodological problems of evaluating the nature of plant-plant interactions is revealed by the problems posed in answering component questions, each themselves involving complex interactions. Does the interaction take place above or below the soil surface? Does the interaction take place because a plant changes the environment of its neighbors? Does the interaction take place because a plant deprives its neighbors of resources? Does the interaction take place because a plant produces toxic chemicals (or toxic chemical conditions) that harm or kill its neighbors? Conversely, does it take place because a plant can resist or tolerate a toxic chemical?

**9.2.1.2 Modal patterns of neighbor interaction in the community**. It is possible is to organize interactions between weeds and crops in a local field into several general categories of observable patterns of coexistence, the effects neighbors have on each other. These categories are based on the effects caused by the interaction and explicitly indicate the quality of benefit or harm in terms of fitness experienced by its participants. There are several possible combinations, ranging from mutually beneficial through neutral to mutually harmful interactions. The level of benefit or harm is continuous and not discrete, such that an interaction may be trivially harmful to deadly. It is important to note that these interactions are not always static. In many cases, two species will interact differently under different conditions. This is particularly true in, but not limited to, cases where species have multiple, drastically different life stages.

The presence of a plant changes the environment of its neighbor, affecting the growth and form of both. Neighbor interactions can take many modal forms, including those of interference and synergy. The variety of patterns of interactions between plants is based on the outcomes of their coexistence in a local community. In simplified form, there are three possible outcomes for an individual species in a mixture: each may gain (+), lose (-) or be unaffected (0). In a two-species mixture both populations may be depressed (- -) (competition), one increase while the other loses (+ -) (antagonism: parasitism, predation, pathogenesis), both may benefit (+ +) (mutualism); the interaction may be neutral for one species and beneficial (+ 0) (commensalism) or detrimental (- 0) (amensalism) to the other; it is also possible for each to be unaffected by the other (0 0) (neutralism). These possible outcomes are presented in figure 9.1.

**9.2.1.3 Neutralism**. Neutralism describes the relationship between two species which interact but do not affect each other (Wikipedia, 12.10). It describes interactions where the fitness of one species has absolutely no effect whatsoever on that of the other. True neutralism is extremely unlikely or even impossible to prove. When dealing with the complex networks of interactions presented by ecosystems, one cannot assert positively that there is absolutely no competition between or benefit to either species. Since true neutralism is rare or nonexistent, its usage is often extended to situations where interactions are merely insignificant or negligible.

**9.2.2 Interference interactions between neighbors**. When neighbors interfere with each other one or both lose. In competitive interactions both lose. In amensal interactions one loses while the other remains unaffected (the amensal). In parasitism one loses and the other gains (the parasite). There are many ways in which these outcomes can come about between neighbors. Interference can be direct (allelopathy, parasitism) or indirect (preferential and early nitrogen absorption). Competitive ability can accrue to the species with superior growth rates, or photomorphogenic internode elongation for enhanced height to overtop neighbors. Interference includes strategies and traits of aggression and offense: luxury nutrient consumption by common lambsquarters (*Chenopodium album*), allelopathy, parasitism by dodder (*Cuscuta* spp.) or *Orabanche* and changes in conditions (e.g., protection from wind, behavior of predators). Resource interference for light includes reducing light intensity and changing light quality. Water interference includes affects on transpiration of limited water and changing the humidity profile. Nutrient interference includes preferential absorption of limited nutrients; sequestering limited nitrogen. Gaseous interference includes reducing $CO_2$ or $O_2$ levels, or their ratios, in the local atmosphere. Interference by modifying pervasive conditions include those to heat, including altering the temperature by protection and exposure. See also ch. 3.4, limiting resources and pervasive conditions in local opportunity.

**Figure 9.1** Two-neighbor interaction space and patterns of mutual effects; neighbors: species 1, 2; interactions: gain (+), unaffected (0), lose (-); interaction effect patterns: facilitation (mutualism, commensalism), neutralism, or interference (antagonism, amensalism, competition)

|  | FACILITATION | | INTERFERENCE | | |
|---|---|---|---|---|---|
| **FACILITATION** | $+_1 +_2$ mutualism | $+_1 \ 0_2$ commensalism | $+_1 -_2$ antagonism | + | |
|  | $0_1 +_2$ commensalism | $0_1 \ 0_2$ neutralism | $0_1 -_2$ amensalism | 0 | Species 1 |
| **INTERFERENCE** | $-_1 +_2$ antagonism | $-_1 \ 0_2$ amensalism | $-_1 -_2$ competition | - | |
|  | + | 0 | - | | |
|  | | Species 2 | | | |

**9.2.2.1 Competition**. Many of the patterns of neighbor interaction are sometimes lumped together into the generic category of competition. The definitions of competition vary depending on whether mechanism of competition or the demographic outcome of neighbor interaction is emphasized.

**competition**:
1: the tendency of neighboring plants to utilize the same quantum of light, ion of a mineral nutrient, molecule of water, or volume of space (Grime, 1979)
2: an interaction between species in a mixture in which each lowers the net reproductive rate of the other (Silvertown and Charlesworth, 2001)
3: the simultaneous demand by two or more organisms or species for an essential common resource that is actually or potentially in limited supply (exploitation competition), or the detrimental interaction between two or more organisms or species seeking a common resource that is not limiting (interference competition) (Lincoln et al., 1998)
4: a mutually detrimental interaction between individuals, populations or species (Wikipedia, 12.10)

As humans we often project our interactions onto plants and conceptualize competition as a contest with a winner and a loser. Competition is almost never a fight, it is a differential, detrimental, seizure and exploitation of opportunity at a particular place and time: utilizing unused resources, accomodating to pervasive conditions, escape-avoiding disturbances. Because communities become crowded quickly with excess members, competition results inevitably with loss by both competing species to differing extents.

#### 9.2.2.2 Amensalism.
It is useful to separate competitive interactions in which both neighbors lose something to those in which one neighbor loses and the other remains unaffected, amensalism.

**amensalism:**
1: an interspecific interaction in which one organism, population or species is inhibited, typically by toxin produced by another (amensal), which is unaffected (Lincoln et al., 1998)
2: an interaction between two species in which one impedes or restricts the success of the other while the other species is unaffected; a type of symbiosis (Wikipedia, 12.10).

Amensalism can be achieved by preferential tolerance to environmental poisons (e.g. differential herbicide resistance; tolerance to allelochemicals) and by growth form (vines exploiting light to the detriment of their architectural hosts), to name but two of many possibilities.

#### 9.2.2.3 Antagonism.
In antagonistic interactions one species benefits at the expense of another: parasitism, predation and pathogenic relationships. Viney plant growth can enable one species to utilize the upright plant architecture of another to gain light while shading the support plant.

#### 9.2.2.3.1 Parasitism.
Parasitism is the form of neighbor interaction in which one organism gains and the other loses.

**parasitism:** an obligatory symbiosis between individuals of two different species, in which the parasite is metabolically dependent on the host, and in which the host is typically adversely affected by rarely killed (Lincoln et al., 1998)

Good examples of plant parasites include dodder (*Cuscuta* spp.) and *Orabanche* spp.

#### 9.2.2.3.2 Predation.
Interactions in which one plant species (weed or crop) loses and another, non-plant, organism gains is found in neighbor interaction patterns of predation by predator or pathogen. Predation is the extreme, fatal form of parasitism in which the host is killed. Parasites feed on living hosts.

**predation:**
1: a biological interaction where a predator (an organism that is hunting) feeds on its prey (the organism that is attacked); predators may or may not kill their prey prior to feeding on them, but the act of predation always results in the death of its prey and the eventual absorption of the prey's tissue through consumption (Begon et al., 1996)
2: an interaction between organisms in which one organism captures biomass from another; widely defined it includes all forms of one organism eating another, regardless of trophic level (e.g. herbivory), closeness of association (e.g. parasitism, parasitoidism) and harm done to prey (e.g. grazing) (Wikipedia, 12.10)

Some examples of predation include phenotypes, and extended phenotypes, that: shelter or exclude predators (or shelter predators of predators); favor or reduce pathogenic activity; encourage defecation or urination that attracts predators into the neighborhood; providing morphological rubbing posts or play objects and so encouraging local trampling.

**predator**
1: the consumption of one animal (the prey) by another animal (the predator) (carnivore)
2: also used to include the consumption of plants by animal (herbivore), and the partial consumption of a large prey organism by a smaller predator (micropredation) (parasites; parasitoids)

Examples of animal predators includes birds eating weed seeds, grazing on weed vegetative and seed tissues (cows, insects), tapeworms, lamprey eels. Susceptibility to predation lowers fitness, resistance to predation increases fitness. There exists a dynamic interaction between individual susceptibilities: changes in resistance (or susceptibility) brings about changes in the forces of natural selection acting on the other.

**9.2.2.3.3 Pathogenic interactions.** Closely related to parasitism and predation is pathogenesis and pathogenic interactions among neighbors.

**pathogen**: parasite which causes disease

**pathogenesis** : the course or development of a disease disease: any impairment of normal physiological function affecting all or part of an organism, esp. a specific pathological change caused by infection, stress, etc. producing characteristic symptoms

Examples of pathogens includes soil microorganisms affecting roots and seeds, and viral or bacterial pathogenic infestations on aerial shoots.

**9.2.2.3.4 Coevolution in parasitism, predation and pathogenesis.**

> "An individual plant that is more prone than its neighbor to attack by pathogens and/or predators will almost inevitably have lowered fitness. The variety of ways in which a host may gain resistance, immunity or a degree of tolerance to such an attack immediately defines equivalent ways in which the predator or pathogen might counter the defence. Just as in the coevolution of competing species at the same trophic level, any change in one component immediately changes the selective forces acting on the other." Harper (1977; p. 768-9)

Pathogen-host interactions have been described as an example of an evolutionary 'arms race'. This race has been compared to that of the Red Queen in Lewis Caroll's "Through the looking-glass and what Alice found there" (Dodgeson, 1871):

> "Alice never could quite make out, in thinking it over afterwards, how it was that they began: all she remembers is that they were running in hand, and the Queen went so fast that it was all she could do to keep up with her: and still the Queen kept crying, 'Faster! Faster!' The most curious part of the thing was, that the trees and the other things round them never changed their places at all: however fast they went, they never seemed to pass anything."
> "… said the Queen. 'Now *here*, you see, it takes all the running *you* can do, to keep in the same place. If you want to get somewhere else, you must run at least twice as fast as that!'"

In this story race the Red Queen needs to keep running just to stay in same place and not fall behind (Ridley, 1995). It is the basis of the Red Queen principle:

**Red Queen principle**: for an evolutionary system, continuing development is needed just in order to maintain its fitness relative to the systems it is co-evolving with (Van Valen, 1973)

The evolutionary relationship of interacting neighbors involves the co-evolution of mechanisms of defense and offense; each elicits response, a new trait to counteract. An apt example began with the introduction of hybrid maize (*Zea mays*) seed in the 20th century, which led to vast improvements in crop productivity and an increase in individual USA farm size. The opportunity to farm larger land areas by a single manager increased after World War 2 with the introduction of the first effective selective herbicide, 2,4-D. Grassy weeds as giant foxtail (*Setaria faberi*) appeared suddenly in midwestern USA crop fields, while susceptible dicot weeds succumbed to 2,4-D. Velvetleaf (*Abutilon theophrasti*) appearance was associated with the vast expansion of soybean (*Glycine max*) production in the USA. This new weed invasion was enabled by the introduction of dinitroaniline herbicides (e.g. Treflan™) in soybean production fields. These weed population shifts in dominant weed species were an early warning of impending change in response to new technological introductions.

The hallmark of the 1950-1980's was the discovery, development and introduction of many new herbicides that vastly increased the selective pressure on weed populations (e.g. atrazine). For a while it looked like humans would eliminate weeds. The resulting response by weed populations was the selection and increased frequency of herbicide resistant weed biotypes. There appeared new and different herbicide resistant species and biotypes with previously unknown detoxification and binding site resistances. In some cases biotypes possessed multiple, preadapted, resistance mechanisms. For a while it looked like the weeds would win (e.g. the Australian experience).

This natural selection-elimination feedback loop that encourages the appearance of herbicide resistance in weeds in turn elicited a response by industrial agriculture to develop and introduce herbicide resistant crops (Dekker and Comstock, 1992). Weed populations then escalated their response by, again, the selection and increased frequency of herbicide resistant weed biotypes in transgenic crop fields. New modes of resistant action were discovered in these new weed biotypes, once again. An indirect consequence of this arms race was the elimination of many unsuccessful herbicide companies and the consolidation of successful companies. Discontinuation of many less effective herbicides relieved many weed populations from the selective pressure of these older chemicals. The increased success and capitalization of the remaining herbicide industry led inexorably to the acquisition of partner seed companies and the monolithic consolidation of the industrial agricultural production enterprise. Currently the 'arms race' continues with consolidation of the agricultural production industries with the food processing and distribution industries in the USA. Undetered by human activity, hubris and capitalization, weed populations continue apace (see unit 2, the evolution of weed populations).

This example of the 'Red Queen principle' is an excellent demonstration of the extended phenotype concept (Dawkins, 1999; see ch. 4.1.1, extended phenotype). Is not the concept of phenotype "extended to include functionally important consequences of gene differences, outside the bodies in which the genes sit ... where the effects influence the survival chances of the gene ..." when observing both weed and human responses to technological innovation?

**9.2.3 Facilitative interactions between neighbors**. Facilitation describes species interactions that benefit at least one of the participants and cause harm to neither (Stachowicz, 2001). Facilitations can be categorized as mutualism (both species benefit) or commensalism (one species benefits, the other is unaffected). Facilitative interactions are cooperative processes.

**cooperation**: the process of working or acting together, intentional and non-intentional; acting in harmony, side by side, or in more complicated ways; the alternative to working separately in competition

Humans benefit from interactions with local ecosystems in a multitude of ways. Collectively, these benefits are becoming known as ecosystem services. Ecosystem services are regularly involved in the provisioning of clean drinking water and the decomposition of wastes. While scientists and environmentalists have struggled with the concept of ecosystem services for decades, the ecosystem services concept itself was popularized by the Millennium Ecosystem Assessment group (MA; 2005) in the early 2000s. Ecosystem services were grouped into four broad categories: provisioning, such as the production of food and water; regulating, such as the control of climate and disease; supporting, such as nutrient cycles and crop pollination; and cultural, such as spiritual and recreational benefits. The concept of ecological services is fraught with anthrocentric value systems, often politicized by assigned economic values to ecosystem services to help inform decision-makers. The scientific benefits of monitizing ecosystems is not apparent.

**9.2.3.1 Mutualism.** A mutualism is an interaction between species that is beneficial to all participants.

**mutualism**: a symbiosis in which both organisms benefit, frequently a relationship of complete dependence; interdependent association; synergism (Lincoln et al., 1998)

An example of a mutualism is the relationship between flowering plants and their pollinators (Boucher et al., 1982; Callaway, 1995). The plant benefits from the spread of pollen between flowers, while the pollinator receives some form of nourishment, either from nectar or the pollen itself.

**9.2.3.2 Commensalism.** A commensalism is an interaction in which one species benefits and the other species is unaffected.

**commensalism**: symbiosis in which one species derives benefit from a common food supply whilst the other species is not adversely affected (Lincoln et al., 1998)

Epiphytes (plants growing on other plants, usually trees) have a commensal relationship with their host plant because the epiphyte benefits in some way (e.g., by escaping competition with terrestrial plants or by gaining greater access to sunlight) while the host plant is apparently unaffected (Callaway, 1995).

## 9.3 Strategic roles and traits of interference and facilitation with neighbors

The adaptive roles individual weed species play in a local community are the evolutionary consequences of complex interactions with other weed and crop plants in local habitats. This section is about the strategic roles individual weed species play in the community, and the traits that underlie these behavioral roles (see ch. 10.4.1, ecological roles-guilds-trades in weed-crop plant communities). Weedy adaptation to neighbor interactions in the local community is the consequence of the three overlapping processes of weed life history, natural selection and invasion biology (table 9.2).

**Table 9.2** Neighbor interactions in vegetative growth: processes of life history, natural selection and invasion biology underlying functional traits of weeds (see table 6.2).

| LIFE HISTORY | NATURAL SELECTION | INVASION BIOLOGY |
|---|---|---|
| **Vegetative growth:**<br>Neighbor Interactions:<br>-Growth-Development<br>-Stress Responses | Pre-Condition 1: excess local phenotypes compete for limited opportunity<br><br>Condition 3: survive to produce the fittest offspring | Colonization & Extinction |

These strategic roles can seen from the point of view of interference or facilitation. Both points of view are matters of degree and a role or trait for one type of interaction can be its apparent opposite depending on the extent/intensity of the behavior, and it's importance at that time in life history. Any interaction can turn from interference to facilitation depending on its effect at the time; these can switch back and forth over time in multiple species interactions. For example a little harmless epiphyte living on the branch of a large tree can be comensal; but an aggressive climbing, choking vine like kudzu can antagonistic in terms of light interference.

**9.3.1 Strategic roles and traits of interference with neighbors.** Interfering with neighbors involves the strategic roles of foraging for local opportunity in space and time.

**foraging:**
1: a manifestation of phenotypic plasticity (or behavior) that increases resource capture in an environment that is spatially heterogeneous for the resource (Silvertown and Charlesworth, 2001)
2: behavior associated with seeking, capturing and consuming food (Lincoln et al., 1998)

**9.3.1.1 Genetic foraging for local opportunity.** Harper (1977) has proposed seven categories of solution to the dilemma of maximizing fitness in a spatially and temporally variable physical-biological environment. These plant qualities can be seen as genetic foraging in space and time to seize and exploit opportunity. He posits five genetic and two phenotypic (gene expression) strategies (table 9.3).

**Table 9.3** Genetic-phenotypic strategic roles and traits to maximize fitness in a spatially and temporally variable habitat.

|  | Strategic roles | Traits |
|---|---|---|
| Genotypic | form locally adapted populations |  |
|  | develop genetic polymorphisms of linked, co-adapted genes | - seed dormancy<br>- phenotypic plasticity<br>- somatic polymorphism |
|  | maintain high population variance | - mating system control of variability<br>- formation of soil seed pools: genetic memory<br>- genetic continuity with species metapopulation |
|  | speciation; formation of breeding groups with species-group and wild-crop-weed groups | speciation by polyploidization |
| Phenotypic | somatic polymorphism | - leaf specialization<br>- seed heteroblasty |
|  | phenotypic plasticity | branching |
| Phenotypic-Genotypic | heterozygotic superiority: variation for changing habitats |  |

**9.3.1.1.1 Genotypic strategies.** Five genetic ways for confronting spatial and temporal diversity include several strategies, at increasing levels of variation.

1: Formation of local specialized races requires inbreeding for local adaptation in the deme.
2: Development of genetic polymorphism with large blocks of linked coadapted genes or supergenes: traits and poly-traits stay together (e.g. complex weed seed germination regulation)
3: Evolution of heterozygotic superiority when in succeeding generations the weed species is confronted by alternating conditions in the community and only the heterozygote can produce progeny with fitness in both conditions.
4: Maintenance of high degree of genetic variance within a population (mating system control of outcrossing rate, variation produced), coupled with longevity of soil seed banks providing memory of past adapted genotypes. This breeding continuity from the plant, to the soil seed pool, also extends to the species' metapopulation gene pool at greater spatial distances.

5: Act of speciation and the formation of distinct breeding groups that can evolve independently. These could include species-groups and wild-crop-weed groups, especially polyploidy speciation and specialization within the broader scale of the metapopulation.

**9.3.1.1.2 Phenotypic strategies.** Two phenotypic (gene expression) ways when confronting spatial diversity include somatic polymorphism and phenotypic plasticity (see ch. 4.3, generate phenotypic variation).

Development of somatic polymorphism is not a response to the local environment. It is where a single genotype develops a variety of phenotypes, each adapted to a different time or space phase of its life history in the community. For example, leaf polymorphisms between seasons or ages of a plant (different times, different needs); cotyledon, unifoliate, trifoliate leaves of soybean; basal rosette (autumn, winter) and elongated primary stem (spring) in winter annuals. For example, seed polymorphism (germination heteroblasty): different seed germination phenotypes shed by one plant into seed bank such as the seed dormancy phenotypes of the cocklebur (*Xanthium sp.*) seed capsule.

Evolution of phenotypic plasticity allows plant changes in response to local conditions. The individual responds directly to the conditions it experiences; gene expression is triggered directly by environment. Phenotypic plasticity is involved with many of the traits involved in performing strategic roles in neighbor interactions described below.

Functional traits fulfill the stragegic roles that have evolved over time for the local adapation of a species with neighbors. The genetic foundation/blueprint, and phenotypic expression, described above for these seven categories are applicable to both spatial and temporal foraging strategies utilized in interference and facilitation interactions with neighbors.

**9.3.1.2 Spatial foraging for local opportunity.**

"... it may be helpful to divide plant growth into two categories: (i) that which has been dominated in its evolution by selective pressures to attain height (and shade neighbours) which leads invevitably to the woody habit; and (ii) that which has been dominated by pressure to expand laterally (and reach limited water and nturients) – this leads equally strongly towards the lateral branching, node rooting habit of clonal species." (Harper, 1977; p.27)

Seed dispersal and germination mechanisms, as well as perennial weed ramet foraging, set the scale to a plant populations spatial heterogeneity. They are the means by which the plant "reaches out" to contact neighbors. They determine the selection pressures the plant will meet in performing strategic roles by means of spatial foraging traits (table 9.4). Seedling recruitment timing and location sets the stage, begins the drama of community assembly, and is developed more fully in ch. 8.4, propagule germination and recruitment.

**Table 9.4** Spatial interference with neighbors: strategic roles and example traits played in competitive, amensal and antagonistic interactions seizing and exploiting local opportunity.

| Strategic roles | Traits |
|---|---|
| Spatial foraging for capture of local opportunity | Annuals & perennials:<br>1. shoot foraging:<br>   -branching, culm tillering architecture<br>   -leaf number and placement<br>   -climbing/prostrate growth habit<br>2. root foraging:<br>   -root branching, architecture<br>   -distribution in soil space |

| | | |
|---|---|---|
| | | 3. whole plant:<br>  -root:shoot ratio partitioning<br>  -rapid growth: C3/C4/CAM leaf photosynthesis |
| | | Annuals:<br>1. seed dispersal mechanisms for location of seedling recruitment |
| | | Perennials:<br>1. genet-ramet independence/physiological dependence<br>2. bud dormancy-germinability<br>3. perinnating shoot (e.g. rhizomes) & rootstock geometry:<br>  -inter-bud length<br>  -branching<br>4. Biomass partitioning:<br>  -perinnating tissue<br>  -food reserve tissue accumulation<br>5. Seed-bud reproductive effort distribution |
| | Spatial offensive attack of neighbors | 1. allelochemical offense (self-tolerance)<br>2. climb on your neighbor (e.g. viney growth habit)<br>3. self-tolerant host of diseases for neighbors |
| | Preempt neighbors ability to spatially seize and exploit opportunity | Nutrients:<br>Luxury nutrient consumption<br><br>Water:<br>Consumption of limited water (e.g. root foraging)<br><br>Light:<br>Architecture, branch/tiller growth for canopy closure: shade your neighbor<br>1. photomorphogenic responses (e.g. internode elongation) |

Spatial foraging is the physical 'reaching out' of a plant to seize and exploit local opportunity. It begins with recruitment when the young seedling extends its axes into the atmosphere and soil and begins acquiring the benefits of that local opportunity: resources, conducive conditions, absence of disturbance and neighbors. Very soon plants contact each other, or begin influencing their neighbors and interfering roles are accomplished with specific traits of the members.

Different plant parts experience different environmental aspects of the habitat; for example, rhizomes and roots in the soil versus lower leaves and stems versus upper leaves and stems. Seed dispersal mechanisms determine the range of environmental heterogeneity that is sampled, e.g. most weeds are dispersed by gravity dispersal. This limited, narrow range of local dispersal may be appropriate for plants that do not need to waste seed exploring variable, unfavorable, nearby habitats. Long range wind dispersal can allow reaching out to a wider range of habitats. Mechanisms of pollen dispersal determines the range of environment that is sampled: pollen travels farther, a larger spatial genetic spread.

**9.3.1.2.1 Perennial bud foraging.** Perennial weed species forage with meristematic buds in space. This type of foraging begins with ramet extension by the parental plant, the genet, and dispersal of its buds in soil (see ch. 5.4.4.4, vegetative clone reproducing species).

A clonal colony or genet is a group of genetically identical individuals (e. g., plants, fungi, or bacteria) that have grown in a given location, all originating vegetatively (not sexually) from a single ancestor. In plants, an individual in such a population is referred to as a ramet. In fungi, "individuals" typically refers to the visible fruiting bodies or mushrooms that develop from a common mycelium which, although spread over a large area, is otherwise hidden in the soil. Clonal colonies are common in many plant species. Although many plants reproduce sexually through the production of seed, some plants reproduce by underground stolons or rhizomes. Above ground these plants appear to be distinct individuals, but underground they remain interconnected and are all clones of the same plant. However, it is not always easy to recognize a clonal colony especially if it spreads underground and is also sexually reproducing (Wikipedia, 5.08).

**9.3.1.2.1.1 Clonal foraging geometry**. Clonal growth can produce linear arrangements, networks or clumped distributions of ramets depending upon the length of connections (or inter-bud spacers) between meristematic nodes/buds, and the frequency of branching (Silvertown and Charlesworth, 2010. There exists a continuum of growth form from the solidly advancing front of ramets, called a *phalanx*, to the production of widely spaced ramets that infiltrate the surrounding vegetation in the *guerilla mode*. The explanation for the variation in clonal morphology probably lies in the benefits and costs associated with two functional aspects of the clonal growth habit: the degree to which ramets are physiologically dependent on each other; and, the degree to which clonal growth helps plants forage. Physiological integration appears to be the rule for clonal plants in nutrient-poor environments (Jonsdottir and Watson, 1997).

**9.3.1.2.1.2 Foraging behavior**. The definition of foraging above includes behaviors with two distinct functions. A *searching function* is triggered in resource-poor patches and improves the probability of encountering a new resource patch. A *holding function* is triggered in resource-rich patches and which improves their utilization and exploitation once found. Many of these foraging behaviors is not limited to clonal plants, they apply to parts of other types of plants, e.g. leaves seeking available light or soil resources. Both the searching and holding functions may be potentially filled by changes in inter-node spacer length affecting whether ramets are locally concentrated or dispersed from the current location, and the holding function by an increase in ramet size or number of rich patches through root and/or shoot proliferation and branching.

**9.3.1.2.2 Offensive attack of neighbors**. Interfering with neighbors includes fulfilling strategic roles with traits for spatial offensive attack: self-tolerant plant production of allelochemicals for offense of neighbors; growth habit supported on neighbor for light capture (e.g. viney habit; *Pueraria* sp. (kudzu), *Vitis* (wild grapes), *Convolvulus arvensis* and *Calystegia sepium* (morning glories), *Fagopyrum esculentum* (wild buckwheat)); self-tolerance of a plant host to diseases for infection of neighbors.

**9.3.1.2.3 Preempt neighbors ability to seize and exploit opportunity**. Interference includes fulfilling strategic roles with traits preempting neighbors from seizing and exploiting local spatial opportunity. Preemption of neighbors includes early growth and development seizing nutrients (luxury consumption), water by root growth monopolizing the soil profile, plastic plant growth of branch and tiller architecture supporting leaf canopy to shade neighbors (reduce photosynthesis, prevent light-stimulated seedling recruitment); photomorphogenic responses such as internode elongation for height to overtop neighbors and preempt light (e.g. see ch. 19, *Abutilon theophrasti* (velvetleaf)).

**9.3.1.2.4 Plant movements for spatial foraging**. Spatial foraging for local opportunity by sessile plants, stationary organisms rooted to a location, involves movement. These movements can be classified as tropic or nastic depending on their directionality in response to stimuli.

**9.3.1.2.4.1 Plant tropisms**. Plant tropisms are phenomena in which plants respond to stimuli in their local environment. A tropism (from Greek τρόπος, tropos, "a turning") is a biological phenomenon, indicating growth or turning movement of a biological organism, usually a plant, in response to an environmental stimulus (Wikipedia, 2.15). In tropisms, this response is dependent on the direction of the stimulus (as opposed to nastic movements which are non-directional responses). Tropisms are usually named for the stimulus involved (e.g. phototropism, reaction to light) and may be either positive (towards the stimulus) or negative (away from the stimulus). The Cholodny–Went model, proposed in 1927, is an early botanical model describing tropism in emerging shoots of monocotyledons, including the tendencies for the stalk to grow towards light (phototropism) and the roots to grow downward (gravitropism). In both cases the directional growth is considered to be due to asymmetrical distribution of auxin, a plant growth hormone

**9.3.1.2.4.1.1 Light tropisms**. Tropic responses to light include phototropism, heliotropism and the exotropic phenomena of photoperiodism and chronobiolgical rhythms.

**Phototropism** is the movement or growth of organisms in response to lights or colors of light. The cells on the plant that are farthest from the light have a chemical called auxin that reacts when phototropism occurs. This causes the plant to have elongated cells on the farthest side from the light. Growth towards a light source is called positive phototropism, while growth away from light is called negative phototropism. Most plant shoots exhibit positive phototropism, and rearrange their chloroplasts in the leaves to maximize photosynthetic energy and promote growth. Roots usually exhibit negative phototropism, although gravitropism may play a larger role in root behavior and growth. Some vine shoot tips exhibit negative phototropism, which allows them to grow towards dark, solid objects and climb them. The combination of phototropism and gravitropism allow plants to grow in an appropriate direction. Phototropism in plants is affected by the wavelength of the light source, and is directed by blue light receptors called phototropins. Other photosensitive receptors in plants include phytochromes that sense red light and cryptochromes that sense blue light. Different organs of the plant may exhibit different phototropic reactions to different wavelengths of light. Stem tips exhibit positive phototropic reactions to blue light, while root tips exhibit negative phototropic reactions to blue light. Both root tips and most stem tips exhibit positive phototropism to red light. Cryptochromes are photoreceptors that absorb blue/UV-A light, and they help control the circadian rhythm in plants and timing of flowering. Phytochromes are photoreceptors that sense red/far-red light, but they also absorb blue light. The combination of responses from phytochromes and cryptochromes allow the plant to respond to various kinds of light. Together phytochromes and cryptochromes inhibit gravitropism in hypocotyls and contribute to phototropism.

**Heliotropism** is movement or growth in response to sunlight. Heliotropism is the diurnal motion or seasonal motion of plant parts (flowers or leaves) in response to the direction of the sun. In floral heliotropism flowers track the sun's motion across the sky from east to west. During the night, the flowers may assume a random orientation, while at dawn they turn again toward the east where the sun rises. The motion is performed by motor cells in a flexible segment just below the flower, called a pulvinus. The motor cells are specialized in pumping potassium ions into nearby tissues, changing their turgor pressure. The segment flexes because the motor cells at the shadow side elongate due to a turgor rise. In general, flower heliotropism could increase reproductive success by increasing pollination, fertilization success, and/or seed development, especially in the spring flowers. Some solar tracking plants are not purely heliotropic: in those plants the change of orientation is an innate circadian motion triggered by light, which continues for one or more periods if the light cycle is interrupted. In *Helianthus* spp. a common misconception is that sunflower heads track the sun across the sky. The uniform alignment of the flowers does result from heliotropism in an earlier development stage, the bud stage, before the appearance of flower heads. The buds are heliotropic until the end of the bud stage, and finally face east. The flower of the sunflower preserves the final orientation of the bud, thus keeping the mature flower facing east. Leaf heliotropism is the solar tracking behavior of plant leaves. Some plant species have leaves that orient themselves perpendicularly to the sun's rays in the morning (diaheliotropism), and others have those that orient themselves parallel to these rays at midday (paraheliotropism). Floral heliotropism is not necessarily exhibited by the same plants that exhibit leaf heliotropism

**Exotropism** is the continuation of growth "outward", in the previously established direction and includes photoperiodism and chronobiology.

**Photoperiodism** is the plant response to the annual seasonal light cycle. Photoperiodism is the physiological reaction of organisms to the length of day or night. Photoperiodism can also be defined as the developmental responses of plants to the relative lengths of light and dark periods. The length of the night period is the controlling factor. Many flowering plants (angiosperms) use a photoreceptor protein, such as phytochrome or cryptochrome, to sense seasonal changes in night length, or photoperiod, which they take as signals to flower. In a further subdivision, obligate photoperiodic plants absolutely require a long or short enough night before flowering, whereas facultative photoperiodic plants are more likely to flower under the appropriate light conditions, but will eventually flower regardless of night length. Biologists believe that it is the coincidence of the active forms of phytochrome or cryptochrome, created by light during the daytime, with the rhythms of the circadian clock that allows plants to measure the length of the night. Other than flowering, photoperiodism in plants includes the growth of stems or roots during certain seasons and the loss of leaves. Artificial lighting can be used to induce extra-long days.

Long-day plants flower when the night length falls below their critical photoperiod. These plants typically flower in the northern hemisphere during late spring or early summer as days are getting longer. In the northern hemisphere, the longest day of the year (summer solstice) is on or about 21 June. After that date, days grow shorter (i.e. nights grow longer) until 21 December (the winter solstice). This situation is reversed in the southern hemisphere (i.e., longest day is 21 December and shortest day is 21 June). Long-day obligate plants include *Dianthus* sp., *Hyoscyamus* sp. and *Avena* sp. Long-day facultative plants include *Pisum sativum*, *Hordeum vulgare* and *Lactuca sativa*.

Short-day plants flower when the night lengths exceed their critical photoperiod. They cannot flower under short nights or if a pulse of artificial light is shone on the plant for several minutes during the night. They require a continuous period of darkness before floral development can begin. Natural nighttime light, such as moonlight or lightning, is not of sufficient brightness or duration to interrupt flowering. In general, short-day (i.e.long-night) plants flower as days grow shorter (and nights grow longer) after 21 June in the northern hemisphere, which is during summer or fall. The length of the dark period required to induce flowering differs among species and varieties of a species. Photoperiodism affects flowering by inducing the shoot to produce floral buds instead of leaves and lateral buds. Short-day facultative plants include *Cannabis* sp., *Gossypium* sp. and *Oryza* sp.

Day-neutral plants do not initiate flowering based on photoperiodism and include cucumbers, roses, and tomatoes. Instead, they may initiate flowering after attaining a certain overall developmental stage or age, or in response to alternative environmental stimuli, such as vernalisation (a period of low temperature).

**Chronobiology** examines periodic (cyclic) phenomena in living organisms and their adaptation to solar- and lunar-related rhythms. These cycles are known as biological rhythms (Koukkari and Sothern, 2007). The related terms chronomics and chronome have been used in some cases to describe either the molecular mechanisms involved in chronobiological phenomena or the more quantitative aspects of chronobiology, particularly where comparison of cycles between organisms is required. The variations of the timing and duration of biological activity in living organisms occur for many essential biological processes. In plants these include leaf movements and photosynthetic reactions (see ch. 18.2, structure-function change in s-triazine resistant plant). The most important rhythm in chronobiology is the circadian rhythm, a roughly 24-hour cycle shown by physiological processes in all these organisms. The circadian rhythm can further be broken down into routine cycles during the 24-hour day: diurnal (organisms active during daytime) and nocturnal (active in the night). While circadian rhythms are defined as endogenously regulated, other biological cycles may be regulated by exogenous signals.

**9.3.1.2.4.1.2 Hydrotropism.** Hydrotropism is movement or growth in response to water. It is a plant's growth response in which the direction of growth is determined by a stimulus or gradient in water concentration. A common example is a plant root growing in humid air bending toward a higher relative humidity level. The process of hydrotropism is started by the root cap sensing water and sending a signal to the elongating part of the root. Hydrotropism is difficult to observe in underground roots, since the roots are not readily observable, and root gravitropism is usually more influential than root hydrotropism. Water readily moves in soil and soil water content is constantly changing so any gradients in soil moisture are not stable.

**9.3.1.2.4.1.3 Thermotropism.** Thermotropism, or thermotropic movement, is the growth in response to changes in temperature, heat-related stimuli. For example, *Rhododendron* sp. leaves curl in response to cold temperatures. *Mimosa pudica* leaf petioles collapse when temperature drops, leading to the folding of leaflets.

**9.3.1.2.4.1.4 Gravitropism.** Gravitropism (or geotropism) is movement or growth in response to gravity. Gravitropism is a turning or growth movement in response to gravity. It is a general feature of all higher and many lower plants as well as other organisms. Charles Darwin was one of the first to scientifically document that roots show positive gravitropism and stems show negative gravitropism. That is, roots grow in the direction of gravitational pull (i.e., downward) and stems grow in the opposite direction (i.e., upwards).

**9.3.1.2.4.1.5 Thigmotropism.** Thigmotropic movement or growth occurs in response to touch or contact stimuli, mechanical signalling. Thigmo mechanisms are adaptations that permit plants to alter growth rates, change morphology, produce tropisms, avoid barriers, control germination, cling to supporting structures, infect a host plant, facilitate pollination, expedite the movement of pollen, spores, or seeds, and capture prey. Usually thigmotropism occurs when plants grow around a surface, such as a wall, pot, or trellis. Climbing plants, such as vines, develop tendrils that coil around supporting objects. Touched cells produce auxin and transport it to untouched cells. Some untouched cells will then elongate faster so cell growth bends around the object. Some seedlings also exhibit triple response, caused by pulses of ethylene which cause the stem to thicken (grow slower and stronger) and curve to start growing horizontally. Roots also rely on touch to navigate their way through the soil. Generally, roots have a negative touch response, meaning when they feel an object, they would grow away from the object. This allows the roots to go through the soil with minimum resistance. Because of this behavior, roots are said to be negatively thigmotropic. Thigmotropism seems to be able to override the strong gravitropic response of even primary roots. Charles Darwin performed experiments where he found that in a vertical bean root, a contact stimulus could divert the root away from the vertical. Tropisms are influenced by the direction of their stimulus, while nastic movements are not.

**9.3.1.2.4.1.6 Chemotropism.** Chemotropism is movement or growth in response to chemicals. Chemotropism is the growth of organisms (or parts of an organism, including individual cells) such as bacteria and plants, navigated by chemical stimulus from outside of the organism or organisms part. The response of the organism or organism part is termed 'positive' if the growth is towards the stimulus or 'negative' if the growth is away from the stimulus. An example of chemo-tropic movement can be seen during the growth of the pollen tube, where growth is always towards the ovules. Conversion of flower into fruit is the example of chemo-tropism. Fertilization of flowers by pollen is achieved because the ovary releases chemicals that produce a positive chemo-tropic response from the developing pollen tube. An example of positive and negative chemo-tropism is shown by a plant's roots; the roots grow towards useful minerals displaying positive chemo-tropism, and grow away from harmful acids displaying negative chemo-tropism. Also, the addition of atmospheric nitrogen, also called nitrogen fixation, is an example of chemo-tropism.

**9.3.1.2.4.1.7 Electrotropism.** Electrotropism is movement or growth in response to an exogenous electric field, usually a cell. Several types of cells such as green algae, spores, and pollen tubes have been already reported to respond by either growing or migrating in a preferential direction when exposed to an electric field. Electrotropism is known to play a role in the control of growth in cells and the development of tissues. By imposing an exogenous electric field, or modifying an endogenous one, a cell or a group of cells can greatly redirect their growth. Pollen tubes, for instance, align their polar growth with respect to an exogenous electric field.

**9.3.1.2.4.1.8 Sonotropism.** Sonotropism is the movement or growth in response to sound.

**9.3.1.2.4.2 Nastic movements.** Nastic movements are non-directional responses to stimuli (e.g. temperature, humidity, light irradiance), and are usually associated with plants. The movement can be due to changes in turgor or changes in growth (i.e. $K^+$ ion concentration controlling movement in plants). Nastic movements differ from tropic movements in that the direction of tropic responses depends on the direction of the stimulus, whereas the direction of nastic movements is independent of the stimulus's position. The rate or frequency of these responses increases as intensity of the stimulus increases. An example of such a response is the opening and closing of flowers (photonastic response). They are named with the suffix "-nasty" and have prefixes that depend on the stimuli.

#### 9.3.1.2.4.2.1 Nastic light movements.

**Photonasty**. Photonasty is the non-directional response to light,

**Nyctinasty**. Nyctinasty is movement at night or in the dark. Nyctinasty is the circadian rhythmic nastic movement of higher plants in response to the onset of darkness. Examples are the closing of the petals of a flower at dusk and the sleep movements of the leaves of many legumes. Nyctinastic movements are associated with diurnal light and temperature changes and controlled by the circadian clock and the light receptor phytochrome. Several leaf-opening and leaf-closing factors have been characterized biochemically. Anatomically, the movements are mediated by pulvini. In the SLEEPLESS mutation of *Lotus japonicus*, the pulvini are changed into petiole-like structures, rendering the plant incapable of closing its leaflets at night

#### 9.3.1.2.4.2.2 Hydronasty. Hydronasty is the non-directional response to water.

#### 9.3.1.2.4.2.3 Thermonasty. Thermonasty is the non-directional response to temperature

#### 9.3.1.2.4.2.4 Nastic growth direction.

**Epinasty**. Epinasty is the downward-bending from growth at the top (e.g. the bending down of a heavy flower). More rapid growth of a plant organ such as a leaf causes unrolling or downward curvature.

**Hyponasty**. Hyponasty is the state of growth in a flattened structure in which the under surface grows more vigorously than the upper. The hyponastic response is an upward bending of leaves or other plant parts, resulting from growth of the lower side. This can be observed in many terrestrial plants and is thought to be linked to the plant hormone ethylene. Submerged plants often show the hyponastic response, where the upward bending of the leaves and the elongation of the petioles might help the plant to restore normal gas exchange with the atmosphere (figure 9.3.1.2.4.2.4, Wikipedia, 5.15). Plants produce ethylene, and normally this dissolves in the air quite easily. But when the plant is submerged ethylene is trapped in the plant.

**Geonasty**. Geonasty (or gravinasty) is movement in response to gravity.

#### 9.3.1.2.4.2.5 Chemonasty. Chemonasty is the non-directional response to diffuse or indirect chemical or nutrient stimuli.

#### 9.3.1.2.4.2.6 Thigmonasty. Thigmonasty (seismonasty; haptonasty) is the nastic response to contact, touch or vibration. Thigmonasty differs from thigmotropism in that nastic motion is independent of the direction of the stimulus. For example, tendrils from a climbing plant are thigmotropic because they twine around any support they touch, responding in whichever direction the stimulus came from. However, the shutting of a venus fly trap is thigmonastic; no matter what the direction of the stimulus, the trap simply shuts (and later possibly opens). Examples include many species in the leguminous subfamily *Mimosoideae*, active carnivorous plants such as *Dionaea* and a wide range of pollination mechanisms. Thigmonasty other than leaf closure occurs in various species of thistles (*Asteraceae*). When an insect lands on a flower, the anthers shrink and rebound, loading the insect with pollen. The effect results from turgor changes in specialized, highly elastic cell walls of the anthers. Similar pollination strategy occurs in *Rudbeckia hirta*. Sensitive leaves also occur in plants of the wood sorrel family (*Oxalidaceae*). Examples include many species of *Oxalis*, *Biophytum sensitivum*, and carambola or star fruit (*Averrhoa carambola*).

#### 9.3.1.3 Temporal foraging for local opportunity. Foraging for local opportunity includes strategic roles and traits to interfere with neighbors over time (table 9.5). These foraging strategies include those to seize and exploit temporal opportunity beginning with appropriate seedling emergence timing. Appropriate recruitment includes early emergence to preempt neighbors and late emergence to seize opportunity after crop production practices cease (e.g. layby) or are much reduced (e.g. harvest, post-harvest times). Rapid rates of early growth, as well as high/plastic/appropriate growth during the growing season, strategic timing of branching and tillering, and scenscence timing are also time foraging traits. Rhythmic behaviors such as diurnal, circadian changes in leaf movements for light capture and differential photosynthetic rate at different times of the day

(e.g. ch. 18, triazine resistant and susceptible plants) are temporal foraging traits. Reproductive traits for temporal interference foraging include the timing of reproductive processes regulated by the seasonal photoperiod, including flowering/anthesis/fertilization timing, seed dormancy induction during embryogenesis, and the timing, frequency and duration of the seed production period.

**Table 9.5** Temporal interference with neighbors: strategic roles and example traits played in competitive, amensal and antagonistic interactions seizing and exploiting local opportunity

| Strategic roles | Traits |
|---|---|
| Temporal foraging for capture of local opportunity | Appropriate seedling emergence timing:<br>1. seed heteroblasty:<br>　-early establishment<br>　-late emergence (e.g. at layby)<br>2. rapid acquisition of seedling autotrophic growth<br>3. rapid early growth<br><br>Vegetative growth:<br>1. high, plastic, appropriate growth rates<br>2. timing of tillering and branching<br>3. scenescence timing<br><br>Rhythmic behaviors:<br>1. circadian, diurnal leaf movements for light capture<br>2. circadian, diurnal photosynthesis/respiration rates (e.g. triazine-resistant/susceptible biotypes)<br><br>Reproduction:<br>1. timing of vegetative-to-reproductive mode induction<br>2. photoperiod signaling/control/stimulation of:<br>　-flowering, anthesis, fertilization<br>　-seed production period time, frequency, duration<br>　-seed dormancy induction during embryogenesis |
| Temporal offensive attack of neighbors | Climb on neighbor's earlier plant architecture growth (viney growth on mature plant) |
| Preempt neighbors ability to temporally seize and exploit opportunity | Nutrients:<br>Early, luxury nutrient consumption<br><br>Water:<br>Early consumption of limited water (e.g. root foraging)<br><br>Light:<br>Architecture growth timing: early branch/tiller growth for canopy closure |

Interfering with neighbors includes fulfilling strategic roles with traits for temporal offensive attack. For example, viney growth habit climbing on taller, more mature plants for light capture. Interference includes fulfilling strategic roles with traits preempting neighbors from seizing and exploiting local temporal opportunity. Temporal preemption of neighbors includes early luxury consumption of nutrients and water by rapid root development exploring the soil profile.

**9.3.2 Strategic roles and traits of facilitation with neighbors.** Facilitation in interactions with neighbors in the local weed-crop community encompasses several strategic roles accomplished by specific traits. The beneficial and neutral effects of species on one another are realized in various ways including escape and avoidance, coexistence, and cooperation with neighbors (table 9.6).

**Table 9.6** Strategic roles and traits played in facilitation interactions with neighbors of by means of commensalism and mutualism.

| Strategic roles | | Traits |
|---|---|---|
| Avoid and escape neighbors | Refuge from neighbors | Physical stress:<br>1. germination in the shadow of nurse plants<br><br>Predation:<br>1. establishment near species more palatable to predators<br>2. minor species avoid density-dependent feeding on more dominant species<br>3. crop mimicry |
| | Defense from neighbors | 1. specialized anti-feeding spines, chemicals<br>2. indigestible seeds for animal predation dispersal |
| | Segregation from neighbors | Spatial segregation:<br>1. species mixtures exploit different light resources with differing leaf canopies<br>2. dominant species leaves unutilized opportunity for other species |
| | | Temporal segregation:<br>1. differential seedling recruitment timing<br>2. different life spans offset life history behaviors: summer and winter annuals, biennials, perennials |
| Co-exist with neighbors | Refuge from competition | Exploit micro-sites and situations with reduced competition |
| | Ecological combining ability: avoid neighbors | Adapt over time with new, improved, modified traits to allow exploitation of alternative opportunity: niche separation & diversification |
| | Tolerate neighbors | Tolerance to herbicides, alleochemicals, shade, drought, salt and mechical injury |
| Co-operate with neighbors | Symbiosis | Symbiotic relationship between endophytic fungi and weed plant |

**9.3.2.1 Escape and avoid neighbors.** Facilitation with neighbors can be accomplished by escape and avoidance, by finding refuge, defense or spatial and temporal segregation from neighbors.

**9.3.2.1.1 Refuge from neighbors.** Escape from neighbors can be accomplished by finding refugia from physical stress and predation.

**9.3.2.1.1.1 Refuge from physical stress.** Facilitation with neighbors can be accomplished by escaping to a safe, or safer, refugia by reducing the negative impacts of a stressful environment. Strict categorization, however, is not possible for some complex species interactions. For example, seed germination and survival in harsh environments is often higher under so-called 'nurse' plants than on open ground (Stachowicz, 2001; Callaway, 1995). A nurse plant is one with an established canopy, beneath which germination and survival are more likely due to increased shade, soil moisture, and nutrients. Thus, the relationship between seedlings and their nurse plants is commensal. However, as the seedlings grow into established plants, they are likely to compete with their former benefactors for resources (Stachowicz, 2001; Callaway, 1995)

#### 9.3.2.1.1.2 Refuge from predation.
Another mechanism of facilitation is a reduced risk of being eaten. Nurse plants, for example, not only reduce abiotic stress, but may also physically impede herbivory of seedlings growing under them (Callaway, 1995). In both terrestrial and marine environments, herbivory of palatable species is reduced when they occur with unpalatable species (Stachowicz, 2001; Callaway, 1995; Bruno et al., 2003). These "associational refuges" may occur when unpalatable species physically shield the palatable species, or when herbivores are "confused" by the inhibitory cues of the unpalatable species (Stachowicz, 2001; Callaway, 1995).

Dominant species can suffer disproportionately from density-dependent feeding, predation, because of their exposure and availability. Some predators or pathogens operate in a density-dependent, or frequency-dependent, manner, culling (e.g. grazing, herbivory) a disproportionately large share from the competitive dominants in a community that delays or prevents competitive exclusion or more minor species.

Density-independent predation can also provide a means of escaping neighbors. For example, vegetation prone to periodic disturbances (e.g. fire, wind) that kills all emerged plants creates opportunity for those able to recruitment the newly cleared locations. The 'intermediate disturbance hypothesis' states that high rates of disturbance will eliminate those species with populations unable to recover quickly, and low rates of disturbance will allow interspecific competition to take its toll; coexistence is favored between extremes of disturbance and competition (Connell, 1978)

Species that are readily predated can adapt their appearance to mimic neighbors that are not predated; for example crop mimicry by barnyardgrass, *Echinochloa crus-galli*, in east Asia can assume the form of rice plants and avoid hand weeding.

#### 9.3.2.1.2 Defense from neighbor stress.
Avoidance of neighbors can be accomplished by utilizing defensive traits for protection from neighbors. These traits include specialized anti-feeding and anti-grazing morphological adaptations such as spines or chemicals to repel, and seed structures to prevent digestion and loss of embryo viability while allowing the predator to disperse the propagule.

#### 9.3.2.1.3 Segregation from neighbors.
Neighbors can facilitate the interactions amongst themselves by spatially or temporally segregated themselves from each other.

#### 9.3.2.1.3.1 Spatial segregation.
Neighbors can facilitate the interactions amongst themselves by spatially segregated themselves from each other. This can be a consequence of clonal growth limitations, limited seed dispersal and subsequent recruitment spatial limitations, or a clumped species spatial distribution.

The movement by animals of items involved in plant reproduction is usually a mutualistic association. Pollinators may increase plant reproductive success by reducing pollen waste, increasing dispersal of pollen, and increasing the probability of sexual reproduction at low population density (Boucher et al., 1982). In return, the pollinator receives nourishment in the form of nectar or pollen. Animals may also disperse the seed or fruit of plants, either by eating it (in which case they receive the benefit of nourishment) or by passive transport, such as seeds sticking to fur or feathers (Bruno et al., 2003).

The effects of facilitation are often observable at the scale of the community, including impacts to spatial structure and diversity. Many facilitative interactions directly affect the spatial structure, distribution, of species. Transport of plant propagules by animal dispersers can increase colonization rates of more distant sites, which may impact the distribution and population dynamics of the plant species (Boucher et al., 1982; Bruno et al., 2003; Tirado and Pugnaire, 2005). Facilitation most often affects distribution by simply making it possible for a species to occur in a site where some environmental stress would otherwise prohibit growth of that species. Many interactions are facilitated in the process of ecological succession. A facilitating species may help drive the progression from one ecosystem type to another, as mesquite apparently does in the grasslands of the Rio Grande Plains (Archer, 1989). As a nitrogen-fixing tree, mesquite establishes more readily than other species on nutrient-poor soils, and following establishment, mesquite acts as a nurse plant for seedlings of other species (Callaway, 1995). Thus, mesquite facilitates the dynamic spatial shift from grassland to savanna to woodland across the habitat. Facilitation affects community diversity (defined in this context as the number of species in the community) by altering competitive interactions. The effect of facilitation on diversity could also be reversed, if the facilitation creates a competitive dominance that excludes more species than it permits (Stachowicz, 2001).

There are situations in which species avoid direct conflict and competing for the same resources at the same time. In the examples below the species (or genus) does better in mixed stands than alone:

- Dioecious species: the different sexes avoid competing with each other.
- Erect (*Polygonum erectum*) and prostrate knotweed (*P. aviculare*): both species of the species-group do better in mixed stands than one species alone; these two species co-exist together on the unpaved edges of high traffic walkways.
- Giant (*S. faberi*), green (*S. viridis*), yellow (*S. pumila*), and knotroot foxtails (*S. geniculata*) in Iowa, USA crop fields: the *Setaria* species-group does better in mixed stands than in single foxtail species alone. They occupy slightly different niches for greater exploitation of field opportunity.
- Velvetleaf in soybeans: *Abutilon theophrasti* does better in soybeans than alone in pure stands where intense self-shading of the weeds results in tall, spindly plants. In soybeans they are able spread their canopy widely and seize light opportunity from the soybean understory.
- Fescue (*Festuca* spp.) populations do better in mixtures with barley than alone.
- Clover (*Trifolium* spp.) in legume-grass associations: locally adapted populations have evolved with neighbor genotypes. If they are placed in new habitats with new neighbor genotypes of the same species they do not do as well.

**9.3.2.1.3.2 Temporal segregation.** Neighbors can facilitate the interactions amongst themselves by temporally segregated themselves from each other. Differences between coexisting species in a locality in terms of life history timing can provide a means for species to facilitate each other. Utilization of local opportunity occurs at different times for different species. Differential life history behavior timing can include:

- Seed dormancy heteroblasty blueprints different recruitment timing patterns
- Seedling recruitment limitation: differences in the soil seed pool quantity or quality between species results in differential seedling recruitment behavior.
- Different life spans of the same or different species in the same community offset their time of seizing and exploiting opportunity. For example: horseweed (*Erigeron* sp.) has summer annual and perennial forms; musk thistle (*Circium vulgare*) has summer annual and biennial forms; dandelion (*Taraxicum officinale*) has summer annual, winter annual, biennial, perennial forms.
- Storage effect of seed pools: differential recruitment from soil propagule pools over time(s); overlapping generations, asynchronous recruitment. Fluctuations in recruitment can promote coexistence if the good years for one species are the bad years for others. Recruits of good years are stored over bad years for the species. For annual weeds this may be the amount of seed stored in soil pools. There must be overlapping generations and weak competition between adults, providing a stage-specific refuge.

**9.3.2.2 Co-exist with neighbors.** Facilitation of interactions in the local community can be accomplished by coexisting, putting up with, neighbors by means of refuge from competition, ecological combining ability, and tolerating neighbors.

**coexistence**: the continuing occurrence of two or more species in the same area or habitat, usually used of potential competitors (Lincoln et al., 1998)

**9.3.2.2.1 Refuge from competition.** Another potential benefit of facilitation is insulation from competitive interactions. For example nurse plants in harsh environments, in a forest, are sites of increased seed germination and seedling survival because the raised substrate of a log frees seedlings from competition with plants and mosses on the forest floor (Harmon et al., 1989).

**9.3.2.2.2 Ecological combining ability.** Selection and adaptation can lead to avoidance of competition between population if enough time elapses: niche separation and diversification. Allard and Adams (1969) indicate that selection in mixed populations favors genotypes with superior ecological combining ability (good neighbors and good competitors). Species should be able to coexist in a community if each inhibits its own population growth more than that of its competitors. One way this can happen is if species have different niches, guilds, roles in the community, selection for ecological combining ability. New niche dimensions are created by evolving plants as they acquire traits that allow them to respond more subtley to environmental and opportunity differences. In a new habitat with the same environment the conditions of life are changed in an essential way. Darwin (1859) also said that the structure of an organism is related in an essential way (although it may be hidden) to that of the other organisms with which it competes or preys upon (neighbors define what you are). Selection can lead to a species adapting to avoid competition. Co-existing forms that are selected together over a long time are less competitive with each other than they are with new unselected variants entering the locality. Selection may favor a divergence in behavior between populations so that each makes less demand on resources needed by the other. Niche separation occurs when intraspecific completion is greater than interspecific due to trait differences causing differential exploitation of opportunity spacetime. Niche separation can reduce the strength of interspecific competition to prevent competitive exclusion. But, this niche separation is not always sufficient, on its own, to prevent competitive exclusion of either species.

Darwin (1859) said competition was greatest within a species; because of the similarity in habits, constitution and structure intra-specific struggle is the most intense. From this one can imagine the most intense competition would exist between siblings, commencing with seedling recruitment (see ch. 3.2.2, niches in the local community). It has been proposed that competition between similar species cannot long endure:

**competitive exclusion principle**:
1: two species competing for the same resources cannot coexist if other ecological factors are constant (Hardin, 1960); complete competitors cannot coexist
2: a theory which states that two species competing for the same resources cannot stably coexist, if the ecological factors are constant; complete competitors cannot coexist. Either of the two competitors will always take over the other which leads to either the extinction of one of the competitors or its evolutionary or behavioural shift towards a different ecological niche (Wikipedia, 5.08)

This proposition is sometimes referred to as Gause's Law of competitive exclusion or just Gause's Law (Gause, 1934). When one species has even the slightest advantage or edge over another, then the one with the advantage will dominate in the long term. One of the two competitors will always overcome the other, leading to either the extinction of this competitor or a behavioral shift towards a different ecological niche.

The nature of weeds is the history of weeds. Mortality dictates natural elimination and selection towards those new phenotypes able to occupy a slightly or greater different role in the weed-crop community, a diversification of function among interacting neighbor species.

**9.3.2.2.3 Tolerate neighbors**. Neighbors can coexist in local communities together by tolerance to many types of inhibitory factors in opportunity spacetime. These facilitative interactions include resistance to herbicides; allelopathy tolerance/specificity; tolerance to resource limitations (shade, drought) and limitations in the pervasive conditions of the environment (heat, high or low; salt); and tolerance to mechanical injury.

**9.3.2.3 Co-operate with neighbors**. Neighbors can cooperate with neighbors, create a synergy in their interactions by improving resource (opportunity) availability. Altruistic interactions are also considered.

**9.3.2.3.1 Improved resource availability**. Facilitation can increase access to limiting resources such as light, water, and nutrients for interacting species. For example, epiphytic plants often receive more direct sunlight in the canopies of their host plants than they would on the ground (Callaway, 1995). Also, nurse plants increase the amount of water available to seedlings in dry habitats because of reduced evapo-transpiration beneath the shade of nurse plant canopies. However, the most familiar examples of increased access to resources through facilitation are the mutualistic transfers of nutrients between symbiotic organisms.

**symbiosis**: a prolonged, close association between organisms

Some examples of mutualistic symbioses include lichens and mychorrhizae. Lichens are associations between fungi and algae, wherein the fungus receives nutrients from the alga, and the alga is protected from harsh conditions causing dessication (Boucher et al., 1982). Mycorrhizae are associations between fungi and plant roots, wherein the fungus facilitates nutrient uptake (particularly nitrogen) by the plant in exchange for carbon in the form of sugars from the plant root (Boucher et al., 1982). There is a parallel example in marine environments of sponges on the roots of mangroves (e.g. genus *Rhizophora*), with a relationship analogous to that of mycorrhizae and terrestrial plants (Callaway, 1995).

Mycorrhizal symbiosis between plants and fungi is one of the most well-known plant–fungus associations and is of significant importance for plant growth and persistence in many ecosystems; over 90% of all plant species engage in mycorrhizal relationships with fungi and are dependent upon this relationship for survival (Bonfante, 2003; Wikipedia, 12.10). The mycorrhizal symbiosis is ancient, dating to at least 400 million years ago (Remy et al., 1994). It often increases the plant's uptake of inorganic compounds, such as nitrate and phosphate from soils having low concentrations of these key plant nutrients (Lindahl et al., 2007; van der Heijden et al., 2006). The fungal partners may also mediate plant-to-plant transfer of carbohydrates and other nutrients. Such mycorrhizal communities are called "common mycorrhizal networks" (Selosse et al., 2006). A special case of mycorrhiza is myco-heterotrophy, whereby the plant parasitizes the fungus, obtaining all of its nutrients from its fungal symbiont (Merckx et al., 2009).

Some fungal species inhabit the tissues inside roots, stems, and leaves, in which case they are called endophytes (Schulz and Boyle, 2005). Similar to mycorrhiza, endophytic colonization by fungi may benefit both symbionts; for example, endophytes of grasses impart to their host increased resistance to herbivores and other environmental stresses and receive food and shelter from the plant in return. For example, dark filamental hyphae of the endophytic fungus *Neotyphodium coenophialum* infiltrate the intercellular spaces of tall fescue leaf sheath tissues (Wikipedia, 12.10).

**9.3.2.3.2 Altruism**. Robert Trivers (1971) demonstrated how reciprocal altruism can evolve between unrelated individuals, even between individuals of entirely different species. The relationship of the individuals involves repeated, or iterated, interactions in which both parties can benefit from the exchange of many seemingly altruistic acts. He said it "take[s] the altruism out of altruism." The premise that self-interest is paramount is largely unchallenged, but turned on its head by recognition of a broader, more profound view of what constitutes self-interest. It does not matter why the individuals cooperate. The individuals may be prompted to the exchange of "altruistic" acts by entirely different genes, or no genes in particular, but both individuals (and their genomes) can benefit simply on the basis of a shared exchange. In particular the benefits of altruism are to be seen as coming directly from reciprocity, and not indirectly through non-altruistic group benefits. Trivers' theory is very powerful, it also predicts various observed behavior, including development of abilities to detect and discriminate against subtle cheaters. Compare Trivers (1971) with Dawkins (1976, 1999); see ch. 4.1.3, genome size, weediness and intra-genomic competition.

## 9.4 The effects of neighbor interactions on plant growth and plant form

"Neighboring plants interfere with each other's activities according to their age, size and distance apart. Such density stress affects the birth rates and death rates of plant parts. As plants in a population develop, the biomass produced becomes limited by the rate of availability of resources so that yield per unit area becomes independent of density - the carrying capacity of the environment. The stress of density increases the risk of mortality to whole plants as well as their parts and the rate of death becomes a function of the growth rate of the survivors. Self-thinning in populations of single species regularly follows a $3/2$ power equation that relates the mean weight per plant to the density of survivors. Density-stressed populations tend to form a hierarchy of dominant and +/- stressed subordinate individuals. The death risk is concentrated within the classes of suppressed individuals." Harper (1977).

"The effects of density do not fall equally on all parts of a plant. In general the size of parts (e.g., leaves or seeds) is much less plastic than the number of parts (e.g., branches). The stress created by the proximity of neighbors may be absorbed in an increased mortality risk for whole plants or their parts, reduced reproductive output, reduced growth rate, delayed maturity and reproduction. The term density is used in a special sense here, to signify the integrated stresses within a community rather than the number of individuals per unit area" Harper (1977).

The nature of the local community includes an explanation defining the interactions among neighbors of the local deme. These interactions are revealed during development when functional traits are expressed that fulfill strategic roles of a particular weed species. The performance of a plant defines the ecological roles, guilds or trades of individual weed species in plant communities (see ch. 10.3.3). An alternative way of explaining the nature of community interactions is a demographic description of the effects of neighbors on plant density, growth, form, reproduction and mortality. Strategic roles and traits provide a qualitative basis to understand the evolving adaption of changing populations in the community. Demographic descriptions of growth, form and mortality can provide a quantitative basis for understanding the nature of weeds in local communities, but inferences derived from demography are confounded.

**9.4.1 Space, neighborhoods and plant density**. Silvertown and Charlesworth (2001) present a demographic representation of neighbor interactions in which the underlying assumption is that the primary resource limiting plant growth is space: plants will fill any available growing space. Space appears in population, genetic, demographic models in two forms:

**demographic space**: demographic space exists in two forms:
1: a particular locality, the specific area over which a plant interacts with neighbors, the ecological and genetic neighborhood
2: plant numerical density, the number of plants per unit area; density is used as an average measure of space occupancy

Density 'stress' is the integrated stresses within a community produced by neighbors on each other, and includes plastic growth (phenotypic plasticity) and well as the altered risk of death. The population-like structure of an individual plant also responds to density stress (unlike animals): varied birth and death rates, and numbers, of plant parts (leaves, branches, flowers, fruits). Compare this concept with that of opportunity spacetime (ch. 3.1.2, seizing and exploiting local opportunity).

**9.4.2 Plant density and productivity per unit area.** There exists a relationship between the yield of dry matter per unit area and the density of plants per unit area (e.g., *Bromus* sp. at three levels of nitrogen fertilization; Donald 1951). Early in growth, and at low numbers of plants per unit area, the number of plants and yield are directly, linearly related. With time, and/or greater numbers of plants per unit area, the yield per unit area becomes independent of plant number: a saturated yield, the holding capacity of that space; the "law of constant yield" (Kira et al., 1953). Variations in sowing density are largely compensated for by the amount of growth made by individual plants.

Plants in a local space are said to obey the law of constant final yield (figure 9.2):

**law of constant final yield**: at high density the total plant yield (biomass, seed weight) of a specified area tends to approach a constant weight

**Figure 9.2.** The law of constant final yield: the relationship between yield (Y; weight per unit area; productivity) and plant density (X; initial plant number per unit area) in a local plant stand.

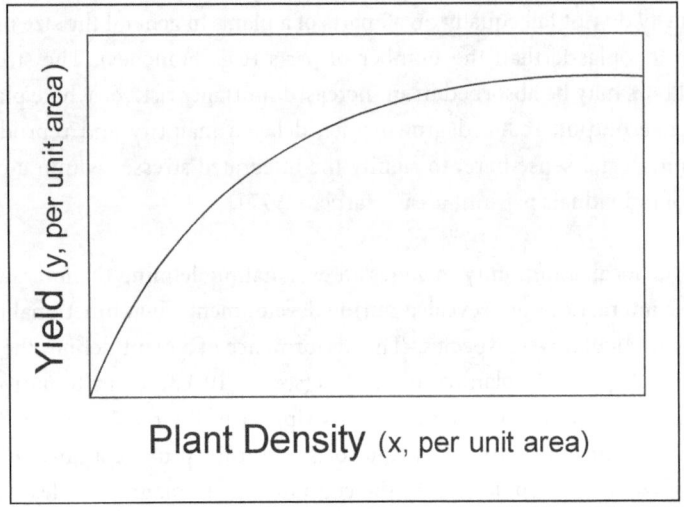

"For stands of annual plants, the 'law of constant yield' (a 'ceiling yield') beyond a certain density will not be quite true, even considering the total biomass production. However, descriptions of the relationships using models based on the assumption of constant ceiling levels will often suffice within limited density ranges." (Hakansson, 2003)

The effects of plant numbers per unit area (density) were described by several Japanese scientists in the 1950's (e.g. Kira et al., 1953; Shinosaki and Kira, 1956). Hakansson (2003) has provided much insight into the competitive effects of plant density on productivity with time in several analyses (e.g. 1988a, 1991, 1997b), summarized in figures 9.3-9.5. For example, the observed relationships of plant density and yield (the "D-Y effect") are generalized in figure 9.3 (compare with figure 9.2). The straight, dotted, lines in figure 9.3 represent density ranges in which intra-specific competition is not apparent (linear interaction of yield and density). The curvilinear, solid, lines in these functions represent non-linear interactions of yield and density. The actual slopes of individual species reaction norms in particular local communities will deviate from this generalized function in figure 9.3.

**Figure 9.3.** Relationships between biomass (Y; yield, weight per unit area; productivity) and plant density (X; initial plant number per unit area) in a single species stand (the D-Y effect). Comparison of yield from 4 stages of plant development and growth: 1, very early; 2-4, increasingly advanced stages (after Hakannson, 2003).

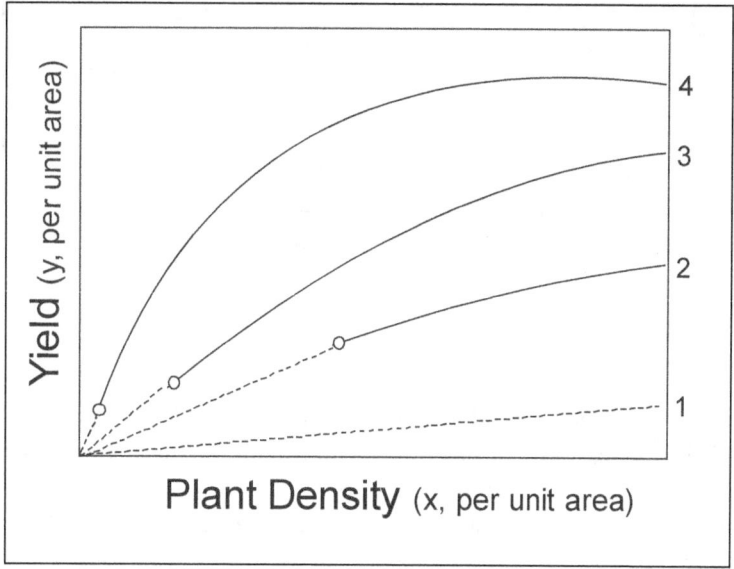

Hakansson provides much insight into the observed effects of plant density with time on yield in figure 9.4 (Hakansson, 2003) in which he summarizes the observations of numerous experiments (e.g. Lang et al., 1956; Bleasdale and Nelder, 1960; Holliday, 1960a, b; de Wit, 1960; Donald, 1963; Bengtsson and Ohlsson, 1966; Willey and Heath, 1969; Bengtsson, 1972; Pohjonen and Paatela, 1974; Hakansson, 1975a, 1979, 1983b, 1988b, 1997a; Ervio, 1983).

Hakansson (2003) distinguishes three density ranges of yield-density relationships. The yield-density relationship is linear at low plant numbers, competitive effects are not apparent (figure 9.4, range 1; see also figures 9.3-9.5, straight dotted lines). Non-linear competitive interactions occur with yield at higher plant densities (figure 9.4, ranges 2-3; see also figures 9.3-9.5, curvilinear lines). These non-linear interactions are complex and are apparent in both crop-weed and intra-specific populations. At mid-level plant densities (figure 9.4, range 2) competitive effects are apparent, the yield-density function is non-linear as it approaches and achieves maximum productivity. Yield increases with increasing plant density to a maximum, at some density in range 2-3 in figure 9.4. The plant density at which this maximum is obtained varied considerably with species, plant part(s) observed and environment (opportunity). Various competitive interactions can occur at plant densities greater than those of maximum yield (figure 9.4, range 3), an apparent violation of the 'law' of constant final yield (figure 9.2) (Hakansson, 2003). Despite this, many models are still based on the assumption of constant final yield (e.g. Shinosaki and Kira, 1956; Holliday, 1960a; de Wit, 1960).

**Figure 9.4.** Generalized relationships of yield (Y; yield, weight per unit area) and density (X; initial plant number per unit area) relationships at a late stage of growth and development of an annual plant stand (the D-Y effect). Comparison of three plant density ranges with indistinct (1-2) and overlapping (2-3) times of occurrence (after Hakannson, 2003).

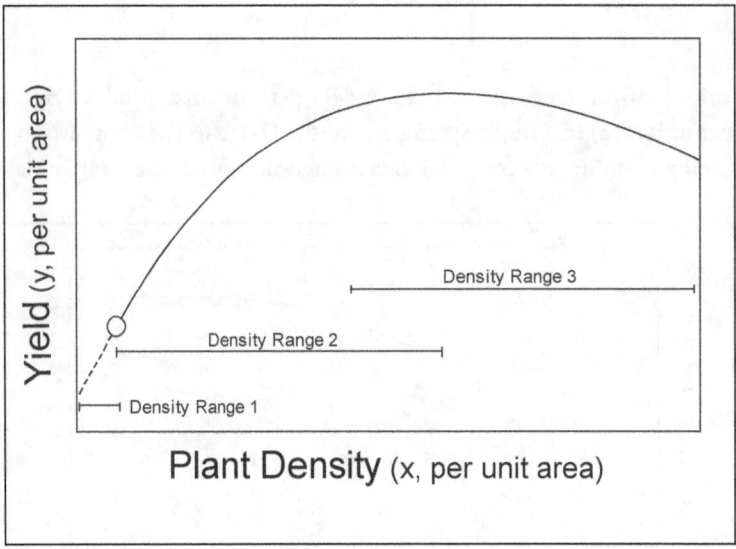

Attempts to define a mathematical function to describe yield-density reaction norms have been confounded by the inherent complexity for plant interactions. Simple, general expressions based on average plant characteristics of a local community (unit area expressions) lose the inherent biodiversity and plasticity of individual weed phenotypes with this statistical homogenization. Hence, Hakansson (2003) concludes:

> "When production-density data represent a wide density range, it is difficult to obtain equally good fits by one and the same equation over the entire range. This is not surprising. The manner of morphological and anatomical adaptation to density differs strongly from one density range to another and seems to differ more or less between species. It is therefore difficult to find simple production-density equations enabling sufficiently good descriptions of the production-density relationship of different plant species in wide density ranges." (Hakansson, 2003)

The slope or form of the generalized D-Y function in figure 9.4 varies with species, plant part considered (e.g. biomass, seed, flower, branch) and opportunity space-time. Biomass functions of entire plants per unit area are sometimes rather flat over a wide range of plant densities. The effects of plant density are strongly influenced by the spatial distribution of crop and weed plants (e.g. crop row spacing). Biomass maxima of generative organs and organs produced late in development (e.g. underground propagules) decrease at lower densities than those of whole plants. The yield and productivity of perennial plant species depends only slightly on the densities of initial, younger plants and is strongly governed by subsequent opportunity space-time.

**9.4.3 Plant density and plant size.** Demographic models utilize the average plant size at several densities to represent competitive interactions among plants in a local community. From the 'law of constant final yield' the competition-density effect (C-D effect), has been obtained: a description of how mean individual plant weight varies at different densities. Individual plant size and yield decreases with increasing plant density per unit area (figure 9.5). This function is the inverse of the yield per area-total plant density relationship in figure 9.4. This inverse demographic function is used to represent competitive interactions among crop and weed plants in a local community.

**Figure 9.5.** Generalized relationships of individual plant size (Y; yield, weight per individual plant) and density (X; initial plant number per unit area) (the C-D effect). Comparison of yield from four stages of plant development and growth: 1, very early; 2-4, increasingly advanced stages (after Hakannson, 2003).

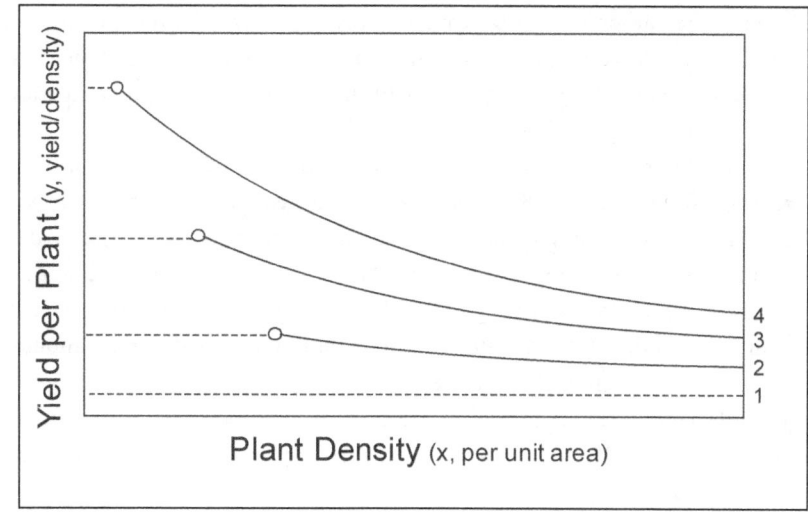

Competition between and among crop and weed species are strongly influenced by the formation of plant size hierarchies in the community, productivity and plant size, and the age- and stage-structure of populations

**9.4.3.1 Plant size hierarchies in the community.** Competition in crowded stands with the growth and development of member plants results in increasing inequality of size among neighbors of the population, producing a plant size hierarchy. Size hierarchies develop when the effect of neighbors on each other is disproportionate to their relative size, relative size asymmetric competition. For example, competition for light can result in larger plants seizing disproportionate shares of radiation as their rate of leaf production exceeds that of shaded neighbors (see case study ch. 16). Relative size symmetry is more typical of soil resource competition, although differential colonization of root systems by mycorrhizas may lead to asymmetric size hierarchies. Asymmetric size competition magnifies small initial differences between crowded seedlings caused by 'priority of emergence', the differential timing of seedling emergence or weeds relative to crop neighbors.

Skewing of the frequency distribution of plant number versus individual plant weight increases with time and with increasing density (plants per unit area). At harvest a hierarchy of individuals is established: typically a few large dominants and a large number of suppressed, small, plants. There exists a danger in assuming that the average plant performance represents the commonest type, or most typical, plant performance: size hierarchies are non-linear.

The place an individual occupies within the hierarchy of a plant population is largely determined in the very early stages of plant establishment and development (the critical role of relative time of emergence). The weight of an individual is a function of:

1: its starting capital (the embryo plus some part of food reserve weight)
2: the relative growth rate of the genotype in the environment provided
3: the length of time for which this growth rate is continued
4: restrictions on the rate or time of growth imposed by the presence, character or arrangement of neighbors in the population.

The "percentage emergence ranking" is an index of the position in emergence ranking an individual occupies in a population "sown" at the same time. The amount of growth made by an individual is more directly determined by its order in the sequence of emergence than by the actual time at which it emerges, or the relative spatial arrangement of neighbors. The greater time that it has been allowed to grow allows it to capture more opportunity space-time. The earlier emerging plant has been able to capture a disproportionate share of environmental resources, with a corresponding deprivation of late emerging plants. Once a difference between neighbors exists, it becomes progressively exaggerated with time (especially when competition for light is the dominant mode of interaction among neighbors).

Intraspecific competition at very high densities leads to density-dependent mortality, or self-thinning, which regulates population size (see ch. 5.2.6, influences of plant density on mortality). Self-thinning leads to the mortality of the smallest (presumably weakest) individuals. Yoda et al. (1963) elucidated a relationship slightly different than the C-D effect among several common annual weed species: the slope of the mean plant weight versus plant density, on log-log scales, was $-{}^3/_2$. Therefore the mean weight per plant is always at the maximum that a locality can support (Dekker, 1983). The authors found widespread applicability of this '$-{}^3/_2$ power law', but concluded its use as a model "could be accepted as a crude approximation".

**9.4.3.2 Productivity and plant size.** Fecundity (numbers of seeds produced) or productivity varies directly with plant size, biomass (Dekker, 2005; Samson and Werk, 1986). The largest plants are capable of exponential seed production, making a vastly disproportionate contribution to the local deme, the soil seed pool. This correlation of biomass to fecundity may not hold true for perennials: seed per unit biomass often increases with plant size, with more seed produced by large plants than small ones (Solbrig et al., 1988).

**9.4.3.3 Age- and stage-structure of populations.** Seedling and propagule emergence timing leads to different age-stage population structure, which determines individual plant competitive advantage, which forms the basis of size frequency distribution hierarchy asymmetry. Age- and stage-structured populations occur as a consequence of differences in seedling emergence, as well as of ecological succession in undisturbed populations (see section 5.2.3, plant age and stage structure). Age- and stage-structured populations also occur in agricultural fields disturbed annually. In populations where recruitment is a infrequent event compared to the lifespan of adults, such as in tree forests, there exists any evident age structure. The presence of an age structure in a population depends on the pattern and periodicity of recruitment; which depends on the periodicity of disturbance and opportunity. At any one time, seeds, seedlings, saplings and adults of various ages may be found. Plants are highly plastic in their rates of growth and development, so that two individuals of the same age but with different local opportunity may be at quite different stages of the life cycle. Therefore, as with an age structure, populations have a stage structure. The size distributions that develop in crowded stands are an example of stage structure. Not all perennial plant populations have an age structure.

**9.4.4 Plant density and plant form.** Plant density influences plant form, reproduction and the diversity of a community.

**9.4.4.1 Influence of plant density on form and reproduction.** Higher plants are plastic in size and form. This plasticity derives from the population-like structure of individuals. The form of repeating units of plant construction (branch/tiller-leaf-flower) is tightly controlled and changes only slightly over a wide range of environments. The number of these units, and thus the size of the whole plant, varies greatly with both age and opportunity.

The dry weight of a plant population compensates more or less perfectly with variations in density, but the parts of individuals are not altered to the same extent. The growth of individuals under density stress results in differential allocation of assimilates between different structures, and resulting differences in the size of those parts. For example, the proportion of seed mass relative to total dry matter weight changes with density stress. High density results in more reproductively inefficient plants. The optimal density for a particular product (seed, storage root, latex, etc.) may be different than that density for dry matter production. For example, the optimal population size for maize (*Zea mays*) seed yield is less than that for silage.

Density stress is generally expressed in a reduction in number of plant parts produced (branches, flowering nodes) and in part by organ abortion (death of old leaves, abscission of flowers and seed pods). An example of this density stress is observed in wheat (*Triticum sp.*). Stable vegetative parts include height, leaf width, stem diameter, and number of spikelets per spike (inflorescence). Plastic parts include branching (tiller shoot formation). Stable reproductive parts include grains per ear and mean weight per grain (seed). Plastic components include fertile tillers per plant (ears, inflorescences).

**9.4.4.2 Plant form and diversity of a community.** One of the elements contributing to the diversity of plant populations, community structure, includes that which arises from the growth form of the individual plant. Harper (1977) said that the "growth pattern of a genet can itself impose order of diversity on a plant community":

"The problems are great: they can be regarded as insoluble and a demography of plants an unattainable ideal, or they can be ignored with a certainty of serious misinterpretation, or they can be grasped and methods, albeit crude, developed to handle the problem."

"The way in which the problem [development of plant demography] can be faced is to accept that there are two levels of population structure in plant communities. One level is described by the number of individuals present that are represented by original zygotes ... independent colonizations ... genets ..."

"Each genet is composed of modular units of construction ... the shoot on a tree ... the ramet of a clone ... the tiller of a grass or the leaf with its bud on an annual ..." Harper (1977)

There exists a population basis to phenotypic plasticity of the individual plant. A single plant is a population of modular units that are attached to, and interact with each other. For example, the above-ground shoot of a higher plant is composed of a repetitive branched system of units each consisting of a segment of stem, a leaf and its axillary bud. The entire shoot system is a population of such units which gains its unity from the possession of a common root system. The branched root system is also a population, a constantly changing number of foraging units. Both root and shoot systems are capable of growing exponentially by increasing the number of parts. The growth rate decreases as the parts make demands on limited carbon fixation resources available to the whole plant (e.g. leaves shade other leaves on the same plant).

The most likely contact between plants in a field, in a local community, is among parts of the same individual. Take the individual plant's point of view. The nearest neighbor of a plant part (a leaf) is another such unit on the same plant (another leaf on the same plant). The greatest contact and interaction is with other parts of same plant self. The next most likely contact is with other plants of same species. This situation is altered in viney, upright, spreading or prostrate plants whose most likely contact is the plant it is climbing over. When different plants grow close to each other we see the influence of one level of population (the number of neighboring plants) on the development of another level of population (the number of structural units that make up an individual plant).

There exists an important difference between animals and plants: the population-like structure of the plant phenotype that responds in a plastic manner to local opportunity space-time. The reliable repeatable unit in plants is the constructional module, not the whole. This kind of plasticity is not found in mature animal morphological development. Birth and death elements of the net reproductive rate of plants (ch. 5.2.1) is readily applied to the leaves or some other modular unit of structure of the individual plant. The plant (quite unlike a higher animal) has an age structure in its organs just like a conventional population of different individual plants.

Examples of plant form diversity include mulberry (*Morus* sp.) trees with two different types of leaves (somatic polymorphism), co-existence and cooperation amongst different plant parts of the same individual. The upper leaves are smaller and lobed allowing more light penetration. The lower leaves are larger, with no lobes allowing more light capture.

Johnsongrass (*Sorghum halpense*) underground organs are rhizomes that have buds, which allow shoot axis emergence for lateral exploitation and foraging for space. These rhizomes are also the root system for nutrient and water absorption.

Soybeans (*Glycine max*) have three different types of leaves for different functions at different times of its early life history: cotyledons, unifoliates, first trifoliate. Weedy grasses typically have several leaf types, beginning with the seed embryo coleoptiles, first true leaf, and the various leaves of the several potential tillers of a plant.

The form of the plant may be such that it avoids contact over time, avoidance of intra-specific competition. For example, winter annuals form rosettes in autumn, then a leafy upright flowering stalk the next spring (e.g. shepard's purse, *Capsella bursa-pastoris*). Biennials act similarly.

Plant community patterns are determined by the morphology of branching in plants. For example, Indian cucumber (*Medeola virginiana*) growth and development gives an intraspecific community (a monoculture) its distinct structure. Rhizome buds emerge at certain angles in the soil, this leads to characteristic, specific patterns of spatial structure and foraging in the field.

**9.4.5 Inference in plant density demograpy**. Several cautions are apparent in demographic representations of weed-crop community interactions (see ch. 15, representation of weed life history). One of the dangers in seeing neighbor interactions from only a demographic perspective arises from the convenience of mathematically grouping plant numbers and productivity on a per-unit-area basis. Mean per unit area behavior becomes inexorably community behavior:

> "The population's apparent behavior is that of an integrated system, reacting independently of individuals, with individual behavior subordinate to that of the population." (Silvertown & Doust, 1993)

This of course is false. Neighbor behavior is a complex emergent property of the community that arises from the heterogeneous collective behaviors of individual plants. The heterogeneity that is homogenized in demography masks the evolutionary process that drives change in local communities: natural selection of the fit, elimination of the less fit. The individuals change, but the numbers remain.

Another danger arises from the use of 'space' being limited to either physical location or as an expression of numerical plant density in that space. A more complete concept of space is found in ch. 3.1 on opportunity space-time. This incomplete recognition of how space is perceived by the individual phenotype is but one of many flawed assumptions within a purely demographic view of the nature of weeds in the community.

Hakansson (2003) concludes in his review of yield-density relationships many of the problems inherent in a purely demographic representation of the effects of neighbor interactions. Once the initial population of seedlings in a crop field begin to interact with each other, behavioral responses become non-linear. Viewed together several key points can be deduced (in opposition to what is too often induced from these functions). Linear growth responses of plants only occur at the lowest densities, early in the season. After plants begin to compete with each other, their growth responses over time and with increasing crowding become non-linear. The functional traits underlying this non-linearity are the plastic growth responses of individual phenotypes to crowding. As crowding and time continue the size distribution of crowded plants forms an asymmetrical hierarchy of a few large dominant individuals, and numerous smaller individuals. Observed differences do not arise from normal distributions, but from reaction norms of more complex interactions of individual growth. As such, these inherently non-linear responses confound statistical analyses based on normally distributed (sized) individuals. As such, these inherently non-linear responses measured experimentally confound statistical analyses based on normally distributed (sized) individuals. Non-linear plant behaviors violate the statistical assumptions that underlie the analysis of variance of normal distributions (see ch.14.2.5.1).

Analysis of variance (ANOVA) statistical techniques are used for resolving the observed variance between data sets into components and determining whether sample differences are explicable as random sampling variation within the same underlying population (Borowski and Borwein, 1991). ANOVA statistical techniques are fundamentally based on an assumption that observations are normally distributed. ANOVA statistical techniques make several crucial assumptions (Kirk, 1982): observations are drawn from normally distributed populations and errors are normally distributed within each treatment population and are independent of each other; and, observations are random samples from the populations and independence of error effects; and, the variances of the populations are homogeneous.

For the empirical scientist these assumptions of normally distributed events in nature present some fundamental problems. There is often good reason to suspect that the additivity of treatment and environmental effects and the independent distribution of experimental errors with a common variance are false (Cochran and Cox, 1957). These confusions are systematically explored in Unit 6 (Representation of weed biology).

Several sources of demographic non-linearity in phenotype reaction norms have been made apparent in this unit. The non-linearity inherent in the phenotypic plasticity of growth and development of plants. For example: branch, leaf, flower number in response to opportunity (e.g. light competition); age and stage (developmental phase of life history) differences in species population and between species. For example the non-linear behaviors arising in the composition and traits of competing species in a local population; differences in annual, biennial and perennial plant species life histories; the non-linearity of the spatial distribution of crop and weed plants in a locality; and differential mortality among and within species in response to opportunity, age, stage, etc.

Demography can reveal quantitative characteristics of the struggle for existence and reproduction during neighbor interactions when statistical assumptions are not violated and phenotypic complexity is embraced. Ecological guilds, roles and functional traits can reveal the qualitative causes of those quantitative changes stimulated by neighbor interactions. Together they might begin to provide a complete explanation of the nature of weed interactions in the local community (see ch. 14, representation of weed life history).

What then is the appropriate utility of demographic techniques in the acquisition of plant knowledge? Knowledge is logically deduced explanations of nature. It is not derived from creative induction. Plant demographic experimentation provides one source of inference about the nature of plants. Demography as discussed in this unit can provide a basis for creative hypothesis and conjecture formulation of future experimental approaches (see U6.1). High inference experimentation testing hypotheses and conjectures may include reaction norm elucidation of key functional traits responsible for competitive behaviors in a local community; for example branch-leaf-flower phenotypic plasticity (see ch. 19, *Abutilon theophrasti*) or seed dormancy heteroblasty (see ch. 17.4.3.1, seed germinability-dormancy heteroblasty). Reaction norm elucidation can provide logical deductions that form an explanation of nature when based on individual phenotypes, the units of natural selection and the sources of local adaptation.

This approach to weed knowledge discovery is revealed in the search for the means by which velvetleaf (*Abutilon theophrasti*) competes with soybeans (*Glycine max*). See this full case study history exemplifying the limits of demographic inference in ch. 19.

I conducted experiments in Michigan in 1978 and 1979 to determine the effect of velvetleaf on soybean productivity. They utilized two complementary, demographic, experimental designs: a weed density-crop yield experiment, and a replacement series experiment with varying proportions of crop and weed at equal population densities. These experiments provided a quantitative estimate of the soybean losses caused by velvetleaf. They also provided insights into their interaction, despite my naïve violation of the statistical assumptions of normality in the analysis. Low mixture infestations of velvetleaf greatly reduced midseason soybean flowering as velvetleaf branching was greatly increased. Velvetleaf branching connotes plant shoot architecture plasticity, the plant body upon which leaves and flowers hang. The former feeds the later, biomass is proportional to seed yield. Low levels of velvetleaf infestations provided opportunity to seize and exploit light above the lower soybean canopy. This provided more space for velvetleaf to branch and form more leaves. The consequential shading increased floral abortion in the soybeans, while its body architecture remained unchanged.

What was the cause of this qualitative interaction among the species gained with quantitative, demographic, observations? What functional, plastic, traits in soybeans and velvetleaf led to this neighborhood interaction? What was the underlying biological explanation for this plastic inhibitory-stimulatory response? Caught up in the interest in allelopathy at the time I inferred from these studies that velvetleaf exuded a toxin that inhibited soybean growth. I was wrong, I had no data to support this erroneous inference (see Gressel and Holm, 1964). Nothing in the yield-density demographic experiments I performed provided a high inference basis of the true competitive interaction mechanism, although I had some strong clues in the branch-leaf-flower data that I (too easily) mis-inferred.

William Akey subsequently conducted several field experiments in 1986-1989. We creatively hypothesized that light was the specific opportunity space-time factor being seized and exploited preferentially by velvetleaf at the expense of soybeans. He clearly demonstrated this was true with detailed light canopy measurements. These involved functional traits for rapid early growth and extensive stem growth leading to tall plant stature. They include a branching architectural pattern that situates much of the velvetleaf leaf area near the top of the plant, above the leaves of shorter competing species, and rapid scenescence of shaded lower leaves. Additionally, active leaf movements are mediated by the pulvinus at the leaf base-petiole terminus which control several advantageous behaviors (water use efficiency during diurnal stress periods; control of light interception; shading of shorter neighboring species).

We learned that quantitative, demographic descriptions of the local agricultural community based in inappropriate mean behaviors could provide some foundation for continuing studies of functional traits. Demography pointed to competitive light interactions, despite its inherent confounding effects of the underlying non-linear complexity of individual behaviors. Creative reflection led to more focused light-trait experimentation from which better explanations were deduced.

## Chapter 10: Weed Community Biodiversity, Structure and Dynamics

"… we may regard a pond, or a forest, as a community with a certain structure, and even a certain stability. But the structure and stability are maintained in the face of a constant turnover of participants. Individuals immigrate and emigrate, new ones are born and old ones die. There is a fluidity, a jumping in and out of component parts, so that it becomes meaningless to try to distinguish 'true' community members from foreign invaders. So it is with the genome. It is not a static structure, but a fluid community." Dawkins, 1999.

**The weeds' story**. I am foxtail.

You humans, you can never fool me, you can never 'out-fox' me. Why? Because I have tricks you find so difficult to discover. Let's start with my family. If you kill me you get 'vendetta', revenge from every place on Earth where the family lives. Kill all of me in a field, what happens next? A giant vaccuum of opportunity for my family. You might as well just hang a sign out on the fence post that says: 'free lunch, enquire within'. You can't stop us.

Let me tell you more about my family, we are different. Sometimes we hang out together in the same field with our crop, 'eater', cousins. Sometimes the big polyploid cousins show up looking for a free meal, a summer picnic in the sun. Plenty of room for everybody. Sometimes its distant relatives that just arrived by boat, train, car, airplane, wind, tractor wheels, you name it. Most of the time it's the older generation that emerge from the same field's dirt. They never ever wait to be invited, they just show up. We owe it all to you. What you call metapopulations we call family, weed mafia, cosa nostra. Wipe out my family? Expect some unwelcome visitors on your doorstep in the near future. It's pay-back time!

Some of you humans think that inviting everyone but us to a field will keep us away. All those nice docile friendly types. Wimps! So you plant lots of different crop species in a field. You rotate them yearly. Encourage lots of wimpy weed species to take up residence. Open the doors to any and all so that the field looks like the United Nations. Good luck humans! Crowds of other plants who don't have the right set of tools are doomed. We will eat them for lunch and take over. So much for your religion of biodiversity. To make it in our neighborhoods you gotta be tough, you gotta have the right moves, the right tools, you gotta dominate everyone in the 'hood. It's what successful gangs always do. You gotta terrorize the locals with all the tricks you have learned from past experience. What you did to kill me in the past makes me stronger than anyone else in the 'hood. We owe it all to you. Thank you humans.

**Summary**. Adaptation in local weed-crop communities is expressed in its biodiversity, community structure and dynamic changes with time in the exploitation of opportunity spacetime. This chapter is about the consequences of ecological evolution of weed populations, adaptation of weed life history from birth through seedling recruitment leading to neighbor interactions that occur in local communities after weeds and crops assemble.

Weed population biodiversity has structure, changing with time in response to the adaptive culminations in the community. The local community population structure at any time is a consequence of past events, and can only be understood in terms of what has happened up until then. Genetic connections, gene flow, between individuals, populations and species exist that provide advantages in the struggle to exploit local opportunity: weed species associations. These include wild-crop-weed complexes, preadaptive colonizing archetypes (generalist-specialist genotypes, reproductive colonizing types) and colonizing species associations (species-groups, polyploid species clusters, aggregate species). Weed community population structure is revealed at several spatial-temporal scales of local opportunity. Larger scale phenomena emerge from smaller scale community structure, from the molecular to the individual: the behavioral traits assume a role, forming a functional guild. This guild is based on clusters of interacting traits that allow them to dominate local opportunity at their neighbors expense in a very particular and specialized manner. Every species in a weed-crop plant community plays a role, fills a niche, utilizes resources and conditions, has a trade or is a member of a guild that functionally defines the species and provides insight into both the opportunity space available in the locality and the predicates of the neighbor interactions that will ensue over the season and life history of the plants.

Weed biodiversity and population structure predicate community dynamics. Weed community dynamics is the formation of a local plant community and the changes that occur within thereafter, including the processes of community assembly and ecological successsion. Weed community assembly begins with the first weed colonization, invasions of local opportunity of an empty local habitat and ends with terminal phases of ecological succession in the fully mature community. Change is constant in weed communities, the action of powerful forces leading to ecological community succession, wherein the current community creates new opportunity for the communities of the future. Community dynamics is an emergent property of interacting phenotypes with successful life history traits. These life history traits allow those populations to play a unique role in exploiting locally available opportunity the community.

Biodiversity is the source of all future local weed populations. Weed biodiversity is the pool of potential candidate populations that might invade, seize and exploit local agricultural opportunity. Biodiversity is the source of novel traits for crop improvement. Local biodiversity is an indicator of locally available opportunity, the predicate and foundation of community dynamics and future evolutionary adaptation by its species. Considerable confusion exists of the nature of biodiversity, complexity and system stability. Weed biodiversity in the community has been linked with the concepts of stability and equilibrium. Biodiversity, stability and sustainability are viewed from different perspectives, with much opinion and some disagreement. It has been conjectured that increased agro-ecosystem biodiversity will result in more stable and sustainable, not to mention more economically viable, crop production systems. There exists no evidence that biodiversity inherently confers stability. Stability is illusory in plant communities. Equilibrium conditions never exist, persist, in an open system such as evolutionary complex adaptive system; open systems require energy to counter entropy

The only stability found in nature is the enduring biodiversity of weed-crop groups, the constancy of human hunger, and the all-consuming human need to manage and control space.

## 10.1 Weedy Community Adaptation

At this point in our weed story we have satisfied the preconditions for the process of evolution to occur: local excess numbers of particular weed species. These weeds have successfully completed the invasion process. Some were naturally selected, others naturally eliminated. The survivors formed communities around the world. They have generated genotypic and phenotypic variation appropriate to the opportunity they have been afforded. The weed species present today have avoided mortality completely, their parentage extends backward, continuous, unbroken in time since the beginning. In that time they have successfully inherited the fittest life history traits in the process of reproduction. As a consequence they have adapted successfully in local demes, in their life histories. This leads us to the weed community, to community biodiversity, which has structure, which continues to change with time. These are the adaptive culminations in the community, the setting of evolution and change, the scene of unceasing seizure and exploitation of opportunity spacetime.

In this chapter we return to the successful weed species first introduced in chapter 1, in tables 1.4 and 1.5. The goal of this chapter is look at the consequences of this ecological evolution of weed populations (unit 2), of the adaptation of weed life history from birth until seedling recruitment (unit 3), and of the neighbor interactions that occur in local communities after weeds and crops assemble (chapter 9). The weed-crop groups of the world today are the plants we have earned for ourselves as humans, the companions we deserve. These specific species are both freed and restrained as weedy organisms because of their biodiversity, their structure in communities, and the consistent dynamics their roles as weeds allow them to perform.

## 10.2 Weed Communities

Understanding weed communities, the populations that live in them and how they change with time, is one of the most important areas of weed biology. Weed communities are the consequence of particular phenotypes seizing and exploiting opportunity. Much remains to be discovered about how weed populations assemble in communities with crops and how they change over time.

John Harper (1977) highlights an important consideration in the quote at the beginning of chapter 12 (Weed-crop communities as complex adaptive systems). He uses metaphors from stories, movies, or the theater to point out the difficulty in studing community dynamics, and how as humans we often look at them in an inappropriate way. There is a tendency to set the scale of observation of the community as system at our convenience, missing the action and failing to observe the plot. The plot in communities is acting out a plants' life history to seize and exploit opportunity at a neighbors' expense. The action is the behavior and timing during that life history, as well as the action of neighbors, disturbance and death. The actors in the story of community are the individual plants, their phenotypes, the traits they display, and the roles they play. The actors also include the dormant members in the soil seed pool as well as the dominating crops of the community. The actors in this story play a role in every community, their guild or trade when the species is observed as a whole. The stage and scenery is the environment and the neighbors of a locality, or a microsite in that locality.

The local community population structure at any time is a consequence of past events, and can only be understood in terms of what has happened up until then. In human terms, this history is filled with tragedy, stress and struggle, in which few survive. Those that do manage to survive are capable of exponential reproductivity. In the study of weed communities it is easy to look at this population structure at times convenient to the observer; e.g. planting, emergence, harvest. But these single 'snapshots' of community tell us little, in the same way that a single frame from a movie reveals little of the story that is unfolding.

All of this raises the question of at what scale (space, time) is it appropriate to describe the system? Based on what? With what criteria? For now, lets set the stage and try to understand who these interesting actors, the weeds, are and what they do over time.

So, given that community structure and dynamics is complex, what insights have been gained to help us understand the weed biology of agricultural communities?

A primary feature of assembly of weeds and crops in agricultural communities is periodic (annual), severe and widespread disturbance that eliminates above-ground vegetation (e.g. winter kill, tillage including seedbed preparation, early season herbicide use). This annual recommencement of community assembly makes studies of agro-communities experimentally more tractable than in more complex, longer-lived plant communities. Initial plant establishment plays a large role in these annually disturbed agricultural fields (Dekker, 2004).

Weed communities will be revealed in this chapter in several ways. First is a discussion of weed biodiversity at several scales and perspectives. Weed biodiversity is the pool of potential candidate populations that might invade, seize and exploit local agricultural opportunity. Second, weed population structure is discussed. Gene flow over historical time has produced the successful weedy phenotypes that exploit opportunity. Although evolution acts at the level of the individual plant, residual genetic connections between individuals, populations and species still exist and provide an often hidden advantage in the struggle to exploit local opportunity. In the final section of the chapter is discussed community dynamics, changes with time. The source of this change is the evolving phenotypes of the community, constantly struggling with the habitat and neighbors. In this relentless adaptation, weed populations often assume a role, form a guild, based on clusters of interacting traits that allow them to dominate in particular localities. Nothing stays the same, change is constant in weed communities. Plant communities, crops and weeds, are typically removed on a regular basis, usually annually. Weed communities thrive because they begin again every year: they are colonizing species. Against this human managed opportunity space-time are powerful forces leading to ecological community succession, wherein the current community creates new opportunity for the communities of the future.

We begin with definitions:

**community**
1: any group of organisms belonging to a number of different species that co-occur in the same habitat or area and interact through trophic and spatial relationships; typically characterized by reference to one or more dominant species (Lincoln)
2: in ecology, an assemblage of populations of different species, interacting with one another; sometimes limited to specific places, times, or subsets of organisms; at other times based on evolutionary taxonomy and biogeography; other times based on function and behavior regardless of genetic relationships (Wikipedia, 6.08)
3: in ecology, assemblages or associations of populations of two or more different species occupying the same geographical area and in a particular time; in its simplest form, groups of organisms in a specific place or time (Wikipedia, 6.14)

**biological community**
1: biocoenosis, biocoenose, biocenose
2: all the interacting organisms living together in a specific habitat (or biotope); biotic community, ecological community; the extent or geographical area of a biocenose is limited only by the requirement of a more or less uniform species composition. (Wikipedia, 6.08)

**local community**: all the interacting demes in a locality

**ecosystem**
1: a community of organisms and their physical environment interacting as an ecological unit; the entire biological and physical content of a biotope
2: an ecosystem is a natural unit consisting of all plants, animals and micro-organisms (biotic factors) in an area functioning together with all of the non-living physical (abiotic) factors of the environment (Wikipedia, 6.08)
3: a biotic community along with its physical environment (Tansley, 1935)

A biological community is characterized by the interrelationships among species in a geographical area. These interactions are as important as the physical factors to which each species is adapted and responding. It is the specific biological community that is adapted to conditions that prevail in a given place. Biotic communities may be of varying sizes, and larger ones may contain smaller ones. The interactions between species are especially evident in their substrate (resources, food or feeding) relationships. Biotic communities can be better understood by observing the utilization of limiting resources, or for animal populations the food network, to identify how species acquire substrate and then determine the system boundary. (Wikipedia, 6.08)

Agricultural communities consist typically of interacting crop species and the weeds that invade that opportunity. Local opportunity demands exploitation by weedy species. Weed biodiversity and population structure predicate community dynamics. Weed populations are preadapted phenotypes that invade opportunity, and successful populations form communities. The structure of communities provides opportunity for certain roles to be played, and not for others. The functional characteristics of those populations, and the time of their expression, are the basis of plant community dynamics. Community dynamics is an emergent property of interacting phenotypes with successful life history traits. These life history traits allow those populations to play a unique role in exploiting locally available opportunity the community. Local opportunity changes over time as a direct consequence of both disturbance and the current roles played by populations in the community. The behavior of populations in an existing community provides opportunity for new community structure (e.g. ecological succession).

Weed biodiversity can be seen as weed-crop community biodiversity. The next two chapter sections cover weed community structure and the strategic role exploitation of opportunity spacetime by means of trait-role guilds. Additionally, also discussed is the tendency for a local community to change, ecological succession, due to the changing of opportunity spacetime created and lost by phases of species-role changes in neighbor interactions,

## 10.3 Weed Community Structure

Weed communities are populated by individual phenotypes that colonize local agricultural opportunity. What is the population structure of those plant communities? Weed community population structure is the emergent property of the evolution of those individuals over time. Gene flow over historical time has produced the successful weedy phenotypes that exploit opportunity. Although evolution acts at the level of the individual plant, residual genetic connections between individuals, populations and species exist that provide advantages in the struggle to exploit local opportunity.

Weed community population structure is revealed at several spatial-temporal scales of local opportunity. At the lowest level of plant system organization, qualities from the molecular to the individual plant reveal the basis for community population structure. Larger scale phenomena are an emergent property arising from smaller scale sources of community structure: the traits that determine the behavior of the phenotype in the community. In the relentless evolutionary process of adaptation, weeds assume a role, they form a functional guild. This guild is based on clusters of interacting traits that allow them to dominate local opportunity at their neighbors expense in a very particular and specialized manner.

Weed communities are populated by individual phenotypes that colonize local agricultural opportunity. Weed community population structure is the emergent property of the evolution of those individuals over time. Weed community population structure is revealed at several spatial-temporal scales of local opportunity. Community structure is apparent at higher levels of plant system organization: global to local community. Observation of this larger scale structure provides insights of the outcome of long-term weedy adaptation: population genetic structure and species associations. Gene flow over historical time has produced the successful weedy phenotypes that exploit opportunity.

**10.3.1 Metapopulation structures**. The spatial distribution of species is patchy at many scales (from local to global) (Silvertown and Charlesworth, 2001). It tends to be particularly fragmentary near the range boundaries, limits of opportunity spacetime, of a species. Regional dynamics describe changes in groups of populations (demes). At scales larger than that of the local deme, populations of the same species are connected in several ways. First, the reciprocal exchange of seeds, propagules and pollen in metapopulations. Second, the one-way migration of propagules to recipient populations that are unable to sustain themselves independently. This probably occurs at the edge of the niche, opportunity, or species range boundaries. Third, remnant populations able persist without immigration through unfavorable periods in the form of soil seed pools.

**metapopulation**:
1: a group of partially isolated populations belonging to the same species; subpopulations of natural populations that are partially isolated one from another and are connected by pathways of immigration and emigration; exchange of individuals occurs between such populations and individuals are able to recolonize sites from which the species has recently become extinct (Lincoln et al., 1998)
2. a network of local populations connected by dispersal, whose persistence depends on reciprocal but unsynchronized migration between populations (Silvertown and Charlesworth, 2001)

**10.3.2 Plant genetic associations.** Although evolution acts at the level of the individual plant, residual genetic connections between individuals, populations and species exist that provide advantages in the struggle to exploit local opportunity: weed species associations. In this section we will observe several of these associations between weed species and crops. These include wild-crop-weed complexes, preadaptive colonizing archetypes (generalist-specialist genotypes, reproductive colonizing types) and colonizing species associations (species-groups, polyploid species clusters, aggregate species). All of these ways of looking at associations of different species overlap. Each provides a different insight into the consequences of evolution in the relentless adaptation to local opportunity that results in the population community structure we see in present day communities. These associations can be viewed by the following criteria of relatedness (table 10.1).

**Table 10.1** Plant genetic associations of closely related species based on genetic evolutionary history, biogeographic distribution and functionally preadapted phenotypes.

| Source | Plant Association Type |
|---|---|
| Genetic evolutionary history | wild-crop-weed complex |
| Genetic evolutionary history & Biogeographic distribution | population genetic structure |
| | species-group |
| | polyploid cluster |
| | aggregate species |
| Genetic evolutionary history | generalist-specialists |
| Functionally preadapted phenotypes | reproductive colonizing types |

These species associations will lead us logically to a discussion of weedy traits, to strategic roles-guilds of trait clusters, and to changes in local communities including weed population shifts and community ecological succession.

**10.3.2.1 The origins of weeds: wild-crop-weed plant complexes.** Where do weeds come from? How long have we had weeds? Wild colonizing plants existed before agriculture, able to seize opportunity when natural disturbances destroyed or altered exisiting plant communities. It was these wild colonizing species that became weeds (and crops) with the advent of agriculture. In many cases, it was these vigorous colonizing species that humans gathered and selected among for the very earliest crops. Refer to chapter 1.5-1.6-1.7 for a more complete development of this important species association.

**10.3.2.2 Biogeographic population genetic structure.** Population genetic structure is the spatial distribution of genotypes (and phenotypes) across the locality, the landscape and around the world. It is biogeography, and is relevant to what species and intra-specific genetic variants occupy a habitat.

**population genetic structure**:
1: the genetic composition and gene frequencies of individuals in a population
2: any pattern in the genetic makeup of individuals within a population which can be characterised by their genotype and allele frequencies (Wikipedia, 12.10)

One of the important aspects of population biology is population genetic structure, which forms the basis of plant spatial and temporal organization. Knowledge of population genetic structure has practical implications (Barrett and Husband, 1990). It can be used to reconstruct the historical process of invasion, migration and colonization, and provide insights into the ecological persistence and evolutionary potential of populations in new habitats, leading to a better understanding of weedy adaptation. It provides a foundation for understanding the spatial structure of individuals within an individual field or local community. Population genetic structure is unique to a species, or species-group.

One of the best ways to understand the population genetic structure and opportunity is to observe it in a widely distributed species-group. The population genetic structure of the weedy *Setaria* species-group is featured in ch. 17.3.2.2, which is a condensed version of the allozyme work of R.L Wang (Wang et al., 1995a, b).

**10.3.2.3 Genotype structuring: species associations for weedy colonization**. Genetic evolutionary history and biogeographic distribution are revealed in species-groups, polyploid clusters and aggregate species.

Weed communities are populated by individual phenotypes that colonize local agricultural opportunity. Weed community population structure is the emergent property of the evolution of those individuals over time. Weed community population structure is revealed at several spatial-temporal scales of local opportunity. Community structure is apparent at higher levels of plant system organization: global to local community. Observation of this larger scale structure provides insights of the outcome of long-term weedy adaptation: population genetic structure and species associations. Gene flow over historical time has produced the successful weedy phenotypes that exploit opportunity. Although evolution acts at the level of the individual plant, residual genetic connections between individuals, populations and species exist that provide advantages in the struggle to exploit local opportunity: weed species associations. In this section we will observe several of these associations between weed species and crops. These include colonizing species associations (species-groups, polyploid species clusters, aggregate species). All of these ways of looking at associations of different species overlap. Each provides a different insight into the consequences of evolution in the relentless adaptation to local opportunity that results in the population community structure we see in present day communities.

**10.3.2.3.1 Species-groups**. Many of the most successful colonizing plant species co-exist together loosely in a species-group. Species-groups provide several advantages in colonization. They provide a more complete exploitation of opportunity niche space and generate variation appropriate to the habitats and landscapes being exploited. Definition:

**species-group**:
1: a group of closely related species, usually with partially overlapping ranges; sometimes used as an equivalent of superspecies (Lincoln et al., 1998).
2: an informal taxonomic rank into which an assemblage of closely-related species within a genus are grouped because of their morphological similarities and their identity as a biological unit with a single monophyletic origin (Wikipedia, 12.10)

**superspecies**: a group of at least two more or less distinctive species with approximately parapatric distributions; a superspecies consisting of two sister [sic] species is called a "species pair" (Wikipedia, 12.10)

One of the best adapted and known weedy species-groups are the foxtails (*Setaria*; see table 10.2; ch. 17.3.2.2.2, species associations; and table 17.3, *Setaria* species associations). Another related way of seeing closely related species group is the polyploid-species cluster (Zohary, 1965), which we will discuss next. The pigweeds (*Amaranthus* spp.) are another infamous species-group. There is considerable diversity within this superspecies, maybe because promiscuous outcrossing between plants (and maybe species) makes firm species identification problematic. Is it any wonder that most weed management tactics miss at least some of these diverse pests? The smartweed (*Polygonum*) species-group consists of several species, including ladysthumb and Pennsylvania smartweed.

Table 10.2. *Setaria* species associations: clade, wild-crop-weed complex, polyploid species cluster (diploid, tetraploid) and species-group.

| Clade | Crop | Wild-Weed | Weed |
|---|---|---|---|
|  | Diploid (2N) | Diploid (2N) | Tetraploid (4N) |
| S. viridis | S. viridis, subsp. italica | S. viridis, subsp. viridis |  |
|  |  |  | S. verticillata |
|  |  |  | S. faberi |
| S. pumila |  |  | S. pumila |
|  |  |  | S. geniculata |
| Clade |  | Species-Group | |
|  | Polyploid Cluster | | |

**10.3.2.3.2 Polyploid species clusters.** Some important colonizing species-groups consist of closely related species consisting of several deriviative polyploids clustered around a diploid ancestor. This species-group/polyploid species cluster is preadapted to exploit opportunity space and is an important type of wild-crop-weed complex, discussed below. The characteristics of a polyploidy species cluster have had a significant impact on the evolution of colonizer species.

The first feature is the occurrence of a diploid "theme", a "common trend of adaptive specialization" of preadaptation to certain habitats and opportunity spaces (Zohary, 1965; p. 418). This pivotal genome is a conservative gene complex that controls the evolutionary theme within a pool of recombinable material in the polyploid, modified, genomes. Thus a dual system is created wherein "the genes controlling the preadaptive "theme" are held together by one part of the chromosomal complement and where wide "variation in theme" is provided by a second part of the chromosomal complement". The result is a species cluster "designed for "canalization" of the evolving population in its preadaptive, basic "theme"" protecting it from haphazard variation from unrestricted recombination.

The *Aegilops-Triticum* is a species-group of aggressive, genetically linked, polyploids preadapted for rapid colonization (Zohary, 1965). This group is more than its component, independent, allotetraploid (amphidiploid) species. Instead they are a genetically linked species cluster, a "compound, loosely interconnected superstructure". The *Aegilops-Triticum* species-group is a "system of compound amphiploidy in combination with a mating system of predominant self-pollination that have apparently enabled the polyploids to build up genetic variation rapidly and evolve successfully as aggressive weedy annuals." This genetic structure facilitates introgression and the creation of large pool of genetic variation by means of a "specific combination of elements: compound amphidiploidy, buffering pivotal genomes, and predominance of self-pollination."

An important contrast in historical gene flow exists between diploids and polyploids within this species-group. The diploid genomic groups are reproductively isolated from one another as opposed to the loose genetic connections between the polyploids. In Zohary (1965, p. 414) we learn, "Another striking contrast between diploids and polyploids (in the historical gene flow within the wheat wild-crop-weed complex) is the reproductive isolation of the diploid genomic groups from one another as compared to loose genetic connections between polyploids. Genetic links and gene flow through occasional spontaneous hybridization and subsequent introgression are apparently a general rule in the case of seven species sharing the pivotal Cu genome."

All this in *Aegilops* is similar to that in the *Setaria* species group: both are searching opportunity space with this type of loosely interrelated and interacting genome systems in the different species. In *Setaria*, green foxtail (*S. viridis*) is diploid theme, giant foxtail (*S. faberi*) and others are the polyploid offshoots.

In the *Setaria* species-group several weedy species often coexist in a single field, each exploiting a slightly different opportunity space or niche, resources left available in somewhat different ways and times (Dekker, 2004b). The geographic distribution of the several weedy foxtail species provides an initial insight into the niches they occupy (Dekker, 2004b). *S. viridis* is the most ancient and widespread weedy foxtail species, and is often the most commonly found species when more than one is found in a single field. The polyploid weedy foxtails are less widely distributed than the diploid *S. viridis*, and are often more specialized. Weedy foxtail biogeographical distribution is associated soils that possess a wide range of moisture, gas, and nutrient regimes with distinctive seasonal and diurnal temperature fluctuations. These seasonal temperature-water cycles are correlated with their cyclical germination behavior (e.g. Dekker et al., 2001; Forcella et al., 1992, 1997). Dekker, 2004b: Speciation by polyploidization (either alloploidy or autoploidy) may be an important means by which new foxtail species were formed. Based on chromosomal data, Khosla and Sharma (1973) speculate polyploidization at various levels has played and active role in speciation and delimiting taxa in *Setaria*. Polyploidization of the more diverse and ancient *S. viridis* may have been the genesis of the specialized and less diverse *Setaria* spp. set (giant, yellow, bristly, knotroot) of specialized and less diverse *Setaria* spp. Allotetraploid forms of *S. faberii* and *S. pumila* have been explained as ancient crosses of *S. viridis* with and unknown diploid species (Li, 1942; Kholsa and Singh, 1971; Till-Bottraud et al., 1992). Li et al. (1942) concluded that *S. faberi* is a product of *S. viridis* and an unknown *Setaria* sp., followed by a chromosome doubling (polyploidization) event, forming an alloploid species. Relatively homogeneous population allozyme data suggest this polyploidization event in *S. faberi* was a relatively recent evolutionary event (Wang et al., 1995b). Genetic analysis has suggested that some variants of *S. verticillata* (n=18) may be the product of chromosome doubling of *S. viridis*, autotetraploidy (Till-Bottraud et al., 1992).

10.3.2.3.3 **Aggregate species**. Species boundaries among several kinds of closely related organisms can be difficult to distinguish morphologically. These groups of organisms are often refered to aggregate or microspecies.

**aggregrate species:**
1: a group of species that are so closely related that they are regarded as a single species.
2: a named species that represents a range of very closely related organisms (Wiktionary, 5.14)
3: a named species consisting of a large group of identified and named microspecies (Wikipedia, 5.14)

Microspecies are usually associated with apomictic plants (ch. 5.4.4.3) which are genetically identical from one generation to the next, each lineage has some of the characters of a true species, maintaining distinctions from other apomictic lineages within the samegenus, while having much smaller differences than is normal between species of most genera. In some genera, it is possible to identify and name hundreds or even thousands of microspecies, which may be grouped together as species aggregates, typically listed in floras with the convention "*Genus species* agg." (such as the bramble, *Rubus fruticosus* agg.). Other examples of aggregate species include *Limonium binervosum* and *Aspergillus niger*; *Chenopodium album*; dandelion (*Taraxicum officinale*); and *Gallium* spp.

A species complex is typically considered as a group of closely related but distinct species that are very similar in appearance to the point that the boundaries between them are often unclear. Terms sometimes used synonymously but with more precise meanings are: cryptic species for two or more species hidden under one species name, sibling species for two cryptic species that are each other's closest relative, and species flock for a group of closely related species living in the same habitat (Wikipedia, 5.16).

**10.3.2.4 Genetic structuring: pre-adaptive colonizing archetypes.** Plant genetic associations of closely related species based on genetic evolutionary history and functionally preadapted phenotypes include generalist-specialists and reproductive colonizing types. These groups clearly show the functional, trait, characteristics as the cause of their association, therefore they are the bridge to traits, roles and guilds developed in the following sections. Another category of weed species associations is based on function within the local community. In both instances these categories are expressions of preadaptation.

**10.3.2.4.1 Generalist-specialist genotypes.** The following sections are taken from Wang, et al. (1995) and provide some insight into generalist and specialist genotypes with the *Setaria* species-group.

Genetic variation and the evolutionary success of colonizers. The population genetic structure of many widely distributed, introduced, self-pollinating, weed species clearly indicates that a high level of genetic variation is not a prerequisite for successful colonization and evolutionary success (Allard, 1965; Barrett and Shore, 1989). Two contrasting, adaptive, strategies are hypothesized to explain weedy adaptation and the success of colonizers: genetic polymorphism with the development of locally adapted genotypes ("specialists"), and phenotypic plasticity for the development of "general purpose" genotypes (generalists) adaptable to a wide range of environmental conditions (Baker, 1965, 1974; Bradshaw, 1965; Barrett and Richardson, 1986). *Setaria* population genetic structure itself allows some insight into the dichotomy of generalism versus specialization. *Setaria* possess both generalists and specific strategy types. A key observation is that although a single, multilocus genotype predominates or is fixed in all local populations, not all multilocus genotypes are equally prevalent within individual weedy *Setaria* species (Wang et al., 1995a, 1995b).

The most striking example of this is in *S. pumila*, where the most common multilocus genotype was found in 53 (of 94 evaluated) accessions from widely separated geographic areas in Europe (Belgium, Czech Republic), Asia (Manchuria, China, Turkey), and North America (Ontario, Canada; eastern U.S.: MD, PA; western US: WA, WY; south U.S.: AK; north and midwestern U.S.: IA, IN, KS, MN, ND, WI) (Wang et al., 1995b). Overlaying this pattern of homogeneity were other, less abundant *S. pumila* genotypes, each with a more narrow geographical and ecological distribution.

The population genetic structure of *S. viridis* suggests that this species also possesses both generally adapted and specially adapted genotypes. Many *S. viridis* populations are strongly differentiated genetically (e. g., northern and southern North America), while other populations remain identical. The most widely distributed *S. viridis* genotype (fixed multilocus genotype) occurred in 25 (of 168 evaluated) accessions from six countries, from both the Old World and the New World (Wang et al., 1995a, 1995b). This common *S. viridis* genotype was found in many geographic locations, in very different ecological habitats: the 270midwestern U.S. Corn Belt areas of Indiana, Minnesota, and North Dakota; the eastern U.S. Corn Belt of Maryland; the western U.S. agricultural valley of Washington; the highlands of Ontario, Canada; the lowlands of Belgium in northwest Europe; the plains of Bohemia, Czechoslovakia in central Europe; the waste areas interspersed between limestone outcroppings in the Akiyoshi Dai National Park, Honshu, Japan; and between cracks in a cement pier over the bay in Yokohama Harbor, Honshu, Japan. Interestingly, this common genotype has not yet been found in Iowa, despite the diversity of *S. viridis* populations found in this state.

These observations reveal a complex hedge-betting strategy by individual foxtail species, a strategy that balances general adaptation with the additional niche opportunities available with specialization: the ratio of general and special genotypes available in a species for invasion. The ratio of generalist to specialist genotypes was quite different within *S. pumila*, *S. geniculata* and *S. viridis* (Wang et al., 1995a, 1995b).

**10.3.2.4.2 Genetic-reproductive colonizing types.** An alternative organization was provided by Ehrendorfer (1965), correlating successful colonizing ability and critical genetic and reproductive characteristics. In some ways like the list of Baker (1965), Ehrendorfer surveyed (interestingly, in the same book as Baker, 1965) a wide range of plant species and collated their traits, looking for patterns of traits in species of particular habitat and behavioral types. This type of analysis is directly relevant to weed roles or guilds in communities, to be discussed later in this chapter.

The traits of interest included life-form (e.g. annual), mode of fertilization (e.g. autogamy), differentiation of seed structure and dispersal mechanism, utilization of vegetative reproduction, and chromosome ploidy condition. Colonizing plant types were also categorized in terms of population structure: mode of reproductive isolation (e.g. floral anatomy), population differentiation (e.g. allopatric), level of hybridization, population variability, and the plant successional habitat colonized.

Three different kinds of plant types were identified, I., II., and III (table 10.3) These types represent variation in genetic and evolutionary strategies of colonization: "The polyploid perennials (I) are "conservative" types. They draw upon genetic diversity mainly built up on the diploid level. In their ecologically more "centrally" located position they operate successfully by mobilizing and recombining genetic material from ecologically divergent parental races. On their more reticulate course of evolution neither rapid nor very decisive changes have occurred.

The dysploid (aneuploid: chromosome number not exact multiple of haploid) annuals (III), on the other hand, can be regarded as evolutionary "avante-garde". Their increased mutation rate is reflected in gross changes of chromosome structure and chromosome number. Their variable offspring often includes rather aberrant new types and has better chances for survival or even expansive colonization in their ecologically more "marginally" located habitats. The structural differentiation of genomes makes swamping of such new types by hybridization less likely. These types then appear as carriers of rapid and divergent evolution. The evolutionary strategy of the autogamous annuals (II) seems to approach partly more type I (e.g. in respect to the occurance of polyploidy), partly more type III (e.g. in respect to the occasional establishment of quite aberrant new lines."

**Table 10.3** Genetic-reproductive colonizing species types (I., II., III.) from Ehrendorfer (1965).

| | GENETIC-REPRODUCTIVE COLONIZING PLANT TYPES | | |
|---|---|---|---|
| TYPE | I. | II. | III. |
| Life-Form | polyploid perennials: -conservative types -ecologically 'central' position | autogamous annuals | dysploid annuals (chromosome number not exact haploid multiple) |
| Course of Evolution/ Reproductive Stragegy | reticulate (net-like): no rapid or decisive changes have occurred | evolutionary strategy: a. partly that of I.: polyploidy b. partly that of III. occassional aberrant line establishment | evolutionary 'avante-garde': a. carriers of rapid and divergent evolution b. variable, aberrant off-spring are new types: enhanced survival-colonizing ability in 'marginal' habitats |
| Genetic Flexibility/ Stability | a. genetic diversity from diploids b. successful by | consequence of: a. short sequence of (annual) generations b. maximum exploitation of gametic mutation rates | |

| | | mobilizing and recombining divergent parental races | c. occasional outcrossing and hybridization (including alloploidy) between selfing lines in type II | c. occasional outcrossing and hybridization (including alloploidy) in the high outcrossing rate in type III<br>a. increased mutation rate: gross chromosome structure-number changes<br>b. swamping by hybrid-ization less likely by new, structurally differentiated, genomes |

The genetic systems in all these types of colonizing species optimize modes of balance between genetic flexibility (selection among highly variable progeny) and stability (fixation and multiplication of successful biotypes) (Stebbins, 1958). Genetic flexibility in type II. and III. annuals is a consequence of the short sequence of generations, maximum exploitation of gametic mutation rates, occasional outcrossing and hybridization (including alloploidy) between selfing lines in type II, and the high outcrossing rate in type III. The outcome of these different genetic and reproductive systems is population genetic structure.

**10.3.3 Ecological roles-guilds-trades in weed-crop communities.** Weed communities are constantly changing. Community dynamics arise and occur as opportunity spacetime changes with time. The source of this change is the evolving phenotypes of the community, constantly struggling with the habitat and neighbors. Nothing stays the same, change is constant in weed communities.

Weed community population structure is revealed at several spatial-temporal scales of local opportunity. In the preceding section community structure was apparent at higher levels of plant system organization: global to local community. Observation of this larger scale structure provided insights of the outcome of long-term weedy adaptation: population genetic structure and species associations. At the lowest level of plant system organization, qualities from the molecular to the individual plant reveal the fundamental bases for community population structure. Larger scale phenomena are an emergent property arising from smaller scale sources of community structure: the traits that determine the behavior of the phenotype in the community.

In the relentless evolutionary process of adaptation, weeds assume a role, they form a functional guild. This guild is based on clusters of interacting traits that allow them to dominate local opportunity at their neighbors expense in a very particular and specialized manner (see ch. 9.3, strategic roles and traits of interference and facilitation with neighbors).

Agricultural plant communities, crops and weeds, are typically removed on a regular basis, usually annually. Weed communities thrive because they begin again every year, they are colonizing species. Against this human managed opportunity spacetime are powerful forces leading to ecological community succession, wherein the current community creates new opportunity for the communities of the future.

**10.3.3.1 Guild structure and community organization**

Every species in a weed-crop plant community plays a role, fills a niche, utilizes resources and conditions, has a trade or is a member of a guild that functionally defines the species and provides insight into both the opportunity space available in the locality and the predicates of the neighbor interactions that will ensue over the season and life history of the plants.

The crop guild is easy to define: its role is to produce biomass, seed for food, fiber, chemicals or intoxicants for human utilization. Every crop species utilizes opportunity spacetime in a locality and leaves some other part of opportunity space unused. It is this unused opportunity space that determines what roles are left unfilled, or what ecological trade of guild a weed species might occupy that will make it successful in seizing and occupying that opportunity space.

Guilds are defined:

**ecological guild**
1: a group of species having similar ecological resource requirements and foraging strategies, and therefore having similar roles (niches) in the community
2: a group of species that exploit the same class of environmental resources in a similar way (Root, 1967)
3: groups of species that exploit resources in a particular way (Silvertown, 2001)
4: groups of functionally similar species in a community

Root (1967) coined the term "guild" to describe groups of functionally similar species in a community. In competitive communities, guilds would represent arenas with the potential for intense interspecific competition, with strong interactions within guilds but weaker interactions with the remainder of their community. A guild is a group of species separated from all other such clusters by an ecological distance greater than the greatest distance between the two most disparate members of the guild concerned. This conservative definition allows complex hierarchical patterns of nesting of smaller guilds within larger ones.

Another very useful technique that depicts some of a community's "connectedness" involves ranking each species' neighbors in niche space from the nearest to the most distant (Inger and Colwell 1977). When overlap is plotted against such nearness ranks in niche space, very similar species (such as those belonging to the same guild) fall out together, whereas species on the periphery of niche space have low overlap with the remainder of the community and tend to fall well below other species.

A group of species that exploit the same class of environmental resources in a similar way (Root, 1967); e.g a group of species having similar requirements and foraging habits and so having similar roles in the community. Species that act as herbivores, omnivores, carnivores and detrivores are examples of guilds. The concept's early roots lie in plant and animal ecology, when ecologists recognized the organization of trophic groups called "Genossenschaften" (Schimper, 1903); see guild and "Synthropia" (Balogh & Loksa, 1956). An assumption derived from competition theory is that species within guilds are most likely competitors, therefore guilds are suggested to form the basis of community organization (Uetz, Halaj & Cady, 1999).

**10.3.3.2 Parameters of weed species ecological role and niche.** The ecological role of a particular weed species in a particular agro-ecosystem is more problematic to define. No one has attempted to define ecological guilds of weeds in agriculture, but herein we make a first approximation. What dimensions, or parameters, would define ecological guild/agro-community structure?

The role, trade, guild or ecological niche that a particular weed species plays and occupies is a function, fundamentally, of its developmental life cycle and life history processes and traits.

Individual weed species guilds are the fulfillment of specific roles they play at different times in their life history. These roles are accomplished by means of specific traits expressed during life history. The processes of life history, natural selection and invasion biology underlie functional traits arising from the fulfillment of these roles. The general roles and traits that form the basis of individual weed guilds can be found in the process and trait tables preceeding each of the life history phases in which weed adaptation occurs (table 10.4).

**Table 10.4** Plant morphological structures, developmental (physiological, morphogenic) life history activities, ecological traits and functional roles that occur during the plant life history processes of agro-community assembly, interactions with neighbors, reproduction and dispersal over generational, and ecological successional, time.

| MORPHOLOGY | LIFE HISTORY PROCESS | DEVELOPMENTAL ACTIVITY | ECOLOGICAL TRAIT | ECOLOGICAL, FUNCTIONAL ROLE |
|---|---|---|---|---|
| seedling | Agro-Community Assembly | seedling recruitment, emergence | •seed dormancy, heteroblasty <br> •seedling emergence time, pattern <br> •relative seedling emergence order | •assembly in community with neighbors |
| vegetative plant | Neighbor Interactions | opportunity space exploitation: | Spatial foraging: <br> •weed shoot and root architecture and function <br> •vegetative bud: <br>   -bud morphogenesis <br>   -plant bud architecture <br>   -bud foraging from parental ortet | •resource allocation <br> •spatial foraging |
| flowering plant | Reproduction | •flowering <br> •pollen dispersal and spatial foraging <br> •birth of zygote: <br>   -fertilization <br>   -embryogenesis <br>   -dormancy induction <br> •meristem architecture and morphogenesis <br> •abscission from parent <br> •time of senescence | •mating system <br> •time and duration of reproduction | •multiplication of parent <br> •display of genotypic variation, novelty |
| seed | Dispersal | Spatial dispersal: <br> •seed movement: local, distant <br> Temporal dispersal: <br> •soil seed pool formation | Propagule qualities: <br> •seed mobility <br> •seed dormancy, heteroblasty | •dispersal in space and time <br> •invasion and colonization <br> •persistence, enduring occupation of locality |
| all plant forms | Multi-Year Processes: <br> •generation periodicity <br> •ecological succession | life cycle | life cycle timing: annual, biennial, perennial | •multiplication, colonization, persistence, enduring occupation of locality |

Certain functional traits have a disproportionate influence on subsequent life history events in the same manner that certain species in a local community have a disproprotionate influence on dynamic interactions among neighbor species.

**keystone species:**
1: species whose importance is not obvious from its size or abundance; a weed that alters nutrient cycles, soil properties or provides food for invasive animals (Booth and Swanton, 2002)
2: a keystone species has a disproportionate effect on local community function relative to it biomass (Paine, 1966, 1969)
3: some species and functions are extremely important to the total ecosystem function (Walker, 1992)

**keystone trait:**
a functional phenotypic trait possessed by an organism disproportionately large effect on its life history events and individual fitness

For example, the traits responsible for seedling emergence timing relative to other crop and weed species in an agroecosystem has a profound effect on subsequent fitness (see ch. 8.4 propagule germination and recruitment). The induction of variable dormancy states among individual seeds of a single parent plant (seed heteroblasty) provides the 'blueprint' for fine scale seedling emergence timing and pattern to maximize exploitation of local opportunity spacetime. As such, traits responsible for dormancy-germinability regulation can be considered 'keystone' traits.

**10.3.3.3 Trait Guild: Relative seedling/bud emergence order.** The time and pattern of seedling emergence of a weed relative to that of the crop is the primary determinate of both plant community assembly and structure, and the ultimate fate of that individual. As such, many traits are responsible for controlling the relative emergence time of a species in a locality. It is these traits that determine the ecological role that that species will play in the community in the seizure of opportunity space and its fitness. Relative emergence time is a very good place to start in understanding the ecological role, or guild, that a species occupies in the crop-weed plant community.

Iowa maize and soybean crop production fields may be a good agro-ecosystem to begin this process of identifying the ecological role of a weed species, and the collection of ecological guilds that comprise an agro-community. Iowa is in many ways monolithic and homogeneous: 21 million acres of row crops, a one- (maize) or two-crop (maize, soybeans) crop rotation system is utilized on the vast majority of this land. This simple rotational crop production system has been in use for some decades (e.g. 1950's till now). The communities that are established on these Iowa crop fields can provide our first view of ecological roles filling opportunity space. The keystone trait guild formed by the relative seedling emergence time and sequence order for common weeds of summer annual crops in Iowa and adjacent areas was revelated in table 8.7, ch. 8.4.5.5 relative emergence order.

## 10.4. Weed Community Dynamics: Community Assembly and Ecological Succession

**10.4.1 Weed community dynamics.** Weed community dynamics is the formation of a local plant community and the changes that occur within thereafter. Weed community dynamics, assembly, begins with the first weed colonization (see ch. 3.5, the nature of plant invasions of local opportunity; ch. 9.1.2, recruitment: community assembly) of an empty local habitat and ends with terminal phases of ecological succession in the fully mature community (see ch. 10.4.4, plant community ecological succession). Most plant communities lie somewhere between these two extremes. Weed community dynamics includes the processes of community assembly and ecological successsion.

Plant communities are organs of the much larger global system of flow driven by the second law of thermodynamics. The biosphere, atmosphere, and hydrosphere are not separate entities but interlocking environments that flow and design themselves together. This is the physical basis for the emergence of hierarchy in nature that guides community assembly and ecological succession (see ch. 11.1.8 plant community structure and assembly).

**10.4.2 Seizing and exploiting local opportunity**. The successful occupation of a locality by a plant species is predicated on a habitat that can provide opportunity over time that can be seized and utilized to allow a complete life cycle of establishment, growth and reproduction of excess individuals of that species (see ch. 3.1.2 seizing and exploiting local opportunity). Weedy plant community dynamics is a consequence of natural selection and reproductive success among excess variable phenotypes (and their functional traits) in response to the structure, quality and timing of locally available opportunity spacetime. The nature of a particular locality is defined by the opportunity spacetime it affords the weed population to survive and reproduce: the local niche. Locally habitable opportunity spacetime for an organism includes its available resources (e.g. light, water, nutrients, gases), pervasive conditions (e.g. heat, terroir), disturbance history (e.g. tillage, herbicides, frozen winter soil), and neighboring organisms (e.g. crops, other weed species, diseases) with which it interacts, all at a particular time. The successful weed species must possess functional traits expressed at appropriate times over the growing season to exploit that the particular opportunity spacetime available in that local habitat. The newly invading species must possess these traits before the invasion process begins in order to take advantage of local opportunity: it must be preadapted.

Plant invasions are formative events in the ecology of community assembly and ecological succession, as well as in the evolution of niche differentiation by speciation. How local opportunity space-time is seized and exploited is revealed in the multi-step process of plant invasion. The plant invasion process is the mass movement or encroachment of an organism from one area to another. An individual invading species possessing certain preadapted traits and a vulnerable local opportunity spacetime must successfully survive three processes: dispersal of that species into that locality, followed by colonization and enduring occupation of that habitat (see ch. 3.5 the nature of plant invasions of local opportunity). Most invading species probably fail to complete all three steps, and all local plant populations become extinct eventually.

**10.4.3 Propagule recruitment: community assembly**. The plant community begins to be assembled with the emergence of seedlings of annual species previously dispersed into a vulnerable local habitat, or with the growth of vegetative meristems of perennial species (see ch. 9.1.2 recruitment: community assembly). With community assembly begins interaction with neighbor plants and other organisms in the local community. Interacting neighbor plants impact each other as the community develops. Inhibitory priority effects occur when a species that arrives first at a site negatively impacts a species that arrives later by reducing the availability of space or resources (e.g. competition). Facilitative priority effects occur when a species that arrives first at a site alters abiotic or biotic conditions in ways that positively impact a species arriving later (e.g. ecological succession). The most typical community that weeds invade is the agricultural community. The hallmark of annual crop production fields is the elimination of all plants that might compete with the yearly crop being established. As such weeds find enormous opportunity (exercising the inhibitory priority effect) to seize and exploit in these managed, fertile, habitats. Neighbors begin with the crop species, and include other competing weed species as the season progresses (see ch. 8.4.5.5, relative emergence order). Crop fields are the beginning, the starting point, of plant community ecological succession.

**10.4.4 Plant community ecological succession**. Ecological succession, a fundamental concept in ecology, refers to more-or-less predictable and orderly changes in the composition or structure of an ecological community (Wikipedia, 6.08). Succession may be initiated either by formation of new, unoccupied habitat (e.g., a lava flow or a severe landslide) or by some form of disturbance (e.g. fire, severe windthrow, logging, agriculture) of an existing community. The former case is often referred to as primary succession, the latter as secondary succession.

**succession**: a directional, progressive, orderly change in vegetation that would ultimately converge on a stable, predictable climax community (Clements, 1916, 1936)

**ecological succession**
1: the chronological distribution of organisms within an area
2: the geological, ecological or seasonal sequence of species within a habitat or community
3: the development of plant communities leading to a climax
4: a process of continuous colonization of and extinction on a site by species populations (Bazzaz, 1979)

Primary succession is a series of changes that occur in an area where no soil or organisms live. Secondary succession is a series of changes occurring in an area where the ecosystems been disturbed, but soil and organisms are still existing. Each is a sequence of species moving from colonizing to density-stressed conditions. Plant succession changes the fitness of a species: later successional species are adapted to density-dependent competition. Colonizing plants avoid successional changes and disturbances by many of their key life history traits: disperse to, invade, disturbed sites; escape succession and go elsewhere to colonized new area: continuous return to beginning of successional cycle. Seed and bud bank buffer against short term changes.

The trajectory of ecological change can be influenced by site conditions, by the interactions of the species present, and by more stochastic factors such as availability of colonists or seeds, or weather conditions at the time of disturbance. Some of these factors contribute to predictability of successional dynamics; others add more probabilistic elements. In general, communities in early succession will be dominated by fast-growing, well-dispersed species (opportunist, fugitive, or r-selected life-histories). As succession proceeds, these species will tend to be replaced by more competitive (k-selected) species.

**successional trajectory:** the likely sequence of plant and animal communities predicted to occur in a habitat following a particular type of disturbance

All plant communities began with seizing and exploiting unoccupied opportunity space. In each new community a species seizes the opportunity created by what the existing species didn't exploit. Subsequently, neighbors compete for occupied space. Earlier successional species facilitate invasion of later successional species by the opportunity spacetime they create in the habitat by their own life cycle completion (exercising facilitative priority effects). As such, each established community provides the opportunity space of the next successional community, with the wild card of disturbance to reset the community structural sequence again in patchy spatial patterns.

**relay floristics**:
1: earlier successional species prepare the environment for later species (Egler, 1954)
2: vegetation in each successional pathway alters the environment and ameliorates it for later invading species, early species facilitate the invasion of later species (Clements, 1916, 1936)

**chaperone species**: a species that may facilitate the invasion of another species either directly assisting it or by inhibiting a third species (Kelly, 1994)

Trends in ecosystem and community properties in succession have been suggested, but few appear to be general. For example, species diversity almost necessarily increases during early succession as new species arrive, but may decline in later succession as competition eliminates opportunistic species and leads to dominance by locally superior competitors. Net Primary Productivity, biomass, and trophic level properties all show variable patterns over succession, depending on the particular system and site.

Ecological succession was formerly seen as having a stable end-stage called the climax (see Frederic Clements), sometimes referred to as the 'potential vegetation' of a site, shaped primarily by the local climate. This idea has been largely abandoned by modern ecologists in favor of non-equilibrium ideas of how ecosystems function. Most natural ecosystems experience disturbance at a rate that makes a "climax" community unattainable. Climate change often occurs at a rate and frequency sufficient to prevent arrival at a climax state. Additions to available species pools through range expansions and introductions can also continually reshape communities.

The development of some ecosystem attributes, such as pedogenesis and nutrient cycles, are both influenced by community properties, and, in turn, influence further community development. This process may occur only over centuries or millennia. Coupled with the stochastic nature of disturbance events and other long-term (e.g., climatic) changes, such dynamics make it doubtful whether the 'climax' concept ever applies or is particularly useful in considering actual vegetation.

The idea of ecological succession goes back to the 19$^{th}$ Century. In 1860 Henry David Thoreau read an address called "The Succession of Forest Trees" in which he described succession in an Oak-Pine forest (Wikipedia, 6.08). Henry Chandler Cowles, at the University of Chicago, developed a more formal concept of succession, following his studies of sand dunes on the shores of Lake Michigan (the Indiana Dunes). He recognized that vegetation on sand-dunes of different ages might be interpreted as different stages of a general trend of vegetation development on dunes, and used his observations to propose a particular sequence (sere) and process of primary succession. Understanding of succession was long dominated by the theories of Frederic Clements, a contemporary of Cowles, who held that successional sequences of communities (seres), were highly predictable and culminated in a climatically determined stable climax. Clements and his followers developed a complex taxonomy of communities and successional pathways. A contrasting view, the Gleasonian framework, is more complex, with three items: invoking interactions between the physical environment, population-level interactions between species, and disturbance regimes, in determining the composition and spatial distribution of species. It differs most fundamentally from the Clementsian view in suggesting a much greater role of chance factors and in denying the existence of coherent, sharply bounded community types. Gleason's ideas, first published in the early 20$^{th}$ century, were more consistent with Cowles' thinking, and were ultimately largely vindicated. However, they were largely ignored from their publication until the 1960s. Beginning with the work of Robert Whittaker and John Curtis in the 1950s and 1960s, models of succession have gradually changed and become more complex. In modern times, among North American ecologists, less stress has been placed on the idea of a single climax vegetation, and more study has gone into the role of contingency in the actual development of communities.

**10.4.5 Weed population shifts and ecological succession.** Ecological succession in disturbed agricultural communities often appears stuck in the early, colonizing, phases of plant species invasions. Annual cropping systems reset the potential sequence of species change each year with soil tillage, seedbed preparation, planting, weed control (typically herbicidal) and harvesting. Within this narrower range of opportunity spacetime change does occur, often dramatically. These changes is floral and species composition in the agricultural community are called weed population shifts.

Weedy microevolution includes the occurrence of population shifts in agricultural fields and local adaptation by individuals and populations to those localities. Weed population shifts are the changes in the individual organisms that make up the population of a locality, often caused by changes in weed management practices. The bottom line for a farmer, and an individual field, is that when things change the weeds adapt. Those practices that control weeds in a field this year probably won't provide acceptable weed control in future years. Why? Because what you do today kills those weeds susceptible to your management practices. The few weeds that survive today's management practices are the parents of the weed problems you will have in the future: leaner, meaner, better able to handle what you dish out. This process has been going on for thousands of years. Every crop grown by humans, in every field, in every year for the history of agriculture has resulted in the weeds you have in your field today. Agronomic practices (e.g. mechanization, herbicides, new crops and crop varieties) and an increase in the size and structure of land holdings (decreasing pest scouting, decrease in secondary tillage, decrease in fences and fence line field continuity, increase in contiguous field spatial arrangement) can change the weed species present in a field (interspecific population shifts), as well as the genotypes of a single species present in a local agroecosystem (intraspecific population shifts). See also the population shift stories by species in Hakansson (2003; ch.13, pp.222-247).

**10.4.5.1 Inter-specific population shifts.** The weed species present in a local weed community can change in response to changes in environment, neighbors, cropping practices and disturbance. Many weed species have undergone very large extensions of the ranges (e.g. *Setaria* spp.-gp.) they thrive in while others formerly widespread have all but disappeared.

**10.4.5.1.1 Grassy weed tolerance to 2,4-D.** Before WWII, the early 1940's, selective herbicides were not used on weeds in Iowa corn fields. The weed communities were different than they are today. At that earlier time broadleaf weeds were considered the major weed problem (farmers often grouped plants in their fields into "weeds" and "grasses"). With the introduction of 2,4-D, this selective weed control killed the broadleaf weeds in most cases, but did little to the grasses. Grassy weed species suddenly became a much larger weed problems than before. *Setaria faberi* (giant foxtail) was largely unknown to midwestern US growers before 1950. Since that time, population shifts have occurred as a response to the introduction of a new herbicide mode of action. For example, in the 1980-1990's the appearance of *Panicum mileaceum* (wild proso millet) and *Eriochloa villosa* (woolly cupgrass), weeds largely unknown previously, have spread across the Midwestern US.

**10.4.5.1.2 Herbicide-induced life cycle shifts.** Populations of summer annuals subjected to herbicide selection pressure has resulted in a population shift to winter annual life cycle or to seed that germinated much later in the season in England.

**10.4.5.2 Intra-specific population shifts.** Changes in crop fields can favor particular variants within a species. The selection for new and better adapted genotypes and phenotypes of a species can cause shifts in the composition of a population of an individual species in a field or area. Just as herbicides selected and shifted different species in a fields weed community, herbicides and other disturbance changes have caused shifts within populations of one species to better adapted variants.

**10.4.5.2.1 Early flowering.** Population shifts to early flowering forms of wild oat (*Avena fatua*) and *Arabidopsis thaliani* were selected for by changes in weed seed cleaning and crop harvesting practices.

**10.4.5.2.2 Crop mimicry.** Barnyardgrass has a variant (*Echinochloa crus-galli* (L.) Beauv. var. *oryzicola* (Vasing) Ohwi.) that mimics cultivated rice so effectively it escapes handweeding. Its growth habit and appearance are more similar to rice than some of rice's closer relatives. The main selective force favoring this intra-specific variant has been intensive handweeding in Asian rice crops.

**10.4.5.2.3 Dwarf variants.** Dwarf variants appeared in populations of *Aethusa cynapium* and *Torilis japonicum* in English cereal crops [were selected by the introduction of harvesting equipment]. The equipment selected against the taller variants, but the dwarf variants thrived.

**10.4.5.2.4 Biodiversity shifts.** Herbicide use led to a decrease in the diversity of genotypes present in a weed species population in California. [NoDak study of oat fields and herbicide history: increase in intra-specific Avena fatua heterogeneity in local populations.]

**10.4.5.3 Herbicide-induced population shifts.** Some of the most dramatic and widespread population shifts in agricultural communities occurs in response to the introduction of herbicide use in a crop. Natural selection quickly selects for resistance, whether caused by alterations to the molecular target site in the plant or by altered metabolism and detoxification of the herbicide. Survivors rapidly exploit the opportunity space left by their dead neighbors.

**10.4.5.3.1 Herbicide resistance genotype shifts: altered herbicide target variants.** Herbicide resistance can be conferred in a plant by many mechanisms. These resistance mechanisms include a change in the target site (usually a protein) that the herbicide binds to, causing death or injury. Below is a partial list of some of these types of population shifts caused by herbicide use that kills the susceptible variants and allows the resistant survivor variants to thrive. Altered herbicide binding (target) site variants include atrazine-resistant *Chenopodium album* biotypes (altered psbA chloroplast gene product and binding site); ALSase-inhibitor resistance (altered acetolactate synthase gene in *Amaranthus rudis* (common waterhemp) and *Kochia scoparia*); ACCase inhibitor resistance (altered acetyl CoA carboxylase gene in *Avena fatua* (wild oat) and *Setaria viridis* (green foxtail)).

**10.4.5.3.2 Herbicide resistance genotype shifts: enhanced herbicide metabolism variants**. Herbicide resistance can also be conferred on variants within a species that detoxify herbicides more rapidly than susceptible genotypes. Repeated herbicide applications kill off the more susceptible and leave the enhanced detoxifying variants to thrive.

Selection for variants of *Tripleurospermum inodorum* (scentless mayweed) with incrementally enhanced herbicide metabolism resulted from repeated applications of several herbicides, including simazine. *Avena fatua* (wild oat) herbicides (barban, difenzaquat) led to an increase in variants with better metabolic detoxification systems. Populations of *Alopecurus myosuroides* Huds. (slender foxtail or blackgrass) in the UK, and *Lolium rigidum* (rigid ryegrass) populations in Australia, have been identified in agricultural fields with high levels of metabolic cross-resistance to several different chemical families of herbicides. In the later instance in Australia, some populations were found to be resistant to all herbicides available for weed control in those crops in which it appears. Several populations of *Abutilon theophrastiI* (velvetleaf) in Maryland and Wisconsin in the US have been identified as resistant to s-triazine herbicides. Resistance was not due to altered triazine binding (psbA gene mutants), but due to enhanced metabolism in those biotypes.

## 10.5 Weed Community Biodiversity

Biodiversity is a broad topic, as broad as the totality of earth's biology in one sense, as narrow as many have re-defined it, re-conceptualized it. And it is of crucial importance to human society. This chapter will not provide a comprehensive view of biodiversity, but will attempt to suggest several different ways of seeing weed biodiversity within the broader views.

Biodiversity is the source of all future local weed populations. It is the source of novel traits for crop improvement. Local biodiversity is an indicator of locally available opportunity, the predicate and foundation of community dynamics and future evolutionary adaptation by its species. Biodiversity can reveal the potential effects (e.g. extinction) of human disturbances on organic life, and global changes in opportunity spacetime (e.g. global warming). Biodiversity of weedy roles enables discovery of what opportunity spacetime is available: the source of future opportunity; what is being lost or gained or unchanged; how it is changing and how it might be managed. Explanations of weed biodiversity could provide the basis for management of opportunity spacetime.

Weed communities are revealed by biodiversity at several scales and perspectives. Weed biodiversity predicates community dynamics. Weed biodiversity is the pool of potential candidate populations that might invade, seize and exploit local agricultural opportunity. The basis of community is the behavior, the life history, of an individual plant. Individuals form local populations in communities. Dispersal of pollen, seeds and other propagules across the landscape link local communities into meta-communities and ecosystems. The behavior at all these scales is an emergent property of the cumulative individual behaviors.

Biodiversity is defined:

**biodiversity**
1: the variety of organisms considered at all levels, from genetic variants of a single species through arrays of species to arrays of genera, families and still higher taxonomic levels
2: the variety of ecosystems, which comprise both the communities of organisms within particular habitats and

the physical conditions under which they live
3: the totality of biological diversity

Biodiversity is an emergent property of the local population, the deme, of the evolution of individuals interacting with neighbors. Biodiversity is the emergent properties of the guild-roles of different species. Biodiversity is not predictable. The second definition shows biodiversity, like the phenotype, like opportunity spacetime, consists of all of biology plus the habitats and environments they exist in, this is the most accurate representation (G x E in a global sense). It is the one that comes closest to representing Harper's view of biodiversity seen from the perspective of the individual plant (the unit of evolution) interacting with neighbors, the precise perspective from which evolution operates (ch. 10.5.2).

This section on biodiversity is organized into several sections: the scales on which weedy biodiversity exists; Harper's perspective on biodiversity as that encountered by interacting neighbors; and the stability of community biodiversity. The later of socio-political interest in conservation biology and sustainable agriculture (stability, community equilibrium and sustainability).

**10.5.1 Scales of weedy biodiversity.** At what scale of time and space should we use to describe weedy biodiversity? What biology occurs within each of these scales? Weed genetic biodiversity exists at several levels of organization, each in its own scale. Insight into a complete representation might be gained by using different, overlapping criteria of the scale of diversity: biological scale (genotype) within spatial (biogeographical) and temporal scales (table 10.5):

**Table 10.5** Scalar space-time opportunity for genetic and plant system biodiversity as a basis/foundation of local plant community structure and dynamics.

| SCALE | LEVEL |
|---|---|
| GENOTYPE | gene/allele |
|  | trait |
|  | individual plant phenotype |
|  | population/deme |
|  | species |
|  | species associations |
| SPATIAL | molecular |
|  | microsite |
|  | locality |
|  | landscape |
|  | continent |
|  | global |
| TEMPORAL | instantaneous |
|  | diurnal |
|  | seasonal |
|  | ecological succession |
|  | long-term |

Of course these are general categories, and much finer or coarser scale divisions can be utilized for many purposes of understanding weedy biodiversity. These scales are overlapping in time and are intimately related. Expressions of weed biodiversity at the species and higher levels are incomplete, they do not include the fundamental units of natural selection (the phenotype) or evolution (the deme).

**10.5.2 Biodiversity encountered by interacting neighbors**. Harper (1977; p. 707-8) defines biodiversity in a different, and more comprehensive, way. He describes it in terms of the levels of diversity that an individual plant may meet when it interacts with its neighbors. A more complete development of this broad view of weed biodiversity in found in Dekker (1997). Harper posits five ways individual plants sense, perceive, or are stimulated by their immediate neighbors in the local biological community, elements contributing to population diversity.

**10.5.2.1 Perception of neighbor morphology**. In the first instance, individuals meet neighbors in the somatic polymorphism of their plant parts, shoots or organs occupying space over time that interact directly with both neighbors of other parts of the same individual (e.g. leaves, roots, other shoots on the same plant). In the community this can occur between sexually (plant from a single seed) or asexually (perennating vegetative meristems) reproduced genets and ramets.

**10.5.2.2 Perception of neighbor development**. The second level of biodiversity that an individual plant may meet when it interacts with its neighbors is the diversity of age-states within the community. In a broader sense, this includes developmental, phenological or life history time.

**10.5.2.3 Perception of neighbor genotype/phenotype**. The third level of biodiversity an individual in a community will encounter is the more common expression of biodiversity, the genetic variants (intra-specific, inter-specific) of neighbor species.

**10.5.2.4 Perception of microhabitat opportunity**. The fourth expression of biodiversity is that of microsites within the habitat that may provide opportunity spacetime for the weed.

**10.5.2.5 Perception of neighbor diversity**. The fifth level of biodiversity that an individual in a community may meet in its interactions with neighbors is groupings at a higher level than the species (e.g. species-groups, polyploidy species clusters). This interesting, and little studied, topic is developed in ch. 10.3 weed population structure.

**10.5.3 Weed community biodiversity: complexity, stability and equilibrium**. Considerable confusion exists of the nature of biodiversity, complexity and system stability. The following quotations indicate some of the differences in perspective taken by scientists.

> "[Complex] systems are open and it may be difficult or impossible to define system boundaries. Complex systems operate under far from equilibrium conditions, there has to be a constant flow of energy to maintain the organisation of the system." Wikipedia, 12.10

> "There is no comfortable theorem assuring that increased diversity and complexity beget enhanced community stability; rather, as a mathematical generality, the opposite is true. The task, then, is to elucidate the devious strategies which make for stability in enduring natural systems. There will be no simple answer to these questions." May, R.M, 1973

> "The general conclusion of these papers has been that diversity and stability may be related but there are enough problems with experimental design and semantics that conclusive evidence about the exact nature of the relationship between diversity and stability is still lacking." Booth and Swanton, 2002 (summarizing multiple papers on diversity's role in ecosystems).

> "Biodiversity at larger scales (community, ecosystem, landscapes, biomes) becomes and important possible indicator of whether an ecosystem is functioning well or not." Booth and Swanton, 2002

> "... diversity is too broad a concept to use as a good indicator [of ecosystem stability], but it continues to have political and illustrative appeal and power." Angermeier and Karr, 1994; Schlapfer, 1999

"The two main areas where the effect of biodiversity on ecosystem function have been studied are the relationship between diversity and productivity, and the relationship between diversity and community stability. More biologically diverse communities appear to be more productive (in terms of biomass production) than are less diverse communities, and they appear to be more stable in the face of perturbations." Wikipedia, 12.10

"Elton (1958) was among the first natural scientists to articulate a link between biological diversity – the number and variety of native species in an ecosystem – and ecological health. Greater diversity conveys a degree of "biotic resistance," he argued, which helps preserve the integrity of an ecosystem over time. A natural, undisturbed ecosystem could be thought of as an immunologic system; invasion a disease. A recent issue of National Geographic described ecological invasion as a "green cancer". The disease metaphor is compelling. There's just one problem. Fifty years of research by invasion biologists around the world have failed to confirm it."

"But biological diversity per se –the number of species in an ecosystem – provides no shield against invasions. In a 1999 issue of the journal Biological Invasions, Daniel Simberloff (Simberloff and Von Holle, 1999), a prominent ecologist at the Institute for Biological Invasions at the University of Tennessee at Knoxville, writes simply, "It seems clear to me that there is no prima facie case for the biotic resistance paradigm.""
Burdick, 2005

For a larger perspective on stability in complex systems see ch. 11.2 fragile and antifragile systems.

**10.5.3.1 Community complexity**. Before prediction and management is the need to explain weed community complexity. The first, primary, over-arching need is to explain, to understand community biodiversity as a complex adaptive system, and all that that implies. For a larger perspective on complexity see unit 5 complex adaptive systems.

**10.5.3.2 Community stability**. Weed biodiversity in the community has been linked with the concepts of stability and equilibrium. Does relatively greater weed-crop biodiversity in an agroecosystem, a cropping system, inherently lead to a more stable, sustainable agroecosystem? What is the relationships between biodiversity, ecosystem or community stability and sustainability? These are important questions to human society. It has been argued that agriculture itself is not sustainable and its stability rests on the continuity of human management. Human management is susceptible to human actions and long-term human social stability.

Biodiversity, stability and sustainability are viewed from different perspectives, with much opinion and some disagreement. Therefore, first are defined some terms to ensure conceptual clarity:

**ecological stability**:
1: connoting a continuum, ranging from resilience (returning quickly to a previous state) to constancy (lack of change) to persistence (simply not going extinct); the precise definition depends on the ecosystem in question, the variable or variables of interest, and the overall context
2: in conservation ecology, populations that do not go extinct; in mathematical models of systems, Lyapunov stability (dynamical system that start out near an equilibrium point stay there forever) (Wikipedia, 6.08)

**community stability**: how communities resist change in response to disturbance or stress (Booth and Swanton, 2002)

Local stability indicates that a system is stable with relatively small, short-lived disturbances, while global stability indicates a system highly resistant to change in species composition and/or food web dynamics. For these definitions of stability it is important to remember that natural selection and elimination is always ongoing, that history is irreversible, and they depend on the perspective from which the system is observed (including the genotypic, spatial and temporal scale of observation, table 10.5). Due to the inconsistent usage of the term stability in ecological literature, specific terms of the types of ecological and community stability have been proposed:

**ecological stability-constancy and persistence:**
1: living systems that can remain unchanged in observational studies of ecosystems (Wikipedia, 6.08)
2: the ability of a community to remain relatively unchanged over time (Barbour et al., 1999)

**ecological stability-resistance and inertia (or persistence):**
1: a system's response to some perturbation (disturbance; any externally imposed change in conditions, usually happening in a short time period)
2: resistance is a measure of how little the parameter of interest changes in response to external pressures
3: inertia (or persistence) implies that the living system is is able to resist external fluctuations (Wikipedia, 6.08)

**ecological stability-resilience, elasticity and amplitude:**
1: the tendency of a system to return to a previous state after a perturbation
2: 'engineering resilience' is the time required for an ecosystem to return to an equilibrium or steady-state following a perturbation; 'engineering resilience' (Hollings, 1973)
3: 'ecological resilience' is the capacity of a system to absorb disturbance and reorganize while undergoing change so as to still retain essentially the same function, structure, identity, and feedbacks; it presumes the existence of multiple stable states or domains (Walker et al., 2004)
4: elasticity and amplitude are measures of resilience; elasticity is the speed with which a system returns; amplitude is a measure of how far a system can be moved from the previous state and still return (Wikipedia, 6.08)
5: the ability of a community to return to its original state following stress or disturbance (Barbour et al., 1999)
6: the speed at which the system returns to its former state following a perturbation (Putnam, 1994)

      The concept of resilience in ecological systems was first introduced in order to describe the persistence of natural systems in the face of changes in ecosystem variables due to 'natural' or anthropogenic causes (Hollings, 1973). In ecology, resilience is one possible ecosystem response to a perturbation or disturbance. A resilient ecosystem resists damage and recovers quickly from stochastic disturbances such as fires, flooding, windstorms, insect population explosions, and human activities such as deforestation and the introduction of exotic plant or animal species. Disturbances of sufficient magnitude or duration can profoundly affect an ecosystem and may force an ecosystem to reach a threshold beyond which a different regime of processes and structures predominates (Folke et al., 2004). Human activities that adversely affect ecosystem resilience such as reduction of biodiversity, exploitation of natural resources, pollution, land-use, and anthropogenic climate change and are increasingly causing regime shifts in ecosystems, often to less desirable and degraded conditions in the opinion of some (Folke et al., 2004; Peterson et al., 1998). Interdisciplinary discourse on resilience now includes consideration of the interactions of humans and ecosystems via socio-ecological systems, and the need for shift from the 'maximum sustainable yield" paradigm to environmental management which aims to build ecological resilience through "resilience analysis, adaptive resource management, and adaptive governance" (Walker et al., 2004). (Wikipedia, 12.10) We await.

The concept of ecological stability is often associated with that of sustainability. To sustain has two, almost opposite, but often confused, meanings (Deutsch, 2011):

**sustain**:
1: to provide someone with what they need
2: to prevent things from changing

**sustainability**:
1: the ability to sustain
2: the ability to provide someone with what they need
3: the ability to prevent things from changing
4: humanity's investment in a system of living, projected to be viable on an ongoing basis that provides quality of life for all individuals of sentient species and preserves natural ecosystems
5: a characteristic of a process or state that can be maintained at a certain level indefinitely.
6: environmental, the potential longevity of vital human ecological support systems, such as the planet's climatic system, systems of agriculture, industry, forestry, fisheries, and the systems on which they depend
7: how long human ecological systems can be expected to be usefully productive; emphasis on human systems and anthropogenic problems (Wikipedia, 6.08)

You have to live the solution to problems, and to set about solving new problems that this creates. Finding new solutions is unsustainable. Evolution does not sustain static populations structure.

On what time and spatial scales do we assess stablility and sustainability? How exactly do ecological stability and sustainability relate to each other? It has been conjectured that increased agro-ecosystem biodiversity will result in more stable and sustainable, not to mention more economically viable, crop production systems. This view takes the form of the:

**biotic resistance hypothesis**: the biological diversity (number and variety of species) in a community or an ecosystem, conveys a degree of "biotic resistance" that preserves the integrity of an ecosystem over time, a protection against invasions; more biologically diverse communities are more stable in the face of perturbations

**species diversity-stability hypothesis**:
1: species diversity allows more ecosystem stability; complex trophic structures result in more stable communities (MacArthur, 1955)
2: human-disturbed communities wherein many species have gone extinct are prone to pest outbreaks and unpredictable fluctuations in population; there exists a negative relationship between native species richness, biodiversity, and community invasibility; communities with many species would be invasion resistant, while species poor communities will be highly invasible (Elton, 1958)

No evidence exists that 'biotic resistance' is enhanced by the presence of more species in deme. There exists no evidence that biodiversity inherently confers stability. Stability is illusory in plant communities.

### 10.5.3.3 Equilibrium in the community.

> "[Complex] systems are open and it may be difficult or impossible to define system boundaries. Complex systems operate under far from equilibrium conditions, there has to be a constant flow of energy to maintain the organisation of the system." (Wikipedia, 12.10)

Equilibrium conditions never exist, persist, in an open system such as evolutionary complex adaptive system; open systems require energy to counter entropy (see Unit 5 complex adaptive weed systems). The idea of equilibrium separates the two different worlds of the noncomplex and complex. In a nonorganic, noncomplex, system an object in equilibrium occurs in a state of inertia. In a complex, organic, system equilibrium only happens in death (Kaufman, 1995).

**equilibrium**: the condition of a system in which competing influences are balanced

**homeostasis**:
1: the property of a system (typically a living organism), either open or closed, that regulates its internal environment and tends to maintain a stable, constant condition; multiple dynamic equilibrium adjustment and regulation mechanisms make homeostasis possible (Wikipedia, 12.10)
2: the maintenance of metabolic equilibrium within an animal by a tendency to compensate for disrupting changes (Collins English Dictionary, 1979; Collins, Glasgow, UK)
3: the tendency for the internal environment of an organism to be maintained constant (Chambers Sci Dictionary)

Weed biodiversity in the community has been linked with the concepts of stability and equilibrium. Does relatively greater weed-crop biodiversity in an agroecosystem, a cropping system, inherently lead to a more stable, sustainable agroecosystem? The stability and equilibrium of biodiversity is an important area of interest, especially in the areas of conservation biology and sustainable agriculture. Much experimental research is conducted at biodiversity in terms of species richness.

**10.5.3.3.1 Biodiversity as species richness.** Species richness is considered by many as an appropriate metric with which to assess the homogeneity of an environment. Typically, species richness is used in conservation studies to determine the sensitivity of ecosystems and their resident species. The actual number of species calculated alone is largely an arbitrary number. These studies, therefore, often develop a rubric or measure for valuing the species richness number(s) or adopt one from previous studies on similar ecosystems.

**species richness**: the number of different species in a given area

**biodiversity (species richness)**: the number and relative abundance of species in an area or community (Booth and Swanton, 2002)

**biodiversity (species evenness)**: abundance of each species in a community (Booth and Swanton, 2002)

In contrast to this perspective is the species redundancy hypothesis which argues that species are often similar, redundant, such that elimination of most of them would not affect ecosystem function (Schulze and Mooney, 1993). Most species are interchangeable and that conservation of most species is not necessary in terms of ecological function (Walker, 1992).

In ecology, a species-area curve is a relationship between the area of a habitat, or of part of a habitat, and the number of species found within that area. Larger areas tend to contain larger numbers of species, and empirically, the relative numbers seem to follow systematic mathematical relationships (Preston, 1962). The species-area relationship is usually constructed for a single type of organism, such as all vascular plants or all species of a specific trophic level within a particular site. It is rarely, if ever, constructed for all types of organisms if simply because of the prodigious data requirements.

Whether species richness is a metric for environmental homogeneity or ecosystem sensitivity remains unclear. Whether representations of habitats based on species of a specific trophic level is realistic remains doubtful. Whether a metric is helpful that avoids representing all types of organisms in a habitat because the data is to prodigious is also suspect. Expressions of weed biodiversity at the species and higher levels inherently avoid the units (gene, phenotype) from which that biodiversity evolved. Considered from an evolutionary, rather than this ecological demographic, perspective this way of viewing biodiversity may be erroneous and lead to incomplete explanations of the nature of weed communities.

The conceptual history of equilibrium in the community is developed in the following two sections. The equilibrium theory of island biogeography attempts to establish and explain the factors that affect the species richness of natural communities (MacArthur and Wilson, 1967). The unified neutral theory of biodiversity and biogeography (Hubbell, 2001) is a hypothesis which aims to explain the diversity and relative abundance of species in ecological communities with the assumption that the differences between members of an ecological community of trophically similar species are "neutral," or irrelevant to their success. These two topics are viewed from different perspectives, with much opinion and some disagreement.

**10.5.3.4 Equilibrium theory of island biogeography.** Island biogeography attempts to establish and explain the factors that affect the species richness of natural communities (Wikipedia, 12.10). The field was started in the 1960s, coined the term theory of island biogeography, as this theory attempted to predict the number of species that would exist on a newly created island (MacArthur and Wilson, 1967). The theory was developed to explain species richness of actual islands. Now it is used in reference to any ecosystem surrounded by unlike ecosystems. For biogeographical purposes, an "island" is any area of suitable habitat surrounded by an expanse of unsuitable habitat. While this may be a traditional island—a mass of land surrounded by water—the term may also be applied to many untraditional "islands", such as the peaks of mountains, isolated springs in the desert, expanses of grassland surrounded by highways or housing tracts, or natural habitats surrounded by human-altered landscapes. Additionally, what is an island for one organism may not be an island for another: some organisms located on mountaintops may also be found in the valleys, while others may be restricted to the peaks. The theory of island biogeography proposes that the number of species found on an undisturbed island is determined by immigration, emigration and extinction. Additionally, the isolated populations may follow different evolutionary routes, as shown by Darwin's observation of finches in the Galapogos Islands and the diversification of reptiles after rainforests collapsed 300 million years ago. Immigration and emigration are affected by the distance of an island from a source of colonists (distance effect). Usually this source is the mainland, but it can also be other islands. Islands that are more isolated are less likely to receive immigrants than islands that are less isolated. The rate of extinction once a species manages to colonize an island is affected by island size (area effect or the species-area curve). Larger islands contain larger habitat areas and opportunities for more different varieties of habitat. Larger habitat size reduces the probability of extinction due to chance events. Habitat heterogeneity increases the number of species that will be successful after immigration. Over time, the countervailing forces of extinction and immigration result in an equilibrium level of species richness (Wikipedia, 12.10).

Defining and studying biodiversity from the single perspective of species numbers in a locality obscures the underying complexity and diversity at other scales of observation (e.g. table 10.5).

**10.5.3.5 Unified neutral theory of biodiversity and biogeography.** The "Unified Theory" is a hypothesis and the title of a monograph (Hubbell, 2001; Wikipedia, 12.10). The hypothesis aims to explain the diversity and relative abundance of species in ecological communities, although like other neutral theories of ecology, Hubbell's hypothesis assumes that the differences between members of an ecological community of trophically similar species are "neutral," or irrelevant to their success. This implies that biodiversity arises at random, as each species follows a random walk (McGill, 2003). The hypothesis has sparked controversy, and some authors consider it a more complex version of other null models that fit the data better.

Neutrality means that at a given trophic level in a food web, species are equivalent in birth rates, death rates, dispersal rates and speciation rates, when measured on a per-capita basis (Hubbell, 2005). This can be considered a null hypothesis to niche theory. Hubbell built on earlier neutral concepts, including the theory of island biogeography (MacArthur and Wilson, 1967) and Gould's concepts of symmetry and null models (Hubbell, 2005).

An ecological community is a group of trophically similar, sympatric species that actually or potentially compete in a local area for the same or similar resources (Hubbell, 2001). Under the Unified Theory, complex ecological interactions are permitted among individuals of an ecological community (such as competition and cooperation), provided that all individuals obey the same rules. Asymmetric phenomena such as parasitism and predation are ruled out by the terms of reference; but cooperative strategies such as swarming, and negative interaction such as competing for limited food or light are allowed (so long as all individuals behave in the same way).

The Unified Theory also makes predictions that have profound implications for the management of biodiversity, especially the management of rare species. The theory predicts the existence of a fundamental biodiversity constant, conventionally written $\theta$, that appears to govern species richness on a wide variety of spatial and temporal scales.

The reader is left free to evaluate the insight this theory provides in our explanation of the nature of weeds, a set of species whose success at a similar trophic level in the community appear to be something other than neutral.

**10.5.4 Weed communities: conclusions**. If Robert May (1973) is correct in the quote above, that it is a mathematical certainty that increased biodiversity threatens community stability, then the challenge faced by human disturbance is to "elucidate the devious strategies which make for stability in enduring natural systems".

At the end of the day what is stable? Weed-crop groups (tables 1.4 and 1.5) and the constancy of the human need for food have been stable for millennia. Human needs and technology, environment and opportunity, these all change with time. Therefore weed communities change. If we can say anything about stability and biodiversity it is the enduring nature of the world weed-crop groups found in table 1.4. Their stable endurance is underwritten by humans need for food, and by the human need to manage and control space (see section 1.3, weeds and human nature). The 'devious strategies' which have allowed weed populations to exploit opportunity, and colonize disturbed habitats, were discussed in units 3 (adaptation in weed life history) and 4 (adaptation in local plant communities). It is certainly true that weedy populations of several important current weed species and species-groups existed before the advent of agriculture some 10,000 years before present. This long-term stability is most apparent in weed species with crop relatives (see tables 1.4 and 1.5; *Amaranthus*, *Brassica*, *Chenopodium*, *Setaria*, *Solanum* and *Oryza* to name but a few). Is this not the ultimate form of plant community stability?

Much of this chapter has focused on the contributions made by ecology to understanding weed community biodiversity. It has been a story of controversy and belief, incomplete knowledge and insubstantial explanation. In unit 5 that follows the reader is exposed to a larger perspective of weeds and weed communities. We enter the world of complex adaptive systems in the hope that a larger perspective, beginning with first principles of thermodynamics, will aid in our understanding of the nature of weeds.

# UNIT 5: COMPLEX ADAPTIVE WEED SYSTEMS

## U5.1 The Nature of Weeds

The nature of weeds is fully revealed in the weed-crop groups presented in chapter 1, table 1.4. These are the consequences of 12,000 years of human agriculture and land management. The weed and crop species cannot be meaningfully separated, they are as one. The nature of weeds is that of the extended phenotype of *Homo sapiens*. As the beaver builds a dam and the pond and lodge become its home, so as we sow do we reap.

We began by understanding the processes of evolution that acts specifically on those plant species humans favored for food, fiber, fuel and intoxication. Local populations were systematically and predictably disturbed by cropping practices which contained resources and specific conditions that provided opportunity for a community of plants to thrive in. The fit became fitter, the less fit were eliminated ruthlessly. The fit dispersed, colonized and endured throughout the world. The fit adopted breeding practices amongst themselves that ensured appropriate novelty and conservation of genetic variation. They continued to breed at slower rates across the world through a vast network of metapopulations, bringing in new novelty, sharing the best they had to offer with their distant relatives. They divided and specialized. They came, they saw, they conquered. They gathered in unbelievably great numbers, hidden in the soil, waiting for the exact right time to resume reproduction. They interacted with each other with such finesse and specificity that they dominated the particularly rich, vast opportunities that the billions and billions of humans created for them.

## U5.2 The Nature of Complexity

Thus far in this book we have developed the evolutionary ecology of weeds incrementally in four units: the nature of weeds in unit 1, the evolution of weed populations in unit 2, adaptation in weed life history in unit 3, and adaptation in local plant communities in unit 4. Evolutionary ecology has provided us the foundation for understanding the nature of weeds. It is the logically deduced explanation of weedy biology from published observations. But understanding the nature of weeds relies on deeper insights than those provided by observations of weed biology. There exist larger forces in nature that need to be understood if evolution is the make sense. We can characterize these larger forces of nature as those that give rise to complex adaptive systems. Therefore, Unit 5 is a synopsis of complexity, complex adaptive systems and their relationship to the evolutionary nature of weeds of which we are part.

## U5.3 Formation of Complex Adaptive Systems

The constructal law guides the formation of complex systems. The thermodynamic flows of water and heat through plants are the drivers, engines, through which weeds exploit opportunity space-time; flows that ultimately determine community structure. The Earth water cycle drives resource seizure by plants in the flow of dissolved nutrients and gases caught in evapo-transpiration streams of the xylem. Trees and forests, plants and their communities, are pumping stations operating all the time to move water from the ground to the air. Flow systems construct their own flow architectures and body rhythms that enable them to move more easily. Everything that flows and moves generates designs that evolve to survive (to live). The constructal law commands that currents in nature move in configurations that flow more easily over time. Self-organization of the flow system occurs with the construction of flow patterns that ease movement. With self-organization a treelike pattern emerges throughout nature, vascularization. The hierarchical structure of design in nature is the concise name for what others describe as the emergence of "complex" design and "hierarchy".

Things in nature can be differentiated by their responses to disturbances, in the long run: things that like, are neutral, or dislike, disturbances. Fragile things are vulnerable to the disorder, volatility, uncertainty. Antifragile things response to a stressor with positive sensitivity to increases in volatility, variability, stress or uncertainty. Antifragile systems are those that change with time (e.g. evolution). Randomness is the 'disorder cluster': disorder, volatility, variability, stress, dispersion of outcomes, uncertainty, error, or risk. Randomness is information fueling antifragility. Phenomena differ between simple, complicated and complex systems. Simple and complicated systems have no interdependencies between component parts. Complex systems have severe interdependencies between component parts. Emergent properties in complex systems are the nonlinear result of adding units: the sum becomes increasingly different from the parts. In a nonlinear system adding two elementary actions to one another can induce dramatic new effects, unexpected structures and events. Simplification fails when something nonlinear is substituted with the linear. Simplifying scientific models mistakes a function of a variable for the variable itself. Nonlinear responses in weed biology include plastic phenotypic growth to local opportunity spacetime. Crowded weed communities form a nonlinear hierarchy of plant size with development: a few large plants and many small plants. The average of something has little to do with this skewed distribution of plant sizes.

## U5.4 Complex Adaptive Systems

Weed-crop communities are complex adaptive systems. A complex adaptative system is a dynamic network made up of a large number of active interacting adaptive agents, acting in parallel, and diverse in both form and capability. The activity of a CAS system of interacting elements results in the emergence of system order, and an anticipatory-nonpredictive strategy for adaptation to the environment. Weed-crop communities are examples of complex adaptative systems: the global metapopulation of a single species, the local deme of that weed species, the seed portion of a weed's life cycle, the plastic life history of an individual phenotype. CAS behavior and evolution arise from agents which sense their environment and develop schema representing rules of action. In CAS, large-scale behaviors emerge from the aggregate interactions of less complex agents forming the hierarchical system structure. Complex adaptive systems are living, adaptable, changeable systems in which complexity, emergence, self-similarity and self-organization occur. Emergence is the arising of novel (even radical) and coherent structures, patterns, properties during process of self-organization in complex systems. In complex adaptive systems an anticipatory-nonpredictive strategy for dealing with an uncertain future, and adaptation to the environment, emerges.

## U5.5 Soil-Seed Communication Systems

Seedling recruitment is a complex adaptive system: the *Setaria* spp.-gp. seed-soil environment communication system. *Setaria* seed heteroblasty places many individual bets on future success: seeds with slightly different germination requirements. Weedy foxtails communicate with the soil to time their departure from seed dormancy to the emerged seedling. Evolution embodies an ongoing exchange of information between organism and environment. What evolves is information. Information is physical. Biology is physical information with quantifiable complexity. Weedy *Setaria* spp. seeds with heterogeneous dormancy states are recruited from soil pools when sufficient oxygen, water and heat accumulate over time. The *Setaria* seed is constructed in such a way as to be receptive to specific soil signals: oxy-hydro-thermal time. A communication system of any type must contain five elements: information source, transmitter, channel, receiver and destination. The source of the message affecting weedy *Setaria* seed is the soil surrounding the individual seed in a particular locality. The message is heat and oxygen dissolved in soil water the *Setaria* embryo requires for germination. The transmitter is the water film adhering to soil particles adjacent to, and continuous with, seed exterior surfaces. The *Setaria* communication channel is the continuous water film connecting soil particles with the seed exterior and interior that transmits the oxygen-water-heat signal to the living seed interior receiver and the embryo destination. The signal transmitted from exterior water films are received by the seed embryo, the destination. Information is physical: memory resides in several locations in the *Setaria* seed. Memory resides in the functional-adaptive traits that regulate all seed behaviors by encoding and decoding soil signals. The message is remembered: plants pass on these functional traits to their polymorphic progeny.

## Chapter 11: Complex Adaptive Systems: Formation and Nonlinear Response

**Summary.** Plant and weed morphological structure and function can be explained partially from first principles, the second law of thermodynamics: the flows of water, pressure and heat through organisms. The thermodynamic flows of water and heat through plants are the drivers, engines, through which weeds exploit opportunity space-time; flows that ultimately determine community structure. The flow of heat in Earth systems both drives the water cycle and plant metabolism. The Earth water cycle drives resource seizure by plants in the flow of dissolved nutrients and gases caught in evapo-transpiration streams of the xylem. Trees and forests, plants and their communities, are pumping stations operating all the time to move water from the ground to the air. All natural designs are engines (of heat, fluid, or mass) driven by useful energy derived from the sun: nature organizes itself to move more easily. Flow systems construct their own flow architectures and body rhythms that enable them to move more easily. Everything that flows and moves generates designs that evolve to survive (to live). The constructal law commands that currents in nature move in configurations that flow more easily over time. Everything that moves, animate or inanimate, is a flow system. All flow systems generate shape and structure in time in order to facilitate this movement. Design emerges because things flow better with configuration. Self-organization of the flow system occurs with the construction of flow patterns that ease movement. With self-organization a treelike pattern emerges throughout nature because it is an effective design for facilitating point-to-area and area-to-point flows. Wherever you find such flows, you find a treelike structure. The treelike structure of flow is vascularization, the interdependency of life, the spread of nourishing currents across an area or throughout a volume. Everything that moves and morphs in order to flow and persist is alive. The hierarchical structure of design in nature, 'few large and many small', is the concise name for what others describe as the emergence of "complex" design and "hierarchy".

Things in nature can be differentiated by their responses to disturbances, in the long run: things that like, are neutral, or dislike, disturbances. Fragile things are vulnerable to the disorder, volatility, uncertainty. The fragile depend on things following the exact planned course, with as few deviations as possible: deviations are more harmful than helpful. Robust things are strong in constitution. Robust things can resist change without adapting its initial stable configuration under perturbations or unusual or conditions of uncertainty. Antifragile things response to a stressor or source of harm with positive sensitivity to increases in volatility, variability, stress or uncertainty. Antifragile systems are those that change with time (e.g. evolution). The antifragile are things that benefit, get better, from shocks: the exact opposite of fragile. The concept of random is complex, having several connotations depending on context. There is no functional difference between these types of randomness, we can never know enough to make distinctions. Randomness is the 'disorder cluster': disorder, volatility, variability, stress, dispersion of outcomes, uncertainty, error, or risk. Randomness is information fueling antifragility. Phenomena differ between simple, complicated and complex systems. Simple and complicated systems have simple responses, there are no interdependencies between component parts. Complex systems is severe interdependencies between component parts. Emergent properties in complex systems are the nonlinear result of adding units: the sum becomes increasingly different from the parts. Causation is either nearly impossible to detect or not really defined because of these complex interactions. Antifragility in biology works in hierarchical layers, from the global to molecular. As such, there exists a tension between nature and individuals in a natural system (e.g. natural selection). Some internal parts of the system (individuals) may be required to be fragile in order to make the system antifragile as a result. Nature in the aggregate survives and is antifragile by the contributions and fragility of species. Species in turn survive and are antifragile by the contributions and fragility of its individual organisms. The random element in nature uses error as a source of information. There exists danger in stability. There is no stability without volatility. In the linear world observed effects are linked to the underlying causes by a set of laws reducing for all practical purposes to a simple proportionality. In a linear system the combined action of two different causes is the superposition of the effects taken individually. In a nonlinear system adding two elementary actions to one another can induce dramatic new effects, unexpected structures and events in the form of abrupt transitions, a multiplicity of states, pattern formation, or an irregular markedly unpredictable evolution in space and time referred to as deterministic chaos. Simplification fails when something nonlinear is simplified with the linear as a substitute. Simplifying scientific models squeeze information into a 'Procrustean bed', mistaking something with the function of something: mistaking a function of a variable for the variable itself. The function of a thing has different properties than the thing itself. The function of something becomes different from the something under nonlinearities; with significant asymmetry they may have nothing to do with each other. With nonlinearity, the function depends more on the volatility around the average. The more nonlinear the response, the less relevant the average, and the more relevant the stability around the average. The average does not matter to the operation of the nonlinear system, the variability of the system is more important, the convexity effect. There exist two kinds of nonlinear responses ($f(x)$) by a variable ($x$). Concave functions curve inward, with more losses than gains with change in the variable ($x$), the fragile. Convex functions curve outward, with more gains than losses with change in the variable ($f(x)$), the antifragile. Nonlinear responses in weed biology are common, notably phenotypic plasticity in growth, development and reproductive responses to local opportunity spacetime. Crowded weed communities tend to form a nonlinear hierarchy of plant size with development: a few large plants and many small plants. The average (function) of something (plant size) has little to do with this skewed distribution of plant sizes. This tendency to fit observations into a 'Procrustean bed' underlies the profound flaws inherent in many demographic representations of weed biology. For the fragile demographer these nonlinearities are fatal. For the antifragile evolutionist the asymmetrical, nonlinear uncertainty of life events are liberating.

**11.1 Formation of Complex Systems: The Constructal Law.** Plant and weed morphological structure and function can be explained partially from first principles, the second law of thermodynamics. As such, plant-weed community structure can also be partially explained by the flows of water, pressure and heat through their member organisms.

The thermodynamic flows of water and heat through plants are the drivers, engines, through which weeds exploit opportunity space-time. The Earth water cycle drives resource seizure by plants in the flow of dissolved nutrients and gases caught in evapo-transpiration streams of the xylem. The flow of heat in Earth systems both drives the water cycle and plant metabolism. Plants, weeds, channel these Earth water and heat flows to seize and exploit opportunity space-time resources and conditions. Resource acquisition is driven by water flow through the plant xylem: nutrients and gases dissolved in soil water. Local conditions provide heat driving plant metabolism: the growth and development of the plant. Light resource is captured by the plant shoot architecture consequential to this growth and development.

A constructal law approach predicts the essential features of vegetation design. This tells us that there is a direction to evolution. It is not the story of random events but the unfolding saga of the emergence and evolution of design for better and better flow in time. The constructal law places a physics principle behind Darwin's ideas about evolution. It tells us why certain changes are better than others and shows that those changes do not arise by accident but through the generation of design. The constructal law also expands our understanding of evolution, showing that the natural tendency of biological change is the same tendency that shapes the inanimate world.

**11.1.1 First principles: the thermodynamics of flow.** The nature of complex systems derives at the outset from first principles of physics. Thermodynamics is the branch of physics concerned with heat and temperature and their relation to energy and work (Wikipedia, 1.16). The behavior of those variables is subject to general constraints that are common to all materials. These general constraints are expressed in the four laws of thermodynamics. Thermodynamics describes the bulk behavior of the body, not the microscopic behaviors. Thermodynamics rests on two laws, both first principles:

1: conservation of energy
2: tendency of all currents to flow from high (temperature, pressure) to low; the "one-way flow" principle

The second law of thermodynamics indicates the irreversibility of natural processes, and, in many cases, the tendency of natural processes to lead towards spatial homogeneity of matter and energy, and especially of temperature. In a natural thermodynamic process, the sum of the entropies of the interacting thermodynamic systems increases.

All natural designs are engines (of heat, fluid, or mass) driven by useful energy derived from the sun: nature organizes itself to move more easily. The movement towards equilibrium, the pull of gravity, puts things in motion, and nature generates configurations to facilitate this flow that has direction in time. In order to get from here to there, everything must create a path. Flow systems construct their own flow architectures and body rhythms that enable them to move more easily. Everything that flows and moves generates designs that evolve to survive (to live). The more efficient the pathway, the more likely is its persistence, whatever the mechanism behind that persistence.

**11.1.2 The constructal law.**

"… a constructal law approach predicts the essential features of vegetation design."
"This tells us that there is a direction to evolution. It is not the story of random events but the unfolding saga of the emergence and evolution of design for better and better flow in time."
"The constructal law places a physics principle behind Darwin's ideas about evolution. It tells us why certain changes are better than others and shows that those changes do not arise by accident but through the generation of design. The constructal law also expands our understanding of evolution, showing that the natural tendency of biological change is the same tendency that shapes the inanimate world." (Bejan and Zane, 2012)

The constructal law commands that currents in nature move in configurations that flow more easily over time (Bejan and Zane, 2012).

**the constructal law**: for a finite-size flow system to persist in time (to live), its configuration must evolve in such a way that provides easier access to the currents that flow through it (given freedom from constraints)

This thermodynamic law is a physical law, a physical phenomenon, a natural tendency, that governs the emergence of macroscopic shape and structure in nature. It can be used to understand why designs emerge and to predict how they will evolve in the future. It is a tendency that unites everything that moves.

**11.1.3 Flow systems.** Everything that moves, animate or inanimate, is a flow system. All flow systems generate shape and structure in time in order to facilitate this movement across the landscape filled with resistance (e.g. friction) (Bejan and Zane, 2012). Designs in nature are not the result of chance. They arise naturally, spontaneously, because they enhance access to flow in time.

**system:** the region in space or the quantity of mass selected by an observer for analysis and discussion defined by a sharp and precise boundary around the entity.

Boundaries impermeable to the flow of mass are closed systems; open systems allow masses to cross boundaries containing inlets and outlets.

**flow (flow systems):**
1: flow represents the movement of one entity (in the channel) relative to another (the background)
2: flow is described by what the flow carries (fluid, heat, mass, information), how much it carries (mass flow rate, heat current, traffic, etc.) and where the stream is located

Flow systems have two basic properties:

      1: the current that is flowing (e.g. fluid, heat, mass, or information)
      2: the design through which the current flows

Every flow system is part of a bigger flow system, shaped by and in service to the world around it.

Flow systems require imperfections, without resistances they would accelerate continuously, out of control. Easier flowing consists of balancing each imperfection, resistance, against the rest. All components of the system collaborate, working to create a whole that is less and less imperfect with time. The distributing and redistributing of imperfection through the complex flow system are accomplished by making changes in the flow architecture, design. The prerequisite is for the flow system to be free to morph. The emerging flow architecture is the means by which the flow system achieves its objective under constraints.

Understanding flow systems in nature consists of three components. First, motion is the cause of every life. All flow systems have the tendency to endow themselves with design: their configuration (architecture, geometry, shape, structure) and their rhythm (the predictable rate at which it pulses and moves). This second understanding requires recognition of what flows through the system and what shape and structure should emerge to facilitate that flow. The configuration is the defining characteristic. The third understanding is that flow system designs evolve, they configure and reconfigure themselves over time. This evolution occurs in only one direction: flow designs get measurably better, moving more easily and farther if possible. Important flow systems herein are preexisting and evolving.

Design emerges because things flow better with configuration. Self-organization of the flow system occurs with the construction of flow patterns that ease movement. With self-organization a treelike pattern emerges throughout nature because it is an effective design for facilitating point-to-area and area-to-point flows. Wherever you find such flows, you find a treelike structure.

**11.1.4 Evolution of flow system structure and design.** The constructal law dictates that flow systems should evolve over time, acquiring better and better configurations to provide more access for currents that flow through them (Bejan and Zane, 2012). Design generation and evolution are macroscopic physical phenomena that arise naturally to provide better and better flow access to the currents that run through them. Complexity arises as result of the constructal design that emerges. The universality and reach of the principle is that it occurs at every scale. The myriad of scaling laws found in nature are only surface reflections of a far deeper harmony, balance and oneness in nature.

The constructal law provides an overarching direction to evolution: design and evolution toward greater "access" for all components in a "finite-size" system. It predicts that evolution should occur because of the tendency of all flow systems to generate better and better designs for the currents that flow through them. All flow designs arise, evolve and compound themselves because of the directional tendency towards enhancing movement. Life is movement and the constant morphing of the design to this movement. To be alive is to keep on flowing and morphing. Evolution occurs for "better" physical adaptations: change in design that facilitates faster, easier movement. This is direction and physical evolution without intention.

The constructal law also recasts organic evolution as a dynamic process that generates better designs. There is a large volume of imperfection involved in biological evolution (e.g. genetic drift, selection on linked alleles, extinction, dispersal limitation, environmental heterogeneity in space and time) as well as idiosyncratic variation. But the central tendency is the selection of characteristics that ease the flow. This definition removes the concept of life from the specialized domain of biology. It aligns it, juxtaposes it, with the physics concept of the dead state: equilibrium with the environment. In thermodynamics a system at equilibrium has the same pressure, the same temperature as its surroundings. Hence, a system in which nothing moves. Uniformity is death; equilibrium means uniformity in every respect.

"Evolution" means design modifications over time. How these changes are happening are mechanisms, which should not be confused with the principle, the constructal law. In the evolution of biological design, the mechanism is mutations, biological selection, and survival. In geophysical design, the mechanism is soil erosion, rock dynamics, water-vegetation interaction, and wind drag. What flows through a design that evolves is not nearly as special in physics as how the flow system acquires and improves its configuration in time. The 'how' is the physics principle, the constructal law. The 'what' are the currents and the mechanisms, and they are as diverse as the flow systems themselves. The 'what' are many and the 'how' is one. Hierarchy more simple than this does not exist. The constructal law advances our understanding of evolution by proclaiming that design should emerge across nature to facilitate flow. It also holds that these configurations should morph with a clear direction in time: to provide better and better flow access. The vascular, hierarchical designs we find throughout nature strike a balance between the speeds of their currents by generating multiscale channels. Each of which selects the mode of flow, slow and fast, that works best for them.

**11.1.5 Flow system structure and movement.** Every flow structure consists of two major (non-static) elements that cover its entire area (Bejan and Zane, 2012). Every tree-shaped flow architecture is defined by these two regimes:

1: the channels
2: the finite sized spaces between adjacent channels (interstices)

There exist two shape-generating ways of flowing. Striking the balance between them is the hallmark of natural design:

1: slow and short
2: fast and long

In all designs currents move more slowly over short distances through interstices and faster over longer distances through channels. This gives rise to pulsing movement of current flows. Vascular flow systems generate multiscale channels because this is a good design for spreading a current from a point to an areas or an area to a point. The emergence of multiscale design hinges on the balancing of these two flow regimes. The key design principle is this: the time to move fast and long should be roughly equal to the time to move slow and short. When this occurs, currents flow with ease over the area inhabited by the entire flow structure. This is the foundation of all constructal designs.

Design starts to appear when the surface slow-short coalesces to form a rivulet. In many cases this involves the transition from laminar to turbulent flow in water, air or other fluid. This transition to a new way of flowing provides the contrast, the essence of the design. The transition is the birth of design. Design emerges over the entire area. In order to move more easily the flow system acquires geometry, design. Two natural phenomena are part of this flow design. First, gravity pulls the flow down. Second, the constructal law accounts for the fact that the flow will form the first channel (rivulet) when the flow becomes rapid enough. When the flow is slow and short, diffusion is the way to go. When the flow intensifies an organized structure with streams and channels is better. Various sized channels emerge in response to the specific resistance encountered by the flow. The overall balance of the system is achieved through a universal design balance: the resistance to moving slow and short should be comparable with the resistance moving fast and long. The time spent flowing slow and short should be roughly equivalent to the time spent flowing long and fast in the channels. The bifurcated design distributes resistances so that globally the flow system becomes less and less imperfect. The constructal law is a principle of physics making us think holistically. Nothing moves in isolation; every flow is part of other flow systems. There exists a basic tendency to balance thermodynamic imperfections, to generate configurations that balance resistances and reduce their combined effect. Good design involves the nearly uniform distribution of imperfection throughout the entire flow system

This structure of channels and interstices can be seen in fitness landscape metaphors. In evolutionary biology, fitness landscapes or adaptive landscapes are used to visualize the relationship between genotypes and reproductive success (fitness) (Wikipedia, 1.16). Fitness landscapes are often conceived of as ranges of mountains with intervening valleys. Fitness is the "height" of the landscape. The set of all possible genotypes, their degree of similarity, and their related fitness values is then called a fitness landscape (Wright, 1932). An evolving population may climb uphill in the fitness landscape, by a series of small genetic changes, until a local optimum is reached. Wright visualized a genotype space as a hypercube; fitness as a function of allele frequency. No continuous genotype "dimension" is defined. Instead, a network of genotypes are connected via mutational paths. In a phenotype fitness landscape, each dimension of the hypercube represents a different phenotypic trait (reaction norm).

**11.1.6 Why do weeds exist?** The constructal law shows us that tree, plant, shape and structure can be predicted from the universal tendency to facilitate flow access (Bejan and Zane, 2012). It's hard to recognize terrestrial plants as flow systems: water is flowing. Terrestrial plants happen because that is where the water is and must flow (upward), not because 'plants like water'. Trees, and all plants, grow in great abundance where there is more water in the ground than in the air. Rivers, trees and plants are designs that emerge to handle the currents that flow through them and along with them. They do not exist in service to themselves but in service to the global thermodynamic flow. They work with all that flows around them, evolving to enhance the movement of everything on Earth. The constructal law integrates trees and plants as part of an immense global flow architecture – along with all river basins, raindrops, and atmospheric and oceanic circulation – that facilitates cyclical flow of water in nature: the Earth water cycle (see ch. 3.3.1, Earth's physical geography). Because of the second law of thermodynamics, water is governed by the natural tendency to equilibrate all the moisture in the environment. Because of the constructal law, a wide range of morphing and mating flow designs have emerged to facilitate that movement. The terrestrial plant is a design for moving water. Only a small fraction is consumed by the plant; 99-99.5% is lost by transpiration. The bulk of it is pumped back into the atmosphere. Trees and forests, plants and their communities, are pumping stations operating all the time to move water from the ground to the air. Beginning with the roots that pull water from the surrounding area, to the stem that conveys water to the branches, to the leaves that release it when they open their pores (stomates) to capture carbon dioxide and sunlight for photosynthesis, the design of the tree is geared toward performing this work efficiently. The second law of thermodynamics proclaims that nature should manifest the tendency to move water from wet to dry both locally and globally. The constructal law teaches us that trees and forests, plants and their community, occur to survive in order to facilitate rapid transfer of water from the air to the ground. It improves on the Darwinian view that cast trees as individuals and separate species competing against their neighbors to survive: a projection of human values on evolution, a struggle for life with neighbors.

**11.1.7 How the constructal law predicts the design of weed**. The constructal law predicts the design of a tree, a plant, from its roots and stems to its branches and leaves (Bejan and Zane, 2012). The plant has to handle two types of flow:

1: movement of water from the ground to the air
2: flow of stresses caused by the wind

Therefore it should have a special architecture that provides access for water coursing through it and mechanical strength against the winds that buffet it. How should the root handle water flow? It must have a porous body to allow for two types of water flow:

1: transversal, from all sides, so that water can enter the system from various depths
2: longitudinal, so the water can move up from the ground

The longitudinal flow (the through channel) has less resistivity than the transversal flow. The constructal law predicts the entire root system should have a treelike architecture because this is a good way to provide flow access from a volume (soil) to a single point (at ground level).

    Mechanical strength is determined by the stresses that flow through an object. If a rod has a uniform cross section, the stresses flow through the entire body without strangulations because this is an efficient way to use the volume of solid material to house the stresses in the smallest, lightest body. The wind sends shivers throughout the tree, terrestrial plant, introducing stresses that flow through it. The plant's design distributes these stresses uniformly, spreading the highest stresses (and the chance to fracture) throughout, so that each part withstands the maximum allowable stress, giving each part a maximum chance at survival. Every part is at equal risk and is equally protected. The round cross sections and branching structure of the root system that facilitate the flow of water also steady the tree. Wind blows through the trees from all directions, so the round cross section is best for distributing the highest stresses among the fibers as they are subjected to bending. Like strong fingers of a hand encircling a lamppost to steady the body in a strong wind, the plant roots grab hold of the earth.

    Where roots collect water and move it up, the trunk disperses it up and out through the branches and leaves. Which is the best shape for the two flows that inhabit the trunk, with water moving up and stresses flowing to and from the ground from the wind? It is broad at the bottom and narrows as it rises because the higher we go, the less water we find as it is dispersed to the lower branches. This design extends to the branches. We find more numerous smaller branches the higher we go in the tree, plant because this is an efficient design for releasing water back into the air. In the real world a host of environmental factors affect branches. Strong winds bend them and shear them away. The lack of sunlight in a crowded forest, or on the north slope of a hill, imperils lower branches. Local environmental conditions are one of many factors that create variations in the shapes and structures of individual trees, plants.

    The constructal design of trees demands a spiral arrangement (the Fibonacci series) of branches around the trunk. For trees and plants to move water efficiently from the ground to the air, each of their branches should grow laterally into the space that is least affected by its neighbors (or itself). This is because all the branches are dumping water into the air, so the driest air will be the one farthest away from other water-emitting branches. Each branch grows and moves toward the space that contains the least humid air. The need to reduce interference between branches is a restatement of the constructal law, that is, the tendency to morph in order to have greater flow access for water from the ground to the air, that is, from the base of the trunk (the point) to the entire tree (the volume).

**11.1.8 Plant community structure and assembly**. The driving forces of Earth thermodynamics creates flows that ultimately determine community structure. Plant community assembly begins with recruitment (see ch. 9.1.2). With community assembly begins interaction with neighbor plants and other organisms in the local community. During plant community ecological succession (see ch. 10.4.2.2) earlier successional species facilitate invasion of later successional species by the opportunity spacetime the create in the habitat (see ch.

3.1.1; ch. 3.5). The distribution of plant (tree) sizes of the forest (the local plant community) seems hodgepodge. We see smaller trees here and there, filling in gaps amidst their larger cousins, finding niches where they can. Just as each tree is an individual pumping station whose design facilitates the flow of water from the ground to the air, the entire forest is a giant pumping station that mixes the number of large and small trees to achieve this on a grand scale, the local community scale (Bejan and Zane, 2012). We take a larger leap in understanding by acknowledging that trees are inextricably linked to, and shaped by, all the other trees and forms of vegetation as well as the environment around them. Trees are components of a larger flow system, the forest, so their size and distribution should be determined by the forest's tendency to generate design that eases flow. Each tree and each forest is a single entity, but all are fed and shaped by one another. The best pumping design, the best distribution of trees, has a few big trees and many smaller ones: a hierarchical relationship among all trees of varying sizes. This was a good way to cover the entire forest floor with vegetation to move water. Despite any Darwinian struggle for survival that might occur among all forms of vegetation on the forest floor, the design is known in advance. This is the physics basis for the emergence of hierarchy in nature that guides community assembly.

Forests, plant communities, are organs of the much larger global system of flow driven by the second law of thermodynamics. This integrative approach reveals that the biosphere, atmosphere, and hydrosphere are not separate entities but interlocking environments that flow and design themselves together. From this idea we can predict their designs and the designs of everything that flows and moves. This insight also challenges the Darwinian concept of winners and losers. In time, some species do flourish and others wither away. It even appears some thrive precisely because they are able to crowd out their "competitors", that is, their neighbors. The constructal law teaches us to see all flow systems as components of a single organism, the entire globe, which evolves its design to enhance its flow. They are not competing against each other but working together. The idea of winners and losers might make sense if evolution were a zero-sum game with no direction in time. But because flows morph to increase flow access for the whole, the whole becomes the winner.

**11.1.9 Hierarchy: the treelike structure of flow.** The treelike structure of flow is vascularization, which captures the central idea of interdependency of life (Bejan and Zane, 2012). The tree suggests the connection between point and area or volume. Vascularization contains the pivotal idea of life-giving flow and of a body (volume, area) filled with life. Design arises in order to spread often-nourishing currents across an area or throughout a volume. Everything that moves and morphs in order to flow and persist is alive. Everything evolves in order to provide greater access to the life-sustaining currents that run through its vasculature. Hierarchy is a cornerstone characteristic of natural design (Bejan and Zane, 2012). It is essential to good design. It arises naturally because it benefits the entire flow system. Hierarchy evolves because good flow often involves multi-scale architectures, channels of varying sizes. All are necessary to spread the current that runs through it to every destination, providing access to the whole area.

The hierarchical structure of design in nature, 'few large and many small', is the concise name for what others describe as the emergence of "complex" design and "hierarchy". The plant, tree, has one main channel, the trunk, a few main branches, and many roots, stems, and leaves. All are necessary for the efficient flow of water from the ground to the air. Every multiscale flow system exhibits a hierarchical structure composed of a few large channels and many smaller ones. The constructal law also predicts that the rigid hierarchy will give way in time to a freely morphing hierarchy. Hierarchy is generated at every step in the formation of a flow system, not just at the end. While all multiscale, point-to-area flow systems generate easier flowing configurations that have hierarchy, they do not simply reiterate the simplest design (fractal) into ever more complex patterns. The smallest detail is not simply a miniature version of the largest drawing. Just as one size never fits all, neither does one design. Flow systems generate just enough complexity for the size of the territory they bathe with current. They create architectures that work for them. The complexity of each architecture is modest, finite. The phenomenon of design in nature is not one where complexity increases in time. Each flow system evolves to acquire the right level (kind) of complexity to flow, to live. A seminal aspect of hierarchy is the interconnectedness and interdependence of every component of the flow system. Hierarchy emerges because all flow systems use the right combination of components of varying sizes to efficiently move the currents that flow on the same, finite territory. Hierarchy arises because it is good for every component of the global flow system. This is the integrative aspect of design, the balance that naturally emerges among all it flow components.

Flow designs are global engines that have arisen to enable currents that flow through them to move easily across the landscape (Bejan and Zane, 2012). The source of almost all movement, all life, on Earth is the sun. For all the diversity we find in nature, the history of our planet is the unfolding story of the interaction between solar energy and the mass it sets in motion (see ch. 3.4.1, thermodynamic Earth).

## 11.2 Fragile-Antifragile Systems

"When a random event happens, it is already too late to react ..." "Post-event adaptation, no matter how fast, would always be a bit late." "... the (post-event) adaptability criterion is innocent of probability ..." The property in a stochastic process of not seeing at any time period $t$ what would happen in time after $t$, this is, any period higher than $t$, hence reacting with a lag, an incompressible lag, is called nonanticipative strategy, a requirement of stochastic integration. The incompressibility of the lag is central and unavoidable. Organisms can only have nonanticipative strategies, hence nature can only be nonpredictive." Taleb, 2010

How do complex adaptive systems respond to an unpredictable future? How do nonlinear asymmetric weed behavioral traits respond to variable and unpredictable changes in local opportunity spacetime? How do weeds anticipate an uncertain future? The information in chapter 10 featured the current approach to weed community structure, dynamics and biodiversity concepts apparent in the ecological literature. The key concepts and criteria in ecological systems include stability, resilience, equilibrium and sustainability. These concepts are related to human needs, anthropogenic means of judging the value of nature to human consumers and scientists (e.g. 'ecological services'). Taleb (2012) has provided a more comprehensive insight, a concept for understanding complexity and complex adaptive systems: the fragile and antifragile. The fragile and anti-fragile are understood in the related concepts of the random, the complex, the nonlinear and the optional.

**11.2.1 The fragile, robust and antifragile**. Things in nature can be differentiated by their responses to disturbances, in the long run: things that like, are neutral, or dislike, disturbances (Taleb, 2012).

**fragile**:
1: vulnerability of something to the disorder, volatility, uncertainty of things that affect it with time
2: easily broken, damaged, or destroyed; physically weak; frail; delicate; tenuous; flimsy
3: a concave sensitivity to stressors, leading to a negative sensitivity to increases in volatility (or variability, stress, dispersion of outcomes, or uncertainty: the "disorder cluster") (Taleb, 2012)

Shocks bring more harm to the fragile as their intensity increases, up to a certain level. For the fragile the cumulative effect of small shocks is smaller than the single effect of an equivalent single large shock: extreme events cause more harm to the fragile. The fragile depend on things following the exact planned course, with as few deviations as possible: deviations are more harmful than helpful. The fragile needs to be very predictive in its approach; conversely, predictive systems cause fragility.

**robust**:
1: strong in constitution; hardy; vigorous; sturdily built; requiring or suited to physical strength
2: immunity from one's external circumstances, good or bad, and an absence of fragility; resilience, resists shocks but remains the same

3: the ability of a system to resist change without adapting its initial stable configuration; the persistence of a system's characteristic behavior, or a characteristic or trait (canalization), under perturbations or unusual or conditions of uncertainty (Wikipedia, 6.14)

**antifragile**:
1: a convex response to a stressor or source of harm (for some range of variation), leading to a positive sensitivity (positive convexity effects) to increases in volatility (or variability, stress, dispersion of outcomes, or uncertainty: the "disorder cluster"); shocks bring more benefits (less harm) as their intensity increases (up to a point) (Taleb, 2012)

Antifragile systems are those that change with time: evolution, culture, ideas, revolutions, political systems, technological innovation, cultural and economic success, corporate survival, good recipes. The antifragile determines the boundary between what is living and organic (or complex) and what is inert. The antifragile are things that benefit, get better, from shocks: the exact opposite of fragile. They are things that thrive and grow when exposed to volatility, randomness, disorder and stressors. They are things that love adventure, risk and uncertainty. Antifragile systems have the singular property of being able to deal with the unknown, the ability to do things without understanding them, and to do them well. The relation between fragility, convexity, and sensitivity to disorder is mathematical, obtained by theorem, not derived from a priori empirical data mining or some historical narrative.

Risk and exposure are the over-riding considerations in assessing fragility. How much of something (e.g. seedlings; $x$) did you invest (your exposure)? What is the risk (potential gains and losses) of that exposure? For example, the fragility of a weed species investment establishing a population in the spring could be assessed by the number of seedlings emerging from a larger dormant seed pool in the soil. How do weeds go about exposing (when, where, what, how) and risking in the process of local adaptation? They do it by trial and error search mechanisms to maximize seizing and exploiting opportunity spacetime.

**11.2.2 The random.** The concept of random is complex, having several connotations depending on context: non-order or non-coherence; a lack of predictability (statistics); random selection (evolution and games); epistemic randomness (knowledge); ontological randomness (complexity and chaos); and algorithmic randomness (information and computation) (Taleb, 2010). A truly random system is in fact random (non-order) and does not have predictable properties. Randomness can arise from lack of knowledge of system (epistemic, algorithmic) or because of the dynamic complexity of the system (ontological). There is no functional difference between these types of randomness, we can never know enough to make distinctions (see ch. 13.1, randomness, uncertainty and information). Randomness in life is not randomness in the casino, of throwing dice or flipping a coin. Randomness is the 'disorder cluster': disorder, volatility, variability, stress, dispersion of outcomes, uncertainty, error, or risk. Randomness is information fueling antifragility.

**11.2.3 The complex.** Phenomena differ between simple, complicated and complex systems (Taleb, 2010). Simple and complicated systems have simple responses, there are no interdependencies between component parts. The hallmark of complex systems is severe interdependencies between component parts. Emergent properties in complex systems are the nonlinear result of adding units: the sum becomes increasingly different from the parts. Causation is either nearly impossible to detect or not really defined because of these complex interactions. Causal relationships cannot be isolated in a complex system. For the complex a state of normalcy requires a certain degree of volatility, randomness, the continuous swapping of information, and stress. Complex systems are harmed when deprived of volatility. The idea of equilibrium separates the two different worlds of the noncomplex and complex. In a nonorganic, noncomplex, system an object in equilibrium occurs in a state of inertia. In a complex, organic, system equilibrium only happens in death (Kaufman, 1995).

Complex systems subjected to randomness, unpredictability, build a mechanism beyond the robust to opportunistically reinvent themselves each generation. This results in continuous change in a local population and a species. For example, in biology, natural selection provokes evolution and adaptation

Evolution benefits from randomness and volatility. The more noise and disturbances in the system the more the effect of the reproduction of the fittest and that of random mutations will play a role in defining the properties of the next generation (barring extreme shocks leading to extinction). The benefits derived from randomness come by two different routes. They both act in a similar way to cause changes in the traits of the surviving next generations: randomness in genotypes (e.g. mutations, sexual recombination); randomness in the environment (opportunity spacetime).

Antifragility in biology works in hierarchical layers, from the global to molecular. As such, there exists a tension between nature and individuals in a natural system: the rivalry between individual organisms contributes to evolution (natural selection). Some internal parts of the system (individuals) may be required to be fragile in order to make the system antifragile as a result. There exists a tension in the trade-off of fragilities in nature. Nature in the aggregate survives and is antifragile by the contributions and fragility of species. Species in turn survive and are antifragile by the contributions and fragility of its individual organisms. Individual organisms are not completely fit for all future random events, they are not preadapted for all such events. Therefore they are fragile for randomness, the uncertainty of future events. Every random event will bring its own antidote, via natural selection and elimination, in the form of ecological variation.

The random element in trial and error uses error as a source of information. Every trial tells you what doesn't work, you start zooming in on the solution, every attempt becomes more valuable, and you make discoveries along the way. Stressors, error and volatility are information allowing opportunistic adjustment. There exists danger in stability. The avoidance of small mistakes makes the large mistakes more severe and vulnerabilities accumulate. There exists a need to see second steps, chains of consequences, and side effects. There is no stability without volatility.

**11.2.4 The nonlinear: concave-convex effects.** Much of science is built on the idea that a natural system subjected to well-defined external conditions will follow a unique course and that a slight change in these conditions will likewise induce a slight change in the system's response. Owing undoubtedly to its cultural attractiveness, this idea, along with its corollaries of reproducibility and unlimited predictability and hence of ultimate simplicity, has long dominated our thinking and has gradually led to the image of a linear world: a world in which the observed effects are linked to the underlying causes by a set of laws reducing for all practical purposes to a simple proportionality (Nicolis, 1995).

A striking difference between linear and nonlinear laws is whether the property of superposition holds or breaks down. In a linear system the ultimate effect of the combined action of two different causes is merely the superposition of the effects of each cause taken individually. But in a nonlinear system adding two elementary actions to one another can induce dramatic new effects reflecting the onset of cooperativity between the constituent elements. This can give rise to unexpected structures and events whose properties can be quite different from those of the underlying elementary laws, in the form of abrupt transitions, a multiplicity of states, pattern formation, or an irregular markedly unpredictable evolution in space and time referred to as deterministic chaos. Nonlinear science is, therefore, the science of evolution and complexity (Nicolis, 1995).

Simplification fails when something nonlinear is simplified with the linear as a substitute. Simplifying scientific models squeeze information into a 'Procrustean bed'. Procrustes was a sadistic thug from Greek mythology who abducted travelers and forced them to lie in a special bed. He was a rogue smith and bandit from Attica who physically attacked people by stretching them or cutting off their legs, so as to force them to fit the size of an iron bed. In general, when something is Procrustean, different lengths or sizes or properties are fitted to an arbitrary standard: simplifications are not simplifications (Taleb, 2010; Wikipedia, 6.14).

There exists a 'Procrustean' tendency in humans to mistake something (a perception, idea, theory) with the function of something: conflation of event and exposure. It is common to mistake a function of a variable for the variable itself. The function of a thing has different properties than the thing itself. The function of something becomes different from the something under nonlinearities. The more asymmetrical they are the more different they are. With significant asymmetry they may have nothing to do with each other.

The more nonlinear the more the function of something divorces itself from the something. The properties of $f(x)$ and those of $x$ become divorced from each other when $f(x)$ is nonlinear. The more volatile the something, the more uncertainty, the more the function divorces itself from the something. With nonlinearity, the function depends more on the volatility around the average. The more nonlinear the response, the less relevant the average, and the more relevant the stability around the average.

The consequences of nonlinear, asymmetrical events in nature in response to volatility differ for the fragile and antifragile. An unfavorable asymmetry exists with exposure to volatility if you have more to lose than benefit from that source: you are fragile and thrive with a lack of volatility and stressors. A favorable asymmetry occurs with exposure to volatility if you have more to gain than lose: you are antifragile, and you may be harmed by a lack of volatility and stressors. This is the fundamental asymmetry in the antifragile: something has more upside than downside in certain situations and tends to gain from volatility, randomness, errors, uncertainty, stressors and time (and the reverse). The first order effect of nonlinearity is that the average does not matter to the operation of the system. The variability of the system is more important than the average. The average is not significant when one is fragile to variations. The second order effect of nonlinearity is that the dispersion of possible outcomes matter more than the average outcome. This is the convexity effect.

There exist two kinds of nonlinear responses ($f(x)$) by a variable ($x$) (figure 11.2): concave negative asymmetry and the convex positive asymmetry. Concave functions curve inward, with more losses than gains with change in the variable ($x$) (figure 11.2, top). This negative asymmetry defines the fragile. Convex functions curve outward, with more gains than losses with change in the variable ($f(x)$) (figure 11.2, bottom). This positive asymmetry defines the antifragile. For convex functions, the antifragile, the average of the function of something ($x$) is going to be higher than the function of the average of something ($f(x)$); and the reverse when the function is concave, the fragile.

Nonlinear responses in weed biology are common, notably phenotypic plasticity in growth, development and reproductive responses to local opportunity spacetime. For example, in many weed species asymmetrical, non-linear responses to variable local opportunity spacetime include branching, flowering and seed fecundity (yield per individual plant), and individual plant size in response to density (plant number per unit area). Compare these nonlinear responses to Malthusian postulates (ch. 2.3.1), assumptions underlying the Verhulst-Pearl logistic equation of population growth in a limited environment (ch. 2.4.1), phenotypic plasticity (ch. 4.3.1), somatic polymorphism (ch. 4.3.2), and the effects of neighbor interactions on plant growth and plant form (ch. 9.4). For example, crowded weed communities tend to form a nonlinear hierarchy of plant size with development: a few large plants and many small plants (see ch. 9.4.3.1, plant size hierarchies in the community). The average (function) of something (plant size) has little to do with this skewed distribution of plant sizes. This tendency to fit observations into a 'Procrustean bed' underlies the profound flaws inherent in many demographic representations of weed biology (see ch. 14.2, the ecological demography of plant population life history dynamics). For the fragile demographer these nonlinearities are fatal. For the antifragile evolutionist the asymmetrical, nonlinear uncertainty of life events are liberating (see ch. 14.3, evolutionary, trait-based, weed life history population dynamics).

**11.2.5 The optional.** Nonlinear asymmetry results in two biases: mistaking an average of a function for the function of the average. The convexity bias is the exact measure of benefits derived from nonlinearity or optionality: the difference between $x$ and a convex *function of x*. This convexity bias comes from a mathematical property called 'Jensen's inequality'. Jensen's inequality relates the value of a convex function of an integral to the integral of the convex function (1906). In its simplest form the inequality states that the convex transformation of a mean is less than or equal to the mean after convex transformation; its simple corollary, the opposite is true of concave transformations (Wikipedia, 6.14). These biases represent optionality: in convex functions, the potential gains outweigh the potential losses; in concave losses outweigh gains. For favorable asymmetries (positive convexity), you will do reasonably well in the long run, outperforming the average in the presence of uncertainty. The more the uncertainty the more role for optionality to help, the more you will outperform. Favorable optional outcomes is the central property of life, antifragility.

**Figure 11.2.** Asymmetrical, nonlinear concave (top; the 'fragile') and convex (bottom; the 'antifragile) effects of changes in variable $x$. Increases in variable $x$ from 'start' result in larger gains ($f(x)$) from positive asymmetry (bottom) than from negative asymmetry (top). Conversely, decreases in variable $x$ from 'start' result in smaller losses ($f(x)$) from positive asymmetry (bottom) than from negative asymmetry (top).

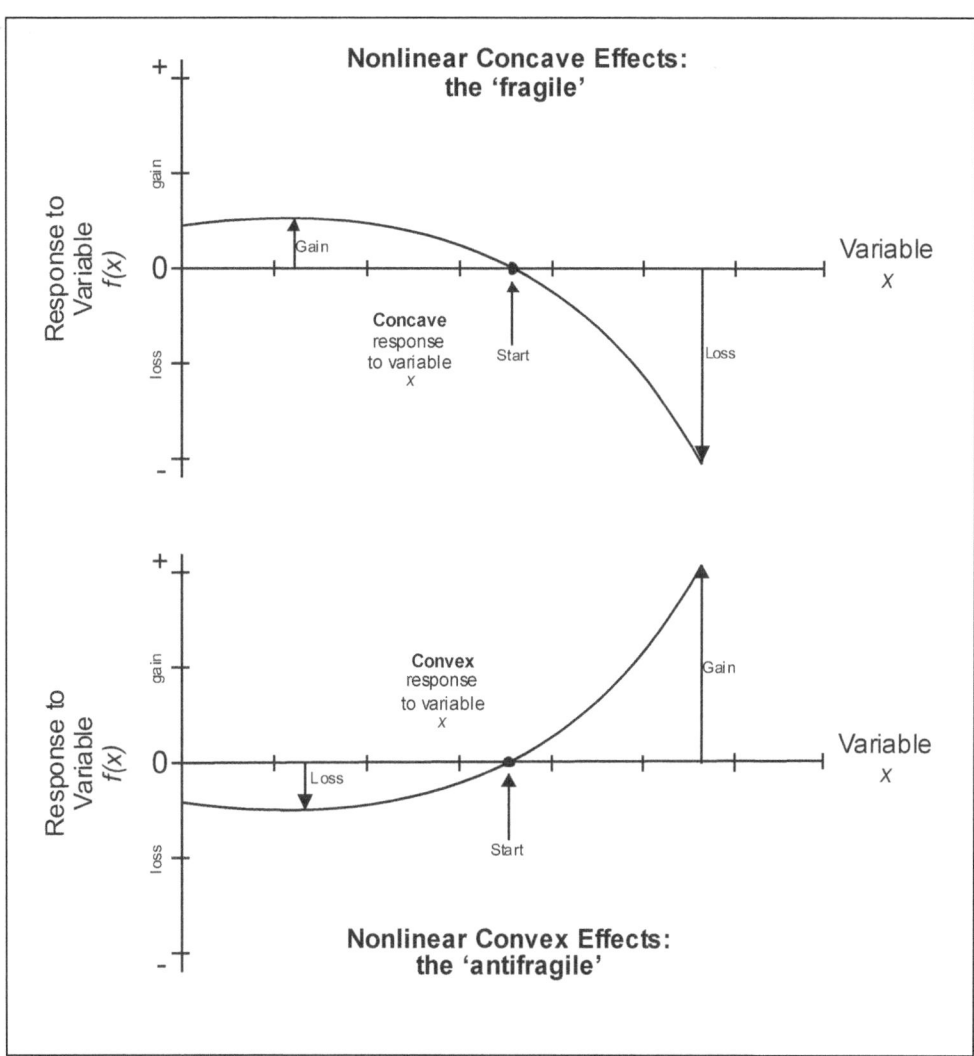

## Chapter 12: Weed-Crop Communities as Complex Adaptive Systems

"The plant population that is found growing at a point in space and time is the consequence of a catena of past events. The climate and the substrate provide the scenery and the stage for the cast of plant and animal players that come and go. The cast is large and many members play no part, remaining dormant. The remainder act out a tragedy dominated by hazard, struggle and death in which there are few survivors. The appearance of the stage at any moment can only be understood in relation to previous scenes and acts, though it can be described and, like a photograph of a point in the performance of a play, can be compared with points in other plays. Such comparisons are dominated by the scenery, the relative unchanging backcloth of the action. It is not possible to make much sense of the plot or the action as it is seen at such a point in time. Most of our knowledge of the structure and diversity of plant communities comes from describing areas of vegetation at points in time and imposing for the purpose a human value of scale on a system to which this may be irrelevant." J.L. Harper, 1977

**Summary.** A complex adaptative system (CAS) is a dynamic network made up of a large number of active interacting adaptive agents, acting in parallel, and diverse in both form and capability. The activity of a CAS system of interacting elements results in the emergence of system order, and an anticipatory-nonpredictive strategy for adaptation to the environment. The overall system behavior is a result of a huge number of decisions, made every moment, by many individual agents acting and reacting in competition and cooperation to what other agents are doing, in which control is highly dispersed and decentralized. The amount of information necessary to describe the behavior of CAS is a measure of its complexity. Examples of CAS include the stock market, social insect and ant colonies, the biosphere and the ecosystem, the brain and the immune system, the cell and the developing embryo. Weed-crop communities are examples of CAS: the global metapopulation of a single species, the local deme of that weed species, the seed portion of a weed's life cycle, the plastic life history of an individual phenotype.

Complexity arises in a system composed of many elements in an intricate arrangement. The numerous elements interact with each other by which numerous relationships are formed among the elements. CAS behavior and evolution arise from agents (e.g phenotypes, strategic traits, cells) which scan/sense/perceive their environment (opportunity spacetime) and develop schema/plans/traits representing interpretive and action rules (e.g. seed heteroblasty). Elements respond only to what is available locally: they are ignorant of system behavior. These schema/plans/traits are subject to change and evolution.

CAS consists of an array of morphologies organized into an elaborate hierarchy. In CAS, large-scale behaviors emerge from the aggregate interactions of less complex agents forming the hierarchical system structure. CAS dynamic interactions are non-linear: small causes can have large effects. CAS hierarchical organization occurs when agents aggregate together to form meta-agents, in turn aggregating to form meta-meta-agents. Aggregate tagging is the mechanism of CAS that facilitates selective interactions among agents, hierarchical organization, and boundary formation. Tagged aggregate behavior is complex, unpredictable, nonlinear.

Complex adaptive systems are living, adaptable, changeable systems in which complexity, emergence, self-similarity and self-organization occur. Emergence is the arising of novel (even radical) and coherent structures, patterns, properties during process of self-organization in complex systems. CAS behavior and evolution is emergent, not predetermined. Emergent properties manifest not so much the material bases of compounds as how the material is organized. Complex systems and patterns emerge out of a multiplicity of relatively simple interactions among agents. Systems emerge from the moment-to-moment decisions made by many players choosing among very many options at each time. Self-organization, self-assembly in complex adaptive systems is the process wherein a structure or global pattern emerges in a system solely from numerous local parallel interactions among lower level components.

The history of CAS behavior and evolution is irreversible. The past is co-responsible for its present behavior. The system future is unpredictable. How then do organisms anticipate an uncertain future? They do so by the emergence of internal models that sense their environment. These internal models are composed of hierarchically arranged building blocks. A framework for representing their adaptive agents in a complex adaptive system consists of a performance system, credit-assignment to consequential behaviors, and a rule discovery algorithm. From this performance system a an anticipatory-nonpredictive strategy for dealing with an uncertain future, and adaptation to the environment, emerges.

## 12.1 Complexity and Complex Adaptive Systems.

A complex adaptative system (CAS) is a dynamic network made up of a large number of active interacting adaptive agents, acting in parallel, and diverse in both form and capability (Axelrod and Cohen, 1999; Holland, 1995). The overall system behavior is a result of a huge number of decisions, made every moment, by many individual agents acting and reacting in competition and cooperation to what other agents are doing, in which control is highly dispersed and decentralized.

**complex**:
1: made up of various interconnected parts; composite; intricate or involved; a whole made up of interconnected or related parts
2: a whole that comprehends a number of intricate parts, especially one with interconnected or mutually related parts

**complexity**:
1: the state or quality of being intricate; something intricate; complication
2: something with many parts where those parts interact with each other in multiple ways
phenomena which emerge from a collection of interacting objects

**12.1.1 Composite systems**. All known stable matter in the universe is made up of three kinds of elementary particle coupled via four kinds of fundamental interaction (Auyang, 1998). The makeup of the infinite diversity and complexity of things we see around us derive from the homogeneity and simplicity of these particles and fundamental interactions. Composition is not mere congregation. The constituents of a compound interact. The interaction generates complicated structures, complex composite systems. How do we explain and represent the complexity of composition?

Large scale composition produces high complexity and limitless possibility. Myriad individuals organize themselves into a dynamic, volatile, and adaptive system that, although responsive to the external environment, evolves mainly according to its intricate internal structure generated by the relations among constituents. Large composite systems are variegated and full of surprises. Despite small scale complexity, sometimes they crystallize into large-scale patterns, meaningful pictures, that can be conceptualized rather simply. These salient patterns are the emergent properties of compounds. Emergent properties manifest not so much the material bases of compounds as how the material is organized. Belonging to the structural aspects of the compounds, they are totally disparate from the properties of the constituents, and the concepts about them are paradoxical when applied to the constituents.

**12.1.2 Complex domains.** Functionally, a complex domain is characterized by a great degree of interdependence between its elements, both temporal (a variable depends on its past changes), horizontal (variables depend on one another), and diagonal (variable A depends on the past history of variable B). As a result of this interdependence, mechanisms are subjected to positive, reinforcing feedback loops, which cause "fat tails", the extreme ends of frequency distributions (Taleb, 2012). That is, they prevent the working of the Central Limit Theorem (see ch. 13.2.5) that establishes thin tails in normal distributions, "convergence to the Gaussian". In complex domains moves are exacerbated over time instead of being dampened by counterbalancing forces. Finally we have nonlinearities that accentuate the fat tails. Complexity implies remote events play a disproportionate role in system behavior.

**12.1.3 Characteristics of complex systems.** Complex adaptive systems possess several central characteristics, key principles, basic properties and mechanisms common to all. All CAS possess a system of interacting elements, agents, whose activity exhibits the emergence of system order, and an anticipatory-nonpredictive strategy for adaptation to the environment.

The ideas and models of complex adaptative systems are essentially evolutionary, grounded in modern chemistry, biological views on adaptation, exaptation and evolution and simulation models in economics and social systems (Wikipedia, 11.10). Examples of complex adaptive systems include the stock market, social insect and ant colonies, the biosphere and the ecosystem, the brain and the immune system, the cell and the developing embryo, manufacturing businesses and any human social group-based endeavor in a cultural and social system such as political parties or communities. The study of complex adaptive systems is the interdisciplinary study of systems using more than one theoretical framework (e.g. evolution, sociobiology, demography, anthropology).

Weed-crop communities are examples of complex adaptive systems. The global metapopulation of a single member species of that local community, as well as the local deme of that weed species, the seed portion of that weed species' life cycle, and the plastic life history of an individual phenotype member of that deme, all are component complex adaptive systems in their own right.

What explanations can complex adaptive system theory provide for understanding the nature of weed interactions with their neighbors, and the nature of the local weed community? Understanding how complex adaptive systems work can provide some insight into understanding how plant communities assembly, how crop and weed neighbors interact within the community, and how crucial traits of individual phenotypes lead to successful reproduction in the deme as it evolves and adapts locally in the seizure and exploitation of opportunity spacetime.

### 12.2 System of Interacting Elements, Agents

Complexity arises in a system composed of many elements in an intricate arrangement. The numerous elements interact with each other by which numerous relationships are formed among the elements. The relationship among system parts is differentiated from other parts outside of system.

Complex adaptive systems are composites made up of various interconnected parts, domains of interconnected or related parts. The number of elements in a CAS is large such that a conventional description is impractical, and does not assist in understanding the system (e.g. Colbach et al., 2001a, b, 2008). Complexity-simplicity in the system is relative, and changes with time.

CAS are composed of large numbers of interacting agents, diverse in both form and capability (Holland, 1995). The agents' behavior is determined by a collection of rules (e.g. stimulus-response). By looking at these rules acting in sequence we arrive at possible behaviors open to an agent. A major part of the environment of any given adaptive agent consists of other adaptive agents, so that a portion of any agent's efforts at adaptation is spent adapting to other adaptive agents.

A key principle of complex adaptive system behavior and evolution is that the basic system units are agents (e.g phenotypes, strategic traits, cells) which scan/sense/perceive their environment (opportunity spacetime) and develop schema/plans/traits representing interpretive and action rules (e.g. seed heteroblasty). Elements respond only to what is available locally: they are ignorant of system behavior (see ch. 13, weed seed-soil communication systems). These schema/plans/traits are subject to change and evolution. Agents are a collection of properties, strategies and capabilities for interacting with local opportunity and other agents. Agents and the complex system are adaptable, with a high degree of adaptive capacity providing resilience to disturbance.

An essential property of complex adaptive systems is the aggregation of agent (elements; organisms) interactions. Large-scale behaviors emerge from the aggregate interactions of less complex agents forming the hierarchical system structure. The tagging of aggregates is the mechanism of CAS that facilitates selective interactions among agents. Tags are the ultimate mechanism behind CAS hierarchical organization. Tagging is a mechanism for aggregation and boundary formation in CAS. Tagged aggregate behavior is complex, unpredictable nonlinear. Complex adaptive systems possess the property of flow. A property of complex adaptive systems consists of an array of morphologies organized into an elaborate hierarchy. Diversity arises when the spread of an agent opens a new niche that can be exploited by modifications of other agents. The diversity of CAS is a dynamic pattern, often persistent and coherent despite changes in agents.

**12.2.1 Aggregation of agent interactions**. A key component of CAS is the property of aggregation of agent (elements; organisms) interactions (Holland, 1995). In CAS large-scale behaviors emerge from the aggregate interactions of less complex agents forming hierarchical system structure (see ch. 11.1.9, Hierarchy: the treelike structure of flow). Similar things aggregate into categories forming hierarchical structure. Aggregates can in turn act as agents at a higher level of organization (meta-agents).

**12.2.2 Tagging for aggregation**. Complex adaptive systems use the mechanism of tagging of aggregates to enable observation and manipulation of properties hidden by symmetries (Bar-Yam, 1992; Holland, 1995). Tagging is a mechanism for aggregation and boundary formation in CAS. Tags provide aggregates with coordination and selectivity, they facilitate selective interaction. Tag-based interactions provide the basis for filtering, specialization and cooperation. Tags are the ultimate mechanism behind CAS hierarchical organization.

**12.2.3 Non-linear interactions**.

"It is little known outside the world of mathematics that most or our mathematical tools, from simple arithmetic through differential calculus to algebraic topology, rely on the assumption of linearity. Roughly linearity means that we can get a value for the whole by adding up the values of the parts. More carefully, a *function* is *linear* if the value of the function, for any set of values assigned to the arguments, is simply a weighted sum of those values." "It is so much easier to use mathematics when systems have linear properties that we often expend considerable effort to justify an assumption of linearity. Whole branches of mathematics are devoted to finding linear functions that are reasonable approximations when linearity cannot be established. Unfortunately, none of this works well for CAS." (Holland, 1995)

Tagged aggregate behavior is complex, unpredictable: the property of nonlinearity (Holland, 1995; see ch. 11.2.4). Nonlinear interactions almost always make behavior of the aggregate more complicated than would be predicted by summing or averaging.

**12.2.4 System network flows.** Complex adaptive systems possess the property of flow (Holland, 1995; see ch. 11.1, Formation of complex systems: the constructal law). A CAS flows over a network of nodes (processors, agents), connectors (designation of possible interactions) and resources. CAS networks vary with time: nodes, connectors appear/disappear as agents adapt/fail to adapt. Neither flows nor networks are fixed in time: they are patterns that reflect changing adaptations as time elapses and experience accumulates.

Tags define the network by delimiting the critical interactions, the major flow connections. The adaptive processes that modify CAS select *for* tags that mediate useful interactions and *against* tags that cause malfunctions. Examples of CAS flows include ecosystems: species, foodweb and biochemical interactions.

CAS flows possess two important properties, the multiplier and recycling effects. The multiplier effect occurs with the injection of additional resource at some node. This produces a chain of changes as the network is transformed, compromising long-range predictions based on simple trends. The recycling effect is a consequence of cycles in networks. Recycling the same raw input in a network produces more resource as each node. For example, the ability of a forest to recapture and recycle critical resources (e.g. mineral nutrients in fallen leaves)

Agents that participate in cyclic flows cause the system to retain resources. The resources so retained can be further exploited: they offer new niches to be exploited by new kinds of agents. Parts of a CAS that exploit these possibilities will thrive, particularly parts that further enhance recycling. Parts that fail to do so will lose their resource to those that do. This is natural selection writ large. It is a process that leads to increasing diversity through increasing recycling. The recycling of resources by the aggregate behavior of a diverse array of agents is much more than the sum of the individual actions. For this reason it is difficult to evolve a single agent with the aggregates's capabilities. Such complex capabilities are more easily approached step by step, using a distributed system. CAS do not settle on a few highly adapted types that exploit all opportunities.

**12.2.5 Hierarchical diversity-variability of structures and agents.** A property of complex adaptive systems consists of a complete array of morphologies organized into an elaborate hierarchy (Bar-Yam, 1992; Holland, 1995). A hierarchical organization of CAS occurs when agents aggregate together to form meta-agents, in turn aggregating to form meta-meta-agents. The persistence of an agent (e.g. organism), or a CAS as an hierarchical array of structures and agents, depends on the context provided by other agents (e.g. other organisms; neighbors). A structural niche in a CAS is defined by neighbor agent interactions. Each kind of agent fills a niche that is defined by the interactions centering on that agent. If we remove one kind of agent from the system, creating a hole, the system typically responds with a cascade of adaptations resulting in a new agent filling that hole: unoccupied opportunity spacetime demands filling. This phenomenon is akin to biological convergence: the same trait fills the same niche in different physiologies (weed species), a niche is determined by the interactions the trait must provide.

Diversity arises with a change in opportunity spacetime (see ch. 10.5 weed community biodiversity). Diversity arises when the spread of an agent opens a new niche that can be exploited by modifications of other agents: opportunities for new interactions (Holland, 1995). The diversity of CAS is a dynamic pattern, often persistent and coherent despite changes in agents (e.g. weed species). In a CAS a pattern of interactions that has been disturbed by the extinction of component agents (species) often reasserts itself, though the new agents may differ in detail from the old (e.g. ecological succession of a local plant community; plant invasions). Extinction disturbance allows weed species to arise to fill the opportunity spacetime vacated. CAS patterns evolve. The diversity observed in a CAS is the product of progressive adaptations. Each new adaptation opens the possibility for further interactions and new niches.

**12.3 Emergence of System Order**

Complex adaptive systems are living, adaptable, changeable systems in which complexity, emergence, self-similarity and self-organization occur. A CAS is a complex, self-similar collectivity of interacting adaptive agents with emergent and macroscopic properties. Complex adaptive system order, behavior and evolution, is emergent, not predetermined. Emergence is the way complex systems and patterns arise out of a multiplicity of relatively simple interactions. Emergence is the arising of novel (even radical) and coherent structures, patterns, properties during process of self-organization in complex systems. Complex adaptive systems are open, defining system boundaries is difficult or impossible. CASs operate far from equilibrium conditions, a constant flow of energy is needed to maintain the organization of the system.

### 12.3.1 Self-similarity and self-organization

"… [there exists a] distinction between planned architecture and *self-assembly*." "The key point is that there is no choreographer and no leader. Order, organization, structure – these all *emerge* as by-products of rules which are obeyed *locally* and many time over, not globally. And that is how embryology works. It is done by local rules, at various levels but especially the level of the single cell. No choreographer. No conductor of the orchestra. No central planning. No architect. In the field of development, or manufacture, the equivalent of this kind of programming is *self-assembly*." (Dawkins, 2009)

The second law of thermodynamics and the constructal law provide the fundamental forces that drive design in nature (see Unit 5.3, formation of complex systems). All natural designs are engines (of heat, fluid, or mass) driven by useful energy derived from the sun: nature organizes itself to move more easily. The movement towards equilibrium, the pull of gravity, puts things in motion, and nature generates configurations to facilitate this flow that has direction in time. In order to get from here to there, everything must create a path. Flow systems construct their own flow architectures and body rhythms that enable them to move more easily. Everything that flows and moves generates designs that evolve to survive (to live). The more efficient the pathway, the more likely is its persistence, whatever the mechanism behind that persistence. Flow system designs evolve, they configure and reconfigure themselves over time. This evolution occurs in only one direction: flow designs get measurably better, moving more easily and farther if possible. Design emerges because things flow better with configuration. Self-organization of the flow system occurs with the construction of flow patterns that ease movement. With self-organization a treelike pattern emerges throughout nature because it is an effective design for facilitating point-to-area and area-to-point flows. Wherever you find such flows, you find a treelike structure.

A complex adaptive system is a complex, self-similar, self-organizing collectivity of interacting adaptive agents with emergent and macroscopic properties. Self-organization, self-assembly in complex adaptive systems is the process wherein a structure or global pattern emerges in a system solely from numerous local parallel interactions among lower level components. Self-organization is achieved by elements distributed throughout the system, without planning imposed by a central authority or external coordinator. The rules specifying interactions among components are executed using only local information (without reference to global patterns). It relies on multiple interactions exhibiting strong dynamical non-linearity. It may involve positive and negative feedback, and a balance of exploitation and exploration. Self-organization is observed in coherence or correlation of the integrated wholes within the system that maintain themselves over some period of time. There exists a global or macro property of "wholeness" (Corning, 2002).

Complex adaptive weed systems exhibit self-similarity. It occurs when a component, or an object of the system is exactly or approximately similar to a part of itself. In complex adaptive weed systems self-similarity is revealed in several ways. In weeds and plants, similar modular units are repeated at several levels of spatial organization. For example, the individual plant shoot is composed of repeating shoot units: leaves and tillers in grasses, branches and leaves in dicots (see table 16/7?.1; ch. 16.3.1, plant morphological structure). The plant inflorescence is composed of repeating units from shoot, to tiller, to tiller inflorescence. In grassy weeds like *Setaria* spp.-gp. the inflorescence consists of a hierarchy of repeating tissues from fascicle to spikelet shoot to floret or bristle shoot. The *Setaria* seed is series of repeating units, differentiated floral organs, from seed hull to caryopsis to embryo and endosperm (e.g. Haar and Dekker, 2011).

Phenotypic plasticity in an individual plant expresses self-similarity in response to local opportunity. Increasing opportunity stimulates growth of the individual plant by means repeating the numbers and architecture of the plant body: shoot branches with modular leaves and photosynthesizing cells, root branches with modular roots, root hairs and absorptive cells.

Self-similar phenotypes are expressed in the temporal population structure of a species in a local deme, as well as in the continuous interacting local-to-global populations of the species' metapopulations. Phenotypic traits in the individual are self-similar in variants produced by parent plants to achieve slightly different strategic roles in local adaptation. These strategic traits in turn are self-similar as preadaptations, exaptations, when those phenotypic traits confront new evolutionary situations in which function can change.

**12.3.2 Scale invariance and self-similarity.** In nature there are many processes forming structures in which, at decreasing scales, similar principles of formation persist. For example, branching processes in biology happen step by step in increasing detail. The principles of construction remain the same throughout: at very small scales the object has similar properties as the whole. This is scale invariance and self-similarity (Schroeder, 1991; Stoyan and Stoyan, 1994; see ch. 13/14?.13.2.6, scalable probability distributions and inference).

**self-similarity:** an invariance with respect to scaling; invariance not with additive translations, but invariance with multiplicative changes of scale

A self-similar object appears unchanged after increasing or shrinking it size. For example, in turbulent flows, large eddies beget smaller ones, and these spawn smaller ones still, etc. Self-similarity is akin to periodicity on a logarithmic scale.

**12.3.3 Emergence.** A key principle of complex adaptive system behavior and evolution is that system order is emergent, not predetermined. Emergence is the way complex systems and patterns arise out of a multiplicity of relatively simple interactions among agents. Systems emerge from the moment-to-moment decisions made by many players choosing among very many options at each time (Waldrop, 1994). A complex system is a system formed out of many components whose behavior is emergent: the behavior of the system cannot be simply inferred from the behavior of its components (Bar-Yam, 1992). The amount of information necessary to describe the behavior of such a system is a measure of its complexity. Emergence is the arising of novel (even radical) and coherent structures, patterns, properties during process of self-organization in complex systems. Synergistic effects are produced by organized emergent systems: emergence itself has been the underlying cause of the evolution of emergent phenomena. Emergent properties found at all levels, from agent to system include adaptation or homeostasis, communication, cooperation, specialization, spatial and temporal organization, and reproduction.

Several qualities of these dynamic interactions are discernible. Interactions are non-linear: small causes can have large effects (see catastrophe theory in ch. 3.3.4; ch. 10.4.5.3, herbicide-induced population shifts; ch. 18, triazine resistant/susceptible pleiotropy). Interaction is physical or involves exchange of information. Interactions are rich: any system element is affected, and affects several other systems. Interactions are primarily with immediate neighbors, but not exclusively. Elements respond only to what is available locally: they are ignorant of system behavior (Camazine et al., 2003) (see concept of terroir, ch. 3.4.3.2). Several rules of interaction characterize emergent phenomena. These rules or laws of emergence or evolution have no causal efficacy, they do not generate anything, they only describe regularities and consistent relationships. The underlying causal agencies of emergent phenomena must be specified separately. Rules cannot predict history because the system involves more than rules. The feedback, recurrency rule states that any interaction can feed back on itself directly, or after a number of intervening steps, and the feedback quality can vary. Emergent phenomena, organized purposeful activity, are shaped by feedback driven influences (e.g. pleiotropy, see ch. 18, triazine-resistant *Brassica*). Causation is iterative: effects are also causes (see Harper, ch. 10.5.2, biodiversity encountered by interacting neighbors); see extended phenotype, ch. 4.1.1). Another rule, a key principle of CAS behavior and evolution, is that history irreversible. All CASs have a history, they evolve, their past is co-responsible for their present behavior. Therefore the system future is unpredictable.

"For everyone who has will be given more, and he will have an abundance. Whoever does not have, even what he has will be taken from him." Matthew 25:29; The holy bible.

Nature is a complex, recursive environment. The world we live in has an increasing number of feedback loops, causing events to be the cause of more events. This generates snowballs and arbitrary and unpredictable plant-wide winner-take-all-effects. In nature information flows rapidly, accelerating epidemics, generating contagions.

**contagious distribution**: a distribution pattern in which values, observations or individuals are more aggregated or clustered than in a random distribution, indicating that the presence of one individual or value increases the probability of another occurring nearby (Lincoln et al., 1998)

A preferential attachment process is any of a class of processes in which some quantity is distributed among a number of individuals or objects according to how much they already have, so that those who are already wealthy receive more than those who are not (Wikipedia, 2.16). "Preferential attachment" is only the most recent of many names that have been given to such processes. They are also referred to under the names "Yule process" (Yule, 1925), "cumulative advantage", "the rich get richer", or the "Matthew Effect". Preferential attachment can, under suitable circumstances, generate power law distributions, where a relative change in one quantity results in a proportional relative change in the other quantity, independent of the initial size of those quantities: one quantity varies as a power of another (see ch. 9.4, effects of neighbor interactions on plant growth and plant form).

Emergent phenomenon in generated systems are, typically, persistent patterns with changing components (Holland, 1998). For example, in a standing wave that forms in front of a rock in a fast-moving stream the flowing water changes but patterns persist; or the turn-over of atoms and molecules in long-lived organisms. The context in which a persistent emergent pattern is embedded determines its function; e.g. multi-functionality in biological systems. Interactions between persistent patterns add constraints and checks that provide increasing 'competence" as the number of such patterns increase; e.g. DNA redundancy. Nonlinear interactions, and the context provided by other patterns, both increase this competence. The number of possible interactions, and hence the possible sophistication of response, rises extremely rapidly, factorially, with the number of interactants

## 12.4 Adaptation in Complex Adaptive Systems: Non-anticipatory-nonpredictive Strategy

The history of complex adaptive system behavior and evolution is irreversible. The system's past is co-responsible for its present behavior: the system future is unpredictable. How then do organisms anticipate an uncertain future? They do so by the emergence of internal models that sense their environment. These internal models are composed of hierarchically arranged building blocks. A framework for representing their adaptive agents in a complex adaptive system consists of a performance system, credit-assignment to consequential behaviors, and a rule discovery algorithm. From this performance system a strategy for dealing with an uncertain future environment emerges.

**12.4.1 Internal models sense environment**. A hallmark of complex adaptive systems is that they anticipate the future by means of an internal model (see schema, Gell-Mann, 1994; Holland, 1995). This internal model is the mechanism for anticipation of an uncertain and unpredictable future. These models are interior to the agent (organism). They sense specific environmental signals. The CAS agent must select patterns in the torrent of input it receives from the environment and then must convert those patterns into changes in its internal structure. The basic maneuver for constructing internal models is to eliminate details so that selected patterns are emphasized. The changes in structure, the model, must enable the agent (organism) to anticipate the consequences that follow when that pattern (or one like it) is again encountered. A fully developed weed species example is presented in ch. 13, weed seed-soil environment communication systems (see ch. 13.3.1.3, schema of intrinsic *Setaria* sp. seed traits). There exist two kinds of internal model. A tacit model prescribes a current action, under an *implicit* prediction of some desired future state (e.g. the *Setaria* sp. seed morphology). An overt model is used as a basis for *explicit*, but internal, explorations of alternatives, a process often called *lookahead* (e.g. the mammalian brain).

A model allows us to infer something about the thing modeled. A given structure in an agent (organism) is an internal model if we can infer something of the agent's environment merely by inspecting that structure. We can infer a great deal about the environment of any organism by studying relevant pieces of morphology and biochemistry (see *Setaria* sp. seed morphology, ch. 17.3.1.1). The structure from which we infer the agent's environment is required to also actively determine the agent's behavior. If the resulting actions anticipate useful future consequences then the agent has an effective internal model. For example, the seed hull of *Setaria faberi* is structured to accumulate soil water and oxygenate it by means of increased air-water surface area, enabling dissolved oxygen stimulated germination in anticipation of favorable seedling emergence timing. Despite differences between tacit (e.g. *Setaria* sp. seed morphology) and overt (e.g. mammal brains) models, there are commonalities. In both cases the organism's chances of survival are enhanced by anticipating uncertain futures. Variants of the model are subject to selection and progressive adaptation.

**12.4.2 Building blocks of internal models.** An internal model is generated by specific building blocks from environmental experience (Holland, 1998). They provide a mechanism for systems exhibiting emergence to extract repeatable features from the changing environment. An internal model must be based on limited samples of a perpetually novel environment: specific, useful signals among the broad signal noise of the environment. The model can only be useful if there is some kind of repetition of the situations modeled. Building blocks range from mechanisms in physics to the way we parse the environment into familiar objects.

The future cannot be predicted. A list of rules cannot be prepared for all possible situations. There exists a need to decompose the situation confronted, evoking rules from an agents (organisms) repertoire of everyday component building blocks of the system. Each of these building blocks has been practiced and refined, adapted by natural selection, in many different situations. When a new situation is encountered, the organism combines relevant, tested building blocks to model the situation in a way that suggests appropriate actions and consequences. We parse a complex scene by searching for elements already tested for reusability by natural selection and learning. The component parts are not arbitrary, they can be used and reused in a great variety of combinations. An example of the mechanism of using building blocks to anticipate an uncertain future in weed species is the behaviors of functional traits responding to specific environmental cues. For example, *Setaria* sp. seed morphological structures respond to soil oxy-hydro-thermal time in the soil which is used in many ways to signal many different kinds of seed behaviors (ch. 13). Branching growth in Abutilon theophrasti responds to ambient intercepted light in many ways (phenotypic plasticity; ch. 19).

The use of building blocks to generate internal models is a pervasive feature of complex adaptive systems. If model making encompasses most of scientific activity, then the search for building blocks becomes *the* technique for advancing that activity. In weed science the utility of demographic building blocks in model building compares unfavorably with functional traits responding to specific environmental signals (see ch. 15, representation of weed life history).

**12.4.3 Framework for representing adaptive agents.** A framework for representing adaptive agents in a CAS consists of a performance system, credit-assignment to behaviors, and a rule discovery algorithm. Agents in a complex adaptive system are formed by aggregation (Holland, 1995). A CAS is described by the evolving interactions of those agents. Tags facilitate and direct these transactions. The diversity of tags highlights the variety of interactions possible in complex flows. This provides short- and long-term coherence to CAS despite diversity, change, and lack of central direction. Nonlinearities are responsible for this coherence (see antifragile convex effects of nonlinearity, ch. 11.2.4). Nonlinearities are embodied in internal models that drive agent interactions. There exists a need to discover building blocks that are combined and recombined to determine CAS 'outward appearance'. There exists a need for a common representation for agents, organisms, of a complex adaptive system.

The formation of a framework for representing adaptive agents involves several steps or stages, a procedure for rule discovery (Holland, 1995). The elucidation of a performance system involves looking for a way to represent the capabilities of different kinds of agents at a particular time, without any concern for changes that might be produced by adaptation. In the next stage, the agent's successes or failures are used to assign credit or blame to parts of the performance system. The last stage concerns making changes in the agent's capabilities, replacing parts assigned low credit with new options.

**12.4.3.1 Agents performance system**. The elucidation of an anti-fragile performance system involves looking for a way to represent the capabilities of different kinds of agents at a particular time, without any concern for changes that might be produced by adaptation (Holland, 1995). Adaptive agents are formally described by rules. For these rules to be a successful unifying device they must meet three criteria. The rules must use a single syntax to describe all CAS agents: systematic rules governing the systems proper formulation. The rule syntax must provide for all interactions among agents. The existence of an acceptable procedure for adaptively modifying the rules.

An agent's (organism's) performance system specifies the agent's capabilities at a fixed point in time, what it would do in the absence of any further adaptation. The three basic elements of the performance system are a set of detectors (agent's capabilities for extracting information from the environment), a set of IF/THEN rules (agent's capabilities for processing that information), and a set of effectors (agent's ability to act on its environment) (figure 12.1). The three elements lose details in their abstraction of the mechanisms employed by the different kinds of agents. The agent is described as a collection of message-processing rules. A common description of agents uses formal rules with single syntax, whatever the agents outward form, useful for all situations: it is adaptable. IF/THEN, stimulus-response rules depend critically on way agent interacts with environment. The agent must filter torrent of input environmental information: detector turns on/off in response to particular property of environment. The EFFECTOR output is an inversion of the detector. An appropriate message activates an elementary effect on the environment: effectors decode standardized messages to cause actions in the environment.

Messages differ. Detector-originating messages have a built-in meaning determined by the environmental properties detected. These rule-originating messages have no assigned meaning except when they are used to activate effectors. The agent can combine several tested, non-contradicting, active rules simultaneously to describe a novel situation. Rules become building blocks; in turn activating appropriate effector sequences.

Performance systems adapt. Rules are hypotheses undergoing testing and confirmation. The objective is to provide contradictions, not to avoid them. There exist alternative rules, competing hypotheses. When one fails competing rules are waiting in the wings to be tried. *Setaria* sp. seed heteroblasty provides a good example of a complex adaptive performance system searching for appropriate local seedling emergence times (ch. 13.3).

**Figure 12.1** Schematic diagram of a complex adaptive performance system: detectors and effectors of environmental messages processes by agents (organisms). See Holland (1995, figure 2.2). Compare to ch. 13, figures 13.1 and 13.3, Shannon communication system in *Setaria* sp. DETECTOR: specific environmental signal; PERFORMANCE SYSTEM: rules; EFFECTOR: response behaviors.

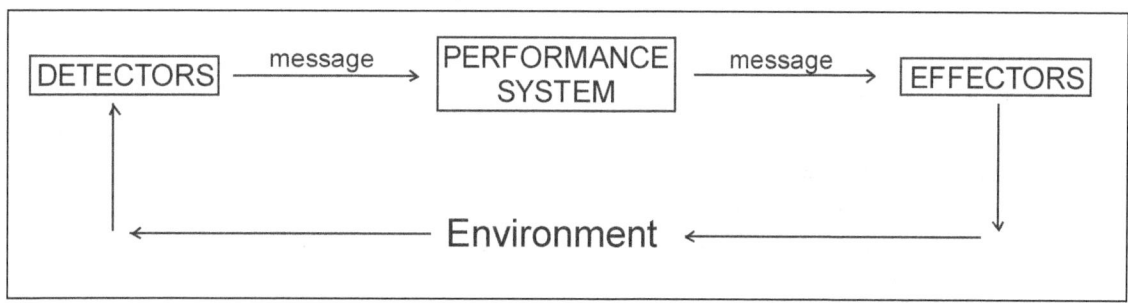

of a complex adaptive performance system searching for appropriate local seedling emergence times (ch. 13.3).

**12.4.3.2 Adaptation by credit assignment.** The agent's successes or failures are then used to assign credit or blame to parts of the performance system: a credit assignment algorithm (Holland, 1995). The rules ability to win competition among rule-hypotheses is based on past utility. Each rule is assigned strength reflecting rules utility to system over time. Modifying strength on the basis of experience over time is credit assignment. The essence of credit assignment is to provide the system with hypotheses that anticipate future consequences, strengthening rules that set the stage for later, overtly rewarding activities. Credit assignment depends on previous, stage-setting, actions that make possible later outcomes useful. The credit assignment procedure strengthens rules that belong to chains of action terminating in rewards.

The building blocks of a complex adaptive performance system are defined by their position in a sequence of events is a schema. More complicated building blocks (schemata) are usually formed by combining simpler building blocks. The result is a hierarchy wherein building blocks at one level are combined to form building blocks at the next level. Adaptive agents can learn strategies. Evolution favors combinations of building blocks that increase fitness, those that have survived in the contexts in which they have been tested. These contexts are provided by other building blocks and the environmental niches the species inhabits (local opportunity spacetime). Building blocks range from mechanisms in physics to the way we parse the environment into familiar objects. They provide a way of extracting repeatable features from the perpetual novelty that attends systems exhibiting emergence (Holland, 1998).

**12.4.3.3 Adaptation by rule discovery algorithm.** The last stage concerns making changes in the agent's capabilities, replacing parts assigned low credit, and the discovery of a rule algorithm that provides utility to the agent (organism) in future activity (see ch. 13.3.1, forecasting *Setaria* seed behavior). The generation of plausible hypotheses centers on the use of tested building blocks incorporating past experience with broad amplitude innovation. Tags play an important role in coupling rules and providing sequential activity because they too have building blocks. Tags are schemata that appear in both condition and action parts of rules. Established tags found in strong rules spawn related tags, providing new couplings, new clusters, and new interactions

**12.4.4 Strategy and environment.** How do organisms anticipate an uncertain future? Complex adaptive systems maintain coherence under change by means of conditional action and anticipation without central direction. CAS have lever points wherein small amounts of input produce large, directed changes (Holland, 1995).

"When we desire a motion to follow a given pattern, the difference between this pattern and the actually performed motion is used as a new input to cause the part regulated to move in such a way as to bring its motion closer that given by the pattern." (Wiener, 1948)

Homeostasis is the knack for self-regulation by an agent, an organism:

**homeostasis**:
1: the property of a system (typically a living organism), either open or closed, that regulates its internal environment and tends to maintain a stable, constant condition; multiple dynamic equilibrium adjustment and regulation mechanisms make homeostasis possible (Wikipedia, 12.10)
2: the maintenance of metabolic equilibrium within an animal by a tendency to compensate for disrupting changes (Collins English Dictionary, 1979; Collins, Glasgow, UK)
3: the tendency for the internal environment of an organism to be maintained constant (Chambers Sci Dictionary)
Dictionary)

One way nature deals with unexpected, unpredictable events (both positive and negative) is redundancy, of three different types (Taleb, 2010).

Defensive redundancy is survival insurance under adversity due to the availability of spare parts. This redundancy incurs costs of maintaining these spare parts and the energy need to keep them around, available for adversity. Simple models of behavior fail when disturbance and perturbation modify some assumption, when you change one parameter, or take a parameter heretofore assumed to be fixed and stable and make it random. Redundancy is the opposite of naïve optimization of one strategy for dealing with the future. Nature usually avoids overspecialization, it limits evolution and weakens organisms by increased exposure to risk.

Size redundancy arises because Nature does not like anything too big. Nature does not limit the interactions between entities; it does limit the size of its units. For example, weed phenotypic plasticity relies on relatively small unit sizes (e.g. branches, leaves) for sensitive responsiveness by the plant to changing opportunity spacetime in a locality. Economies of scale that are stimulated by optimization, efficiency, and the illusion of stability, lead to vulnerability to outside contingencies, unpredictable events. A certain class of unforeseen errors and random shocks hurts large organisms vastly more than small ones.

Connective redundancy arises because Nature does not like too much connectivity and globalization. Larger environments are more scalable than smaller ones. This allows the biggest to get even bigger, at the expense of the smallest, through the mechanism of preferential attachment.

Structure-function redundancy in Nature enables same function to be performed by identical elements, and the same function by two different structures.

Preadaptation, exadaptation redundancy allows an element to be used to perform a certain function that is not its current central one. A certain adaptation leads to a new function in the right environment. Anything that has a secondary use, and one you did not pay for, will present an extra opportunity should a heretofore unknown application emerge or a new environment appear. The organism with the largest number of secondary uses is the one that will gain the most from environmental randomness.

Many examples of redundancy are apparent in weed species. Multiple, overlapping dormancy-germinability mechanisms exist in weed seeds (e.g. 3-4 in *Setaria*, see ch. 16 ; 4-5 in *Chenopodium*, Altenhofen, 2009, Altenhofen and Dekker, 2014), preadaptations highly sensitive to future environmental randomness. The evolution of in *psbA* chloroplast weed mutants with altered photosynthesis conferred preadaptation for herbicide resistance prior to the introduction of atrazine (see ch. 17).

When a plant, a weed, has several functional redundancies, environmental randomness helps on balance. The low/no cost of many redundancies allow the plant to benefit from randomness more than you can be hurt by it. This is the 'convexity to uncertainty' argument, a disproportionate, favorable nonlinear response (see ch. 11.2.4, the non-linear, concave-convex effects).

## Chapter 13  Weed Soil Environment-Seed Communication Systems

**Summary**: A keystone, threshold event in a weed's life history is germination, emergence from dormancy and assembly with other plants in the local community. Seedling recruitment is a complex adaptive system in its own right. This chapter is about one such well-characterized complex adaptive weed system, the *Setaria* spp.-gp. seed-soil environment communication system. The *Setaria* spp.-gp. seed has acquired a keystone trait, seed heteroblasty, allowing it to place many individual bets on future success: seeds with slightly different germination requirements. Seed germination is carefully timed by sensing signals in the soil. This chapter is the story of how the weedy foxtails communicate with the soil to time their departure from the relative safety of seed dormancy back into the land of the living, the emerged seedling and plant.

Information and probability can be envisioned in the transmission and reception of opportunity spacetime by an organism leading to successful adaptation in that locality. The nature of weeds is an environment-biology communication system. Biology is information. Evolution embodies an ongoing exchange of information between organism and environment. What evolves is information. Information is physical. Biology is physical information with quantifiable (Kolmogorov) complexity. The gene is not the information-carrying molecule, the gene is the information. An important problem in science is to discover the language of information in biological systems; the predictable developmental events of weed life history.

The nature of the weedy *Setaria* is revealed in the physical (morphological and genetic spatial structures) and the phenomenal (life history behavior instigated by functional traits). The environment-biological informational system with which weedy *Setaria* life history unfolds is represented in the seed-soil communication system. Weedy *Setaria* spp. seeds with heterogeneous dormancy states are recruited from soil pools when sufficient oxygen, water and heat accumulate over time. Weed seeds like *Setaria* require only three resources/conditions (signals) from their immediate soil environment to germinate and emerge as seedlings: temperatures (T) favorable for germination and seedling emergence (thermal-time), adequate but not excessive soil moisture (hydro-time), and oxygen dissolved in imbibed water to support germination metabolism (oxy-hydro time). When any of these three signals is insufficient or excessive, a living seed in the soil will remain dormant. When all three signals are adequate germination will occur.

The nature of weedy *Setaria* seed-seedling life history is a complex adaptive, soil-seed communication system arising from its component functional traits. Functional traits controlling seed-seedling behavior are physical information that has evolved in ongoing communication between organism and environment leading to local adaption. Weedy *Setaria* seed life history behaviors are controlled by environmental information (signals) flowing from the soil to the seed embryo. The *Setaria* seed is constructed in such a way as to be receptive to specific signals contained within the entire soil information available to it. The specific signal to which *Setaria* is tuned affecting seed behavior in the soil is the amount of oxygen and heat (T, thermal) in soil water over time, oxy-hydro-thermal time ($O_2$-$H_2O$-T-Time). The message that directly controls *Setaria* seed behavior is oxygen and heat accumulating in the embryo modulating seed respiration.

Information theory was developed by Claude E. Shannon to find fundamental limits on signal processing operations such as compressing data and on reliably storing and communicating data. A communication system of any type must contain the following five elements: information source, transmitter, channel, receiver and destination.

The source of information generating the message affecting weedy *Setaria* seed life history development is the soil surrounding the individual seed in a particular locality. This information flows constantly within the soil. The soil-borne seed responds to only a limited portion of the entire soil information, the message controlling behavior. The message is heat and oxygen dissolved in soil water that the *Setaria* embryo requires for continued development, whether continued quiescence or germination. The message was naturally selected during *Setaria* evolution from among the larger entire set of soil information as a dependable signal stimulating-inhibiting seed germination-dormancy behaviors. The specific signal that affects *Setaria* behavior is the amount of heat and oxygen dissolved in soil water films connecting soil particles with the seed with time, oxy-hydro-thermal-time ($O_2$-$H_2O$-heat-time).

The transmitter converts/encodes/changes the message to produce a suitable signal. The transmitter is the water film adhering to soil particles adjacent to, and continuous with, seed exterior surfaces (hull, placental pore). The message sent is heat and oxygen dissolved in soil water, information is physical. The transmitter modulates the heat and oxygen content of water films that physically connect soil particles with seed surfaces and that are received by the receiver, the living tissues of the seed interior. The transmitter changes, encodes, the amount of oxygen dissolved in soil water by formation of continuous water films physically connecting soil particle-seed surfaces. Signal encoding is the change in the water film message by the morphology of the seed surface. *Setaria* seed hull surface morphology and topography accomplish these changes that encode the signal, information is physical.

The *Setaria* communication channel is the continuous water film connecting soil particles with the seed exterior (hull, placental pore) and interior (TACL, endosperm, embryo). This water film communication channel transmits the oxygen-water-heat signal to the living seed interior receiver and the embryo destination.

The signal transmitted from exterior water films (terminating with the placental pore) are received by the seed interior by TACL membrane transport and diffusion equilibrating heat and dissolved gases between the exterior-interior seed compartments. Water oxygenated by the seed surface water films is the signal that is decoded by the seed interior receiver in the form of the free oxygen message available to stimulate respiration and hence embryo germination (the destination).

The destination is the embryo. The resumption of life history growth and development by the embryo is stimulated by the accumulation of adequate interior oxygen leading to germination and seedling emergence.

Information is physical: memory resides in several locations in the *Setaria* seed. *Setaria* seed memory has both short and long term expressions. In the short term, memory is expressed by the amount of oxygen accumulated in the seed interior. *Setaria* seed memory is the current germination-dormancy state of each living seed in a local soil pool. Memory is expressed in the long-term by responsiveness to $O_2$-$H_2O$-heat messages as determined by the morpho-physiological soil-seed communication system (hull, TACL membrane, scavenger protein). Memory resides in the functional-adaptive traits that regulate all seed behaviors by transducing/transforming/encoding and decoding inorganic soil $O_2$-$H_2O$-heat signals over time: the three morpho-physiological mechanisms forming the soil-seed communication system. The message is remembered: plants arising from seed utilizing this communication system pass on these functional traits to their polymorphic progeny, each of which in turn passes on the ability to generate its own range of heteromorphic seeds appropriate to continuing, successful local adaptation.

## 13.1 *Setaria* spp.-gp. Seed-Soil Communication System

How should we, can we represent complex adaptive weed systems? This chapter is about one such well-characterized complex adaptive weed system, the *Setaria* spp.-gp. seed-soil environment communication system (Dekker, 2013, 2014). A keystone event in life history of the *Setaria* spp.-gp. is the transition of a dormant seed to a germinating seed in the soil. Seedling emergence is a singularly crucial commitment by the plant, an all-or-nothing bet that a miniscule amount of stored energy in the seed will be enough to allow it to seize and exploit the local opportunity spacetime available, leading to reproductive success. This commitment is not random. Over evolutionary time the *Setaria* spp.-gp. seed has acquired a keystone trait, seed heteroblasty, allowing it to place many individual bets on future success. All these bets, all these seeds with slightly different germination requirements, are closely timed by means of signals in the soil. This is the story of this chapter: how the weedy foxtails communicate with the soil to time their departure from the relative safety of seed dormancy back into the land of the living, the emerged seedling and plant.

### 13.1.1 Communication and information theory.

"... information must not be confused with meaning." (Shannon and Weaver, 1949)

**communication**: the activity of conveying information through the exchange of messages, or information, as by signals or behavior (Wikipedia, 10.12)

**information**:
1: message being conveyed; a sequence of symbols that can be interpreted as a message, recorded as signs, or transmitted as signals (Wikipedia, 10.12)
2: any kind of event that affects the state of a dynamic system (Wikipedia, 10.12)
3: negative entropy, choice and uncertainty, surprise, difficulty; anything that changes probabilities or reduces uncertainty (Gleik, 2011)
4: a mathematical abstraction of the content of any meaningful statement or data, enabling the study of the most efficient way of recording or transmitting them (Borowski and Borwein, 1991)
5: also called **uncertainty**; formally, a real-valued function of events in a probability space that depends only on the probability of events, and is such that events of probability one have zero uncertainty, that uncertainty increases as probability drops, and that the uncertainty of the simultaneous occurrence of two independent events is the sum of their individual uncertainties (Borowski and Borwein, 1991)
6: knowledge acquired through experience or study; knowledge of specific and timely events or situations; news (Hanks et al., 1979)

The lack of information is uncertainty. Randomness is a fundamental concept in making inferences about uncertain events. A lack of predictability of events can arise from a system's incoherence, a lack of knowledge of the relevant phenomena, because of the dynamic complexity of the system, or the lack of information in the measurements we are able to make of system behavior. Often we are unable to distinguish between these types of randomness preventing us from quantifying the probability of an event occurring. Worse, our inability to understand the nature of randomness can lure us into making incorrect assumptions of event uncertainty, and therefore inappropriate event probability distributions.

**information theory**:
1: information theory is a branch of probability theory that deals with uncertainty, accuracy, and information content in the transmission of messages; random signals (noise) are often added, altering the message during transmission; information theory determines the probability that a sent signal is the same as that received; redundancy, repeating the signal, is needed to overcome limitations in the transmission system; the statistics of choosing a message out of all possibilities determines the amount of information it contains (Daintith and Clark, 1999)
2: a collection of mathematical theories, based on probability theory, that are concerned with methods of coding, decoding, storing and retrieving information, and with the likelihood of a given degree of accuracy in the transmission of a message through a channel that is subject to probabilities of failure that are described by a probability law (Borowski and Borwein, 1991)

Information theory was developed by Claude E. Shannon (Shannon and Weaver, 1963) to find fundamental limits on signal processing operations such as compressing data and on reliably storing and communicating data. Since its inception it has broadened to find applications in statistical inference, networks (e.g. evolution and function of molecular codes, model selection in ecology) and other forms of data analysis. Closely related to these concepts of algorithmic randomness is the entropy of a random variable in information theory:

**random variable**:
1: formally, a measurable function defined on a probability space and having range in the interval [0,1] (Borowski and Borwein, 1991)
2: a quantity that may take any of a range of values (either continuous or discrete) that cannot be predicted with certainty but only described probabilistically (Borowski and Borwein, 1991)
3: (Statistics) a variable that can assume any of a given set of values with assigned probability (Lincoln, et al., 1998)
4: assignment of a numerical value to each possible outcome of an event space (Wikipedia, 3.12)

**entropy**:
1: lack of pattern or organization; disorder (Hanks et al., 1979)
2: the macroscopic variable representing the degree of disorder with a system (Borowski and Borwein, 1991)
3: (Statistical mechanics) the increase in entropy in a closed system to a maximum at equilibrium is the consequence of the trend from a less probable to a more probable state (Walker, 1988)
4: the entropy of a system is a measure of its degree of disorder. The total entropy of any isolated system can never decrease in any change; it must either increase (irreversible process) or remain constant (reversible process). The total entropy of the *Universe* therefore is increasing, tending towards a maximum, corresponding to complete disorder of the particles in it (assuming that it may be regarded as an isolated system). (Uvarov et al., 1982)
5: a measure of the efficiency of a system, such as a code or language, in transmitting information (Hanks et al., 1979)
6: (information theory) a measure of the uncertainty associated with a random variable; Shannon entropy (Wikipedia, 3.12)

**Shannon entropy** (Shannon and Weaver, 1963):
1: quantity of the expected value of the information contained in a message (a specific realization of the random variable), usually in units such as bits
2: a measure of the average information content one is missing when one does not know the value of the random variable
3: an absolute limit on the best possible lossless compression of any communication

Entropy is a key measure of information, which is usually expressed by the average number of bits needed to store or communicate one symbol in a message. Entropy quantifies the uncertainty involved in predicting the value of a random variable. For example, specifying the outcome of a fair coin flip (two equally likely outcomes) provides less information (lower entropy) than specifying the outcome from a roll of a die (six equally likely outcomes).

Information and probability can also be envisioned in the discovery of nature: uncertainty, accuracy and knowledge derived from empirical observations. It can also be envisioned in the transmission and reception of opportunity spacetime by an organism leading to successful adaptation in that locality. These two utilizations of information come together in the concept of making inferences about uncertain knowledge of weed evolution, and how phenotypic consequences can be empirically described within the random noise generated by limitations of our observational, logical and statistical tools.

### 13.1.2 Shannon communication system.

"The fundamental problem of communication is that of reproducing at one point either exactly or approximately a message selected at another point." (Shannon and Weaver, 1949).

Claude Shannon communication theory bridges information and uncertainty; information and entropy; information and chaos. Shannon defined the measure of information ($H$) as the measure of uncertainty: "of how much 'choice' is involved in the selection of the event or how uncertain we are of the outcome." (Shannon, p.18; 1949).

**measures of information (H):**
- $H$ = measures of information
- = measures of choice and uncertainty (number of choices)
- = entropy of a message
- = Shannon entropy
- = the information

**noise source:** everything that corrupts the signal

**transform:** to change greatly the appearance or form of; to change the nature, condition or function of (Wikipedia, 10.12)

**transduction:**
1: (biophysics) conveyance of energy from a donor electron to a receptor electron, during which the class of energy changes; a committed process, no return path
2: (physiology) the transportation of stimuli to a biological receptor; conversion of a stimulus from one form to another.
3: (biology) a mechanical/physical stimulus alerting events is converted into an action potential which is transmitted along a biological channel where it is integrated

**signal transduction:**
1: signal transduction is any biological process by which a transducer converts one kind of signal/energy/stimulus into another (Wikipedia, 10.12)
2: the original form of a signal is converted to another form of energy using a transducer, implying the use of a sensor/detector; a transducer is a device that converts one form of energy to another (electrical, mechanical, electromagnetic (including light), chemical, acoustic or thermal energy) (Wikipedia, 10.12)

### 13.1.3 Communication and biological complexity.
**13.1.3.1 Message communication: usefulness and meaning.** The concept of logical depth is a measure of value that captures the usefulness of a message in a particular domain (Bennett, 1988). Information per se is not a good measure of message value, the amount of work it takes to compute something. The value in a message lies in its buried redundancy, the parts predictable only with difficulty, things the receiver could find in principle have figured out with being told, but only at considerable cost in money, time, computation. Logical depth and self-organization is how complex structures develop in nature. Evolution starts with simple initial conditions,

complexity arises apparently building on itself. The basic processes of self-organization and evolution resemble computation.

The meaning of a message is made in reference to some system with certain physical or conceptual entities. Meaning results when a cognitive agent (e.g. a human, a seed) takes a signal and turns it into information; information is in the head of the receiver. Meaning is evident in system behavior, "any change in an entity with respect to its surroundings".

**13.1.3.2 Complexity of information**. A dynamical system produces information. If it is unpredictable it produces a great deal of information.

**Kolmogorov complexity**:
1: the complexity of a number, message, set of data is the inverse of simplicity and order, and it corresponds to information
2: calculating complexity in terms of algorithms; the complexity of an object is the size of the smallest computer program needed to generate it;
3: complexity as the amount of information, the degree of randomness, in a message; information = randomness = complexity

## 13.2 Biological Communication

> "What lies at the heart of every living thing is not a fire, not a warm breath, not a 'spark of life'. It is information, words, instructions ... If you want to understand life, don't think about vibrant, throbbing gels and oozes, think about information technology." (Dawkins, 1986).

The nature of weeds is an environment-biology communication system (see ch. 3.4.4.1 signal space dimensions). Biology is information. Evolution itself embodies an ongoing exchange of information between organism and environment. For biology, information comes via evolution; what evolves is information in all its forms and transforms. Information is physical. Biology is physical information with quantifiable (Kolmogorov) complexity. The gene is not the information-carrying molecule, the gene is the information.

Information in biological systems can be studied. An important problem in science is to discover another language of biology, the language of information in biological systems. "The information circle becomes the unit of life. It connotes a cosmic principle of organization and order, and it provides an exact measure of that." (Loewenstein, 1999). This information circle for weedy plants is the predictable developmental events of the annual life history.

Information theory was developed by Claude E. Shannon (Shannon and Weaver, 1949) to find fundamental limits on signal processing operations such as compressing data and on reliably storing and communicating data. Since its inception it has broadened to find applications in statistical inference, networks (e.g. evolution and function of molecular codes, model selection in ecology) and other forms of data analysis. A communication system of any type (e.g. language, music, arts, human behavior, machine) must contain the following five elements (E) (figure 13.1):

A Shannon communication system includes these elements, as well as the concepts of message and signal. A message is the object of communication; a vessel which provides information, it can also be this information; its meaning is dependent upon the context in which it is used. A signal is a function that conveys information about the behavior or attributes of some phenomenon in communication systems; any physical quantity exhibiting variation in time or variation in space is potentially a signal if it provides information from the source to the destination on the status of a physical system, or conveys a message between observers, among other possibilities.

**Figure 13.1.** Schematic diagram of Shannon information communication system (Shannon and Weaver, 1949).

| | Communication Element | Description |
|---|---|---|
| 1 | Information source | entity, person or machine generating the message (characters, math function) |
| 2 | Transmitter | operates on the message in some way: it converts (encodes) the message to produce a suitable signal |
| 3 | Channel | the medium used to transmit the signal |
| 4 | Receiver | inverts the transmitter operation: decodes message, or reconstructs it from the signal |
| 5 | Destination | the person or thing at the other end |

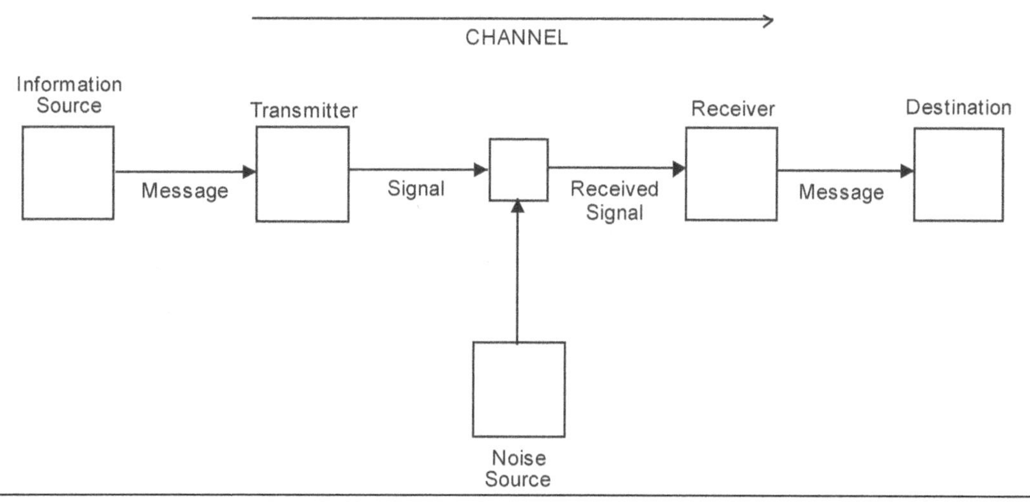

### 13.3 *Setaria* Life History as a Complex Adaptive Soil-Seed Communication System

The nature of the weedy *Setaria* is revealed in the physical (morphological and genetic spatial structures) and the phenomenal (life history behavior instigated by functional traits). The nature of *Setaria* is the consequence of structural self-similarity and behavioral self-organization. Structural self-similarity in morphology was revealed in seed envelope compartmentalization and individual plant tillering; in genetics by local populations and *Setaria* species-associations forming the global metapopulation. Behavioral self-organization was revealed in the self-pollenating mating system controlling genetic novelty, seed heteroblasty blueprinting seedling recruitment, and phenotypic plasticity and somatic polymorphism optimizing seed fecundity.

The consequence of structural self-similarity and behavioral self-organization has been the evolution of a complex adaptive seed-soil communication system. Weedy *Setaria* life history is represented in algorithmic form as FoxPatch, a model to forecast seed behavior. The environment-biological informational system with which weedy *Setaria* life history unfolds is represented in the seed-soil communication system.

**13.3.1 Forecasting *Setaria* seed behavior: FoxPatch.** Weedy *Setaria* spp. seeds with heterogeneous dormancy states are recruited from soil pools when sufficient oxygen, water and heat accumulate over time (Dekker, 2004). The variable timing of this recruitment is the first step in assembly with crops and other weeds in annually disturbed agricultural communities. The timing of seedling emergence is the single most important factor in determining subsequent weed control, competition, crop losses and replenishment of the soil seed pool.

FoxPatch is a computational tool to predict the behavior of individual weedy *Setaria* spp. seeds, including after-ripening, dormancy re-induction, germination and seedling emergence from the soil (Dekker et al., 2003; Dekker, 2009, 2011a). It is based on intrinsic morpho-physiological traits in the seed (the germinability-dormancy state) at abscission responding to extrinsic oxy-hydro-thermal time signals it receives from the environment (Dekker et al., 2003). Three rules of behavior define the interaction of individual seeds with signals from their immediate environment, and thus determine patterns of seed-seedling life history behaviors.

Algorithmically, FoxPatch consists of a schema derived from the morpho-physiological traits in the seed determining its dormancy state at any time in its life history. This schema responds to soil signals in predictable ways. These predictable responses are the foundation for three seed behavior rules which together define the seed-seedling portion of the weedy *Setaria* sp. life history. Forecasting *Setaria* seed behavior relies on three separate sources of information: dormancy state heterogeneity, environmental signals and patterns of seedling emergence.

**13.3.1.1 Seed state and process model.** FoxPatch is a predictive tool, a formalization of what we empirically know about the *Setaria* spp. seed traits (schema) and the life history behaviors that arise from the interaction of these traits with the environments in which they thrive (behavior rules). The life history of weedy *Setaria* sp. seed is schematically represented in figure 13.2.

**Figure 13.2.** Schematic diagram of weedy *Setaria* sp. soil seed pool behavior based on the life history of the seed (seed states and transitions between states). Life history states (terminal, boxes; flux, circles): [1] dormant seeds (DORM), [2] germination candidate seeds (fully after-ripened; CAN), [3] germinated seeds (GERM), [4] seedlings (SEEDLING), and [5] dead seeds or seedlings (DEAD). Life history processes (arrows), or transitions between states: [A] induction of dormancy and input (dispersal) of dormant seed at abscission (the seed rain), [B] after-ripening of dormant seed (AR), [C] re-induction of (2°) dormancy in germination candidates (DORM RE-INDUCT), [D] germination growth (GROW), [E] seedling emergence (EMERGE), and [F: $F_1$-$F_4$] death (from states 1-4).

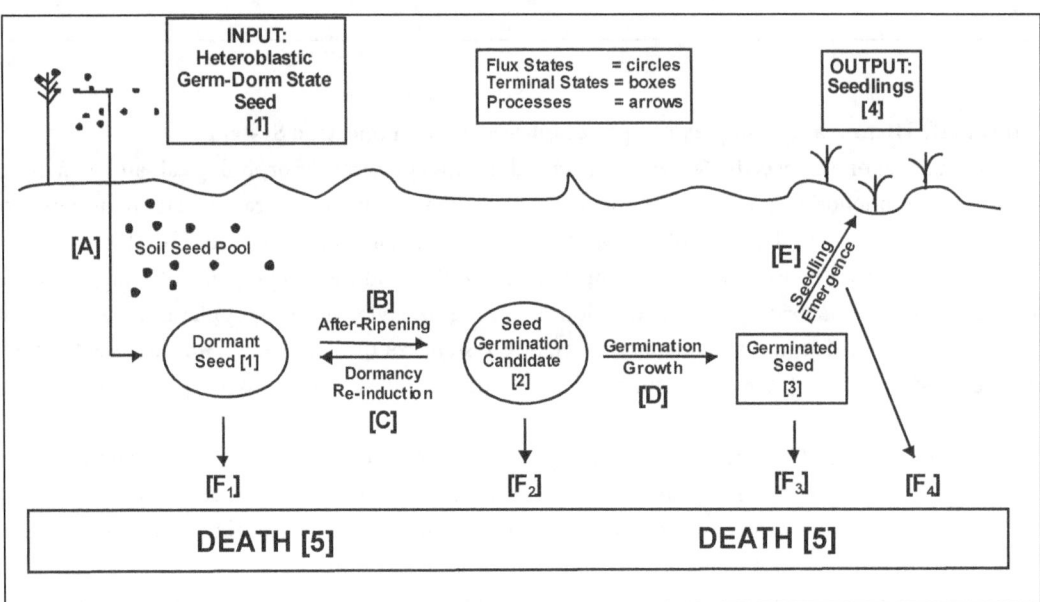

The seed portion of the life history can also be represented by the behavioral processes leading to transitions between seed states:

$$DORM_{AB} \xrightarrow{AR} \underset{DORM\ RE-INDUCT}{\longleftarrow} CAN \xrightarrow{GROW} GERM \xrightarrow{EMERGE} SEEDLING$$

Where: $DORM_{AB}$, dormancy state at abscission.

**13.3.1.2 Soil signal controlling seed behavior: oxy-hydro-thermal time.** Weed seeds like *Setaria* require only three resources/conditions (signals) from their immediate soil environment to germinate and emerge as seedlings: temperatures (T) favorable for germination and seedling emergence (thermal-time), adequate but not excessive soil moisture (hydro-time), and oxygen dissolved in imbibed water to support germination metabolism (oxy-hydro time). For each of these three resources/conditions there is a threshold amount an individual seed needs to germinate. Each of these three thresholds has a minimum and maximum value range within which seeds germinate. When any of these three signals is insufficient or excessive, a living seed in the soil will remain dormant. When all three signals are within their thresholds germination will occur. The conditions within all three thresholds can also be conceptualized as the germination "window" through which a seed must pass to germinate and emerge as a seedling, germination phenotype space.

FoxPatch can be expressed quantitatively by defining the environmental signal required by an individual to complete each of the component processes, given the initial dormancy state of individuals. All viable weedy *Setaria* spp. seed in the soil obey the:

General seed behavior rule: The behavior of an individual weedy *Setaria* seed in the soil is regulated by the amount of oxygen dissolved in water (the oxygen-water signal) that accumulates in the seed symplast, and temperatures favorable (or not) to germination growth (the germination temperature signal), over some time period (cumulatively oxy-hydro-thermal time).

(Atchison, 2001; Dekker, 1999, 2000, 2003, 2004a; Dekker *et al.*, 1996, 2001; Dekker & Hargrove, 2002; Donnelly *et al.*, 2012; Sareini, 2002).

All FoxPatch predictions are based on the assumption that when an individual, dormant, foxtail seed receives the minimum environmental signal required for after-ripening and germination in the soil it will emerge (barring mortality). Therefore, the parameters critical to predicting the seedling emergence of an individual seed are [1] the soil signals (T, $H_2O$, $O_2$) required by an individual *Setaria* sp. seed to overcome its inherent level of dormancy to after-ripen and germinate (the schema), and [2] the signals that an individual seed in the soil actually experiences and utilizes (soil signal). Provided these two critical pieces of information are available, seedling emergence predictions in FoxPatch are based on the following:

General prediction algorithm: An individual weedy foxtail seed will change state when the minimum inherently-required oxy-hydro-thermal time signal is received from its realized environment (plus signals not causing an effect due to inefficient transduction or insensitivity).

There are two signals in the soil that interact to control weedy *Setaria* seed behavior. Both are based on temperature and moisture. The oxygen signal is OxSIG, oxy-hydro-time. The heat signal is TSIG, germination thermal time. The combined influence of these two signals on foxtail seed behavior is the oxy-hydro-[thermal]-time signal.

**13.3.1.2.1 The oxy-hydro-time signal: OxSIG.** One of the most striking features of the soil in the temperate regions of the globe is the diurnal fluctuation in soil temperatures that occur every day, and the ample moisture in springtime soils contains the seasonally maximum amount of oxygen immediately after thawing. This diurnal phenomena (the "big spring temperature oscillator") plays a crucial role in after-ripening and germination (Dekker et al., 2001). During the cool night period water entering the seed placental pore brings with it much oxygen due to its high solubility. During the mid-day, temperatures rise and $O_2$ solubility decreases causing a de-gassing effect. The water is purged, sending a pulse of free oxygen into the seed symplast because diffusion out of the narrow pore opening is restricted. The contribution of both chronic $O_2$ availability (based on solubility alone) and acute $O_2$ availability (based on degassing with temperature change) are the two sources of the oxy-hydro signal to which *Setaria* spp. have adapted and synchronized their life history in the soil.

The signal controlling *Setaria* seed after-ripening, 2° dormancy re-induction, and the continuing supply of $O_2$ during the germination processes is OxSIG: the oxygen mass per water volume of seed symplast (caryopsis) per time period per seed. An individual *Setaria* seed experiences two different OxSIG's (chronic, acute) from the soil environment depending on the diurnal pattern of temperature (constant, alternating, or both). The OxSIG affecting any single seed is the summation over time of the $O_2$ received:

$$OxSIG_{Total} = [(OxSIG_{Chronic}) + (OxSIG_{Acute})]$$

The chronic $O_2$ signal ($SIG_{Chronic}$) is that amount of dissolved $O_2$ (g L$^{-1}$) that is diffuses into the seed symplast when the ambient temperature is constant for some time period (g L$^{-1}$ Time$^{-1}$):

$$OxSIG_{Chronic} = (OXY_{Temp°C} * VH_2O_{Seed}) * (TIME_{Temp°C}) \text{ seed}^{-1}$$

Where: $OXY_{Temp°C}$, $O_2$ *solubility (g per l) at a particular temperature;* $VH_2O_{Seed}$, $H_2O$ *volume (g or ml) of a single seed symplast;* $TIME_{Temp°C}$, *time (h) at a particular temperature.*

We have experimentally determined the seed symplast water volume of a locally (Ames, IA, USA) adapted *S. faberi* population (author lot # 3781; K99-40) to vary from 0.00036 to 0.00050 ml (360-500 nanoliters) per seed, depending on the immediate microsite moisture availability and temperature (Atchison & Dekker, unpublished data).

The acute $O_2$ signal ($SIG_{Acute}$) is the difference in $O_2$ (g) in solution that is de-gassed (lost from solution) when the symplast $H_2O$ changes from a low to a high temperature, a process that occurs every day of the year in the soil:

$$OxSIG_{AR\ Acute} = (OXY_{LOW\ Temp°C} * VH_2O_{Seed}) - (OXY_{HIGH\ Temp°C} * VH_2O_{Seed})$$

Where: $OXY_{LOW\ Temp°C}$ *and* $OXY_{HIGH\ Temp°C}$ *are the oxygen solubilities in $H_2O$ at the low and high temperatures, respectively.*

**13.3.1.2.2 The thermal signal: TSIG.** Temperature (T) plays two different roles as a signal controlling *Setaria* seed behavior. The ambient temperature determines $O_2$ solubility (inversely related to T) as well as the germination growth environment. Temperature's two roles often have opposite effects on an individual seed: cool provides more $O_2$, while inhibiting germination; warm provides less $O_2$ but promotes embryo growth.

**13.3.1.2.3 Oxy-hydro-[thermal]-time: OxSIG & TSIG.** The combined influence of these two signals on foxtail seed behavior is the oxy-hydro-[thermal]-time signal (chronic and acute OxSIG & TSIG). Experimentally, local soil moisture and temperature are measured at depth in the soil hourly, continuously, for accurate signal quantification.

**13.3.1.3 Schema of intrinsic *Setaria* sp. seed traits.** Each individual *Setaria* sp. seed possesses inherent traits, morphological and physiological, that determine how it will behave in a variety of environmental conditions it might encounter. These traits of the seed are inherited from the parent plant at abscission, and can be conceptualized as a plan for future action, a schema. The schema is a blueprint constraining an individual's unique set of reactions to local environmental conditions (biotic and abiotic), the realized phenotype. This schema is biologically complete in the sense it predicates all the behaviors an individual will perform such as after-ripening and germination. What is the schema that weedy *Setaria* sp. seed possess? In its most basic computational form it is the amount of oxy-hydro-thermal time signal required to germinate an individual seed. Experimentally this is determined by germination assays at seed abscission under optimal, controlled conditions (see dormancy induction below).

**13.3.1.4 Germinability-dormancy induction.** The local environment surrounding an individual *Setaria* panicle during embryogenesis, and the physiology of dormancy induction traits in developing embryos, conspire to produce a heterogeneous collection of seeds, each with potentially different dormancy states at abscission, the time they first become independent of the parent plant, seed germinability-dormancy heteroblasty. The amount of dormancy in each seed, and the variability in dormancy quantity among individuals, is a direct display of the future timing of recruitment. The consequence of past selection and local adaptation determines the "germinability-dormancy heterogeneity bandwidth" among a cohort of seeds produced by a single successful parent plant panicle. The precise mechanisms and environmental conditions leading to particular dormancy states in individual seeds are unknown, but the role of light may be important (Atchison, 2001; Dekker, 2003, 2004a; Dekker et al., 1996). When this process is understood the definition of a dormancy induction rule during embryogenesis may be possible. Until that time, the quantity of dormancy induced in an individual seed at the time of abscission ($DORM_{AB}$) can be experimentally determined in optimal conditions:

Initial individual seed dormancy state:   The inherent dormancy state of an individual foxtail seed at abscission is defined as the minimum oxy-hydro-thermal time signal required to stimulate germination.

The initial dormancy state is experimentally determined by the minimum amount of after-ripening (AR) required for a dormant seed (DORM) to become a germination candidate (CAN),
and for that CAN to germinate (GROW), in favorable (controlled, ideal) environmental conditions (e.g. AR: 4° C, moist, dark; GROW: 15-35° C alternating diurnal, moist, light). The population data is presented in frequency or cumulative distribution diagrams of AR dose versus percent germination. The signals required to after-ripen and germinate individual foxtail seeds can be expressed in a number of ways, including AR time and oxy-hydro-thermal time units.

**13.3.1.5 Rules for individual weedy *Setaria* sp. seed behavior.** The rules that control individual weedy *Setaria* sp. seed behavior are illustrated in its life history (Figure 13.2). *Setaria* sp. seed behavior is the consequence of the interaction between extrinsic local environment influences and intrinsic seed traits (the schema). These rules of behavior are the inevitable consequence of past selection and adaptation molding the *Setaria* sp. phenotypes we see in the field today. They pertain to living processes and changes in state, as opposed to the more stochastic and unpredictable factors involved in mortality.

FoxPatch is based on three rules of individual weedy *Setaria* sp. seed behavior. These rules are defined by field observations of dormancy state heterogeneity among seeds at abscission from an individual panicle, as well as by after-ripening, dormancy re-induction, and germination. What particular behavior occurs is dependent on the initial state of the seed (the schema) and the rate and amount of signal accumulated. Each of these behavioral transitions is governed by a specific foxtail seed behavior rule, presented here in their qualitative forms.

**13.3.1.5.1 The after-ripening rule, and its inverse the dormancy re-induction rule.** From the time a *Setaria* spp. seed is shed (typically August-November in North America) it is exposed to moisture-temperature conditions that either after-ripen (AR) the seed (generally cool, moist) or maintain (or induce) dormancy (generally hot, dry). When sufficient AR has occurred, and the seed symplast accumulates enough oxygen (dissolved in water) over a period of time, it becomes a germination candidate.

After-Ripening (see figure 13.2): $$DORM_{AB} \xrightarrow{AR} CAN$$

After-ripening rule: An individual dormant foxtail seed (DORM) will after-ripen when the rate of oxygen dissolved in water entering the symplast exceeds the capacity to remove that oxygen for some time period, during which the minimum inherently-required amount of oxygen has accumulated in the symplast (CAN) allowing germination (GERM) to occur at some temperature.

After-ripening prediction algorithm: An individual dormant foxtail seed (DORM) will after-ripen to a germination candidate (CAN) when the minimum inherently-required oxy-hydro time signal is received from its realized environment (plus signals not causing an effect due to inefficient transduction or insensitivity) allowing germination (GERM) to occur at some temperature.

When conditions in the field are generally hot and dry for some time (e.g. summer) the oxygen and water levels in the seed decrease and dormancy is re-induced as the effects of previous after-ripening are lost.

Dormancy re-induction (see figure 13.2): $$DORM_{AB} \xleftarrow{DORM\ RE-INDUCT} CAN$$

Dormancy re-induction rule: An individual after-ripened foxtail seed (CAN) will reacquire dormancy when the rate of oxygen dissolved in water entering the symplast is less than the capacity to remove oxygen for some time period, during which the minimum inherently-required amount of oxygen is no longer present.

Dormancy re-induction algorithm: An individual foxtail seed germination candidate (CAN) will have dormancy re-induced when the minimum inherently-required oxy-hydro time signal is not received from its realized environment (plus signals not causing an effect due to inefficient transduction or insensitivity) preventing germination from occurring at some temperature.

Experimentally we determine AR and DORM by removing seed from the field at intervals through the year and determine the number of viable (tetrazolium assay) seeds that immediately germinate (GERM) or not (DORM) under optimal conditions.

**13.3.1.5.2 The germination candidate threshold state**. Once a seed becomes fully after-ripened, saturated with sufficient dissolved oxygen in the symplast, it is poised on the threshold of germination, a germination candidate (CAN), awaiting only favorable temperature conditions to germinate. It is from the after-ripened seed pool (CAN) in the soil that seedlings are recruited. It is the least dormant seed in the soil that after-ripen the earliest in a season, and therefore they compose the most likely germination candidates for recruitment.

**13.3.1.5.3 The seed germination rule**. When a fully after-ripened weedy *Setaria* sp. seed, a germination candidate, is exposed to some temperature (usually higher than that for AR) for some time it germinates.

Germination (see figure 13.2):   $CAN \xrightarrow{GROW} GERM$

Seed germination rule: An individual foxtail germination candidate (CAN; an after-ripened seed with sufficient free symplastic oxygen for sustained respiration) will germinate when exposed to minimally favorable temperatures for some time period.

Seed germination algorithm: An individual foxtail germination candidate seed will germinate when exposed to minimally favorable temperatures from its realized environment (plus signals not causing an effect due to inefficient transduction or insensitivity) for some time period.

Minimally favorable temperature often is a narrow temperature range that increases as the oxygen in the symplast increases (after-ripening depth).

The most complex part of FoxPatch at this time is elucidation of a quantitative function that predicts germination of a germination candidate based on both oxy-hydro (OxSIG) and thermal (TSIG) time. Two opposing forces are at work at this time: lower temperatures (e.g. 0° C liquid water) possess high oxygen solubility and inhibit germination; higher temperatures (35° is more favorable than 15° C) stimulate germination but have low oxygen solubility. After-ripening is favored by cool, moist conditions. Germination is favored by relatively high symplast oxygen saturation and high temperatures. The correct function to predict these opposing forces will be a computational trade-off weighing their relative influences and interaction. These opposing elements provide a very powerful, yet sensitive, environment-sensing mechanism allowing *Setaria* spp. to detect very precisely favorable conditions leading to recruitment and subsequent success.

Forecasting the seed behavior of weedy *Setaria* spp. seeds in agricultural soils based on intrinsic dormancy qualities, environmental oxy-hydro-thermal-time signals, and consistent seasonal patterns of seedling emergence could provide us with a robust tool of immense value, an algorithm of the first assembly step of disturbed agricultural communities.

### 13.3.2 *Setaria* soil-seed communication system.

**13.3.2.1 Shannon communication system for weedy *Setaria* seed-seedling.** The nature of weedy *Setaria* seed-seedling life history is a complex adaptive, soil-seed communication system arising from its component functional traits. Functional traits controlling seed-seedling behavior are physical information that has evolved in ongoing communication between organism and environment leading to local adaption. A Shannon communication system of any type must contain the five elements (E) presented in figure 13.1, as well as message and signal. (Shannon and Weaver, 1949)

Weedy *Setaria* seed life history behaviors are controlled by environmental information (signals) flowing from the soil to the seed embryo. The *Setaria* seed is constructed in such a way as to be receptive to specific signals contained within the entire soil information available to it. The specific signal to which *Setaria* is tuned affecting seed behavior in the soil is the amount of oxygen and heat (T, thermal) in soil water over time, oxy-hydro-thermal time ($O_2$-$H_2O$-T-Time). The message that directly controls *Setaria* seed behavior is oxygen and heat accumulating in the embryo modulating seed respiration.

The Shannon environmental-biological communication system between the soil and the *Setaria* seed contains the five elements (E) and components as schematically organized in figure 13.3.

**Figure 13.3.** Schematic diagram of Shannon environment-biology information-communication system for weedy *Setaria* seed-seedling life history development. Communication elements (E): E1, information source, soil; E2, transmitter, soil particle contact with seed surface water films; E3, channel, continuous soil particle-seed surface water films; E4, receiver, living seed interior from TACL membrane to aleurone layer to embryo; E5, destination, embryo. The signal is soil $O_2$-$H_2O$-T-Time; the message is $O_2$-$H_2O$-T stimulating embryo respiration.

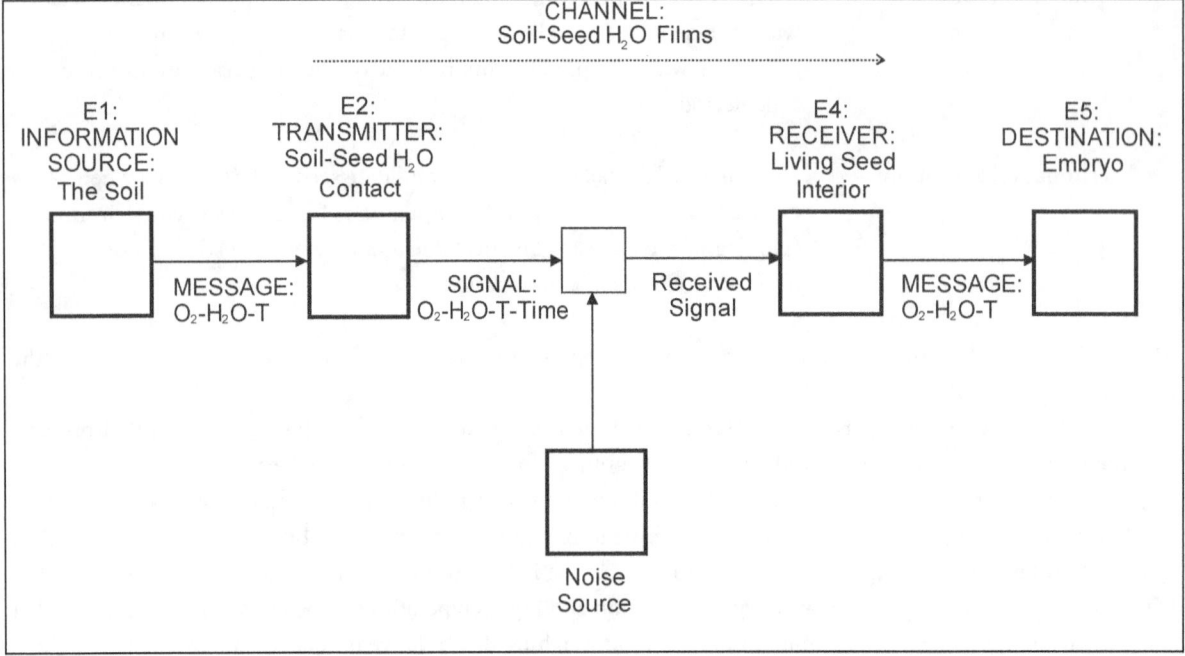

### 13.3.2.1.1 Information source (E1): the soil adjacent to the *Setaria* seed.

"The *information source* selects a desired *message* out of a set of possible messages" (Shannon and Weaver, 1949)

The source of information generating the message affecting weedy *Setaria* seed life history development is the soil surrounding the individual seed in a particular locality. There are many sources of information impinging on the seed, including changing cycles of oxygen-water-heat in seasonal climate and environment, site latitude-longitude-elevation-slope-aspect, soil quality (tilth, structure, texture, compaction, organic matter, chemicals), neighboring seeds and plants, microbial and animal activity, and agricultural and other human disturbances. This information flows constantly within the soil. The soil-borne seed responds to only a limited portion of the entire soil information, the message controlling behavior. The message is heat and oxygen dissolved in soil water that the *Setaria* embryo requires for continued development, whether continued quiescence or germination. The growth and development behaviors of the embryo include 'remain dormant' or 'resume growth and development'. The message was naturally selected during *Setaria* evolution from among the larger entire set of soil information as a dependable signal stimulating-inhibiting seed germination-dormancy behaviors. The specific signal that affects *Setaria* behavior is the amount of heat and oxygen dissolved in soil water films connecting soil particles with the seed with time, oxy-hydro-thermal-time ($O_2$-$H_2O$-heat-time).

### 13.3.2.1.2 Transmitter (E2): water film contact between soil-seed.

"The *transmitter* changes this *message* into the *signal* which is actually sent over the *communication channel* from the transmitter to the *receiver*." (Shannon and Weaver, 1949)

The transmitter converts/encodes/changes the message to produce a suitable signal. The transmitter is the water film adhering to soil particles adjacent to, and continuous with, seed exterior surfaces (hull, placental pore). The message sent is heat and oxygen dissolved in soil water, information is physical. The specific signal transmitted is the amount of heat and oxygen dissolved in soil water films connecting soil particles with the seed interior-exterior with time (oxy-hydro-thermal-time) received by the embryo-endosperm affecting *Setaria* behavior. The transmitter modulates the heat and oxygen content of water films that physically connect soil particles with seed surfaces and that are received by the receiver, the living tissues of the seed interior. The transmitter changes, encodes, the amount of oxygen dissolved in soil water by formation of continuous water films physically connecting soil particle-seed surfaces. Signal encoding is the change in the water film message by the morphology of the seed surface. *Setaria* seed hull surface morphology and topography accomplish these changes that encode the signal, information is physical.

**13.3.2.1.3 Channel (E3): soil-seed water films.** The *Setaria* communication channel is the continuous water film connecting soil particles with the seed exterior (hull, placental pore) and interior (TACL, endosperm, embryo). This water film communication channel transmits the oxygen-water-heat signal to the living seed interior receiver and the embryo destination. The capacity of this water film channel can be described in terms of the water-oxygen information it transmits per unit time from the adjacent, continuous, soil source.

**13.3.2.1.4 Noise source.** Introduction of noise in the communication channel means the received message exhibits increased uncertainty. The received signal is selected out of a more varied set than the transmitted set. There can be several sources of noise between the transmitter and receiver, everything that corrupts the water-oxygen signal: water film discontinuities, physical interference from soil biota and chemicals, soil structural changes, seed damage, localized areas of anomolous gas and water content/exchange.

**13.3.2.1.5 Receiver (E4): the seed interior.**

"The *receiver* is a sort of inverse transmitter, changing the transmitted signal back into a message, and handling this message on to the destination." (Shannon and Weaver, 1949)

The signal transmitted from exterior water films (terminating with the placental pore) are received by the seed interior by TACL membrane transport and diffusion equilibrating heat and dissolved gases between the exterior-interior seed compartments. Water oxygenated by the seed surface water films is the signal that is decoded by the seed interior receiver in the form of the free oxygen message available to stimulate respiration and hence embryo germination (the destination).

**13.3.2.1.6 Destination (E5): the embryo.** The resumption of life history growth and development by the embryo is stimulated by the accumulation of adequate interior oxygen leading to germination and seedling emergence. Insufficient oxygen maintains or stimulates dormancy and embryo quiescence. This message is 'remembered' by the weedy Setaria seed in the process of evolutionary adaptation. Successfully surviving and reproducing plants arising from seed utilizing this communication system pass on these functional traits to their polymorphic progeny, each of which in turn passes on the ability to generate its own range of heteromorphic seeds appropriate to continuing local adaptation

**13.3.2.2 *Setaria* soil-seed communication transmission algorithm.** The *Setaria* soil-seed communication system seed behavior described above (Figure 13.3) can be also expressed as operations (processes) computed by seed algorithms (Table 13.1).

**13.3.3 Seed memory and adaptive evolution.** Information is physical: memory resides in several locations in the *Setaria* seed. *Setaria* seed memory has both short and long term expressions. In the short term, memory is expressed by the amount of oxygen accumulated in the seed interior. *Setaria* seed memory is the current germination-dormancy state of each living seed in a local soil pool. The seed germination-dormancy capacity is 'hard-wired' (intrinsic) to seed at birth, embryogenic seed dormancy induction. Information storage is an important component of communication. The ability of *Setaria* seeds to accumulate oxygen in the seed interior provides a message storage capability for a time interval determined by the seed's germination-dormancy capacity. Memory is expressed in the long-term by responsiveness to $O_2$-$H_2O$-heat messages as determined by the morpho-physiological soil-seed communication system (hull, TACL membrane, scavenger protein). Memory resides in the functional-adaptive traits that regulate all seed behaviors by transducing/transforming/encoding and decoding inorganic soil $O_2$-$H_2O$-heat signals over time: the three morpho-physiological mechanisms forming the soil-seed communication system. The message is remembered: plants arising from seed utilizing this communication system pass on these functional traits to their polymorphic progeny, each of which in turn passes on the ability to generate its own range of heteromorphic seeds appropriate to continuing, successful local adaptation.

**Table 13.1**. Seed algorithms, operations (developmental processes), computed by weedy *Setaria* seeds during life history development by means of the Shannon soil-seed communication system.

| Soil-Seed Communication System | |
|---|---|
| Algorithmic Operations Computed | Life History Developmental Processes |
| Step 1: Acquire information from soil water source adjacent to seed (E1) | |
| Step 2: Transmit oxygen-water signal to seed exterior encoder and antenna | Step 2A: Assemble transmitter encoder and antenna by adhesion of continuous water films in contact with soil particles and seed exterior |
| | Step 2B: Encode transmitter signal by amplifying soil-seed water film oxygen content. Change oxygen content of water films by contact with seed hull surface (e.g. increased surface area exposed to dissolve soil atmosphere gases) |
| | Step 2C: Seed exterior antenna transmits signal by diffusive equilibration of oxygen content throughout continuous soil-seed water film communication channel (hull, placental pore) connected to TACL membrane receiver |
| Step 3: Transmit oxygen-water signal to seed interior receiver | Step 3A: Form water films continuous with hull, placental pore and TACL membrane entrance to seed interior. |
| | Step 3B: Oxygen-water signal movement across seed surface facilitated by hull topography channeling to basal placental pore entrance |
| | Step 3C: Oxygen-water signal physically filtered by apoplastic placental pore tissues |
| | Step 3D: Oxygen-water signal entrance into seed interior receiver is physically restricted (tuner, resistor) to TACL membrane by continuous surrounding gas- and water-tight caryposis coat surrounding the embryo and endosperm compartment |
| | Step 3E: Equilibrate oxygen-water content throughout soil-TACL channel by water-gas film diffusion |
| | Step 3F: Equilibrate seed exterior-interior oxygen-water content by membrane-regulated diffusion |
| Step 4: Receive oxygen-water signal in seed interior | Step 4A: Accumulate oxygen-water message information in embryo-endosperm capacitor |
| | Step 4B: Remove oxygen in embryo-endosperm water by action of interior oxygen scavenger protein |

| Step 5: Provoke embryo behavior with oxygen-water message at seed interior destination | Step 5A: Stimulate-inhibit respiration and embryo germination with adequate oxygen-water accumulation |

# UNIT 6: REPRESENTATION OF WEED BIOLOGY

"In theory there is no difference between theory and practice; in practice there is." Yogi Berra, 2010

" 'You don't see something until you have the right metaphor to let you perceive it' [Robert Stetson] Shaw said, echoing Thomas S Kuhn." Gleick, 1987

"Algorithms are an exacting standard. It's often said that you don't really understand something until you can express it as an algorithm. (As Richard Feynman said, "What I cannot create, I do not understand.") "In any area of science, if a theory cannot be expressed as an algorithm, it's not entirely rigorous." Domengos, 2015

## U6.1 Finding Knowledge

How can weed biology be represented with the understandings developed in previous units on the nature of weeds, the evolution and adaptation of populations and local communities, and of complex adaptive weed systems? How can knowledge of weeds be discovered, and then represented? The real source of scientific theories is not from sensory experience, but from conjecture (Deutsch, 2011). The real source of our knowledge is conjecture alternating with criticism. The role of experiment and observation is to choose between existing theories, not to be the source of new ones. The most important source of variation in explanatory theories is creativity. How do we acquire knowledge? Not by imitation. Scientific knowledge is impossible without pre-existing knowledge about what to look at, what to look for, how to look, and how to interpret what one sees. Therefore, theory has to come first. It has to be conjectured, not derived. No idea can be represented entirely explicitly. An explicit conjecture has an inexplicit component whether we are aware of it or not. So does all criticism.

**conjecture:**
1: the formation of conclusions from incomplete evidence; guess; the inference or conclusions so formed
2: In mathematics, a conjecture is a conclusion or proposition based on incomplete information, for which no proof has been found (Wikipedia, 6.16)

**knowledge:**
1: the facts, feelings, or experiences known by a person or group of people
2: awareness, consciousness, or familiarity gained by experiene or learning
3: specific information about a subject
4: familiarity, awareness or understanding of someone or something, such as facts, information, descriptions, or skills, which is acquired through experience or education by perceiving, discovering, or learning; a theoretical or practical understanding of a subject. (Wikipedia, 6.16)

**creativity:**
1: the capacity to create new explanations

2: a phenomenon whereby something new and somehow valuable is formed; the created item may be intangible (such as an idea, a scientific theory, a musical compostion or a joke) or a physical object (such as an invention, a literary work or a painting) (Wikipedia, 6.16)
3: a process of becoming sensitive to problems, deficiencies, gaps in knowledge, missing elements, disharmonies, and so on; identifying the difficulty; searching for solutions, making guesses, or formulating hypotheses about the deficiencies: testing and retesting these hypotheses and possibly modifying and retesting them; and finally communicating the results (Wikipedia, 6.16, Torrance, P.)

The substance of scientific theories is explanation. Explanation *of errors* constitutes most of the content of the design of any non-trivial scientific experiment. Good explanations are explanations that are hard to vary in the sense that changing the details would ruin the explanation.

**explanation**:
1: statement about what is there, what it does, and how and why
2: a set of statements constructed to describe a set of facts which clarifies the causes, context, and consequences of those facts (Wikipedia, 6.16)

An explanation may establish rules or laws, and may clarify the existing ones in relation to any objects, or phenomena examined. An explanation is often underpinned by an understanding that is represented by different media (e.g. music, text, and graphics). Thus, an explanation is subjected to interpretation, criticism and discussion. In scientific research, explanation is one of several "purposes" for empirical research. Explanation is a way to uncover new knowledge, and to report relationships among different aspects of studied phenomena. Explanation attempts to answer the "why" question. Explanations have varied explanatory power. The formal hypothesis is the theoretical tool used to verify explanation in empirical research.

Phenomenon can possess the quality of reach. When the knowledge in a gene happens to have reach, it will help the individual to help itself in a wider range of circumstances, and by more, than the spreading of the gene strictly requires (possibly sources of preadaption, exadaptation).

**reach**:
1: the ability of some explanations to solve problems beyond those that they were created to solve
2: an intrinsic attribute of an explanation, not an empirical or inductive assumption
3: the reach of ideas into the world of abstractions is a property of the knowledge that they contain (not of the brain in which they may happen to be instantiated)
4: reach always has an explanation, the explanation is not yet known

**jump to universality**: the tendency of gradually improving systems to undergo a sudden large increase in functionality (reach), becoming universal in some domain

Darwinian evolution is the creation of knowledge through alternating variation and selection. The central idea of neo-Darwinism is that evolution favors the genes (the information) that spread best through the population. Knowledge embodied in genes is knowledge of how to get themselves replicated at the expense of their rivals, often by imparting useful functionality to their organism (e.g. functional traits). A population of replicators subject to variation will be taken over by those variants that are better than their rivals at causing themselves to be replicated. Both human knowledge and biological adaptations are abstract replicators: forms of information which, once they are embodied in a suitable physical system, tend to remain so while most variants of them do not. For example, *Setaria* seed morphology as memory and knowledge (see ch. 13.3.3). Knowledge is information which, when it is physically embodied in a suitable environment, tends to cause itself to remain so. Knowledge and biological adaptations are hard to vary.

All knowledge-creation depends on, and physically consists of, emergent phenomena. Deutsch (2011) provides an alternate definition (see ch. 12.3, emergence of system order):

**emergence**:
1: class of high-level phenomena that can be well explained in terms of each other alone, with no direct reference to anything at (the atomic) (lower) level or below: the behavior of that whole class of high-level phenomena is quasi-autonomous (almost self-contained)
2: resolution into explicability of phenomena at a higher (quasi-autonomous) level

There exist explanations at several levels of emergence independent of each other, sets of phenomena that can be explained well in terms of each other without analyzing them into their constituent (lower level) entities (such as atoms). There is no inconsistency in having multiple explanations of the same phenomenon, at different levels of emergence. Explanations at any level of emergence can be fundamental, they do not form a hierarchy with the lowest being the most fundamental. This perspective on emergent phenomena in nature bring into question two misconceptions: reductionism and holism.

**reductionism**: the misconception that science must or should always explain things by analyzing them into components (and hence that higher-level explanations cannot be fundamental)

**holism**: the misconception that all significant explanations are of components in terms of wholes rather than vice versa

Related to the misconceptions of reductionism and holism is the misconception of inductivism (see ch. 14.1.3, knowledge and induction).

**inductivism**: the misconception that scientific theories are obtained by generalizing or extrapolating repeated experiences, and that the more often a theory is confirmed by observation the more likely it becomes

**induction**: the non-existent process of 'obtaining' referred to in inductivism

**principle of induction**: the idea that 'the future will resemble the past' combined with the misconception that this asserts anything about the future

Inductive inference (low-weak; high-strong) can be a stimulus for conjecture. Logical deduction provides the basis of explanation, knowledge.

There exists in all these understandings a deep hunger, an unfilled need, for a theory of weed science.

## U6.2 Representation, uncertainty and cultural relativity in weed biology

Knowledge of weeds requires consideration of the randomness, uncertainty and probability of complex systems in the way we represent their life history. The lack of information is uncertainty. Whether inductive inferences are justified, or under what conditions, is the problem of induction. Inductive inferences cannot be logically justified, they may be false with more observations. The deductive method of is based on observations of nature in which it is logically possible to conclusively decide truth and falsity. Mathematical probability definitions are not defined in terms of the probability of unknown events not previously observed. It is not possible to observe all conceivable events of a phenomenon. The scientist must reason from particular cases to wider generalities, a process of uncertain inference.

Statistics is the science of techniques to evaluate the uncertainty of inductive inferences. The normal distribution is the basis of most statistics. Much of biology is nonlinear: there is reason to suspect the additivity of treatment and environmental effects and the independent distribution of experimental errors with a common variance are false. Other distributions appropriate to represent complex natural phenomena include exponential distributions of events from Poisson processes, skewed distributions, and scalable distributions including those following power laws. These 'long-tailed' event probabilities require an appreciation of the role unpredictable, consequential events play in complex adaptive systems. Unpredicable events with significant consequences have been called 'Black Swans'.

There exist several different ways of describing, or representing, plant life history and population dynamics: ecological demography and the evolution of selectable life history characteristics of phenotypes in the deme, the local population. Experimentally they approach plant life history from very different perspectives.

The basis of most current weed models is quantitative and demographic: comparisons over years of the numbers and sizes of plants per unit area at different times of their life history, the fundamental unit of description of the organism. The demographic form of a life history model is quantifiable, with measurements of changes in phase state pool number and sizes with time, often expressed on a unit area basis. The phenotypic membership of the local deme and soil seed pool changes as natural selection favors some and eliminates others. The consequence of phenotypic plasticity is that plants growing under density stress typically have a skewed, nonlinear distribution.

Evolutionary models represent weed life history dynamics in a local population by capturing the physical and behavior information contained in functional traits of the individual weed phenotype that respond to specific environmental signals, opportunity spacetime, in a manner that optimizes their fitness in terms of survival and reproduction.

Representation of the nature of weed biology is confounded by the human scientists that make them, their beliefs, values and models. What we see, what we want to see, what is expected of us by our education and job, all these put limitations on our ability to observe, describe and represent. Human perception structures natural laws. For scientific history there are always conflicting views, uncertain outcomes, unpredictable developments; often due to the role of the emotions, the limits of imagination, the conservatism of institutions.

The socio-political cultural relativity of scientific rationality and model paradigms is revealed in the nature of scientific revolutions. Scientific theories, paradigms, tell scientists what nature contains and doesn't contain, which provides a map which scientists elucidate. When paradigms change, there are usually significant shifts in the criteria determining the legitimacy both of problems and of proposed solutions. A paradigm crisis can occur: contradictions between theory and observation accumulate. A new paradigm may appear that explains these anomalies. A new paradigm alone will not change accepted theory.

Change can be dificult for scientists, for humans, to do: belief and faith are broken.

## Chapter 14: Randomness, Probability and Inference in Weed Biology

"Today's statistician, applying the laws of chance, is able to place a precise and objective measure on the uncertainty of his inference." Steel and Torrie, 1960

**Summary**. Knowledge of weeds depends on the way we represent their life history. Making these inferences has relied on statistics that assume normal frequency distributions of observations, distributions that fail to consider randomness, uncertainty and probability in complex systems. Herein we consider aspects of randomness, uncertainty and inductive inferences relevant to any attempt to represent weed life history, ch. 15. The concept of random is complex, having several connotations depending on context. There is no functional difference between these types of randomness, we will never know enough to to make distinctions. The lack of information is uncertainty.

The question whether inductive inferences are justified, or under what conditions, is known as the *problem of induction*. Inductive inferences cannot be logically justified regardless of how many observations they are based on because any conclusion they lead us to may turn out to be false with yet more observations. The *deductive method* of critically testing theories based on observations of nature in the empirical sciences must be conclusively decidable with respect to their truth and falsity. To verify them and to falsify them must both be logically possible. If a theory withstands these tests and is not superseded by another theory we say it has been corroborated by past experience. The probability of an event is a measure of the likelihood or frequency that it will occur, from impossible to certainty. Mathematical probability definitions are not defined in terms of the probability of unknown events which have not been previously observed. It is not possible to observe all conceivable events of a phenomenon. The scientist must reason from particular cases to wider generalities, a process of uncertain inference.
This inductive process enables one to disprove incorrect hypotheses, but it does not permit one to prove hypotheses that are correct. Subjective interpretation of probability theory is indicated by the frequent use of psychologistic expressions: mathematical 'expectation' or 'normal' law of 'error'.

Statistics is the science, pure and applied, of creating, developing, and applying techniques such that the uncertainty of inductive inferences may be evaluated. The normal distribution is the basis of most statistics. ANOVA statistical techniques make several crucial assumptions: observations are drawn from normally distributed populations and errors are normally distributed within each treatment population and are independent of each other; observations are random samples from the populations and independence of error effects; and, the variances of the populations are homogeneous. There is often good reason to suspect that the additivity of treatment and environmental effects and the independent distribution of experimental errors with a common variance are false.

There exist other distributions more appropriate to represent complex natural phenomena. These include exponential distributions of events from Poisson processes, skewed distributions, and scalable distributions including those following power laws. Scalable phenomena and power law distributions are common in nature. These 'long-tailed' event probabilities require an appreciation of the role unpredictable, consequential events play in complex adaptive systems. There exists a fragility and vulnerability of the Gaussian in the estimation of tail events in which extreme deviations decrease at an increasing rate. Measures of uncertainty that are based on the bell curve simply disregard the possibility, and the impact, of sharp jumps or discontinuities and are therefore inapplicable in complex natural systems. Appreciation of the extreme event should be the starting point, not the exception to be ignored, in empirical studies. Unpredicable events with significant consequences have been called 'Black Swans'. The existence of Black Swan events raises questions about making inductive predictions based on historical observations. How can we know the future, given knowledge of the past? How can we determine properties of the (infinite) unknown based on the (finite) known? This is the problem of inductive knowledge (see U6.1, finding knowledge). These caveats provide the basis and setting to look at how weeds are represented in ch. 15.

## 14.1 Randomness, Uncertainty and Information

The ability of acquire knowledge of the nature of weeds is founded on how we represent weed life history. Making inferences about weed life history behavior, adaptation and evolution has historically relied on statistical models that assume normal frequency distributions of observations. The ability of Gaussian distributions to describe weed phenotype norms of reaction is mitigated by failing to consider the properties of randomness, uncertainty and probability in complex systems. Herein we consider aspects of randomness and event uncertainty, problematic inductive inferences compromising the ability to make inferences about future weed behavior. These considerations will frame any attempt to represent weed life history, ch. 15.

### 14.1.1 Randomness.

"... to come up with an appropriate definition [of randomness] one has no choice but to consider issues of perception and analysis." Wolfram, 2002

The concept of random is complex, having several connotations depending on context: non-order or non-coherence; a lack of predictability (statistics); random selection (evolution and games); epistemic randomness (knowledge); ontological randomness (complexity and chaos); and algorithmic randomness (information and computation). Clarity in understanding these matters can be provided by consideration of the various meanings attributed to the key concepts of randomness and probability.

**random (non-order):**
1: haphazard, no recognizable pattern (Lincoln et al., 1998)
2: lacking any definite plan or prearranged order; haphazard (Hanks et al., 1979)
3: having no definite aim or purpose; not sent or guided in a particular direction; made, done, occurring, etc., without method or conscious choice; haphazard (OED)
4: a non-order or non-coherence in a sequence of symbols or steps, such that there is no intelligible pattern or combination (Wikipedia, 3.12)

**random (unpredictable):**
1: (Statistics) having a value that cannot be determined before the value is taken, but only described probabilistically (Borowski and Borwein, 1991)
2: (Statistics) with an *a priori* probability of occurrence other than zero or unity (Lincoln et al., 1998)
3: a lack of predictability, but admitting regularities in the occurrences of events whose outcomes are not certain; situations in which the certainty of the outcome is at issue and notions of haphazardness are irrelevant (Wikipedia, 3.12)

**epistemic randomness (unknown):**

1: random because my knowledge about causation is incomplete, not necessarily because the process has truly unpredictable properties; uncertainty due to incomplete information; what I cannot guess (Taleb, 2010)
2: randomness is unknowledge, the world is opaque, appearances fool us (Taleb, 2010)
3: something in which none of our standard methods of perception and analysis can detect regularities; something in which no simple program can detect regularities (Wolfram, 2002)

**ontological randomness (ahistorical)**: the type of randomness where the future is not implied by the past (or not even implied by anything); with ontic uncertainty the future is created continuously by the complexity of ongoing actions; ontic uncertainty is much more fundamental than epistemic which comes from imperfections in knowledge (Taleb, 2010)

Algorithmic randomness, or an algorithmically random sequence is an infinite sequence of binary digits that appears random to any computable algorithm. There are three equivalent definitions of algorithmic randomness (Wikipedia, 3.12):

**algorithmic randomness (Kolmogorov complexity)** (Schnorr 1973, Levin 1973):
1: a random sequence is incompressible: no prefix can be produced by a program much shorter than the prefix; a measure of the computational resources needed to specify the object
2: a string (usually of bits) is Kolmogorov random if and only if it is shorter than any computer program that can produce that string; a random string is also an "incompressible" string, in the sense that it is impossible to give a representation of the string using a program whose length is shorter than the length of the string itself

**algorithmic randomness (constructive null covers)** (Martin-Löf 1966):
1: a random real number should not have any property that is "uncommon"
2: a random sequence is not contained in any constructive null cover set

**algorithmic randomness (constructive martingales)** (Schnorr 1971):
1: no effective procedure should be able to make money betting against a random sequence.
2: a martingale is a betting strategy; the martingale characterization says that no betting strategy implementable by any computer can make money betting on a random sequence.
3: in probability theory, a martingale is a model of a fair game where no knowledge of past events can help to predict future winnings; a martingale is a sequence of random variables (a stochastic process) for which, at a particular time in the realized sequence, the expectation of the next value in the sequence is equal to the present observed value even given knowledge of all prior observed values at a current time; martingales exclude the possibility of winning strategies based on game history, and thus they are a model of fair games.

A truly random system is in fact random (non-order) and does not have predictable properties (Taleb, 2010). A chaotic system has entirely predictable properties, but they either are hard to know or unknowable. Randomness can arise from lack of knowledge of system (epistemic, algorithmic) or because of the dynamic complexity of the system (ontological). There is no functional difference between these types of randomness, we will never know enough to to make distinctions.

**14.1.2 Uncertainty and information**. The lack of information is uncertainty. Randomness is a fundamental concept in making inferences about uncertain events in weed life history behavior, adaptation and evolution (see ch. 13.1.1 communication and information theory). Chance and accident are the opposite of rationality. They are not knowledge but an acknowledgement of its absence. Chance is a code word for saying there is too much conflicting data, too many variables for us to make sense of the whole. It is an admission that we cannot see the pattern, which is the opposite of randomness and noise (Bejan and Zane. 2012).

### 14.1.3 Knowledge and induction.

**14.1.3.1 The problem of induction: the White Swan.** The question whether inductive inferences are justified, or under what conditions, is known as the *problem of induction*, posed by Karl Popper in 1935 (2009). Observations of nature in the empirical sciences rely on *inductive methods* to generate knowledge. Inductive inferences are derived from particular, *singular statements* of results from observations and experiments. These singular statements pass to form *universal statements* (e.g. hypotheses, theories), potentially new knowledge. These universal statements cannot be logically justified or inferred from their singular statements, regardless of how many observations they are based on because any conclusion they lead us to may turn out to be false with yet more observations: "no matter how many instance of white swans we may have observed, this does not justify the conclusion that all swans are white".

**14.1.3.2 Deductive testing of theories.** The *deductive method* of critically testing theories based on observations of nature in the empirical sciences is deductive. Easily testable or applicable singular statements (predictions) are deduced from previously accepted statements of a theory. These and other derived statements are compared with the results of experiments and practical applications to obtain a decision about the theoretical prediction. If the conclusions reached about the singular statement are positive, verifiable, then the theory is acceptable, there is no reason to discard it. If the conclusions reached about the singular statement are negative, if it has been falsified, then this falsification also falsifies the theory from which the prediction has been logically deduced. As long as a theory withstands detailed and severe tests and is not superseded by another theory in the course of scientific progress we may say it has been *corroborated* by past experience. This deductive procedure does not resemble anything in the inductive procedure. All meaningful statements of empirical science "must be capable of being finally decided, with respect to their truth *and* falsity ... they must be *conclusively decidable* ... their form must be such that *to verify them and to falsify them* must both be logically possible."

### 14.2 The Probability of Events

"There is no other reason for introducing the concept of probability than the incompleteness of our knowledge." (Waismann, 1930)

**14.2.1 Probability.** The probability of an event is a measure of the likelihood or frequency that it will occur, from impossible to certainty:

**probability**:
1: (statistics) a measure of the relative frequency or likelihood of occurrence of an event; values lie between zero (impossibility) and one (certainty) and are derived from a theoretical distribution or from observations (Hanks et al., 1979)
2: the likelihood of a given event occurring (Daintith and Clark, 1999)
3: a measure or estimate of the degree of confidence one may have in the occurrence of an event, measured on a scale from 0 (impossibility) to 1 (certainty) (Borowski and Borwein, 1991)
4: (mathematics) the proportion of favourable outcomes to the total number of possible outcomes if these are indifferent (Borowski and Borwein, 1991)
5: the likelihood that events will occur in random, stochastic phenomena (Gullberg, 1997)

**14.2.2 Probability interpretations.** The term probability has been used in a variety of ways since its first use in games of chance (Wikipedia, 4.12).

"... to solve the following philosophical problem. *Is the degree of corroboration or acceptability of a theory a probability ... Does it obey the rules of the probability calculus?*" Popper, 2009

There are two broad categories of probability interpretations which can be called "physical" and "evidential" probabilities. We interpret the probability values of probability theory when we answer questions such as "does probability measure the real, physical tendency of something to occur?", or "is it just a measure of how strongly one believes it will occur?". Physical probabilities (objective or frequency probabilities) are associated with random physical systems such as roulette wheels, rolling dice and radioactive atoms. In such systems, a given type of event (e.g. the dice showing a six) tends to occur at a persistent rate, or "relative frequency", in a long run of trials. Physical probabilities either explain, or are invoked to explain, these stable frequencies. Thus talk about physical probability makes sense only when dealing with well-defined random experiments. Probability does not have a single definition. There exist several broad categories of probability interpretations, which possess different often conflicting views about the fundamental nature of probability.

**14.2.2.1 Frequency probability.** Frequency probability is the interpretation that defines an event's probability as the limit of its relative frequency in a large number of trials. The probability of a random event denotes the *relative frequency of occurrence* of an experiment's outcome, when repeating the experiment. This occurs only with experiments that are random and well-defined. Probability is the relative frequency "in the long run" of outcomes (Hacking, 1965).

**frequency probability**: the relative frequency of a certain class of occurrences within a certain other class (axiomatically, the probability of $x_1$ with regard to $x_2$) (Popper, 2009)

The set of all possible outcomes of a random experiment is called the sample space of the experiment. An event is defined as a particular subset of the sample space that you want to consider. For any event only one of two possibilities can happen; it occurs or it does not occur. The relative frequency of occurrence of an event, in a number of repetitions of the experiment, is a measure of the probability of that event. The *critical region* of a hypothesis test is the set of all outcomes which cause the null hypothesis to be rejected in favor of the alternative hypothesis.

An objective interpretation of probability, very closely related to frequency theory, is the propensity theory (Popper, 1959, 2009):

"... probability theory can best be interpreted as *a theory of the propensities of events* to turn out one way or another." "I regard the propensity interpretation as a conjecture about the structure or the world."

The propensity theory is one interpretation of the concept of probability. This interpretation views probability as a physical propensity, or disposition, or tendency of a given type of physical situation to yield an outcome of a certain kind, or to yield a long run relative frequency of such an outcome. This kind of objective probability is sometimes called *chance*. Propensities, or chances, are not relative frequencies, but purported *causes* of the observed stable relative frequencies. Propensities are invoked to *explain why* repeating a certain kind of experiment will generate a given outcome type at a persistent rate (Wikipedia, 4.12).

**14.2.2.2 Subjective probability.** Subjective probability is ordinarily used to describe an attitude of mind towards some proposition of whose truth we are not certain (Stuart and Ord, 2009). The proposition of interest is usually of the form "will a specific event occur?". The attitude of mind is of the form "how certain are we that the event will occur?". Numbers are assigned per subjective probability, i.e., as a degree of belief (de Finetti, 1970). The certainty adopted can be described in terms of a numerical measure. This number, between 0 and 1, is called probability (Jaynes, 1986). The higher the probability of an event, the more certain we are that the event will occur. Thus, probability in an applied sense is a measure of the confidence a person has that a (random) event will occur. (Wikipedia, 4.12)

Bayesian, epistemic (knowledge of causation), probability provides probability a subjective status by regarding it as a measure of the 'degree of belief' of the individual assessing the uncertainty of a particular situation.

**Baysian probability**: an abstract concept, a quantity that we assign theoretically, for the purpose of representing a state of knowledge, or that we calculate from previously assigned probabilities (Jaynes, 1986; Wikipedia, 4.12)

Epistemic or subjective probability is sometimes called *credence*, as opposed to the term *chance* for a propensity probability. Bayesian approaches include expert knowledge as well as experimental data to produce probabilities. The expert knowledge is represented by a prior probability distribution. The data is incorporated in a likelihood function. The product of the prior and the likelihood, normalized, results in a posterior probability distribution that incorporates all the information known to date (Hogg et al., 2004).

### 14.2.3 The fundamental problem of the theory of chance.

"It is unanimously agreed that statistics depends somehow on probability. But, as to what probability is and how it is connected with statistics, there has seldom been such complete disagreement and breakdown of communication since the Tower of Babel". (Savage, 1954)

**14.2.3.1 The logical problem of subjective probability.** When studying a phenomena, it is not possible to observe all conceivable events. Variation will exist among those observations since exact sources of cause and effect are unknown. The scientist must then reason from these particular cases to wider generalities, a process of uncertain inference. This inductive process enables one to disprove hypotheses that are incorrect, but it does not permit one to prove hypotheses that are correct (Popper, 2009). Observing only part of the possible information necessarily leads only to uncertain inference. The cause of the uncertainty arises because chance is involved in supplying the information. If observations fail to disprove a hypothesis, one may wonder if the theory does not embrace facts beyond the scope of inference of the experiment or if, with modification, it cannot be made to embrace such facts (Steel and Torrie, 1960). Application of probability theory to random' events is characterized by a peculiar kind of incalculability. After many unsuccessful attempts this can lead one to believe that all rational methods of prediction must fail in these cases: "only a prophet could predict them." (Popper, 2009). Subjective interpretation of probability statements about single events as indefinite predictions are not objectionable. They are confessions of deficient knowledge about particular events. They are not objectionable as long as it is clearly recognized that objective frequency statements are fundamental, since they alone are empirically testable. Interpretation of these formally singular subjective probability statements (indefinite predictions) are not statements about an objective state of affairs. They are only statements about their internal objective statistical state of affairs (Popper, 2009).

**14.2.3.2 Objective system of probability.**

"There does not exist a formal, objective mathematical system of probability for empirical science." (Popper, 2009)

Mathematical probability definitions are not defined in terms of the probability of unknown events which have not been previously observed. They are stated for cases of either the number of possible and equally likely outcomes; or, as probabilities of different possible events not already known but that have occurred in some other number of observation trials. There is a lack of a consistent definition of probability: "we still lack a satisfactory axiomatic system for the calculus of probability." (Popper, 2009):

**axiomatic probability**:
1: a study of *probability* in terms of *probability measure*
2: the formal study of random or chance events, usually in terms of probability measure, independent random variables and their generalizations (Borowski and Borwein, 1991)
3: a primitive statement of the rules of a *deductive formal system* for the symbolic calculation of probability

**probability measure**:
1: a mapping, P, of a sample space, X, into the interval [0, 1] subject to the conditions that P maps the whole space to 1; P is then the probability density function (a function representing the relative distribution of the frequency of a continuous random variable) (Borowski and Borwein, 1991)
2: a mapping of the entire sample space of the relative frequency distribution of a continuous random variable

**14.2.3.3 Subjective system of probability.** Subjective interpretation of probability theory is indicated by the frequent use of psychologistic expressions: mathematical 'expectation' or 'normal' law of 'error'. The degree of probability is treated as a measure of the feelings of certainty or uncertainty, of belief or doubt, which are aroused in the observer by certain assertions or conjectures.

> "Although probability statements play such a vitally important role in empirical science, they turn out to be in principle *impervious* to strict falsification." "How can we explain the fact that from incalculability – that is, from ignorance – we may draw conclusions which we can interpret as statements about empirical frequencies, and which we then find brilliantly corroborated in practice?" (Popper, 2009)

Subjective approaches are able to give consistent solutions to the problem of how to decide probability statements. But these statements are non-empirical, tautologies: propositions that are always true whether its component terms are true or false (something either is or isn't). Subjective probability generates non-empirical statements of events, tautologies that are immune to falsification. Paradoxically, this incalculability can lead us to conclude that probability calculations can be applied to these events. This paradox disappears when we accept the subjective theory of probability. The subjective theory is a method for carrying out logical transformation of what we already know; what we do not know. It is just when we lack knowledge that we carry out these transformations. This conception dissolves the paradox. But it does not explain how a statement of ignorance, interpreted as a frequency statement, can be empirically tested and corroborated.

**14.2.4 Frequency distributions of events.** Statistics is the science, pure and applied, of creating, developing, and applying techniques such that the uncertainty of inductive inferences may be evaluated (Steel and Torrie, 1960). One of the most important methods in empirical science analysis is to organize the data in a frequency distribution of the observed variables:

**frequency distribution**: (statistics) the function of the distribution of a sample that corresponds to the probability density function of the given underlying population and tends to it as the sample size increases; the set of relative frequencies of the sample points falling within given intervals of the range of the random variable (Borowski and Borwein, 1991)

**variable:**
1: characteristics which show variability or variation; chance variables, random variables; quantitative or qualitative; continuous or discrete (Steel and Torrie, 1980)
2: a property with respect to which individuals within a sample differ in some discernible way (Lincoln, et al., 1998)
3: an expression that can be assigned any of a set of values (Borowski and Borwein, 1991)
4: the actual property, character, measured by individual observations in an experiment (Sokal and Rohlf, 1981)
5: a changing quantity that might have any one of a range of possible values (Daintith and Clark, 1999)

**random variable:**
1: (Statistics) a variable that can assume any of a given set of values with assigned probability (Lincoln et al., 1998)
2: chance variable; *stochastic* variable; a quantity that can take any one of a number of unpredicted values. A *discrete random* variable has a definite set of possible values with corresponding probabilities. A *continuous random* variable can take any value over a continuous range, the probabilities of a particular value occurring is given by a probability density function. (Daintith and Clark, 1999)
3: a quantity that may take any of a range of values (either continuous or discrete) that cannot be predicted with certainty but only described probabilistically. Formally, a measurable function defined on a probability space and having range in the interval [0, 1] (Borowski and Borwein, 1991)
4: the mathematical representation of a variate associated with a stochastic phenomenon (Walker, 1988)

**independent variable:**
1: a variable that has no dependent relationship on another variable (Daintith and Clark, 1999)
2: a variable in a mathematical equation or statement whose value determines that of the dependent variable; the argument (Borowski and Borwein)
3: (Statistics) the variable that defines a set of distinct experimental conditions, or that an experimenter deliberately manipulates in order to observe its relationship with some other quantity; the predictor (Borowski and Borwein)

**dependent variable:**
1: a variable whose value depends on another chosen variable's value (Daintith and Clark, 1999)
2: a variable whose value is determined by that taken by the independent variables (Borowski and Borwein
3: (Statistics) a variable whose values are observed for different values of the independent variable; response variable; predicted variable (Borowski and Borwein)

Several types of statistics are calculated to measure the dispersion of observations around the sample mean:

**variance:**
1: a measure of the dispersion of a statistical sample (Daintith and Clark, 1999)
2: a measure of the dispersion of the distribution of a random variable, obtained by taking the expected value of the square of the difference between the random variable and its mean; the square of the standard deviation (Borowski and Borwein, 1991)

**standard deviation:**
1: a measure of the dispersion of a statistical sample, equal to the square root of the variance (Daintith and Clark, 1999)

2: a measure of the dispersion of a distribution, given by the square root of the variance; the unit whose multiples are used to describe the divergence of a random variable from its mean; the interval within one *sd* of the mean contains approximately 68% of the population (Borowski and Borwein, 1991)

**analysis of variance:**
1: any of a number of techniques for resolving the observed variance between sets of data into components, especially to determine whether the difference between two or more samples is explicable as random sampling variation with the same underlying population (Borowski and Borwein, 1991)
2: a statistical test whether two or more sample means could have been obtained from populations with the same parametric mean (Sokal and Rohlf, 1981)

Explaining random sampling variation with a population is facilitated by statistical test theory and the formulation of empirically testable null and alternative hypothesis. Analysis of variance provides an insight into the nature of variation of natural events and nature itself. However, it can create constructions of nature in the mind of the scientist that give rise to misleading or unproductive conclusions (Lewontin, 1974; Sokal and Rohlf, 1981). Statistically, these measures of uncertainty of events are termed 'errors':

**error:**
1: the uncertainty in a measurement or estimate of a quantity (Daintith and Clark, 1999)
2: the difference between some quantity and an approximation to or estimate of it, often expressed as an absolute or relative range (Borowski and Borwein, 1991)
3: variation produced by disturbing factors, both known and unknown; important effects man be wholly or partially obscured by experimental error; conversely, the experimenter may be misled into believing in effects that do not exist (Box, et al., 1978)

There are two basic types of error:

**random error:**
1: any deviation, the magnitude and direction of which cannot be predicted (Lincoln et al., 1998)
2: an error that occurs in any direction, cannot be predicted, and cannot be compensated for; it includes the limitations in the accuracy of the measuring instrument and the limitations in reading it (Daintith and Clark, 1999)

**systematic error:** error arising from the faults or changes in conditions that can be corrected for (Daintith and Clark, 1999)

How can you predict the qualities and quantities of random errors, especially in a complex system? Logical conclusions made from experimental observations can result two general classes of inductive errors:

**type I error:**
1: error of rejecting a true hypothesis (false positive) [but not explicitly stating it is false?]
2: false positive
3: a true null hypothesis was incorrectly rejected

**type II error:**
1: error of not rejecting a false hypothesis [ but not explicitly stating it is true?]
2: false negative

3: one fails to reject a false null hypothesis

In statistical test theory an unambiguous statement of a null hypothesis (a default state of nature, event) is compared to an alternative hypothesis (negation of the null hypothesis) using the concept of statistical error (Wikipedia, 4.12). The statistical test may negative, relative to null hypothesis (not observed) or positive (observed). If the result of the test corresponds with reality, then a correct decision has been made; if not, then an error has occurred. A type I error is a wrong decision made when a test rejects a true null hypothesis, an error of excessive scepticism. It is comparable with a so called false positive in other test situations. A type II error is a wrong decision made when a test fails to reject a false null hypothesis, an error of excessive credulity. It is comparable with a so called false negative in other test situations. A type I or type II error depends directly on the null hypothesis. Negation of the null hypothesis causes type I and type II errors to switch roles. Logically this can lead to four types of errors depending on the truth or falsity of the statement, hypothesis:

| ↓ TRUE/FALSE    OUTCOME → | Reject Null Hypothesis | Fail to Reject Null Hypothesis |
|---|---|---|
| True Null Hypothesis | false positive  *type I error* | true negative  *correct outcome* |
| False Null Hypothesis | true positive  *correct outcome* | false negative  *type II error* |

The goal of the test is to determine if the null hypothesis can be rejected. A statistical test can either reject (prove false) or fail to reject (fail to prove false) a null hypothesis, but never prove it true. Failing to reject a null hypothesis does not prove it true. In many practical applications type I errors are more delicate than type II errors. A false positive (with null hypothesis of no effect) in scientific research suggest an effect, which is not actually there, while a false negative fails to detect an effect that is there. The only way to minimize both types of error, without just improving the test, is to increase the sample size, and this may or may not be feasible.

In the following two sections are presented two differing types of frequency and probability distributions that may be utilized in making inferences about empirical events.

**14.2.5 Normal probability distributions and inference.**

"Most biological data when plotted in a frequency curve, closely fit a mathematically defined curve called the normal frequency curve." (Little and Hills, 1978)

"Many biological phenomena result in data which are distributed in a manner sufficiently normal that this distribution is the basis of much of the statistical theory used by the biologist. In fact, the same is true in many other fields of application." (Steel and Torrie, 1960)

The normal distribution is the basis of most statistics. Introductory statistics textbooks begin with the assumption that the normal distribution is the appropriate statistical model for representing most biological observations and for making inferences about knowledge-uncertainty.

**normal distribution**:
1: Gaussian distribution; the type of statistical distribution where the variation of a quantity $x$ about its mean value of the same measurement taken several times is entirely random; the graph of the normal distribution is bell-shaped and symmetrical about the mean. (Daintith and Clark, 1999)
2: Gaussian distribution; a distribution that is continuous and symmetrical with mean, median and mode coincident; many quantitative measurements appear to be approximately normally distributed; this is in part a consequence of the *central limit theorem* (Borowski and Borwein, 1991)

3: an exceptionally' thin tail distribution'

**central limit theorem**: if sufficiently many samples are successively drawn from any population, the sum or mean of the sample values can be thought of, approximately, as an outcome from a normally distributed random variable (Borowski and Borwein, 1991)

"If a random event occurs only about one time in twenty, we agree to label the event "unusual. ... this definition is easily the most common." (Steel and Torrie, 1960)

The Gaussian approach has much utility with variables for which there is a rational reason for the largest not to be too far away from the average. These include variables with physical limitations preventing large observations. They also include situations where there are strong forces of equilibrium bringing things back rather rapidly after conditions diverge from equilibrium (Taleb, 2010). Rare events in a complex biological population could raise doubts of the appropriateness of the central tendency theorem to represent many phenomena.

### 14.2.5.1 Assumptions in the analysis of variance of normal distributions.

"Readers not so inclined must take it on faith that the model postulated above, the result of many factors acting independently and additively, will approach normality. However, at the beginning of this procedure we made a number of severely limiting assumptions for the sake of simplicity." "Lifting these restrictions makes the assumptions leading to a normal distribution compatible with innumerable biological situations. It is therefore not surprising that so many biological variables are approximately normally distributed." (Sokal and Rohlf, 1981)

Analysis of variance (ANOVA) statistical techniques are used for resolving the observed variance between data sets into components. These statistical methods are especially useful in determining whether sample differences are explicable as random sampling variation within the same underlying population (Borowski and Borwein, 1991). ANOVA statistical techniques are fundamentally based on an assumption that observations are normally distributed.

Normal frequency distributions of observed events arise in nature from many single or composite factors; which occur independently of each other; and are independent in effect (their effects are additive); and these factors equally contribute to the variance (a measure of the dispersion of the distribution of a random variable) (Sokal and Rohlf, 1981). ANOVA statistical techniques make several crucial assumptions (Kirk, 1982):

1: Observations are drawn from normally distributed populations and errors are normally distributed within each treatment population and are independent of each other;
2: Observations are random samples from the populations and independence of error effects; and, the variances of the populations are homogeneous.

Assumptions central to the normal distributions, mild randomness, made in 'coin-flip' games are (Taleb, 2010):

1: That the coin flips are independent of each other, the coin has no memory;
2: There occur no wild jumps, no uncertainty to the step size in a random walk.

The step size is the building block of the basic random walk; and it is always known, only one step. If either of these two central assumptions are not met, your moves, coin tosses, will not lead to the bell curve; they can lead to wild style scale-invariant randomness (Mandelbrotian).

For the empirical scientists these assumptions of normally distributed events in nature present some fundamental problems. There is often good reason to suspect that the additivity of treatment and environmental effects and the independent distribution of experimental errors with a common variance are false (Cochran and Cox, 1957). When observations deviate from the normal distribution the discussion in most introductory statistics textbooks focus on mathematical remedies to make observations consistent with these assumptions. For example (Cochran and Cox, 1957):

"The same problem may arise because the experimental errors follow a distribution that is decidedly skew." "If the nature of the functional relationship is known, a transformation can be found that will …"

"The most serious disturbances appear to arise when the experimental error variance is not constant over all observations." "Where this type of disturbance is suspected, the remedy is to divide the error …"

The appropriateness of these mathematical transformations is problematic.

**14.2.5.2 Inference in normal distributions.** Most observations in normal, Gaussian distributions occur near the average. The odds of a deviation decline exponentially as you move away from that average: 68% of the observed events fall within one standard deviation of the mean. There exists a dramatic increase in the speed in decline in the odds as you move away from the average. This rapid decline in odds allows you to ignore outliers, outlying events in a normal distribution. Assuming normal distributions may not present problems if you're dealing with qualitative inference, yes/no answers to which magnitudes don't apply. The impact of the improbable cannot be too large. No single observation, but itself, can disrupt the overall findings. If you're dealing with aggregates, where magnitudes do matter, then you will have a problem and get the wrong distribution if you use the Gaussian, as it does not belong here; one single number, observation, can disrupt all your averages (Taleb, 2010).

There exists a need to know the bell curve intimately and identify where it can and cannot hold. Empirical natural scientists, users of the bell curve, need to justify its use, not the opposite. The Gaussian, and such as the Poisson law, are the only class of distributions that are sufficiently described by the standard deviation and the mean. Standard deviations are a measure of the degree of risk and randomness. Standard deviation is just a number that you scale things to, a matter of mere correspondence if phenomenon are Gaussian. Standard deviations do not exist outside the Gaussian; if they do exist they do not matter or explain much. Other statistical notions that do not have significance outside the Gaussian include *correlation* and *regression*. A single measure for randomness, standard deviation, cannot be used to describe risk. A simple characterization of uncertainty does not exist in complex natural systems. There is a need to understand randomness fully.

**14.2.6 Scalable probability distributions and inference.** Despite the ubiquity and emphasis placed on normal probability distributions in empirical sciences, there exist other distributions which may provide a more appropriate representation of complex natural phenomena. These include exponential distributions of events from Poisson processes, skewed distributions, and scalable distributions including those following power laws.

**14.2.6.1 Exponential frequency distributions.** The exponential distribution is a family of continuous probability distributions in probability theory and statistics. The exponential distribution is a member of a larger class of probability distributions that includes the normal distribution, binomial distribution, gamma distribution, Poisson distribution (Poisson, 1837), and many others. Exponential distributions describes the time between events in a Poisson process, i.e. a process in which events occur continuously and independently at a constant average rate. A Poisson process is a stochastic process which counts the number of events and the time that these events occur in a given time interval. The time between each pair of consecutive events has an exponential distribution. Each of these inter-arrival times is assumed to be independent of other inter-arrival times. In a stochastic or random process there is some indeterminacy. Even if the initial, starting condition is known, there are several (often infinite) directions in which the process may evolve. Contrasted to this, the exponential distribution describes the time for a continuous process to change state. In real-world scenarios, the assumption of a constant rate (or probability per unit time) is rarely satisfied (Wikipedia, 4.12).

**14.2.6.2 Skewed frequency distributions.**

"On the other hand, when we find departure from normality, this may indicate certain forces, such as selection, affecting the variable under study." "… skewness …" "… bimodality …" (Sokal and Rohlf, 1981)

Uncertainty in empirical science is found in the occurrence of rare events, confounding the representation of natural phenomena in a probability distribution. There is no reliable way to compute these small probabilities, the odds of rare events (Taleb, 2009). They can be described using measures called kurtosis and skew representing how 'fat' the tails of a distribution are, how much rare events play a role.

**kurtosis:**
1: (statistical) a measure of the concentration of a distribution about its mean (Hanks et al., 1979)
2: departed from a normal distribution; 'peakedness' of the distribution; *leptokurtic*, more observations near the mean and the tails and less in the intermediate regions relative to a normal distribution; *platykurtic*, fewer observations at the mean and the tails but with more in the intermediate regions relative to a normal distribution (Sokal and Rohlf, 1981)

**skewness:**
1: (statistical) a measure of the symmetry of a distribution about its mean (Hanks et al., 1979)
2: observed asymmetric frequency distribution departed from a normal distribution in which one tail of the distribution is more drawn out more than the other; mean and median do not coincide (Sokal and Rohlf, 1981)

In computing these small probabilities the sampling error is too large for any statistical inference about how non-Gaussian something is. If you miss a single number you miss the whole thing. The instability of kurtosis implies that a certain class of statistical measures should be totally disallowed: standard deviation, variance, least square deviation (Taleb, 2009).

**14.2.6.3 Scalable phenomena: fractal randomness.** Benoît Mandelbrot has argued that normal distributions do not properly capture empirical and "real world" distributions. There are other forms of randomness that can be used to model extreme changes in risk and randomness. He proposed seven states of randomness in probability theory, extending the concept of the normal distribution to include aspects of real world turbulence in complex systems. The seven states are: proper mild randomness; borderline mild randomness; slow randomness with finite delocalized moments; slow randomness with finite and localized moments; pre-wild randomness; wild randomness; and extreme randomness (Mandelbrot, 1997). He observed that randomness can become quite "wild" if the requirements regarding finite mean and variance are abandoned. Wild randomness corresponds to situations in which a single observation, or a particular outcome can impact the total in a very disproportionate way. Traditional normal distributions are at the mild end of the scale within this categorization. Wild randomness has been used in the analysis of turbulent situations (Holmes et al., 2008). Traditional "bell curves" are inadequate for measuring risk, such curves disregard the possibility of sharp jumps or discontinuities. Traditional approaches are based on random walks, contrasting with a world primarily driven by random jumps. Tools designed for random walks address the wrong problem (Mandelbrot and Taleb, 2006).

Mandelbrotian randomness is fractal randomness, uncertainty. It links randomness to the geometry of nature: fractal geometry. Fractality is the repetition of geometric patterns at different scales, revealing smaller and smaller versions of themselves. Small parts resemble, to some degree, the whole.

**fractal**:
1: geometry of the rough and broken; from Latin *fractus* (Taleb, 2009)
2: (Math) a set with non-integral Hausdorff dimension (Borowski and Borwein, 1991)
3: a curve or surface that has a fractal dimension and is formed by the limit of a series of successive operations; generated by an iterative process and that a small part of the figure contains the information that could produce the whole figure; self-similar (Daintith and Clark, 1999)
4: mathematical models for very irregular, very detailed sets, their degree of 'wildness' or 'roughness' characterized by fractal dimension (higher is rougher) (Stoyan and Stoyan, 1994)
5: (Math) a set such that two standard definitions of its dimension give different answers; a curve is one-dimensional according to one definition but a curve with arbitrarily small irregularies may have a larger dimension by another definition, and it is then called a fractal curve (Walker et al., 1998)

Significant differences in the definition of an objects dimension connote a high degree of structural complexity. There are many processes in nature forming structures in which, at decreasing scales, similar principles of formation persist (e.g. branching processes in biological growth occur stepwise in increasing detail). Parts of an object has similar properties as the whole, the property of scale invariance and self-similarity (Stoyan and Stoyan, 1994). Fractality is the repetition of geometric patterns at different scales, revealing smaller and smaller versions of themselves. Small parts resemble, to some degree, the whole (Taleb, 2009). There is no qualitative change when an object changes size.

Fractals generate pictures of ever increasing complexity by using a deceptively minuscule recursive rule; a rule that can be reapplied to itself infinitely. The fractal has numerical and statistical measures that are (somewhat) preserved across scales; the ratio is the same, unlike the Gaussian. The shapes are never the same, yet they bear an affinity to one another, a strong family resemblance. Self-affinity might be a more appropriate term for natural complexity than strictly mathematically self-similar. This character of self-affinity implies that one deceptively short and simple rule of iteration can be used, either by a computer or, more randomly, by Mother Nature, to build shapes of seemingly great complexity. It is how nature works.

### 14.2.6.4 Scalable phenomena: power laws in biology.

"The problem of explaining the observed statistical features of complex systems can be phrased mathematically as the problem of explaining the underlying power laws, and more specifically the values of the exponents." (Bak, 1996)

Fractals are characterized by power law distributions (Bak, 1996). Unlike the exponential decline from the mean found in normal distributions, the odds of a deviation from the average declines geometrically. The probabilities do not drop faster, they do not accelerate as you move away from the mean. It is impossible to use fractal distributions to provide acceptably precise probabilities because a very small change in the "tail exponent" coming from observation error would result in very large changes in probability. Fractal randomness is a way to reduce the surprise of rare, consequential events. But it does not yield precise answers, it is about unknown unknowns (Taleb, 2009).

A power law is a special kind of mathematical relationship between two quantities. When the frequency of an event varies as a power of some attribute of that event (e.g. the independent variable, size), the frequency is said to follow a power law (figure 14.1).

**Figure 14.1** A power-law graph, size versus frequency; left (A), few large sizes dominating the distribution (the '80-20' rule); right (B), the 'long tail' of increasingly rare event frequencies; areas of regions A and B are equal.

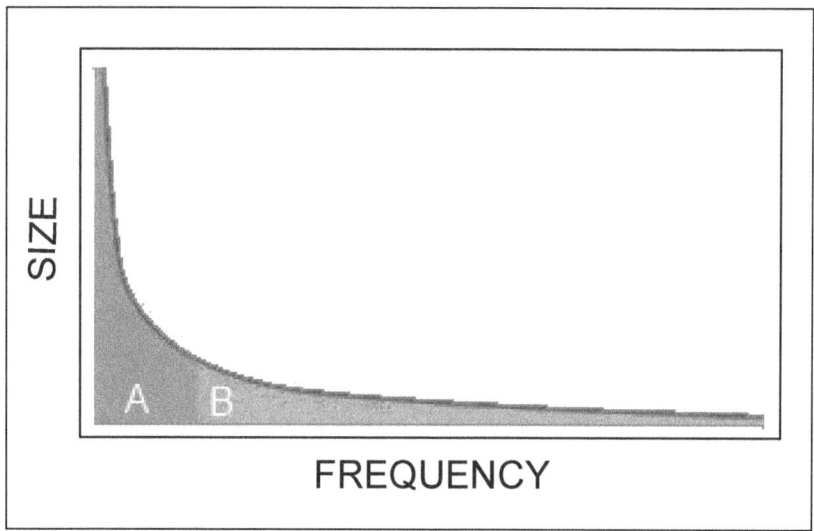

The distributions of a wide variety of physical, biological, and man-made phenomena follow a power law, including the foraging pattern of various species, the species richness in clades of organisms, and many other quantities. Skewed distributions dominated by the few are common in biology (see ch. 9.4.3.1, plant size hierarchies in the community; ch. 11.1.9, hierarchy: the treelike structure of flow).

In "long-tailed" distributions a high-frequency or high-amplitude population is followed by a low-frequency or low-amplitude population which gradually "tails off" asymptotically. The events at the far end of the tail have a very low probability of occurrence. 'Fat-tailed', as well as 'heavy-tailed" and 'long-tailed', distributions are probability distributions that exhibit extremely large skewness or kurtosis. Long-tail distortions arise with the inclusion of some unusually high (or low) values which increase (decrease) the mean, skewing the distribution to the right (or left). In fat-tailed probability distributions a larger share of population exists within the tail than observed under a "normal" or Gaussian distribution. The ubiquitous normal distribution possesses an exceptionally thin-tailed distribution (as well as to the exponential distribution). The tail of a fat-tail distributions has a power law decay, but do not necessarily follow a power law at all scales. The Pareto principle (the 80–20 rule; the law of the vital few; the principle of factor sparsity) states that, for many events, roughly 80% of the effects come from 20% of the causes. For such population distributions the majority of occurrences are accounted for by a minority of items in the distribution. What is unusual about a long-tailed distribution is that the most frequent events represent less than 50% of occurrences; the least frequent are more important as a proportion of the total population. Power law distributions or functions characterize an important number of natural behaviors, arousing scientific interest in such distributions, and the relationships that create them (Wikipedia, 4.12).

The relationships among different components and processes in a physical or biological system must be adjusted as size changes so that the organism can continue to function. Many anatomical and physiological attributes of organisms change with size so that they remain self-similar. These relationships among critical structural and functional variables are maintained over a wide range of scales (many orders of magnitude). Such self-similarity is fractal. The relationships among the variables can be described by a fractal dimension or a power function (Brown et al., 2000). The use of power functions to characterize scaling laws, allometric equations, has a venerable and well established history (e.g. Thompson, 1917). They take the form of equation 1:

$$Y = Y_0 M^b \qquad \text{(Eq. 1)}$$

Where: Y = some dependent variable; $Y_0$ = some normalization constant; M = some independent variable (e.g. body mass); b = the scaling exponent. Biological scaling relationships are called allometric, because the exponent, b, typically differs from unity. If b = 1, the relationship is called isometric (plots as a straight line on linear and logarithmic axes; e.g. change in independent variable body mass linear with change in dependent variable). If b ≠ 1, the relationship is called allometric (it plots as a curve on linear axes, but plots linear on logarithmic axes; therefore: b is the slope of the linear plot on logarithmic axes) (Brown et al., 2000).

An example of allometry in geometric scaling is found in spheres (or any object of self-similar shape) in which one can describe changes as a function of a linear dimension. These equations can also be applied to biological organisms. It might be expected that organisms of the same general body plan would scale geometrically, e.g. the of size-nested Russian Matryoshka wooden dolls. If organisms preserve self-similar shapes as they vary in size, then their linear dimensions should vary as the 1/3, and their surface areas as the 2/3, powers of their body mass. But, biological organisms usually do not scale geometrically. There exist powerful constraints on structure and function that do not allow organisms to main the same geometric relationships among their components as size changes over several orders of magnitude. For example, as tree size increases the stem cross-sectional and leaf surface area increase more rapidly ($M^{3/4}$) than expected with geometric scaling ($M^{2/3}$): increased mechanical trunk resistance to stress; greater laminar gas exchange to support greater biomass (Brown et al., 2000).

Organismal structure-function relationships cannot simply be scaled geometrically, with exponents that are simple multiples of 1/3. But, quarter-power scaling relationships in biology are ubiquitous. Because biological time scales as $M^{1/4}$, most biological rate processes scale as the reciprocal of time, $M^{-1/4}$. Two reasons are proposed why biological scaling exponents characteristically take on values that are simple multiples of ¼. Together they suggest the existence of a common scaling mechanism linking most anatomical, physiological, and ecological characteristics of organisms so that their allometric exponents are predicted to be simple multiples of ¼. The first reason is the fundamental importance of metabolic rate. Metabolic rate limits all biological processes at all levels of organization from molecules to cells to individuals and populations. The rates of uptake and release of all metabolites are characterized by similar scaling exponents. As such, many biological rate processes scale as $M^{-1/4}$, and most biological times scale as $M^{1/4}$. The second reason is that even the simplest organism is an extremely complex system. It depends on the integrated performance of its component structures and functions for its existence and reproduction. These structures and functions, and their integration, must be preserved with narrow limits as organisms vary in size. The component systems cannot be optimized separately. Because of their interdependence, all must be balanced with each other and scaled in some integrated way. Biological scaling relationships are manifestations of a single underlying scaling process based on quarter powers and to be unique to living things (Brown et al., 2000).

The problems associated with functioning over a wide range of sizes has led to the evolution of a three-part solution in biological design. First, some of the most basic component structures and functions are invariant (invariant quantities; see ch. 7.4.4.5, seed size plasticity and stability). Second, others are constrained to obey a common set of scaling laws (interrelated scaling exponents). Third, natural selection imposes an economy of design (symmorphosis) influencing the coevolution of all traits. These three principles are fundamental to scaling the anatomical, physiological, and ecological characteristics of organisms.

Biologists traditionally study scaling relationships at three levels: within individual organisms, among different individual organisms of varying size, and within assemblages of multiple individuals or species of organisms. For molecules, macromolecules and cells to function as an integrated whole within an individual organism these invariant structural units must be connected by transport, communication, protection, and structural support systems. Natural selection for efficient design of these systems has resulted in the evolution of networks with self-similar, hierarchically scaled architectures (e.g. plant branching patterns). Allometry is applied over many orders of magnitude in body size variation among individual organisms. Phenotypic plant plasticity encompasses many of these growth phenomena (see ch. 4.3.1). Scaling relationships expressed at levels of biological organization greater than individual organisms includes populations, species-groups, communities (assemblages of multiple species). Few have been reported, but these hint at potentially important features of ecology and evolution. The best example is may be found in Yoda et al., 1963. In this work a plant 'thinning' law describes the relationship between individual plant size and density (plant number per unit area) in single- or mixed-species stands. It presents a geometric -3/2 power law relating the spacing of competing plants to the volumes their leaf canopies occupy. Other work indicates a non-geometric, -4/3 power law, may fit better than Yoda's -3/2 geometric law.

Mathematical scaling laws may be able to describe how the sizes and shapes of structures and the rates of processes vary with body size. These scaling laws often account a large proportion of the variation in attributes when organisms vary in size over many orders of magnitude. The geometric, physical, and biological constraints that organisms must obey, as well as the opportunities that organisms have exploited leading to diverse size, are reflected in these scaling laws. The allometric equations used to characterize these scaling laws are power functions in which the magnitude of a trait is expressed as a function of body mass raised to some exponent. A unique, pervasive feature of biological scaling relationships is that nearly all of them are fit by allometric equations which have exponents that are multiples of ¼. A common basis for these biological scaling laws explaining their origin is suggested by so many biological attributes that scale as quarter powers of body mass. The origin of this commonality is based on three apparently universal principles of biological scaling: an integrated set of scaling laws with quarter-power exponents of body mass; some invariant quantities that do not scale with body size ($M^0$); and, natural selection imposes an economy of design so that the magnitudes of structures and functions tend to just meet maximum demands (Brown et al., 2000).

## 14.3 The Structure of Randomness and Probability

Scalable phenomena and power law distributions are common in nature. These 'long-tailed' event probabilities require an appreciation of the role unpredictable, consequential events play in complex adaptive systems.

### 14.3.1 Unpredictable, consequential events: The Black Swan.

"No matter how many instance of white swans we may have observed, this does not justify the conclusion that all swans are white" (Popper, 2009)

There exists a fragility and vulnerability of the Gaussian in the estimation of tail events in which extreme deviations decrease at an increasing rate. Measures of uncertainty that are based on the bell curve simply disregard the possibility, and the impact, of sharp jumps or discontinuities and are therefore inapplicable in complex natural systems. Unpredictable, large deviations cannot be dismissed because they are rare; cumulatively, the impact of outliers can be dramatic. The world is dominated by these extreme, unknown, and very improbable events. As such, appreciation of the extreme event should be the starting point, not the exception to be ignored, in empirical studies. Unpredicable events with significant consequences have been called 'Black Swans' (Taleb, 2010), and event with three attributes:

**Black Swan event:**
1. unpredictability; it is an outlier, it lies outside the realm of regular expectations because nothing in the past can convincingly point to its possibility
2. consequences; it carries and extreme impact
3. retrospective explainability; in spite of its outlier status, human nature makes us concoct explanations for its occurrence after the fact, making it explainable and predictable

The Black Swan is about consequential epistemic limitations, both psychological (hubris and biases) and philosophical (mathematical) limits to knowledge. Something that is not expected is always something consequential and unpredictable by a particular observer: empirical relativity.

The existence of Black Swan events raises questions about making inductive predictions based on historical observations. How can we know the future, given knowledge of the past? How can we determine properties of the (infinite) unknown based on the (finite) known? This is the problem of inductive knowledge, the white swan (Popper, 2009).

"It is perfectly possible to calculate fitness for any given strategy using the method of "forward iteration". "The problem is that the total number of possible strategies is enormous, so enormous that this approach is not feasible even with a large computer. The backward method, by contrast is extremely efficient. … the entire computation for the problem described here takes only a second or two on a typical desktop computer. Backward iteration [calculating fitness by working backwards in time] is fast, efficient and accurate." (Clark and Mangel, 2000)

Naively mistaking past observations as something definitive or representative of the future is the one and only cause of our inability to understand the Black Swan. A series of corroborative facts is not necessarily evidence. Negative instances get closer to the truth, not by verification. It is misleading to build a general rule from observed facts. Popper (2009) generated a large-scale theory around the asymmetry of having a large quantity of corroborative instances and a small amount of negative instances. The theory was based on an empirical and logical technique called 'falsification', to prove wrong). You know what is wrong with a lot more confidence than you know what is right. Popper proposed making conjectures and refutations: you formulate a (bold) conjecture and seek observation that would prove you wrong (a refutation). Frequently scientists only search for confirmatory instances (the confirmation bias) (Taleb, 2010). Ontological randomness is where the future is not implied by the past, it is not implied by anything, it is ahistorical. With ontic uncertainty the future is created continuously by the complexity of ongoing actions. Ontic uncertainty is much more fundamental than epistemic which comes from imperfections in knowledge (Taleb, 2010).

The more random information is, the greater its dimensionality, and the more difficult to summarize. 'Kolmogorov complexity' defines the degree of randomness, how much pattern we see in information. The more you summarize, the more order you bring, the less randomness. Simplification leads us to see the world as less random than it actually is. Black Swan events are what we leave out of this process of simplification.

**14.3.2 The domain of unpredictable, consequential events**. Are all natural systems capable of generating Black Swans, or are there particular domains in which complex or simple events are more probable? Taleb in his book on Black Swans (2009) defined two conceptual domains in which either the simple or the complex were more likely to occur: extremistan and mediocristan; the domains of the extreme and routine. Understanding these two domains provides crucial insight into how nature works, about how to understand and study complex adaptive systems such as weeds. These two domains share much with what was discussed previously of normal (Gaussian), as opposed to scalable (power law dominated), event probability distributions.

Extremistan produces Black Swans, the domain of extreme events. It is the province where the total can be conceivably impacted by a single event. In this domain inequalities exist such that one single observation can disproportionately impact the aggregate, or the total: the tyranny of the singular, the accidental, the unseen, and the unpredicted must be endured. Large disturbances are everything in Extremistan. Probability distributions of events are fractal, scalable, dominated by power laws; corresponding generally to numbers. In extremistan, matters are subjected to wild/superwild type II Mandelbotian randomness. In this scalable domain, an arbitrarily large number is possible; inequalities should not stop above some know maximum bound. Inferences can be made about things that do not appear in data, these things still are possible. Event transitions into crises and contagions are unpredictable. In this complex domain traditional statistics is not applicable, appropriate. Statistical methods applicatble to normal distributions in mediocristan are too often used inappropriately for events in extremistan as an 'approximation'. Statistics too readily allows 'absence of proof' to be mistaken with 'proof of absence'. With this asymmetric logic, one exception allows rejection of the Gaussian, but millions of observations do not fully confirm their validity. The Gaussian disallows large deviations. The tools of extremistan do not disallow long quiet stretches, long-tailed distributions, anticipating those rare consequential events.

Mediocristan is dominated by the mediocre, with few extreme successes or failures; the domain of the routine. The aggregate is not meaningfully affected by a single event: the tyranny of the collective, the routine, the obvious, and the predicted must be endured. No single instance will significantly change the aggregate or the total when the sample is large (central limit theorem). The probability of events are non-scalable, events distributed by Gaussian bell curve functions: corresponding generally to physical quantities. In mediocristan, matters subjected to mild type I Mandelbotian randomness. The bell curve is grounded in mediocristan. There is a qualitative difference between Gaussian probability and scalable laws.

Classical thermodynamics produces Gaussian variations in mediocristan. Informational variations and informational randomness are from extremistan. In statistics empirical scientist are confronted with differing approaches to randomness and uncertainty. The pervasiveness of Gaussian thinking in statistical models leads inexorably to mediocristan as a starting point. The assumption that ordinary fluctuations are the dominant source of randomness, therefore probabilities are easily computable. Embracing nature as extremistan recognizes Black Swans as the dominant source of randomness, probabilities are not easily computable.

The more random information is, the greater its dimensionality, and the more difficult to summarize. 'Kolmogorov complexity' defines the degree of randomness, how much pattern we see in information. The more you summarize, the more order you bring, the less randomness. Simplification leads us to see the world as less random than it actually is. Black Swan events are what we leave out of this process of simplification.

"In addition to independent and dependent variables, all experiments include one or more nuisance variables ... undesired sources of variation in an experiment that may affect the dependent variable ... the effects of nuisance variable are of no interest per se." (Kirk, 1982)

There is a strong impulse in empirical studies to simplify, to generalize reality and make it easily understood. The greek philosopher Plato saw nature in idealized form, the idealized representation of nature. With this view of the world the ideal Platonic representation enters into contact with messy complex reality, and the side effects become apparent in our models of nature. The gap between what you know and what you think you know becomes dangerously wide; it is where the rare and consequential is produced. This gap between our empirical idealizations and messy, Black Swan-ridden, complexity is the 'Platonic fold' (Taleb, 2009).

Platonicity focuses on pure, well-defined, and easily discernible objects at the cost of ignoring those objects of seemingly messier and less tractable structures. It prefers crisp constructs and ideas over less elegant objects. It encourages us to think we know more than we do, but in specific unpredictable aspects of our understanding. To idealize in this way is to mistake the map for the territory. It is not known beforehand where the map wrongly represents the territory, which can lead to severe consequences. It is thinking that is top-down, formulaic, closed-minded, commoditized. Thinking that is a-Platonic is bottom-up, open-minded, skeptical, empirical. An excellent example in weed biology is the concept of species. Taxonomy and phylogeny enable large perspectives on community dynamics. Probing beneath the seemingly monolithic and homogeneous surface of a particular weed species reveals large areas of heterogeneity and imprecision.

"... "all models are wrong, but some are useful" reveals not understanding that the real problem is that "some are harmful" ..." (Taleb, 2010)

"Taleb: Assume that a coin is fair, i.e., has an equal probability of coming up heads or tails when flipped. I flip it 99 times and get heads each time. What are the odds of my getting tails on my next throw?

Statistician: Trivial question. One half, or course, since you are assuming 50% odds for each and independence between draws.

Taleb: What do you say Tony?

Fat Tony: I'd say no more than 1%, of course.

Taleb: Why so? I gave you the initial assumption of a fair coin, meaning that it was 50% either way.

Fat Tony: You are either full of crap or a pure sucker to buy that "50 pehcent" business. The coin gotta be loaded. It can't be a fair game.

(Translation [from Brooklynese]: It is far more likely that your assumptions about the fairness are wrong than the coin delivering 99 heads in 99 throws.)

Taleb: But the statistician said 50%.

Fat Tony (whispering in his ear): I know these guys with the nerd examples from the bank days. They think way too slow. And they are too commoditized. You can take them for a ride." (Taleb, 2010)

The 'ludic fallacy' (*ludus*, latin for games) is the manifestation of the Platonic fallacy in the study of uncertainty. It arises when studies of chance are based on the narrow world of games and dice. A-Platonic randomness has an additional layer of uncertainty concerning the rules of the game in real life. The bell curve (Gaussian) is the application of the ludic fallacy to randomness. The attributes of real life uncertainty have little connection to the idealized ones encountered in exams and games, they ignore a crucial layer of uncertainty. In real life you do not know the odds; you need to discover them, and the sources of uncertainty are not defined.

**14.3.3 Representation of events in the unpredictable, consequential domain.** Predicting uncertain events, and the consequences of decisions guided by those forecasts, can be roughly categorized into four event domains (quadrants) (Taleb, 2010). A prediction map of the four quadrants of uncertain risk, and their consequences, can be made as a function of the complexity of the system in which the events occur (table 14.1).

**Table 14.1.** Four event domains (quadrants) of uncertain risk and consequences of decisions as a function of system complexity.

|  |  | Event Risk & Decision Consequences | |
|---|---|---|---|
|  |  | Simple (binary) | Complex (open-ended, variable) |
|  |  | Simple Payoffs, Consequences | Complex Payoffs, Consequences |
| System Event Complexity | Mediocristan: the collective, routine, obvious; mild randomness | 1st Quadrant: Extremely Safe | 2nd Quadrant: (Sort of) Safe |
| | Extremistan: the complex, fractal; wild randomness | 3rd Quadrant: Safe | 4th Quadrant: Black Swan Domain |

Depending on the complexity of the system in which an event occurs, and the risk associated with those domains, what consequences will model predictions yield? Where will we get hurt by what we don't know? Can systematic limits be set on the fragility of our knowledge, the exact location where these maps no longer yield useful predictions? Can decisions be categorized based on the severity of the potential estimation error: event probability times consequences equals the payoff of decisions? Where will the most harmful errors using models be more severe?

Only one *a priori* assumption need be made. When you look at the generator of events you can tell which environment can deliver large events (extremistan) and which cannot (mediocristan). Two types of decisions emerge from events generated in these domains. The first decision type is simple, binary exposure: is something true or false? They do not depend on high-impact events as their payoff is limited. Binary outcomes are not common, in life payoffs are usually open-ended, variable. The second decision type is more complex and more open-ended. Event frequency or probability are important; but the impact, and the uncertainty of impact, are also crucial.

In the first quadrant simple binary payoffs are generated in mediocristan (e.g. controlled laboratory experiments; games). Predictive models work and forecasting is safe. In the second quadrant complex payoffs are generated in mediocristan. Statistical methods work with some risk. In the third quadrant simple payoffs are generated in extremistan. Little harm results from being wrong because the possibility of extreme events does not impact those payoffs. In the fourth quadrant complex payoffs are generated in extremistan. Problems and opportunities reside in this domain: exposure to positive or negative Black Swans. The consequences or rarer events are more difficult to predict than more frequent ones. Randomness lacks structure. In the first three quadrants you can rely on the best models or theories; doing so in the fourth quadrant is dangerous: no theory or model is just any theory or model. In the fourth quadrant the difference between absence of evidence and evidence of absence becomes acute. Statistical metrics (standard deviation, linear regression, least squares regression) in the fourth quadrant are unstable and do not measure anything. They underestimate the total possible error, and thus overestimate what knowledge one can derive from the data. The fourth quadrant is the domain of Black Swans.

## 14.4 Conclusions

I was misled, we all were profoundly misled, in our college statistics courses to believe that what we were learning would reveal the truth of natural phenomena by using statistically 'normal' probability distributions and statistical tests. They do not, they are tautologies of our own preconceived prejudices, they are logically circular going nowhere. Too little time was spent on violations of assumptions underlying analysis of variance, that weed biology was complex, nonlinear. Evolution, natural selection changes the local deme continuously, inherently violating the covariance structure of your study population. Demographic assumptions of normally distributed phenomena are too often wrong due to phenotypic plasticity and somatic polymorphism, natural selection and elimination, scalable phenomena where power laws dominate. These caveats provide the basis and setting to look at how weeds are represented in ch. 15.

## Chapter 15: Represention of Weed Life History

**Summary.** There exist several different ways of describing, or representing, plant life history and population dynamics: ecological demography and the evolution of selectable life history characteristics of phenotypes in the deme, the local population. Experimentally they approach plant life history from very different perspectives.

The basis of most current weed models is quantitative and demographic: comparisons over years of the numbers and sizes of plants per unit area at different times of their life history, the fundamental unit of description of the organism. The demographic form of a life history model is quantifiable, with measurements of changes in phase state pool number and sizes with time, often expressed on a unit area basis. What this model does not contain are the deterministic biological processes that drive growth and development during life history. These uncharacterized processes are represented as transitions between quantitative state pools. The demographic representation of weed population dynamics is an incomplete abstraction because it ignores the importance of phenotypic variation by averaging behaviors at experimentally convenient times in life history. Measurement of quantities and sizes of uncharacterized phenotypes, and the uncharacterized processes of transitions between life history states, provide little inherent inference of population dynamics. An artifact of demographic representations of populations arises from the changing phenotypic structure of the local community with time. The phenotypic membership of the local deme and soil seed pool changes as natural selection favors some and eliminates others. Natural selection violates this covariance structure by assuming the individual phenotypes are the same at each life history measurement time in the local habitat. This plasticity of form confounds the ability of a purely demographic model to make predictions of population growth rates, biomass and even productivity. The consequence of phenotypic plasticity is that plants growing under density stress typically have a skewed, nonlinear distribution.

Evolutionary models represent weed life history dynamics in a local population by capturing the physical and behavior information contained in functional traits of the individual weed phenotype that respond to specific environmental signals, opportunity spacetime, in a manner that optimizes their fitness in terms of survival and reproduction.

### 15.1 Representing Weed Life History

There exist several different ways of describing, or representing, plant life history and population dynamics: ecological demography (ch. 15.2); and the evolution of selectable life history characteristics of phenotypes in the deme, the local population (ch. 15.3, Dekker, 2011a; ch. 13, Dekker, 2013; ch. 17, Dekker, 2014). Experimentally they approach plant life history from very different perspectives.

Survival, growth and reproduction have been represented in several different ways. The precondition to natural selection is the formation of a local population (deme) in which excess individuals compete for limited local opportunity spacetime. Population growth in a limited environment can be represented by the Verhulst-Pearl logistic equation (see ch. 2.4.1, precondition to natural selection, p.65). Survival and reproduction are represented in demography by the net reproductive rate ($R_o$), the summation of reproduction and survival/mortality (see ch. 5.2.1, demographic survival and reproduction, p.114; ch. 14.2.2, demographic weed life history population dynamics models, p.311). It is one of several ways of viewing survival and reproduction and is widely used in ecology studies (e.g. Silvertown & Doust, 1993, Ch. 9). Population growth and size in a local space obey the law of constant final yield (biomass, seed weight) in which the total plant yield of a high density population in a specified area tends to approach a constant weight (see ch. 9.4.2, plant density and productivity per unit area, p. 214). Much confusion has resulted from the application of these types of growth models to plants (Harper, 1977; see ch. 14.2.2, Demographic weed life history population dynamics models, p. 311). These confusions are systematically explored in this chapter.

The weedy *Setaria* phenotype can be described by its structure and behavior, the seed life history represented as a complex adaptive system. The weedy phenotype can first be described in terms of the structure of its genotypes, the genetic structure of the local population (deme), and morphology. The genetic and morphological structure provides a long-term foundation for the evolving story of life history behavioral adaptation: plant functions, functional traits and regulation of function. The structural and behavioral phenotype can be formally represented as behavioral regulation by morpho-physiological seed compartments, by seed heteroblasty, by environmental seed-soil signal transduction.

Conceptually the study of weed life history will be much improved when a true synthesis in which numerical demography is firmly grounded in a trait-based study of life history behavioral evolution. This then could lead to experimental methodologies able to explain the ecological evolution of weed life histories (Dekker, 2009).

## 15.2 The ecological demography of plant population life history dynamics

"The existence of two levels of population structure in plants makes for difficulties, but the problems are much greater if their existence is ignored. One of the strongest reasons why a population biology of plants failed to develop alongside that of animals was that counting plants gives so much less information than counting animals. A count of the number of rabibits or *Drosophila* or voles or flour beetles gives a lot of information: it permits *rough* predictions of population growth rates, biomass and even productivity. A count of the number of plants in an area gives extraordinarily little information unless we are also told their size. Individual plants are so "plastic" that variations of 50,000-fold in weight or reproductive capacity are easily found in individuals of the same age. Clearly, counting plants is not enough to give a basis for a useful demography. The plasticity of plants lies, howerever, almost entirely in variations of the number of their parts. The other closely related reason why plant demography has been slow to develop is that the clonal spread of plants and the break-up of old clones often makes it impossible to count the number of genetic individuals.

The problems are great: they can be regarded as insoluable and a demography of plants an unattainable ideal, or they can be ignored with a certainty of serious misinterpretation, or they can be grasped and methods, albeit crude, developed to handle the problem.

The way in which the problem can be faced is to accept that there are two levels of population structure in plant communities. One level is described by the number of individuals present that are represented by original zygotes (genets, e.g. seedling, clone; Kays and Harper, 1974). Such units represent independent colonizations. Each genet is composed of modular units of construction – the convenient unit may be a shoot of a tree, a ramet of a clone, the tiller of a grass or the leaf with its bud in an annual."
(Harper, 1977; pp. 25-26)

Weedy and invasive plants perform the plant colonization niche. Weedy plants are the first to seize and exploit the opportunity spacetime created by human disturbance, notably in resource-rich agricultural cropping systems. The urge to understand and predict weed life history behavior with time has provided a strong scientific and practical motivation for the development of these models. Weed models are tools with the potential to provide improved scientific understanding of changing weed populations, including insights into the biological functioning of these plants, and prediction of future life history population dynamics. Weed modeling can also provide practical support for crop management decision making, including evaluation of weed management tactics and strategy, risk, economics and efficacy. Modeling can also be a less expensive means of providing information compared to that of field experimentation. Much progress has been made to realize the potential of weed modeling, but much remains undone.

The basis of most current weed models is quantitative and demographic: comparisons over years of the numbers and sizes of plants per unit area at different times of their life history. The current state of affairs, including limitations of current demographic models, have been featured in two reviews (Holst et al., 2007; Freckleton and Stephens, 2009).

The opportunities and limitations of models arise from the manner in which weeds and their life histories are represented, the inferences that can be derived from the informational content of the models, and the consequences of these factors on the ability of the model to predict future behavior. The purpose of these sections is to assess the limits and potentials of two different, but compatible, types of weed population dynamics models: demographic, and those based on functional phenotypic traits and the biological processes of natural selection, elimination and evolutionary adaptation. Models of both types are assessed in terms of how they represent weed life histories as well as ability to infer and predict future behavior based on their inherent informational content.

**15.2.1 Weed life history models.** A model is a representation of reality. It is inherently an abstraction and a simplification. It is a conceptual framework of a system constructed by indicating which elements should be represented and how these elements are interrelated. This conceptual framework then is translated into algorithms, precisely defined step-by-step procedures by which dynamics are carried out. What elements should be represented in a weed population dynamics model? The first, most important element is a group of plants of one weed species occupying a local habitat. This population is usually isolated to some degree from other populations, but local populations over spatial scales of landscape to global interact (e.g. gene flow) with each other to form metapopulations. It is the local population, the deme, that is the unit of evolution. Populations change with time. Population dynamics are changes in the quality and quantity of member phenotypes, as well as the biological and environmental processes influencing those changes.

What conceptual framework can best represent the interrelationships among members of a population? One crucial component of any conceptual framework of weed models is the life history of the weed species: "The life cycle is the fundamental unit of description of the organism." (Caswell, 2001). Holst et al. (2007) also conclude that "Almost all models consist of a number of life cycle stages, nearly always including at least seeds, seedlings and mature plants." (figure 15.1).

**Figure 15.1** Representation of the an annual weed species life history by discrete life history phases, or states (mature plant, new seeds, seed bank, seedlings) and the metrics used for their measurement (number, size); arrows represent transitions between states; redrawn from Holst et al. 2007.

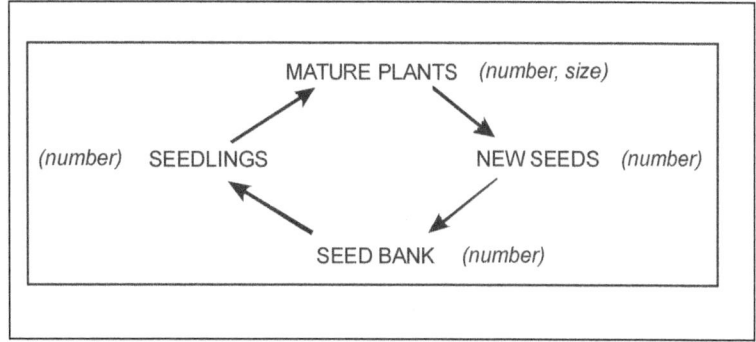

A weed population dynamics model is a representation of the phenomena of a weed population's life history growth and development, from fertilization to death. The model in figure 15.1 represents weed life history phases as discrete phenotypic states of the individual organism with growth. The demographic form of a life history model is quantifiable, with measurements of changes in phase state pool number and sizes with time, often expressed on a unit area basis. What this model does not contain are the deterministic biological processes that drive growth and development during life history. These uncharacterized processes are represented as transitions between quantitative state pools (arrows in figure 15.1).

**15.2.2 Demographic weed life history population dynamics models.** Representations of weed life history population dynamics have largely been accomplished using demographic models:

> "The essence of population biology is captured by a simple equation that relates the numbers per unit area of of an organism $N_t$ at some time $t$ to the numbers $N_{t-1}$ one year earlier." (Silvertown & Doust, 1993, p.1)

From this perspective, weed population dynamics can be represented in its most essential form by this function describing the interrelationships of elements:

$$N_t = N_{t-1} + B - D + I - E$$

Where:
  $N_t$, number per unit area organisms at time $t$;
  $N_{t-1}$, number per unit area organisms one year later;
  B, number of births;
  D, number of deaths;
  I, immigrants in;
  E, emigrants out.

Schematically translating this demographic function onto figure 15.1 reveals four potential life history state phases, or pools, and the relationship of the life history to its complementary metapopulation (I, E), as well as to mortality (D; figure 15.2).

Weed population dynamics are algorithmically represented by the calculation of lambda ($\lambda$), the rate of population size change over one generation.
Where:
  $\lambda = R_0$, net reproductive rate, rate of population increase over a generation;
  $\lambda_t = N_t / N_{t-1}$, annual population growth rate, finite rate of increase.

The finite rate of increase for a population is also expressed as a measure of $W$, so-called Darwinian fitness. The most common formulation of weed population models is as an iterative equation with next year's population calculated from that of the current year (Holst, et al., 2007). The rate of population change over several generations is schematically represented in figure 15.3.

**Figure 15.2** Demographic representation of an annual weed species life history by quantification (number, size) of discrete life history states (mature plant, new seeds, seed bank, seedlings); population size influenced by metapopulaton immigration and emigration dispersal events into and out of the soil seed pool; arrows represent transitions between states.

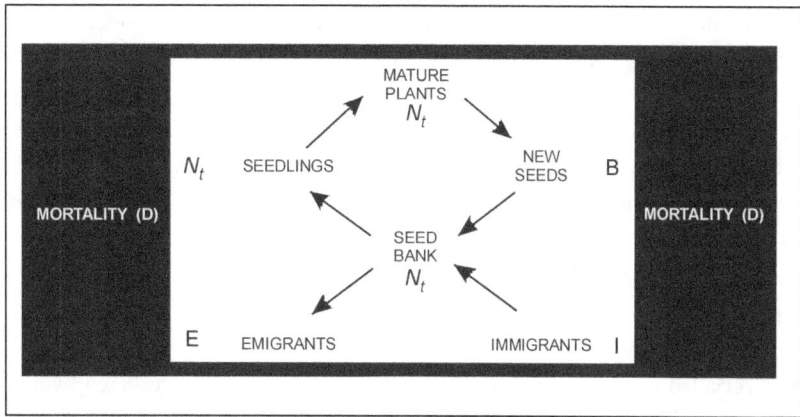

**Figure 15.3** Change in numbers or organisms per unit area ($N_t$) with life history generation time ($N_t$, $N_{t-1}$, etc.) and associated net reproductive rates ($\lambda = R_0$) per generation (1, 2, 3, 4).

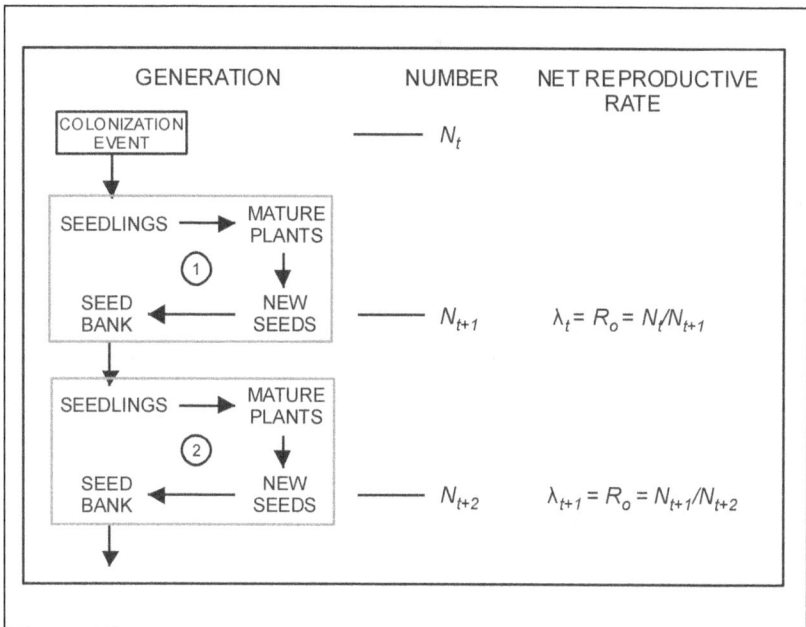

The simplest form of demographic models presented above has been refered to as the "standard population model" (Holst et al., 2007; Sagar and Mortimer, 1976). This simple form has been expanded and extended in several ways (Colbach et al., 2001a, 2001b, 2008). Variation on this basic theme include genetic or spatial components or aspects. The primary amplification of these demographic models occurs in increased attention to life history states: seedling recruitment, mature plants, new seeds and the soil seed bank. Interactions among plants is typically expressed in terms of final biomass of crop and weeds at harvest, with various shortcuts utilized predict the outcome in terms of yield. In most models the outcome of this process is simply reflected in a single competition parameter, which is kept constant over the years. Stochasticity is introduced in some of the models to generate a more irregular population development. Quantification of new seed is based on fecundity per unit area, and is derived in various ways. Little is said in Holst, Rasmussen and Bastiaans (2007) about soil seed pools, which appear to be equated with quantities of old and new seeds, immigrant and emigrant seeds, all with similar or uniform qualities.

Given that the current state of the art of weed population dynamics modeling is represented almost entirely with demographic models, what inherent properties of quantitative models provide the ability to predict future weed behavior?

**15.2.3 Representation and information, inference and prediction.** The opportunities and limitations of models arise from the manner in which weeds and their life histories are represented, the inferences that can be derived from the informational content of the models, and the consequences of these factors on the ability of the model to predict future behavior.

**15.2.3.1 Representation and information.** Information is the meaning given to data (facts, norms) by the way in which it is interpreted. It is a message received and understood.

> "Information is revealed in the correlation between two things that is produced by a lawful process (as opposed to coming about by sheer chance). Information itself is nothing special. It is found wherever causes leave effects. What is special is information processing. We can regard a piece of matter that carries information about some state of affairs as a symbol: it can "stand for" that state of affairs. As a piece of matter, it can do other things as well, physical things, whatever that kind of matter in that kind of state can do according to the laws of physics and chemistry." (Pinker, 1997).

Information can be viewed as a type of input important to the function of the organism (e.g. resources), causal inputs. Information is captured in the physical structure of the organism that symbolize things in nature and that respond to external stimuli allowing causation. Information exists in the physical and physiological structures that capture the way a complex organism can tune itself to unpredictable aspects of the world and take in the kinds of stimuli it needs to function (Pinker, 1997). A good example of biological information taking a physical form is the genome, the informational content of DNA. Closer to home for weed modelers, it is contained in the functional traits of a weed phenotype. Functional traits, such as seed dormancy capacity (e.g. see discussion below on seed heteroblasty), respond to the environment in a particular manner during life history to maximize survival and reproduction (see ch. 13 weed seed-soil environment communication systems). Information can take other forms too.

For weed modelers, what is information? How is information represented in a weed model? Different models contain differing amounts of useful information, the basis of inference and prediction. Inference and prediction are restricted to the informational content of the model.

In the demographic model presented above, the phenotypic identity of weed plants in a local population is represented in their numbers and sizes in spacetime. The informational content derives from the meaning and interpretation given to the numbers and sizes of plants of a particular weed species observed in a particular place, at a particular time (season, life history phase). These metrics provide no information about causation or dynamics. Causation can be inferred indirectly if the same plant in the same location is observed at a later time.

In the evolutionary model presented below, phenotypic identity and informational content are contained in the variation in germination-dormancy capacities induced by a parent plant during embryogenesis, seed heteroblasty. Seed heteroblasty is information physically captured in the structure of the various seeds. It is the behavioral blueprint that responds to specific environmental signals in the soil resulting in seedling emergence carefully timed to the historical occurrence of predictable cropping system disturbances (e.g. herbicide application, tillage, harvesting, frozen winter soil). Seed heteroblasty is the physical information encoded in the morpho-physiology of the seed, it is the cause and determinant of its subsequent life history. It acquired this preadapted physical information by the processes of natural selection-elimination over time. Causation for its life history can be directly inferred from a seed germination assay at seed abscission.

Evolutionary models represent weed life history dynamics in a local population by capturing the physical and behavior information contained in functional traits of the individual weed phenotype that respond to specific environmental signals, opportunity spacetime, in a manner that optimizes their fitness in terms of survival and reproduction.

A complementary way of measuring informational content of a weed model is provided by algorithmic information theory, which measures the information content of a list of symbols based on how predictable they are, or more specifically how easy it is to compute the list through a program. A symbol is an entity with two properties glued together. This symbol carries information, and it causes things to happen. When the caused things themselves carry information, we call the whole system an information processor, or a computer. The processing of symbols involves arrangements of matter that have both representational and causal properties, they simultaneously carry information about something and take part in a chain of physical events. Those events make up a computation (Pinker, 1997). Algorithmically, it is a set of rules that provides an accurate and complete description of a life history capturing its key properties. It provides an algorithmic, or computational, means of forecasting the future behavior of that life history.

Weed phenotypes contain these 'symbols' in the physical form of the functional traits they possess. For example, the DNA coding for the multiple traits of phenotypic plasticity in a weed species is information. It is a physical algorithm computed by the phenotype at every step of its life history that results in its current form and function closely tracking the environmental signals it receives in the local community. Model representations can contain algorithmic forms of this type of information: step-by-step recipes that will, when given an initial state, proceed through a well-defined series of successive states, eventually terminating in an end-state.

A truly dynamic weed population model then would be one that incorporates model algorithms that specify the life history steps a phenotype will go through given an initial state (e.g. seed heteroblasty; time of emergence), as modulated by the specific environment encountered by the individual phenotype in a local population. The biological foundation of these model algorithms is the manner in which specific functional traits are represented.

Evaluating a weed population dynamics model should include a search for its informational content, specifically its representation of the biological traits of the phenotypes of a local population that determine its life history trajectory to survive and reproduce.

**15.2.3.2 Inference**. Inference is the process of reasoning from premises to a conclusion, a deduction. The primary premise, or assumption, of demographic models is that the essence of population biology is captured by a simple equation that relates changes in numbers of organisms per unit area of space with time (Silvertown and Doust, 1993). With this premise, what inferences can, and cannot, be derived from a demographic model of weed population dynamics? A critical review of demographic models reveals a dearth of informational content, flaws in its representation of the deme and life history developmental behavior, and insufficient model formalization.

**15.2.3.2.1 The deme**. Several fundamental flaws are associated with the way the local population is represented in demographic models. The first artifact is the confounding effects of plant, as opposed to animal, population structure. The second derives from how unique individual phenotypes in the local population are represented. The third arises from population membership changes with time that compromise assumptions of covariance structure.

**15.2.3.2.1.1 Population structure.** The local population, the deme, is the setting of biological evolution. The deme consists of unique individual phenotypes, the unit of natural selection. Of crucial importance is how these two components of any weed population dynamic model are represented. Plants respond in a highly plastic manner to locally available opportunity. Unlike animals, plant quantification fails to capture the qualities of the population that drive future dynamics. Demographic weed models that fail to represent this structural nature of plant populations are therefore compromised at conception. John Harper (1977, pp. 25-26) warned of this fatal flaw in weed models:

"The existence of two levels of population structure in plants makes for difficulties, but the problems are much greater if their existence is ignored. One of the strongest reasons why a population biology of plants failed to develop alongside that of animals was that counting plants gives so much less information than counting animals. A count of the number of rabbits or *Drosophila* or voles or flour beetles gives a lot of information: it permits *rough* predictions of population growth rates, biomass and even productivity. A count of the number of plants in an area gives extraordinarily little information unless we are also told their size. Individual plants are so "plastic" that variations of 50,000-fold in weight or reproductive capacity are easily found in individuals of the same age. Clearly, counting plants is not enough to give a basis for a useful demography. The plasticity of plants lies, however, almost entirely in variations of the number of their parts."

Demographic representation of the structure of a local plant population depends therefore on the specification of the number of individuals (level one), and on the number and variability of parts (e.g. leaves, axes) of each individual (level two).

**15.2.3.2.1.2 Individual phenotypic identity.** Natural elimination acts by nonrandomly selecting the fittest individuals in the local population. Changes in demes therefore are an adaptive reflection of the unique biodiverse qualities of those survivors. Weed models are critically evaluated for their ability to represent individual phenotypic identity by means of their functional properties. Weeds assemble in local communities as collections of unique phenotypes. The urge to simplify their representation by categorical or average qualities obscures this biodiversity.

"The assumptions of population thinking are diametrically opposed to those of the typologist. The populationist stresses the uniqueness of everything in the organic world. What is true for the human species – that no two individuals are alike – is equally true for all other species of animals and plants. Indeed, even the same individual changes continuously throughout its lifetime when placed in different environments. All organisms and organic phenomena are composed of unique features and can be described collectively only in statistical terms. Individuals, or any kind of organic entities, form populations of which we can determine the arithmetic mean and the statistics of variation. Averages are merely statistical abstractions, only the individuals of which the population are composed have reality. The ultimate conclusions of the population thinker and of the typologist are precisely the opposite. For the typologist, the type (*eidos*) is real and the variation an illusion, while for the populationist the type (average) is the abstraction and only the variation is real. No two ways of looking at nature could be more different." (Mayr, E. 1959)

Demographic representations of weed populations are limited to the extent that numbers of plants fail to provide information of the qualities of their members. The demographic representation of weed population dynamics is an incomplete abstraction because it ignores the importance of phenotypic variation by averaging behaviors at experimentally convenient times in life history. Measurement of quantities and sizes of uncharacterized phenotypes, and the uncharacterized processes of transitions between life history states, provide little inherent inference of population dynamics.

**15.2.3.2.1.3 Local population dynamics.** The third, and possibly the most telling, artifact of demographic representations of populations arises from the changing phenotypic structure of the local community with time. Natural selection eliminates lesser fit individuals to the enrichment of the survivors. As such the phenotypic-genotypic composition of the deme is constantly changing with time. During the growing season mortality alters the composition of the population. The population genetic structure of a local soil seed pool is different every year with the addition of offspring from those favored individuals. As such, demographic models represent populations as constant qualitative entities. Causation cannot be inferred from plant numbers that consist of different individual phenotypes. Inferences derived from them are incomplete as the assumptions underlying population covariance structure are violated. Covariance is a measure of how much two variables change together, for example plant number with time. The phenotypic membership of the local deme and soil seed pool changes as natural selection favors some and eliminates others. Natural selection violates this covariance structure by assuming the individual phenotypes are the same at each life history measurement time in the local habitat.

**15.2.3.2.2 Life history development and behavior.** Strong inferences of weed population dynamics can be made when the functional traits driving individual phenotypic behavior in local population are represented. Some of the most important functional traits of weeds are found in individual plant polymorphism and plasticity.

**15.2.3.2.2.1 Life history states and processes.** Demographic models represent weed populations by quantifying numbers of plants, their sizes and their density per unit area of space at discrete times in their life history. Lacking is a representation of the developmental growth processes that cause transitions between life history states to occur (arrows, figure 15.1). Demographic models do not embrace the dynamic processes causing weed life history, despite the claim that:

> "Processes governing the transition from one stage to the other, like germination and seed production and processes responsible for the losses that occur throughout, like seed mortality, plant death and seed predation, are included." (Holst et al., 2007).

Process is indirectly inferred, a surrogate derived from computational number-size frequency transitions between discrete life history times. Mature plant number and size are not the competitive processes of interaction among neighbors. Soil seed pool numbers are not the motive forces driving the processes of germination, dormancy reinduction and seedling recruitment. New seed numbers do not reveal adaptive changes in these new phenotypes caused by natural selection and elimination: changes in the genetic-phenotypic composition of the local population. Life history developmental behavior is motivated by specific environmental signals stimulating functional traits inherent in the phenotype. Weed population dynamics come about as a direct adaptive consequence of generating phenotypic-trait variation among excess progeny in the deme, followed by the survival and reproduction of the fittest phenotypes among those offspring with time.

**15.2.3.2.2.2 Polymorphism and plasticity.** Individual weed phenotypes derive fitness from their heterogeneity by exploiting local opportunity. Weed population structure is difficult to model unless somatic polymorphism and phenotypic plasticity are represented, inherent functional traits that control life history behavior as well as allow the individual to assume a size and function appropriate to its local opportunity spacetime. Somatic polymorphism is the production of different plant parts, or different behaviors, within the individual that is expressed independently of its local environment. Seed heteroblasty is an example of parentally-induced dormancy heterogeneity among offspring that provides strong inferences of future behavior. Phenotypic plasticity is the capacity of a weedy plant to vary morphologically, physiologically or behaviorally as a result of environmental influences on the genotype during the plant's life history. Experimentally capturing this level of population structure entails measuring the population of phenotypes expressed by a single genotype when a trait changes continuously under different environmental and developmental conditions: the reaction norm. The reaction norm in population structure is expressed by number and size of constituent leaf, branch, flower and root modules of the individual plant that vary in response to the locally available opportunity spacetime. This plasticity of form confounds the ability of a purely demographic model to make predictions of population growth rates, biomass and even productivity. The consequence of phenotypic plasticity is that plants growing under density stress typically have a skewed distribution of individual plant weights, especially when they compete for light. Skewing of the frequency distribution (numbers of plants versus weight per plant) increases with time and with increasing density (plants per unit area). Typically at harvest a hierarchy of individuals is established: a few large dominants and a large number of suppressed, smaller, plants. The individual weeds in the hierarchical population structure possess the potential for explosive, nonlinear exponential growth and fecundity. Individual weed plants have the potential to produce a very large range of seed numbers depending on their size. The range in reproductive capacities of plants extends from 1 to $10^{10}$ (approaching infinity for vegetative clone propagule production; Harper, 1977). There exists a danger in assuming that the average plant performance represents the commonest type, or most typical, plant performance (Dekker, 2009).

**15.2.3.2.3 Model formalization and measurement metrics.**

**15.2.3.2.3.1 Hypotheses of local weed population dynamics.** Any model of weed behavior must be preceded by an experimental hypothesis of how population dynamics comes about: to what is the deme adapted? It should be a statement of an overarching intuition of how the biological system works, or the primary forces driving its expression. Such a hypothesis should appropriately begin with the intelligent designer of the system: human agricultural activity. Such a hypothesis could provide a tool to realistically guide the mathematical, algorithmic and statistical formalization of model components, metrics and output. No hypothesis of this type has been proposed for demographic models.

**15.2.3.2.3.2 Mathematical, algorithmic, statisitical model formalization and component description.** A model is a representation of reality. It is inherently an abstraction and a simplification. It is a conceptual framework of a system constructed by indicating which elements should be represented and how these elements are interrelated. This conceptual framework then is translated into algorithms, precisely defined step-by-step procedures by which dynamics are carried out. Many models are published without "... a complete description of the model logic and mathematics, including the parameter values." Of the 134 papers reviewed in this review, 16-19% were not open for re-use or even critique. (Holst et al., 2007). Inference in simple and complex systems and models derives from the definition of model parameter space and algorithmic solutions of population dynamics:

> "An intelligent system, then, cannot be stuffed with trillions of facts. It must be equipped with a smaller list of core truths and a set of rules to deduce their implications." (Pinker, 1997)

> "The real issue here is the apparent reduction in simplicity. A skeptic worries about all the information necessary to specify all the unseen worlds. But an entire ensemble is often much simpler than one of its members. The principle can be stated more formally using the notion of algorithmic content."
> "... the whole set is actually simpler than a specific solution ..."
> "The lesson is that complexity increases when we restrict our attention to one particular element in an ensemble, thereby losing the symmetry and simplicity that were inherent in the totality of all the elements taken together."
> (Tegmark, M. 2009)

> "Spatially explicit models tend to get complex, or mathematically demanding, like the model of neighbourhood interference between *Abutilon theophrasti* and *Amaranthus retroflexus* (Pacala & Silander, 1990). Another hindrance to fully grasp these models is that they may contain so many details, that it makes a full description of the model in scientific journals impossible, e.g. the within-field model of Richter *et al.* (2000) or the landscape model of Colbach *et al.* (2001b). To counteract this inherent complexity in spatial processes, one can reduce the complexity of the weed model itself.

But this makes for very abstract models which can be difficult to relate to real weed population dynamics (e.g. Wang et al., 2003)." (Holst et al., 2007)

"... the danger that the model develops into a monstrous specimen covering far too many facets and bearing an enormous parameter requirement. Collecting relevant parameters then becomes a time consuming exercise or might even develop into an objective on its own, putting the focus on analysis, rather than on synthesis of knowledge. Additionally, models containing too many parameters are often characterized by enormous error margins, and often lose their robustness." (Holst et al., 2007)

**15.2.3.2.3.3 Random-nonrandom processes.** Any model of weed population dynamics must accurately represent both random and nonrandom processes. Holst et al. (2007) indicate that stochastic models can be used to explain past population dynamics. If successful, stochastic models gain credibility as predictive tools of long-term population dynamics. The authors indicate that stochastic models are a tool to handle the uncertainty of future conditions. This review classifies environmental unpredictability, agricultural practice (cropping disturbances) and statistical error in model parameter estimates as random, unpredictable, and stochastic. Classification of some of these experimentally tractable phenomena (e.g. cropping disturbance; survival and reproduction) as random is inappropriate. Significantly, they classify natural demographic variation in reproduction and mortality as stochastic. Apparently Charles Darwin's contributions (1859) are underappreciated by demographic weed modelers. Variational evolution of a population or species occurs through changes in its members by natural selection, the processes of nonrandom elimination and nonrandom sexual selection (Mayr, 2001).

## 15.2.3.3 Predicting weed population dynamics.

"It's hard to make predictions, especially about the future." (Yogi Berra, 2010).

Two recent reviews of weed modeling have come to a similar conclusion (Holst et al., 2007; Freckleton and Stephens, 2009). Predicting the future is problematic (see ch. 14.1 randomness, uncertainty and information). Future behavior is an emergent property of the inherent biological information contained within the individual weed phenotype (and its traits) as it accomplishes its life history survival and reproduction. Demographic models inherently do not contain this biological information. Limitations in the inferences that demographic models possess render them of limited utility in predicting future behavior. The predictability of a model is based in its complexity. The work of nobel laureate F.A. Hayek (1974) is revealing. He distinguished the capacity to predict behavior in simple systems and those in complex systems through modeling. Complex biological phenomenon could not be modeled effectively in the same manner as those that dealt with essentially simple phenomena like physics. Complex phenomena, through modeling, can only allow pattern predictions, compared with precise predictions made of non-complex phenomena. How then is it possible to predict weed population dynamics? What is missing in demographic population models is the biological information contained in weedy traits whose expression drives the missing deterministic processes, processes incorrectly attempted to be replaced by stochastic probabilities of knowable weed phenomena (Holst et al., 2007). What then are the "...smaller list of core truths and a set of rules to deduce their implications." (Pinker, 1997) that will simplify weed population models and allow strong inference and predictability? Intuitively, these core truths most come from the inherent biological traits of the weeds themselves. It is to this that evolutionary models are directed.

## 15.5 Evolutionary, Trait-based, Weed Life History Population Dynamics

"As the famous geneticist T. Dobzhansky has said so rightly, "Nothing in biology makes sense, except in the light of evolution." Indeed, there is no other natural explanation than evolution for [biological phenomena]." (Mayr, 2001).

An evolutionary model of population changes based on the actions of functional traits might be guided by the following hypothesis. Weedy and invasive plants perform the plant colonization niche. Weedy plants are the first to seize and exploit the opportunity spacetime created by human disturbance, notably in resource-rich agricultural cropping systems. Local opportunity spacetime is the habitable space available to an organism at a particular time which includes its resources (e.g. light, water, nutrients, gases) and conditions (e.g. heat, climate, location), its disturbance history (e.g. tillage, herbicides, winter), and neighboring organisms (e.g. crops, other weed species). Therefore, it is hypothesized that weedy plant life history behavior in a deme is a consequence of natural selection and reproductive success among excess variable phenotypes (and functional traits) in response to the structure, quality and timing of locally available opportunity spacetime. Evidence in support of this hypothesis is revealed in ch. 13, weed seed-soil environment communication systems; ch. 17, weedy *Setaria* species-group; and ch. 18, triazine resistant and susceptible *Brassica napus*; ch. 19, *Abutilon theophrasti*).

What alternative is there to quantitative demographic life history models to represent weed population dynamics? How can the limitations and artifacts of quantitative demographic models be overcome? How is the essence of population biology captured in a life history representation? The thesis of this section is that understanding population dynamics in agroecosystems requires a qualitative evolutionary representation of local populations based upon the two component processes of natural selection and elimination resulting in weedy adaptation. Evolutionary models based on the two component processes of natural selection (generation of variation, selection and elimination) are discussed in terms of these same critical factors.

The essence of population biology is captured by a weed life history representation stated in the form of the processes of natural selection: the fittest parents generate phenotypic variation in their offspring that preferentially survive and reproduce in the local deme. Figure 15.1 can be redrawn (figure 15.4) to represent this in a much simplified form:

**Figure 15.4** Representation of an annual weed species life history in terms of the two component processes of natural selection and elimination: step 1, production of phenotypic variation by the fittest parent plants; step 2, survival and reproduction of the fittest phenotypes, elimination of the others.

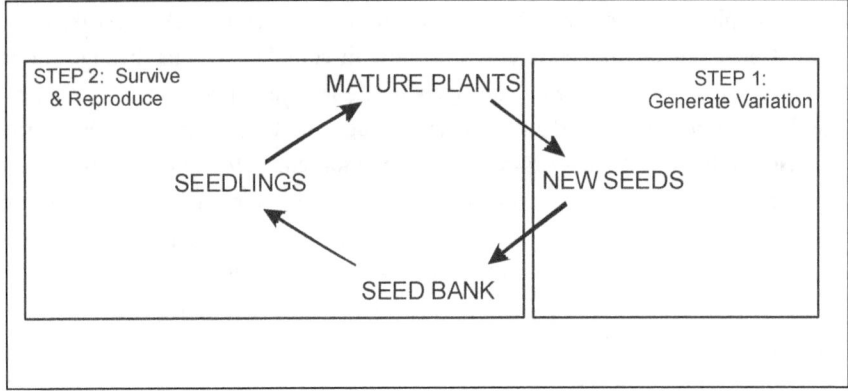

In each generation, new seed dispersed into the local deme comes from the fittest parent plants of the previous generation. In this view the phenotypic composition of the local deme is constantly changing during life history. Life history is not a repetitive cycle, but a spiral of overlapping life histories of changing individuals better adapted to the local habitat (figure 15.5).

**Figure 15.5** Schematic representation of the adaptive changes in the local population of an annual weed species through several generations (life cycles) as a consequence of the two processes of natural selection and elimination.

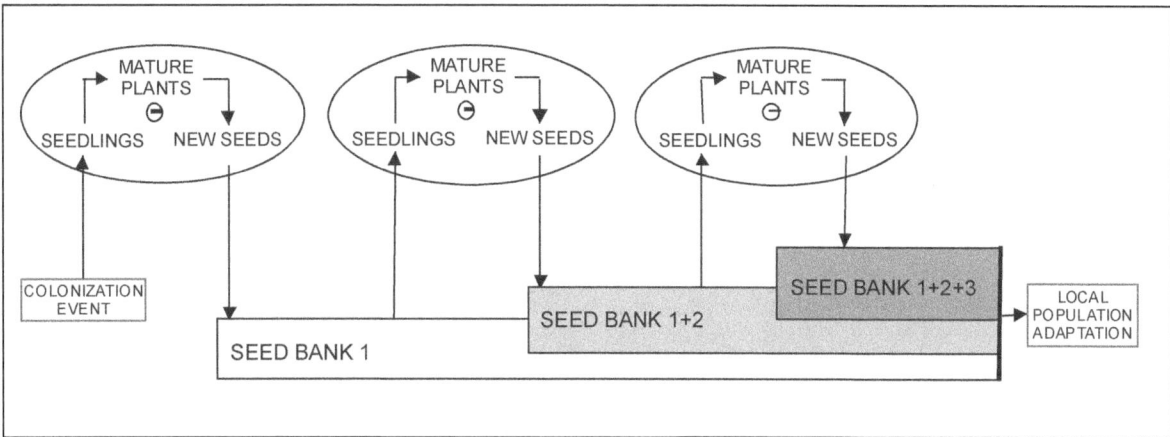

Life history does not begin at the same starting condition with each new generation. New surviving seed join the preexisting seed pool in the soil to form the new local population every winter in a sexually reproducing annual weed species. The local population is dynamic, its phenotypic composition (the plant communities of the future) changes with new addition and loss. It is an expanding spiral for growing populations, a constricting spiral for dying populations. Therefore both the quantity and quality (traits) of the individual phenotypes in the deme change with time: adaptation to the local habitat. This evolutionary adaptation is the most dynamic element of the weed population that any realistic model must represent.

The phenotypic composition of each new local population changes with the recruitment of seedling from the seed pool. The composition of the seed pool is the dynamic element of local adaptation: the progeny of the fittest individuals selected from the previous generations. The representation of this changing seed pool is most challenging element in the formalization of a realistic life history model.

Sexually reproducing, annual, weed population dynamics are the adaptive consequence of natural selection and elimination of excess individuals in the local deme. This evolutionary process is represented by two processes and 5 conditions which are presented fully herein unit 2, ch. 2-5. Several weed species have provided examples of this evolutionary model system: *Setaria* spp.-gp. (ch. 13, 17), herbicide resistant *Brassica napus* (ch. 18), and *Abutilon theophrasti* (ch. 19).

# Chapter 16: Cultural Relativity of Rationality

"There are some people who, if they don't already know, you can't tell 'em". Yogi Berra, 2010

**Summary**. Representation of the nature of weed biology is confounded by the human scientists that make them, their beliefs, values and models. What we see, what we want to see, what is expected of us by our education and job, all these put limitations on our ability to observe, describe and represent. Human perception structures natural laws.

The modern misconception of scientific progess is one with a straight-forward logic of discovery: that one discovery leads almost automatically to another. Instead, the modern understanding of scientific history is that there are always conflicting views, uncertain outcomes, unpredicable developments. There are psychological and cultural factors working against innovation. In order to make sense of these delays we should not look to the inflexible logic of discovery, but to other factors: the role of the emotions, the limits of imagination, the conservatism of institutions.

The socio-political cultural relativity of scientific rationality and model paradigms is revealed in the nature of scientific revolutions. Scientific theories, paradigms, tell scientists what nature contains and doesn't contain, which provides a map which scientists elucidate. Paradigms give form to the scientific life. They provide a vehicle for scientific theory, telling the scientist about the entities that nature does and does not contain and about the ways in which those entities behave. Therefore, when paradigms change, there are usually significant shifts in the criteria determining the legitimacy both of problems and of proposed solutions.

Once learned, accepted and adopted by a scientific group a paradigm remains the way to represent nature until, and only until several things happen. There must occur a paradigm crisis: anomalies, contradictions between theory and observation accumulate that can no longer be avoided, when compared to the existing paradigm. There must appear a new paradigm that explains these anomalies, but a new paradigm alone will not change accepted theory. A change will only occur with the successful completion of three paradigm-nature comparisons: between the existing paradigm and nature; between the new paradigm and nature; and then, between the existing and new paradigm. This change is hard for scientists, for humans, to do: belief and faith are broken.

## 16.1 Introduction to Cultural Relativity of Rationality

Representation of the nature of weed biology is confounded by the human scientists that make them, their beliefs, values and models. The goal of this chapter is to provide a connection between weed biology and how I represented them in my scientific career, the content of unit 7, weed case history. For each weed species I studied I was confronted with the mystery of unknown biology and the 'normal science' that preceeded my work. The choice of subject and experimental implementation occurred within the university institutional cultural tradition. With every weed species I studied I discovered cultural resistance. In addition to elucidating weed biology I learned of the cultural relativity of scientific rationality. What we see, what we want to see, what is expected of us by our education and job, all these put limitations on our ability to observe, describe and represent. Human perception structures natural laws.

**culture:**
1: a set of ideas that cause their holders to behave alike in some ways; ideas are information that can be stored in people's brains and can affect their behavior
2: the way of life, the ideas, the general customs and beliefs, and social behavior of a particular group of people or society at a particular time
3: that complex whole which includes knowledge, belief, art, morals, law, custom and any other capabilities and habits acquired by man as a member of society
4: a social domain that emphasizes the practices, discourses, and material expressions, which, over time, express the continuities and discontinuities of social meaning of a life held in common

"Up to now it has been assumed that all our cognition must conform to the objects; but ... let us once try whether we do not get farther with the problems of metaphysics by assuming that the objects must conform to our cognition." (I. Kant, in Rohlf, 2010)

Our understanding of the external world had its foundations not merely in experience, but in both experience and *a priori* knowledge (Wikipedia, 6.14). The external world provides things we sense. It is our mind that processes this information about the world and gives it order, allowing us to comprehend it. Our mind supplies the conditions of space and time to experience objects. The concepts of the mind and the perceptions or intuitions that garner information from phenomena are synthesized by comprehension. Without the concepts, intuitions are nondescript; without the intuitions, concepts are meaningless:

"Thoughts without content are empty, intuitions without concepts are blind." (Kant, 1781).

**16.1.1 History of scientific progress.** The history of science is not one of progress but of delay. A history not of events, but of non-events. A story not of an inflexible logic but of a sloppy logic; not of overdetermination, but of underdetermination. These cases of delay, non-events, sloppy logic and underdetermination are the norm, not the exceptions in science (Wootton, 2006).

The modern misconception of scientific progess is one with a straight-forward logic of discovery: that one discovery leads almost automatically to another, that one researcher picks up where another has left off, as if passing a baton in a relay race ("... dwarfs standing on the shoulders of giants ..."; Bernard of Chartres).

Instead, the modern understanding of scientific history is that there are always conflicting views, uncertain outcomes, unpredicable developments. There has been progress; but not nearly as much as most of us believe. Why was progress so slow? There were psychological and cultural factors working against innovation. What was lacking was a clear model of how to proceed. There existed only one model, the concept of 'normal' as opposed to 'revolutionary' science."

**16.1.2 Scientific institutions.**

"... Iowa State Agricultural College, the state's lethargic land grant institution ..." "In 1885, at nineteen, Harry (H.C. Wallace, father) headed off to Iowa State Agricultural College in Ames. The experience was not invigorating. "The college was nominally an agricultural college but very little agriculture was taught,", Henry (H. Wallace, grandfather) wrote." "By 1906 the humble agricultural college at Ames had begun to think of itself in grander terms. It was called the Iowa State College of Agricultural and Mechanical Arts, and its president declared it was "perhaps without peer in the country" in the field of agricultural education. But the institution Henry A. Wallace entered that fall remained a backwoods "cow college" in many respects. Its buildings were ramshackle; it had no gymnasium or auditorium or even a permanent home for its library." "The teachers he most admired had about them a "quality of enthusiasm" that sparked in students a desire to learn, a characteristic, Wallace observed, that made them beloved by students and "nearly always distrusted by their fellow faculty members. Others he considered merely intellectual hacks. He had already proven to himself the uselessness of corn shows, but his professors continued to prattle about the pedigrees and aesthetics of plants and animals." Culver and Hyde, 2000

In order to make sense of these delays we should not look to the inflexible logic of discovery, but to other factors: the role of the emotions, the limits of imagination, the conservatism of institutions (Wootton, 2006). What psychological, cultural, or institutional factors represented an obstacle to science developing as it might have? Scientific institutions have a life of their own. In certain circumstances a scientific institution will implement policies that no individual person within the institution thinks are good. It can happen if an outside agency controls the institution's funding and requires that it meet certain criteria that nobody within the institution actually believes in. A situation often found in contemporary universities and institutions that rely on government funding. It was the norm under communism. It is the norm under religion. It is becoming more prevalent in US public universities adopting monetary goverance policies and top-down research priorities. There exist circumstances in which an institution can take a decision, or pursue a policy not from any individual within the institution. Institutions can thus take on a life of their own (e.g. Dekker, 1999, 2005; Dekker and Comstock, 1992).

**16.1.3 Intellectual coherence and truth**. Historically, the test of truth within the scholastic intellectual world of the universities was intellectual coherence, not practical effectiveness (Wootton, 2006). As long as science met that test it was subject to no other, it was a monopoly. Because it was a monopoly there was no need for it to prove its superiority by comparing alternatives. Sometimes there existed disagreement over knowledge, about what the relevant information was, and about how to interpret it. When points of view were radically different, a choice had to be made.

**16.1.4 Scientific crisis and change**. Kuhn (1962) sharply distinguished 'normal science' and science during periods of intellectual crisis (Wootton, 2006). He argued that major intellectual advances take place only in the context of a crisis within existing ways of thinking and doing. For long periods 'normal science' went largely unquestioned. Real progress began only when that assumption began to be questioned. Science was held together by what he called 'paradigms':

"... a laboratory activity, learnt by generations of students ..."; "... a model solution to a problem ..."; "... the epitome of something you needed to know to belong to an intellectual community ..."; "Thus one could give an account of a paradigm that related it to a practice, a theory, or a sociological community."

## 16.2 The Nature of Scientific Revolutions

The socio-political cultural relativity of scientific rationality and model paradigms is revealed in the nature of scientific revolutions (Kuhn, 1996). The basis of scientific change in weed science is revealed the nature of scientific problems and paradigms and how they change.

**16.2.1 Scientific problems**. Paradigms exist where problems are solved within a scientific research tradition that is constrained by rules, preconceptions or established viewpoints (belief and faith) (Kuhn, 1996). If a problem is to be classified as a scientific puzzle it must be characterized by more than an assured solution. There must also be rules that limit both the nature of acceptable solutions and the steps by which they are to obtained. The use of the term 'rule' must be broad, one in which it can occasionally be equated with 'established viewpoint' or with 'preconception'.

**16.2.2 The nature of scientific paradigms.** Scientific theories, paradigms, tell scientists what nature contains and doesn't contain, which provides a map which scientists elucidate (Kuhn, 1996). Because nature is very complex this map guides continued elucidation and experimentation to further develop the paradigm. Paradigms also provide the scientist with directions for map-making; theory, methods and standards are inextricable. As paradigms change, in order for paradigms to change, shifts in the criteria to determine the legitamacy of both problems and solutions change.

Paradigms give form to the scientific life. They provide a vehicle for scientific theory, telling the scientist about the entities that nature does and does not contain and about the ways in which those entities behave. That information provides a map whose details are elucidated by scientific research. This map is as essential as observation and experiment to science's continuing development because nature is too complex and varied to be explored at random. Paradigms provide scientist not only with a map but also with some directions essential for map-making. In learning a paradigm the scientist acquires theory, methods, and standards together, usually in an inextricable mixture. Therefore, when paradigms change, there are usually significant shifts in the criteria determining the legitimacy both of problems and of proposed solutions.

**16.2.3 Paradigm: disciplinary matrix.** Scientists share a paradigm. Members (of a particular community of specialists) share a paradigm or set of paradigms that accounts for the relative fullness of their professional communication and the relative unanimity of their professional judgements (Kuhn, 1996). A disciplinary matrix refers to the common possession of the practitioners of a particular discipline which is composed of ordered elements of various sorts, each requiring further specification. Paradigms are also called the disciplinary matrix, which has at least four components.

**16.2.3.1 Symbolic generalizations.** Formal and readily formalizable components of the disciplinary matrix: symbolic generalizations, expressions, deployed without question or dissent by group members, which can be readily cast in a logical form.

**16.2.3.2 Commitments to particular models.** Shared commitments and beliefs in particular models determining acceptable explanations ('metaphysical paradigms'), that supply the group with preferred or permissible analogies and metaphors that help determine what is an accepted explanation and as a puzzle-solution.

**16.2.3.3 Shared values.** Widely held values shared by all that provide a sense of community to natural scientists as a whole; usually more widely shared among different communities than either symbolic generalizations or models.

**16.2.3.4 Exemplars of nature.** Paradigm 'exemplars' are the means by which common interpretations of observations of nature are achieved. Exemplars are shared symbolic generalizations illustrated by technical problem-solutions found in the periodical literature to achieve common interpretations of observations of nature. They are concrete problem-solutions that students encounter from the start of their scientific education, whether in laboratories, on examinations, or at the ends of chapters in science texts. Exemplars are shared examples that show how the job is done.

**16.2.4 Scientific paradigm crisis and change.** Once learned, accepted and adopted by a scientific group (discipline, subset of specialists) a paradigm remains the way to represent nature until, and only until several things happen (Kuhn, 1996). First, there must occur a paradigm crisis: anomalies between theory and observation. A large body of contradictions, anomalies, to the paradigm accumulate that can no longer be avoided, when compared to the existing paradigm. Second, there must appear a new paradigm that explains these anomalies. The new paradigm allows comparisons to be made. A new paradigm alone will not change accepted theory. A change will only occur with the successful completion of three paradigm-nature comparisons: between the existing paradigm and nature; between the new paradigm and nature; and then, between the existing and new paradigm.

The consequences of a paradigm shift is a change in the criteria used to determine the legitimacy of problems and their solutions. This change is hard for scientists, for humans, to do: belief and faith are broken.

When scientists are confronted by crises caused by the emergence of novel theories, even severe and prolonged anomalies, there is one thing they never do. Though they may consider alternatives, they do not renounce the existing paradigm that has led them to the crisis. They do not, that is, treat anomalies as counterinstances, though that is what they are. Once a scientific theory has achieved the status of paradigm it is declared invalid only if an alternative candidate is available to take its place. In the history of science no process has been revealed whereby an existing theory is falsified by direct comparison with nature. Scientific judgement rejecting a previously accepted theory is always based upon more than a comparison of that theory with the world. The decision to reject one paradigm is always simultaneously the decision to accept another, and the judgement leading to that decision involves the comparison of both paradigms with nature *and* with each other.

Human psychology contributes to the relativity of rationality. Some people, otherwise intelligent, have a deficiency of that human ability to impute to others knowledge that is different from their own (also Asperger syndrome, mild form of autism). Their personal knowledge is the same as that of others. This is known as the problem of underdevelopment of a human faculty called "theory of mind" or "folk psychology" (Taleb, 2009). You cannot do anything with knowledge unless you know where it stops. What you know and what you don't know allows you to determine the costs of using that knowledge. Science should provide maps showing what current knowledge and current methods do not do for us.

Stimuli are processed by perception and only the 'best' interpretations, sensations, are useful for future scientists; the truth is tacit. Paradigms change when anomalies in observation lead to better paradigms, which must be compared to observation and the old paradigm to cause a revolution in a science.

**16.2.5 Knowledge embedded in stimuli-sensation**. Among the things we know about our awareness of 'something' are that very different stimuli can produce the same sensations; that the same stimulus can produce very different sensations; and that the route from stimulus to senation is in part conditioned by education (Kuhn, 1996). One of the fundamental techniques by which the members of a group (e.g. an entire culture; a sub-community of specialists within it) learn to see the same things when confronted with the same stimuli is by being shown examples of situations that their predecessors in the group have already learned to see as like each other and as different from other sorts of situations.

Adoption of stimuli-to-sensation perceptions are evolutionary: selection and elimination among competing ideas (memes; Dawkins, 1989) in a population leads over time to more useful theories. Having very few ways of seeing are exactly the theories that have withstood the tests of group use, and are therefore worth transmitting from generation to generation. Because they have been successfully selected over historic time the experience and knowledge of nature is embedded in the learned stimulus-to-sensation route.

Built into the neural process that transforms stimuli to sensations are the following characteristics: they are transmitted through education; they have been found to be more effective by trial than historical competitors in a groups's current environment; they are subject to change both through further education and through the discovery of misfits with the environment.

Those are characteristics of knowledge. It is strange usage, for one other characteristic is missing. We have no direct access to what it is we know, no rules or generalizations with which to express this knowledge. Rules which could supply that access would refer to stimuli not sensations. Stimuli we can know only through elaborate theory. In it absence, the knowledge embedded in the stimulus-to-sensation route remains tacit.

**16.3 Cultural Relativity of Rationality in Weed Science**

Some of the concepts presented in this book are not fully embraced in the current practice of weed science today. Ecological thinking in weed science has increased in the last decades, but demographic representation still dominates the dialogue. Plant invasions are seen as an independent category of organisms and the dialogue politicized. Most prominent is the resistance to evolutionary thinking (especialy apparent in the USA), to the central role of functional traits and the phenotype, and the importance of complexity and uncertainty in knowledge creation.

# UNIT 7: WEED CASE HISTORY

"Past the flannel plains and blacktop graphs and skylines of canted rust, and past the tobacco-brown river overhung with weeping trees and coins of sunlight through them on the water downriver, to the place beyond the windbreak, where untilled fields simmer shrilly in the A.M. heat: shattercane, lamb's-quarters, cutgrass, sawbrier, nutgrass, jimsonweed, wild mint, dandelion, foxtail, muscadine, spine-cabbage, goldenrod, creeping Charlie, butter-print, nightshade, ragweed, wild oat, vetch, butcher grass, invaginate volunteer beans, all heads gently nodding in a morning breeze like a mother's soft hand on your cheek. An arrow of starlings fired from the windbreak's thatch. The glitter of dew that stays where it is and steams all day. A sunflower, four more, one bowed, and horses in the distance standing rigid and still as toys. All nodding. Electric sounds of insects at their business. Ale-colored sunshine and pale sky and whorls of cirrus so high they cast no shadow. Insects all business all the time. Quartz and chert and schist and chondrite iron scabs in granite. Very old land. Look around you. The horizon trembling, shapeless. We are all brothers." David Foster Wallace, in his unfinished novel The Pale King, 2011

"So far we have considered only the theoretical branch, which has almost monopolized the philosophy of biology. Thoughtful biologists, however, are increasingly emphasizing the historicity of evolution and the role of narrative in explaining evolutionary events. They realize that historical narratives, which differ in aim, method, and conceptual emphasis from therortetical models, are not only more appropriate but almost mandatory for explaining specifc events in the evolution of life on earth. Theoretical generalization is powerful, but generalization exacts a heavy price, which may destroy the concreteness of history." Auyang, 1998.

**Summary**. This unit is the story, or stories, of three weeds I have come to know personally. My first, an annual dicot, velvetleaf (*Abutilon theophrasti*), was my first experience with an individual species. With my Ph.D. velvetleaf experience I came to know an individual species in some depth. It isn't until you come to know an organism when it's at home as a unitary individual as a scientist that you develop empathy for your subject of observation. And it is then that you can begin to understand biology, for me weeds, in much greater depth. This deeper insight arises because I began to see my more closely focused, specialized, observations through the lens of the entire individual. It is the individual in a population that evolves in nature. It is the individual that reveals the beautiful complexity of life as a whole.

So I came to know and learn from my first weed. Velvetleaf is so entirely selfish with the sun, never letting anyone get in its way. Reaching for the sky if need be, putting all its close neighbors in the shade, stretching up as far as necessary to ensure the first place in the sun. Never ever learning to share.

My second species was the annual dicot triazine resistant *Brassica napus*, rapeseed, a half-wild crop with an incompletely domesticated nature, a model for the triazine resistant weeds species. I began uncovering *B. napus* in my time in Ontario, and continued seriously when I emigrated to Iowa in 1985. At my new home there I began by documenting an emerging resistance problem: *Kochia scoparia*, common lambsquarters

(*Chenopodium album*), Pennsylvania smartweed (*Polygonum pensylvanicum*), and then finally my favorite closest friend giant foxtail (*Setaria faberi*). All herbicide resistant.

It was *Brassica* that first showed me how truly weird weeds can be, the discovery of a chronomutant. Here was a weed that turned water and sun into sugar at the oddest times of day, a trait that arose from a minute alteration in one of its genes hiding in that eukaryotic hitchhiker the chloroplast. This tiny little lesion led to an explosion of compensating changes in its body, in its life. It was more than just doing photosynthesis at weird times of the day, but changing its membranes, keeping its leaf stomata holes open all the time, more dormant seeds, on and on ... This was an adventure until forced to be more 'practical'. I needed to keep my job.

My third species became during the 1990's my closest weed friend of all, the annual grass species that allowed me to 'pull back the curtain' and peer deeply into its secrets, the weedy foxtails. This was primarily giant foxtail, but also its close cousins and 'comrades-in-arms' green (*S. viridis*) and yellow foxtail (*S. pumila*). With time, and a bit of coaxing, I also made acquaintance with knotroot (*S. geniculata*) and bristly foxtail (*S. verticilata*). It was in the observation of his-her seeds that the charms of giant foxtail were revealed to me. It was the discovery of that lovely, tiny, 'nano-thermos bottle' that led to so many deeply satisfying epiphanous moments during my career: oxygen solubility in water is *inversely* related to temperature! No wonder their biggest flush is in the spring. Discovering a myserious protein inside the nano-thermos seed: an oxygen 'sponge'! Seed hull topography was an oxygenator! That these little seeds were a Shannon communication system (soil-seed contact) living in dirt! Foxtail mothers produce many babies, each with a unique self-timer to germinate when opportunity spacetime is maximal! This was pure fun, and practical at that. I kept my job.

In this unit I want to connect concepts in the first units with these empirical, experimental experiences as examples of a broader perspective on the evolutionary ecology, the nature of weeds. Linkages are provided throughout the text to these examples. Alternatively, table U7.1 is provided to facilitate linkage of weed species case histories to a sample of concepts presented in this book.

**Table U7.1** Evolutionary ecological concept examples apparent in weed species case histories presented in Unit 7: keystone functional traits; representation of weedy nature; book concepts and relevant book chapter sections.

| Weed Species | Keystone Weedy Functional Traits | Representation of Weedy Nature | EEW Concepts (text) |
|---|---|---|---|
| *Setaria* species-group | Seed dormancy heteroblasty via regulation of seed interior oxygen: 1] hull surface topography oxygenation 2] placental pore control of $H_2O$-$O_2$ diffusion 3] caryopsis exclusion of $H_2O$-$O_2$ diffusion 4] seed interior oxygen-scavenger protein | Complexity, complexity hierarchies and emergent behavior: 1] seed dormancy heteroblasty as seedling emergence blueprint 2] seed-soil environment communication system | 1] wild-crop-weed complexes (1.5.4; 10.3.2) 2] limiting resources (3.4.2.4) 3] signal spacetime (3.4.4) 4] speciation (4.2.2.3.6) 5] somatic polymorphism (4.3.2.2) 6] embryo adaptation (7.3) 7] dispersal in time (8.3) 8] seedling recruitment (8.4; 8.4.6.1; 8.4.7) 9] emergence (12.3) 10] seed-soil communication (ch. 13) 11] trait-based representation (15.3) |
| TR/S *Brassica napus* | Diurnal phase shift in carbon assimilaton: 1] altered photosynthesis | Pleiotropic complexity, complexity hierarchies and emergent behavior: | 1] limiting resources (3.4.2.1) 2] neighbor competition (9.2.2.1) 3] emergence (12.3) |

|  |  |  |  |
|---|---|---|---|
|  | 2] altered stomatal function<br>3] altered chloroplast-membrane structure-function<br>4] s-triazine resistance | 1] trait basis of carbon fixation in stressful, shaded habitats<br>2] pleiotropic emergence | 4] demographic representation (15.2) |
| *Abutilon theophrasti* | Leaf canopy light interception:<br>1] phototropism: inter-node elongation in shade<br>2] rapid shaded lower leaf scenescence | Trait basis of light competition<br><br>Limitations of demographic inference | 1] limiting resources (3.4.2.1)<br>2] neighbor competition (9.2.3; 9.3.2.2; 9.4)<br>3] population shifts (10.4.5.3) |

## Chapter 17: Weedy *Setaria* Species-Group (Foxtails)

**17.1 The General Nature of Weeds, The Specific Nature of Weedy *Setaria***

The nature of communities is revealed with a complete phenotype life history description of each plant species in a local community. Local plant community structure and behavior is an emergent property of its component species. Community dynamics emerges from the interacting life histories of these self-similar, self-organizing, specific components. Understanding an individual weed species in detail can provide the basis of comparison among and within weed species of a local plant community. The challenge is to discover the qualities of each member of the weed-crop community, the nature and variation of species traits used to exploit local opportunity. These insights provide a deeper, specific, understanding of biodiversity responsible for community assembly, structure and (in)stability.

It is in the nature of weedy *Setaria* to process ambient environmental information as a communication system in the process of life history seizure and exploitation of local opportunity (Dekker, 2013, 2014). The setting of weed evolution, the stage upon which diversifying evolution takes place, is a local population of variable phenotypes of a weed species in a particular locality. The nature of a particular locality is defined by the opportunity spacetime available to the weed population to seize and exploit, to survive and reproduce. Opportunity spacetime for a population is the locally habitable space for an organism at a particular time. It is defined by its available resources (e.g. light, water, nutrients, gases), pervasive conditions (e.g. heat, terroir), disturbance history (e.g. tillage, herbicides, frozen winter soil), and neighboring organisms (e.g. crops, other weed species) with which it interacts (Dekker, 2011B). It is the local niche, the niche hypervolume (Hutchison, 1957).

The weedy *Setaria* species-group provides an exemplar of widely distributed species whose complex life history behavior arises from multiple interacting traits. The specific nature of *Setaria* seed-seedling biology provides strong inferences of the nature of all weeds, elucidating the range of adaptations used by individual species to seize and exploit opportunity in agro-communities.

**17.2 The Nature of Weedy *Setaria* Seed-Seedling Life History**

The nature of weedy *Setaria* seed-seedling life history can be described as a complex adaptive, soil-seed communication, system arising from its component functional traits. Complex seed structures and behaviors emerge as a consequence of self-organization of self-similar parts forming this adaptive soil-seed communication system. Functional traits controlling seed-seedling behavior are physical information that have evolved in ongoing communication between organism and environment leading to local adaption.

The nature of weedy *Setaria* life history emerges when self-similar plant components self-organize into functional traits possessing biological information about spatial structure and temporal behavior. Spatial plant structure extends through the morphological (embryo to plant) and genetic (plant to metapopulation). Temporal life history behavior begins in anthesis/fertilization and embryogenesis; development continues with inflorescence tillering, seed dispersal in space and time, and resumption of embryo growth with seedling emergence. The interaction of self-similar plant components leads to functionally adapted traits, self-organization. These heritable functional traits are the physical reservoirs of information guiding life history development, emergent behavior. Information contained in structural and behavioral traits is communicated directly between seed and soil environment during development. The specific nature of *Setaria* is elucidated in table 17.1.

### 17.2.1 Seed-seedling life history.

**17.2.1.1 Threshold Events.** Several discrete, threshold events characterize *Setaria* summer annual life history. These threshold events provide time points allowing individual comparison in the elucidation of development. The threshold life history events begin and end with successful fertilization, continues with seed abscission, germination, and seedling emergence when the new vegetative plant develops to fertilization of new progeny.

**Table 17.1.** Plant structure (spatial, temporal) and emergent behavior examples of self-similar components, self-organization functional traits, and communication-information in the life history of weedy *Setaria*.

| | | Self-Similar Components | | Self-Organization: Functional Traits | Biological Information | Emergent Behavior |
|---|---|---|---|---|---|---|
| **SPATIAL STRUCTURE** | | | | | | |
| Plant Structure | Morphological | embryo | | somatic polymorphism | environment-plant-seed communication system | induction of differential dormancy-germinability capacity in individual seeds of inflorescence |
| | | seed envelopes | aleurone layer | | | |
| | | | caryopsis | | | |
| | | | coat | | | |
| | | | hull | | | |
| | | | glumes | | | |
| | | seed | | • somatic polymorphism<br>• phenotypic plasticity | | light capture architecture |
| | | inflorescence | spikelet | | | |
| | | | fascicle | | | |
| | | | panicle | | | |
| | | tiller | primary | | | |
| | | | 2° | | | |
| | | | 3° | | | |
| | Genetic | individual plant | | | | |
| | | local population (deme) | | self-pollination mating system | gene flow communication system | control of genetic novelty |
| | | intra-specific variants | | | | |
| | | species associations | | polyploidization speciation | | trait dispersal for local adaptation |
| | | meta-populations | | trait reservoir | | |
| **TEMPORAL STRUCTURE** | | | | | | |
| Life History Behavior | Seed formation | time of embryogenesis of individual seeds on inflorescence | | seed germination heteroblasty: differential development: hull, placental pore, TACL membrane; oxygen scavenger protein | environment-plant-seed communication system | heterogeneous seed germinability capacity |
| | | time of tillering inflorescence branching | | | | |
| | Seed dispersal | time of abscission of individual seeds | | invasion and colonization | | seed dispersal in space |
| | Seed pool behavior in soil | individual seeds from several inflorescences, plants, years | | • hull topography: water film oxygenation | soil-seed oxygen communication system | • seed dispersal in time |
| | | | | • placental pore: water film channel | | • annual dormancy-germination cycling in soil |
| | | | | • TACL membrane: O$_2$ transport | | |
| | Seedling emergence | | | • oxygen scavenger: embryo O$_2$ regulation | | locally adapted emergence patterns |
| | | **Self-Similar Components** | | **Self-Organization: Functional Traits** | **Biological Information** | **Emergent Behavior** |

**17.2.1.2 Germination and seedling emergence.** One of the most important events in a plant's life history is the time of seed germination and seedling emergence, the resumption of embryo growth and plant development. Emergence timing is crucial, it is when the individual plant assembles in the local community and begins its struggle for existence with neighbors. Resumption of growth at the right time in the community allows the plant to seize and exploit local opportunity at the expense of neighbors, allowing development to reproduction and replenishment of the local soil seed pool at abscission. Soil seed pools are the source of all future local annual weed infestations, and the source of enduring occupation of a locality.

Community assembly of crops and weeds in agroecosystems, and its consequences, is a complex set of phenomena (Dekker et al., 2003). Accurate predictions of the time of weed interference, weed control tactic timing, crop yield losses due to weeds, and replenishment of weed seed to the soil seed pool, require information about how agricultural communities assemble and interact. Despite attempts at description (e.g. Booth and Swanton, 2002), little is known about the rules of community assembly. Their elucidation may remain an empirically intractable problem.

Despite this, there exist two opportunities to understand agroecosystem community assembly during the recruitment phase (e.g. seedling emergence) that predicate future interactions with other plants. The first advantage derives from the annual disturbance regime in agricultural fields that eliminates above ground vegetation (e.g. winter kill, tillage including seedbed preparation, early season herbicide use). Understanding community assembly is most tractable when starting each growing year with a field barren of above-ground vegetation and possessing only dormant underground propagules (e.g soil seed and bud pools), a typical situation in much of world agriculture.

The second advantage derives from the observation that the time of emergence of a particular plant from the soil relative to its neighbors (i.e. crops, other weeds) is the single most important determinate of subsequent weed control tactic use, competition, crop yield losses and weed seed fecundity. Seedling recruitment is the first assembly step in these disturbed agricultural communities, and is therefore the foundation upon which all that follows is based. Information predicting recruitment therefore may be the single most important life history behavior in weed management.

**17.2.2 The nature of *Setaria*: Spatial structure and temporal behavior.** Any complete description of an organism includes the concept of phenotypic function, which consists of two universes: the physical (the quantitative, formal structure) and the phenomenal (qualities that constitute a 'world') (Sacks, 1998). The nature of *Setaria* weed seeds and seedlings is presented in this review as the story of plant spatial structure (genetic, morphological) and temporal life history behaviors, instigated by functional traits, and resulting in local adaptation and biogeography. The weedy phenotype can be described in terms of its morphological structure, the embryo and specialized structures forming the seed, and the self-similar shoot tiller architecture on which reproductive inflorescences arise. The individual, self-similar seeds, form local populations (the deme) which aggregate, self-organize, into the global metapopulation. The genetic and morphological structure provides an enduring foundation for the evolution of life history behavioral adaptation: plant functions, functional traits and regulation of function. The behavioral regulation of the *Setaria* phenotype is an emergent property arising from the interaction of several morpho-physiological seed compartments (embryo, caryopsis, hull, spikelet, fascicle, inflorescence tiller). The emergent property of these interacting, self-organizing, self-similar components is seed heteroblasty: the abscission of individual seeds from a panicle, each with different inherent dormancy-germination capacities. Seed heteroblasty is the physical information forming memory of successful past seedling emergence times appropriate to a locality. It is the hedge-bet for seizing and exploiting future opportunity spacetime.

**17.2.3 *Setaria* model system of exemplar species.**

"One major biological question is how different species become unique organisms. To understand the origins of adaptation ... it is particularly useful to investigate multiple species, especially when they have independently evolved an ability to prosper under similar environmental conditions. With the sequencing of the *Setaria* genome, evolutionary geneticists now have an annual, temperate, $C_4$, drought- and cold-tolerant grass that they can comprehensively compare to other plants that have or have not evolved these adaptations ... particular traits were targeted ... for biotechnical improvement, namely drought tolerance, photosynthetic efficiency and flowering control. With a completed genome sequence, the door is now open for further development of *Setaria* as a model plant. This model can be applied to understanding such phenomena as cell wall composition, growth rate, plant architecture and input demand that are pertinent to the development of biofuel crops. In addition to its use as a panicoid model for switchgrass, pearl millet, maize and *Miscanthus, Setaria* has the model characteristics that will encourage its development as a study system for any biological process, with pertinence to the entire plant kingdom and beyond." (Bennetzen et al., 2012)

The weedy *Setaria* species-group (green foxtail, *S. viridis*; bristly foxtail, *S. verticillata*; giant foxtail, *S. faberi*; yellow foxtail, *S. pumila*; knotroot foxtail, *S. geniculata*) (Rominger, 1962) is presented herein as a weed exemplar of a complex adaptive, soil-seed communication, system. The nature of the weedy *Setaria* species-group as a complex communication system is an exemplar in the sense of Kuhn (1962), "... concrete problem-solutions ...".

*Setaria* provides a model system for seed germination, plant architecture, genome evolution, photosynthesis, and bioenergy grasses and crops. It is used "... as an experimental crop to investigate many aspects of plant architecture, genome evolution, and physiology in the bioenergy grasses ... whose genome is being sequenced by the Joint Genome Institute (JGI) of the Department of Energy.", "... it is closely related to the bioenergy grasses switchgrass (*Panicum virgatum*), napiergrass (*Pennisetum purpureum*), and pearl millet (*Pennisetum glaucum*), yet is a more tractable experimental model because of its small diploid genome ... and inbreeding nature." (Doust et al., 2009). *Setaria* provides a "... potentially powerful model system for dissecting $C_4$ photosynthesis ..." (Brutnell et al., 2010) and "... provide novel opportunities to study abiotic stress tolerance and as models for bioenergy feedstocks." (Li and Brutnell, 2011). A high-quality reference genome sequence has been generated for *Setaria italica* and *Setaria viridis* (Bennetzen et al., 2012).

The weedy *Setaria* species-group (Darmency and Dekker, 2011; Dekker, 2003, 2004a; Dekker et al., 2012a, 2012b), *S. glauca* and *S. verticillata* (Steel et al., 1983), and *S. viridis*, (Douglas et al., 1985) have been reviewed.

The author has collected a very large (more than 3000 accessions) *Setaria* spp.-gp. germplasm collection stored in the Weed Biology Laboratory, Agronomy Department, Iowa State University, Ames. It includes Japanese salt-tolerant germplasm (e.g Dekker and Gilbert, 2008), herbicide resistant biotypes (e.g. Thornhill and Dekker, 1993), systematic pre-transgenic crop era USA (notably Iowa) collections, and north temperate world populations (e.g. Wang et al., 1995a, b).

### 17.3. *Setaria* Spatial Structure

The weedy *Setaria* phenotype can be described in terms of the spatial structure of its seed and plant morphology, and by its genotypes and population genetic structure (Table 17.2). This spatial structure extends from the cells and tissues of the embryo axes, the surrounding seed envelopes, the individual seed and then plant, the local deme, and ending with the aggregation of local populations forming the global metapopulation.

**Table 17.2.** Plant spatial structure and emergent behavior examples of self-similar components, self-organization functional traits, and communication-information in the life history of weedy *Setaria*.

| | | | | SPATIAL STRUCTURE | | |
|---|---|---|---|---|---|---|
| | | **Self-Similar Components** | | **Self-Organization: Functional Traits** | **Biological Information** | **Emergent Behavior** |
| Plant Structure | Morphological | caryopsis | embryo | somatic polymorphism | environment-plant-seed communication system | induction of differential dormancy-germinability capacity in individual seeds of inflorescence |
| | | | endosperm | | | |
| | | placental pore | | | | |
| | | seed envelopes | aleurone layer | | | |
| | | | caryopsis coat | | | |
| | | | hull | | | |
| | | | glumes | | | |
| | | individual seed | | • somatic polymorphism | | |

|  |  |  | spikelet | • phenotypic plasticity |  | light capture architecture |
| --- | --- | --- | --- | --- | --- | --- |
|  |  | inflorescence | fascicle |  |  |  |
|  |  |  | panicle |  |  |  |
|  |  | tiller | primary |  |  |  |
|  |  |  | 2° |  |  |  |
|  |  |  | 3° |  |  |  |
|  | Genetic | individual plant |  |  |  |  |
|  |  | local population (deme) |  | self-pollination mating system | gene flow communication system | control of genetic novelty |
|  |  | intra-specific variants |  |  |  |  |
|  |  | species associations |  | polyploidization speciation |  | trait dispersal for local adaptation |
|  |  | meta-populations |  | trait reservoir |  |  |

Phenotypic traits in the individual are self-similar in variants produced by parent plants to achieve slightly different strategic roles in local adaptation. These strategic traits in turn are self-similar as preadaptations, exaptations, when those phenotypic traits confront new evolutionary situations in which function can change. *Setaria* self-organization arises from the interaction of self-similar plant parts (e.g. seed polymorphism, plant architecture) forming functional traits (e.g. seed heteroblasty, seed tillering infloresences) by means of environment-plant-seed communication (e.g. plastic tiller shoot development, fascicle-spikelet seed number plasticity). These emergent properties include tiller shoot development and growth of shoots forming plant architecture for capturing light for photosynthesis, and for photoperiod-seed signal communication influencing differential dormancy induction in tiller inflorescences. Emergent morphological properties in *Setaria* also include dispersal of heteroblastic seed with differential germinability capacities for fine-scale timing of germination and seedling emergence allowing *Setaria* to seize and exploit local opportunity spacetime at the expense of their neighbors.

**17.3.1 Plant morphological structure.** Plant morphological structure is the emergent property of *Setaria* seed and plant self-organization by means of functional traits for plant phenotypic plasticity (e.g. tillering inflorescence branching, fertile flower number per spikelet) and seed somatic polymorphism (e.g. seed heteroblasty). The *Setaria* seed consists of self-similar structural components nested as integrated, self-organized compartments: hull > caryopsis > embryo (Dekker, 2003; Dekker et al., 1996). The *Setaria* plant consists of self-similar structural parts integrated in a hierarchy forming the architecture of the plant body: shoot branches (tillering panicle inflorescences) > modular leaves and photosynthesizing cells > root branches with modular roots, root hairs and absorptive cells. Plant morphology provides the physical and physiological means for locally adaptation of *Setaria* phenotypes.

**17.3.1.1 Seed morphology.** The life cycle of the *Setaria* seed begins with fertilization and embryogenesis: development of the the embryo, endosperm and living tissues within the enveloping caryopsis; development of the placental pore channel connecting the exterior seed and environment to the living interior; and development of the enveloping hull and glumes of the seed exterior (Dekker, 2000). Self-similar parental tissues surround, protect and modulate the behavior of living interior zygotic tissues. Three separate nuclear genomes interact and produce the tissues that compose the *Setaria* seed. Parental tissues (2N) include the seed glumes, hull (palea, lemma), many of the crushed layers forming the caryopsis coat, and vascular remnants and residual tracheary elements at the basal abscission area, the placental pore (Rost, 1973, 1975). The zygotic tissues include those of the endosperm (3N; aleurone, aleurone transfer cells) and the embryo (2N).

**17.3.1.1.1 Caryopsis.** The caryopsis of *Setaria* seed is enveloped by the hull and consists of dead and living, parental and zygotic, tissues. The caryopsis surrounds the living components of the seed: endosperm, aleurone layer, transfer aleurone cell layer (TACL) membrane and the embryo (Dekker, 2000; Haar et al., 2012). Within the living caryopsis interior is also a putative oxygen-scavenging, heme-containing protein that sequestering oxygen to buffer the embryo against premature germination (Dekker and Hargrove, 2002; Sareini, 2002).

**17.3.1.1.1.1 Embryo.** Within the caryopsis is the embryo: scutellum, coleoptile and coleorhiza (Dekker, 2000; Haar et al., 2012). The exterior placental pore and the interior TACL are located in close proximity to the scutellum and coleorhizal tissues of the embryo at the basal end of the seed. A cementing substance causes the outer epidermis of the coleorhiza and other embryo parts to adhere to the aleurone layer (Rost and Lersten, 1970, 1973). This intimate contact provides a continuous route for the uptake of gas-saturated water from the exterior. The embryo is approximately one half the caryopsis length and is found near the surface of the caryopsis covered only by the caryopsis coat (Haar et al., 2012). The embryo scutellum surrounds the embryonic axis below and along the margins, forming a cuplike structure in which the axis lies (Rost, 1973). Before water imbibition the caryopsis is dry, the caryopsis coat wrinkled and the embryo sunken within the endosperm.

**17.3.1.1.1.2 Endosperm**. The first zygotic tissue that gas-saturated water contacts is the endosperm, which is entirely sealed from the outside by the caryopsis coat, except in the placental pore region (Dekker, 2000).

**17.3.1.1.1.3 Aleurone layer, transfer aleurone cell layer (TACL) membrane**. The outermost layer of the endosperm is the aleurone layer, which is continuous around the caryopsis (Dekker, 2000). It consists of thick tabular cells, each cell is flattened, somewhat rectangular in surface view, and 25-50 μm in length (Rost, 1971a, b). A thick primary cellulose wall encloses each cell. The matrix of the aleurone cells appears as a gray network intermeshed between storage materials. This cell layer has an abundance of protein bodies of various types, lipid bodies (oil droplets or spherosomes), as well as plastids, mitochondria, and other membrane structures, but little starch (Rost and Lersten, 1973). This outermost layer of endosperm is known to produce enzymes, and is the site where germination is first initiated. The aleurone layer is continuous around entire caryopsis, but adjacent to the placental pad the aleurone cells are strikingly different in appearance: the transfer aleurone cells. Transfer aleurone cells occur near base of the caryopsis, adjacent to the end of the coleorhiza where the seed attaches to the parent plant inflorescence. The transfer aleurone endosperm cells are enlarged, approximately columnar, and somewhat elongate perpendicular to the fruit coat (Rost and Lersten, 1970). The thickened portion of the cell wall appears heterogeneous, with that part closest to the middle lamella having a fibrous or porous appearance. These specialized aleurone cells have thick walls bearing ingrowths on the outer radial and outer tangential walls which extend into the cell protoplasm. These ingrowths form a labyrinth, the plasmalemma follows the contours of the ingrowths, thereby significantly increasing the membrane surface area of each cell. Internally the wall has a porous, sponge-like appearance. The wall ingrowths sometimes are very deeply lobed and convoluted. The inner radial and inner tangential parts of the wall lack these ingrowths, and have a middle lamella and typical appearing primary wall. These inner radial and tangential cell walls have little or no ingrowths, indicating they are not receptive to outside solute transport. These aleurone transfer cell wall ingrowths appear similar to those of certain of the transfer cells described by Pate and Gunning (Gunning and Pate, 1969; Gunning, 1977; O'Brien, 1976; Pate and Gunning, 1972). These specialized cells have already been described as playing a role in mature seed of other species (Zee, 1972; Zee and O'Brien, 1971).

**17.3.1.1.1.4 Caryopsis coat**. Immediately beneath the tracheary elements of the placental pad is a thick layer of dark, dense, suberized cells, the caryopsis coat. The caryopsis coat surrounds the embryo and endosperm. It appears as a filmy cuticle layer, shiny, oily to the touch, and gray with dark spots and 3-10 μm thick. Seen in section it is a gossamer-like film, analogous to the cuticle. The coats specked appearance derives from the degradation contents in epidermis pericarp cells. The structure of the coat is a complex of many layers of parental origin formed from crushed cells in various stages of degradation (nucellus, integuments, pericarp). The expansion of the developing caryopsis causes these cells to become crushed, thereby forming the complex caryopsis coat. At maturity the caryopsis coat is water- and gas-impermeable, except at the proximal (basal) end of the caryopsis (the placental pad and pore region). Entry of water and dissolved gases into the caryopsis occurs only through the placental pore and TACL membrane. Entry of water is never restricted through this region of the caryopsis.

**17.3.1.1.2 Placental pad and pore**. At the basal end of the foxtail seed is the site where water and dissolved gases enter, the placental pore. This hull structure is rounded and hard with a soft center consisting of remnants of the parental vascular connection into seed interior, degraded strands of former xylem and phloem tissues (Rost, 1971b). During the ontogeny of the caryopsis a single placental vascular bundle from the parent panicle supplies nutrients and water to the developing ovule. The vascular bundle enters the ovule in a proximal position on the bottom surface of the developing caryopsis. The non-living portion of the placental pore includes residual vascular elements and tracheary elements left from the pedicel connection of the seed to the parental panicle (Rost, 1972). The morphology of the placental pore is identical in dormant and non-dormant caryopsis structures. At the base of the caryopsis is a thickened region called the placental pad. The dark necrotic contents of the placental pad layers make the structure appear as a dark oval-shaped area (about 0.2 mm) when seen from the outside (Dekker et al., 1996). The caryopsis coat in the placental pad and placental pore region is different than that around rest of caryopsis. The transfer aleurone cells rest immediately adjacent to the last layer of pad cells. The placental pad may serve as a second spongy filter, yet allow free entry of gas-saturated water. At caryopsis maturity, a thick, dark, oval pad remains in the position occupied by the placental bundle. In longitudinal section this appears as an elongated multi-layered placental pad. The placental pad of the caryopsis shows evidence of reddish coloration, a morphological indicator of physiological maturity at abscission of the seed (Dekker et al., 1996). The flaired tissue beneath the placental pad is the region where the caryopsis is connected to the palea. This region is the only water-gas entrance to seed (unless physically damaged), and may serve as the first spongy filter of debris, fungal spore, and bacteria entry into the seed. The loose arrangement of the pad cells and the presence of a large number of pits allow for a relatively unimpeded flow of water and solutes into the transfer aleurone cells.

**17.3.1.1.3 Hull**. The seed hull surrounds the caryopsis and consists of the concave lemma and the palea (Haar et al., 2012). Both these non-living structures have ridges on their surface, in some species they are transverse, in others they follow the longitudinal axis. These ridges appear like the drainboard of a sink, and may provide drainage channels for liquids, gases or solid particles in the soil adjacent to the seed. Glume and hull surface ridges may function together to both to mix water and air at, and funnel gas-laden water into, the placental pore (Donnelly et al., 2012). The lemma and palea surround the caryopsis and together form a hard protective covering commonly referred to as the hull (Gould, 1968; Rost, 1975). A door or lid-like structure known as the germination lid is found at the proximal end of the lemma. It is through this structure that the coleorhiza exits the hull during germination. The germination lid is attached, hinged to the lemma on one side. The three unhinged sides lack any physical connection to the adjacent lemma. The hinge provides the only resistance to the opening of the germination lid.

**17.3.1.1.4 Glumes**. The outmost envelopes of the foxtail seed are the papery glumes that partially surround the seed hull (Dekker, 2000). These absorbent structures protect the seed as well as wick and funnel water to the placental pore region, their point of attachment to the seed. Ridges on the glumes are often at right angles to hull (lemma, palea) surface ridges. The fragile glumes detach from the seed sometime after entry into the soil.

**17.3.1.1.5 Seed**. The dispersal unit for *Setaria* is referred to as a seed. The perfect or fertile floret above is often referred to as the fruit, grain or seed. Each seed consists of a single floret subtended by a sterile lemma and two glumes. The fertile floret (seed) of *Setaria* consists of a hull (tightly joined lemma and palea) that encloses the caryopsis (embryo and endosperm surrounded by the caryopsis coat) (Dekker et al., 1996; Haar et al., 2012; Narayanaswami, 1956; Rost, 1973, 1975). The seed is composed of an indurate, usually transversely wrinkled, lemma and palea (hull) of similar marking and texture, which tightly enclose the caryopsis within at maturity. The degree of rugosity of the lemma is a valuable taxonomic diagnostic character. Rugosity varies from smooth and shiny in foxtail millet (*S. viridis*, subspecies *viridis*), to finely ridged in *S. viridis*, to very coarse rugose seed in other *Setaria* species (Rominger, 1962). This rugosity may play a role in soil-seed contact (water, gas exchange) and seed germination. Seed production is an emergent property of *Setaria* reproductive morphology. Seed numbers produced by a *Setaria* are a function of differences in plant architecture: shoot tillering-panicle formation, panicle-fascicle branching, and fascicle spikelet-floret development. Inherent plastic differences in these reproductive structures among *Setaria* species determine the reproductive responses of species and populations to available opportunity in its immediate environment (Haar et al., 2012).

**17.3.1.2 Plant Morphology.** The branching architecture of *Setaria* plants consists of hierarchically organized, nested, structural sets: tiller culm, panicle, fascicle, spikelet and floret, as in most grasses. As such, *Setaria* plant architecture is emergent property of self-organized self-similar plant parts. *Setaria* plant architecture is plastic, and includes the ability to form one or more tillering shoots whose stature and number are precisely sized to local conditions (Dekker, 2003).

**17.3.1.2.1 Panicle inflorescence.** *Setaria* panicle inflorescences develop at the terminal ends of shoot tillers (Haar et al., 2012). The inflorescence in the subgenus *Setaria* is usually narrow, terminal on the culm, very dense, cylindrical, and spicate, with very short branches (fascicles) only a few mm long (Rominger, 1962; Willweber-Kishimoto, 1962; Naryaswami, 1956). The fascicles are spirally arranged around the main axis, each bearing a number of branchlets. *Setaria* spp., panicles are composed of fascicles, which consist of spikelet (with florets) and bristle (seta only) shoots (Narayanaswami, 1956; figure 17.1).

**Figure 17.1.** Schematic diagram of weedy *Setaria* species-group reproductive shoot architecture and panicle structure; tiller and panicle types (1°, primary; 2°, secondary; 3°, tertiary): left; fascicle branching ($F_{1-5}$) on panicle axis: top, right; fascicle structure and arrangement of bristle (seta) shoots (BS) and spikelet shoots (SS) along rachilla axis: *S. pumila* (bottom, middle); *S. viridis* and *S. faberi* (bottom, right).

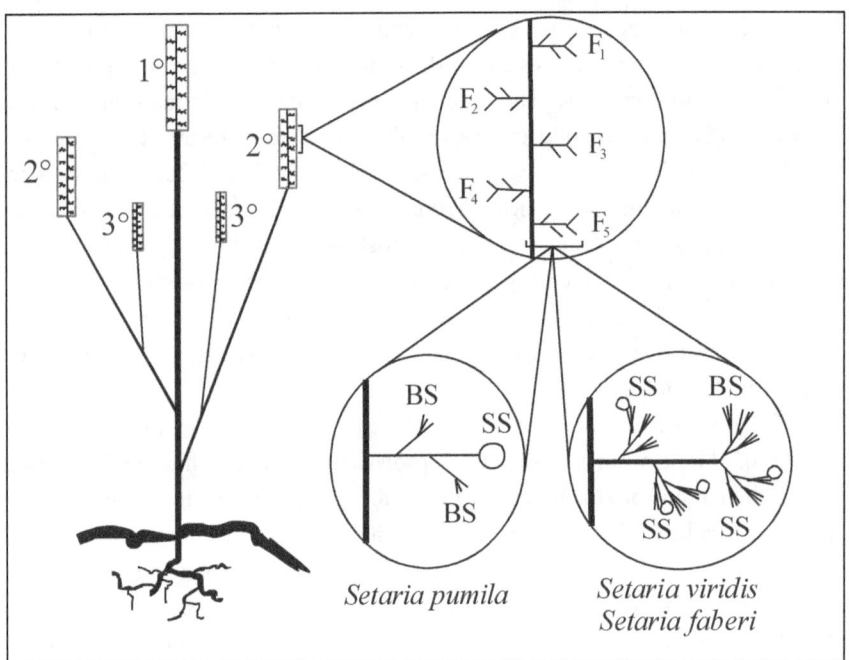

Flowering branches (fascicles) are spirally arranged along the cylindrical inflorescence of the panicle (Dekker et al., 1996). Most fascicles are composed of several branchlets, with fertile and nonfertile spikelets subtended by bristles (setae). In the mature inflorescence, the number of fascicles per unit length of panicle axis decreased from the distal to the basal end (figure 17.2). Flowering branch density along the axis ranged from 16 fascicles/cm at the distal end to 5 fascicles/cm at the basal end of the panicle. This arrangement of fascicles could allow greater light penetration to those flowers at the basal end of the panicle.

**Figure 17.2.** Number of giant foxtail flowering branches (fascicles) per length (cm) of main axis expressed as a proportion (%) of the total panicle length.

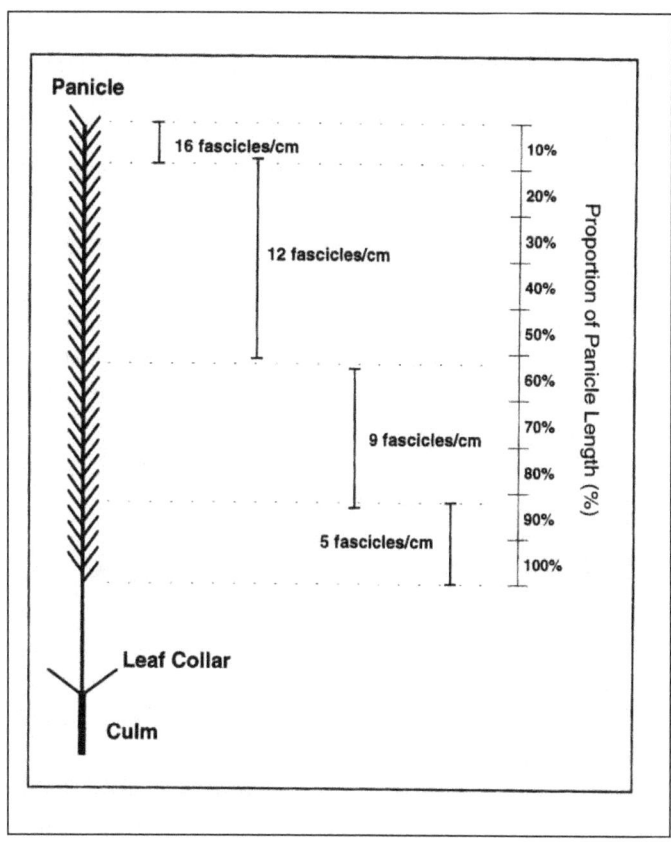

The fascicle-spikelet structure differs among *Setaria* species. In *S. faberi* or *S. viridis* the number of fascicles within each panicle varies, a plastic response to a plant's immediate environment (Clark and Pohl, 1996). Longer, earlier developing *S. faberi* or *S. viridis* panicles often have the most extensive fascicle branching, as well as more spikelets per fascicle (typically 4-6 or more per fascicle). The plastic response of *S. faberi* or *S. viridis* to its immediate environment is also revealed in the number of spikelets-florets that fully mature, which determines the seed number per panicle length (seed density). Under favorable conditions more spikelets are able to develop into seeds, while under unfavorable conditions spikelets may abort. *S. pumila* panicle morphology is different from that of *S. faberi* or *S. viridis*. Only a single, terminal, spikelet-floret is found in each *S. pumila* fascicle (Clark and Pohl, 1996). This stable, fixed morphology limits the ability of *S. pumila* to respond in a plastic way to its environment in terms of seed number relative to that of *S. faberi* or *S. viridis*.

17.3.1.2.1.1 **Spikelet**. The basic unit of the *Setaria* inflorescence is a dorsally compressed, two-flowered spikelet disarticulating below the glumes and subtended by one to several bristle-like setae (Hitchcock, 1971; Rominger, 1962). The spikelet consists of the rachilla, a sterile or staminate floret below, a perfect floret above, and three empty glumes (Li et al., 1935; Willweber-Kishimoto, 1962). For *S. viridis*, *S. faberi* and *S. verticillata* the lower floret is entirely degenerated. In *S. glauca* the lower floret has no pistil, instead it has three well developed anthers which open 3 to 7 days after the upper floret (Willweber-Kishimoto, 1962). *Setaria* spikelets are subtended by one to several setae (stalks of abortive spikelets) that persist after the spikelets disperse (Chase, 1937; Hofmeister, 1868; Prasada Rao et al., 1987). It is these distinctive setae that give the inflorescence of *Setaria* spp. it characteristic appearance (foxtails). *S. faberi* has longer panicles than *S. viridis*. *S. viridis* panicles have a greater number of seed and higher seed density than *S. pumila*. Earlier-developing panicle types were always greater than or similar to the later developing panicle type for each of these parameters (Haar et al., 2012).

**17.3.1.2.1.2 Fascicle.** The fascicle, the panicle branch, consists of 1 to six or more spikelets amid a cluster of setae (bristles) in a complicated system of branching (Naryaswami, 1956; Rominger, 1962). Spikelets are clustered on the rachis that diverges from the main panicle axis. The number of fertile spikets per fascicle within each *S. faberi* or *S. viridis* panicle is variable (e.g. 1 to 3), and plastic in response to environmental conditions (Clark and Pohl, 1996). *Setaria glauca* fascicles, unlike that of *S. viridis* or *S. faberi*, bear only a single spikelet. In the mature inflorescence, the number of fascicles per unit length of panicle axis decreases from the distal to the basal end (Dekker et al., 1996). This arrangement of fascicles is a second mechanism allowing greater light penetration to basal flowers of the panicle.

**17.3.1.2.2 Tiller.** Panicles that develop at the end of the main shoot are referred to as primary panicles (1°; Figure 17.1) (Haar et al., 2012). Secondary panicles (2°) arise at the nodes of the primary tiller, and tertiary panicles (3°) are those that branch laterally from secondary tillers. Developmentally, primary panicles flower first on a plant, followed by secondary, then tertiary.

**17.3.2 Plant genetic structure.** *Setaria* genetic structure includes the genotype structure of individuals within each species as well as the population genetic structure of those individuals aggregated into local populations, species associations and the global metapopulation. The plant genetic structure is an emergent property of *Setaria* plant self-organization by means of the functional trait for a self-pollination mating system controlling gene flow communication between local populations. Self-similar individual plants share genes from the local population to the global metapopulation to express the genetic spatial structure of *Setaria*. Soil seed pools of local populations form a physical memory of past successful genotypes whose functional traits can be shared by gene flow communication between continuous local populations of the larger global metapopulation. Rare (auto-, allo-) polyploidization events have resulting in new, reproductive, species (e.g. *Setaria faberi*) from which self-organization is responsible for wild-crop-weed, species-group and polyploid cluster species associations.

**17.3.2.1 Genotype structure.** *Setaria* genotype structure includes individual plants and variants of each species.

**17.3.2.1.1 Individual plant**. World-wide there are 125 *Setaria* species divided among several subgenera, 74 of these species are from Africa (Hubbard, 1915; Rominger, 1962). The taxonomy of the genus is very complex, and an accurate classification has been confounded by the high degree of overlapping morphological characters both within and between species, and the diverse polyploidy levels within the genus. The genus *Setaria* belongs to the tribe Paniceae, subfamily Panicoideae and family Poaceae (Pohl, 1978). *S. faberi, S. pumila, S. verticillata, S. viridis* (both subspp. *italica* and *viridis*) are Eurasian adventives. The origins of *S. geniculata* are of particular note. *S. geniculata*, a perennial, is the only weedy *Setaria* native to the New World, and very closely resembles *S. pumila*, an annual species of Eurasian origin (Rominger, 1962; Wang et al., 1995b). *Setaria geniculata* and *S. pumila* are frequently misidentified. It has been suggested that the cause of this enigma was an ancient, pre-Columbian, dispersal event westward from the Old to New World (Rominger, 1962). Foxtail millet (*S. viridis* subsp. *italica*) and weedy *S. viridis* subsp. *viridis* are subspecies of *S. viridis* and have continuous and overlapping genetic variation, evidence of the weedy origins of the crop (Prasada Rao et al., 1987; Wang et al., 1995a). Relative genetic diversity within each of the four weedy foxtail species is low to exceedingly low (*S. geniculata* > *S. viridis* > *S. pumila* > *S. faberi*, monomorphic) in comparison to an "average" plant species (Hamrick and Godt, 1990) (Wang et al., 1995a, b). The genus *Setaria* is cosmopolitan, and can be found on every landmass in the world except the polar regions (Prasada Rao et al., 1987; Rominger, 1962). In the western hemisphere, *Setaria* has its center of distribution in Brazil, and radiates poleward north and south (Rominger, 1962). The weedy *Setaria* spp. are primarily temperate species but are widely distributed between 45° S and 55° N latitudes (Holm et al., 1977, 1997; Wang et al., 1995a, b). They are found in every state in the continental U.S. and every province in Canada (Lorenzi and Jeffery, 1987). The Eurasian *Setaria* are rare in the southwest U.S., Mexico and more southerly, tropical regions.

**17.3.2.1.2 Intra-specific variation.** A number of morphological variants of *Setaria* viridis have been named (e.g. *S. viridis* var. major (Gaud.) Posp., giant green; *S. viridis* var. robusta-alba Schreiber, robust white; *S. viridis* var. robusta-purpurea Schreiber, robust purple; *S. viridis* var. pachystachys (Franch et Savat.) Makino et Nemoto; *S. viridis*, var. vivipara (Bertol.) Parl.; (Dore and McNeill, 1980; Hubbard, 1915; Kawano and Miyaki, 1983; Schreiber and Oliver, 1971)), although their taxonomic validity has been questioned (Wang et al., 1995a) (Dekker, 2003). Often the most striking morphological differences among *Setaria* arise from biotypes with similar allelic variation (e.g. compare *S. viridis* subsp. *viridis* var. pachystachys and *S. viridis* subsp. *italica* race maxima), while genotypic variation can be wide within nearly identical morphologies (e.g. compare Old World *S. pumila* and New World *S. geniculata*). Many of the morphological variants described as biotypes are in fact extreme examples of continuous characters (e.g. leaf coloration in *S. viridis* var. robusta-purpurea); Schreiber and Oliver, 1971). One of the most striking observations of *Setaria* behavior is the occurrence of phenotypic heterogeneity, rather than genetic diversity, among the individuals of a population as a means of exploiting a locality (Scheiner, 1993; Wang et al., 1995a, 1995b). This phenotypic heterogeneity takes the form of phenotypic plasticity and somatic polymorphism in many of its most important traits, especially during reproduction. Most of the phenotype variation of *Setaria* has not been characterized. Only the most obvious behaviors and morphologies have been described, those characters most apparent in crop management situations. An increase in herbicide resistant *Setaria* variants has been observed. The resistance mechanisms include those that exclude, as well as metabolically degrade, the herbicides (e.g. Thornhill and Dekker, 1993; Wang and Dekker, 1995). Physiological variation in dormancy and germinability exists (Tranel and Dekker, 2002; Dekker et al., 1996; Norris and Schoner, 1980). Variation in drought tolerance among *Setaria* has been observed (Blackshaw et al., 1981; Manthey and Nalewaja, 1982, 1987; Taylorson, 1986). Potentially salt-tolerant genotypes of *S. viridis* and *S. faberi* been observed along the seacoasts of Japan (Chapman, 1992; Kawano and Miyaki, 1983; author's personal observation, 1992 and 2000, data not reported). There may exist an intimate physiological and morphological relationship between drought tolerance, salt tolerance and seed dormancy (Dekker and Gilbert, 2008). This is most apparent in extreme habitats where foxtail millet remains an important crop and cultivated cereal (e.g. Central Asia including Afghanistan, India, sub-Saharan Africa).

**17.3.2.2 Population genetic structure.** The emergent property of interacting local *Setaria* populations and species associations is the global metapopulation. Self-similar individual seeds and plants aggregate into local soil-seed pools and communities across the global landscape to form a *Setaria* metapopulation. This self-organization occurs by means of a gene flow communication system within the metapopulation. Low diversity species are associated in habitats to exploit overlapping niches, opportunity space-time. Individual demes are highly differentiated among themselves over the landscape and world metapopulation, populations finely adapted to local opportunity. Appropriate levels of *Setaria* biodiversity are maintained by a conservative, self-pollenating mating system with carefully limited outcrossing for the introduction of novel traits anticipating changing habitats.

**17.3.2.2.1 Local population (deme).** The pattern of genetic diversity within an individual weedy *Setaria* sp. is characterized by unusually low intra-population genetic diversity, and unusually high genetic diversity between populations, compared to an "average" plant (Wang et al. 1995a, b). These two patterns of population genetic structure appear to typify introduced, self-pollinated weeds that are able to rapidly adapt to local conditions after invasion and colonization. Although relative genetic diversity within each of the several *Setaria* species is very low, differences between homogeneous populations are high, indicating a strong tendency for local adaptation by a single genotype. Nearly all populations analyzed in America consisted of a single multilocus genotype, while more diversity could be found within European and Chinese populations.

**17.3.2.2.1.1 *Setaria viridis*.** Genetic bottlenecks associated with founder events may have strongly contributed to that genetic structure. The founder effect has been observed in *S. viridis* to a certain degree. *S. viridis* accessions from North America have reduced allelic richness compared to those of Eurasia. Genetic drift probably has occurred in *S. viridis*, as indicated by the many fixed alleles in North American accessions. Multiple introductions of *S. viridis*, in the absence of local adaptation, should have produced a random, mosaic pattern of geographic distribution among North American accessions. Instead, a strong intra-continental differentiation is observed in *S. viridis* populations, both in Eurasia and North America (Jusef and Pernes 1985; Wang et al. 1995a). *S. viridis* populations in North America are genetically differentiated into northern and southern groups separated on either side of a line at about $43.5°$ N latitude. The northern type is less variable than southern type. This regional divergence suggests that natural selection has partitioned *S. viridis* along a north-south gradient. These observations imply that the present patterning among *S. viridis* populations in North America is the consequence of multiple introductions into the New World followed by local adaptation and regional differentiation.

**17.3.2.2.1.2 *Setaria pumila-S. geniculata*.** *Setaria pumila* populations are genetically clustered into overlapping Asian, European, and North American groups (Wang et al. 1995b). *S. pumila* populations from the native range (Eurasia) contain greater genetic diversity and a higher number of unique alleles than those from the introduced range (North America). Within Eurasia, Asian populations have greater genetic diversity than those from Europe, indicating *S. pumila* originated in Asia, not Europe. These observations indicate there have been numerous introductions of *S. pumila* from Eurasia to North America, the majority from Asia. This pattern may also explain the enigma of the origins of *S. pumila* and *S. geniculata*. The pattern of *S. pumila* genetic variability is North American was unexpected: nearly the entire diversity of this species appears to be encompassed by accessions from Iowa, whereas populations collected from other North American locations were nearly monomorphic for the same multilocus genotype (Wang et al. 1995b). In this respect, it is significant or coincidental that this pattern was repeated in the diversity data for *S. viridis*, also a native of Eurasia (Wang et al. 1995a). Iowa possesses a surprising *Setaria* genetic diversity: all five weedy *Setaria* species are present. Typically two or more *Setaria* species occur in the same field at the same time. Iowa is the center of the north-south agro-ecological gradient in North America, perhaps leading to greater environmental heterogeneity. Despite originating on different continents, the genetic diversity patterns for *S. geniculata* parallel those for *S. pumila* and *S. viridis*: greater genetic diversity occurs in accessions from the New World compared to those from the introduced range (Eurasia) (Wang et al. 1995b). This most likely reflects genetic bottlenecks associated with sampling a limited number of founding propagules and the history of multiple introductions from the Americas to Eurasia. The population genetic structure of *S. geniculata* consists of three nearly distinct clusters, groups from Eurasia, northern United States, southern United States. Accessions from Eurasia and North America are approximately equally diverse genetically. Within North America, *S. geniculata* accessions were strongly differentiated into southern and northern groups at about the Kansas-Oklahoma border ($37°$ N latitude); indicating greater genetic differentiation within North American populations than between North American and Eurasian populations.

**17.3.2.2.1.3 *Setaria faberi*.** *S. faberi* contains virtually no allozyme variation. Fifty of the 51 accessions surveyed by Wang et al. (1995b) were fixed for the same multilocus genotype.

**17.3.2.2.1.4 General and specialized genotypes.** The population genetic structure of many widely distributed, introduced, self-pollinating, weed species clearly indicates that a high level of genetic variation is not a prerequisite for successful colonization and evolutionary success (Allard, 1965; Barrett and Shore, 1989). Two contrasting, adaptive strategies are hypothesized to explain weedy adaptation and the success of colonizers: genetic polymorphism with the development of locally adapted genotypes ("specialists"), and phenotypic plasticity for the development of "general purpose" genotypes (generalists) adaptable to a wide range of environmental conditions (Baker, 1965, 1974; Bradshaw, 1965; Barrett and Richardson, 1986). Weedy *Setaria* population genetic structure allows some insight into the dichotomy of generalism versus specialization. *Setaria* possess both generalists and specific strategy types. A key observation is that although a single, multilocus genotype predominates or is fixed in all populations, not all multilocus genotypes are equally prevalent within individual weedy *Setaria* species (Wang et al., 1995a, 1995b). The most striking example of this is in *S. pumila*, where the most common multilocus genotype was found in 53 (of 94 evaluated) accessions surveyed by Wang et al., (1995b) from widely separated geographic areas in Europe, Asia, and North America. Overlaying this pattern of homogeneity were other, less abundant *S. pumila* genotypes, each with a more narrow geographical and ecological distribution. The population genetic structure of *S. viridis* suggests that this species also possesses both generally adapted and specially adapted genotypes. Many *S. viridis* populations are genetically strongly differentiated (e. g., northern and southern North America), while other populations remain identical. The most widely distributed *S. viridis* genotype (fixed multilocus genotype) occurred in 25 (of 168 evaluated) accessions from six countries, from both the Old World and the New World (Wang et al., 1995a, 1995b). Interestingly, this common genotype has not yet been found in Iowa, despite the diversity of *S. viridis* populations found in this state. These observations reveal a complex hedge-betting strategy by individual weedy *Setaria* species that balances general adaptation with the additional niche opportunities available with specialization. The ratio of general to special genotypes in locally adapted populations in a species for invasion is quite different within *S. pumila*, *S. geniculata* and *S. viridis* (Wang et al., 1995a, 1995b).

**17.3.2.2.2 Species associations.** The several self-similar *Setaria* species have self-assembled into several genetic and ecologically adapted associations (table 17.3). *Setaria* is phylogenetically divided into two clades: *S. viridis*, *S. pumila* (Doust and Kellogg, 2002). The diploid parental *S. viridis* precedes the tetraploid *S. verticillata* and *S. faberi* forming a polyploid species cluster (Darmency and Dekker, 2011; Dekker, 2011b). The genus *Setaria* also contains the important world crop foxtail millet (*S. viridis*, subsp. *italica*) whose geographic distribution and evolutionary history is intimately connected with the weedy foxtails. Gene flow communication between millet crop and weeds in agro-ecosystems forms the stable, global, wild-crop-weed complex (De Wet, 1966) in both clades (Dekker, 2003, 2004a). The several weedy *Setaria* frequently exploit the same range, or same field, the emergent property of which is the *Setaria* species-group (a group of closely related species, usually with partially overlapping ranges; Lincoln et al., 1998). Frequently more than one *Setaria* species co-exist together in a locality; possibly allowing a more complete exploitation of resources left available by human disturbance and management. Several weedy foxtail species often coexist in a single field (most commonly with *S. viridis*), each exploiting a slightly different "opportunity space" or niche.

Table 17.3. *Setaria* species associations: clade, wild-crop-weed complex, polyploid species cluster (diploid, tetraploid) and species-group.

| Clade | Crop | Wild-Weed | Weed |
|---|---|---|---|
| | Diploid (2N) | Diploid (2N) | Tetraploid (4N) |
| *S. viridis* | *S. viridis*, subsp. *italica* | *S. viridis*, subsp. *viridis* | |
| | | | *S. verticillata* |
| | | | *S. faberi* |
| *S. pumila* | | | *S. pumila* |
| | | | *S. geniculata* |
| Clade | | Species-Group | |
| | Polyploid Cluster | | |

**17.3.2.2.3 Metapopulations.** Gene flow and seed dispersal linkages form a metapopulation communication system providing continuity and stability to local demes, conservation of adaptive traits (physical knowledge), and a long-term hedge-bet for enduring success of the species-group. The limited diversity contained within weedy foxtails is partitioned across the landscape, continent and world among populations (metapopulation structure) as alternative homozygous genotypes; heterozygosity was rarely detected (Wang et al., 1995a, b). This structure suggests strong inbreeding in nearly all populations of the four species. A common multilocus genotype of both *S. viridis* and *S. pumila* occurred in many accessions from widely separated geographic areas, and indication of general adaptation. Therefore, the geographic distance (from global to local) separating foxtail populations does not indicate the genetic distance separating them. Metapopulation structure provides a means of conserving and preserving useful traits, available as conditions change in local demes. Self-pollination is a conservative mating system, the means by which *Setaria* generates appropriate amounts of variability. Natural selection and stochastic forces both act to keep species genetic diversity low. Each local population is an island of local adaptation, sharing with neighbors at very low rates of gene flow. Rapid and numerous dispersal via human mediation ensures functional traits move between disturbed agro-habitats to ensure fine-scale adaptation in the seizing and exploiting the changing nature of local opportunity spacetime.

### 17.4. *Setaria* Seed-Seedling Life History Behavior

**17.4.1 Life history behaviors: Functions, traits and information.** *Setaria* plant spatial structure is the foundation for emergent life history behavior: self-similar timing of life history processes regulated by functional traits expressed via environment-plant communication (table 17.4).

**Table 17.4.** Plant temporal structure and emergent behavior examples of self-similar components, self-organization functional traits, and communication-information in the life history of weedy *Setaria*.

| | | TEMPORAL STRUCTURE | | |
|---|---|---|---|---|
| Life History Processes | Self-Similar Components | Self-Organization: Functional Traits | Biological Information | Emergent Behavior |
| Life History Behavior — Seed formation | time of embryogenesis of individual seeds on an inflorescence; time of tillering inflorescence branching | seed germination heteroblasty: differential development: hull, placental pore, TACL membrane; oxygen scavenger protein | environment-plant-seed communication system | heterogeneous seed germinability capacity |
| Life History Behavior — Seed dispersal | time of abscission of individual seeds | seed size, number; location | | seed dispersal in space |
| Life History Behavior — Seed pool behavior in soil | time of state changes in individual seeds from several inflorescences, plants, years | • hull topography: water film oxygenation<br>• placental pore: water film channel<br>• TACL membrane: $O_2$ transport | soil-seed oxygen communication system | • seed dispersal in time<br>• annual dormancy-germination cycling in soil |
| Life History Behavior — Seedling emergence | | • oxygen scavenger: embryo $O_2$ regulation | | locally adapted emergence patterns |

*Setaria* life history behavior is a Markov chain of irreversible (dormancy induction; seed dispersal, germination, seedling emergence, neighbor interactions) and reversible (seed after-ripening, dormancy re-induction) processes of seed-plant state changes (flowering plants, dormant seed, seed germination candidate, germinated seed, seedling) regulated by morpho-physiological traits acting through environment-plant communication systems (environment-plant-seed, soil-seed) (figure 17.3).

**17.4.2 Seed formation and dormancy induction.** Individual panicles on a single parent plant produce a diverse array of seeds, each with potentially different after-ripening requirements for germination (Dekker and Hargrove, 2002). The production of seeds with different levels of dormancy (experimentally revealed by the after-ripening dose (e.g., time at 4° C, moist, dark) required for germination) is a function of plant architecture. Earlier fertilized seeds (both intra- and inter-panicle) are relatively more dormant than later developing seeds. The first seeds on an individual panicle were shown to possess relatively greater dormancy (greater after-ripening requirement) than the last seeds maturing on the same panicle (Haar, 1998). Additionally, primary (1°) panicles produce seeds with relatively greater dormancy than those produced on secondary (2°), and again on tertiary (3°), panicles of the same parent plant. Significant heterogeneity in dormancy states among seeds shed by a single plant allow these species to emerge at appropriate times within a cropping season and in different years (Dekker et al., 1996; Forcella et al., 1997). Soil seed banks consisting of diverse foxtail species and genotypes, each contributing a heterogeneous collection of dormancy phenotypes, reveal a hedge-betting strategy for adaptation to changing conditions within agroecosystems (e.g. Cohen, 1966; Philippi and Seger, 1989).

**Figure 17.3.** Schematic diagram of weedy *Setaria* sp. seed-seedling life history behavior in soil pools: plant/seed state pools (1-5) and processes (A-G; C-D are reversible).

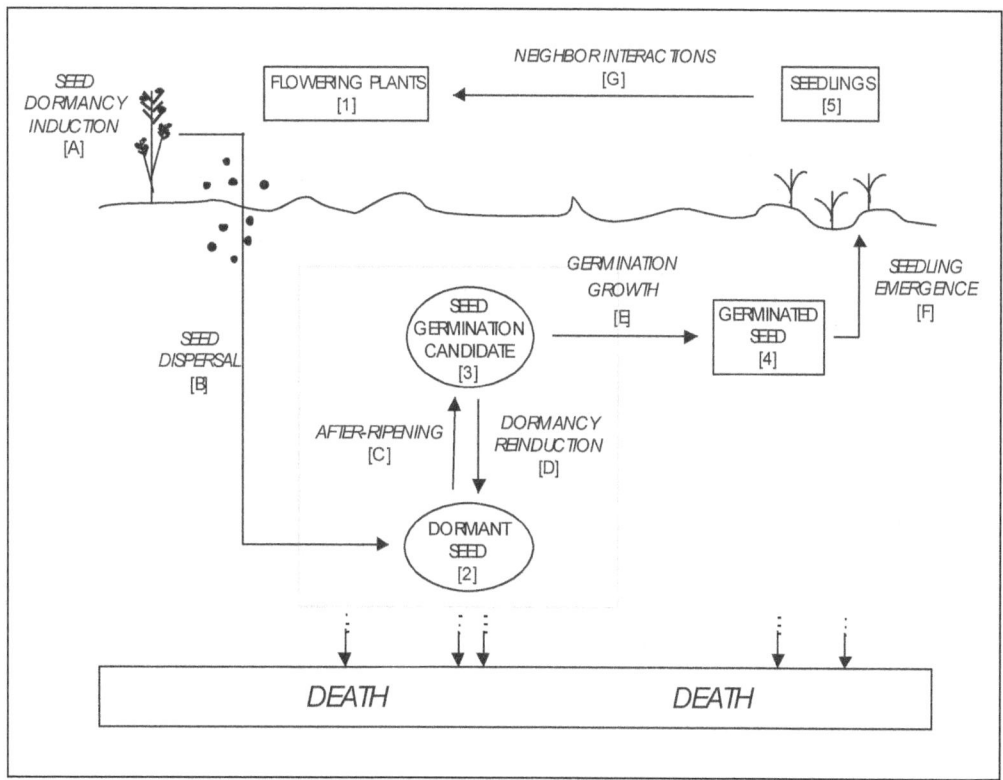

**17.4.2.1 Inflorescence and flowering.** The branching architecture of *Setaria* plants is plastic, an ability to form one or more tillering shoots whose stature and number are precisely sized to local conditions. A complex pattern of branching, from plant to spikelet, provides diverse microenvironments within which different levels of dormancy are induced in individual seeds of the panicle, and among panicles on a common plant (Dekker, 2003).

The pattern of flowering and seed fertilization is a complex process occurring on several different spatial and temporal scales: within and among individual panicles on the same plant, during the annual growing season, and during a daily or diurnal period (Darmency and Dekker, 2011; Dekker, 2003).

**17.4.2.1.1 Flowering pattern on panicle main axis.** The first externally visible event in the reproductive phase is the emergence of the immature panicle from the subtending leaf sheath of its tiller. Once flowering has proceeded basipetally to the bottom of the panicle the culm elongates and extends out of the leaf collar. Two patterns of flowering along the inflorescence were observed (Dekker et al., 1996). A consistent pattern of flowering of the first spikelet in each fascicle along the axis was observed. The flowering of these first spikelets of the fascicle was first apparent in the distal 30-40% section of the panicle axis (figure 17.4). The second group of spikelets then flowered in two directions at the same time, in one direction toward the proximal end and in another direction toward the distal end of the axis. Flowering of these first spikelets reached completion at the distal end first, followed by those at the basal end. A second pattern of flowering was observed among the other fertile spikelets within individual fascicles along the axis. This second pattern was very complex and no consistent observations of its nature along the axis were made in this study.

**Figure 17.4.** Pattern of *S. faberi* flowering spikelets on the panicle main axis over time with development expressed as a proportion (%) of the total panicle length; circled numbers indicate axis area of temporal progression of flowering: 1, first axis zone of flowering; 8, last area of axis to flower.

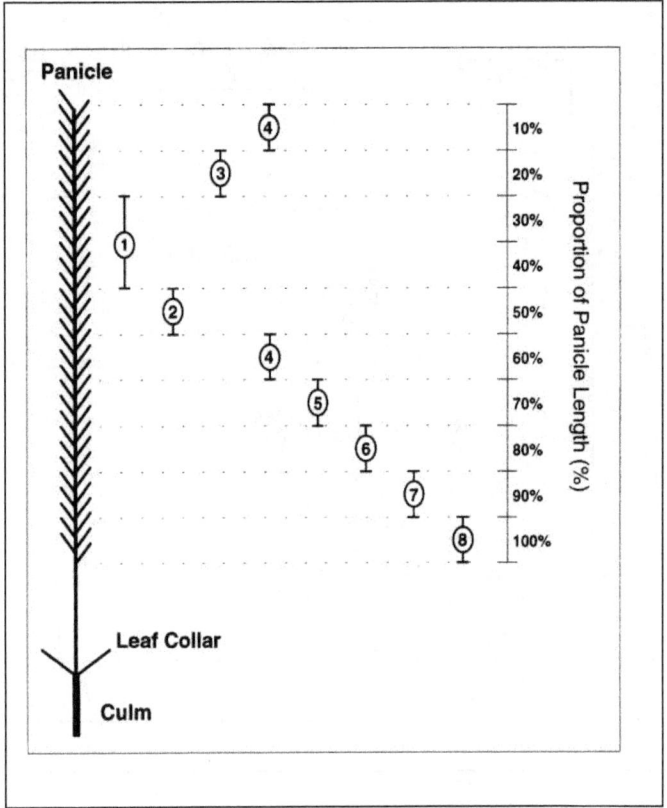

**17.4.2.1.2 Elongation pattern of panicle and culm axis.** A consistent pattern of panicle and culm elongation was observed; namely, elongation ceased distal to the appearance of the first flower in a section of the axis (figure 17.5) (Dekker et al., 1996). Axis elongation then occurred basipetally to that flower towards the culm and leaf collar. As new flowers appeared along the axis toward the basal end, elongation distal to these flowers also ceased. The culm supporting the panicle did not elongate until flowering had proceeded to the basal end of the inflorescence. At that time the culm elongated and extended out from the leaf collar. The time from the appearance of the first flower on the axis until the time when the first spikelet of the most basipedal fascicle flowered was about 9 d. In *S. pumila* the lower floret has no pistil, instead it has three well developed anthers which open 3 to 7 days after the upper floret (Willweber-Kishimoto, 1962). Elongation of the inflorescence and culm could also allow greater light penetration into the panicle as it rises above its nearby competitors (intraplant, intraspecies, interspecies). Both fascicle spacing and axis elongation could serve as mechanisms differentially modifying the light microenvironment of individual spikelets along the axis during flowering and embryogenesis. This differential light reception may be a mechanism by which different seed germinability phenotypes are shed from a panicle. Differential induction of germinability in *S. faberi* seed as a response to variable field light conditions has been observed previously in giant foxtail (Schreiber, 1965a).

**Figure 17.5.** Pattern of *S. faberi* panicle and culm axis elongation growth (cm) overt time (9d) with development; A-G represent the location of the first flowering spikelet to appear on the individual day indicated, with the zone of elongation indicated below it; BS, base of inflorescence defined by most basal spikelet; LC, leaf collar area subtending the panicle; the inflorescence peduncle (culm) is that distance between BS and LC.

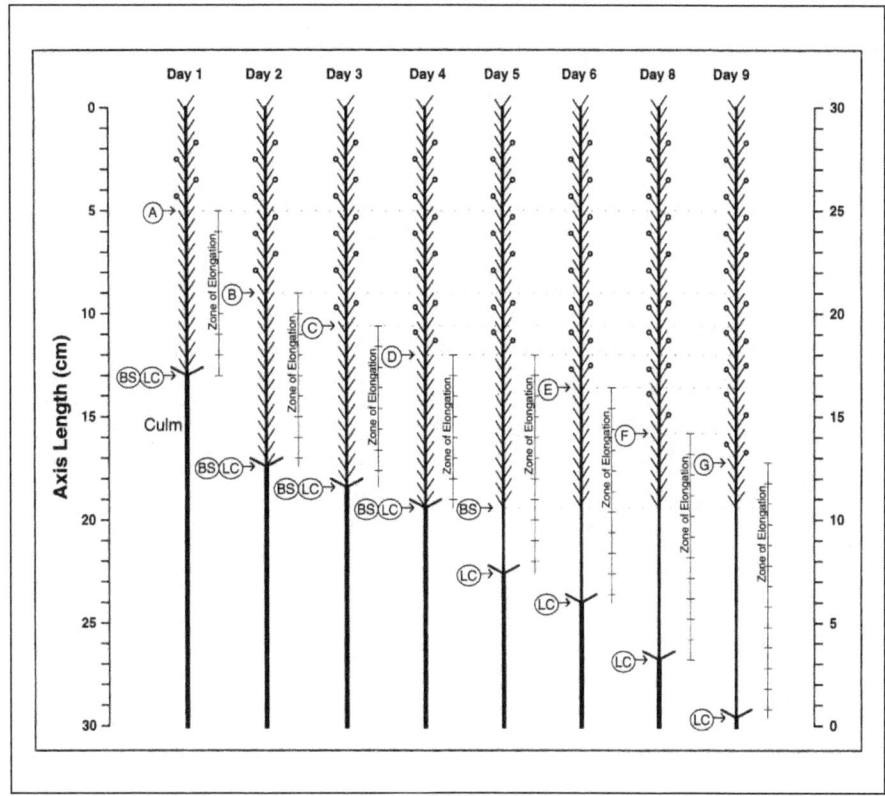

**17.4.2.1.3 Seasonal flowering pattern.** Panicles usually emerge after the summer solstice, and flowering commences and continues after that time into the autumn. There is considerable variation in the patterns of panicle tiller emergence, flowering, embryogenesis and seed abscission among and within weedy *Setaria* spp. inflorescences. Time of seedling emergence, as well as the photoperiod and temperature coinciding with an individual plant's seasonal growth period, are major determinants of these flowering phenomena (Haar, 1998; Stevens, 1960). The time from seedling emergence to the appearance of the first *S. faberi* and *S. viridis* panicle is highly variable and is dependent on the diurnal photoperiod (Dekker et al., 1996; Nieto- Hatem, 1963; Schreiber and Oliver, 1971; Schreiber, 1965a; Stevens, 1960). *Setaria* appear to initiate flowering and fertilization in response to the shortening photoperiod (lengthening dark period) after the summer solstice. Long daylengths appear to prolong these reproductive responses, while continuous light markedly delays panicle production and inhibits flowering, in all the weedy and crop *Setaria* spp. (Fabian, 1938; King, 1952; Peters and Yokum, 1961; Peters et al., 1963; Santelman et al., 1963; Schreiber, 1965a; Vanden Born, 1971). Triazine resistant biotypes of *S. viridis* flower earlier than susceptible variants (Ricroch et al., 1987). Blooming in wild type biotypes is negatively correlated with temperature and positively with relative humidity (Li et al., 1935). The response of flowering to photoperiod in weedy *Setaria* spp. may be hastened by higher temperatures (Ricroch et al., 1987; Schreiber and Oliver, 1971).

**17.4.2.1.4 Diurnal flowering pattern.** *Setaria* flowering usually occurs during specific time periods of the day, depending on local environmental conditions, and is an endogenously regulated rhythm with two daily maxima peaks (i.e. 0400 to 0800 h and 1900 to 0000 h) in the dark (Dekker et al., 1996; Li, 1935; Rangaswami Ayyangar et al, 1933; Willweber-Kishimoto, 1962). The average time between flower opening and closing is 70 min. (Kishimoto, 1938).

**17.4.2.2 Mating and Fertilization**

**17.4.2.2.1 Hybridization.** The weedy *Setaria* are primarily a self-pollenated species (Mulligan and Findlay, 1970; Pohl, 1951). Wind pollenation (anemophily) is the mode in those rare circumstances of outcrossing (Pohl, 1951; Nguyen Van and Pernes, 1985). Pollen and gene flow in the weedy *Setaria* can be intra-specific (autogamy, self-fertilization; allogamy, outcrossing) or inter-specific hybridization (introgression between different *Setaria* species) (Dekker, 2003). Although the anthers are visible at flowering, and then feathery stigmates also exert from the glumes, the fertilization principally occurs within the flower. Stigmate maturity is generally synchronous with anther dehiscence, which results in a high probability of self-pollination. Some rare cases of protogyny, or in contrast of protandry, can be observed according to the genotype and the environmental conditions, but in both cases the more proximate pollen belongs to the other spikelets of the same panicle, thus leading again to self-pollination. Wind pollination (anemophily) is the mode in those rare circumstances of outcrossing (Pohl 1951), with pollen probably moving at most a few dozen meters (Wang et al. 1997). (Darmency and Dekker, 2011). Natural outcrossing between *Setaria* plants of the same species, intraspecific hybridization, does occur and is an important source of new variants. In *S. viridis*, spontaneous outcrossing rates among plants in the field have been observed to be between 0 and 7.6% of the autogamous rates (Darmency et al., 1987a, b; Prasado Rao et al., 1987; Till-Bottraud, 1992; Takashi and Hoshino, 1934; Li, 1934; Li et al., 1935; Macvicar and Parnell, 1941). Till-Bottraud et al. (1992) found 0.74% outcrossing for *S. viridis* plants spaced every 0.25 m. Similarly, a selfing rate exceeding 99% was reported in Jasieniuk et al. (1994) for *S. viridis*. Using a dominant herbicide-resistance marker, Volenberg and Stoltenberg (2002) found an outcrossing rate ranging from 0 to 2.4% for *S. faberi* planted at 0.36 m interval. Introgression of pollen between weedy species, interspecific hybridization, is a rare event. When it does occur it can have very important consequences, although the progeny are almost always sterile (Clayton, 1980; Li et al.,1942; Osada, 1989; Poirier-Hamon and Pernes, 1986; Stace,1975; Till-Bottraud et al., 1992; Willweber-Kishimoto, 1962).

**17.4.2.2.2 Asexual reproduction.** Asexual (meiotic) reproduction is not a common mode of reproduction in weedy *Setaria*, and is probably limited to *S. geniculata*, a perennial species with short, branched, knotty rhizomes (Rominger, 1962). Agamospermy (seed formation without fertilization) in *S. viridis* has been noted (Mulligan and Findlay, 1970), and has also been suggested for *S. glauca* and *S. verticillata* (Steel et al., 1983). Apomixis has also been reported in other *Setaria* species within the *Setaria* subgenus (*S. leucopila*, *S. machrostachya* and *S. texana*; Chapman, 1992; Emery, 1957). *Setaria faberi* and *S. pumila* tillers readily root in soil when cut, separated from the plant, and buried in moist soil, an important trait allowing weedy *Setaria* to reestablish themselves after cultivation and mowing (Barrau, 1958; Santlemann et al., 1963; Schreiber, 1965b)

**17.4.2.2.3 Speciation and reproductive barriers between *Setaria* species.** New *Setaria* species can be formed by several processes (e.g. mutation, hybridization, polyploidization, etc.), but such events are very rare. Partial reproductive barriers exist between the *Setaria* species, a genetic condition favoring introgression and gene flow at very low levels within the species-group and wild-crop-weed complex (Darmency et al., 1987a, b; de Wet, 1966; Harlan, 1965; Harlan et al., 1973). Panicle height differences, and times of fertilization (pollen shed; stigma receptivity), both prevent hybridization events from occurring between many populations (Willweber-Kishimoto, 1962). Reproductive barriers between the *Setaria* species does not occur at the level of pollen germination and pollen tube growth in the stigma in any combination of pollen and stigma among *S. italica*, *S. viridis*, *S. faberi*, *S. pumila*, and *S. pallide-fusca* (Willweber-Kishimoto, 1962). Polyploidization (either alloploidy or autoploidy) has played an active role in speciation and delimiting taxa in *Setaria* (Khosla and Sharma, 1973). Polyploidization of the more diverse and ancient *S. viridis* may have been the genesis of the specialized and less diverse *Setaria* spp. set (*S. faberi*, *S. pumila*, *S. verticillata* and *S. geniculata*). Allotetraploid forms of *S. faberi* and *S. pumila* have been explained as ancient crosses of *S. viridis* with an unknown diploid species (Li, 1942; Kholsa and Singh, 1971; Till-Bottraud et al., 1992). This polyploidization in *S. faberi* may be a relatively recent evolutionary event (Wang et al., 1995b). *S. verticillata* (n=18) may be the product of chromosome doubling of *S. viridis*, autotetraploidy (Till-Bottraud et al., 1992). Geographical barriers to inter-species hybridization in current times are much less important now than in the past.

**17.4.2.3 Seed formation and embryogenesis.** The life cycle of foxtail seed begins with seed development, the development of outer seed envelopes, the inner endosperm and the embryo (Dekker, 2003, 2004). Embryogenesis in *Setaria* begins with anthesis and fertilization of the new embryo, and ends when the embryo, enveloped with parental tissues (i.e. lemma, palea, caryopsis coat), becomes physiologically separated from the parent plant (abscission). Three separate nuclear genomes interact and produce the tissues that compose the foxtail seed. The parent contributes sporophytic tissues to the seed that nurture and protect the developing embryo, as well as provide significant contributions to dormancy and control of post-abscission germination timing (e.g. seed envelopes such as the caryopsis coat), and physical protection in the soil. The duration of seed development and embryogenesis of an individual *S. faberi* seed grown in controlled environmental conditions is 8 to 15 days (Haar, 1998; Dekker et al., 1996).

The morphology of foxtail seeds also provides important clues about what environmental factors limit germination and maintain dormancy (Dekker and Hargrove, 2002). Morphologically, the foxtail seed caryopsis (embryo, endosperm, and aleurone layer) is surrounded by several enveloping layers which control its behavior (Dekker et al., 1996). The caryopsis coat is filmy, oily to the touch, water- and gas-tight, and continuous except at the placental pore opening on the basal end of the seed. Although the mature foxtail seed is capable of freely imbibing water and dissolved gases, entry is restricted and regulated by the placental pore tissues, where there is membrane control by the transfer aleurone cell layer (TACL). Gases entering the moist seed interior must be dissolved in the imbibed water passing through the narrow placental pore and TACL. The importance of gases entering dry seed is unknown. The morphology of foxtail seeds strongly suggests that seed germination is restricted by water availability in the soil and by the amount of oxygen dissolved in water reaching the inside of the seed to fuel metabolism. Oxygen solubility in the water entering the seed interior is an inverse function of the diurnally, seasonally, and annually changing soil temperature. The control by this oxygen-limited, gastight morphology is supported by observations of increased germination when foxtail seed envelopes, including the caryopsis coat, are punctured (Stanway, 1971; Dekker, et al., 1996). Based on these observations, biogeographic distribution of weedy *Setaria*, as well as individual seed behavior, is regulated by the amount of oxygen dissolved in water taken into the seed over time. When adequate amounts of water and oxygen reach the embryo, sufficient energetic equivalents are generated to support germination metabolism. When inadequate amounts of oxygen reach the embryo, dormancy is maintained or secondary dormancy is induced.

An inherent problem with seed germination only being regulated by a restriction of $O_2$-$H_2O$ entry (TACL regulation) (Dekker, 2000) is the observation that foxtail seed germination in natural soil habitats occurs several weeks after both favorable temperatures and $O_2$-$H_2O$ are available in the spring (Fig. 1; Forcella et al., 1992, 1997). What is the basis of delayed weed seed germination timing in the soil after adequate temperature and resource conditions are obtained? Hiltner (1910) suggested that inside dormant grass seeds there exists an oxygen-absorbing substance that would prevent $O_2$ from reaching the embryo. A robust seed regulatory system could result from the interaction of the TACL which restricts $O_2$-$H_2O$ entry into the seed, and an oxygen scavenging system which delays the transduction of $O_2$ in the symplast. These restrictions in uptake and $O_2$ scavenging could act together to prolong the time before $O_2$ stimulates the first metabolic events of seed germination. Heterogeneous foxtail seeds with differing levels and combinations of placental pore (TACL) size and $O_2$ absorption could also provide the means by which the seedling emergence of different individuals from the same parent plant would occur over the course of an individual growing season and also over many years. Heterogeneous collections of foxtail seeds in the soil would provide a hedge-betting strategy well suited to the disturbances typical of agroecosystems.

**17.4.2.3.1 Hull development.** The size and shape of most of the panicle tissues of parental origin do not change appreciably after emergence from the culm, and before seed fertilization commences (culm length and the caryopsis coat are notable exceptions). Several qualitative changes occur in the hull and the caryopsis with development from anthesis to after abscission (Dekker et al., 1996). The seed hull structures are present at anthesis, but change during seed ontogeny from open, soft tissues to a hard, enclosing structure at anthesis (Dekker et al., 1996). The seed hull color and hardness change with development as well as the shape of the palea and its position in relationship to the enclosing lemma. Caryopsis size and shape change with development as did the endosperm, embryo, caryopsis coat, and the placental pad. The size of the hull (parental tissue) changed relatively little compared to the caryopsis. The length of the hull remained relatively constant with development, ranging from 2.4 to 2.8 mm. The width of the hull ranged from 1.4 to 1.7 mm. The color of the hull changed with development from light green to dark green and dark brown. The hull began development as a soft and rubbery structure, progressing to hardness. The relative position of the palea to the enclosing lemma changed with development. At anthesis the palea was sunken below (inward toward the seed interior) the edge, or lip of the lemma, and had a concave surface (curved sightly inward to the cavity formed by the two hull structures). This relation between lemma and palea changed until the palea was flat, even with lemma lip edge. Later the palea shape ranged from flat to convex. The palea became approximately even with the lemma lip edge with time, and then protruded beyond the lemma lip.

**17.4.2.3.2 Placental pore and pad development.** During the ontogeny of the caryopsis a single placental vascular bundle from the parent panicle supplies nutrients and water to the developing ovule. As the caryopsis matures, a thick, dark, oval pad remains in the position occupied by the placental bundle. The point of attachment of the caryopsis (zygotic) with the palea (parental) is the placental pad. The placental pad was first visible as a dark circle at the base of the caryopsis on the side appressing the palea. At excision of the caryopsis from the hull, this area exhibited a distinctive color that changed with age. The color of this structure was white from two to ten DAA. From nine DAA and thereafter, the placental pad began to change color. The first color was a very pale pink (9-10 DAA) followed by a deepening color from pale pink to rust-red. The necrotic contents of the layers of the placental pad (Rost, 1973) and the red coloration in this area of the caryopsis (Narayanaswami, 1956) have been noted previously. It is suggested that this red coloration is a morphological marker in S. *faberii* indicating physiological maturity (Dekker et al., 1996). The formation of this abscission layer or zone is directly analogous to the black layer formation in *Zea mays* L., an indicator of physiological maturity (Daynard and Duncan, 1969; Sexton and Roberts, 1982).

**17.4.2.3.3 Caryopsis development.** The size and the shape of the caryopsis (COP) and its component structures were tubular early in development, and only partially filled the internal cavity formed by the lemma and palea (Dekker et al., 1996). The caryopsis matures and expands, filling the hull width by day 6, and the length by abscission. During this early time the remnant style and anther stalks were also present in the cavity. The caryopsis coat becomes fully developed at approximately day 4, and appears shiny, oily, and gray with dark spots. Sometime after day 4 the caryopsis coat becomes gas impermeable, except at the proximal (basal) end of the caryopsis (the placental pad and pore region). Entry of water into the caryopsis occurs only through the placental pore after day 4. Entry of water is never restricted through this region of the seed from day 4 until germination occurs (Rost, 1971b, 1973, 1975). With development the COP became ovoid and later plump, filling the hull cavity. The endosperm developed during this time, beginning with only a few cells, then becoming more visible as a clear liquid. Thereafter the endosperm became firmer, lost moisture and changed from liquid, liquid dough, soft dough, crumbly or granular dough, and finally to a firm and hard dough. The COP (mostly zygotic tissues) grew, progressively filling the hull cavity with development. At anthesis it was 0.75 mm, and increased to 2.2 mm just after abscission. After that time, the COP length remained constant or decreased slightly, possibly as a consequence of desiccation. The COP width increased from 0.35 mm just after anthesis, to a maximum of 1.5 mm just prior to abscission. It remained relatively constant after that time. The smallest difference between the hull and caryopsis length (0.35-0.5 mm) occurred between 11 and 15 DAA, then increased slightly after that time due to COP desiccation. The difference between the hull and caryopsis widths did not change (0.1-0.25 mm) after 6-7 DAA, apparently filling the hull cavity in that dimension at that time. Although the caryopsis appears to fill the cavity formed by the hull completely, arguments implicating the hull in physical restraint of embryo germination are incomplete (Simpson, 1990). Although germination must occur through the germination lid of the lemma (Rost, 1973), some space is apparent inside the seed cavity: the embryo and endosperm both shrink somewhat within the cavity after 16 DAA (Fig. 4). Several seeds dissected in the course of these experiments (Dekker et al., 1996) had partially germinated coleoptiles and coleorhizas (never both in the same seed), indicating arrested viviparous (precocious) germination.

**17.4.2.3.4 Embryo development**. The embryo first became visible at 2 days after anthesis (DAA) as a faint, darkish spot and ring at the basal end on the lemma side of the COP. The remainder of the COP at this time was clear. Later the coleorhizal end of the embryo axis became prominent and began to protrude from the globular-shaped caryopsis. This coleorhizal protrusion progressed from a rounded shape to a boxy, fully formed structure that filled the germination lid cavity on the basal end of the lemma. The embryo shrank below the endosperm surface (14 DAA and thereafter) as the seed became dehydrated.

**17.4.2.3.5 Dormancy induction**. Variable germinability-dormancy capacity is induced in developing seeds during embryogenesis (Dekker, 2003, 2004). *Setaria faberi* embryos become capable of independent germination at about day 6 of embryogenesis (Dekker et al., 1996). At that time embryo germination is very high (i.e. 95%), and occurs in both axes (coleoptile, coleorhiza). By day 8 embryo germination decreases, becomes more variable, and axis-specific germination first appears (germination of only one axis). This embryo dormancy induction period occurs from day 8 through anthesis, when embryo germination is very low (i.e. 10%). The variability in embryo and caryopsis germination among individual seeds shed by a panicle increases from when dormancy is induced until after abscission. Four qualitatively different types of embryo germinability were observed (Dekker et al., 1996). The four germination types were mutually exclusive; only one type appeared in any individual seed. These different germination types were correlated with the age of the embryo. Immature embryos exhibited three apparently different types of germinative growth.

The first type was observed in caryopses 0-5 DAA and appeared as disorganized, undifferentiated callus growth at the basal, coleorhizal end of the caryopsis. Germinability was evidenced as a pointed protrusion of cellular growth overlying the site of the excised pedicel, immediately basipetal to the placental pad. Also, a slight swelling of the basal, coleorhizal end of the caryopsis was often associated with this type of germination. It was unclear whether this type of germination arose from vascular (parental) or coleorhizal (zygotic) tissue. Although this type of germination was observed in many very young COPs, it may be an artifact associated with the excision of the vascular bundle to the pedicel. A similar type of germinability, disorganized proliferation of the radicle (callus), was noted previously in wheat *(Triticum aestivum)* in which the pericarp was removed (Symons et al., 1983).

The second type of germination of immature embryos was observed from 5-7 DAA, the time when the embryo was first large enough for excision from the caryopsis. At that time almost all embryo coleoptiles and coleorhizas appeared to be differentiating and growing. This germination took the form of shortening and thickening of those two axes. The growth never progressed very far, but each shortened and thickened axis appeared as rounded "balls" of tissue within the underlying scutellar depression (the scutellar "cup").

The third type of germination involved germinative growth of the immature (7-11 DAA) scutellum (modified cotyledonary structure of the embryo, intercalated into the endosperm in the caryopsis). Infrequently cellular growth and swelling would occur in areas of the scutellum appressed to the endosperm, usually at the coleoptile end of the embryo. Other types of germinative growth did not occur when this scutellar germination was observed.

More typically, germination and growth of the coleoptile and coleorhiza occurred in embryos aged seven DAA and older. Several different combinations of this type of axis-specific germinability of S. *faberi* embryos were observed. Germination of only the coleoptile axis of the embryo took several different forms. This type of germination first took the form of swelling of the basal area immediately adjacent to the coleoptile itself. When the embryo was part of the caryopsis, the first indication of this type of germination was a cracking of the overlying caryopsis coat. While some embryos did not continue to germinate after this, some coleoptiles continued to grow. The next step observed was the growth of the coleoptile itself causing it to increase in length and for the axis to curve and lift away from the scutellar "cup" it was enclosed by before germination. Some portion of the embryos observed continued to germinate, as evidenced by continued elongation and extension of the axis followed by the emergence of the first true leaf. Axis-specific germinability also was observed in which only the coleorhiza grew. The first indication of this type of germination was a swelling and protrusion of the coleorhiza leading to its extension beyond the end of the caryopsis. In some embryos this would appear as a pointed protrusion from the axis end, in others as the entire blunt, rounded coleorhiza end swelling and growing. Growth in some of these embryos would not proceed beyond this stage. In still others continued germination was evidenced by the appearance and extension of the trichomes. If germination proceeded, the radicle would extend from the end of the coleorhiza. All of these different axis-specific germinability states were observed in S. *faberi* embryos, as well as germination in which both axes germinated. Axis-specific germinability in wheat has been reported previously (Mitchell, 1980; Symons et al., 1983).

Germination of embryos differed not only with stage of development, but also as a function of the tissues surrounding them: hull plus non-embryo caryopsis, caryopsis, coleorhlza trlchomes (endosperm, caryopsis coat), and the isolated embryo. Quantitative and axis-specific differences in germinative capacity among embryos were also observed.

**17.4.2.3.5.1 Seed germinability**. Seed were completely nongerminable from fertilization to abscission and shortly thereafter. No seed germinated in these environmental conditions until after-ripening had occurred.

**16.4.2.3.5.2 Caryopsis germinability**. Three different, axis-specific types of caryopsis germinability were observed: coleoptile (CPT) only, coleorhiza (CRZ) only, coleoptile plus coleorhiza. Caryopses first became germinable at 8 DAA. This was primarily due to coleoptile germinability from eight to 16 DAA. Germinability of caryopses, as well as variability among similar-aged COPs, increased during the second half of embryogenesis, primarily as axis-specific coleoptile germination. Apparently other COP tissues are positively influencing the embryo relative to its behavior in isolation, delaying the onset of full developmental arrest. Coleorhizal, as well as CPT plus CRZ, germinability was very low from 8 to 16 DAA. After-ripened caryopses were highly germinable, primarily due to CPT plus CRZ germinability, although some germinability was restricted to single axes at that time. Variability of all types of caryopsis germinability was greatest just after abscission (12-16 DAA). Germinability of caryopses was very high after after-ripening had occurred.

**17.4.2.3.5.3 Embryo germinability**. The same three types of axis-specific germinability observed in caryopses were also observed in isolated embryos. Embryos first become germinable at 6 DAA, most of which was CPT plus CRZ "ball" type germination. Germinability among embryos was very high at that time. More normal axis germinability was observed at 2-8 DAA. After six DAA embryo germinability decreased until after abscission (12 DAA). During this time both CPT plus CRZ and coleorhiza-only germinability decreased, while coleoptile germinability increased, then decreased. Coleorhiza-only germination remained very low at all ages. This decrease in embryo germinability in the second half of embryogenesis has been reported for other species (e.g., sunflower: Le Page-Degivry et al., 1990). At 12 DAA, just after abscission, embryo germinability of all types was low. After that time embryo CPT plus CRZ germinability and variability increased, while CPT-only and CRZ-only germination remained low. Variability in embryo germinability increased with development from 6 DAA until 16 DAA. After-ripened embryos were highly germinable, primarily due to CPT plus CRZ germinability, but some low levels of single-axis germinability were observed.

**17.4.2.3.5.4 Multiple germinability phenotypes.** There is no single, qualitative state of germinability into which all seeds must enter, remain, and exit (Simpson, 1990). Instead, there are multiple states of germinability arising from the wide array of environmental conditions that these plants face, changing sensitivities of seed organs from anthesis through after-ripening to these conditions, different roles played by the inflorescence, fascicle, spikelet, and seed components, and a wide variation in genetic constitutions. Although seed germination is a singular (threshold) event in the foxtail seed life cycle, there are many different developmental paths by which it can be attained. The foxtail seed is a plastic, dynamic biological entity situated at all times in a complex, changing environment. These selection pressures over evolutionary time have led to multiple germinability phenotypes within a single genotype. A single foxtail plant shedding seed with different germination capacities has the ability to ensure its enduring occupation of a locality with changing environmental conditions. Giant foxtail seed rain consisting of multiple germinability phenotypes is a consequence of influences exerted at the level of the inflorescence, the fascicle, the spikelet, and compartments within the seed. Germinability in S. *faberi* embryos is achieved quickly after anthesis. This germinability has four different germination forms, but the importance of some of them is not apparent. Embryo germination can take place in either or both axes at any time from the middle of embryogenesis through after-ripening. The germinability of isolated embryos is not necessarily an indication of their germinability when enclosed in the caryopsis or hull. The embryo can respond differently in each seed compartment (hull, caryopsis, embryo), and each compartment probably exerts a regulatory influence independent (as well as dependent) of the others. This compartment-specific regulation changes with development. For example, at eight DAA the surrounding caryopsis tissues inhibited the enclosed embryo relative to the high germination of isolated embryos of that age. Conversely, the surrounding caryopsis tissues enhanced the germination of the enclosed embryo at 12 DAA, just after abscission, relative to the low germination of the naked embryo at that time. Seed germination does not occur until each compartment independently has achieved a significant amount of embryo germinability in both axes, especially in the coleorhiza. The lack of seed germination may be a direct consequence of asynchrony among these three seed compartments, with seed germination occurring when these three seed compartments act synchronously. The times of seed abscission, dissemination from the parent panicle, and entry into the soil seed bank may have an important evolutionary and ecological significance to the species. At these times seed leaving the parent plant are non-germinable, and isolated embryos are in their least germinable state. At abscission both caryopses and isolated embryos possess their maximum variation in germinability as a population, either in terms of percent germinability or axis-specific germinability. The population of S. *faberi* seed leaving a panicle after abscission expresses a large, possibly maximal, number of germinability phenotypes in its life cycle. This may be an important adaptive strategy of foxtail to ensure enduring survival as it enters the hostile soil seed bank, as the chances of survival of the population is increased by disseminating the maximum number of germination options for an uncertain future.

**17.4.2.3.5.5 Light and dormancy induction.** Atchison (2001) observed that seed maturing early in the growing season in partial shade are considerably more dormant than those on nearby plants in full sun. Axis elongation and fascicle spacing are two mechanisms differentially modifying the light microenvironment of individual spikelets (Dekker et al., 1996). This differential light reception may be a mechanism by which different levels of dormancy are induced in individual florets of a panicle: relatively greater light levels are associated with relatively reduced dormancy in later maturing seeds on a panicle. Greater light penetration is correlated with relatively less dormancy in the more widely spaced basal florets, while the greater height and light above the soil surface (and competitors) is correlated with less dormant florets maturing later on an individual panicle (Atchison, 2001; Dekker et al., 1996; Haar, 1998).

**17.4.2.3.5.6 Morpho-physiological dormancy traits.** The germination of giant foxtail embryos is modified by the presence of several seed envelopes, notably the hull (glumes, lemma, palea) and the caryopsis coat (Rost, 1973, 1975). Induction of dormancy in caryopses and seeds coincides with the sealing of the caryopsis by the caryopsis coat, and the maturation of the hull. The variability in germination among individual caryopses and seeds increases with time after dormancy induction. These events may describe a "phenocritical" period in seed genesis when dormancy is induced (Christianson and Warnick, 1984). Compartment germination differences provide evidence of multiple seed tissue control of germination.

**17.4.2.3.5.7 Compartmentalization of seed germinability.** The germinability state of an individual seed at anyone instance in its life cycle is a function of the germinability of its component compartments, the embryo, and the enclosing and associated structures of the seed (hull, caryopsis) (Dekker et al., 2012). Several different germinability states were observed in S. *faberii* embryos as they developed. The hull was observed to exist in two states, dependent on the resistance of the germination lid (Rost, 1973) to internal pressure and opening by the coleorhiza. The embryo, whether isolated or in the caryopsis, exists in four different states: both axes nongerminable, both axes germinable, coleoptile germinable only, or coleorhiza only germinable. The germinability states of S. *faberi* seed can be modeled based on the independent, and at other times dependent, relationship between the three structural compartments. From the observed germinability states, 32 hypothetical qualitative states can be envisioned. Twenty eight of the 32 states were observed at some time during development. The common feature of these four missing states is that they all have open hulls (H +), and the isolated embryo has nongerminable coleoptiles and germinable coleorhizas. Either these states do not exist or they were not revealed in the times during the life cycle we studied. We propose that dormancy in S. *faberi* seed can be defined by the asynchronous germinability of the embryo within each of the three compartments of the seed (embryo, caryopsis, and hull), and germination can be defined by the synchronous germinability of the embryo within each of the three compartments. From these studies we hypothesize a dynamic, developmental basis of germinability regulation resulting in multiple germinability phenotypes at all stages of development, particularity when seed is shed from the parent plant and the soil seed bank is replenished.

**17.4.2.3.6 Seed abscission.** Weedy *Setaria* spp. seeds become physiologically independent when the abscission layer in the pedicel forms, indicated by the red coloration of the placental pad on the caryopsis (Dekker et al., 1996). Abscission is a singular, threshold event in the life history of a new *Setaria* plant, but it is only a partial separation of the new and parental generations: parental, tissues surround the embryo, affecting its ability to germinate. The basic unit of the *Setaria* inflorescence is a dorsally compressed, two-flowered spikelet, which disarticulates below the glumes, and is subtended by one to several bristle-like setae (Hitchcock, 1971; Rominger, 1962). *Setaria faberi* seeds absciss from the parent plant almost entirely dormant, a threshold life history event (Dekker, 2003, 2004).

**17.4.2.4 Seed fecundity and plasticity.** Seed production is an emergent property of *Setaria* reproductive morphology. Seed numbers produced by a *Setaria* are a function of differences in plant architecture: shoot tillering-panicle formation, panicle-fascicle branching, and fascicle spikelet-floret development. Inherent plastic differences in these reproductive structures among *Setaria* species determine the reproductive responses of species and populations to available opportunity in its immediate environment.

*Setaria* panicle inflorescences develop at the terminal ends of shoot tillers. Developmentally, primary panicles flower first on a plant, followed by secondary, then tertiary. *Setaria spp.*, panicles are composed of fascicles, which consist of spikelet (with florets) and bristle (seta only) shoots. The fascicle-spikelet structure differs among *Setaria* species. In S. *faberi* or S. *viridis* the number of fascicles within each panicle varies, a plastic response to a plant's immediate environment (Clark and Pohl, 1996). Longer, earlier developing S. *faberi* or S. *viridis* panicles often have the most extensive fascicle branching, as well as more spikelets per fascicle (typically 4-6 or more per fascicle). The plastic response of S. *faberi* or S. *viridis* to its immediate environment is also revealed in the number of spikelets-florets that fully mature, which determines the seed number per panicle length (seed density). Under favorable conditions more spikelets are able to develop into seeds, while under unfavorable conditions spikelets may abort. S. *pumila* panicle morphology is different from that of S. *faberi* or S. *viridis*. Only a single, terminal, spikelet-floret is found in each S. *pumila* fascicle (Clark and Pohl, 1996). This stable, fixed morphology limits the ability of *S. pumila* to respond in a plastic way to its environment in terms of seed number relative to that of *S. faberi* or *S. viridis*.

*Setaria* seed rain exhibited some stable, and many more plastic, responses. *S. faberi* panicles were consistently longer than those of *S. viridis*. *S. viridis* parameters were greater than *S. pumila*. Earlier panicles were greater than, or similar to, later ones for all parameters. More typically, tillers and panicles responded to local conditions in a plastic way, confounding the formulation of seed production generalizations. In *S. faberi* and *S. viridis* no consistent relationship between seed number and panicle length was observed among different tiller types. A more consistent relationship between parameters was observed for *S. pumila* compared to the others, making prediction possible for this species. The stability and plasticity of these relationships is partially due to the differences in *S. faberi* and *S. viridis* panicle, fascicle and spikelet morphology compared to *S. pumila*. These stable and plastic responses provide fine-scale adjustment to a locality, maximizing exploitation of local opportunity.

The numbers and yield of seed produced by a weedy *Setaria* plant is highly plastic, and strongly dependent on biomass accumulation, plant tillering and fascicle architecture (Haar, 1998; Haar and Dekker, 2011; Kawano and Miyake, 1983). *Setaria* plant biomass is strongly dependent on available resources, competition from neighbors, and the time of seedling emergence (Dekker, 2004b). In general, plant size and seed number is greater for plants emerging earlier in the growing season than later (Schreiber, 1965b). Reports of seed productivity therefore vary from 1 to 12,000 seeds/plant (Peters et al., 1963; Rominger, 1962; Slife, 1954; Steel et al., 1983; Vanden Born, 1971). Seed number per panicle, panicle length and seed number per unit panicle length (seed density) were found to vary among *Setaria* species (*S. faberi*, *S. viridis*, *S. pumila*), panicle types (1°, 2°, 3°) and field locations (Haar,1998). For example, Haar (1998) found seed number per panicle varied by *Setaria* species, location and panicle type: *S. faberi*, 165 to 2127; *S. viridis* subsp. *viridis*, 144 to 725; and *S. pumila*, 54 to 213. Because the weedy *Setaria* spp. seed rain is continuous over a weeks-to-months period, productivity can be underestimated by infrequent collection methods and inefficient capture techniques. Attempts to provide experimental seed productivity estimates (e.g. relationship between panicle length and seed number per unit length measurements) have produced inconclusive results (Barbour and Forcella, 1996; Defelice et al., 1989; Fausey et al., 1997; Haar, 1998).

**17.4.3 Seed rain dispersal.** *Setaria* seed is dispersed in both time (seed germinability-dormancy heteroblasty) and space (invasion and colonization, formation of soil seed pools).

**17.4.3.1 Seed germinability-dormancy heteroblasty: Dispersal in time**.

"The discovery of clines in seed polymorphism ... suggests that seed polymorphism may be a most sensitive indicator of evolution in weedy species." (Harper, 1964)

"Seed polymorphisms seem particularly likely to be sensitive indicators of evolutionary change in alien invaders." (Harper, 1964)

Weedy *Setaria* seed is shed from the parent plant primarily in a nongerminable, but viable, state of developmental arrest. Studies have concluded that all seed shed from the parent plant is dormant (*S. pumila* (Rominger, 1962): Povalitis, 1956; Peters and Yokum, 1961; Rost, 1975; *S. verticillata*: Lee, 1979), "almost completely" dormant (*S. pumila*: Taylorson, 1966; Kollman, 1970; *S. faberi*: Stanway, 1971), or shed with variable germinability (*S. pumila*: Norris and Schoner, 1980; *S. viridis*: Martin, 1943). Morphological variants of *S. viridis* (robust white, robust purple and giant green) have been reported indicating they do not possess seed dormancy (Schreiber, 1977). Triazine resistant *S. faberii* seed were reported to be inherently more dormant than susceptible biotypes (Tranel and Dekker, 1996, 2002), consistent with observations in other species (Mapplebeck et al., 1982). Vivipary (premature, precocious germination) in *S. viridis* plants have been observed (Dore and McNeill, 1980; Haar, 1998; Hubbard, 1915).

Individual *Setaria* panicles on a single parent plant produce a diverse array of seeds, each with potentially different dormancy states (seed heteroblasty) (Atchison, 2001; Dekker et al., 1996; Moore and Fletchall, 1963; Haar, 1998; Kollman, 1970; Martin, 1943; Nieto-Hatem, 1963; Schreiber, 1965a; Stanway, 1971; Taylorson, 1986; Vanden Born, 1971). Heterogeneity of dormancy among seed from a single parent plant, seed heteroblasty, is one of several types of heteroblastic development of repeating, self-similar, plant units (metamers) during plant ontogeny (Gutterman, 1996; Jones, 1999). This phenomenon has been characterized in other plant species (e.g. *Xanthium pensylvanicum*, Weaver & Lechowicz, 1983; *Chenopodium album*, Williams and Harper, 1965). Systematic, quantitative characterization of heteroblasty among individuals within a locally adapted population, between local populations, and their relationship to subsequent life history behaviors has not been reported previously.

Evidence has been provided for a complex model of germinability regulation based on the independent, asynchronous actions of the embryo, caryopsis, and hull compartments, as well as on their dependent, synchronous action (Dekker et al., 1996). These studies provide evidence for a dynamic, developmental model of *S. faberi* germinability regulation resulting in phenotypes with a wide range of germinability shed from an individual panicle. These diverse germinability phenotypes are found at all stages of development, but particularly when the seed is shed and the soil seed pool is replenished.

An individual weedy *Setaria* spp. plant synflorescence produces individual seeds whose germination requirements vary. This seed heteroblasty occurs in any parental environment, even those within controlled, constant conditions for the duration of its life history (Dekker, et al., 1996; Harr, 1998). Seed heteroblasty is therefore a constitutively expressed genetic (or epigenetic) trait. The dormancy capacity of an individual *S. faberi* seed refers to its dormancy state at abscission, as well as the quantity of environmental signals (oxy-hydro-thermal time; Dekker et al., 2003) required to stimulate a change in state (e.g. after-ripening duration) (Dekker, et al., 1996; Dekker & Hargrove, 2002; Sareini, 1970). Germination capacity is an inherent quality of a *Setaria* seed which is retained for its entire life.

Changes in germination were correlated with tiller development and relative time of seed maturity within a panicle (Haar and Dekker, 2012). Seed produced on tillers that developed earlier (panicle age; PA, time of first flower of panicle) were more likely to be dormant than seed from later-developing tillers (Figure 17.6). Seed that matured later on a panicle (panicle position; PP, PDP) were more likely to germinate than seed that matured earlier on the same panicle. A consistent trend toward later maturing seed (calendar date; CD) having less dormancy was found for seed grown under different environments which implies an inherent or parental source for variation in giant foxtail seed dormancy. The variation in percentage germination at abscission and following stratification after-ripening treatments (4° C, dark, moist) indicates that the *S. faberi* seed rain consists of individual seeds, possibly each with a different degree of dormancy.

Studies were conducted to determine the relationship between weedy *Setaria faberi* seed dormancy and subsequent behaviors in the soil culminating in the timing of seedling recruitment (Atchison, 2001; Dekker et al., 2012b; Jovaag, 2006; Jovaag et al., 2011, 2012a, b). A robust characterization of seed heteroblasty at the time of dispersal for 39 locally adapted *S. faberi* populations, as influenced by parental genotype (time of embryogenesis) and environment (year, location) has been provided. The heteroblastic structure of each population was revealed by the germination response to increasing durations of after-ripening (in "ideal" conditions). The production of seeds with different germination requirements (germinability-dormancy) is a function of genotype, plant architecture and the environmental conditions during ca. 12 day embryogenic period. Hence, individual *Setaria* plant populations are most accurately defined as a unique combination of species, individual parent plant location and time of abscission (calendar Julian week (JW) and year).

**Figure 17.6** Schematic description of *Setaria faberi* seed cohorts from primary (1°), secondary (2°) and tertiary (3°) panicle types: calendar date (time) of seed abscission (CD); seed position on panicle at time of abscission (PP, PDP); cohort of seed from entire panicle based on age (PA, time of first flower of panicle).

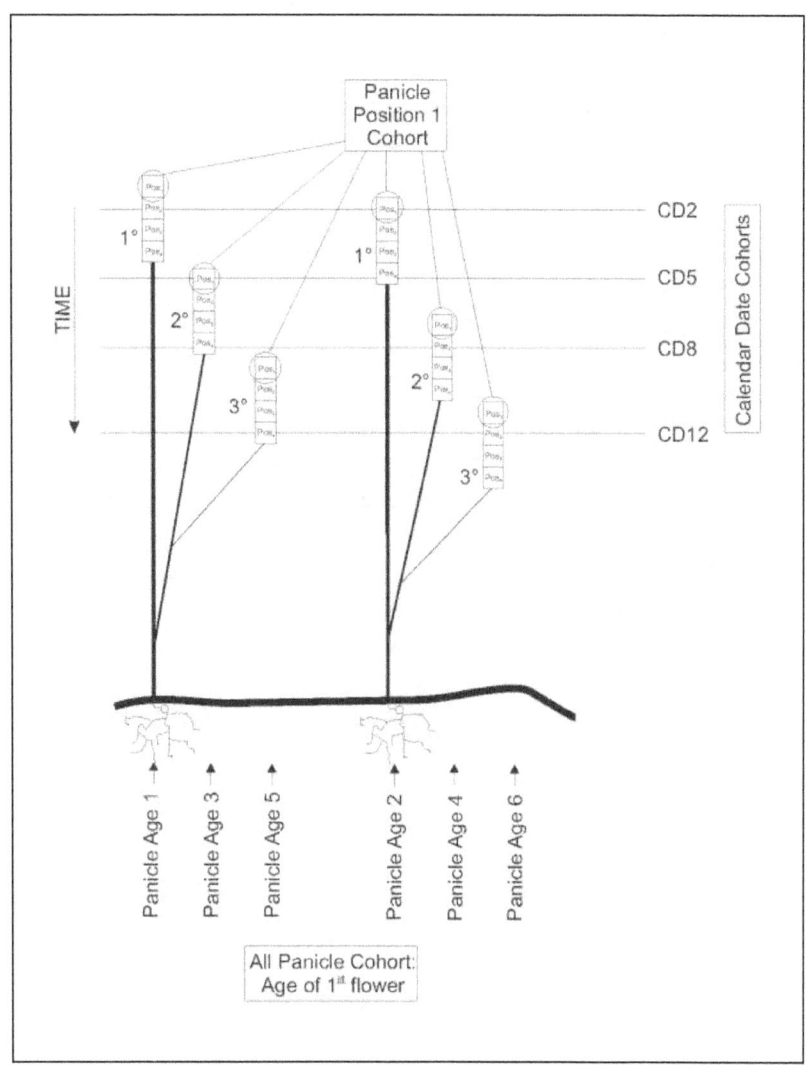

For *Setaria*, considerable heterogeneity in germination requirements existed among and within all populations tested. Year, seasonal time of abscission, and location all proved to be important factors influencing the germination heterogeneity structure of *Setaria* populations as revealed by after-ripening. These observed heteroblastic differences within and among populations revealed a fine scale adaptation to local space and time conditions. Earlier harvested seed was consistently less germinable than later harvested seed. Earlier fertilized seeds produced on *Setaria* 1° synflorescences (typically August in the midwest U.S.) are relatively more dormant than those from 2° panicles (ca. September), which are in turn more dormant than seed on 3° panicles (ca. October) and subsequent panicles. This is consistent with earlier studies (Dekker, 2004; Dekker et al, 1996) suggesting the importance of photoperiod, which is longest while the early (JW32, August) seeds are developing, then decreases through the middle (JW36, September) and late (JW40, October) periods of the seed rain.

Germination responses of seeds to after-ripening were used to visualize the general patterns phenotype space. A schematic diagram of this generalized germination reaction norm is given in figure 17.7. The majority of the populations were differentiated from each other; this variation indicated a fine scale adaptation to different local environments. Taken together, the 39 responses represented *Setaria's* "seed dormancy phenotype space" and revealed three different generalized dormancy patterns. The first pattern, low dormancy populations, had high initial germination in response to low doses of after-ripening. The second, high dormancy populations, had no or low initial germination with little additional response to increased after-ripening. Most populations had the third pattern, intermediate to the others, with low initial germination and increasing germination with increasing after-ripening dose. The third, most common pattern, occurred within the B1 and B2 regions (Figure 17.7). These regions defined a pattern of low (<10%) initial germination, then increasing germination with increasing AR, followed by a plateau at about 40-50 days of after-ripening when the germination response to added after-ripening was saturated (at 70-100% germination).

These studies emphasize the crucial role that environment (year, season) at the time of seed formation and embryogenesis plays in the dormancy phenotype produced by plants in a locality. The location itself (soil properties, fertility, past management history, etc.), the 'terroir', can also influence the quality of plant phenotypes produced. The germination patterns observed and dormancy phenotype space occupied, defined the heteroblastic hedge-bet structure at abscission. The B1/B2 regions were the most commonly occupied phenotype spaces, suggesting a population's "best bet" generally laid within this region. Variation among populations within this region indicated a fine scale adaptation to the local environment. Regions A and C were less commonly occupied, suggesting they were "long shots" at the edge of local adaptation.

**Figure 17.7.** Schematic diagram of the hedge-bet structure within the heteroblasty phenotype space. A: High initial germinability, rapid germination response to after-ripening (AR). B1: Low initial germinability, increasing germinability with increasing AR duration until a plateau of high germinability is reached. B2: Increasing germinability with increasing AR, but less overall germination than in B1. C: Little or no early germinability, but some increase with long AR durations. D: "Perfect" crop response—all seeds immediately germinable.

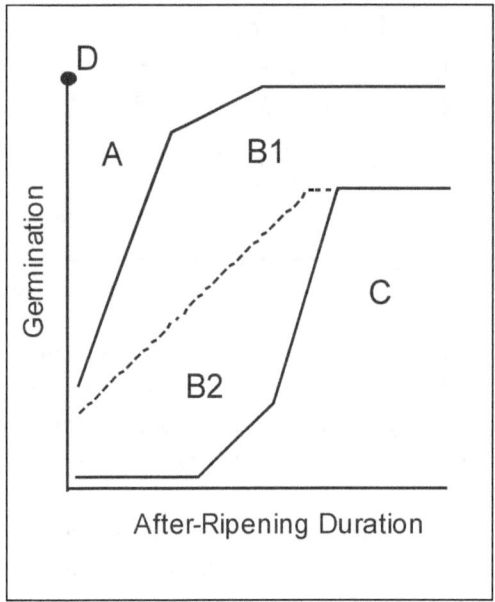

Germination responses were also used to rank populations based on their dormancy level to facilitate later comparisons with emergence behavior. Weedy *Setaria* seed dormancy capacity heterogeneity at abscission (seed heteroblasty) provides a "blueprint" for subsequent behaviors in the soil culminating in seedling recruitment.

**16.4.3.2 Soil seed pool formation: Dispersal in space.** Seed heteroblasty inevitably leads to formation of long-lived seed pools in the soil. A seed pool begins with heterogeneous dormant seed dispersed into the soil from local or distant plants. One of the most important traits weedy *Setaria* spp. possess for invasiveness, colonizing ability and enduring occupation of a locality is their ability to form long-lived soil seed pools, the source of all future annual weed infestations (Buhler et al., 1997a; Dekker, 1999). Dormancy inherent in *Setaria* seeds inevitably leads to formation of soil seed pools. Significant heterogeneity in dormancy states among weedy *Setaria* seeds results in precisely timed germination over large temporal scales (hours to decades; Atchison, 2001; Dekker et al., 1996; Forcella et al., 1992, 1997). Soil seed pools consisting of a diverse collection of *Setaria* species, genotypes and dormancy phenotypes reveals a hedge-betting strategy for adaptation to changing conditions within agroecosystems (e.g. Cohen, 1966; Philippi and Seger, 1989). The reasons why *Setaria* spp. are so successful in disturbed habitats derives from their biodiversity (Wang and Dekker, 1995; Wang et al., 1995a, 1995b; Dekker, 1997), especially their ability to produce diverse phenotypes from a single parent plant (e.g. Dekker et al., 1996) and their ability to time germination precisely. This results in individual seedlings emerging from the seedbank at different times over large temporal scales (hours to decades; Forcella et al., 1992, 1997).

The life of an individual weedy *Setaria* parental plant ends with reproduction, seed abscission and dispersal replenishing the soil seed pool. (Dekker, 2003). Weedy *Setaria* spp. seeds separate (shatter) from the parent plant soon after abscission, but the time of dispersal varies based on pedicel deterioration and wind (Dekker, 2003). Seeds of the crop foxtail millet have been selected to resist shattering, allowing near complete harvest by the grower. *Setaria* seed dispersal occurs by gravity, animal (e.g. birds), wind, water and human (e.g. farm machinery) vectors (Bor, 1960; Ridley, 1930; Steel et al., 1983; Wilson, 1980). The retrorsely barbed setae of *S. verticillata* stick easily to animals and humans facilitating their dispersal (Bor, 1960). The *Setaria* seed rain period in the north temperate regions of the world occur generally from July through December, depending on seasonal environmental conditions (notably the time of soil freezing and killing frost at the end of the season) and harvesting (Atchison, 2001; Haar, 1998; Peters et al., 1963; Stevens, 1960). The time of seed rain is continuous, with the seeds from 1° panicles occurring first, followed by 2° and subsequent tillers. On individual panicles, the seed rain progresses in approximately the order in which they flower (Dekker et al., 1996).

**17.4.3.3 Dispersal and local adaptation: Evolutionary history of *Setaria* invasion.** Local adaptation of weedy *Setaria* spp. today is the evolutionary and ecological consequence of past dispersal events: invasion, colonization and enduring occupation. The weedy *Setaria* are a very ancient group of plants, and their evolutionary history is characterized by invasion and colonization followed by selection and local adaptation for enduring occupation of a site (Dekker, 2004). There appears to be five major phases, or waves, of *Setaria* invasiveness in earth's history. The tropical genus *Setaria* dispersed to Eurasia from its probable first home in Africa. The first invading species was probably a *S. viridis* or *viridis*-like diploid annual (Rominger, 1962; Stapf and Hubbard, 1930).

In the second phase, an *S. viridis* weedy progenitor species spread over Eurasia as a wild colonizing species. With domestication it then spread both as a weed (Li et al., 1942; Li et al., 1945; Wang et al., 1995a, b) and a crop (foxtail millet, *S. viridis*, subsp. *italica*; Wang et al., 1995a, b; de Wet, 1995; de Wet et al., 1979; Fukunaga et al., 1997, 2002; korali, *S. pumila*). Foxtail millet was a crop in China 6000 years B.P. (Cheng, 1973; Gao and Cheng, 1988; Ho, 1975; Kawase and Sakamoto, 1984; Li and Wu, 1996), while its use as a crop in Europe dates to 3500 B.P. (Dembinska, 1976; de Wet and Harlan, 1975; Kuuster, 1984; Neuweiler, 1946). *S. pumila* seed were gathered from wild plants and later cultivated in India (de Wet, 1992). Other weedy foxtail species arose from *S. viridis*-like ancestors (Li et al., 1942; Li et al., 1945; Rominger, 1962; Willweber-Kishimoto, 1962; Williams and Schreiber, 1976; Prasado Rao et al., 1987). These polyploid specialists probably included *S. pumila* and *S. verticillata*.

In the third wave, the weedy foxtails spread to the New World at two different times: pre-Columbian (before ca. 1500 A.D.) and post-Columbian invasions (after ca. 1500 A.D.). The pre-Columbian origins of *S.*

*geniculata*, the only weedy foxtail native to the New World, suggest an ancient dispersal event westward from the Old World (Africa) to New World (Rominger, 1962). A species of *Setaria* was the oldest cultivated cereal in the Americas, its origins dating to 8000 B.P. (Callen, 1965, 1967; Smith, 1968). Prehistoric migration of *Setaria* could date from when the ancestors of contemporary native Americans first crossed the Bering Strait land connecting Asia and North America (Beringia; approximately 10,000 B.P.). The *S. pumila* invasion of the Americas came primarily came from Asia (Wang et al., 1995b). The post-Columbian dispersal of weedy foxtails by Eurasian human emigration over the last 500 years is most likely the major source of the weedy foxtail floral introductions we see in North American agroecosystems today (Wang et al., 1995a, 1995b). For example, *S. viridis* had invaded Canada by 1821 (Rousseau and Cinq-Mars, 1969). The foxtails have expanded their range since their introduction to North America in the last several hundred years as a consequence of global trade and human-mediated dispersal (Hafliger and Scholz, 1980).

The fourth wave may have been determined recently by an extremely rare allo(tetra)polyploidization event giving rise to a fertile hybridization product in southern China, *S. faberi* (Pohl, 1951; Pohl, 1966; Wang et al., 1995b). *S. faberi* was first introduced to North America in the 1920's near New York City (Fairbrothers, 1959). Soon thereafter it was found in Philadelphia (1931) and Missouri (1932). From this early colonization *S. faberi* spread rapidly in the eastern and midwestern states (Fernald, 1950; Rominger, 1962; Reed, 1970). More recently, *S. faberi* has spread northward into Ontario in the 1970's (Dore and McNeill, 1980; Warwick, 1990). The largest increase and spread of *S. faberi* occurred across the North American corn and cereal growing regions in post-WWII era. The introduction and adoption of the herbicide 2,4-D to control dicotyledonous weeds created an opportunity space for, or population shift to, grassy weeds like the foxtails (Pohl, 1951; Slife, 1954). Introduction of other new selective herbicides, as well as expansion of maize and soybean production in North America, accelerated these trends (Warwick, 1990). In little over a half century, *S. faberi* has colonized a large part of the most fertile, disturbed areas of North America (Hafliger and Scholz, 1980).

Currently we are experiencing a fifth invasive phase, from the late 1970's through the 2000's, with the appearance of many different types of herbicide resistant biotypes of foxtail (e.g. Holt and LeBaron, 1990; LeBaron and Gressel, 1982; Ritter et al., 1989; Stoltenberg and Wiederholt, 1995; Thornhill and Dekker, 1993). Concurrent with this is the incremental spread of *S. faberi* and *S. pumila* in the northern prairies of North America.

Geographic invasion can simultaneously be viewed from the perspective of genetic invasion: *Setaria* polyploidization. Polyploid foxtails more specifically adapted to niche opportunities in agroecosystems include those four weedy foxtails with more narrow geographic distributions than diploid *S. viridis*: knotroot, yellow, *S. verticillata*, and more recently *S. faberi*. The 4$^{th}$ geographic invasive wave was also the 4$^{th}$ genetic-polyploidization, weed speciation event.

**17.4.4 Seed behavior in the soil.** Weedy *Setaria* behavior in the soil is regulated by the amount of oxygen dissolved in soil water taken into the seed over time.

There exists a cascade of three morpho-physiological mechanisms that act together to modulate received oxygen, water and heat and thereby constrain and control *Setaria* spp. seed embryo behavior. The seed hull and outer envelopes act to attract and accumulate water, enhance gas solubility in surface water film by means of surface rugosity, and channel that gas-laden water to the placental pore. The transfer aleurone cell layer (TACL) is a membrane whose diameter and function regulates diffusion of gas-laden water in and out of the living seed interior. An oxygen-scavenging heme-containing protein in the living interior acts to buffer the seed against premature germination by sequestering oxygen. This morphology strongly suggests that seed germination is restricted by water availability in the soil, and by the amount of oxygen dissolved in water reaching the inside of the seed symplast to fuel metabolism.

Soil water, temperature and oxygen play a unifying role in regulating both the global biogeographic distribution of weedy *Setaria* spp., as well as the responses of individual seeds in a soil microsite. Environment and morpho-physiological traits interact to stimulate weedy *Setaria* seed behaviors: induction of embryogenic dormancy, induction of summer dormancy, after-ripening, germination and seedling emergence. The signal stimulating all these *Setaria* behaviors is the quantity of oxygen within the water content of the living seed caryopis interior per time unit, oxy-hydro-thermal-time: mass $O_2^{-1}$ volume $H_2O^{-1}$ time$^{-1}$ seed$^{-1}$.

*Setaria* seed behavior in the soil is predicated on dormancy capacity heterogeneity at abscission and modulated by the seasonal environmental conditions experienced in the field: seed germinability-dormancy heteroblasty is the blueprint for all post-abscission behaviors in the soil.

**17.4.4.1 Control of seed germinability.** Seed germination includes all types of embryo growth under any condition. Germinability refers to the capacity of an embryo to germinate under some set of conditions. After-ripening refers to the period after dispersal when the seed cannot germinate, even under favorable conditions, and during which changes occur allowing it to germinate. Dormancy is avoided except when referring to specific published reports that use the term (Dekker et al., 1996).

**17.4.4.1.1 Regulation of weedy *Setaria* seed behavior.** The wide geographic range of adaptation, heterogeneity in dormancy phenotypes, and genotypic diversity raise the question of what mechanisms in weedy *Setaria* spp. seeds drive seed behaviors (e.g., induction of dormancy, after-ripening, germination, induction of summer dormancy, and seedling emergence) (Dekker, 2003). There is evidence that soil water, temperature and oxygen play a unifying role in regulating both the global biogeographic distribution of weedy *Setaria* spp., as well as the responses of individual seeds in a soil microsite (Dekker, 2000; Dekker and Hargrove, 2002; Dekker et al., 2001).

The effects that naturally occurring gases (oxygen, nitrogen, carbon monoxide) may cause in dormant giant foxtail (*Setaria faberi*) seed germination under favorable temperature and moisture conditions were investigated (Dekker and Hargrove, 2002). The germination responses to gas mixtures supported the hypothesis that *S. faberi* germination behavior is regulated by the amount of oxygen taken into hydrated seed over time. *S. faberi* seed germination was markedly affected by $O_2$ concentration (in $N_2$) above and below that of air (20% $O_2$): the largest increase in germination (from 37 to 60%) occurred between 20-25% $O_2$; between 0-10% $O_2$ germination increased from 0-30%; and surprisingly germination at 10 and 20% $O_2$ was similar. These observations reveal an asymmetrical response to incremental changes in $O_2$ above and below that typically found in agricultural soils. Carbon monoxide had opposite effects on *S. faberi* germination in air depending on concentration, stimulation and inhibition: germination increased from 37 to 56% with the addition of 1% CO, but decreased from 37 to 14% with 75% added CO. An explanation may be that there are two separate effects of CO, each occurring in different physiological systems of dormant seeds at the same time. At high concentrations (75%) in air CO inhibited seed germination, probably by inhibiting mitochondrial respiration. But low CO concentrations (0.1 or 1%) in air stimulated seed germination. It was not apparent which physiological system(s) CO and $O_2$ affected. It seems unlikely that CO-stimulated germination arises from effects on the respiratory apparatus, but may be a consequence of CO interactions with an as yet uncharacterized physiological factor in the seed (Sareini, 2002).

The morphology of weedy *Setaria* spp. seeds provides clues about which environmental factors limit germination and maintain dormancy. This morphology strongly suggests that seed germination is restricted by water availability in the soil, and by the amount of oxygen dissolved in water reaching the inside of the seed interior to fuel metabolism. Oxygen solubility in the water entering the interior is an inverse function of the diurnally, seasonally, and annually changing soil temperature. The control by this oxygen-limited, gas-tight morphology is supported by observations of increased germination when *Setaria* seed envelopes, including the caryopsis coat, are punctured (Dekker, et al., 1996; Stanway, 1971), as well as by the stimulatory effects of oxygen and carbon monoxide (Dekker and Hargrove, 2002).

There exists a cascade of three morpho-physiological mechanisms that act together to modulate received oxygen, water and heat and thereby constrain and control *Setaria* spp. seed embryo behavior. The first is the seed hull and outer envelopes that act to attract and accumulate water, enhance gas solubility in those water film by means of its surface rugosity, and channel that gas-laden water to the placental pore (the basal opening to the seed interior; the only water entry point in the seed) (Donnelly *et al.*, 2012). The placental pore (PP) terminates with the transfer aleurone cell layer (TACL), a membrane whose diameter and function regulates diffusion of gas-laden water in and out of the interior (Rost, 1971a, b, 1972, 1973, 1975; Rost and Lersten, 1970, 1973). The foxtail seed is gas- and water-tight except at this pore due to the enveloping caryopsis coat (CC). Together the PP, TACL and CC provide the second controlling mechanism. The third element is an oxygen-scavenging heme-containing protein ("X", figure 17.8) in the interior that acts to buffer the seed against premature germination by sequestering oxygen (Dekker and Hargrove, 2002; Sareini, 2002). These controlling mechanisms are represented schematically in figure 16.8.

**Figure 17.8.** Schematic diagram of the *Setaria* sp. seed, surrounding soil particles and oxygen dissolved in water ($H_2O$-$O_2$). The seed interior (aleurone, TACL, endosperm, $O_2$-scavenging protein (X), embryo) is surrounded by the non-living glumes, hull, placental pore and the gas- and water-impermeable caryopsis coat.

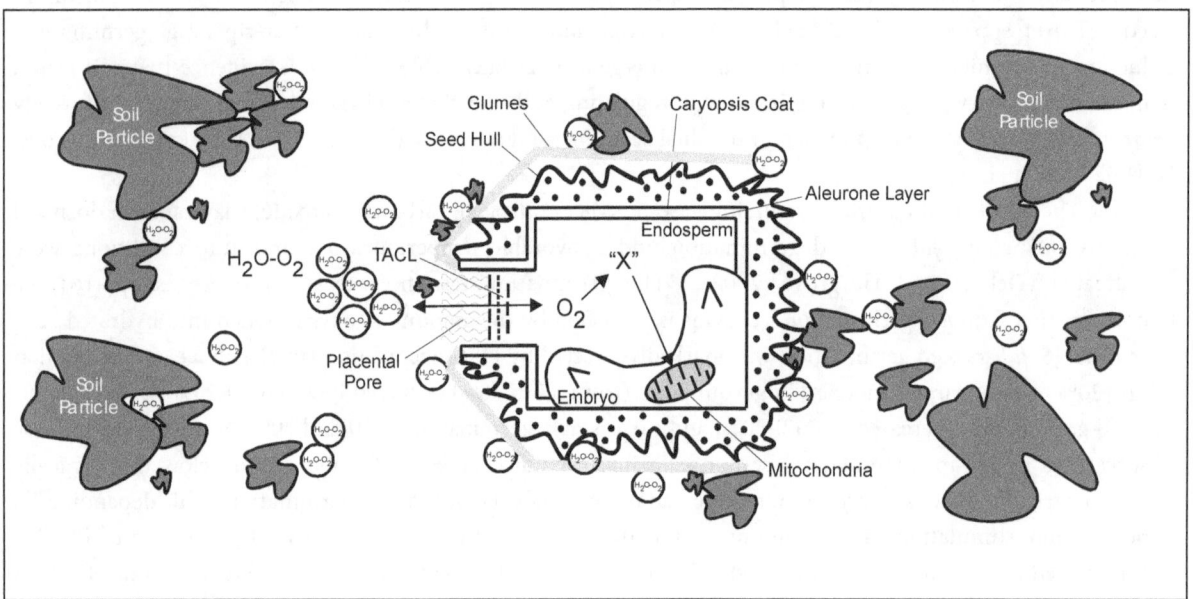

The interaction of these three mechanisms in any individual seed defines its unique dormancy state at abscission, which interacts with its immediate environment to define its subsequent behavior. Environmental signals change the initial dormancy state of an individual seed over the course of its life history in the soil. The quantity of these signals required to change behavior is an intrinsic quality of the individual and is retained for the entire life of the seed. The dormancy state an individual living seed inherits from its parent plant is never lost or "broken". The behavior of an individual seed in the soil during its life history is always a consequence of its current intrinsic state responding to its current extrinsic conditions, a Markov chain process.

#### 17.4.4.1.2 Movement of water-gas between soil and seed.

> "There is increasing evidence that specialized properties of the seed, its shape, its size and form, and the nature of its surface determine the type of contact the seed makes with the soil and on this depends the ability of the seed to germinate successfully." (Harper, 1964)

Seed morphology therefore seems to be a ripe area for investigation, in particular the morphology of seed surfaces as the interface, the 'antenna', linking the soil-borne weed with its immediate microsite environment (Dekker and Luschei, 2009; Donnelly et al., 2012). The seed morphological structures that water and gases encounter as they enter the seed from the soil environment extend from the glumes to the embryo: glumes-hull-placental pore-placental pad-caryopsis coat-TACL membrane-aleurone layer-endosperm-embryo. The environmental signal regulating weedy *S. faberi* behavior is the amount of oxygen dissolved in water received by the embryo over time, and temperatures conducive to both growth and oxygen solubility (i.e. oxy-hydro-thermal-time, Dekker et al., 2003). These controlling mechanisms are represented schematically in figure 17.9.

**Figure 17.9.** Schematic diagram of the *Setaria* sp. seed and the morpho-physiological components responsible for seed dormancy and behavior. The interior (aleurone layer, transfer aleurone cell layer (TACL), endosperm, oxygen-scavenging protein (X), and the embryo) is surrounded by two enclosing structures, the hull (lemma, palea) and the gas- and water-tight caryopsis coat. Water and dissolved gas entry into the interior is restricted to the placental pore at the basal end of the seed.

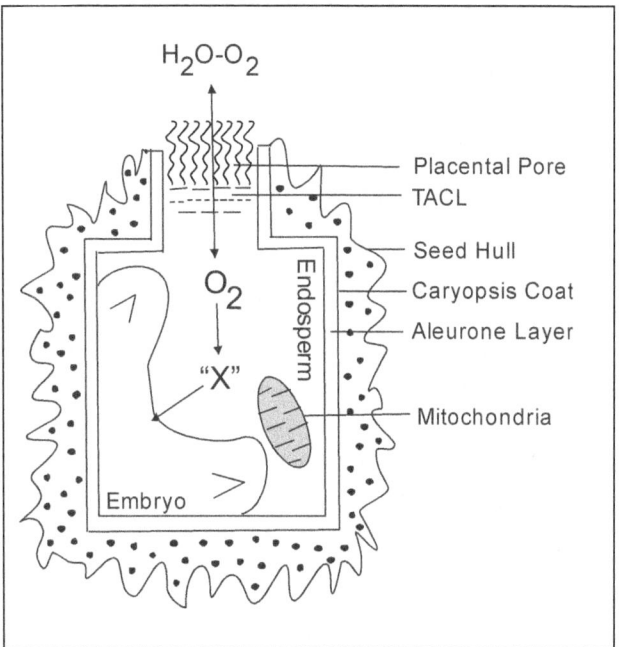

Temperature, oxygen and moisture signals are exchanged between three continuous partitions: (1) the soil matrix, (2) the surface of the seed (hull) which includes the placental pore, and (3) the interior of the seed (figure 17.10). The soil matrix in contact with the seed surface is the local source of moisture and gases required by the seed to initiate germination. The re-supply rate of this external water/gas pool varies over time in delayed response to macroscopic hydrological and thermal cycles (weather). Germination requires conditions favorable for rehydration and sustained metabolism. For a variety of plant species, lower temperatures, causing increased dissolved oxygen content and lower metabolic demands in seeds, can play an important part in seasonal germination timing (Dekker and Hargrove, 2002). It is therefore logical that the presence of surface films or excessive moisture as supplied by the environment can alter the degree to which a given internal moisture content and temperature are favorable for embryonic germination.

**Figure 17.10.** Schematic diagram of the transduction of water-gas (e.g. $H_2O$-$O_2$) and temperature (HEAT) signals from the soil matrix to the *Setaria* sp. seed symplast. Soil water-gas mixtures can be present in three (3) continuous partitions (gray shaded areas): soil matrix water-gas (top), seed exterior (hull) and apoplast (bottom left gray), and the seed symplast (bottom right gray). Heat from the environment affects oxygen solubility in water and metabolic behaviors (e.g. germination).

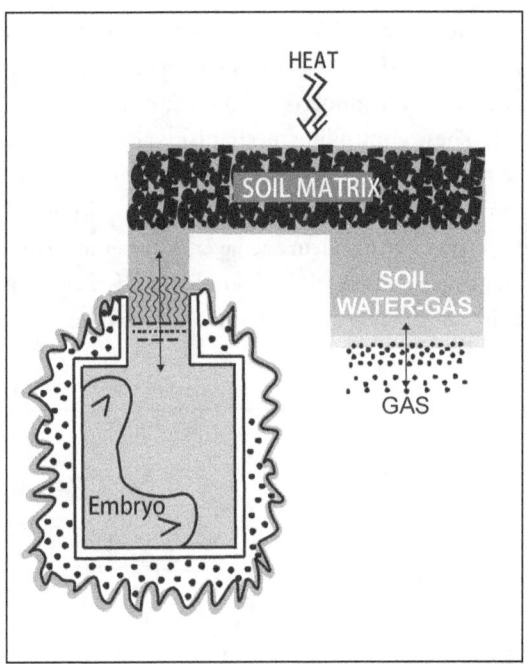

It is hypothesized that the change in moisture content of the *Setaria* spp. seed is driven by the dynamics of two anatomical structures: the hull and the caryopsis. Water saturates the hull by adhesion and the formation of surface films. In contrast, for caryopses with intact suberized coats, water enters and hydrates the caryopsis only by passing through the placental pore. Thus the anatomy of the seed hull may function as more than a simple protective structure for the *Setaria* seed; it may represent a mechanism by which the seed has tuned its germination timing by indirectly altering the frequency of its internal moisture and gas availability.

Water saturates the hull by adhesion and the formation of surface films (Dekker and Luschei, 2009). In contrast, water typically enters and hydrates the gas- and water-tight caryopsis only by passing through the placental pore. By analyzing the rate of water loss during a drying period, it was possible to determine the initial moisture content of the two anatomical structures and to reveal how the loss of water from the hull (partition 2) and caryopsis (partition 3) varied with moisture availability and thermocycle conditions in the local environment (partition 1). The drying procedure removed water in three distinct phases. In the first, water was lost quickly from the hull surface and more slowly from the caryopsis via the narrow placental pore. In the second, water continued to be lost from the caryopsis alone. After some time the interior water was tightly bound in the seed, the final phase. The moisture incubation experiments demonstrated that *S. faberi* seeds did not absorb water in the manner expected if it was a homogeneous material. Hydration of dry *S. faberi* seeds began with preferential partitioning of water to the interior seed caryopsis and the embryo. With additional local moisture availability the hull and caryopsis compartments absorbed moisture in a similar manner until the moisture content becomes sufficiently high to saturate the caryopsis. The caryopsis saturated at a lower local water availability content than did the seed hull. With saturation of the caryopsis, the hull preferentially absorbed the excess moisture. Hull water was spread uniformly over the seed surface forming a water film (partition 2). It is speculated that the water film changes the manner in which the seed interfaces with the external soil microsite environment (partition 1), the locus of oxygen exchange. The delivery rate of oxygen to the embryo is a function of its phase, partial pressure gradient and diffusion rate to the embryo (partition 3). The thickness of the seed surface water film is therefore a powerful mechanism regulating the pool of dissolved oxygen pool (partition 2) directly available to the seed embryo (partition 3), hence its crucial role in determining subsequent seed behaviors. Evidence that hull surface water quantity, in the physical form of boundary layer film thickness, controls germination was provided by the observation that maximum seed germination occurs in conjunction with intermediate levels of moisture availability in the soil environment. Soil moisture levels above and below this optimal hydration level resulted in lower germination. This observation provides a strong inference that hull surface water quantity in the physical form of boundary layer film thickness controls germination. Consistent with the theory of oxygen-water control of *S. faberi* germination (e.g. Dekker and Hargrove, 2002), these observations support the concept that optimal water film thicknesses on the seed surface are those that maximize the trade-off between hull surface area available for oxygen diffusion and water availability to support and sustain germination metabolism. Relatively low hull surface water contents provide thinner water films for soil atmosphere gases to penetrate, and therefore support greater gas diffusion but provide insufficient moisture for germinative growth. Thicker surface films provide greater resistance to gas diffusion but provide ample moisture for germination. Intermediate water availability optimizes the trade-off between water film gas diffusion (the dissolved oxygen pool) and moisture for germination.

Alternating diurnal thermocycles stimulated greater seed germination than did constant temperatures for all moisture contents greater than 0.25 ml, given the same heat units in each condition. These results are consistent with the hypothesis that alternating diurnal temperature conditions provide additional oxygen to the seed embryo and interior caryopsis due to the de-gassing effect that occurs when water temperature is increased. When temperatures increase in a water solution with oxygen at equilibrium with the local atmosphere, gas solubility decreases and excess oxygen is forced out of solution as a gas to the adjacent environment as it reaches the new equilibrium. This pulse of gaseous oxygen is immediately available to the seeds' metabolic physiology, the acute oxy-hydro-thermal time signal (Dekker et al., 2001, 2003).

These experiments indicate a robust, dynamic system of germination control by the transduction of external moisture and thermocycle signals from the soil to the seed embryo as modulated by the characteristics of the seed hull surface and the placental pore. The seed in unable to control its temperature, but the morphology of its exterior surface provides a means by which the seed is able to regulate the other two requisites for germinative growth, moisture and oxygen.

In a study by Donnelly (Donnelly et al., 2012), evidence was provided that seed shape in the genus *Setaria* is a product of the tradeoff between the constraints of phylogeny and local adaptation. Although there is a general "foxtail shape" (i.e. *Setaria* seeds from any of the taxa examined are recognizably similar), phylogenetic relatedness alone cannot explain the pattern of shape differences found in this species-group. Green foxtail is more similar in shape to the taxa that share its ecological niche (invasive-colonizing-weedy) than to the crops which are conspecific with it. Evidence was found of selection for size, surface to volume ratio, shape, seed surface rugosity, and seed hull permeability to water and gasses. From these results the null hypothesis was rejected, that phylogeny alone is sufficient to explain the pattern of variation in shape between taxa in *Setaria*. Because there was a clear divide between taxa of differing life histories (weeds versus crops), the possibility that the differences in shape between taxa are random (neutral drift) was also rejected. Thus, it is likely there is a functional explanation to explain the pattern of variation observed.

The results can be explained by the hypothesis that seed shape in *Setaria* is adapted to transduce an environmental signal consisting of oxygen, water, temperature, and time (oxy-hydro-thermal time) (Dekker and Hargrove 2002b, Dekker 2003). This signal is used by the seed interior to regulate seed behavior, particularly germination. Moisture availability plays a key role in regulation of germination in *S. faberi* (Dekker and Luschei 2009), with too much or too little water inhibiting germination (Dekker and Luschei 2009). With the sensitivity *Setaria* exhibit to temperature fluctuations, oxygen levels (Dekker and Hargrove 2002), and volume of water forming a film on the surface of the seed (Dekker and Luschei 2009), we conclude that transducing an oxy-hydro-thermal time signal is a crucial role of the seed exterior. To transduce the signal, the outer surfaces of the seed must accumulate soil water, passively oxygenate the accumulated water, and channel it to the seed interior via the placental pore. In other words, the lemma and palea of a *Setaria* seed act as an antenna to receive, modulate, and transduce information about the environment that the seed interior can use to determine germination.

The efficiency of this antenna is affected by several features of the seed's exterior, size, surface to volume ratio, rugosity, and the permeability of the lemma, palea, and lemma/palea suture to water. The differences we found between taxa show how each group has evolved outer seed morphology to increase the signal reception. *Setaria viridis* and *S. verticillata* have very small seeds, which are more elongate to increase surface area relative to seed volume. Elongation in these species is likely a result of the tradeoff between size and the surface-to-volume ratio (S:V). *Setaria faberi*, with slightly larger seeds, has also evolved elongation which has resulted in a higher S:V than *S. viridis* and *S. verticillata*. The seeds of *S. geniculata* are larger and the most elongate with a moderately high S:V. *Setaria pumila* has the highest S:V despite being the least elongate and having the largest seed size of the weedy taxa. It may be that life history differences between *S. pumila* and *S. geniculate* (*S. geniculata* was the only perennial species in the study) explain the differences in elongation between them. As a perennial, *S. geniculata* has multiple years for reproduction and the potential to reproduce vegetatively (Rominger 1962). This could reduce selection pressure for increased seed volume as found in *S. pumila*. Despite their differences, all the weeds show a distinct selective pressure acting on seed shape. The seeds are elongate to at least a limited extent and all have relatively flat paleas. Foxtail millets' (*Setaria viridis*, subsp. italic; *S. italica*) domestication has resulted in much rounder (less elongate), moderately-sized seeds with an intermediate S:V. The palea is extremely rounded in foxtail millet, likely due to selection for maximizing starch content.

Rugosity is likely to amplify the differences seen in S:V in the weedy taxa. *S. pumila* and *S. geniculata*, the taxa with the highest S:V, are also the most visibly rugose. *Setaria faberi* has more rugose seed surfaces than *Setaria viridis* or *S. verticillata* and also a greater S:V. However, S:V in foxtail millet will likely decrease in relation to the weeds when rugosity is factored in. The seeds of foxtail millet are relatively smooth in comparison to their weedy relatives (Pohl 1951, Rominger 1962, Pohl 1978).

Although it is hard to separate the crops out from the weeds by looking at seed size and S:V, there is a large difference between the taxa in terms of water permeability along the lemma/palea (LP) suture and in seed surface fragility. Because the LP suture is completely fused in the weedy taxa, water must be channeled to the seed interior via the placental pore (abscission point). This is not necessary in foxtail millet because the LP suture is not completely sealed at the distal end and the lemma and palea of foxtail millet are thin and fragile (Pohl 1951, Rominger 1962).

Another difference between the weeds and crops is germination behavior. The weeds germinate throughout the growing season, adjusting to local agricultural practices (Atchison 2001, Jovaag et al. 2012c). Foxtail millet, on the other hand, has been bred to exhibit nearly simultaneous germination. If seed shape in the weedy foxtails is adapted to transduce the environmental signal the seed interior uses for germination, it is likely the differences in seed shape and water permeability in foxtail millet are due to the release of the selection pressures acting on seed shape. In addition, selective breeding for increased starch volumes in foxtail millet is likely the reason foxtail millets are much more spherical than the weedy taxa (maximizing volume while minimizing surface area).

With the results of this study, the oxy-hydro-thermal time signal regulating seed behavior, sensitivity to water film thickness, and the importance of the outer seed surface as the primary interface between the seed interior and its environment, the null hypothesis that seed shape differences between the taxa in the genus *Setaria* can be explained by phylogenetic relatedness is rejected. It is also likely that Harper's (1964) supposition that seed shape is sensitive to environmental differences is correct for this weedy/invasive/colonizing species-group.

#### 17.4.4.2 Seed behavior.
#### 17.4.4.2.1 Annual seed-dormancy-germinability cycling in the soil. The fate of heteroblastic *Setaria faberi* seed entering the soil post-abscission was elucidated by Jovaag (Jovaag et al., 2011). Introduction of four

populations of *S. faberi* seeds with heterogeneous dormancy capacities into the soil of a no-till *Glycine max* field resulted in the formation of enduring pools with varying seasonal cycles of dormancy, after-ripening, germination, dormancy re-induction and death (figure 17.11).

The buried seed rain of highly dormant seed after-ripened with time and became highly germinable, awaiting favorable temperature and moisture conditions: the heterogeneous germination candidate pool. As this pool was depleted in the spring and early summer by seedling emergence and death, dormancy was re-induced in the living seeds that remained in the soil. The seeds remained dormant throughout the summer, then resumed after-ripening during late autumn. This dormancy-germinability cycle exhibited complexity both within and among *S. faberi* populations. Seed heteroblasty within *S. faberi* populations was retained, and germinability responses to the yearly seasonal environment varied among *S. faberi* populations. Further, local adaptation was shown by the differential germinability responses among *S. faberi* populations in common location agricultural nurseries. Seed mortality patterns also exhibited complexity within and among populations. Within an individual *S. faberi* population, mortality patterns changed as seeds aged in the soil. Among *S. faberi* populations differential mortality responses were observed in response to yearly seasonal environments and common nurseries.

**Figure 17.11.** Schematic diagram of weedy *Setaria* spp. soil seed pool behaviour based on the life history of the seed (seed states and processes (transitions between states)). Life history states: Dormant seeds, seed germination candidate, germinated seeds, seedlings, and dead seeds or seedlings. Life history processes: induction of dormancy and input (dispersal) of dormant seed at abscission (seed rain), after-ripening of dormant seed, dormancy re-induction of germination candidates, germination and growth of germination candidates, seedling emergence, and death.

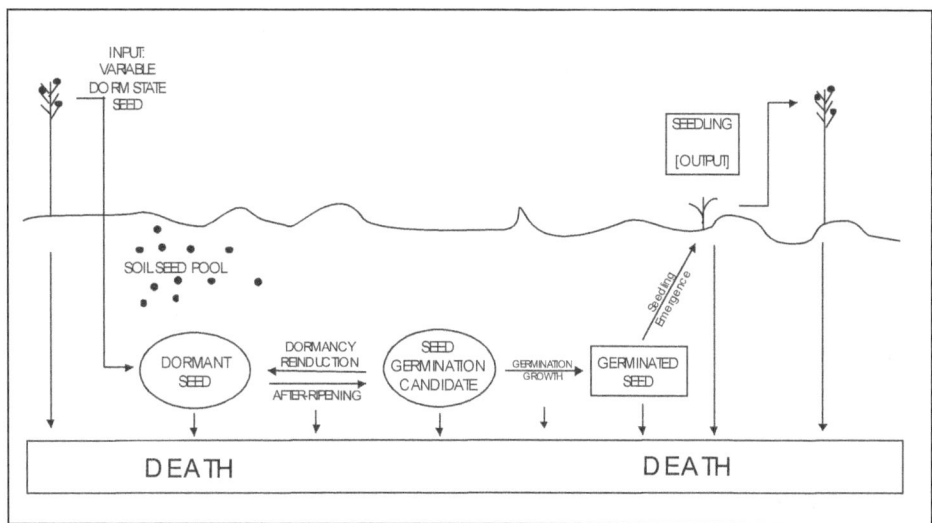

These observations clearly indicate *S. faberi* seeds have an annual dormancy-germinability cycle similar to that of the summer annual weeds reported by Baskin and Baskin (1985): high spring germinability, summer dormancy re-induction, higher autumn germinability. Unlike that previous report, evidence is now provided that individual *S. faberi* seed populations retained heteroblastic qualities throughout this annual cycle, that cyclic responses to the yearly seasonal environment varied among populations, and that adaptation to local conditions occurred in the germinability of an individual population. The retention of heteroblasty within a population is evident in the variation of the proportions of candidate seeds within a single population and time, and the heterogeneous decline in germinability, from early spring to early summer. The Lorentzian function model provided a sensitive method to distinguish this heteroblastic variation in the proportion of germination candidates over seasonal time among populations. Germinability responses to yearly environmental conditions were also shown. Local adaptation was evident in the differential germinability responses among *S. faberi* populations to the common nursery environments (i.e. a phenotype by environment interaction). This differential response occurred in both years. These observations provide evidence that the heterogeneous dormancy capacity among individual seeds remained, consistent with the hypothesis that dormancy capacity is an inherent quality of the seed and dormancy capacity heterogeneity of the population remains over time.

**17.4.4.2.2 Seed longevity in the soil.** The length of time that weedy *Setaria* spp. seeds are able to survival in the soil is quite variable, but mortality decreases with depth of burial (Dekker, 2003). The maximum period of survival in the soil varies by species: *S. pumila*: typical, 13 years; maximum, 30 years (Darlington; 1951; Dawson and Bruns, 1975; Kivilaan and Bandurski, 1973; Toole and Brown, 1946); *S. viridis*: maximum 10 to 39 years (Burnside et al.; 1981; Dawson and Bruns, 1975; Thomas et al., 1986; Toole and Brown, 1946); *S. verticillata*: maximum, 39 years (Toole and Brown, 1946). More typically, the majority of weedy *Setaria* spp. seed live in the soil for much shorter time periods (Atchison, 2001). In general, *S. viridis* and *S. pumila* seed on the surface lose viability sooner than buried seed (Banting et. al., 1973; Thomas et al., 1986). The rate of survival of *Setaria* seed is longer under uncultivated conditions (Stoller and Wax, 1974; Waldron, 1904). Burial of weedy *Setaria* spp. seed increases both the level of dormancy, viability, and the longevity, of those seeds, possibly due to decreased oxygen at greater depths or within soil aggregates (Banting et. al., 1973; Dekker and Hargrove, 2002; James, 1968; Pareja and Staniforth, 1985; Pareja et al., 1985; Stoller and Wax, 1974). Dawson and Bruns (1975) showed precipitation had no effect on *S. viridis* seed longevity in the soil.

**17.4.5 Seedling recruitment.** Weedy *Setaria* spp. seeds with heterogeneous dormancy states are recruited from soil pools when sufficient oxygen, water and heat accumulate over time (Dekker et al., 2003). The variable timing of this recruitment is the first step in assembly with crops and other weeds in annually disturbed agricultural communities. The timing of seedling emergence is the single most important factor in determining subsequent weed control, competition, crop losses and replenishment of the soil seed pool.

**17.4.5.1 Seedling emergence and community assembly.** Seedling emergence is the birth of a *Setaria* plant, its entrance into a local community, the first committed step in assembly with neighbors. Soil seed pool spatial structure, seed dormancy, and environment determine where and when seedlings emerge. Successful assembly of *Setaria* spp. in local communities with other species is determined by key traits typifying their annual life history. The first assembly step in many agricultural fields is the emergence of seedlings or herbaceous foliage of perennial species. Successful foxtail seedlings then interact with other species in the locality. The competitive phase concludes with reproduction and entry of new seed into the soil. Seedling emergence prediction guides early season risk management and the use of weed tactics. Time of seedling emergence is the single most important factor determining subsequent yield losses from, and competitiveness of, a weed. A weed's competitive success determines yield losses, seed bank replenishment and future weed infestations.

**17.4.5.1.1 Seedling emergence.** In general, weedy *Setaria* spp. seeds are capable of germinating and emerging anytime the soil is unfrozen (Dekker, 2003). Most *Setaria* seedlings emerge in the early part of the growing season of a locality (e.g. April-June in the northern hemisphere; Atchison, 2001; Buhler et al., 1997b; Forcella et al., 1992, 1997): *S. viridis* (Banting et al., 1973; Chepil, 1946; Dawson and Bruns; 1962); *S. pumila* (Dawson and Bruns, 1962, 1975; Manthey and Nalewaja,1987; Martin, 1943; Sells, 1965; Steel et al., 1983; Stoller and Wax, 1974); *S. verticillata* (Martin, 1943). With knowledge of dormancy states and environmental $O_2$-$H_2O$ signals in a local seed pool it may be possible to predict *Setaria* emergence in a locality with precision (Dekker et al., 2001).

**17.4.5.1.2 Seed germination depth from the soil.** Weedy *Setaria* spp. seedlings emerge from relatively shallow depths in the soil, usually between 1 to 5 cm depth (Dekker, 2003). Emergence is greatest at 1.5 to 2.5 cm and declines with depth to a maximum of about 7.5 to 14 cm (Buhler, 1995; Buhler and Mester, 1991; Dawson and Bruns, 1962; Gregg, 1973; Waldron, 1904). Emergence of weedy *Setaria* spp. does not usually occur on the soil surface except in the debris of no-till systems, an indication that light may not play an important role in seed germination. The soil depths from which the maximum *S. faberi* emergence occurs is a function of the tillage system, the greater the depth of tillage the deeper and more varied the seed placement and seedling emergence: 0 to 2 cm (no-till); 1 to 3 cm (disk); 1 to 4 cm (chisel plow); 1 to 5 cm (moldboard plow) (Mester and Buhler, 1986; Yenish et al. 1992). Atchison (2001) observed that *S. faberi* in the first year after dispersal emerged from shallower depths (0 to 5 vs. 5 to 10 cm), but emergence of older, more dormant seed was similar from these two depths. Others have correlated depth of emergence and seed weight, surmising that greater seed weight indicates more seed reserves (Dawson and Bruns; 1962; King, 1952). Oxygen content decreases with soil depth, inhibiting after-ripening and germination at greater depths and enforcing dormancy (Dekker and Hargrove, 2002; James, 1968).

**17.4.5.2 Patterns of seedling emergence.** Seedling recruitment of heteroblastic *Setaria faberi* seed entering the soil post-abscission was elucidated (Jovaag et al., 2012). Evidence was provided that weedy *Setaria* seedling recruitment behavior is predicated on dormancy state heterogeneity at abscission (seed heteroblasty), as modulated by environmental signals. Complex oscillating patterns of seedling emergence were observed during the first half of the growing season in all 503 soil burial cores of the 39 populations studied. These patterns were attributed to six distinct dormancy phenotype cohorts arising from inherent somatic polymorphism in seed dormancy states. Early season cohorts were formalized using a mixture model consisting of four normal distributions. Two, numerically low, late season cohorts were also observed. Variation in emergence patterns among *Setaria* populations revealed a fine scale adaptation to local conditions. Seedling recruitment patterns were influenced by both parental-genotypic (time of embryogenesis) and environmental (year, common nursery location, seed age in the soil) parameters. The influence of seed heteroblasty on recruitment behavior was apparent in that *S. faberi* populations with higher dormancy at the time of dispersal had lower emergence numbers the following spring, and in many instances occurred later, compared to those less dormant. Heteroblasty was thus the first determinant of behavior, most apparent in recruitment number, less so in pattern. Environment modulated seedling numbers, but more strongly influenced pattern. The resulting pattern of emergence revealed the actual "hedge-bet" structure for *Setaria* seedling recruitment investment, its realized niche, an adaptation to the predictable mortality risks caused by agricultural production and interactions with neighbours. These complex patterns in seedling recruitment behavior support the conjecture that the inherent dormancy capacities of *S. faberi* seeds provides a germinability 'memory' of successful historical exploitation of local opportunity, the inherent starting condition that interacts in both a deterministic and plastic manner with environmental signals to define the consequential heterogeneous life history trajectories.

*Setaria faberi* is a very successful invasive agricultural weed because of its ability to form long-lived soil seed pools of heterogeneous seed (Jovaag, Dekker & Atchison, 2009a, b) that cycle annually between dormancy and germination candidacy until environmental conditions allow germination and emergence to occur. Seedling recruitment is an irreversible, committed step in the life history of a plant. It is the time when the annual *S. faberi* plant resumes autotrophic growth, assembles in local agro-communities, and begins interactions with neighbors. As such, its timing is crucial to subsequent behaviors, including survival and reproductive success. Seedling emergence timing is a trade-off between the risks of mortality from disturbance and competition with neighbours and exploitation of opportunity space-time to accumulate biomass and produce seed. *S. faberi* has evolved in agro-ecosystems characterized by annual disturbance and selection by predictable farming practices (e.g., planting, tillage, herbicides, harvesting). The variable risks in these agricultural habitats have not led to a single, best time to germinate. Is there a pattern to the periodicity of *S. faberi* seedling recruitment? Past research only reveals a qualitative pattern: an early flush of seedling emergence in the spring (April-June) which decreases as temperatures warm, followed by low numbers during the late season (e.g., Forcella, et al., 1992, 1997). Is it possible to discover a more precise, quantitative pattern to seedling emergence? If so, does it bear a relationship to seed heteroblastic patterns among the same seeds at abscission? A successful survival strategy for a prolific seed producing species like *Setaria* may be to disperse heteroblastic seeds able to take advantage of as many recruitment opportunities as are available, in both seasonal time and field space. The resulting pattern of seedling emergence timing may reveal the 'hedge-bet' for individual *Setaria* fitness. It is hypothesized that there exists a pattern to *Setaria* seedling recruitment timing which is predicated in the first instance on the dormancy state heterogeneity (seed heteroblasty) at abscission. This pattern in the second instance is modulated by the environment (year, common nursery location, seed age in the soil). Thus, patterns in *Setaria* emergence will reflect the historically advantageous recruitment times (opportunity) enabling survival and reproductive success.

**17.4.5.2.1 General seedling recruitment pattern.** A complex pattern of *S. faberi* seedling emergence was observed consisting of six discrete cohorts during the growing season: Julian week (JW) 16-49 (Jovaag et al., 2012). The majority were recruited during four early season (JW16-31) emergence periods, and a small minority during two later season (JW32-49) emergence flushes. This complex pattern of discrete, discontinuous seedling emergence was unexpected. *A priori* expectations were of a large, continuous flush of emergence in early season followed by low, sporadic numbers during the late season, as reported previously (e.g., Forcella et al 1992, 1997).

**17.4.5.2.2 Early season recruitment.** A complex oscillating pattern of *S. faberi* seedling emergence was observed in the first half of the growing season (JW16-31) in all 503 cores of the 39 populations studied, and formalized using a mixture model consisting of four normal distributions which consistently fit data from different populations and years (figure 17.12). The consistent appearance of these four normal distributions provides strong evidence of the existence of discrete recruitment cohorts arising from inherent somatic polymorphism in seed dormancy states (seed heteroblasty). Changes in the recruitment of early season *S. faberi* cohorts in response to several factors were reflected in their timing, the proportional distribution among the individual seasonal cohorts, and the total number of seedlings that emerged.

**17.4.5.2.3 Parental and environmental influences on recruitment.** Fine scale changes in seedling emergence patterns in response to changes in parental qualities (time of abscission) and environmental conditions (year, common nursery, seed age in the soil) indicate that *S. faberi* has the ability to adapt its behavior on a fine scale to a locality by generating multiple, well-timed and numbered, pulses of seedlings over the course of the growing season.

**Figure 17.12.** *S. faberi* seedling emergence (proportion of total) with time (Julian week, JW) during the spring and early summer of: the first year after burial for all populations (top panel); year of emergence (year 1, 1998; year 2, 1999; year 3, 2000) of all populations buried in 1997 (middle panel); the first year after burial of populations with a common time of seed abscission (August, JW 32; September, JW 36; October, JW 40) (bottom panel). Bars: relative frequency; solid line: mixture model estimate (4 normal components with equal variance); dashed lines: model's 4 components weighted by the mixing proportions (Jovaag et al., 2012).

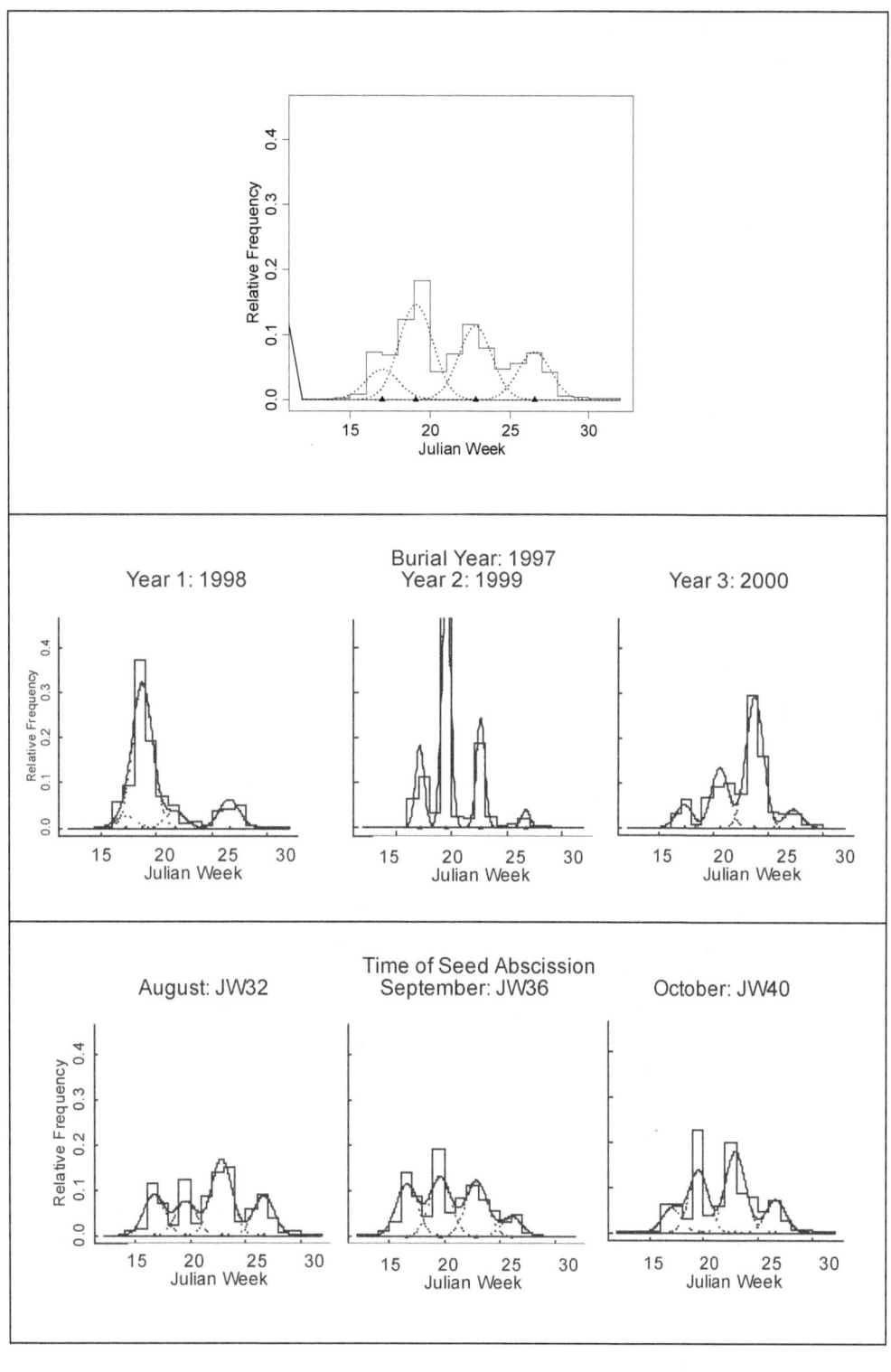

**17.4.5.2.3.1 Parental influences.** Seed heteroblasty is induced during embryogenesis and modulated by the specific seasonal conditions experienced by the parental plant (Dekker, 2003). The influence of seasonal environment on dormancy induction is apparent at the time of seed abscission from the parent plant: the earliest maturing (August) seed was more dormant than middle (September) seed, which in turn was more dormant than late (October) seed (Jovaag, Dekker & Atchison, 2012a). The influence of seed heteroblasty was more apparent in recruitment number, less so in pattern. Total seedling recruitment in was often inversely related to the time of seed maturity from August to October. This inverse relationship between the time of abscission and the number of seeds subsequently emerging during the spring and early summer was observed in 1999 and the second half of 2000. Further, mid-summer to autumn emergence during both 1999 and 2000, while low, came only from middle and late maturing populations. During the first half of 2000, there was generally no difference in the emergence numbers, proportions or timing among seeds with different abscission times. This similarity indicated that environmental conditions dominated emergence patterns in 2000. The greater variation in emergence proportion and number during 1999 indicated that environmental conditions were less dominant, and parental variability more influential, than in 2000. Seed heteroblasty had some influence on emergence pattern: less dormant populations sometimes emerged earlier than more dormant. In years producing more germinable seeds, the majority generally emerged earlier compared to those years producing relatively more dormant seeds.

**17.4.5.2.3.2 Year.** Seasonal conditions (year) had the greatest influence on *S. faberi* recruitment relative to other environmental parameters studied.

**17.4.5.2.3.3 Seed age in the soil.** There was no consistent relationship of seed age in the soil with the timing and proportion of emergence, evidence of the inherency of seed dormancy capacity. Recruitment numbers (as well as the soil seed pool size) declined rapidly during the early season (JW16-31) as seed age in the soil increased from one to three years.

**17.4.5.3 Seed germinability-dormancy heteroblasty blueprints seed behavior in the soil.**

**17.4.5.3.1 Relationship of seed heteroblasty and seedling emergence.** Fine scale differences in seed dormancy capacities (heteroblasty; Jovaag et al., 2012a), dormancy cycling in the soil (Jovaag et al., 2011), and seedling emergence behavior (Jovaag et al., 2012b) were observed among *S. faberi* populations. Further, a clear relationship between these behaviors of a *S. faberi* population during its seed-seedling life history stages was observed, evidence consistent with the hypothesis that seedling recruitment is predicated at the outset by seed heteroblasty, and subsequently modulated by the environment.

The life history trajectory of the individual *S. faberi* seed, from anthesis and embryogenesis to germination and seedling emergence was determined in the first instance by the inherent, parentally-induced, dormancy capacity of the seed, but was sensitive to the environmental conditions confronted on different time scales. This, consequential, life history trajectory was first apparent in dormancy-germinability cycling in the soil (Jovaag et al., 2011). At the widest temporal scale studied, heteroblastic qualities largely determined the numbers of seeds that are recruited in a single year and locality. On a finer temporal scale (weekly or daily), the distribution and timing (pattern) of seedling emergence was strongly influenced by the unfolding seasonal environment. The life history trajectory is therefore both deterministic and plastic, conditional on starting conditions (parentally-induced dormancy capacity) and modulated by environmental conditions confronted subsequently. The resulting seedling recruitment in a locality is an emergent property of these complex interacting factors.

A relationship between seed heteroblasty at abscission and its subsequent behavior in the soil was observed. Individual *S. faberi* seed populations retained heteroblastic qualities throughout their annual soil germinability cycle (high spring germinability, summer dormancy re-induction, increased autumn germinability). Retention of heteroblasty was evident in the variable proportions of candidate seeds within a single population in the soil at individual seasonal times. The influence of heteroblasty was also apparent in the heterogeneous decline in germinability of seed in the soil from early spring to early summer. Further, the germinability pattern of the candidate pool in the soil (high in spring, low in summer, somewhat increased in autumn) was consistent with the emergence pattern (high in spring and early summer, low in mid-summer, none in late summer-early autumn, low in mid-late autumn).

A relationship between seed heteroblasty at abscission from the parent plant and its subsequent seedling recruitment behaviour was also observed. This influence, modulated by environment, was apparent in both seedling recruitment numbers and temporal emergence patterns. Evidence of this relationship between heteroblasty and emergence numbers was provided by the positive Spearman correlation between dormancy capacity at abscission and the cumulative number of seeds emerged during the first year after burial for both the 1998 and 1999 *S. faberi* populations. Additionally, more dormant populations had lower emergence numbers during the first year after burial than less dormant populations. This was more apparent during the 1999 season than the 2000 season. Further evidence was provided by the inverse relationships between time of abscission and subsequent emergence numbers during 1999 and the second half of 2000. Early maturing seed was the most dormant and had the least number of seeds emerging. Seed maturing late in the season was the least dormant and had the greatest number of seeds emerging. During the first half of 2000, environmental conditions strongly dominated emergence numbers and patterns such that no relationship with time of abscission was observed. The 1998 seed dormancy heteroblasty (after-ripening, AR) groups (Jovaag et al., 2011) also provided evidence of this relationship: the most dormant seed (AR group 3) had the least emergence, while the least dormant seed at abscission (AR group 1) subsequently had the greatest number of seeds emerging. Further, the least dormant seed generally emerged earlier: AR Group 1 had the most early spring emergence, and the least early summer emergence.

### 17.4.5.3.2 Seedling emergence hedge-betting for assembly in agro-communities.
These studies by Atchison and Jovaag (Jovaag et al., 2011, 2012a, b) provide insights about weedy hedge-betting strategies of *S. faberi* for invasion, colonization and enduring occupation of central Iowa agricultural fields. Fitness in *S. faberi* is conferred by strategic diversification of seedling recruitment when confronted by risks from agricultural disturbance and heterogeneous, changeable habitats. Evolutionarily, hedge-betting is a strategy of spreading risks to reduce the variance in fitness, even though this reduces intrinsic mean fitness. Hedge-betting is favored in unpredictable environments where the risk of death is high because it allows a species to survive despite recurring, fatal, disturbances. Risks can be spread in time or space by either behaviour or physiology. Risk spreading can be conservative (risk avoidance by a single phenotype) or diversified (phenotypic variation within a single genotype) (Cooper & Kaplan, 1982; Philippi & Seger, 1989; Seger & Brockman, 1987).

Hedge-betting of seedling recruitment time and pattern reduces temporal variance in fitness of *S. faberi* in agroecosystems characterized by annual cycles of disturbance and unpredictable climatic conditions. Earlier emergence allows for greater biomass and potentially explosive seed production at season's end (Dekker 2004b), but also has a high risk of death from agricultural practices. Later emergence has a lower potential for seed production, but also a lower risk of death. Thus variation in germination time results in both a decrease in the maximum and an increase in the minimum potential seed production, reducing fitness variability and ensuring enduring occupation of the locality. Variation in germination time has the additional benefit of reducing sibling competition, which enhances individual fitness (Cheplick 1996).

Hedge-betting by *S. faberi* begins with the birth of seeds and the induction of heteroblastic differences that are subsequently echoed in seedling emergence behavior. The complex pattern of emergence observed in this study (Jovaag et al., 2012b) reveals the actual hedge-bet structure for central Iowa *S. faberi* seedling recruitment, its realized niche. The relative seedling investment made by *S. faberi* at different times during the season is a consequence of the changing balance between the risk of mortality and the potential fecundity during those periods.

There exists some predictability to the primary sources of mortality and recruitment opportunities for *S. faberi* in Iowa agroecosystems over the course of their annual life histories (tables 17.5, 17.6). The majority of seedlings are recruited in the spring when the risk of mortality is very high from crop establishment practices (seedbed preparation, planting, weed control) and the rewards are the greatest. Weed seedlings emerging early have the greatest time available for biomass accumulation and competitive exclusion of later emerging neighbours. Subsequent fitness devolves on those individual *S. faberi* plants that escape these disturbances.

Seedling recruitment in the late spring and early summer is affected by the cessation of all agricultural disturbances and weed control tactics, a time termed 'layby' in the Midwestern US. Seeds emerging after layby avoid mortality from cropping disturbances, and the risk of death shifts to that from interactions with neighbours. Freed from cropping disturbance risk, seedlings emerging post-layby face increasing competition from crop and weed neighbours and decreasing time for biomass production. Subsequent fitness devolves on those plants freed from human intervention that compete effectively with neighbours until harvest, an advantageous opportunity space-time apparent from the seedling investment observed.

**Table 17.5.** *S. faberi* seedling recruitment cohort (time, Julian week (JW)) exploitation of changing opportunity spacetime in Iowa, USA, maize-soybean cropping fields (Jovaag et al., 2012b).

| COHORT | 1 | 2 | 3 | 4 | 5 | 6 |
|---|---|---|---|---|---|---|
| SEASON | EARLY | | | | LATE | |
| | Early Spring | Mid-Spring | Late Spring | Early Summer | Summer | Autumn |
| TIME (JW) | 16-18 | 18-20 | 22-24 | 26-28 | 32-35 | 45-49 |
| Fecundity Potential | very high | very high | high | medium | low | very low |
| Mortality Risk | very high | very high | high | low | low | low |
| Source(s) of Risk | crop disturbance | crop disturbance | crop disturbance | neighbors | neighbors | crop disturbance; climate |
| Weed Strategy | escape cropping | escape cropping | escape; post-layby opportunity | post-layby opportunity | post-layby opportunity | post-harvest opportunity |
| Seedling Investment | 12% | 38% | 30% | 20% | 0.2% | 0.1% |

Very small *S. faberi* seedling investments are made late in the season when the risks of mortality are much lower. Seedlings recruited late experience a senescing crop light canopy. Those emerging during or after harvesting experience mortality risks only from the approaching winter climate. Floral induction in *S. faberi* is increasingly stimulated by decreasing photoperiod, allowing a seedling with as few as 2-3 leaves to produce a few seeds very late in the season (e.g. Oakwood site, 1998; data not reported). This justifies a small hedge-bet investment yielding low numerical returns but ensuring some potential increase in soil seed pool size for future infestations.

The time of seedling emergence determines what individuals assemble in plant communities and how they interact with neighbors. Recruitment timing therefore determines how much biomass successful plants can accumulate, which directly predicates seed productivity (Dekker, 2004b). Fitness devolves on pre-adapted individual *S. faberi* plants that anticipate the unfolding opportunity space-time. The hedge-betting structure of *S. faberi* revealed in these papers supports the conjecture that the inherent variation in seed dormancy capacities among seeds dispersed by a parent plant provides a germinability 'memory' of historically successful seedling recruitment responses to local disturbance allowing exploitation of local opportunity in space and time (Trewavas, 1986).

**Table 17.6.** Calendar of historical, seasonal times (Julian week, month) of agricultural field disturbances (seedbed preparation; planting; weed control, including tillage and herbicides; time after which all cropping operations cease, layby; harvest and autumn tillage), and seedling emergence timing for central and southeastern Iowa, USA, *Setaria faberi* population cohorts (all *S. faberi* combined; Jovaag et al., 2012b)

# Evolutionary Ecology of Weeds

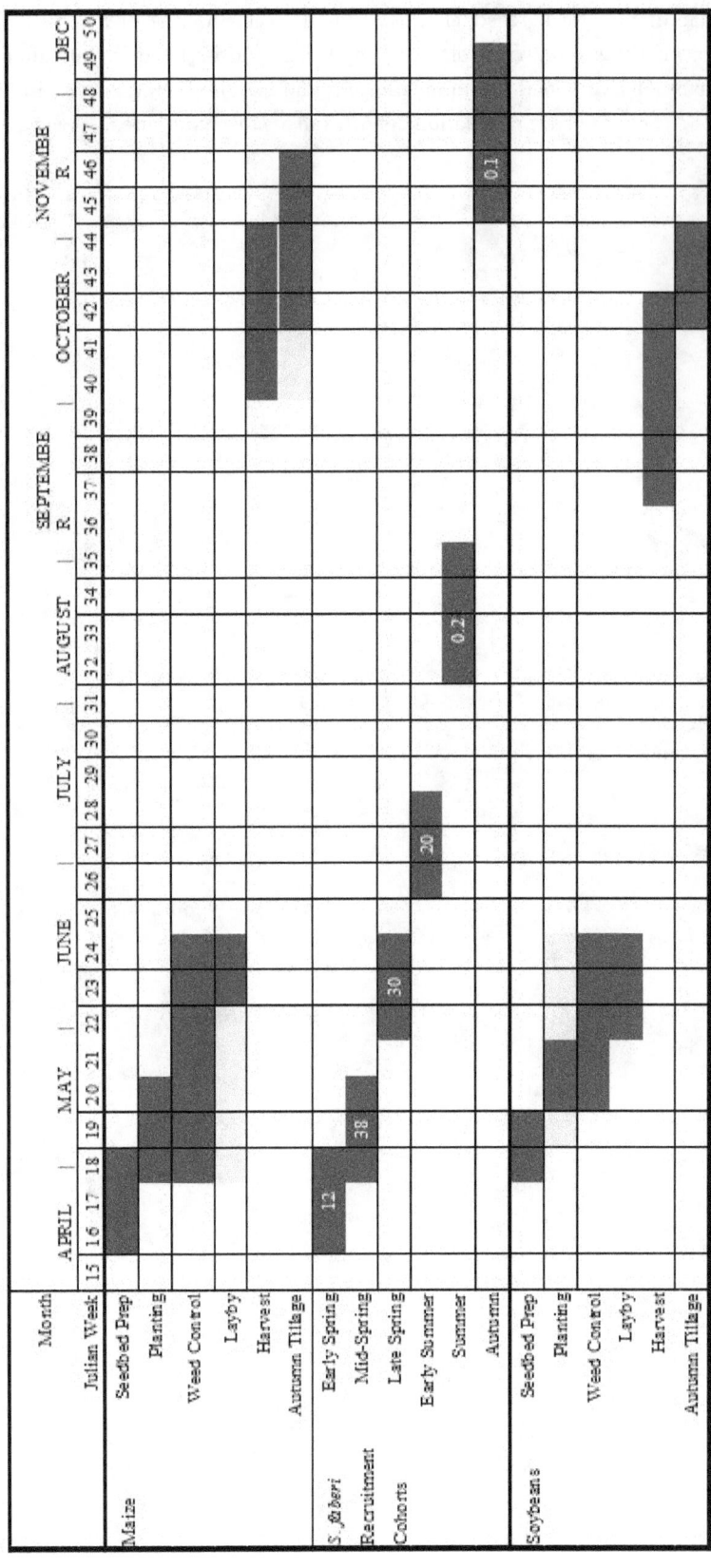

## Chapter 18: Triazine Resistant and Susceptible *Brassica napu*s

**Summary.** The nature of s-triazine resistant plants (R) is a complex adaptive photosynthetic system arising as a consequence of a single base pair lesion in the chloroplast *psbA* gene. The D-1 protein product of *psbA* is a key element of photosystem II electron transport. Altered electron transport in R causes a pleiotropic cascade of self-reorganization of interacting, interdependent, functional traits. Order in this highly conserved photosynthetic system is emergent: a new homeostatic equilibrium among the plastids, cells and organs of the whole plant R phenotype emerges.

The nature of R plants is how local photosynthetic opportunity spacetime is seized and exploited relative to that of the susceptible (S) wild type. The altered D-1 protein product of the *psbA* gene has been regarded as less photosynthetically efficient in the R biotypes of several species. Past studies have shown lower carbon assimilation (A) and whole plant yields in R relative to S; in other comparisons R was greater than S; in still others R and S were comparable. The equivocal nature of our understanding of how photosynthesis differs between R and S, and how it is regulated, led us to focus experimental efforts on the chronobiology of both phenotypes.

A differential pattern of leaf disc chlorophyll fluorescence (LCF) between S and R *Brassica napus* was observed: R is a chronomutant. A phase shift in LCF maxima occurred in the daily light-dark cycle, supporting the hypothesis that alterations in chloroplast structure conferring s-triazine resistance imply altered temporal behavior of photosynthetic activity.

Studies were conducted observing diurnal patterns of A during development. Younger R plants had greater photosynthetic rates early and late in the diurnal, while those of S were greater during midday as well as the photoperiod as a whole. As *B. napus* plants aged, differences in A between biotypes increased: only during the late diurnal period was R greater. As *B. napus* began reproductive development, a reversal of photosynthetic differences occurred. R carbon assimilation was greater than S early, midday, and for the whole day period; S was superior to R only late in the day. These results indicate a more complex pattern of photosynthetic carbon assimilation than previously reported. The photosynthetic superiority of a biotype is a function of the time of day and the age of the plant.

R plants also differ in stomatal function from S plants. Total conductance to water vapor and intercellular $CO_2$ partial pressure in R was either equal to, or greater than, S over the lifetime of those plants, with the possible exception of some atypical episodes late in ontogeny and late in the light period. As a consequence of these phenomena, leaves of R plants were either the same temperature, or cooler, than leaves of S plants for the entire lives of both biotypes. At lower leaf and air temperatures R and S leaves function in a similar way. As air temperature increases, their responses diverge, allowing R plants compensate for their sensitivity to high temperature. R leaves generally are cooler and total conductance to water is greater than in S, probably due to greater stomatal aperture size. As a result, at higher air temperatures R leaves photosynthesize at cooler leaf temperatures closer to the optimal for both biotypes. In addition to other pleiotropic effects, R plants appear to be stomatal mutants. They constitute a model system to study regulation of stomatal function and the relationship between environmental cues and stomatal behavior.

Studies were conducted to test if there is any regulatory role for ribulose-1,5-bisphosphate carboxylase/oxygenase (Rubisco) in photosynthetic carbon assimilation in R and S *Brassica napus*. Rubisco percent activation and initial activity may account for R and S carbon assimilation differences during midday. Differences in A early and late in the day were not accounted for by differences Rubisco (initial, total, % activation).

Studies were conducted to determine the response of R and S to different temperatures and gas atmospheres with infrared gas analysis and pulse amplitude modulated chlorophyll fluorescence techniques. Photosynthetic regulation can be separated into three categories based on these studies. The first category is Rubisco-limited photosynthesis. When carbon assimilation was Rubisco-limited there was little difference between R and S biotypes. A second category, feedback-limited photosynthesis, was most evident at 15°C, less so at 25°C. The third category, photosynthetic electron transport-limited photosynthesis, was evident at 25 and 35° C. R exhibited much more electron transport-limited carbon assimilation at 25°C and 35°C than did S. At 15°C neither R nor S exhibited electron transport limited carbon assimilation. The primary limitation to photosynthesis changes with changes in leaf temperature: electron transport limitations in R may be significant only at higher temperatures.

The reorganized R plant interacts with the environment in a different manner than does S. Under environmental conditions highly favorable to plant growth, S often has an advantage over R. Under certain less favorable conditions to plant growth, stressful conditions, R can be at an advantage over S. It can be envisioned that there were environmental conditions prior to the introduction of s-triazine herbicides in which R had an adaptive advantage over the more numerous S individuals in a population of a species. Under certain conditions R might have exploited a photosynthetic niche under-utilized by S. These conditions may have occurred in less favorable environments and may have been cool (or hot), low light conditions interacting with other biochemical and diurnal plant factors early and late in the photoperiod, as well as more complex physiological conditions late in the plant's development. R survival and continuity could have been favored under these conditions.

## 18.1 The Nature of s-Triazine Resistant Plants

The nature of s-triazine resistant plants (R) is a complex adaptive photosynthetic system. Complex photosynthetic adaptation of the whole plant phenotype arises in the dynamic network of interacting components behaving in parallel: the psbA gene, the D-1 protein, chloroplast ultrastructure, the R cell, and leaf tissues and organs. Order in this system is emergent: coherent structures and functional traits appear during the process of self-organization of the entire photosynthetic process. The nature of R plant self-organization is a pleiotropic consequence of a single base pair lesion in the chloroplast *psbA* gene. This single mutation affects multiple phenotypic traits in a cascade from gene to global R metapopulation behavior. The R product of the *psbA* gene, the D-1 protein, is a key functional element of photosystem II electron transport. Altered electron transport in R leads to a profound re-organization of interacting, interdependent, functional units in the chloroplast. Change in the highly conserved photosynthetic apparatus leads to a new homeostatic equilibrium among the plastids, cells, tissues, organs and the whole plant R phenotype. The nature of R plants is the pleiotropic consequence of how and when the new phenotype seizes and exploits local photosynthetic opportunity spacetime relative to that of the susceptible (S) wild type.

## 18.2 Structure-Function Change in s-Triazine Resistant Plants

**18.2.1 Structural change in s-triazine resistant plants.** s-triazine resistance in higher plants is due to a single base pair mutation to the *psbA* chloroplast gene (Hirschberg and McIntosh, 1983). The R variant has a single nucleotide base pair mutation of guanine for adenine. This mutation results in substitution of a serine codon (AGT) for a glycine codon (GGT) at position 264 in the polypeptide product. The codon 264 change in the maternally inherited, cytoplasmic, *psbA* gene causes a change in its product, the trans-membrane D-1 protein, a key functional element in PSII electron transport. This single base pair mutation results in profound decrease in the ability of the protein to bind s-triazines that induce toxic responses.

Several changes in thylakoid lipid chemical composition in R have been observed (Pillai and St. John, 1981). R phospholipids had higher linolenic acid concentrations and lower levels of oleic and linoleic fatty acids (Dekker, 1993; Lemoine et al., 1986). R plants overall were richer in unsaturated fatty acids, had higher proportions (and quantitatively greater amounts of) monogalactosyl diglyceride and phosphatidyl choline, and had lower proportions of digalactosyl diglyceride and phosphatidyl choline, than in S (Burke et al., 1982). As a consequence of a higher proportion of appressed thylakoids, R plants had a greater proportion of $\Delta 3$-trans-hexadecenoic acid in phosphatidylglycerol (Boardman, 1977).

Although the leaf anatomy of R and S is similar, many plastid ultrastructural characters are different (Dekker, 1993). R has decreased plastid starch content and increased grana stacking (and the associated characters of lower chl $a/b$ ratio, increased chl $a/b$ light-harvesting complex, and lower P700 chl $a$ and chloroplast coupling factor amounts) (Mattoo et al., 1984). The resistant mutant had a larger volume of the chloroplast as grana lamellae and more thylakoids per granum. Vaughn reported that several triazine-resistant species, including *Brassica napus* have a shade-adapted chloroplast ultrastructure (Vaughn and Duke, 1984; Vaughn, 1986). These plastid ultrastructural changes in R are similar to those found in shade-adapted leaves (Boardman, 1977; Dekker, 1993; Dekker and Burmester, 1992).

A single base-pair substitution at codon 264 in the *psbA* chloroplast gene in the highly conserved photosynthetic apparatus (PS) leads to a cascade of changes in the plants morphology, physiology and ecological reaction to its immediate environment, a non-intuitive result. With the genetic lesion, a dynamic re-organization of interacting, interdependent, plant functional units occurs, a new homeostatic equilibrium among parts. This pleiotropic cascade of phenotypic changes that are stimulated by *psbA* gene codon 264 mutation is not predictable from genomic information.

### 18.2.2 Functional change in s-triazine resistant plants.

**18.2.2.1 Resistance to s-triazine herbicides.** The herbicide atrazine was introduced commercially around 1960 for broadleaf weed control in maize (*Zea mays*) production fields. S-triazine resistance (R) in higher plants was first discovered in common grounsel (*Senecio vulgaris*) in 1968 in response to protracted exposure to simazine and atrazine in nursery stock (Ryan, 1970). Since that time s-triazine resistance has been confirmed in many species throughout the world. The rapid increase in R weed species and locations has been associated with sole reliance on s-triazine herbicides for weed control, simple selection by a single herbicide. In Iowa, for example, four species rapidly appeared: kochia (*Kochia scoparia*; Dekker et al., 1987), giant foxtail (*Setaria faberi*; Thornhill and Dekker, 1993), common lambsquarters (*Chenopodium album*; Dekker and Burmester, 1989) and Pennsylvania smartweed (*Polygonum pensylvanicum*; Dekker et al., 1991).

Herbicide resistance is conferred by a marked decrease in atrazine binding to the D-1 protein encoded by the chloroplast (maternally inherited) *psbA* gene mutant (Hirschberg and McIntosh, 1983). Studies revealed that this type of herbicide resistance was a consequence of reduced herbicide binding to the 32kD chloroplast protein, D-1, in the resistant (R) biotype (Pfister et al. 1981; Steinback et al. 1981). S-triazine herbicides act by binding to the D-1 protein in susceptible (S) species in the chloroplast thylakoid membranes and thereby inhibiting photosystem II (PS II) electron transport. The s-triazine resistance mechanism is not physiological; R is not a function of differential uptake, translocation, metabolism, accumulation or membrane permeability (Radosevich, 1970).

**18.2.2.2 Differences in carbon assimilation efficiency.** Since the first discovery of s-triazine resistance (R) in higher plants, the altered D-1 protein product of the psbA gene has been regarded as less photosynthetically efficient in those R biotypes of the species (Dekker, 1993). s-triazine resistant plants have been shown to have a decreased quantum efficiency of $CO_2$ assimilation compared to s-triazine susceptible plants (S) (Holt et al., 1981); and have been generally regarded as less fit than S plants (Gressel and Segel, 1978). This decreased quantum efficiency of $CO_2$ assimilation in R compared to the S wild type is credited to an altered redox state of PSII quinone acceptors and a shift in the equilibrium constant between $Q_A-$ and $Q_B$ in favor of $Q_A-$ (Arntzen et al., 1979).

Several studies have shown lower photosynthetic carbon assimilation rates (A) in R *Amaranthus hybridus* (Ahrens and Stoller, 1983) and *Senecio vulgaris* (Holt et al., 1981; Ort et al., 1983) relative to S. Beversdorf et al. (1988) found lower R whole plant yields in field evaluations of *Brassica napus*. Additionally, van Oorschot and van Leeuwen (1984) found A in S *Amaranthus retroflexus* was greater than that in R; while R and S A were comparable in *Chenopodium album*, *Polygonum lapathifolium*, *Poa annua*, *Solanum nigrum* and *Stellaria media*. R biotypes of *Phalaris paradoxa* have been found to be photosynthetically superior to their S counterparts (Schonfeld et al., 1987). Jansen et al., (1986) observed that R *Chenopodium album* chloroplasts had lower electron transport rates between water and plastoquinone compared to S; yet no differences were found in the rate or quantum yield of whole-chain electron transport, or in A, between R and S. These inconsistent responses by R and S biotypes have led several to conclude that the change conferring R is not necessarily directly linked to inferior photosynthetic function (Holt and Goffner, 1985; Jansen et al., 1986; Schonfeld et al., 1987).

An assessment of these and other studies reveals several possible reasons why different responses may have been observed. They include pleiotropic reorganization of the R chloroplast and the dynamic interelationship between components of photosynthesis; the role of environment in altering responses; genetic factors, such as differences between biotypes and genome interactions within a biotype; and the possibility of an unnamed factor controlling photosynthesis (Arntzen and Duesing, 1983; Duesing and Yue, 1983). Comparison of different photosynthetic responses by R is further complicated by the many different environmental and biological conditions under which they were conducted. These include changes in plant species, plant age, plant uniformity, plant tissue, temperature, photosynthetic photon flux density (PPFD), the degree of experimental environmental control (field, glasshouse or controlled environment chambers) and the diurnal light period length and variation of conditions. Few systematic studies have been conducted to separate these factors.

A mitigating factor in comparing R and S photosynthetic responses has been the use of model systems in which other genes besides the mutation to *psbA*, have differed (e.g. Holt et al., 1981). Inferences from these studies have been confounded because they relied on a non-isogenic model system. McCloskey and Holt (1990) has suggested nuclear genome differences may compensate for differences in productivity between non-isogenic R and S selections, and that detrimental effects may be attenuated by interactions of plastid and nuclear genomes (Stowe and Holt, 1988). Many of these limitations may have been overcome in studies with the nearly isonuclear biotypes of *Brassica napus* (Beversdorf and Hume, 1984; Dekker and Burmester, 1990; Jursinic and Pearcy, 1988; Vaughn, 1986).

**18.2.2.3 Complex *psbA* mutant phenotype.** Decreases in electron transport function in the chloroplast have been believed to be the cause of decreased carbon assimilation rates and plant productivity in many reports (Dekker, 1993). What is less clear in the literature is whether this change in D-1 structure and electron transport function directly modifies whole-leaf photosynthesis and plant productivity or only indirectly influences these functions (Holt and Goffner, 1985). The dynamic nature of these responses have led several to conclude that the primary effect of R is complex, involves more than one aspect of photosynthesis, and can be mitigated by other processes in the system (Dekker and Sharkey, 1992; Ireland et al., 1988; McClosky and Holt, 1990). For example, it has been pointed out that decreased $Q_A$- to $Q_B$ electron transport in R is more rapid than the normally rate limiting oxidation of plastoquinol (Barber, 1983; Ort et al., 1983), while others studies indicate this step may be rate limiting (Jursinic and Pearcy,1988).

**18.2.2.4 *psbA* mutant pleitropic reorganization.** The genetic change in R plants leads to a profound, pleiotropic, reorganization of structural and functional units in the chloroplast (Dekker and Burmester, 1993) (table 18.1). This adaptive reorganization of photosynthetic components in the chloroplast may be a compensatory mechanism to maintain a functional interaction of the PS II complex lipids and proteins (Pillai and St. John, 1981). This pleiotropic cascade includes both structural (Burke et al., 1982; Mattoo et al., 1984) and functional changes (Arntzen et al., 1979; Pillai and St. John, 1981; Tranel and Dekker, 2002; Vaughn, 1986). The dynamic nature of the chloroplast to reach a new, markedly different, structural and functional equilibrium in response to the mutation of a key plastidic gene has been observed previously (Hugly et al., 1989; Kunst et al., 1989). This profound pleiotropic cascade of functional and structural changes conferred by changes in the D-1 protein could imply that the amino acid substitution is close to a primary functional and structural source of photosynthetic regulation. Mattoo (Mattoo et al., 1984) has suggested that the rapid anabolism-catabolism rate of the D-1 protein could serve as a signal resulting in the reorganization of membranes around the PSII complex. This dynamic reorganization has consequences for evaluating and understanding regulatory effects of electron transport in carbon assimilation (Dekker, 1993).

**Table 18.1** Cascade of pleiotropic effects at the plastid, cell, tissue, organ, plant, population and community levels of plant organization consequential to the mutation of the *psbA* chloroplast gene.

| Gene | Plastid/Cell | Tissue/Organ | Plant | Population | Community |
|---|---|---|---|---|---|
| psbA chloroplast gene condon 264 mutation | altered chloroplast D-1 protein: structural changes in PSII electron transport | | s-triazine resistance | enrichment of R biotypes | enrichment of R species |
| | | differential R/S diurnal pattern of photosynthetic carbon assimilation and chlorophyll fluorescence | higher R carbon assimilation early and late in the day | enhanced intra-specific photosynthetic phenotype diversity | enhanced inter-specific photosynthetic niche diversity |
| | | | higher carbon assimilation in older R plants | | |
| | | increased stomatal aperture opening | cooler leaf temperature | | |
| | increased thylakoid membrane stacking | shade-adapted leaf morphology | enhanced resistance to low temperature stress | | |
| | | | enhanced photosynthetic efficiency in lower light conditions | | |
| | increased thylakoid grana lamellae fatty acid unsaturation | greater low temperature lipid fluidity | increased seed dormancy | | |

The equivocal nature of our knowledge of how photosynthesis differs between R and S, and how it is regulated, led us to focus experimental efforts on chronobiological understandings of the R phenotype under more dynamic, but closely controlled, growth conditions (Dekker and Burmester, 1993). In particular we have observed other pleiotropic effects in R: differential patterns of both carbon assimilation (A) and chlorophyll fluorescence over the course of the light-dark diurnal cycle. Subsequently we observed that R plants have markedly different leaf and stomatal responses to temperature.

### 18.3 Chlorophyll fluorescence in R and S

Variations in leaf disc chlorphylll fluorescence (LCF; Chl *a*; terminal, $F_t$) intensity with time of day were studied to test the hypothesis that alterations in chloroplast structure that confer s-triazine resistance (R) may also imply altered temporal organization of chlorophyll fluorescence activities (Dekker and Westfall, 1987). Two periods of reduced photosynthetic efficiency occurred in the daily light-dark cycle in both biotypes (figure 18.1). The times these occurred during the diurnal differed between biotypes. A phase shift in LCF maxima occurred between R and susceptible (S) biotypes. The R biotype was less photosynthetically efficient than S early and late in the light period. The S biotype was less efficient in the middle of the light and dark periods of the diurnal. This differential pattern of LCF supports the hypothesis that s-triazine resistance chloroplast alterations also imply an alteration in the temporal organization of chloroplast physiological function: the s-triazine resistant phenotype is a chronomutant.

**Figure 18.1** Diurnal oscillations in s-triazine resistant and susceptible *Brassica napus* leaf disc chlorphyll fluorescence; relative daily maxima and minima combined over three days of a controlled environment experiment; r, resistant means; s, susceptible means; variations in LCF amplitude removed.

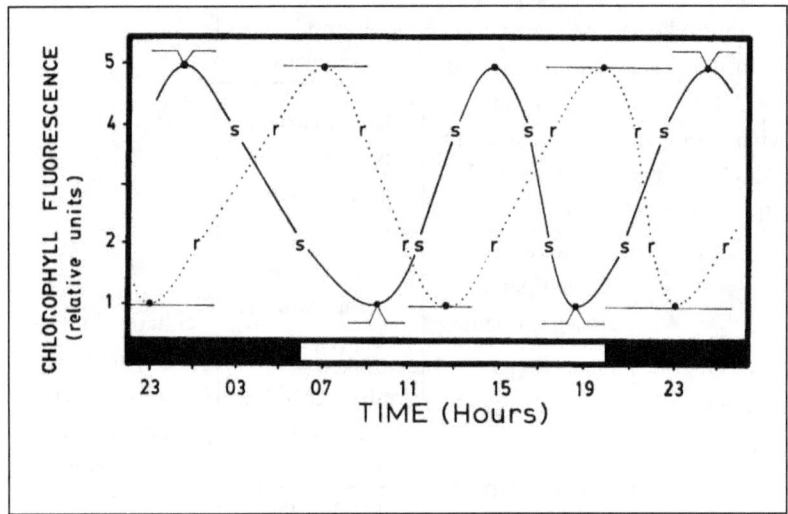

### 18.4 Carbon assimilation in R and S

Insights about the chronobiology of R and S chlorophyll fluorescence stimulated observations of how carbon assimilation changes during the light period, and how it changes over the life of the two biotypes. Studies were conducted testing the hypothesis that the mutation to the *psbA* plastid gene that confers s-triazine resistance also results in an altered diurnal pattern of photosynthetic carbon assimilation relative to that of the susceptible biotype.

Studies were conducted observing changes in the diurnal patterns of photosynthetic carbon assimilation (A) in R relative to that of the susceptible (S) wild-type over the ontogenetic development of the plant (Dekker and Burmester, 1992). Carbon assimilation rates (A) approximately tracked the increasing and decreasing diurnal light levels (figure 18.2).

Younger (3 to 4 leaf) R plants had greater photosynthetic rates early and late in the diurnal light period, while those of S were greater during midday as well as the photoperiod as a whole. A differential pattern of A was observed in 3 to 3 $1/2$ leaf plants: early and late in the photoperiod R carbon assimilation exceeded that of S, while S carbon assimilation exceeded that of R during the midday period (figure 18.3, table 18.2A). The same differential pattern of A between R and S was observed in 4 leaf plants.

**Figure 18.2.** Changes in photosynthetic photon flux density (PPFD; μmol quanta m$^{-1}$ period$^{-1}$) with time (hour of the day) in separate experiments on 3-9$^{1}/_{2}$ leaf *Brassica napus* plants; periods of the diurnal: early (0060-1000 h), midday (1000-1800 h) and late (1800-2150 h) (Dekker and Burmester, 1992).

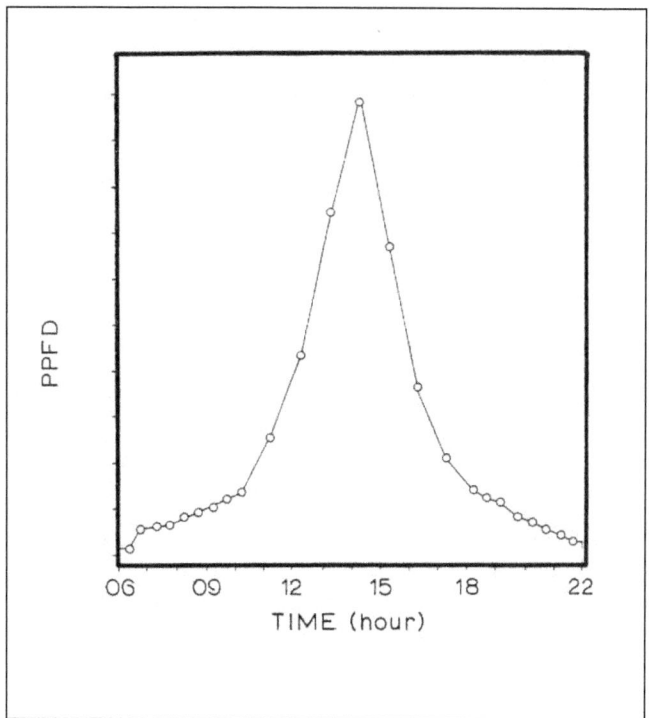

As *B. napus* plants aged, differences in A between S and R became greater. At the 5 $^{1}/_{2}$ to 6 leaf growth stage, S leaf A exceeded R during midday, was similar to R early, while only during the late period was R greater (figure 18.4, table 18.2A). For both biotypes, maximum leaf A was reached during this stage of development.

S continued to assimilate more carbon than R in 6 $^{1}/_{2}$ to 7 $^{1}/_{2}$ leaf plants, plants near the end of the vegetative phase of development, for most times of the day (figure 18.5, table 18.2A). A in R equalled that in S late in the day.

As *B. napus* began the reproductive phase of development, 8 $^{1}/_{2}$ to 9 $^{1}/_{2}$ leaf plants, a nearly complete reversal of photosynthetic differences observed in previous growth stages occurred. R carbon assimilation was greater than S early, midday, and for the whole day period while S (for the first time in its development) was superior to R later in the day (figure 18.6, table 18.2A). An erratic pattern of A, especially in S plants, was observed late in the photoperiod. A increased for a time while PPFD was incrementally decreasing. It is unclear why this occurred. Overall, A at this stage of plant development was considerably less (S ca. 65% less; R ca. 45% less) than the vegetative phase plants of one week previous, and was the lowest A in their ontogeny. Also, this erratic pattern could be a function of plants overcoming limitations imposed during the preceding high light, midday, period; or erratic responses to the onset of senescence.

These relative photosynthetic relationships change in several ways with ontogeny, and are summarized in table 18.2A. These results support a generalized model of carbon assimilation during the diurnal light period with changes in development (figure 18.7, table 18.2A).

**Figure 18.3** Changes in carbon assimilation (A) ($\mu mol\ CO_2\ m^{-2}\ s^{-1}$) in 3 to $3^1/_2$ leaf s-triazine resistant (R) (•) and susceptible (S) (o) *Brassica napus* plants with time (hour of the day); leaf temperature 25° C; S.E.=+/- standard error of the mean, n=9.

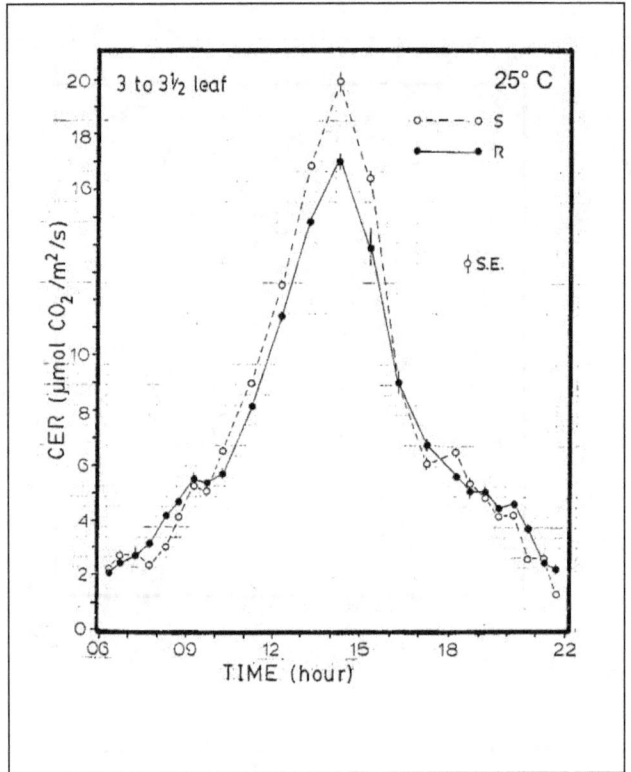

**Table 18.2A** Summary of ontogenic (3-9½ leaf plants) effects on photosynthetic carbon assimilation (A; $\mu mol\ CO_2\ m^{-2}\ s^{-1}$), leaf temperature (°C), Total conductance to water vapor (g; $mmol\ m^{-2}s^{-1}$), and leaf intercellular $CO_2$ partial pressure ($C_i$; $\mu L$) in s-triazine resistant (R) and susceptible (S) biotypes at 25° C air temperature with changes within a diurnal photoperiod (early, 0600-1000h; midday, 1000-1800h; late, 1800-2150h; and all day, 0600-2150h) (Dekker and Burmester, 1992); table 17.2B organized by leaf stages within photosynthetic parameters.

| Leaf Stage | Early 0600-1000h | Midday 1000-1800h | Late 1800-2150h | All Day 0600-2150h |
|---|---|---|---|---|
| Photosynthetic carbon assimilation (A) ($\mu mol\ CO_2\ m^{-2}s^{-1}$) | | | | |
| $3-3^1/_2$ | **R > S** | S > R | **R > S** | S > R |
| 4 | **R > S** | S > R | **R > S** | S > R |
| $5^1/_2-6$ | S = R | S > R | **R > S** | S > R |
| $6^1/_2-7^1/_2$ | S > R | S > R | S = R | S > R |
| $8^1/_2-9^1/_2$ | **R > S** | **R > S** | S > R | **R > S** |
| Leaf temperature (°C) | | | | |
| $3-3^1/_2$ | S = R | S = R | S = R | S = R |
| 4 | S = R | S > R | S = R | S > R |
| $5^1/_2-6$ | S > R | S > R | S = R | S > R |

| | | | | |
|---|---|---|---|---|
| $6^1/_2$-$7^1/_2$ | S > R | S > R | S = R | S > R |
| $8^1/_2$-$9^1/_2$ | S > R | S > R | S = R | S > R |
| Total conductance to water vapor (g) (mmol m$^{-2}$s$^{-1}$) | | | | |
| 3-$3^1/_2$ | S = R | S = R | R > S | R > S |
| 4 | R > S | R > S | S = R | R > S |
| $5^1/_2$-6 | R > S | S = R | R > S | R > S |
| $6^1/_2$-$7^1/_2$ | S = R | S = R | S > R | S = R |
| $8^1/_2$-$9^1/_2$ | R > S | R > S | S > R | R > S |
| Leaf intercellular CO$_2$ partial pressure (C$_i$) (μL) | | | | |
| 3-$3^1/_2$ | S = R | S = R | S = R | S = R |
| 4 | R > S | R > S | S = R | R > S |
| $5^1/_2$-6 | R > S | R > S | S = R | R > S |
| $6^1/_2$-$7^1/_2$ | S = R | S = R | S > R | S = R |
| $8^1/_2$-$9^1/_2$ | S = R | R > S | S > R | S = R |

**Figure 18.4** Changes in carbon assimilation (A) (μmol CO$^2$ m$^{-2}$ s$^{-1}$) in 5 $^1/_2$ to 6 leaf s-triazine resistant (R)(·) and susceptible (S)(o) *Brassica napus* plants with time (hour of the day); leaf temperature 25° C; S.E.=+/- standard error of the mean, n=9.

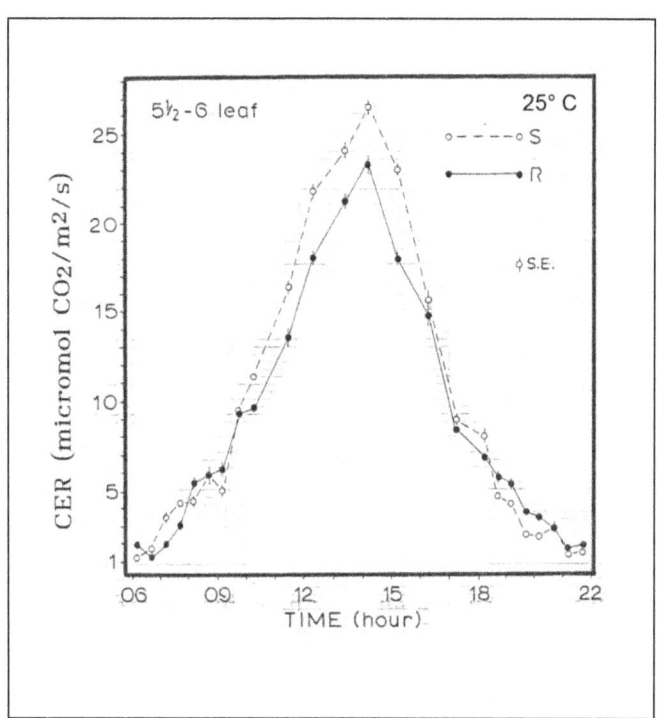

**Figure 18.5.** Changes in carbon assimilation (A) (μmol $CO^2$ $m^{-2}$ $s^{-1}$) in 6 $\frac{1}{2}$ to 7 $\frac{1}{2}$ leaf s-triazine resistant (R)(·) and susceptible (S)(o) *Brassica napus* plants with time (hour of the day); leaf temperature 25° C; S.E.=+/- standard error of the mean, n=9.

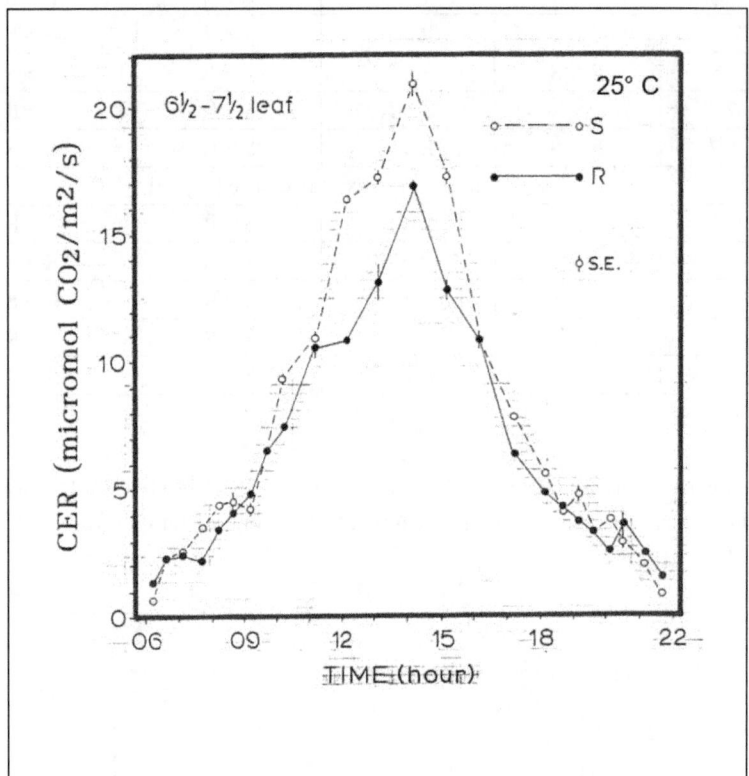

As the plants age during the vegetative phase of development, S gradually assimilates more than R in the early ("A" of figure 18.7), and then in the late ("C" of figure 18.7), part of the light period. At the end of the vegetative phase of development, R is photosynthetically inferior to S at most times of the day, and is never better. This relationship between the two biotypes dramatically changes with the onset of the reproductive phase (8 1/2 to 9 1/2 leaf) of plant development: R is now superior to S during all periods ("B" of figure 18.7) of the diurnal light period with the exception of the late part of the day.

As both R and S plants developed from 3 to 9 1/2 leaves, they assimilated more carbon on an all-day basis until the 5 1/2 to 6 leaf stage when A was the greatest, and then A declined through the reproductive stage when the oldest (8 1/2 to 9 1/2 leaf) plants had the lowest A ("D" of figure 18.7). Differences observed in A at that later developmental stage could be a function of R losing photosynthetic competence at a slower rate than S. This quality of R could have important implications in agriculture and the critical seed development period: possibly greater A late in development could result in greater accumulated carbon for seed development, a factor that might partially overcome reduced photosynthetic carbon assimilation earlier in the vegetative phase of development.

These studies indicate a more complex pattern of photosynthetic carbon assimilation than previously observed. The photosynthetic superiority of one biotype relative to the other was a function of the time of day and the age of the plant. The lower rate of R photosynthetic carbon assimilation can be overcome in some instances by certain environmental and developmental conditions. These results support the hypothesis tested: *psbA* plastid gene mutation conferring R also confers a different diurnal pattern of photosynthetic function than that in S. This work is also consistent with the pattern of differential chlorophyll fluorescence ($F_t$) during the diurnal previously reported (Dekker and Westfall, 1987).

**Figure 18.6** Changes in carbon assimilation (A)($\mu$mol $CO^2$ $m^{-2}$ $s^{-1}$) in 8 $^1/_2$ to 9 $^1/_2$ leaf s-triazine resistant (R)(·) and susceptible (S)(o) *Brassica napus* plants with time (hour of the day); leaf temperature 25° C; S.E.=+/- standard error of the mean, n=9.

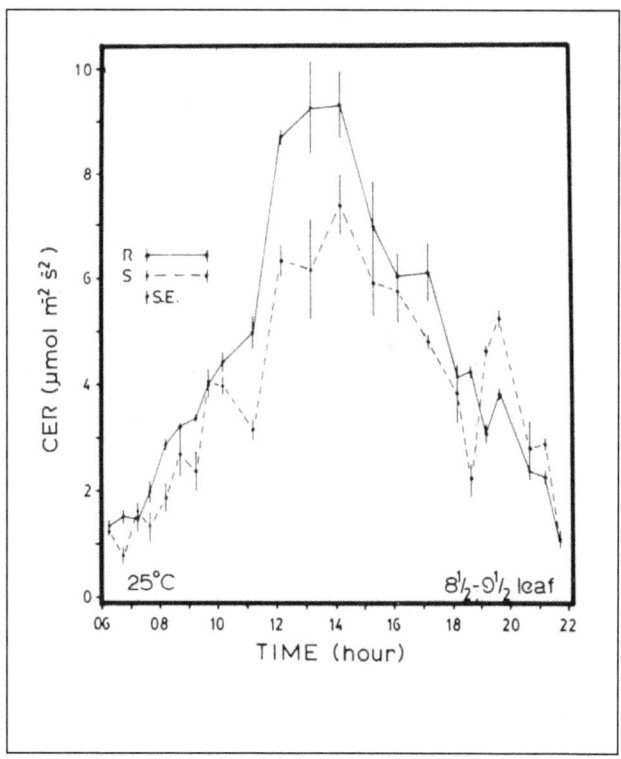

## 18.5 Temperature effects on photosynthetic function in R and S

**18.5.1 Diurnal temperature effects.** The diurnal pattern of carbon assimilation differs between R and S at different air temperatures and stages of development: R is greater than S at higher air temperature, and late in ontogeny (table 18.2A). When carbon assimilation (A) was greater in R: S leaf temperatures were similar to R (except $8^1/_2$-$9^1/_2$ leaf); and S stomatal conductance (g) and $C_i$ were less than or equal to R (table 18.2B) (Dekker and Burmester, 1992). When carbon assimilation was greater in S: R leaf temperatures and stomatal conductances were less than or similar to S (except $8^1/_2$-$9^1/_2$ leaf, late); and S Ci were less than or equal to R (except $8^1/_2$-$9^1/_2$ leaf, late). Regardless of differences or similarities in A between the two biotypes: S leaf temperatures were greater than or equal to R (R was never greater than S); and R stomatal conductances and Ci were greater than or equal to S (except $6^1/_2$-$7^1/_2$ and $8^1/_2$-$9^1/_2$ leaf, late).

S plants assimilated more carbon over the entire day than R plants at 10°, 15° and 25° C air temperatures, but assimilated less than R at 35° C air temperatures (Dekker and Burmester, 1993) (Table 18.3, 18.4; figures 18.8, 18.11). During the early part of the light period, A in R plants was either similar (15° C air temperature), or greater (25° and 35° C) compared to that in S. During the midday part of the light period, A in S was greater than that at 15° and 25° C air temperatures, while A in R was greater than that in S at 35° C. A in R was greater than in S at all 3 air temperatures at some times of the diurnal.

**Figure 18.7** Model of the relationship between relative carbon assimilation rate (A) and the time of the day in s-triazine resistant (R) and susceptible (S) *Brassica napus*; "A" indicates changes in relative A during the early portion of the diurnal, "B" indicates changes in relative A during the midday portion of the diurnal, "C" indicates changes in relative A during the late portion of the diurnal, "D" indicates changes in relative A during ontogeny.

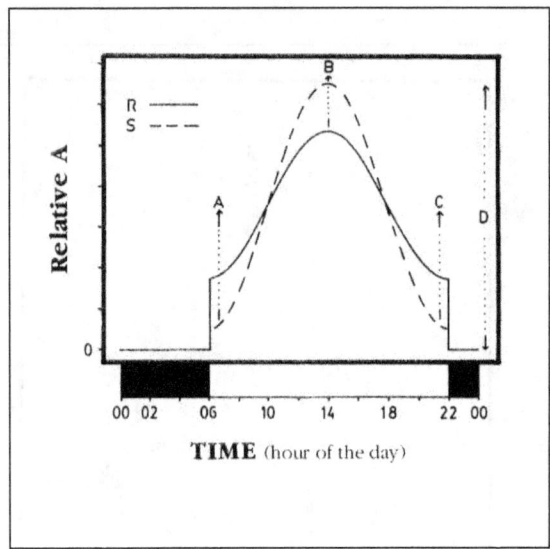

**Table 18.2B** Summary of ontogenic (3-9½ leaf plants) effects on photosynthetic carbon assimilation (A; μmol $CO_2$ $m^{-2}$ $s^{-1}$), leaf temperature (°C), Total conductance to water vapor (g; mmol $m^{-2}s^{-1}$), and leaf intercellular $CO_2$ partial pressure ($C_i$; μL) in s-triazine resistant (R) and susceptible (S) biotypes at 25° C air temperature with changes within a diurnal photoperiod (early, 0600-1000h; midday, 1000-1800h; late, 1800-2150h; and all day, 0600-2150h) (Dekker and Burmester, 1992); table 18.2A organized by photosynthetic parameters within leaf stage.

| Leaf Stage | Parameter | Early 0600-1000h | Midday 1000-1800h | Late 1800-2150h | All Day 0600-2150h |
|---|---|---|---|---|---|
| 3-3½ | A | R > S | S > R | R > S | S > R |
|  | Leaf Temp | S = R | S = R | S = R | S = R |
|  | g | S = R | S = R | R > S | R > S |
|  | $C_i$ | S = R | S = R | S = R | S = R |

| Leaf Stage | Parameter | Early 0600-1000h | Midday 1000-1800h | Late 1800-2150h | All Day 0600-2150h |
|---|---|---|---|---|---|
| 4 | A | R > S | S > R | R > S | S > R |
|  | Leaf Temp | S = R | S > R | S = R | S > R |
|  | g | R > S | R > S | S = R | R > S |
|  | $C_i$ | R > S | R > S | S = R | R > S |

| Leaf Stage | Parameter | Early 0600-1000h | Midday 1000-1800h | Late 1800-2150h | All Day 0600-2150h |
|---|---|---|---|---|---|

| Leaf Stage | Parameter | Early 0600-1000h | Midday 1000-1800h | Late 1800-2150h | All Day 0600-2150h |
|---|---|---|---|---|---|
| $5^{1}/_{2}$-6 | A | S = R | S > R | **R > S** | S > R |
| | Leaf Temp | S > R | S > R | S = R | S > R |
| | g | **R > S** | S = R | **R > S** | **R > S** |
| | $C_i$ | **R > S** | **R > S** | S = R | **R > S** |

| Leaf Stage | Parameter | Early 0600-1000h | Midday 1000-1800h | Late 1800-2150h | All Day 0600-2150h |
|---|---|---|---|---|---|
| $6^{1}/_{2}$-$7^{1}/_{2}$ | A | S > R | S > R | S = R | S > R |
| | Leaf Temp | S > R | S > R | S = R | S > R |
| | g | S = R | S = R | S > R | S = R |
| | $C_i$ | S = R | S = R | S > R | S = R |

| Leaf Stage | Parameter | Early 0600-1000h | Midday 1000-1800h | Late 1800-2150h | All Day 0600-2150h |
|---|---|---|---|---|---|
| $8^{1}/_{2}$-$9^{1}/_{2}$ | A | **R > S** | **R > S** | S > R | **R > S** |
| | Leaf Temp | S > R | S > R | S = R | S > R |
| | g | **R > S** | **R > S** | S > R | **R > S** |
| | $C_i$ | S = R | **R > S** | S > R | S = R |

In most instances, leaf temperatures and stomatal conductance (g) was similar in R and S at 15° C air temperatures. At 25° C, S leaf temperatures were greater than, or equal to, R. At 35° C, R leaf temperatures were less than those of S during all parts of the light period (table 18.3; figure 18.9). At 25° and 35°C, g in R plants was almost always greater than that of S (table 18.3, figure 18.10).

**Table 18.3.** Changes in carbon assimilation rate (A; $\mu mol\ CO_2\ m^{-2}\ period^{-1}$) and conductance to water vapor (g; $mol\ m^{-2}\ s^{-1}$) in 4 leaf s-triazine resistant (R) and susceptible (S) *Brassica napus* plants grown at 15°, 25° and 35° C air temperatures within a diurnal period (early, 0600-1000h; midday, 1000-1800h; late, 1800-2150h; and all day, 0600-2150h).

| Air Temp | Early 0600-1000h | Midday 1000-1800h | Late 1800-2150h | All Day 0600-2150h |
|---|---|---|---|---|
| Photosynthetic carbon assimilation (A) ($\mu mol\ CO_2\ m^{-2}s^{-1}$) | | | | |
| 15°C | R = S | S > R | **R > S** | S > R |
| 25°C | **R > S** | S > R | **R > S** | S > R |
| 35°C | **R > S** | **R > S** | **R > S** | **R > S** |
| Leaf Temperature (°C) | | | | |
| 15°C | R = S | **R > S** | R = S | R = S |
| 25°C | R = S | S > R | R = S | S > R |
| 35°C | S > R | S > R | S > R | S > R |
| Total conductance to water vapor (g) ($mol\ m^{-2}s^{-1}$) | | | | |
| 15°C | R = S | R = S | **R > S** | R = S |
| 25°C | **R > S** | **R > S** | R = S | **R > S** |
| 35°C | **R > S** | **R > S** | **R > S** | **R > S** |

**Table 18.4.** Comparison of carbon assimilation (CER; μmol $CO^2$ $m^{-2}$ $s^{-1}$) in s-triazine resistant (R) and susceptible (S) Brassica napus plants at 3-4 (figure 18.2) and 8 ½-9 ½ leaf (figure 18.5) at 25° C, and 3-4 leaf at 35° C (figure 18.9) air temperatures.

| Temp | Leaf Stage | Early 0600-1000h | Midday 1000-1800h | Late 1800-2150h | All Day 0600-2150h |
|---|---|---|---|---|---|
| 25° C | 3-4 | R > S | S > R | R > S | S > R |
| | 8 ½-9 ½ | R > S | R > S | S > R | R > S |
| 35° C | 3-4 | R > S | R > S | R > S | R > S |

**Figure 18.8.** Changes in carbon assimilation (CER; μmol $CO^2$ $m^{-2}$ $s^{-1}$) in 3 to 4 leaf s-triazine resistant (R) and susceptible (S) *Brassica napus* plants with time (hour of the day); leaf temperature 35° C; S.E. = $^+/_-$ standard error of the mean, n=9.

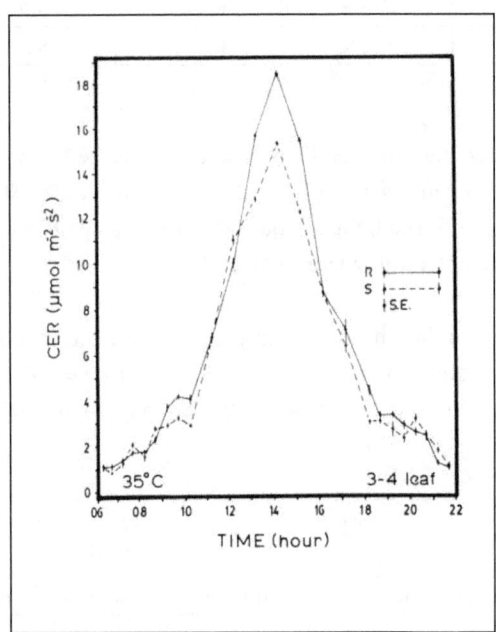

Carbon assimilation in R was greater than that of S at higher temperature, and late in ontogeny. This contrasts with other studies in which S carbon assimilation was greater than R when leaf temperatures where held constant. When leaf temperature is not controlled, greater stomatal conductance and leaf cooling in R can overcome these disadvantages.

In addition to the differences in carbon assimilation, R plants also differ in stomatal function from S plants (Dekker and Burmester, 1992; Dekker et al., 1990). Total conductance to water vapor and intercellular $CO_2$ partial pressure in R was either equal to, or greater than, S over the lifetime of those plants, with the possible exception of some atypical episodes late in ontogeny and late in the light period. As a consequence of these phenomena, leaves of R plants were either the same temperature, or cooler, than leaves of S plants for the entire lives of both biotypes. The physiological or biochemical linkage connecting the primary R defect and stomatal function is not apparent, but could provide an interesting experimental model system. These results on altered, cooler, leaf temperatures and stomatal function have been also observed in previous studies under quite different conditions (Dekker and Sharkey, 1992). In all instances in these experiments, both R and S plants were functioning at or near (26-29°C) their photosynthetic temperature optima of ca. 28°C (Dekker and Sharkey, 1992).

**Figure 18.9** Change in leaf temperature at constant 35° C air temperature in s-triazine resistant (R) and susceptible (S) *Brassica napus* plants with time (hour of the day); S.E. = $^+/_-$ standard error of the mean.

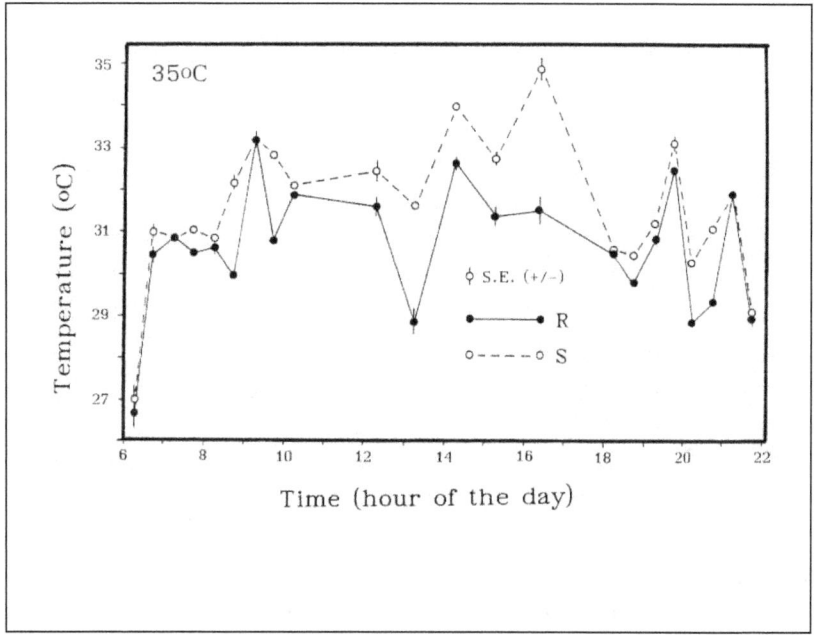

**Figure 18.10** Stomatal conductance of water vapor versus diurnal time of R and S at 35° C air temperature.

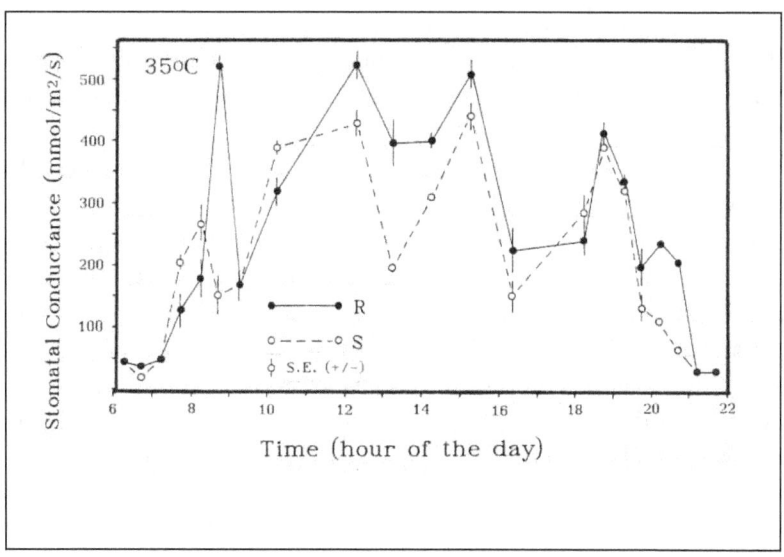

**Figure 18.11** Changes in carbon assimilation (CER; μmol $CO^2$ $m^{-2}$ $s^{-1}$) in s-triazine resistant (R) and susceptible (S) *Brassica napus* plants with time (hour of the day); leaf temperature 10° C; S.E. = $^+/_-$ standard error of the mean.

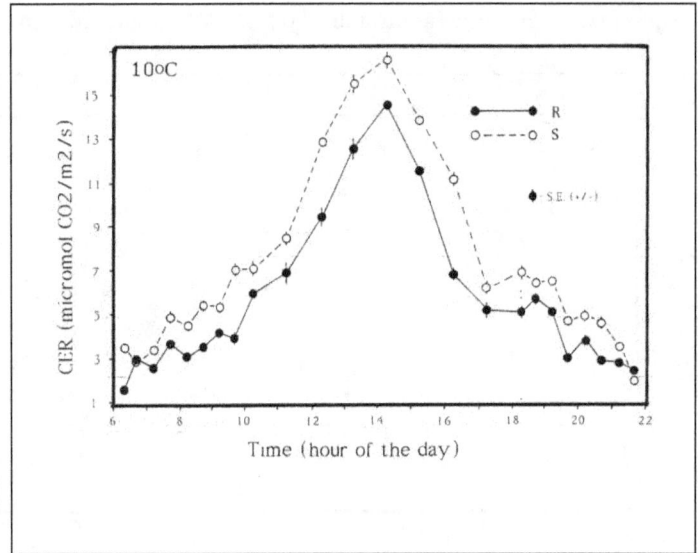

**18.5.2 Controlled leaf temperature effects.** At 15°C A in R and S was similar, but as leaf temperature increased to 35°C S assimilated increasing more carbon than R (figure 18.12) (Dekker and Burmester, 1993). For both biotypes, the optimal temperature for A was ca. 26-27°C. Associated with increases in leaf temperature was a greater total conductance to water vapor in R relative to S under similar temperature conditions (figure 18.13) (Dekker and Burmester, 1993).

At lower leaf and air temperatures (15°C), R and S leaves function in a similar way (Dekker and Burmester, 1993). But, as the temperature of their environment increases, their responses diverge considerably. If leaf temperature is controlled externally, R leaves assimilate considerably less carbon than S leaves, especially at hyper-optimal temperatures (e.g. 35°C).

If leaf temperature is not directly controlled, and both R and S are immersed in an identical air temperature environment, there responses to the same air temperature are different (Dekker and Burmester, 1993). R leaves generally are cooler and total conductance to water is greater than in S, probably due to greater stomatal aperture size. The consequence of this is that at higher air temperatures (e.g. 35°C) R leaves photosynthesize at cooler leaf temperatures, leaf temperatures closer to the optimal for both biotypes. In this way R plants compensate for their high temperature sensitivity.

In addition to other pleiotropic effects of the mutation to the psbA gene in R, R plants appear to be stomatal mutants. R and S *Brassica napus* biotypes may constitute a good model system to study regulation of stomatal function and the relationship between environmental cues and stomatal behavior.

**Figure 18.12** Carbon exchange rate (CER; μmol $CO_2$ $m^{-2}$ $s^{-1}$) versus leaf temperature (°C) in s-triazine resistant (R) and susceptible (S) *Brassica napus*.

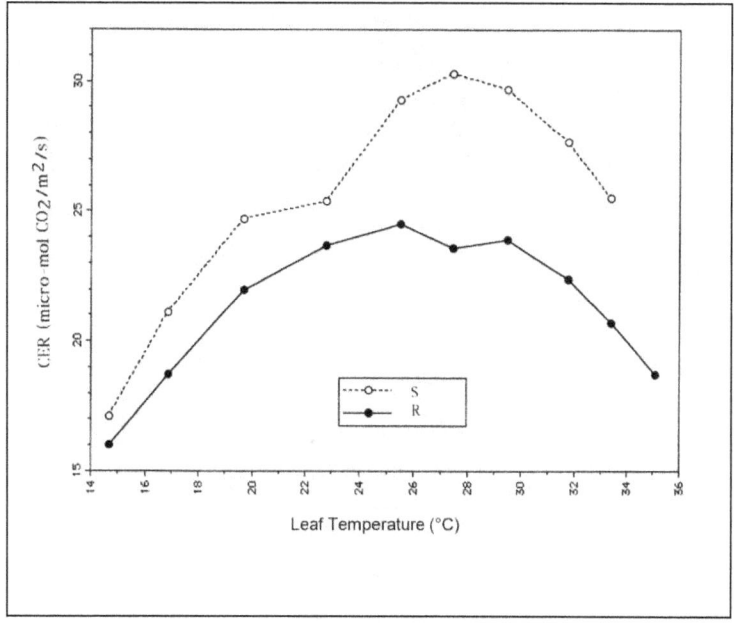

**Figure 18.13** Stomatal conductance to water vapor (mol $m^{-2}$ $s^{-1}$) versus leaf temperature (°C) in s-triazine resistant (R) and susceptible (S) *Brassica napus*.

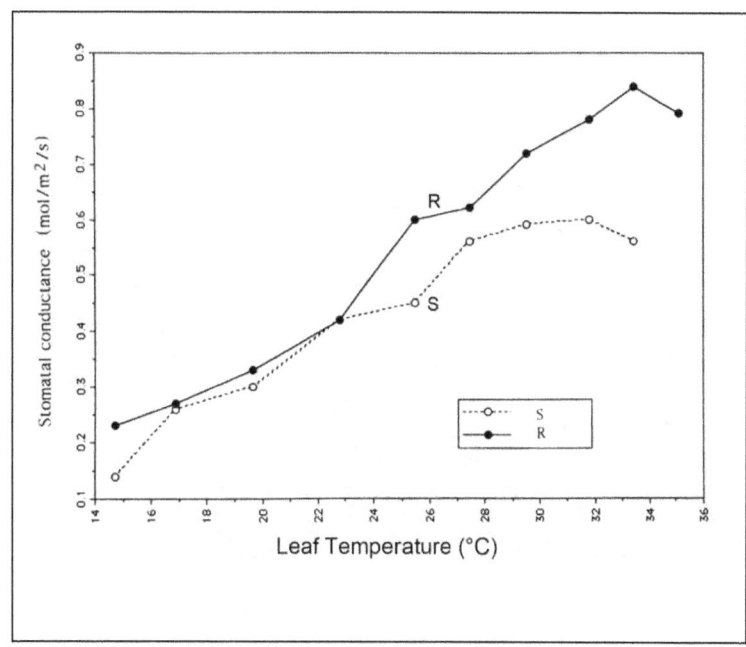

## 18.6 Rubisco Activity in R and S

Studies were conducted to test if there is any regulatory role for ribulose-1,5-bisphosphate carboxylase/oxygenase (Rubisco) in the differential photosynthetic carbon assimilation previously reported between the R and S *Brassica napus* (Dekker et al., 1990; Mazen and J. Dekker, 1990). Diurnal oscillations in initial, total and percent Rubisco activation (initial/total) were studied in two biotypes. Plants were evaluated under dynamic, changing light, conditions representative of a typical daily light cycle in the field (figure 18.1).

Initial activity started to increase early in the day and reached its peak at midday, reaching a similar rate as that of the total (figures 18.14, 18.15; table 18.5). By late afternoon initial activity declined to values similar to those of the morning and remained constant overnight. Total Rubisco activity for both R and S remained constant, or decreased slightly, during the light period (figure 18.15; table 18.5). At the very early and late portion of the day, and overnight, total activity was about twice that of the initial activity at those times.

**Figure 18.14.** Carbon exchange rate (CER; µmol $CO^2$ $m^{-2}$ $s^{-1}$) and initial Rubisco activity (relative units) versus diurnal time (hour of the day).

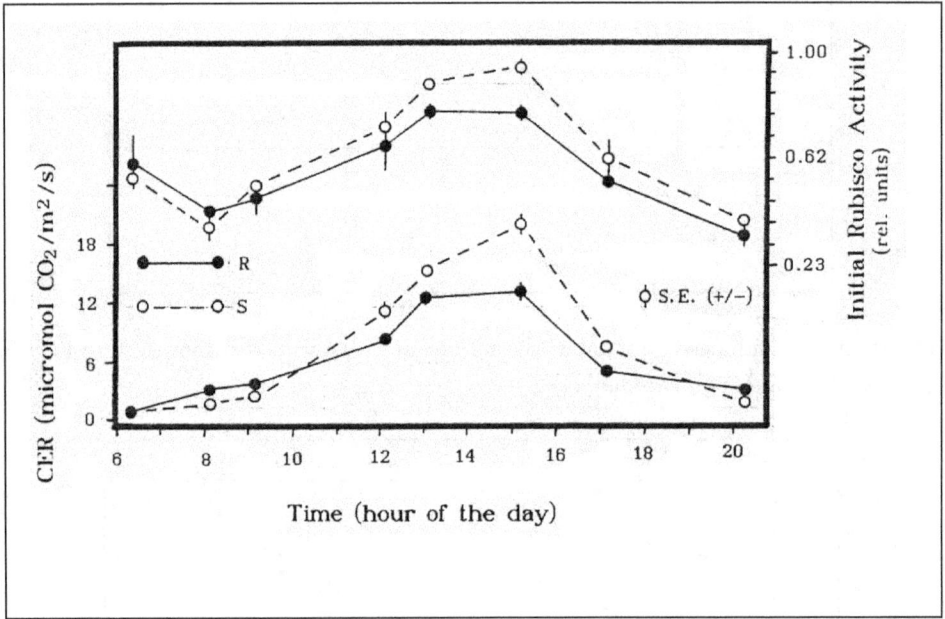

Rubisco percent activity (% activation; initial/total) (figure 18.16) and initial activity (figure 18.14) may account for the carbon assimilation differences observed between R and S during the midday (high light) period: their diurnal rise and fall corresponded to CER at those times (table 18.5). CER differences early and late in the day (low light) were not accounted for by differences in any of the Rubisco activity indices (initial, total, % activation).

## 18.7 Photosynthetic regulation in R and S

Studies were conducted to determine the response of two biotypes of *Brassica napus* (triazine resistant (R) and susceptible (S)) to different temperatures and gas atmospheres (Dekker and Sharkey, 1990, 1992). The response of photosynthetic carbon assimilation and chlorophyll fluorescence quenching to changes in intercellular $CO_2$ partial pressure ($C_i$), $O_2$ partial pressure, and leaf temperature (15-35° C) in R and S were examined to determine the effects of the changes in the resistant biotype on the overall process of photosynthesis in intact leaves. Observations included photosynthetic carbon assimilation (A) using infrared gas analysis techniques and pulse amplitude modulated chlorophyll fluorescence. Plants were evaluated with strict leaf temperature control and leaf gases ($N_2$, $CO_2$, $O_2$, $H_2O$).

**Figure 18.15.** Carbon exchange rate (CER; μmol $CO^2$ $m^{-2}$ $s^{-1}$) and total Rubisco activity (relative units) versus diurnal time (hour of the day).

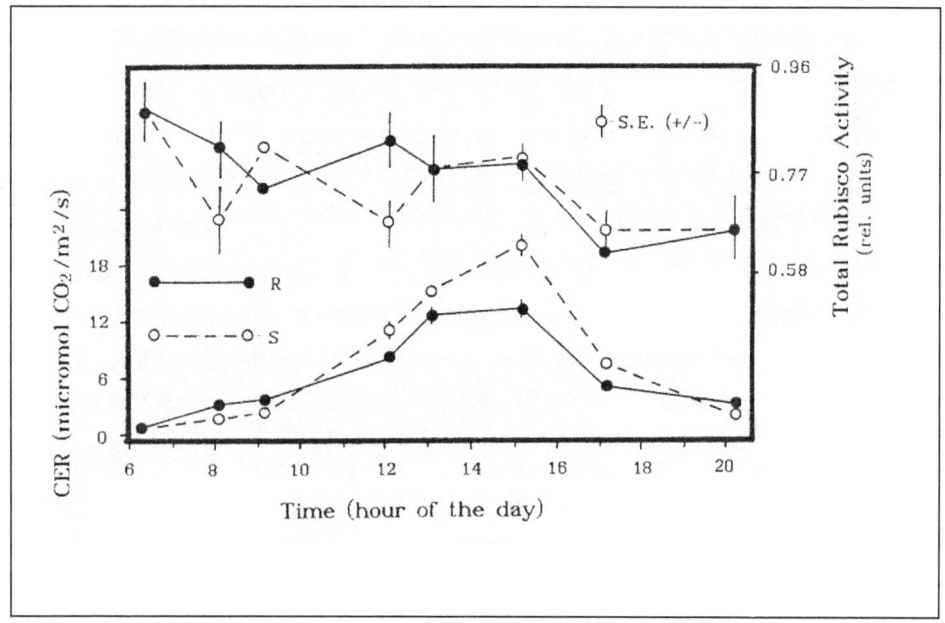

**Figure 18.16** Carbon exchange rate (CER; μmol $CO^2$ $m^{-2}$ $s^{-1}$) and % Rubisco activitation (initial/total; relative units) versus diurnal time (hour of the day).

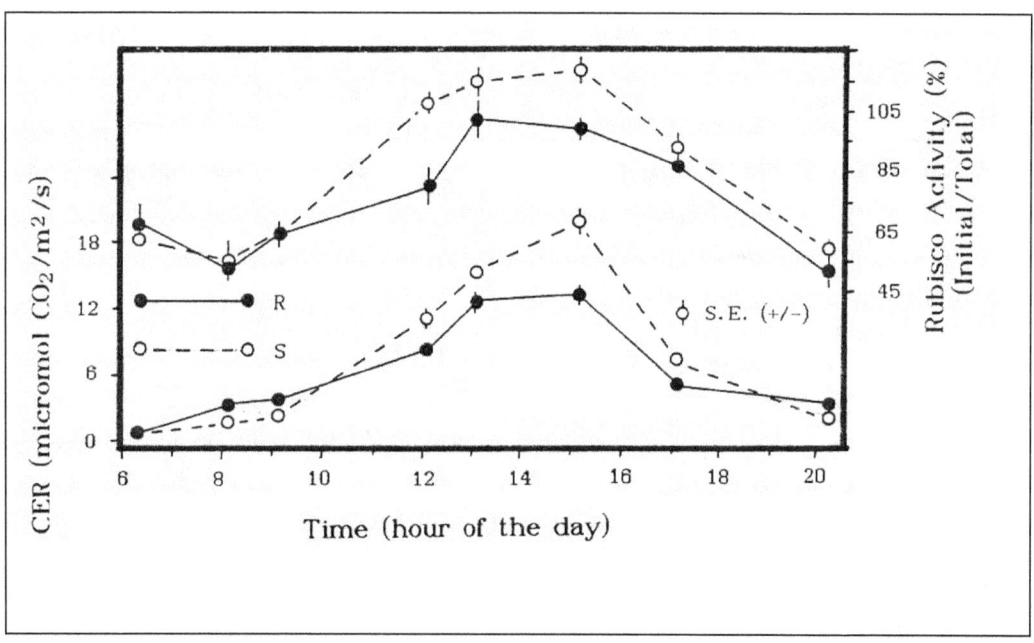

**Table 18.5** Rubisco (initial; total; % activation, initial/total) and carbon assimilation (µmol $CO^2$ $m^{-2}$ $s^{-1}$) of R and S *Brassica napus* during the daily diurnal period (early, midday, late) (figures 18.13, 18.14, 18.15; Mazen and J. Dekker, 1990).

|  | Early | Midday | Late |
|---|---|---|---|
| Rubisco: Initial | R = S | S > R (1300-1400h) | R = S |
| Rubisco: Total | no consistent relationship between R and S total activity | | |
| Rubisco: % Activation | R = S | S > R (1100-1600h) | R = S |
| Carbon Assimilation | R = S | S > R (1200-1700h) | R = S |

Several observations were made. Both variants responded to increasing availability of $CO_2$ (figure 18.17). S was insensitive to very low atmospheres of $O_2$ at all temperatures evaluated (15-35°C) (figure 18.17). S [$O_2$] insensitivity could indicate A in that biotype is limited by feedback from sucrose synthesis. R was insensitive to very low $O_2$ atmospheres at 15°C, but was very sensitive to low $O_2$ at 25°C. R is limited by electron transport limitations at 25°C and therefore responds to low [$O_2$] (removal of limitations imposed by photorespiration). Regulation in R changes markedly at 15°C, where the limitation to photosynthesis becomes similar to that of S, feedback limitation from sucrose synthesis.

Based on these observations photosynthesis can be separated into three categories based on chlorophyll fluorescence quenching and responses to intercellular partial $O_2$ pressure (p($O_2$)), and $C_i$ (Dekker and Sharkey, 1992) (table 18.6). The first category of photosynthetic response occurs when photosynthetic $CO_2$ assimilation increases with $C_i$ and $q_P$ (photochemical chlorophyll fluorescence quenching) increases or $q_N$ (non-photochemical chlorophyll fluorescence quenching) decreases indicating increasing electron transport with $C_i$. This category is called Rubisco limited photosynthesis. When carbon assimilation was Rubisco limited there was little difference between the resistant and susceptible biotypes. Rubisco activity parameters were similar between the two biotypes.

A second category, called feedback-limited photosynthesis, was evident at 15 and 25° C above 300 microbars $C_i$. Characteristics of this category include constant or declining carbon assimilation rates as $C_i$ increased, no p($O_2$) response or $O_2$ stimulation of the carbon assimilation rate, and decline of $q_P$ with increasing $C_i$. At 25°C $q_N$ also increased with $C_i$. This condition has been called feedback limited photosynthesis (Sharkey, 1990) and was most evident at 15°C, less so at 25°C. The importance of feedback at low temperature has been reported before (Sage and Sharkey, 1987).

The third category, photosynthetic electron transport-limited photosynthesis, was evident at 25 and 35° C at moderate to high $CO_2$. Characteristics of this category include the carbon assimilation rate increasing with $C_i$ consistent with $C_i$ suppression of photorespiration, $O_2$ inhibition of carbon assimilation, and $q_P$ and $q_N$ independent of $C_i$. This is believed to represent photosynthetic electron transport limited photosynthesis (Sharkey, 1990). The resistant biotype exhibited much more electron transport limited carbon assimilation at 25°C and 35°C than did the susceptible biotype. At 15°C neither resistant nor susceptible biotype exhibited electron transport limited carbon assimilation.

These observations reveal the increasing importance of photosynthetic electron transport in controlling the overall rate of photosynthesis in the resistant biotype as temperature increases. These observations indicate that the primary limitation to photosynthesis changes with changes in leaf temperature, and that electron transport limitations in R may be significant only at higher temperatures (Dekker and Sharkey, 1990, 1992). At low temperatures (e.g. 15° C) R carbon assimilation rates responded to changing temperatures much quicker than S; a form of cold tolerance in R conferred by increased lipid fluidity and polarity. At low temperature, when the response curves of carbon assimilation to $C_i$ indicated little or no electron transport limitation, carbon assimilation was similar in the resistant and susceptible biotype. With increasing temperature more and more electron transport limited carbon assimilation was observed and a greater difference between resistant and susceptible biotypes was observed. These observations are consistent with previous reports that resistant biotypes are more sensitive to high temperature (Ducruet and Ort, 1988; Havaux, 1989). Our interpretation of our results is that this is caused by the increasing importance of photosynthetic electron transport in controlling the overall rate of photosynthesis as temperature increases. In summary, the temperature sensitivity of the effect of triazine resistance in *B. napus* is accounted for by the temperature dependence of electron transport limitations to carbon assimilation, with no need to invoke an additional temperature sensitivity in the resistant biotype.

**Figure 18.17** Carbon exchange rate (CER; micromol $CO_2$ $m^{-2}$ $s^{-1}$) versus intercellular $CO_2$ partial pressure ($C_i$) at 15° C (top) and 25° C (bottom) in s-triazine resistant (R) and susceptible (S) *Brassica napus* leaves; low $[O_2]$, 700 μbar intercellular $O_2$ partial pressure.

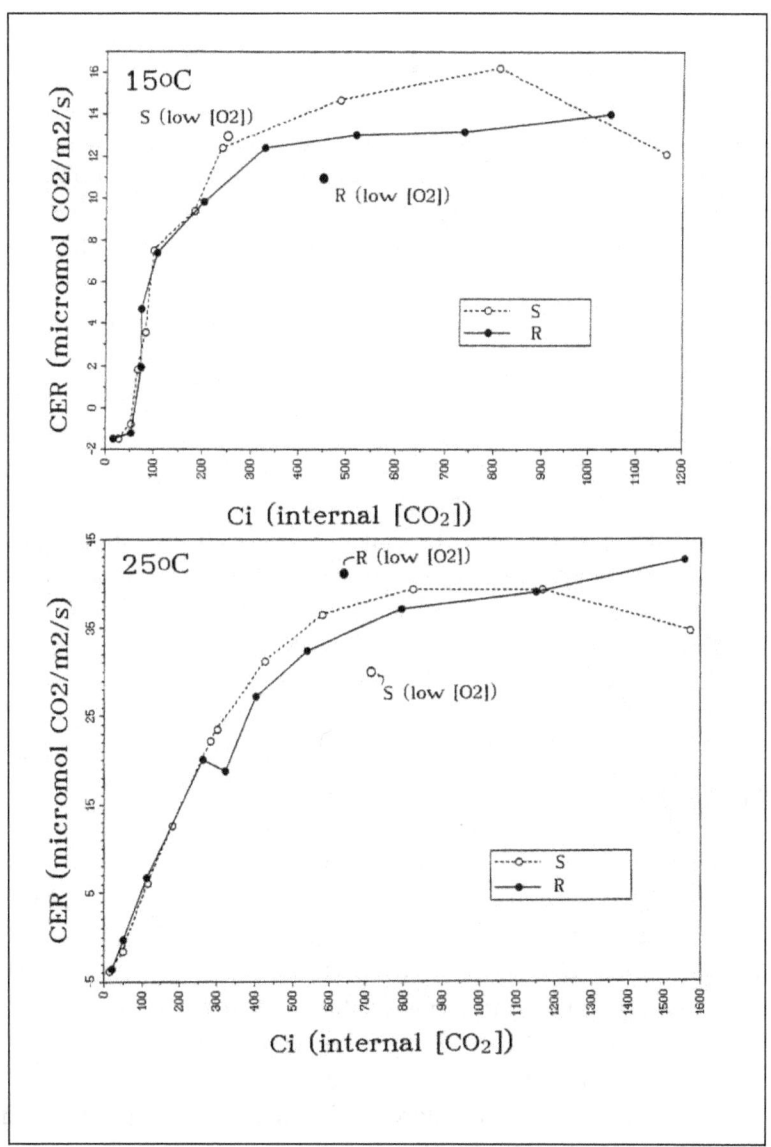

**Table 18.6** R and S carbon assimilation rate (A) limitations and regulation change with leaf temperature changes:

| Leaf Temp | Carbon Assimilation | $O_2$ Sensitivity | Limitations-Regulation |
|---|---|---|---|
| 25-35° C | S > R | | electron transport rate is limited |
| 25° C | | S [$O_2$] insensitive | Rubisco feedback inhibition from sucrose synthesis |
| | | R responsive to [$O_2$] | electron transport is limited |
| 15° C | | S [$O_2$] insensitive | Rubisco feedback inhibition from sucrose synthesis |
| | | R [$O_2$] insensitive | |
| | R = S | | electron transport rate is not limited |

## 18.8 Evolutionary ecology of s-triazine resistant plants

**18.8.1 Pleiotropy in R.** A single base-pair substitution at codon 264 in the *psbA* chloroplast gene in the highly conserved photosynthetic apparatus leads to a cascade of changes in the plants morphology, physiology and ecological reaction to its immediate environment, a non-intuitive result. With the genetic lesion, a dynamic re-organization of interacting, interdependent, plant functional units occurs, a new homeostatic equilibrium among parts. The equivocal nature of our understanding of how photosynthesis differs between R and S, and how it is regulated, has led us to focus experimental efforts on chronobiological, environmental and regulatory understandings of R under static and dynamic, closely controlled, growth conditions.

Many plant species exhibit an endogenous rhythm of carbon assimilation and stomatal function once entrained in a photoperiod. This rhythm is regulated to some extent independently of the plant's direct response to PPFD (Browse et al., 1981). Work in our laboratory indicated a consistant, differential, pattern of Chl *a* fluorescence ($F_t$), carbon assimilation, leaf temperature, total conductance to water vapor (g), and leaf intercelluar $CO_2$ partial pressure ($C_i$) (Dekker, 1993; Dekker and Burmester, 1992; Dekker and Westfall, 1987 a, b) between S and R *Brassica napus* over the course of a diurnal light period: R is a chronomutant. Pleiotropic changes in R result in chloroplast and whole leaf morphology similar to that of low light, dark-adapted, plants. The acquisition of shade-adapted morphology in R is not a plastic response of the phenotype to environment. A consistent, differential, pattern of many photosynthetic functions was observed between R and S *Brassica napus* over the course of a diurnal light period. Photosynthetic superiority of one biotype to another was a function of the time of day, the age of the plant and the temperature of the environment

R plants varied in their relative advantage (or disadvantage) over S in terms of carbon assimilation as they aged, and in response to temperature. Carbon assimilation in R was much lower than that in S at high leaf temperatures (e.g. 35° C) when the leaf temperature was closely controlled. These results are consistent with those of others (Ducuret and Ort, 1988; Havaux, 1989; Gounaris and Barber, 1983). When leaf temperature was not directly controlled, but air temperature was, R carbon assimilation exceeded that of S at relatively high temperatures (e.g. 35° C air temperature) (Dekker and Burmester, 1990, 1992). In both experimental conditions R leaf stomatal conductances were usually greater than in S. R possesses greater heat tolerance than S due to leaf cooling from greater stomatal conductance and leaf intercellular $CO_2$ partial pressure. As a consequence, at all important physiological temperatures (10-35° C) R leaves are cooler than S leaves. Stomatal function differentially regulates carbon assimilation in these two biotypes.

R also possesses tolerance to cool temperatures, and to changes in temperature, which is conferred by pleiotropic membrane lipid changes. It is hypothesized that these ontogenetic and diurnal patterns of differential photosynthesis may be a consequence of correlative diurnal fluctuations in fatty acid biosynthesis and the dynamic changes in membrane lipids over the course of the light-dark daily cycle, changes in leaf membrane lipids with age (Lemoine et al., 1986), or microenvironmental temperature influences (Ireland et al., 1988; Dekker and Burmester, 1990; Ducruet and Lemoine, 1985). At relatively cool temperatures it has been hypothesized that the change in lipid saturation of chloroplast membranes could confer cold tolerance to R plants, resulting in greater carbon assimilation rates in R under those conditions (Pillai and St. John, 1981).

R biotypes may have a selective, adaptive, advantage over S in certain unfavorable ecological niches independent of the presence of s-triazine herbicides: in cool, low-light environments early and late in the day; at high temperatures; and late in the plant's development.

**18.8.2 R adaptation to the environment and regulation of carbon assimilation**. Regulation of photosynthesis in R and S are controlled by many different factors. Limitations in electron transport in R is not the only critical factor in yield losses at the whole plant level. The pleiotropic effects observed in R result in a new equilibrium between functional and structural components. It is this new dynamic pleiotropic reorganization that regulates carbon assimilation in R. Electron transport limitations are only one possible regulatory point in the photosynthetic pathway leading from light-harvesting and the photolysis of water, through ribulose bisphosphate carboxylase/oxygenase, to starch/sucrose biosynthesis, translocation, and utilization. Carbon flux through the leaf is regulated at many points. Electron transport, even in R, is not the only critical regulatory step. In fact, Dekker and Sharkey (1992) have shown that the primary limitation to photosynthesis changes with changes in leaf temperature, and that electron transport limitations in R may be significant only at higher temperatures.

The reorganized R plants interact with the environment in a different way than does S. It is this that causes the functional result observed. Under environmental conditions highly favorable to plant growth, S often has an advantage over R. Under certain less favorable conditions to plant growth, stressful conditions, R can be at an advantage over S.

It can be envisioned that there were environmental conditions prior to the introduction of s-triazine herbicides in which R had an adaptive advantage over the more numerous S individuals in a population of a species. Under certain conditions R might have exploited a photosynthetic niche under-utilized by S. These conditions may have occurred in less favorable environments and may have been cool (or hot), low light conditions interacting with other biochemical and diurnal plant factors early and late in the photoperiod, as well as more complex physiological conditions late in the plant's development. Under these conditions R survival and continuity could have been ensured at a higher frequency of occurrence than that due to the mutation rate of the *psbA* plastid gene alone, independent of the existence of a postulated plastome mutator (Duesing and Yue, 1983).

## Chapter 19: *Abutilon theophrasti* (Velvetleaf)

**Summary.** I conducted experiments in Michigan in 1978 and 1979 to determine the effect of velvetleaf on soybean productivity. To answer this question I used two complementary, demographic, experimental designs: a weed density-crop yield experiment, and a replacement series experiment with varying proportions of crop and weed at equal population densities.

At the conclusion of these experiments I had a quantitative estimate of the soybean losses caused by velvetleaf in these Michigan conditions. I also had gained a big clue about their interaction: low levels, or mixture proportions, of velvetleaf greatly reduced midseason soybean flowering; while at the same time velvetleaf branching was greatly increased, much more than was expected when it was alone in monoculture. Velvetleaf branching connotes plant shoot architecture plasticity, the plant body upon which leaves and flowers hang. The former feeds the later, biomass is proportional to seed yield. What I observed was that when low numbers of velvetleaf infested a soybean field there was greater opportunity to seize and exploit light above the lower crop canopy. This led to two things: more room for velvetleaf to branch out and seize opportunity in the form of more leaves; the consequential shading increased floral abortion in the soybeans, while its body architecture remained unchanged.

What was the cause of this qualitative interaction among the species gained with quantitative, demographic, observations? What functional, plastic, traits in soybeans and velvetleaf led to this neighborhood interaction? What was the underlying biological explanation for this plastic inhibitory-stimulatory response? I needed to know more if I was to really understand what was going on. I needed to know a lot more if I was to be able to predict future soybean-velvetleaf interactions in any field, in any year. It was William Akey who subsequently brought light to our understanding.

William Akey conducted several field experiments in 1986-1989 for his Ph.D. thesis research to address these questions. We hypothesized that light was the specific opportunity spacetime being seized and exploited preferentially by velvetleaf at the expense of soybeans. He clearly demonstrated this was true. A combination of traits appears to give velvetleaf an advantage over soybean in competition for light. These functional traits include a rapid early growth rate; including rapid, extensive stem growth leading to tall plant stature. They also include a branching architectural pattern that situates much of the leaf area near the top of the plant, above the leaves of shorter competing species, and rapid scencence of lower leaves when they become shaded. Additionally, active leaf movements are mediated by the pulvinus at the leaf base-petiole terminus which control several advantageous behaviors. Active movements optimize water use efficiency by regulating leaf angle exposure during diurnal stress periods. Control of diurnal leaf angle also controls velvetleaf light interception, photosynthetic activity and concomitant shading of shorter neighboring species.

These studies impressed on us that we could derive more inferences of the causes of plastic weed and crop behavior and adaptation in the field from research focused on qualitative functional traits relative to those derived from only a quantitative, demographic description of the local agricultural community.

Stimulated by questions of weed control failures due to velvetleaf escaping herbicides in the shallow soil surface layer, I studied the biology of velvetleaf seedling emergence as a minor component of my Ph.D. thesis studies. This was my first venture into a weed species' seed life history, a theme of later foxtail (*Setaria spp.*) research. What I found was that large numbers of velvetleaf seedlings emerged early in the season, with very reduced numbers after that time. Velvetleaf seedlings emerged from successively shallower depths in the soil as the growing season progressed. I found that velvetleaf seedling emergence depth may be stimulated by periods of shallow soil drying. It is hypothesized, therefore, that shallow soil drying in some manner opens the chalazal slit of the intact seed, allowing water entry past the hard coat, stimulating velvetleaf to germinate.

Heather MacKenzie, and eager and bright undergraduate student at the University of Guelph, Ontario, Canada discovered that velvetleaf seedlings accumulated less biomass as the soil pH decreased. This seedling growth reduction was associated with interference of nutrient uptake and toxicity. These nutrient imbalances also led to enhanced injury by several common herbicides.

### 19.1 *Abutilon theophrasti* Taxonomy

Velvetleaf, *Abutilon theophrasti*, is a relatively homogeneous genetically, self-pollinated annual weed species that reproduces strictly by seed. It is native to south Asia, India, and infests considerable parts of agricultural North America (Gleason and Cronquist, 1963; Stubbendieck et al., 1995).

The plant can attain a wide range in height, 0.2-2 m, with a taproot root system. It has erect stems and is sparingly branched. The alternate branchs are smooth, and covered with short, velvety pubescence. The alternate leaves are also pubscent, softly hairy, broad, cordate, 10-15 cm long and wide. The petioles are about 10-15 cm in length.

Velvetleaf has solitary, yellow, perfect (hermaphroditic, self-pollinating) flowers in the upper axils, about 1.5-2.5 cm in diameter, with 5 petals. It typically flowers from July-October in North America. The peduncles joined above the middle, and elongated to 2-3 cm at maturity.

The velvetleaf fruit is cup-shaped, a schizocarp, 1.5-3 cm in diameter. This fruit is composed of 10-15 densely hairy carpels united in a ring around a central axis, with conspicuous, horizontally spreading beaks pointing upward. The fruit contains 3-15 ovules/seeds per carpel. At maturity the carpels dehiscent across the top, eventually falling from the branch axis. Velvetleaf seed are rounded to triangular, 3-4 mm long, notched, flattened, grayish-brown to black.

### 19.2 Velvetleaf Polymorphism & Plasticity

I began my graduate research career in weed science at Michigan State University in East Lansing with Bill Meggit in the Crop Science Department in 1977. My primary non-degree responsibility to him was to coordinate the herbicide trials in soybeans across Michigan. I chose to study velvetleaf, *Abutilon theophrasti*, for my Ph.D. studies because it was difficult to control with the herbicides available at that time. Therefore it seemed velvetleaf biology was a suitable, practical, topic to understand in more depth for my degree.

I began my formal Ph.D. graduate research career asking the question 'what effect does VLF have on soybean productivity?' I was strongly influenced by Alan Putnam in the Horticulture Department at that time, by his interests and teaching in the then emerging areas of allelopathy (e.g. Putnam and Duke, 1978) and weed ecology. I immersed myself in the demographic literature, notably in C.T. de Wit's 'On Competition" (1960) and the scientific inferential power of the replacement series experimental design. The experimental approach offered the chance to control changes in population density (plants/m$^2$) and observe the plastic responses of the plant (e.g. numbers and mortality, growth, yield). I was also strongly influenced by several workers from the 1950's and 1960's: Hozumi, Kira, Shinosaki and Yoda (e.g. Hozumi et al., 1968; Shinozaki and Kira, 1956, 1961; Yoda et al., 1957, 1963) to name but a few. Much of this density-yield experimental design work was rehashed later by weed scientists, with a significant loss in fidelity from these original Japanese pioneers.

The velvetleaf work was completed in 1980, but I was left with a lingering hunger for an understanding of what caused the observed neighbor interactions and soybean yield losses. These questions were continued with the excellent subsequent work by William Akey, soon after my move in 1985 to Iowa State University in Ames. At the end of the velvetleaf work I was getting the first glimmers of the importance of functional traits to explain the often highly variable demographic responses we observed, the limitations pointed out by Harper (1977) missed by so many others (Dekker, 2011).

So here is the story of my first weed affair, the first time a weed shared its secrets with me. The first time I asked.

### 19.3     Velvetleaf Competitive Plasticity

**19.3.1  Branching and flowering plasticity predicate dry weight and seed yield plasticity.** I conducted experiments in Michigan in 1978 and 1979 to determine the effect of velvetleaf on soybean productivity (Dekker, 1980; Dekker and Meggit, 1983a, 1983b; Dekker et al., 1983; see Abstracts Cited, *A. theophrasti*). To answer this question I used two complementary, demographic, experimental designs: a weed density-crop yield experiment, and a replacement series experiment with varying proportions of crop and weed at equal population densities.

What I discovered was that low infestations of *A. theophrasti* inhibited soybean branching, flowering, dry weight and seed yield. These inhibitory plastic growth responses were most apparent in seed yield and flowering, and least apparent in dry matter accumulation. The effect of velvetleaf as a neighbor was most apparent in mixtures with high and equal proportions of soybeans. During development in these mixtures, midseason velvetleaf branching stimulation and soybean flowering inhibition resulted in velvet leaf dry matter and seed weight stimulation and soybean dry matter and seed yield inhibition at season's end. Velvetleaf preferentially seized more locally available opportunity stimulating its plastic branching architecture in these mixtures with soybeans. This enabled velvetleaf to accumulate more dry matter and hence more seed at the expense of soybeans. Soybean flowering decreased, resulting in dry matter and seed weight inhibition in these mixtures with velvetleaf. The consequence of this asymmetrical interaction is a skewed distribution of branch numbers: velvetleaf preferentially seizes light and leaf production accelerates, while simulateously soybean is shaded and growth decelerates.

At the conclusion of these experiments I had a quantitative estimate of the soybean losses caused by velvetleaf in these Michigan conditions. I also had gained a big clue about their interaction: low levels, or mixture proportions, of velvetleaf greatly reduced midseason soybean flowering; while at the same time velvetleaf branching was greatly increased, much more than was expected when it was alone in monoculture. Velvetleaf branching connotes plant shoot architecture plasticity, the plant body upon which leaves and flowers hang. The former feeds the later, biomass is proportional to seed yield (Dekker, 2004b). What I observed was that when low numbers of velvetleaf infested a soybean field there was greater opportunity to seize and exploit light above the lower crop canopy. This led to two things: more room for velvetleaf to branch out and seize opportunity in the form of more leaves; the consequential shading increased floral abortion in the soybeans, while its body architecture remained unchanged.

What was the cause of this qualitative interaction among the species gained with quantitative, demographic, observations? What functional, plastic, traits in soybeans and velvetleaf led to this neighborhood interaction? What was the underlying biological explanation for this plastic inhibitory-stimulatory response? These studies, and these two demographic experimental designs, were not capable of answering these more fundamental questions. Additionally, these soybean yield losses in response to velvetleaf infestations were highly variable within the years studied in Michigan, and were not applicable to other years and locations with any degree of agriculturally-meaningful precision.

Naively, in the flush of my early scientific enthusiasm, I stated in my thesis "Allelopathy by means of water relationship interference is implicated." This was wrong. There existed no implication. I needed to know more if I was to really understand what was going on. I needed to know a lot more if I was to be able to predict future soybean-velvetleaf interactions in any field, in any year. It was William Akey who subsequently brought light to our understanding.

**19.3.1.1 Introduction.** What effect do velvetleaf and soybeans have on each other in agricultural neighborhoods in Michigan fields? What demographic interactions occur between these two species in terms these highly plastic growth parameters as their annual life history progresses from midseason branching and flowering nodes to season-end dry matter accumulation and seed yield? To answer these questions for my Ph.D. research I used two different experimental designs, one to estimate actual agricultural production practices (density-yield) wherein population density changed with weed infestation, and a companion replacement series design to untangle plastic neighbor responses due to species mixture proportion and population density.

**19.3.1.2 Experimental methods.** Studies were conducted on the Michigan State University Agricultural Experiment Station, East Lansing, in 1978 and 1979 to assess the mutual interactions between *A. theophrasti* and soybean neighbors. Experimental neighborhood designs were of two complementary types. A density-yield design was used to estimate growth and yield plasticity of both species in typical agricultural conditions of a Michigan soybean field: 20-22 soybean/m$^2$, with additional infestations of 0-25 *A. theophrasti* plants/m$^2$, both in rows. A replacement series design was also used to estimate the effects of mixtures of these two species at a single constant population density similar to that of field soybeans: monocultures and 25-75% of each species in mixtures.

> "A count of the number of plants in an area gives extraordinarily little information unless we are also told their size. Individual plants are so "plastic" that variations of 50,000-fold in weight or reproductive capacity are easily found in individuals of the same age. Clearly, counting plants is not enough to give a basis for a useful demography. The plasticity of plants lies, howerever, almost entirely in variations of the number of their parts." (Harper, 1977)

With the above caveat on demographic approaches to these studies, the parmeters I observed (per unit area; per plant) were plant number, branch number, flower number (midseason, harvest), dry weight, seed weight.

**19.3.2 Branching.** The presence of low (e.g. 3/m$^2$) infestations of *A. theophrasti* reduced soybean branching at midseason (figure 19.1, left). The highest infestations reduced soybean branching by almost 50%. In mixtures, soybean responses to velvetleaf were similar to those in monoculture (figure 19.1, right). In mixtures, velvetleaf responses to soybeans were similar to those in monoculture except in minority mixtures, when they were greater. This resulted in greater total branching in those mixtures.

**19.3.3 Flowering.** The presence of low (e.g. 2-4/m$^2$) infestations of *A. theophrasti* reduced soybean flowering at midseason (figure 19.2, left). This was more apparent in higher velvetleaf densities and in 1978. In mixtures, soybean flowering decreased in all mixtures with velvetleaf compared to those in monoculture (figure 19.2, right). In mixtures, velvetleaf responses to soybeans were similar to those in monoculture. This resulted in decreased total branching in mixtures, entirely due to soybean flowering inhibition.

**19.3.4 Dry matter accumulation.** The presence of low (e.g. 3-5/m$^2$) infestations of *A. theophrasti* reduced soybean dry weight at seasons end (figure 19.3, left). Higher velvetleaf densities (7-25/m$^2$) did not result in appreciably more soybean dry weight losses. In mixtures, soybean dry weight decreased in most mixtures with velvetleaf compared to those in monoculture, especially those with velvetleaf in the minority (figure 19.3, right). Velvetleaf dry weight in equal proportioned mixtures with soybeans were greater than those in monoculture; but similar in other mixtures. This resulted in decreased total dry weight in low velvetleaf mixtures due to soybean dry weight inhibition. In equal proportion mixtures, total dry weight was similar to that expected in monocultures due to the offsetting consequences of soybean inhibition and velvetleaf stimulation.

**Figure 19.1** Effect of various *A. theophrasti* densities on soybean branching nodes/m² (expressed as a percent of the weed-free control) in a constant density soybean stand in 1979 (left); replacement series diagram of soybeans and *A. theophrasti* branching nodes/m² at midseason 1979 at a constant typical soybean field density (right).

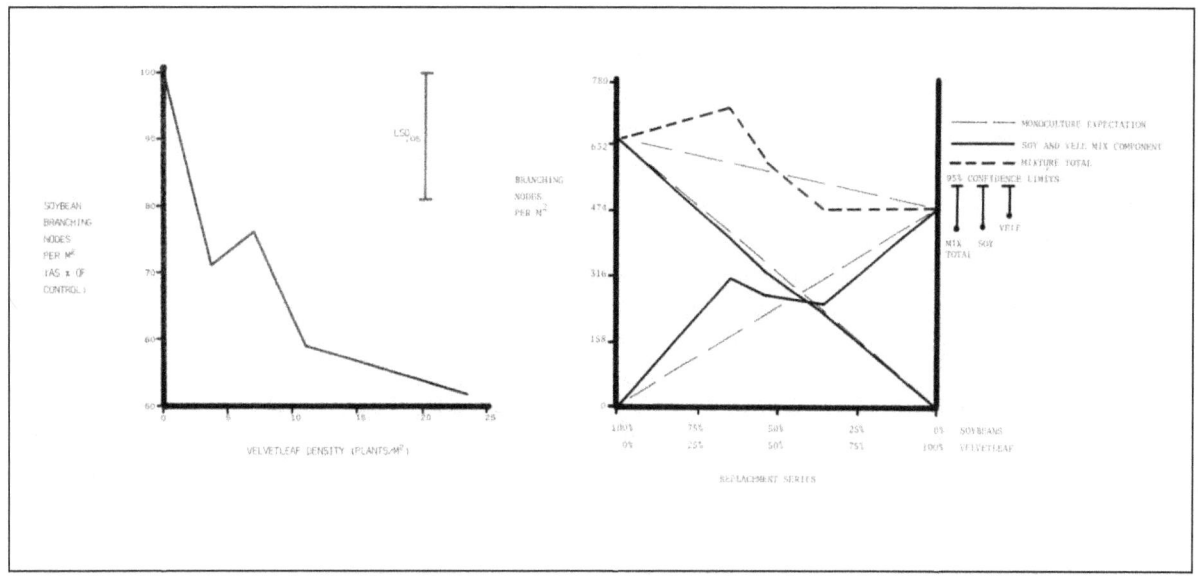

**Figure 19.2** Effect of various *A. theophrasti* densities on soybean flowering nodes/m² at midseason (expressed as a percent of the weed-free control) in a constant density soybean stand (left); replacement series diagram of soybeans and *A. theophrasti* flowering nodes/m² at midseason 1979 at a constant typical soybean field density (right).

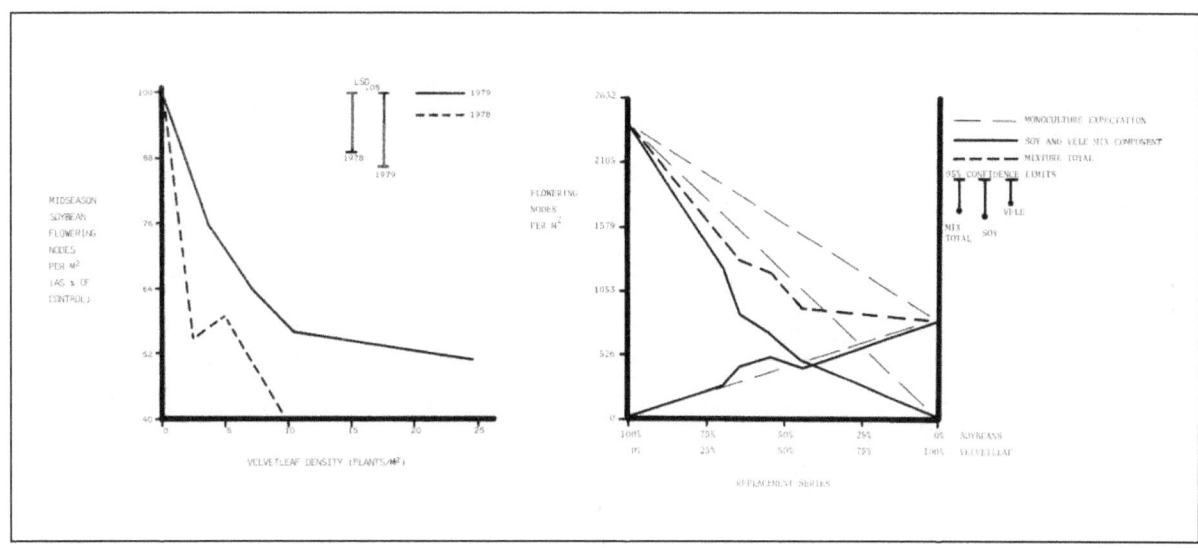

**Figure 19.3** Effect of various *A. theophrasti* densities on soybean dry weight/m$^2$ (expressed as a percent of the weed-free control) in a constant density soybean stand (left); replacement series diagram of soybeans and *A. theophrasti* dry weight/m$^2$ at midseason 1978 at a constant typical soybean field density (right).

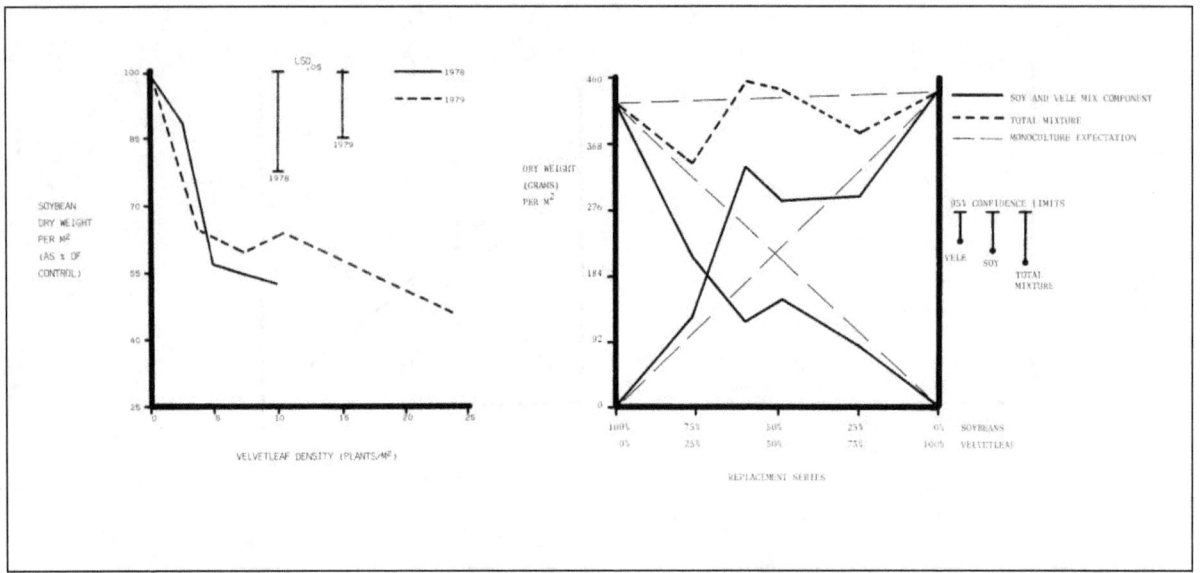

**19.3.5 Seed weight at harvest.** The presence of low (e.g. 5/m$^2$) infestations of *A. theophrasti* reduced soybean seed weight at harvest (figure 19.4, left). Higher velvetleaf densities (7-25/m$^2$) resulted in greater soybean seed yield losses (e.g. 75% loss at 25 velvetleaf/m$^2$). In mixtures, soybean seed weight responses were similar to those for dry weight. Soybean seed yield decreased in most mixtures with velvetleaf compared to those in monoculture, especially those with velvetleaf in the minority (figure 19.4, right). Velvetleaf seed weight in low proportion mixtures with soybeans was greater than those from the monoculture expectation; and greater or similar in other mixtures. This resulted in different total seed yield responses depending on mixture proportions. In low velvetleaf proportions total seed yield was depressed, in equal proportions it was similar to monoculture expectations, in both cases due the offsetting consequences of soybean inhibition and velvetleaf stimulation.

## 19.4 Velvetleaf Photomorphogenic Plasticity

**19.4.1 Phototropic and photomorphogenic plasticity of velvetleaf leaf architecture predicates soybean yield losses.** Earlier studies provided the clue that low numbers of velvetleaf infesting a soybean field caused it to branch more than when it grew alone in monoculture. While velvetleaf got bigger, soybean floral numbers decreased. At season's end velvetleaf dry and seed weight were greater, and soybean less, than expected when each was grown in monoculture. This insight raised several questions. What was the underlying biological explanation for this plastic inhibitory-stimulatory response? What functional, plastic, traits in soybeans and velvetleaf led to this neighborhood interaction?

William Akey conducted several field experiments in 1986-1989 for his Ph.D. thesis research to address these questions. We hypothesized that light was the specific opportunity spacetime being seized and exploited preferentially by velvetleaf at the expense of soybeans. He clearly demonstrated this was true.

Analysis of results from a deWit replacement series experimental design indicated that interference between velvetleaf and soybean did not occur early in the growing season when individual plants were fairly widely separated. Velvetleaf and soybean did not significantly interfere with each other early in the growing season, but velvetleaf gained resources at the expense of soybean during the middle and later parts of the season.

As canopy closure occurred, velvetleaf gained resources at the expense of soybean and, thereby, reduced the vegetative and reproductive growth of the crop.

**Figure 19.4** Effect of various *A. theophrasti* densities on soybean seed weight yield/m² (expressed as a percent of the weed-free control) in a constant density soybean stand (left); replacement series diagram of soybeans and *A. theophrasti* seed weight yield/m² 1978 at a constant typical soybean field density (right).

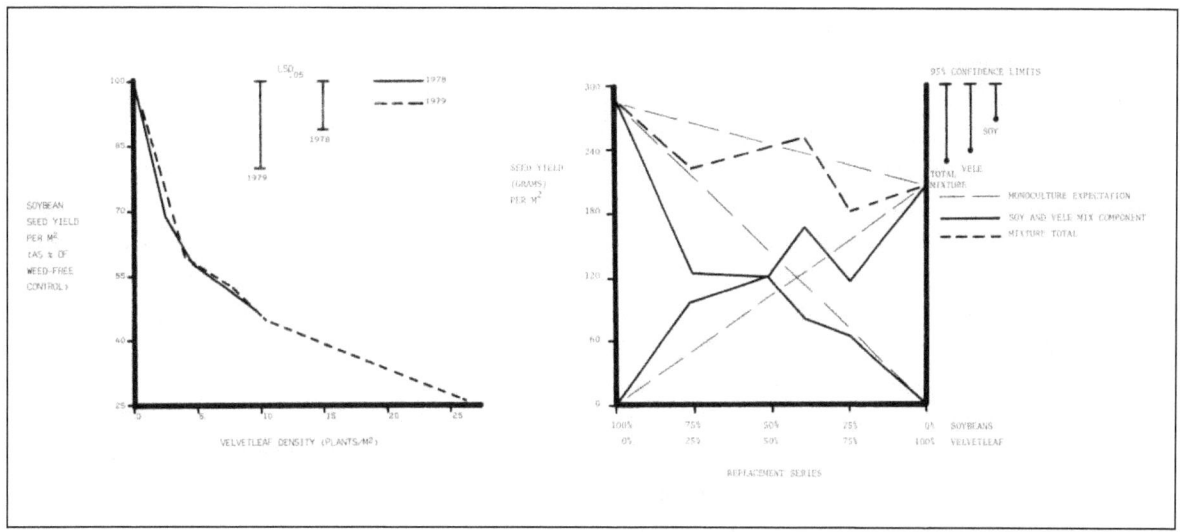

Velvetleaf was taller than soybean over most of the growing season and had more branches, especially at the top of the canopy. Velvetleaf had greater light interception than soybean in the upper part of the canopy, particularly early and late in the day, although total light interception by canopies of the two species was similar on most sample dates. Velvetleaf had higher light utilization efficiency, or conversion of intercepted light into dry matter, in the middle and later part of the growing season.

Relative above ground dry weight of soybean in mixtures was higher than expected from monoculture values early in the growing season but lower than expected later in the season, whereas velvetleaf had higher than expected values later in the season. Velvetleaf depressed seed yield of soybean in all mixtures; relative seed yield of velvetleaf was greater in all mixtures than in monoculture.

A combination of traits appears to give velvetleaf an advantage over soybean in competition for light. These functional traits include a rapid early growth rate; including rapid, extensive stem growth leading to tall plant stature. They also include a branching architectural pattern that situates much of the leaf area near the top of the plant, above the leaves of shorter competing species, and rapid scencence of lower leaves when they become shaded. Additionally, active leaf movements are mediated by the pulvinus at the leaf base-petiole terminus which control several advantageous behaviors. Active movements optimize water use efficiency by regulating leaf angle exposure during diurnal stress periods. Control of diurnal leaf angle also controls velvetleaf light interception, photosynthetic activity and concomitant shading of shorter neighboring species.

These studies impressed on us that we could derive more inferences of the causes of plastic weed and crop behavior and adaptation in the field from research focused on qualitative functional traits relative to those derived from only a quantitative, demographic description of the local agricultural community.

**19.4.2 Introduction.** From 1986 through 1989 William Akey conducted several field experiments for his Ph.D to understand the interaction between velvetleaf and soybeans (Akey, 1989; Akey et al., 1990, 1991; see Abstracts Cited, *A. theophrasti*) based on the clues obtained from previous work. The clue was that when low numbers of velvetleaf infested a soybean field its branching was greater than when grown alone in monoculture. While velvetleaf got bigger, soybean floral numbers decreased. At season's end velvetleaf dry and seed weight were greater, and soybean less, than expected when each was grown in monoculture. What functional, plastic, traits in soybeans and velvetleaf led to this neighborhood interaction? What was the underlying biological explanation for this plastic inhibitory-stimulatory response? We hypothesized that light was the specific opportunity spacetime being seized and exploited preferentially by velvetleaf at the expense of soybeans. William Akey clearly demonstrated this was true.

## 19.5 Velvetleaf Seedling Emergence

**19.5.1 Seedling emergence plasticity in space and time.** Stimulated by questions of weed control failures due to velvetleaf escaping herbicides in the shallow soil surface layer, I ventured into the biology of velvetleaf seedling emergence as a minor component of my Ph.D. thesis studies. This was my first venture into a weed species' seed life history, a theme of later foxtail research.

What I found was that large numbers of velvetleaf seedlings emerged early in the season, with very reduced numbers after that time. Velvetleaf seedlings emerged from successively shallower depths in the soil as the growing season progressed at two Michigan locations in 1978, but not 1979. Average emergence depth was 28 and 31 mm (late June) to 9 mm (mid-September). Emergence depth ranged from the surface to 50-60 mm at all locations. Interestingly, no velvetleaf germinated at less than 10 mm depth in June, and there always occurred some surface germination from July through October sample dates. Emergence depth may decrease with decreasing soil moisture, as the soil surface layers dry with increasing heat and water use. Additionally, germination flushes of velvetleaf occur after extended periods in which rainfall events are less than 125 mm. Velvetleaf seedling emergence depth may be stimulated by periods of shallow soil drying. It is hypothesized, therefore, that shallow soil drying in some manner opens the chalazal slit of the intact seed, allowing water entry past the hard coat, stimulating velvetleaf to germinate.

Later, as a young assistant professor in the Crop Science Department at the University of Guelph, Ontario, Canada, I had an opportunity to look at velvetleaf seedling growth with Heather MacKenzie, and eager and bright undergraduate student. What she discovered was that velvetleaf seedlings accumulated less biomass as the soil pH decreased. This seedling growth reduction was associated with interference of nutrient uptake and toxicity. These nutrient imbalances also led to enhanced injury by several common herbicides.

**19.5.1.1 Introduction.** I conducted a study in 1978 and 1979 at part of my Ph.D. thesis work on the timing and depth of velveltleaf seedling emergence from the soil (Dekker and Meggitt, 1986). Researchers at American Cyanamide Corporation interested me in determining what depth velvetleaf emerged from because of weed control failures by their dinitroaniline herbicide pedimethalin. They thought velvetleaf might be germinating under their 0-4 cm soil layer of herbicide, and thus seizing and exploiting non-corporate opportunity spacetime. Corporate curiosity was all it took for me to conduct investigate velvetleaf seedling emergence, although I never utilized the results reported here to actually study herbicide uptake from soil depths. It was to be my first adventure into the seed portion of a weed's life history, a major theme in my future foxtail research reported later in this book in much greater detail. For that, I am grateful to them.

**19.5.1.2 Methods.** Several agricultural sites in Michigan were choosen because they had along history of heavy natural infestations of velvetleaf distributed in the plow sole layer, 0-23 cm. Once every 3 weeks I counted velvetleaf seedlings in the experimental plots and cut them off at the soil line. Using a soil sampling probe I extracted the seedling to 15 cm. Carefully removing the soil around the seedling I and measured the distance from the soil surface to the most proximal root hairs, the location of the original germinating seed. I then used paraquat to remove all existing vegetation, ensuring that all the seedlings at the next sampling time had emerged only in that time interval. Thus, information was not cumulative, but represented seedling recruitment in that discrete period.

**19.5.2 Seedling emergence depth.** Velvetleaf seedlings emerged from successively shallower depths in the soil as the growing season progressed in both locations evaluated in 1978: averaging 28 and 31 mm (late June) to 9 mm (mid-September) at those two locations. At the one location studied in 1979, there was no consistent change in emergence depth with time: 31 mm (late June) to 35 mm (early Octotober). The range of emergence depth was from the surface to 50-60 mm at all locations in both years. Interestingly, no velvetleaf germinated at less than 10 mm depth in June, and there always occurred some surface germination from July through October sample dates, in all locations in all years. These two observations support the conclusion that emergence depth decreases with decreasing soil moisture, as the soil surface layers dry with increasing heat and water use by plants with seasonal time. Additionally, the data also suggest that flushes of germinating velvetleaf are associated with extended periods during which individual rainfall events do not exceed 125 mm. These observations suggest that velvetleaf seedling emergence depth (and possibly number) may be stimulated by seasonal periods of shallow soil drying. It is hypothesized, therefore, that shallow soil drying in some manner opens the chalazal slit, the hilum-like opening of the intact (hard-coated) seed to which water entry into the interior embryo is restricted. Opening of the hard coat may stimulate velvetleaf seed germination.

**19.5.3 Seedling emergence numbers and pattern**. Large numbers of velvetleaf seedlings (14-88/$m^2$) emerged early in the season (June-early July), with very reduced numbers (0-8/$m^2$) after that time (late June-early October), at all five Michigan locations in both 1978 and 1979. No apparent relationship existed between monthly rainfall at the several locations and the depth or numbers of seedlings emerging through the growing season.

**19.5.4 Seedling growth and soil pH**. Velvetleaf seedlings accumulated less biomass as the soil pH decreased from neutral (pH 6.9) to acidic (ph 4.3) reaction (Dekker, et al., 2006). This seedling growth reduction was associated with interference of uptake and incorporation, as well as toxicity caused by excessive amounts of, of several nutrients in seedling tissue. These nutrient imbalances also led to enhanced injury by several common herbicides used for the control of *A. theophrasti*.

## SELECTED READINGS

Baker, H.G. and G.L. Stebbins (editors). 1965. Genetics of colonizing species. Academic Press, New York, NY.

Darwin, C. 1859. On the origin of species by means of natural selection, or the preservation of favoured races in the struggle for life. John Murray, London.

Dawkins, R. 1976. The selfish gene. Oxford University Press, NY.

Dawkins, R. 1999. The extended phenotype - the long reach of the gene. Oxford University Press, UK.

Harper, J. 1977. Population biology of plants. Academic Press, NY.

Mayr, E., 2001. What evolution is. Basic Books, New York, 318 p.

Popper, K. 2009. The logic of scientific discovery. Routledge Classics, Oxon, UK.

Ridley, M. 1995. The Red Queen: Sex and the evolution of human nature. Penguin Books, UK.

Schopenhauer, A. 1969. The world as will and representation; vol. 1-2. Dover Publications, NY.

Shannon, C.E. and W. Weaver. 1963. The mathematical theory of communication. University of Illinois Press, Chicago, IL.

Taleb, N.N. 2010. The Black Swan, The Impact of the Highly Improbable, 2$^{nd}$ edition. Random House Trade Paperbacks, NY.

Taleb, N.N. 2012. Antifragile: things that gain from disorder. Random House, NY.

# REFERENCES CITED

Ahrens, W.H., E.W. Stoller. 1983. Competition, growth rate and $CO_2$ fixation in triazine-susceptible and resistant smooth pigweed (*Amaranthus hybridus*). Weed Science 31:438-444.

Akey, William. 1989. The role of competition for light in interference between velvetleaf (*Abutilon theophrasti*) and soybean (*Glycine max*). Ph.D. Thesis, Crop Production and Physiology, Agronomy, Iowa State University.

Akey, W. C., T. W. Jurik and J. Dekker. l990. Competition for light between velvetleaf (*Abutilon theophrasti*) and soybean (*Glycine max*). Weed Research 30:403-411.

Akey, W. C., T. W. Jurik and J. Dekker. l991. A replacement series evaluation of competition between velvetleaf (*Abutilon theophrasti*) and soybean (*Glycine max*). Weed Research 31:63-72.

Alex, J.F. and C.M. Switzer. 1976. Ontario weeds. Ontario Ministry of Agriculture and Food, Agdex 640, Toronto, Ontario, Canada.

Allard, R.W. 1965. Genetic systems associated with colonizing ability in predominantly self-pollinated species. In: Genetics of colonizing species, H.G. Baker and G.L. Stebbins, editors. Academic Press, New York, NY.

Allard, R.W. and J. Adams. 1969. Population studies in predominantly self-pollinating species. XIII. Intergenotypic competition and population structure in barley and wheat. American Naturalist 103(934):621-645.

Altenhofen, L. 2009. The effects of light, temperature, after-ripening, nitrate and water on *Chenopodium album* seed germination. M.S. Thesis, Ecology and Evolutionary Biology Inter-departmental Graduate Program and Agronomy Department, Iowa State University, Ames.

Altenhofen, L. and J. Dekker. 2014. The effects of light, temperature, after-ripening, nitrate and water on Chenopodium album seed germination. Environment and Ecology Research 2(2):80-90.

Anderson, R.N. 1968. Germination and establishment of weeds for experimental purposes. W.F. Humphrey Press, Geneva, NY.

Anderson, E. 1952. Plants, man and life. Little, Brown, Boston, MA.

Anderson, E. 1956. Man as a maker of new plants and plant communities; In: Man's role in changing the face of the earth, W.L. Thomas, Jr. (Ed.), p. 363-377; University of Chicago Press, IL.

Andujar, J.G., D.L. Benoit, A. Davis, J. Dekker, F. Graziani, A. Grundy, L. Karlsson, A. Mead, P. Milberg, P. Neve, I.A. Rasmussen, J. Salonen, B. Sera, E. Sousa, F. Tei, K.S. Torresen, J.M. Urbano. 2013. Continental diversity of *Chenopodium album* seedling recruitment. arXiv:1310.0483[q.bio.PE]. http://arxiv.org/abs/1310.0483

Angermeier, P.L. and J.R. Karr. 1994. Biological integrity versus biological diversity as policy directives. BioScience 44:690-697.

Anonymous. 1970. Selected weeds of the United States. Agricultural Research Service, U.S. Department of Agriculture, Agriculture Handbook No. 366.

Anonymous. 1981. Weeds of the North Central states. North central regional research publication no. 281. University of Illinois bulletin 772.

Anonymous. 1984. The holy bible – New international version. Harper Paperbacks, NY, NY.

Anonymous. 1994. Herbicide handbook, 7th edition. Weed Science Society of America, Champaign, IL.

Anonymous. 1999. Executive Order 13112 of February 3, 1999—Invasive Species. Federal (U.S.) Register: February 8, 1999 (Volume 64, Number 25)

Anonymous. 2004. Invasivespecies.gov: A gateway to Federal and State invasive species activities and programs. National (U.S.) Agricultural Library; National (U.S.) Invasive Species Council; http://www.invasivespecies.gov.

Anonymous. 2005. Millennium Ecosystem Assessment (MEA). Ecosystems and Human Well-Being: Synthesis. Island Press, Washington.

Anonymous. 2008, 2010. Wikipedia: http://en.wikipedia.org/wiki/Main_Page

Archer, S. 1989. Have southern Texas savannas been converted to woodlands in recent history? American Naturalist 134: 545-561.

Arntzen, C.J, D.L. Ditto and P.E. Brewer. 1979. Chloroplast membrane alterations in triazine-resistant *Amaranthus retroflexus* biotypes. Proceedings of the National Academy of Science USA 76:278-282.

Arntzen, C.J., J.H. Duesing. 1983. Chloroplast-encoded herbicide resistance. *In:* Advances in gene technology: molecular genetics of plants and animals, pp 273-294. K. Downey, R.W. Voellmy, F. Ahmad, J. Schultz, (Eds.), Academic Press, New York.

Atchison B., 2001. Relationships between foxtail (*Setaria* spp.) primary dormancy at abscission and subsequent seedling emergence. M.S. Thesis, Iowa State University, Ames, USA.

Austin, D.F. 2006. Foxtail millets (*Setaria*: Poaceae)- abandoned food in two hemispheres. Economic Botany 60:143-158.

Auyang, Sunny. 1998. Foundations of complex-system theories. Cambridge University Press, NY.

Axelrod, R. and M.D. Cohen. 1999. Harnessing complexity. The Free Press, Simon & Schuster, NY, NY.

Babcock, E.B. and G. L. Stebbins. 1938. The American Species of *Crepis*: their interrelationships and distribution as affected by polyploidy and apomixis. Carnegie Inst. Wash. Publ. no. 504.

Bailey, L.H. and E.Z. Bailey. 1941. Hortus the second. Macmillan, NY.

Bak, P. 1996. How nature works. The science of self-organized criticality. Springer-Verlag, NY.

Baker, H.G. 1965. Characteristics and modes of origin of weeds. In: Genetics of colonizing species; H.G. Baker and G.L Stebbins (Eds.).; pp. 147-172. Academic Press, NY.

Baker, H.G. 1974. The evolution of weeds. Annu. Rev. Ecol. Syst. 5:1-24.

Ballare, C.L. and J.J. Casal. 2000. Light signals perceived by crop and weed plants. Field Crops Research 67(2):149-160.

Balogh, J. and I. Loksa. 1956. Untersuchungen u¨ber die Zoozo¨nose des Luzernenfeldes. Acta Zool. Hungarica, 2:17–114.

Banting, J. D., E. S. Molberg and J. P. Gebhardt. 1973. Seasonal emergence and persistence of green foxtail. Can. J. Plant Sci. 53:369-376.

Barber, J. 1983. Photosynthetic electron transport in relation to thylakoid membrane composition and organization. Plant, Cell and Environment 6:311-322.

Barbour, J.C. and F. Forcella. 1993. Predicting seed production by foxtails (*Setaria* spp.). Proc. North Cent. Weed Sci. Soc. 48:100.

Barbour, M.G., J.H. Burk, W.D. Pitts, F.S. Gillam and M.W. Schwartz. 1999. Terrestrial plant ecology, 3$^{rd}$ ed. Benjamin-Cummings, Menlo Park, CA.

Barrau, J. 1958. Subsistence agriculture in Melanesia. Bernice P. Bishop Museum, Hawaii
Bull. 219.

Barrett, S. C. H., and B. J. Richardson. 1986. Genetic attributes of invading species, p.21-33. *In* R. H. Groves and J. J. Burdon [eds.], Ecology of biological invasions. Australian Academy of Science, Canberra.

Barrett, S. C. H., and J. S. Shore. 1989. Isozyme variation in colonizing plants, p.106-126. In D. E. Soltis and P. S. Soltis [eds.], Isozymes in plant biology. Dioscorides, Portland, OR.

Bar-Yam, Y. 1992. Dynamics of complex systems. Perseus Books, Reading, MA.

Baskin, J.M., Baskin, C.C. 1985. The annual dormancy cycle in buried weed seeds: a continuum. Bioscience 35(18):492-498.

Baskin, C.C. and J.M. Baskin. 1998. Seeds – Ecology, biogeography, and evolution of dormancy and germination. Academic Press, NY.

Baskin, J. and C.C. Baskin. 2004. A classification system for seed dormancy. Seed Science Research 14:1-16.

Barker, A.V. and D.J. Pilbeam. 2007. Handbook of plant nutrition. CRC Press, Boca Raton, FL.

Barrett, S.C.H. 1992. Gender variation in *Wurmbea dioica* (Liliaceae) and the evolution of dioecy. Journal of Evolutionary Biology 5:423-444.

Bazzaz, F.A. 1979. The physiological ecology of plant succession. Annual Review of Ecological Systems 10:351-371.

Begon, M., Townsend, C., Harper, J. 1996. Ecology: individuals, populations and communities (3$^{rd}$ edition). Blackwell Science, London, UK.

Behrendt, S. and M. Hanf. 1979. Grass weeds in world agriculture. BASF Aktiengesellschaft, Ludwigshafen am Rhein, Federal Republic of Germany.

Bejan, A. and J.P. Zane. 2012. Design in nature: How the constructal law governs evolution in biology, physics, technology, and social organization. Anchor Books, Random House Inc., NY.

Bengtsson, A. 1972. Radavstand och utsadesmangd for vavete och korn. Lantbrukshogskolan Meddelanden A 160. Uppsala, Sweden.

Bengtsson A. and I. Ohlsson, 1966. Utsadesmangdsforsok med varsad. Lantbrukshogskolan Meddelanden A 43. Uppsala, Sweden.

Bennett, C.H. 1988. Logical depth and physical complexity; In: The universal Turing Machine: a half-century survey, Rolf Herken (Ed.). Oxford University Press, UK.

Bennetzen, J.L. et al. 2012. Reference genome sequence of the model plant *Setaria*. Nature Biotechnology 30:555-561.

Bernoulli, J. 1713. Ars conjectandi, opus posthumum. Accedit Tractus de seriebus infinitis, et epistola gallicé scripta de ludo pilae reticularis. Basel: Thurneysen Brothers.

Berra, Y., 2010. The Yogi Book. Workman Publishing, NY.

Beversdorf, W.D., D.J. Hume. 1984. OAC Triton spring rapeseed. Canadian Journal of Plant Science 64:1007-1009.

Beversdorf, W.D., D.J. Hume, M.J. Donnelly-Vanderloo. 1988. Agronomic performance of triazine-resistant and susceptible reciprocal spring canola hybrids. Crop Science 28: 932-934.

Bewley, J.D. and M. Black. 1994. Seeds: germination, structure, and composition. Springer, US. [ch8]

Black, J.N. 1959. Seed size in herbage legumes. Herbage Abstracts 29:235-241.

Blackshaw, R. E., E. H. Stobbe, C. F. Shayewich and W. Woodbury. 1981. Influence of soil temperature and soil moisture on green foxtail (*Setaria viridis*) establishment in wheat (*Triticum aestivum*). Weed Sci. 29:179-184.

Blatchley, W.S. 1912. The Indiana weed book. Nature Publications Company, Indianapolis, IN.

Bleasdale, J.K.A. and J.A. Nelder, 1960. Plant population and crop yield. Nature 188:342.

Bonfante, P. 2003). Plants, mycorrhizal fungi and endobacteria: a dialog among cells and genomes. The Biological Bulletin 204(2):215-220.

Booth, B.D. and C.J. Swanton. 2002. Assembly theory applied to weed communities. Weed Science 50:2-13.

Bor, N.L. 1960. Grasses of Burma, Ceylon, India and Pakistan. Int. Monogr. Pure and Appl. Biol. Vol. 1. Pergamon Press, London.

Boardman, N.K. 1977. Comparative photosynthesis of sun and shade plants, Annual Review of Plant Physiology 28:355-377.

Borowski, E.J. and J.M. Borwein (Eds). 1991. The HarperCollins Dictionary of Mathematics. Harper Collins Publishers, NY.

Boucher, D. H., S. James, and K. H. Keeler. 1982. The ecology of mutualism. Annual Review of Ecology and Systematics 13: 315-347.

Box, G.E.P., W.G. Hunter, and J.S. Hunter. 1978. Statistics for experimenters. John Wiley & Sons, NY.

Bradford, K.J. and H. Nonogaki. 2007. Annual plant reviews. Vol. 27. Oxford, Blackwell, Ames, IA. p. 367.

Bradshaw, A. D. 1965. Evolutionary significance of phenotypic plasticity in plants. Advances in Genetics 13: 115-155.

Brown, D.E. 1991. Human universals. McGraw-Hill, NY.

Brown, J.H., G.B. West, and B.J Enquist. 2000. Scaling in biology: patterns and processes, causes and consequences. In: Scaling in biology, J.H. Brown and G.B. West (eds.), Santa Fe Institute studies in the science of complexity, Oxford University Press, NY.

Browse, J., P.G. Roughan and C.R. Slack. 1981. Light control of fatty acid synthesis and diurnal fluctuations of fatty acid composition in leaves. Biochemical Journal 196:347-354.

Brenchley, W.E. 1920. Weeds of farmland. University Press, Aberdeen; Longmans, Green and Co., London; UK.

Bruno, J. F., J. J. Stachowicz, and M. D. Bertness. 2003. Inclusion of facilitation into ecological theory. TREE 18: 119-125.

Brutnell, T.P., L. Wang, K. Swartwood, A. Goldschmidt, D. Jackson, X.-G. Zhu, E. Kellogg and J. Van Eck. 2010. *Setaria viridis*: a model for $C_4$ photosynthesis. The Plant Cell 22(8):2537-2544.

Buchholtz, K.P. 1967. Report of the terminology committee of the Weed Science Society of America. Weeds 15:388-389.

Buhler, D.D. 1995. Influence of tillage systems on weed population dynamics and management in corn and soybeans in the central USA. Crop Sci. 35:1247-1258.

Buhler, D.D., R.G. Hartzler, and F. Forcella. 1997a. Implications of weed seed and seed bank dynamics to weed management. Weed Sci. 45:329-336.

Buhler, D.D., R.G. Hartzler, F. Forcella and J.L. Gunsolus. 1997b. Relative emergence sequence for weeds of corn and soybeans. Iowa St. Univ. Extension Publ. SA-11.

Buhler, D.D., and T.C. Mester. 1991. Effect of tillage systems on the emergence depth of giant foxtail (*Setaria faberi*) and green foxtail (*Setaria viridis*). Weed Sci. 39:200-203.

Bunting, A.H. 1960. Some reflections on the ecology of weeds; In: The biology of weeds: A symposium of the British Ecological Society, J.L. Harper (Ed.), p. 11-26; Blackwell Scientific Publications, Oxford, UK.

Burdick, A. 2005. The truth about invasive species. Discover magazine, May.

Burke, J.J., R.F. Wilson and J.R. Swafford. 1982. Characterization of chloroplasts isolated from triazine-susceptible and triazine-resistant biotypes of *Brassica campestris* L. Plant Physiology 70:24-29.

Burnside, O.C., C.R. Fenster, L.L. Evetts, and R.F. Mumm. 1981. Germination of exhumed weed seed in Nebraska. Weed Sci. 29:577-586.

Callaway, R. M. 1995. Positive interactions among plants (Interpreting botanical progress). The Botanical Review 61: 306-349.

Callen, E.O. 1965. Food habits of some Pre-Columbian Mexican Indians. Econ. Bot. 19:335-343.

Callen, E.O. 1967. The first New World cereal. American Antiquities 32(4):535-538.

Camazine, S., Deneubourg, Franks, Sneyd, Theraulaz, Bonabeau. 2003. Self-Organization in Biological Systems, Princeton University Press, NJ.

Casper, B.B. 1990. Seedling establishment from one- and two-seeded fruits of *Cryptantha flava*: a test of parent-offspring conflict. American Naturalist 136:167-177.

Caswell, H. 2001. Matrix population models, 2$^{nd}$ edition. Sinauer, Sunderland, MA.

Cavalier-Smith, T. 1978. Nuclear volume control by nucleoskeletal DNA, selection for volume and cell growth rate, and the solution of the DNA C-value paradox. Journal of Cell Science 34:247-278.

Cavers, P.B. and J.L. Harper. 1966. Germination polymorphism in *Rumex crispus* and *Rumex obtusifolius*. The Journal of Ecology 1:367-382.

Chambers scientific and technology dictionary. 1988. W & R Chambers-Cambridge University Press, UK.a [7.3.1]

Chapman, G.P. 1992. Domestication and its changing agenda. In: Grass evolution and domestication, G.P. Chapman (Ed.). Cambridge Univ. Press, Cambridge, UK. pp. 316-337.

Chase, A. 1937. First book of grasses. San Antonio, TX: 2nd ed.

Cheng, K. 1973. Radio carbon dates from China: some initial interpretations. Current Anthropology 14: 525-528.

Chepil, W.S. 1946. Germination of weed seeds. I. Longevity, periodicity of germination and vitality of seed in cultivated soil. Sci. Agric. 26:307-347.

Christianson, M.L., and D.A. Warnick. 1984. Phenocritical times in the process of in vitro shoot organogenesis. Developmental Biology 101(2):382–390.

Clack, T., S. Mathews and R.A. Sharrock, 1994. The phytochrome apoprotein family in Arabidopsis is encoded by five genes: the sequences and expression of PHYD and PHYE. Plant Molecular Biology 25:413-427.

Clark, C.W. and M. Mangel. 2000. Dynamic state variable models in ecology. Oxford University Press, NY.

Clark, L.G. and R.W. Pohl. 1996. Agnes Chase's first book of grasses, 4th ed. Smithsonian Institution Press, Washington and London.

Clayton, W. D. 1980. *Setaria.*. In: Flora Europa; T.G. Tutin, V.H. Heywood, N.A. Burges, D.M. Moore, D.H. Valentine, S.M. Walters and D.A. Webb (Eds.). Vol. I., pp. 263-264; Vol. 5, Cambridge University Press. Cambridge, U.K.

Clements, F.E. 1916. Plant succession. Publication number 242. The Carnegie Institution, Washington, D.C.

Clements, F.E. 1936. The nature and structure of climax. Journal of Ecology 22:9-68.

Clements, F.E. and V.E. Shelford. 1939. Bio-ecology. Chapman and Hall, London, UK.

Cochran, W.G. and G.M. Cox. 1957. Experimental designs, 2nd Edition. John Wiley & Sons, NY.

Cockburn, A. 2015. Weed whackers: Monsanto, glyphosate, and the war on invasive species. Harper's Magazine Vol. 331, No. 1984, September, p.57-63.

Cohen, D. 1966. Optimising reproduction in a randomly varying environment. Journal of Theoretical Biology 12:119-129.

Colbach, N., C. Durr, S. Gruber and C. Pekrun. 2008. Modelling the seed bank evolution and emergence of oilseed rape volunteers for managing co-existence of GM and non-GN varieties. European Journal of Agronomy 28(1):19-32.

Colbach, N., C. Clermont-Dauphin, and J.M. Meynard. 2001a. GENSYS: a model of the influence of cropping system on gene escape from herbicide tolerant rapeseed crops to rape volunteers. I. Temporal evolution of a population of rapeseed volunteers in a field. Agriculture, Ecosystem and Environment 83:235-253

Colbach, N., C. Clermont-Dauphin, and J.M. Meynard. 2001b. GENSYS: a model of the influence of cropping system on gene escape from herbicide tolerant rapeseed crops to rape volunteers. II. Genetic exchanges among volunteer and cropped populations in small regions. Agriculture, Ecosystem and Environment 83:255-270

Cole, L.C. 1954. The population consequences of life history phenomena. Quarterly Review of Biology 29:103-137.

Connell, J.H. 1978. Diversity in tropical rainforests and coral reefs. Science 199:1302-1310.

Cooper, W.S. and R.H. Kaplan. 1982. Adaptive "coin-flipping": a decision-theoretic examination of natural selection for random individual variation. Journal of Theoretical Biology 94, 135-151.

Corning, P.A. 2002. The re-emergence of "emergence": a venerable concept in search of a theory. Complexity 7(6): 18–30

*Cowles, Henry Chandler. "The ecological relations of the vegetation of the sand dunes of Lake Michigan". 1899. Botanical Gazette

Crosby, A.W. Jr. 1972. The Columbian Exchange: Biological and cultural consequences of 1492. Greenwood, Westport, CN.

Culver, J.C. and J. Hyde. 2000. American Dreamer – A life of Henry A. Wallace. W.W. Norton, NY, NY.

Daintith, J. and J. Clark (Eds.). 1999. The Facts on File Dictionary of Mathematics, 3rd Edition. Markethouse Books Ltd., NY.

Darlington, H. T. 1951. The seventy year period of Dr. Beal's seed vitality experiment. Am. J. Bot. 38:379-381.

Darmency, H. and J. Dekker. 2011. *Setaria*. In: Wild crop relatives: Genomic and breeding resources, Millets and grasses, C. Kole (ed.); Wealth of Wild Species: Role in Plant Genome Elucidation and Improvement, Vol 2.; Chapter 15, pp. 275-296;. Springer-Verlag Berlin, Heidelberg.

Darmency, H., C. Ouin, and J. Pernes. 1987a. Breeding foxtail millet (*Setaria italica*) for quantitative traits after interspecific hybridization and polyploidization. Genome 29:453-456.

Darmency, H., G.R. Zangre, and J. Pernes. 1987b. The wild-weed-crop complex in *Setaria*: a hybridization study. Genetica 75:103-107.

Darwin, C. 1859. On the origin of species by means of natural selection, or the preservation of favoured races in the struggle for life. John Murray, London.

Dawkins, R. 1976. The selfish gene. Oxford University Press, NY.

Dawkins, R. 1986. The blind watchmaker. Norton, New York.

Dawkins, R. 1999. The extended phenotype - the long reach of the gene. Oxford University Press, UK.

Dawkins, R. 2004. The ancestor's tale –a pilgrimage to the dawn of evolution. Mariner Books, Houghton Mifflin, New York, NY.

Dawkins, R. 2009. The greatest show on earth. Black Swan edition, Transworld Publishers, London, UK.

Dawson, J. H. and V. F. Bruns. 1962. Emergence of barnyardgrass, green foxtail and yellow foxtail seedlings form various soil depths. Weeds. 10:136-139.

Dawson, J. H. and V. F. Bruns. 1975. Longevity of barnyardgrass, green foxtail and yellow foxtail seed in soil. Weed Science. 23:437-440.

Daynard, T.B. and W.G. Duncan. 1969. The black layer and grain maturity in com. Crop Science 9: 473-476.

Defelice, M.S., W.B. Brown, R.J. Aldrich, B.D. Sims, D.T. Judy, and D.R. Guethle. 1989. Weed control in soybeans (*Glycine max*) with reduced rates of postemergence herbicides. Weed Sci. 37:365-374.

de Finetti, B. 1970. Logical foundations and measurement of subjective probability. Acta Psychologica 34:129–145.

Dekker, J. 2010. Seed heteroblasty: how weeds exploit local opportunity. Proceedings of the15th European Weed Research Society Symposium:58; Kaposvar, Hungary.

Dekker, J. 1980. Interference between velvetleaf (*Abutilon theophrasti* Medic.) and soybeans (*Glycine max* (L.) Merrill). Ph.D. Thesis, Michigan State University.

Dekker, J. 1991. An overview of new techniques and advances in weed physiology and molecular biology. Weed Science 39: 480-481.

Dekker, J. 1992. Pleiotropy and photosynthetic regulation in triazine resistant *Brassica napus*. Photosynthesis Research 34:235.

Dekker, J. 1993. Pleiotropy in triazine resistant *Brassica napus*: Leaf and environmental influences on photosynthetic regulation. Zeitschrift Naturforschung 48c:283-287.

Dekker, J. 1997. Weed diversity and weed management. Weed Science 45:357-363.

Dekker, J. 1999a. Ethics and choice of research topics in the public sector. Weed Science 47:629-630.

Dekker, J. 1999b. Soil weed seed seed banks and weed management. Journal of Crop Production 2(1):139-166. Simultaneously published in book format as "Expanding the Context of Weed Management", D. Buhler (Ed.), Haworth Press Inc. New York.

Dekker, J. 2000. Emergent weedy foxtail (*Setaria* spp.) seed germinability behavior. In: Seed Biology: Advances and Applications (M. Black, K.J. Bradford, and J. Vasquez-Ramos, Eds). CAB International International, Wallingford, UK, pp. 411-423.

Dekker, J. 2003. The foxtail (*Setaria*) species-group. Weed Science 51:641-646.

Dekker, J. 2004a. The evolutionary biology of the foxtail (*Setaria*) species-group. In: Weed Biology and Management; Inderjit (Ed.). Kluwer Academic Publishers, The Netherlands. Pp. 65-113.

Dekker, J. 2004b. Invasive plant life history determinants of yield in disturbed agro-ecosystems. XII[th] International Conference on Weed Biology, Dijon, France. Association Française pour la Protection des Plantes Annales: 5-14.

Dekker, J. 2005. Biology and anthropology of plant invasions. In: Invasive plants: ecological and agricultural aspects; Inderjit (Ed.). Birkhäuser Verlag BioSciences, Basel. Pp. 235-250.

Dekker, J. 2009. FoxPatch: An evolutionary model system for weedy *Setaria* spp.-gp. seed life history dynamics. XIII[th] International Conference on Weed Biology, Dijon, France. Association Française pour la Protection des Plantes Annales: 117-127.

Dekker, J. 2011a. Evolutionary models of weed life history population dynamics. In: Advances in environmental research, Volume 10, pp. 143-165; J.A. Daniels (Ed.). Nova Science Publishers, Hauppauge, NY.

Dekker, J. 2011b. Evolutionary ecology of weeds. http://www.agron.iastate.edu/~weeds/AG517/517Course/EEWbook.html

Dekker, J. 2013. Soil-weed seed communication systems. Universal Journal of Plant Science 1(1):9-20.

Dekker, J. 2014. Seed dormancy, germination and seedling recruitment in weedy *Setaria*. In: Handbook of Plant and Crop Physiology, 3rd Edition, Chapter 2, pp. 33-102; M. Pessarakli, Editor. Taylor and Francis Group, CRC Press, Boca Raton, FL; 1031 pp.

Dekker, J., D. Adams, A. van Aelst, B. Dekker, J. Donnelly, and M. Haar, M. Hargrove, H. Hilhorst, C. Karssen, J. Lathrop, E. Luschei, and D. Todey. 2012a. Weedy *Setaria* seed germination-dormancy behavior: Regulatory compartmentalization. Lambert Academic Publishing, Saarbrücken, Germany. ISBN: 978-3-659-24309-7

Dekker, J., B. Atchison, M. Haar, and K. Jovaag. 2012b. Weedy *Setaria* seed life history: Heterogeneous seed rain dormancy predicates seedling recruitment. Lambert Academic Publishing, Saarbrücken, Germany. ISBN: 978-3-8454-7859-3

Dekker, J, B. Atchison and K Jovaag. 2003. *Setaria* spp. seed pool formation and initial assembly in agro-communities. Aspects of Applied Biology 69:247-259

Dekker, J. and R. G. Burmester. 1989a. Differential diurnal carbon exchange and photoinhibition in a psbA plastid gene chronomutant of *Brassica napus*. *In*: Current Research in Photosynthesis, M. Baltscheffsky (Ed.), Kluwer Academic Publishers, Dordrecht, The Netherlands, Vol. IV(16):263-266.

Dekker, J. and R. G. Burmester. 1989b. Differential diurnal photosynthesis in a psbA gene chronomutant of *Brassica napus*. Physiologia Plantarum 76:182.

Dekker, J. and R. Burmester. 1989c. Differential diurnal photosynthetic function and regulation in a psbA plastid gene chronomutant of *Brassica napus*. Chronobiologica 16:127.

Dekker, J. and R. Burmester. 1989d. Mutant weeds of Iowa: s-triazine resistant plastids in *Chenopodium album* L. Journal of the Iowa Academy of Science 96:61-64.

Dekker, J. and R. G. Burmester. 1990a. Differential diurnal photosynthetic function in a psbA plastid gene chronomutant of *Brassica napus* L. *In*: Chronobiology: Its role in clinical medicine, general biology and agriculture; Pt. B., Cellular and Molecular Mechanisms. Eds. D. Hayes, J. E. Pauly and R. J. Reiter. Pp. 243-251. Wiley-Liss, N.Y., N.Y.

Dekker, J. and R. G. Burmester. 1990b. Differential pleiotropy in a psbA gene mutant of *Brassica napus* implies altered temporal photosynthesis and thermal tolerance. Zeitschrift fur Naturforschung 45c:474-477.

Dekker, J.H. and R.G. Burmester. 1992. Pleiotropy in triazine resistant *Brassica napus*: Ontogenetic and diurnal influences on photosynthesis. Plant Physiology 100:2052-2058.

Dekker, J.H. and R.G. Burmester. 1993. Pleiotropy in triazine resistant *Brassica napus*: Differential stomatal and leaf responses to the environment. *In*: Research in photosynthesis, N. Murata (Ed.), Vol. IV:631-634. Kluwer Academic Publishers, Dordrecht, The Netherlands.

Dekker, J., R. Burmester, K. C. Chi and L. Jensen. 1987. Mutant Weeds of Iowa: s-triazine resistant plasids in *Kochia scoparia*. Iowa Journal of Research 62:183-188.

Dekker, J., R. Burmester and T. Sharkey. 1990. Pleiotropy in triazine resistant *Brassica*: IV. Differential responses to temperature. Plant Physiology Supplement 93:209.

Dekker, J., R. Burmester and J. Wendel. 1991. Mutant weeds of Iowa: S-triazine resistant *Polygonum pensylvanicum* L. Weed Technology 5:211-213

Dekker, J. H. and K. Chandler. 1985. Effects of several graminicides on the viability of quackgrass (*Agropyron repens*) rhizome buds. Canadian Journal of Plant Science 65:1057-1064.

Dekker, J. and G. Comstock. 1992. Ethical, environmental and ecological considerations in the release of herbicide resistant crops. Agriculture and Human Values 9(3):31-44.

Dekker J., Dekker B.I., Hilhorst H., Karssen C. 1996. Weedy adaptation in Setaria spp.: IV. Changes in the germinative capacity of S. faberi embryos with development from anthesis to after abscission. American Journal of Botany, 83(8), 979-991.

Dekker, J. and S.O. Duke. 1995. Herbicide resistance in field crops. Advances in Agronomy 54:69-116.

Dekker, J. and J. Gilbert. 2008. Interaction of salt, temperature, light and dormancy affecting giant foxtail (*Setaria faberi*) seed germination. WSSA Abstracts 48:265.

Dekker, J., M. Haar and J. Donnelly. 2012. Weedy *Setaria* Seed Germination-Dormancy Behavior: Regulatory Compartmentalization. Lambert Academic Publishing, Saarbrücken, Germany. (In preparation)

Dekker J., M.Hargrove. 2002. Weedy adaptation in *Setaria* spp.: V. Effects of gaseous environment on giant foxtail (*Setaria faberii* R. Hermm.) (Poaceae) seed germination. American Journal of Botany 89(3):410-416.

Dekker, J., J. Lathrop, B. Atchison, and D. Todey. 2001. The weedy *Setaria* spp. phenotype: How environment and seeds interact from embryogenesis through germination. Proceedings of the Brighton Crop Protection Conference-Weeds 2001:65-74.

Dekker, J., Luschei E.C. 2009. Water partitioning between environment and *Setaria faberi* seed exterior-interior compartments. *Agriculture Journal*, 4(2), 66-76.

Dekker, J., H. MacKenzie and K. Chandler. 2006. The effects of soil pH on *Setaria viridis* and *Abutilon theophrasti* seedling growth and tissue nutrients. Agriculture Journal 1(1):1-4.

Dekker, J., A.M.A. Mazen, R. Burmester and R. Thornhill. 1990. Pleiotropy in triazine resistant *Brassica*: III. Regulation of differential photosynthesis by ribulose bisphosphate carboxylase. Plant Physiology Supplement 93:208.

Dekker, J. H., W. F. Meggitt and A. R. Putnam. 1983. Experimental methodologies to evaluate alleleopathic plant interactions: the *Abutilon theophrasti-Glycine max* model. Journal of Chemical Ecology 9:945-981.

Dekker, J. H. and W. F. Meggitt. 1983a. Interference between velvetleaf (*Abutilon theophrasti* (L.) Medic.) and soybeans (*Glycine max* (L.) Merr.). I. Growth. Weed Research 23:91-101.

Dekker, J. H. and W. F. Meggitt. 1983b. Interference between velvetleaf (*Abutilon theophrasti* (L.) Medic.) and soybeans (*Glycine max* (L.) Merr.). II. Population dynamics. Weed Research 23:103-107.

Dekker, J. H. and W. F. Meggitt. 1986. Field emergence of velvetleaf (*Abutilon theophrasti*) in relation to time and burial depth. Iowa State Journal of Research 61(1):65-80.

Dekker, J. and T. Sharkey. 1990. Pleiotropy in triazine resistant *Brassica*: II. Differential responses to [$CO_2$] AND [$O_2$]. Plant Physiology Supplement 93:207.

Dekker, J. and T. D. Sharkey. 1992. Regulation of photosynthesis in triazine resistant and susceptible *Brassica napus*. Plant Physiology 98:1069-1073.

Dekker, J. and K. C. Vaughn. 1987. An adaptive advantage of triazine resistance: A temporal phase mutation of photosynthetic function. Chronobiologica 14:49.

Dekker, J. and B. Westfall. 1987a. Circadian oscillations in chlorophyll fluorescence in a triazine-resistant chronomutant of *Brassica napus*. J. E. Pauly and L. E. Schering(eds.), *In*: Progress in Clinical and Biological Research. Advances in Chronobiology Pt. A., Pp. 81-93. Alan R. Liss, Inc., New York.

Dekker, J. H. and B. Westfall. 1987b. A temporal phase mutation of chlorophyll fluorescence in triazine resistant *Brassica napus*. Zeitschrift fur Naturforschung 42c:135-138.

Del Castillo, R.F. 1994. Factors influencing the genetic structure of Phacelia dubia, a species with a seed bank and large fluctuations in population size. Heredity 72:446-458.

Dembinska, M. 1976. Wild corn plants gathered in the 9th to 13th centuries in light of paleobotanical materials. Folia Quaternaria 47:97-103.

De Petter, E., L. Van Wiemeersch, R. Rethy, A. Dedonder, H. Fredericq and J. De Greef. 1988. Fluence-response curves and action spectra for the very low fluence and the low fluence response for the induction of Kalanchoë seed germination. Plant Physiology 88:276-283.

Deutsch, D. 2011. The beginning of infinity: Explanations that transform the world. Penquin Books, NY.

De Wet, J.M.J. 1966. The origin of weediness in plants. Proceedings of the Oklahoma Acadamy of Science for 1966.

de Wet, J.M.J. 1992. The three phases of cereal domestication. In: Grass evolution and domestication, G.P. Chapman (Ed.). Cambridge Univ. Press, Cambridge, UK. pp. 176-198.

de Wet, J.M.J. 1995. Foxtail millet *Setaria italica*. In: Evolution of crop plants, 2$^{nd}$ Ed., J. Smartt and N.W. Simmonds (Eds.), pp. 170-172. Longman Scientific and Technical, Essex, UK.

de Wet, J. and J. Harlan. 1975. Weeds and domesticates: evolution in the man-made habitat. Econ. Bot. 29:99-107.

de Wet, J.M.J., L.L. Oestry-Stidd, and J.I. Cubero. 1979. Origins and evolution of foxtail millets. J. Agric. Trop. Bot. Appl. 26:54-64.

de Wit, C.T. 1960. On competition. Versl. Landbouwk. Onderz. 66(8):1-82.

Dobzhansky, T. 1973. "Nothing in biology makes sense except in the light of evolution. The American Biology Teacher 35 (March):125-129.

Dodgeson, C.L (Lewis Carroll). 1865. Alice's adventures in wonderland. Penquin Books, UK.

Dodgeson, C.L (Lewis Carroll). 1872. Through the looking-glass and what Alice found there. Macmilan, UK.

Dodgeson, C.L (Lewis Carroll). 1876. The Hunting Of The Snark: An Agony In Eight Fits. Macmillan, UK.

Domingos, P. 2015. The master algorithm. Allen Lane, UK.

Donald, C.M. 1951. Competition among pasture plants. I. Intra-specific competition among annual pasture plants. Australian Journal of Agricultural Research 2:355-376.

Donald, C.M. 1963. Competition among crop and pasture plants. Advances in Agronomy 15:1-118.

Donnelly, J., D. Adams, and J. Dekker. 2012. Does phylogeny determine the shape of the outer seed hull in *Setaria* seeds? In: Dekker, J., D. Adams, A. van Aelst, B. Dekker, J. Donnelly, and M. Haar, M. Hargrove, H. Hilhorst, C. Karssen, J. Lathrop, E. Luschei, and D. Todey. 2012. Weedy *Setaria* seed germination-dormancy behavior: Regulatory compartmentalization. Lambert Academic Publishing, Saarbrücken, Germany. ISBN: 978-3-659-24309-7

Donnelly, J.L., D.C. Adams and J. Dekker. 2014. Weedy adaptation in *Setaria* spp.: VI. *S. faberi* seed hull shape as soil germination signal antenna. arXiv:1403.7064 [q.bio.PE]. http://arxiv.org/abs/1403.7064

Donohue, K., M.S. Heshel, G.C.K. Chiang, C.M. Butler and D. Barua. 2007. Phytochrome mediates germination responses to multiple seasonal cues. Plant, Cell and Environment 30:202-212.

Dore, W.G. and J. McNeill. 1980. Grasses of Ontario. Res. Branch, Agric. Canada, Monograph 26, Hull, Quebec, Canada.

Douglas, B.J., A.G. Thomas, I.N. Morrison, M.G. Maw. 1985. The biology of Canadian weeds. 70. *Setaria viridis* (L.) Beauv. Canadian Journal of Plant Science 65:669-690.

Doust, A.N., E.A. Kellogg. 2002. Inflorescence diversification in the Panicoid "Bristle Grass" clade (Paniceae, Poaceae): Evidence from molecular phylogenies and developmental morphology. American Journal of Botany 89:1203-1222.

Doust, A.N., E.A. Kellogg, K.M. Devos and J.L. Bennetzen. 2009. Foxtail Millet: A Sequence-Driven Grass Model System. Plant Physiology 149(1):137-141.

Doyle, J. J. 1999. Origins, colonization, and lineage recombination in a widespread perennial soybean polyploid complex. Proceedings of the National Academy of Science 96:10741-10745

Ducruet, J.M. and Y. Lemoine. 1985. Increased heat sensitivity of the photosynthetic apparatus in triazine-resistant biotypes from different plant species. Plant Cell Physiology 26:419-429.

Ducruet, J.M. and D.R. Ort. 1988. Enhanced susceptibility of photosynthesis to high leaf temperature in triazine-resistant *Solanum nigrum* L. evidence for photosystem II D1 protein site of action. Plant Science 56:39-48.

Duesing, J.H., S. Yue. 983. Evidence for a plastome mutator (cpm) in triazine-resistant *Solanum nigrum*. Weed Science Society of America Abstracts 23:191.

Du Rietz, G.E. 1931. Life-forms of terrestrial flowering plants. I. Acta Phytogeographica Suecica 3 (1): 95 pp.

Egler, F.E. 1954. Vegetation science concepts I. Initial floristic composition, a factor in old-field vegetation development with 2 figs. Vegetatio 4:412-417.

Elton, C. 1958. The ecology of invasions by animals and plants. Methuen, London, UK.

Emerson, R.W. 1878. Fortune of the republic. Houghton and Osgood, Boston, MA.

Emery, W.H.P. 1957. A study of reproduction in *Setaria macrostachya* and its relatives in the southwestern United States and Northern Mexico. Bull. Torrey Bot. Club 84:106-121.

Ehrendorfer, F. 1965. Dispersal mechanisms, genetic systems, and colonizing abilities in some flowering plant families. In: Genetics of colonizing species; H.G. Baker and G.L Stebbins (Eds.). Academic Press, NY.

Ervio, L.-R. 1983. Competition between barley and annual weeds at different sowing densities. Annales Agriculturae Fenniae 22:232-239.

Fabian, I. 1938. Beitrage zum Lang- und Kurztagsproblem. Zeitschr Bot 33:305-357.

Fagan, B. 2004. The long summer, how climate changed civilization. Basic Books, NY.

Fairbrothers, D.E. Morphological variation of *Setaria faberii* and *S. viridis*. Brittonia 1959; 11:44-48.

Fausey, J.C., J.J. Kells, S.M. Swinton, and K.A. Renner. 1997. Giant foxtail (*Setaria faberi*) interference in non-irrigated corn (*Zea mays*). Weed Sci. 45:256-260.

Fernald, M.L. 1950. Gray's manual of botany. 8th ed. New York: American Book Co.

Fisher, R.A. 1941. Average excess and average effect of a gene substitution. Annals of Eugenics 11:53-63.

Fisher, R.A. 1958. The genetical theory of natural selection, $2^{nd}$ edition. Dover, NY.

Flematti G.R., E.L Ghisalberti, K.W. Dixon, and R.D. Trengove. 2004. A compound from smoke that promotes seed germination. Science 305(5686):977.

Folke, C., Carpenter, S., Walker, B., Scheffer, M., Elmqvist, T., Gunderson, L., Holling, C.S. 2004. Regime shifts, resilience, and biodiversity in ecosystem management. Annual Review of Ecology, Evolution, and Systematics 35:557-581.

Forcella, F., R.G. Wilson, J. Dekker, R.J. Kremer, J. Cardina, R.L. Anderson, D. Alm, K.A. Renner, R.G. Harvey, S. Clay, and D.D. Buhler. 1997. Weed seed bank emergence across the corn belt. Weed Science 45:67-76.

Forcella, F., R.G. Wilson, K.A. Renner, J. Dekker, R.G. Harvey, D.A. Alm, D.D. Buhler and J.A. Cardina. 1992. Weed seedbanks of the U.S. corn belt: magnitude, variation, emergence, and application. Weed Science 40:636-644.

Ford, R.I. 1985. Prehistoric food production in North American. Anthropological Papers 75, Museum of Anthropology, University of Michigan, Ann Arbor, MI.

Freckleton R.P., Stephens P.A., 2009 - Predictive models of weed population dynamics. *Weed Research*, 49, 225-232.

Fukunaga, K., E. Domon, and M. Kawase. 1997. Ribosomal DNA variation in foxtail millet, *Setaria italica* (L.) P. Beauv., and a survey of variation from Europe and Asia. Theor. Appl. Genet. 95:751-756.

Fukunaga, K., Z. Wang, K. Kato, and M. Kawase. 2002. Geographical variation of nuclear genome RFLPs and genetic differentiation of foxtail millet, *Setaria italica* (L.) P. Beauv. Genet. Res. Crop Evol. 49:95-101.

Gao, M. J., and J. J. Cheng. 1988. Isozymic studies on the origin of cultivated foxtail millet. Acta Agronomica Sinica 14: 131-136

Gause, G.F. 1934. The struggle for existence. Williams and Wilkins, Baltimore, MD.

Gell-Mann, M. 1994. The quark and the jaguar – Adventures in the simple and the complex. W.H. Freeman, NY.

Genty, B., J.M. Briantais, and N.R. Baker. 1989. The relationship between quantum yield of photosynthetic electron transport and quenching of chlorophyll fluorescence. Biochimica et Biophysica Acta 990:87-92.

Gleason, H.A. and A. Cronquist. 1963. Manual of vascular plants of Northeastern United States and adjacent Canada. D. Van Nostrand Co., NY

Gleick, J. 1987. Chaos: Making a New Science, Viking, NY.

Gogol, N.V. 1842. Dead souls. Penquin Books Ltd.,1961, London, UK.

Goldberg, D.E. 1990. Components of resource competition in plant communities. In: Perspectives on plant competition; B. Grace and D. Tilman, editors. Academic Press, San Diego, CA.

Gould, F. W. 1968. Grass systematics, 2nd edition. McGraw-Hill, New York. 382 pp.

Gould, S.J. 1998. An evolutionar perspective on strengths, fallacies, and confusions in the concept of native plants. Arnoldia 58(1):3-10.

Gounaris, K. and J. Barber. 1983. Monogalactosyldiacylglycerol: The most abundant polar lipid in nature. Trends in Biochemical Sciences 8:378-381.

Graham, L.E. 1993. Origin of land plants. John Wiley and Sons, NY, NY.

Grant, V. 1971. Plant speciation. Columbia University Press, New York, NY.

Grass, Günter. 2010. The tin drum. Mariner Books, Houghton Mifflin Harcourt.

Gray, A. 1879. The predominance and pertinacity of weeds. American Journal of Science 118:161-167.

Gregg, W. 1973. Ecology of the annual grass *Setaria lutescens* on old fileds of the Pennsylvania Piedmont. Proc. Nat. Acad. Natural Sci. Philadelphia 124:135-196.

Gressel, J.B. and L.G. Holm. 1964. Chemical inhibition of crop germination by weed seeds and the nature of inhibition by *Abutilon theophrasti*. Weed Research 4(1):44-53.

Gressel, J. and L.A. Segel. 1978. The paucity of genetic adaptive resistance of plants to herbicides: possible biological reasons and implications. Journal of Theoretical Biology 75:349-371.

Grime, P. 1977. Evidence for the existence of three primary strategies in plants and its relevance to ecological and evolutionary theory. American Naturalist 111:1169-1194.

Grime, J.P. 1979. Plant strategies and vegetation processes. Wiley, Chichester, UK.

Gullberg, J. 1997. Mathematics, from the birth of numbers. W.W. Norton & Co., NY.

Gunning, B.E.S., 1977. Transfer cells and their roles in transport of solutes in plants. Sci. Prog., Oxf. 64:539-568.

Gunning, B.E.S. and J.S. Pate. 1969. "Transfer cells" Plant cells with wall ingrowths, specialized in relation to short distance transport of solutes-Their occurrence, structure, and development. Protoplasma 68:107-133.

Gutterman, Y. 1996. Environmental influences during seed maturation and storgage affecting germinability in *Spergularia diandraga* genotypes inhabiting the Negev Desert, Israel. Journal of Arid Environments 34, 313-323.

Haar, Milt. 1998. Characterization of foxtail (*Setaria* spp.) seed production and giant foxtail (*S. faberi*) seed dormancy at abscission. Ph.D. thesis, Interdepartmental Plant Physiology; Iowa State University.

Haar, M., A. van Aelst and J. Dekker. 2012. Micrographic analysis of *Setaria faberi* seed, caryopsis and embryo germination. In: Dekker, J., D. Adams, A. van Aelst, B. Dekker, J. Donnelly, and M. Haar, M. Hargrove, H. Hilhorst, C. Karssen, J. Lathrop, E. Luschei, and D. Todey. 2012. Weedy *Setaria* seed germination-dormancy behavior: Regulatory compartmentalization. Lambert Academic Publishing, Saarbrücken, Germany. ISBN: 978-3-659-24309-7

Haar M.J., A. van Aelst and J. Dekker. 2014. Weedy adaptation in *Setaria* spp.: VIII. Structure of *Setaria faberi* seed, caryopsis and embryo germination. arXiv:1403.7096 [q.bio.PE]. http://arxiv.org/abs/1403.7096

Haar, M.J. and J. Dekker. 2011. Seed production in weedy *Setaria* spp.-gp. Journal of Biodiversity and Ecological Sciences 1(3):169-178.

Haar, M.J. and J. Dekker. 2012. Seed rain dormancy structure in *Setaria faberi*. In: Dekker, J., B. Atchison, M. Haar, and K. Jovaag. 2012. Weedy *Setaria* seed life history: Heterogeneous seed rain dormancy predicates seedling recruitment. Lambert Academic Publishing, Saarbrücken, Germany. ISBN: 978-3-8454-7859-3

Haar, M.J. and J. Dekker. 2014. Weedy adaptation in *Setaria* spp.: VII. Seed germination heteroblasty in *Setaria faberi*. arXiv:1403.7072 [q.bio.PE]. http://arxiv.org/abs/1403.7072

Hacking, I. 1965. The Logic of Statistical Inference. Cambridge University Press, UK.

Hafliger, E., L. Hamet-Ahti, C.D.K. Cook, R. Faden, and F. Speta. 1982. Monocot weeds 3. Documenta CIBA-GEIGY, Basle, Switzerland.

Hafliger, E., Scholz, H. Grass weeds 1-weeds of the subfamily Panicoideae. Basle, Switzerland: Documenta Ciba-Geigy, 1980.

Hafliger, E., and H. Scholz. 1981. Grass weeds 2. Documenta CIBA-GEIGY, Basle, Switzerland.

Hakansson, 1975a. Grodan-vaxtodlingssystemmet och ograset pa akern. Lantbrukshogskolan. Ogras och ograsbekampning. 16:e svenska ograskonferensen, K 1-K 20.

Hakansson, S. 1979. Basic research in crop oroduction. II. Factors of importance for plant establishment, completion and production n crop-weed stands. Sveriges lantbruksuniversitetersitet, Inst. for vaxtodling. Rapporter och avhandlingar 72. Uppsala, Sweden.

Hakansson, S. 1988a. Growth in plant stands of different density. VIIIeme Colloque International sur la Biologie, l'Ecologie et la systematique des mauvaises herbes. EWRS symposium, Dijon, France.

Hakansson, S. 1988b. Competition in stands of short-lived plants. Density effects measured in three-component stands. Crop Production Science 3. Swedish University of Agricultural Sciences, Uppsala.

Hakansson, S. 1991. Growth and competition in plant stands. Crop Production Science 12. Swedish University of Agricultural Sciences, Uppsala.

Hakansson, S. 1997a. Competitive effects and competitiveness in annual plant stands. 1. Measurement methods and problems related to plant density. Swedish Journal of Agricultural Research 27:53-73.

Hakansson, S. 1997b. Competitive effects and competitiveness in annual plant stands. 2. Measurement of plant growth as influenced by density and relative time of emergence. Swedish Journal of Agricultural Research 27:75-94.

Hakansson, S. 2003. Weeds and weed management on arable land-An ecological approach. CABI Publishing, Cambridge, MA, USA.

Hamrick, J. L., and M. J. W. Godt. 1990. Allozyme diversity in plant species. *In* A. H. D. Brown, M. T. Clegg, A. L. Kahler, and B. S. Weir [eds.], Plant population genetics, breeding and genetic resources, 43-63. Sinauer, Sunderland, MA.

Hanf, M. 1974. Weeds and their seedlings. BASF United Kingdom Ltd., Ipswich, UK.

Hanf, M. 1983. The arable weeds of Europe. BASF Aktiengesellschaft, Ludwigshafen, Germany.

Hanks, P., T.H. Long, and L. Urdang (Eds.). 1979. Collins dictionary of the English language. Williams Collins & Sons, Ltd., Glasgow, UK.

Hardin, G. 1960. The Competitive Exclusion Principle. Science 131:1292-1297.

Harker, K. N. and J. Dekker. 1988b. Temperature effects on translocation patterns of several herbicides within quackgrass (*Agropyron repens*). Weed Science 36:545-552.

Harker, K. N. and J. Dekker. 1988a. Effects of phenology on translocation patterns of several herbicides within quackgrass (*Agropyron repens*). Weed Science 36:463-472.

Harlan, J.R. 1965. The possible role of weed races in the evolution of cultivated plants. Euphytica 14:173-176.

Harlan, J.R. 1966. Plant introduction and biosystematics. In: Plant breeding symposium. Iowa State University Press, Ames.

Harlan, J.R. 1971. Agricultural origins: centers and non-centers. Science 174(4008):465-473.

Harlan, J.R. 1982. The origins of indigenous African agriculture; In: Cambridge history of Africa, Volume 1, pp. 624-657. Cambridge University Press, UK.

Harlan, J.R. 1992. Crops and man, 2nd edition. American Society of Agronomy, Inc., Crop Science Society of America, Inc., Madison, WI.

Harlan J.R., and J.M.J. deWet, 1965. Some thoughts about weeds. Economic Botany 19:16-24.

Harlan, J.R., J.M.J. de Wet, and E.G. Price. 1973. Comparative evolution of cereals. Evolution 27:311-325.

Harper, J.L. 1964. Establishment, aggression, and cohabitation in weedy species. In: Genetics of colonizing species; H.G. Baker and G.L Stebbins (Eds.).; pp. 245-286. Academic Press, NY.

Harper, J. 1977. Population biology of plants. Academic Press, NY.

Harper, J.L. 1981. The population biology of modular organisms. In: Theoretical Ecology; R.M. May, editor. Sinauer Associates, Sunderland, MA; pp. 53-77.

Harper, J.L., J.T. Williams, and G.R. Sagar. 1965. The behavior of seeds in the soil: I. The heterogeneity of soil surfaces and its role in determining the establishment of plants. Journal of Ecology 53:272-286.

Harper, R.M. 1944. Geographic survey of Alabama: preliminary report on the weeds of Alabama. Bulletin 53. Wetumpka Printing, AL.

Harmon, M. E. and J. F. Franklin. 1989. Tree seedlings on logs in Picea-Tsuga forests of Oregon and Washington. Ecology 70: 48-59.

Hart J.J., A. Stemler. 1990a. Similar photosynthetic performance in low light-grown isonuclear triazine-resistant and -susceptible *Brassica napus* L. Plant Physiology 94:1295-1300.

Hart J.J., A. Stemler. 1990b. High light-induced reduction and low light-enhanced recovery of photon yield in triazine-resistant *Brassica napus* L. Plant Physiology 94:1301-1307.

Hartzler, R. G., Frank Forcella, and J. L. Gunsolus. 1997. Relative emergence sequence for weeds of corn and soybeans. Iowa State University, University Extension Publication.

Havaux, M. 1989. Comparison of atrazine-resistant and -susceptible biotypes of *Senecio vulgaris* L.: Effects of high and low temperatures on the *in vivo* photosynthetic electron transfer in intact leaves. Journal of Experimental Botany 40:849-854.

Havaux, M., R.J. Strasser, H. Greppin. 1991. A theoretical and experimental analysis of the $q_P$ and $q_N$ coefficients of chlorophyll fluorescence quenching and their relation to photochemical and nonphotochemical events. Photosynthesis Research 27:41-55.

Hayek, F.A. 1974. The pretence of knowledge. Nobel Prize in Economics Lecture, Sweden.

Henderson's dictionary of biological terms. 1979. 9th ed. Van Nostrand Reinhold, NY.

Hendrix, S.D. 1984. Variation in seed weight and its effects on germination in *Pastinaca sativa* L. (Umbelliferae). American Journal of Botany 71:795-802.

Heschel, M.S., J. Selby, C. Butler, G.C. Whitelam, R.A. Sharrock and K. Donahue. 2007. A new role for phytochromes in temperature-dependent germination. New Phytologist 174:735-741.

Hill, R. 2001. Dialogues of the dead. Avon Books, HarperCollins Publishers, New York, NY.

Hiltner, L. 1910. Die Prufung des Saatgutes auf Frische und Gesundheit. Jahresbericht der Vereinigung fur Angewandte Botanik 8: 219.

Hirschberg, J. and L. McIntosh. 1983. Molecular basis of herbicide resistance in *Amaranthus hybridus* L. Science 222:1346-1349.

Hitchcock, A.S. 1971. Manual of the grasses of the United States. 2nd Ed., Vol. 2. New York, NY: Dover Public.

Ho, P.-T. 1975. The cradle of the East. Univ. Chicago Press, Chicago, IL.

Hofmeister, W. 1868. Allgemeine morphologie der Gewachse. Leipzig: Handbuch der Physiologischen Botanik, Vol. 1, Pt. 2.

Hogg, R.V., J.W. McKean, and Craig, A. 2004. Introduction to Mathematical Statistics, 6th ed. Upper Saddle River, NJ; Pearson Educational.

Holland, J.H. 1995. Hidden order – How adaptation builds complexity. Helix Books, Addison-Wesley Publishing Co., Reading, MA

Holland, J.H. 1998. Emergence – From chaos to order. Perseus Books, Reading, MA.

Holliday, R. 1960a. Plant population and crop yield. Nature (London) 186:22-24.

Holliday, R. 1960b. Plant population and crop yield. Field Crop Abstracts 13:159-167.

Holliday, R. 1990. Mechanisms for the control of gene activity during development. Biol Rev Camb Philos Soc 65(4):431–71.

Holling, C.S. 1973. Resilience and stability of ecological systems. Annual Review of Ecology and Systematics Vol 4:1-23

Holm, L., J. Doll, E. Holm, J. Pancho, and J. Herberger. 1997. World weeds – Natural histories and distribution. John Wiley and Sons, NY.

Holm, L., J.V. Pancho, J.P. Herberger, and D.L. Plucknett. 1979. A geographical atlas of world weeds. Wiley-Interscience, John Wiley and Sons, NY.

Holm, L.G., D.L. Plucknett, J.V. Pancho, and J.P. Herberger. 1977. The world's worst weeds – distribution and biology. East-West Center Book, University Press of Hawaii, Honolulu.

Holmes, T.P., J.P. Prestemon, and K.L. Abt. 2008. The economics of forest disturbances: wildfires, storms and invasive species. Springer: Dordrecht, The Netherlands.

Holsinger, K.E. 2000. Reproductive systems and evolution in vascular plants. Proceedings of the National Academy of Sciences of the USA 97:7037-7042.

Holst, N., I.A. Rasmussen and L. Bastiaans L. 2007. Field weed population dynamics: a review of model approaches and applications. Weed Research 47: 1-14.

Holt, J.S. and D.P. Goffner. 1985. Altered leaf structure and function in triazine-resistant common groundsel (*Senecio vulgaris*). Plant Physiology 79:699-705.

Holt, J.S., and H.M. LeBaron. 1990. Significance and distribution of herbicide resistance. Weed Tech. 4:141-149.

Holt, J.S., A.J. Stemler and S.R. Radosevich. 1981. Differential light responses of photosynthesis by triazine-resistant and triazine- susceptible *Senecio vulgaris* biotypes. Plant Physiology 67:744-748.

Hozumi, K., K. Shinozaki, and Y. Tadaki. 1968. Studies on the frequency distribution of the weight of individual trees in a forest stand. I. A new approach toward the analysis of the distribution function and the $-^3/_2$ power distribution. Jap. Jour. Ecol. 18:10-20.

Hubbard, F.T. 1915. A taxonomic study of *Setaria* and its immediate allies. Am. J. Bot. 2:169-198.

Hubbell, S. 2001. The unified neutral theory of biodiversity and biogeography. https://pup.princeton.edu/chapters/s7105.html

Hubbell, S. P. 2005. The neutral theory of biodiversity and biogeography and Stephen Jay Gould. Paleobiology 31:122–123.

Hugly, L. Kunst, J. Browse and C. Somerville. 1989. Enhanced thermal tolerance of photosynthesis and altered chloroplast ultrastructure in a mutant of *Arabidopsis* deficient in lipid desaturation. Plant Physiology 90: 1134-1142.

Hutchison, G.E. 1957. The multivariate niche. Cold Spring Harbour Symposia in Quantitative Biology 22:415-421.

Ireland, J.S., P.S. Telfer, P.S. Covello , N.R. Baker and J. Barber. 1988. Studies on the limitations to photosynthesis in leaves of the atrazine- resistant mutant of *Senecio vulgaris* L. Planta 173:459-467.

James, A. L. 1968. Some influences of soil atmosphere on germination of annual weeds. Ph.D. dissertation. Iowa State University, Ames, Iowa, USA.

Jansen, M.A.K., J.H. Hobe, J.C. Wesselius, and J.J.S. van Rensen. 1986. Comparison of photosynthetic activity and growth performance in triazine-resistant and susceptible biotypes of *Chenopodium album*. Physiologie Vegetale 24:475-484.

Jasieniuk M, Brûlé-Babel A, Morrison I (1994) Inheritance of trifluralin resistance in green foxtail (*Setaria viridis*). Weed Sci 42:123-127

Jaynes, E.T. 1986. Bayesian methods: General background; In: Maximum-Entropy and Bayesian Methods in Applied Statistics, J. H. Justice (Ed.). Cambridge University Press, UK.

Jennnings, G. 1980. Aztec. Forge Book, T. Doherty Assoc., NY, NY.

Jensen, J.L.W.V. 1906. Sur les fonctions convexes et les inégalités entre les valeurs moyennes. Acta Mathematica 30(1):175–193.

Jones, C.S. 1999. An essay on juvenility, phase change, and heteroblasty in seed plants. International Journal of Plant Science 160, S105-S111.

Jonsdottir, I.S. and M.A. Watson. 1997. Extensive physiological integrations: An adaptive trait in resource-poor environments? In: The ecology and evolution of clonal plants; H. de Kroon and J. van Groenendael, editors. Bachhuys, Leiden, NL; pp. 109-136.

Jovaag, Kari Ann. 2006. Life history of weedy *Setaria* species-group. Ph.D. thesis; Interdepartmental Ecology & Evolutionary Biology Program and Statistics Department; Iowa State University.

Jovaag, K., J. Dekker, and B. Atchison. 2011. *Setaria faberi* seed heteroblasty blueprints seedling recruitment: II. Seed behavior in the soil. International Journal of Plant Research 1(1):1-10.

Jovaag, K., J. Dekker, and B. Atchison. 2012a. *Setaria faberi* seed heteroblasty blueprints seedling recruitment: I. Seed dormancy heterogeneity at abscission. International Journal of Plant Research 2(3): 46-56.

Jovaag, K., J. Dekker, and B. Atchison. 2012b. *Setaria faberi* seed heteroblasty blueprints seedling recruitment: III. Seedling recruitment behavior. International Journal of Plant Science 2(6):165-180.

Jursinic, P.A. and R.W. Pearcy. 1988. Determination of the rate limiting step for photosynthesis in a nearly isonuclear rapeseed (*Brassica napus* L.) biotype resistant to atrazine. Plant Physiology 88:1195-1200.

Justus, J. 2006. Ecological and Lyanupov stability.Biennial Meeting of The Philosophy of Science Associaion, Vancouver, Canada.

Justus, J. 2008. Ecological and Lyanupov stability. Philosophy of Science 75(4):421–436
Kawase, M. and S. Sakamoto. 1984. Variation, geographical distribution and genetical analysis of esterase isozymes in foxtail millet, *Setaria italica* (L.) Beauv. Theoretical and applied genetics 67:529-533.

Jusuf, M. and J. Pernes. 1985. Genetic variability of foxtail millet (*Setaria italica* Beauv.) electrophoretic studies of five isozyme systems. Theor. Appl. Genet. 71:385-391.

Kant, I. 1965. Critique of Pure Reason (1781); N.K. Smith (Translator). St. Martins, NY.

Kaufman, S. 1995. At home in the universe: the search for laws of self-organization and complexity. Oxford University Press, UK.

Kawano, S. and S. Miyake. 1983. The productive and reproductive biology of flowering plants. X. Reproductive energy allocation and propagule output of five congeners of the genus *Setaria* (Gramineae). Oecologia 57:6-13.

Kawase, K., and S. Sakamoto. 1984. Variation, geographical distribution and genetical analysis of esterase isozymes in foxtail millet, *Setaria italica* (L.) P. Beauv. Theoretical and Applied Genetics 67: 529-533.

Kawase, M. and S. Sakamoto. 1987. Geographical distribution of landrace groups classified by intraspecific hybrid pollen sterility in foxtail millet, *Setaria italica* (L.) Beauv. Japan Journal of Breeding 37:1-9.

Kays, S. and J.L. Harper. 1974. The regulation of plant and tiller density in a grass sward. Journal of Ecology 50:97-105.

Kegode, G.O., and R.B. Pearce. 1998. Influence of environment during maternal plant growth on dormancy of shattercane (*Sorghum bicolor*) and giant foxtail (*Setaria faberi*) seed. Weed Science 46(3): 322-329.

Kegode, G.O., R.B. Pearce, and T.B. Bailey. 1998. Influence of fluctuating temperatures on emergence of shattercane (*Sorghum bicolor*) and giant foxtail (*Setaria faberi*). Weed Science 46(3):330-335.

Kelly, K. 1994. Out of control: the new biology of machines. Fourth Estate, London.

Khosla, P.K. and M.L. Sharma. 1973. Cytological observations on some species of *Setaria*. The Nucleus 26:38-41.

Khosla, P.K. and S. Singh. 1971. Cytotaxonomical investigations on *S. glauca* Beauv. complex. Abst., 1st All India Cong. Cytol. Genet. 15.

Kimura, M. 1968. Evolutionary rate at the molecular level. Nature 217(5129): 624–626.

Kimura, M. 1983. The neutral theory of molecular evolution. Cambridge University Press, Cambridge, UK.

King, L. J. 1952. Germination and control of the *S. faberi* grass. Contr Boyce Thompson Inst 16:469-487.

King, R.C. and W.D. Stansfield. 1985. A dictionary of genetics; 3rd edition. Oxford University Press, New York, NY.

Kira, T., H. Ogawa, and N. Sakazaki. 1953. Intra-specific competition among higher plants. I. Competition-yield-density interrelationship in regularly dispersed populations. Osaka City University Journal of the Institute of Polytechnics 4:1-16.

Kirk, R.E. 1982. Experimental design procedures for the behavioral sciences. Brooks/Cole Publishing, Montery, CA.

Kishimoto, E. 1938. Chromosomenzahlen in den Gattungen *Panicum* und *Setaria*, part I. Chromosomenzahlen einier *Setaria*-Arten. Cytologia 9:23-27

Kivilaan, A. and R. S. Bandurski. 1973. The ninety year period for Dr. Beal's seed viability experiment. Am. J. Bot. 60:140-145.

Kivilaan, A. and R. S. Bandurski. 1981. The one hundred-year period for Dr. Beal's seed viability experiment. American Journal of Botany 68:1290-1292.

Kollman, G. E. 1970. Germination-dormancy and promoter-inhibitor relationships in *Setaria lutescens* seeds. Ph.D. dissertation. Iowa State Univ. Ames, IA.

Koukkari, W.L. and R.B. Sothern. 2007. Introducing biological rhythms: A primer on the temporal organization of life, with implications for health, society, reproduction, and the natural environment. Springer Science & Business Media, USA.

Kuhn, T.S. 1996. The structure of scientific revolutions, 3rd edition. The University of Chicago Press, IL

Kunst, L., J. Browse, and C. Somerville. 1989. Altered chloroplast structure and function in a mutant of *Arabidopsis* deficient in plastid glycerol-3- phosphate acyltransferase activity. Plant Physiology 90: 846-853.

Kuuster, H. "Neolithic plant remains from Eberdingenhochdorf, southern Germany". In *Plants and ancient man (Studies in paleoethnobotany)*, W.V. Van Zeist and W.A. Caparoe, eds.. Rotterdam: A.A. Bakame, Rotterdam; 1984.

Lang, A.L., J.W. Pendleton and G.H. Dungan. 1956. Influence of population and nitrogen levels on yield and protein and oil contents of nine corn hybrids. Agronomy Journal 48:284-289.

LeBaron, H.M., and J. Gressel. 1982. Herbicide resistance in plants. John Wiley & Sons, Inc., New York.

Lee, S.M. 1979. The distribution and abundance of three species of *Setaria* Beauv. around London, Canada, with particular reference to the effects of shade. M.Sc. Thesis, Univ. of Western Ontario, London, Ont..

Lemoine, Y., J.P. Dubacq, G. Zabulon and J.M. Ducruet. 1986. Organization of the photosynthetic apparatus from triazine-resistant and -susceptible biotypes of several plant species. Canadian Journal of Botany 64:2999-30007.

Leon, R.G. and M.D.K. Owen. 2003. Regulation of weed seed dormancy through light and temperature interactions. Weed Science 51(5):752-758.

LePage-Degivry, M-.T., P. Barthe, and G. Garello. 1990. Involvement of endogenous abscissic acid in onset and release of *Helianthus annuus* embryo dormancy. Plant Physiology 92: 1164-1168.

Levin, L. 1973. On the notion of a random sequence. Soviet Mathematics Doklady 14:1413–1416.

Lewontin, R.C. 1965. Selection for colonizing ability. In: Genetics of colonizing species, H.G. Baker and G.L. Stebbins, editors. Academic Press, New York, NY.

Lewontin, R.C. 1974. The analysis of variance and the analysis of causes. American Journal of Human Genetics 26:400-411.

Li, P. and T.P. Brutnell. 2011. *Setaria viridis* and *Setaria italica*, model genetic systems for the Panicoid grasses. Journal of Experimental Botany 62(9):3031-3037.

Li, C.H., W.K. Pao, and H.W. Li. 1942. Interspecific crosses in *Setaria*. J. Heredity 33:351-355.

Li, H.W. 1934. Studies in millet breeding methods. Bul. Col. Agr., Honan Univ. 2:1-22 (in Chinese).

Li, H.W., C.H. Li, and W.K. Pao. 1945. Cytological and genetical studies of the interspecific cross of the cultivated foxtail millet, *Setaria italica* (L.) Beauv. and the green foxtail millet *S. viridis* L. J. Amer. Soc. Agron. 37:32-54

Li, H.W., Meng, C.J., Liu, T.N. 1935. Problems in the breeding of millet (*Setaria italica* (L.) Beauv.). J. Amer. Soc. Agron. 27:963-970.

Li, Y., and S.Z. Wu. 1996. Traditional maintenance and multiplication of foxtail millet (*Setaria italica* (L.) P. Beauv.) landraces in China. Euphytica 87:33-38.

Lincoln, R., G. Boxshall, and P. Clark. 1998. A dictionary of ecology, evolution and systematic, 2nd Ed. Cambridge University Press, UK.

Lindahl B.D., Ihrmark K., Boberg J., Trumbore S.E., Högberg P., Stenlid J., Finlay R.D. 2007. Spatial separation of litter decomposition and mycorrhizal nitrogen uptake in a boreal forest. New Phytologist 173(3):611–620.

Little, T.M. and F.J. Hills. Agricultural experimentation. John Wiley & Sons, NY.

Loewenstein, W. 1999. The touchstone of life: molecular information, cell communication, and the foundations of life. Oxford University Press, New York.

Lorenzi, H. J., and L. S. Jeffery. 1987. Weeds of the United States and their control, p.78-80. Van Nostrand Reinhold Co., New York.

Luschei, E.C., D.D. Buhler, and J. Dekker. 1998. Effect of separating giant foxtail (*Setaria faberi*) seeds from soil using potassium carbonate and centrifugation on viability and germination. Weed Science 46:545-548.

Lyanpunov, A.M. 1966. Stability of motion, Academic Press, New-York, London, UK.

MacArthur, R.H. 1955. Fluctuations of animal populations and a measure of community stability. Ecology 36:533-536.

MacArthur, R.H. and E.O. Wilson. 1967. The theory of island biogeography. Princeton University Press, NJ.

MacNeish, R.S. 1992. The origins of agriculture and settled life. University of Oklahoma Press, Norman, OK.

MacVicar, R.M. and H.R. Parnell. 1941. The inheritance of plant colour and the extent of natural crossing in foxtail millet. Scient. Agric. 22:80-84.

Malthus, T.R. 1798. An essay on the principle of population. Oxford University Press, Oxford World Classics edition (2008), UK.

Mandelbrot, B. 1997. Fractals and scaling in finance: discontinuity, concentration, risk. Springer.

Mandelbrot, B. and N. Taleb. 2006. A focus on the exceptions that prove the rule, Financial Times, March 23.

Mangelsdorf, 1964. Origins of agriculture in Middle America; In: Natural environments and early cultures, Vol. 1, Handbook of Middle American Indians, R.C. West (Ed.); University of Texas Prss, Austin, TX.

Manthey, D. R. and J. D. Nalewaja. 1982. Moisture stress effects on foxtail seed germination. Proc. North Central Weed Control Conf. 37:52-53.

Manthey, D.R. and J.D. Nalawaja. 1987. Germination of two foxtail (*Setaria*) species. Weed Technology 1:302-304.

Mapplebeck, L.R., V. Sousa Machado, and B. Grodzinski. 1982. Seed germination and seedling growth characteristics of atrazine-susceptible and resistant biotypes of *Brassica campestris*. Can. J. Plant Sci. 62:733-739.

Martin, J. N. 1943. Germination studies of the seeds of some common weeds. Proc. Iowa Acad. Sci.; 50:221-228.

Martin-Löf, P. 1966. The definition of random sequences. Information and Control 9(6): 602–619.

Mathews, S. and R.A. Sharrock. 1997. Phytochrome gene diversity. Plant, Cell and Environment 20:666-671.

Mattoo, A.K., J.P. St. John and W.P. Wergin. 1984. Adaptive reorganization of protein and lipid components in chloroplast membranes as associated with herbicide binding. Journal of Cell Biochemistry 24, 163-175.

May, R.M. 1973. Stability and complexity in model ecosystems. Princeton Univ. Press, NJ

Maynard Smith, J. 1978. The evolution of sex. Cambridge University Press, UK.

Mayr, E. 1959. Darwin and the evolutionary theory in biology. In: Evolution and anthropology: a centennial appraisal, pp. 1-10. Washington, D.C.: Anthropological Society of American:

Mayr, E., 2001. What evolution is. Basic Books, New York, 318 p.

Mazen, A.M.A. and J. Dekker. 1990. Diurnal oscillations in ribulose bisphosphate activity in triazine resistant and susceptible *Brassica napus*. Weed Science Society of America Abstracts 30:163.

McClosky, W.B. and J.S. Holt. 1990. Triazine resistance in *Senecio vulgaris* parental and nearly isonuclear backcrossed biotypes is correlated with reduced productivity. Plant Physiology 92:954-962.

McGill, B. J. 2003. A test of the unified neutral theory of biodiversity. Nature 422(6934):881.

Merckx V., Bidartondo M.I., Hynson N.A. 2009. Myco-heterotrophy: when fungi host plants. Annals of Botany in press (7):1255–1261.

Mester, T. C. and D. D. Buhler. 1986. Effects of tillage on the depth of giant foxtail germination and poplation densities. Proc. North Cent. Weed Control Conf. 41:4-5.

Millennium Ecosystem Assessment (MA). 2005. Ecosystems and Human Well-Being: Synthesis. Island Press, Washington.

Mitchell, B. 1980. The control of germination ability in developing wheat grains. Ph.D. dissertation, University of London, London.

Mithen, S. 2003. After the ice, a global human history 20,000-5000 B.C. Harvard University Press, Cambridge, MA.

Mohler, C.L. 2001. Weed evolution and community structure. In: Ecological management of agricultural weeds, M. Liebman, C.L. Mohler, and C.P. Staver, Eds. Cambridge University Press, UK.

Moore, D.J. and O.H. Fletchall. 1963. Germination-regulating mechanisms of giant foxtail (*Setaria faberii*). Res Bull. Missouri Ag. Exp. Sta., No. 829.

Muenscher, W.C. 1955. Weeds, 2nd Edition. Cornell University Press, Ithaca, NY.

Mulligan, G.A. and J.N. Findlay. 1970. Reproductive systems and colonization in Canadian weeds. Can. J. Bot. 48:859-860.

Murdoch, A.J. 2010. Dormancy and germination of *Chenopodium album* seeds from different latitudes in Europe and North America. 15th EWRS symposium, Kaposvár, Hungary; Proceedings of the European Weed Research Society.

Murray, J.A.H, H. Bradley, W.A. Cragie, and C.T. Onions (Eds.). 1961. Oxford English dictionary. The Clarendon Press, Oxford.

Narayanaswami, S. 1956. Structure and development of the caryopsis in some Indian Millets. VI. *Setaria italica*. Botanical Gazette 118:112-122.

Neuweiler, E. 1946. Nachtrage urgeschichtlicher Pflanzen. Vierteljahrsschr. Naturf. Ges. Zurich 91:122-236.

Nguyen Van, F. and J. Pernes. 1985. Genetic diversity of foxtail millet (*Setaria italica*). In: Genetic differentiation and dispersal in plants, P. Jacquard (Ed.), NATO ASI Series, Vol. G5, pp. 113-128, Springer-Verlag, Berlin.

Nicolis, G. 1995. Introduction to nonlinear science. Cambridge University Press, UK.

Nieto-Hatem, J. 1963. Seed dormancy in *Setaria lutescens* Ph.D. dissertation. Iowa State Univ. Ames.

Norris, R. F., C. A. Schoner, Jr. 1980. Yellow foxtail (*Setaria lutescens*) biotype studies: dormancy and germination. Weed Science. 28:159-163.

Nurse, R.E. and A. DiTommaso. 2005. Corn competition alters the germinability of velvetleaf (*Abutilon theophrasti*) seeds. Weed Science 53:479-488.

O'Brien, T.P. 1976. Transfer cells. In: Transport and transfer processes in plants; Eds. I. F. Wardlaw and J. B. Passioura. Academic Press, Inc. New York. Pp. 59-63.

Op den Kamp, J.A.F., B. Roelofsen and L.L.M. van Deenen. 1985. Structural and dynamic aspects of phosphatidylcholine in the human erythrocyte membrane. Trends in Biological Sciences 10:320-323.

Orgel, L.E. and F.H.C. Crick. 1980. Selfish DNA: the ultimate parasite. Nature 284:604-607.

Ort, D.R., W.H. Ahrens, B. Martin and E. Stoller. 1983. Comparison of photosynthetic performance in triazine-resistant and susceptible biotypes of *Amaranthus hybridus*. Plant Physiology 72:925-930.

Oxford English Dictionary. 1989. Oxford University Press, UK.

Osada, T. 1989. Illustrated grasses of Japan. Heibonsha Ltd. Publishers, Tokyo.

Pacala, S.W. and J.A. Silander. 1990. Field tests of neighbourhood population dynamics models of two annual weed species. Ecological Monographs 60:113-134.

Pareja, M. R., and D. W. Staniforth. 1985. Seed-soil microsite characteristics in relation to weed seed germination. Weed Science 33: 190-195

Pareja, M. R., D. W. Staniforth, and G. P. Pareja. 1985. Distribution of weed seed among soil structural units. Weed Science 33: 182-189.

Paine, R.T. 1966. Food web complexity and species diversity. American Naturalist 110:65-75.

Paine, R.T. 1969. A note on trophic complexity and community stability. American Naturalist 103:91-93.

Pate, J.S. and B.E.S.Gunning. 1972. Transfer cells. Ann. Rev. Plant Physiol. 23:173-196.

Patterson, D.T. 1985. Comparative ecophysiology of weeds and crops. In: Weed physiology, Volume 1: Reproduction and ecophysiology, S.O. Duke, editor; p.101-129. CRC Press, Boca Raton, FL.

Pearl, R. and L.J. Reed. 1920. On the rate of growth of the population of the United States since 1790 and its mathematical representation. Proceedings of the National Academy of Science, U.S.A. 6:275-288.

Peters, R.A., Meade, J.A., Santelmann, P.W. 1963. Life history studies as related to weed control in the northeast. 2. Yellow foxtail and giant foxtail. Agric. Exp. Sta., Pp. 18, Univ. Rhode Island, Kingston.

Peters, R.A., Yokum, H.C. 1961. Progress report on a study of the germination and growth of yellow toxtail (*Setaria glauca* (L.) Beauv.) Proc. Northeast Weed Control Conference 15:350-355.

Peterson, G., Allen, C. R., Holling, C.S. 1998. Ecological resilience, biodiversity, and scale. Ecosystems 1(1):6-18.

Pfister, K., K. Steinback, G. Gardner, and C. Arntzen. 1981. Photoaffinity labeling of a herbicide receptor protein in chloroplast membranes. Proceedings of the National Academy of Science USA 78: 981-985.

Pflanze, O. 1990. Bismarck and the development of Germany; volume 1 (of 3). Princeton University Press, NJ.

Philippi, T. and J. Seger. 1989. Hedging one's bets, revisited. Trends Ecology and Evolution 4: 41-44.

Pianka, E.R. 1970. On r- and K-selection. American Naturalist 104:592-597.

Pielou, E.C. 1991. After the Ice Age: The Return of Life to Glaciated North America. University of Chicago Press, IL.

Pillai P., J.B. St. John. 1981. Lipid composition of chlorplast membranes from weed biotypes differentially sensitive to triazine herbicides. Plant Physiology 68: 585-587.

Pinker, S. 1994. The language instinct. HarperCollins, NY.

Pinker, S. 1997. How the mind works. Penquin Books, London, UK.

Pinker, S. 1999. Words and rules: The ingredients of language. HarperCollins, NY.

Pinker, S. 2002. The blank slate: The modern denial of human nature. Penguin Books, London, UK.

Pohjonen, V. and J. Paatela. 1974. Effect of planting interval and seed tuber size on the gross and net potato yield. Acta Agriculturae Scandinavica 24:126-130.

Pohl, R.W. 1951. The genus *Setaria* in Iowa. IA St. Jour. Sci. 25:501-508.

Pohl, R.W. 1966. The grasses of Iowa. Iowa St. Jour. Sci. 40:341-373.

Pohl, R.W. 1978. How to know the grasses, 3rd Edit., Wm. C. Brown Co. Publishers, Dubuque, IA. 200 pp.

Poirier-Hamon, S. and J. Pernes. 1986. Instabilite chromosomique dans les tissus somatiques des descendants d'un hybride interspecifique *Setaria verticillata* (P. Beauv.), *Setaria italica* (P. Beauv.). R. Acad. Sci. Paris 302:319-324.

Poisson, S.D. 1837. Recherches sur la probabilité des jugements en matière criminelle et en matière civile, précédées des règles générales du calcul des probabilitiés (Research on the Probability of Judgments in Criminal and Civil Matters). Bachelier, Paris, France.

Popovsky, M. 1984. The Vavilov affair. Archion Books, The Shoe String Press, Hamden, CN.

Popper, K. 1959. The propensity interpretation of probability. British Journal for the Philosophy of Science 10:25-42.

Popper, K. 2009. The logic of scientific discovery. Routledge Classics, Oxon, UK.

Povilaitis, B. 1956. Dormancy studies with seeds of various weed species. Proceedings Intl. Seed Testing Assoc.; 21:87-111.

Prasada Rao, K.E., J.M.J. de Wet, D.E. Brink, and M.H. Mengesha. 1987. Intraspecific variation and systematics of cultivated *Setaria italica*, foxtail millet (*Poaceae*). Econ. Bot. 41:108-116.

Preston, F.W. 1962. The canonical distribution of commonness and rarity: Part I. Ecology 43:185-215, 431-432.

Pritchard, T. 1960. Race formation in weedy species with special reference to *Euphorbia cyparissias* L. and *Hypericum perforatum* L.; In: The biology of weeds, J.L. Harper (Ed.), p. 61-66; Blackwell Scientific Publications, Oxford, UK.

Putnam, A.R. and W.B. Duke. 1978. Allelopathy in agroecosystems. Ann. Rev. Phytopathol. 16:431-451.

Putnam, R.1994. Community ecology. Chapman and Hall, London.

Radosevich, S.R. 1970. Mechanism of atrazine resistance in lambsquarters and pigweed. Weed Science 25:316-318.

Rangaswami Ayyangar, G.N., Narayanan, T.R., Seshadri Sarma, P. 1933. Studies in *Setaria italica* (Beauv.), the Italian millet. Part 1. Anthesis and pollination. Indian Jour. Agr. Sci 3:561-571.

Raunkiær, C. 1904. Om biologiske Typer, med Hensyn til Planternes Tilpasninger til at overleve ugunstige Aarstider. Botanisk Tidsskrift 26, p. XIV. Also as Ch. 1: Biological types with reference to the adaption of plants to survive the unfavourable season, in: The Life Forms of Plants and Statistical Plant Geography. Oxford University Press, Oxford; p. 1.

Reed, C.F. Selected weeds of the United States. U.S. Dept. Agric. Agric. Handbook 366, 1970.

Rees, M. 1993. Delayed germination of seeds: a look at the effects of adult longevity, the timing of reproduction, and population age/stage structure. American Naturalist 144:43-64.

Rees, M. 1994. Delayed germination of seeds: a look at the effects of adult longevity, the timing of reproduction, and population age/stage structure. American Naturalist 144:43-64.

Remy, W., Taylor T.N., Hass H., Kerp H. 1994). 4-hundred year old vesicular-arbuscular mycorrhizae. Proceedings of the National Academy of Sciences USA 91(25):11841–11843.

Renfrew, J. 1973. Paleoethnobotany: The prehistoric food plants of the Near East and Europe. Columbia University Press, New York, NY.

Richter, O., U. Bottcher, and P. Zwerger. 2000. Modelling of the spread of herbicide resistant weeds by the linking of population genetic models with cellular automaton models. Zeitschrift fur Pflanzenkrankheiten und Pflanzenschultz, Sonderheft 17:329-336.

Ricroch, A., Mousseau, M., Darmency, H., Pernes, J. 1987. Comparison of triazine-resistant and -susceptible cultivated *Setaria italica*: growth and photosynthetic capacity. Plant Physiol Biochem 25:29-34.

Ridley, H.N. 1930. The dispersal of plants throughout the world. L. Reeve and Co., Pp. 744, Ashford, Kent, U.K.

Ridley, M. 1995. The Red Queen: Sex and the evolution of human nature. Penguin Books, UK.

Ridley, M. 2011. The rational optimist – How prosperity evolves. Harper Perennial, NY.

Ritter, R.L., L.M. Kaufman, T.J. Monaco, and W.P. Novitsky, and D.E. Moreland. 1989. Characterization of triazine-resistant giant foxtail (*Setaria faberi*) and its control in no-tillage corn (*Zea mays*). Weed Sci. 37:591-595.

Roberts, H.A. and P.M. Feast. 1973. Emergence and longevity of seeds of annual weeds in cultivated and undisturbed soil. Journal of Applied Ecology 10:133-143.

Rohlf, M. 2010. "Immanuel Kant"; The Stanford Encyclopedia of Philosophy, E.N. Zalta (ed.).

Rominger, J.M. 1962. Taxonomy of *Setaria* (Gramineae) in North America; Volume 29. Illinois biological monographs, University of Illinois Press, Urbana.

Root, R.B. 1967. The niche exploitation pattern of the blue-gray Gnatcatcher. Ecological Monographs 37:317-350.

Ross, M.A. 1968. The establishment of seedlings and the development of patterns in grassland. Ph.D. thesis, University of Wales, UK.

Rost, T.L. 1971a. Fine structure of endosperm protein bodies in *Setaria lutescens* (*Gramineae*). Protoplasma 73:475-479.

Rost, T.L. 1971b. Structural and histochemical investigations of dormant and non-dormant caryopses of *Setaria lutescens* (*Gramineae*). Ph.D. Thesis, Iowa State University, Ames.

Rost, T.L. 1972. The ultrastructure and physiology of protein bodies and lipids from hydrated dormant and nondormant embryos of *Setaria lutescens* (*Gramineae*). Amer. J. Bot. 59:607-616.

Rost, T.L. 1973. The anatomy of the caryopsis coat in mature caryopses of the yellow foxtail grass (*Setaria lutescens*). Bot. Gaz. 134:32-39.

Rost, T.L. 1975. The morphology of germination in *Setaria lutescens* (Gramineae): The effects of covering structures and chemical inhibitors on dormant and non-dormant florets. Ann. Bot. 39:21-30.

Rost, T.L. and N.R. Lersten. 1970. Transfer aleurone cells in *Setaria lutescens* (*Gramineae*). Protoplasma 71:403-408.

Rost, T.L. and N.R. Lersten. 1973. A synopsis and selected bibliography of grass caryopsis anatomy and fine structure. IA State J. Res. 48:47-87

Rousseau, C., Cinq-Mars, L. Les plantes introduites du Quebec. Jeune Sci 1969; 7:163-222.

Ryan, G.F. 1970. Resistance of common groundsel to simazine and atrazine. Weed Science 18:614-616.

Sacks, Oliver. 1998. The man who mistook his wife for a hat, and other clinical tales. P. 129. Touchstone, NY.

Sagar, G.R. and A.M. Mortimer. 1976. An approach to the study of the population dynamics of plants with special reference to weeds. Applied Biology 1:1-47.

Sage, R.F. and T.D. Sharkey. 1987. The effect of temperature on the occurrence of $O_2$ and $CO_2$ insensitive photosynthesis in field grown plants. Plant Physiology 84:658-664.

Salisbury, E.J. 1942. Reproductive capacity of plants. Bell, London, UK.

Samson, D.A. and K.S.Werk. 1986. Size-dependent effects in the analysis of reproductive effort in plants. American Naturalist 127:667-680

Santleman, P.W., Meade, J.A., Peters, R.A. 1963. Growth and development of yellow foxtail and giant foxtail. Weeds 11:139-142.

Sareini, H T. 2002. Identification and characterization of the major seed peroxidase in *Setaria faberii*. M.S. thesis, Molecular Biology and Biochemistry Department, Iowa State University, Ames.

Savage, L.J. 1954. The foundations of statistics. John Wiley & Sons, Inc., NY.

Scheiner, S.M. 1993. Genetics and evolution of phenotypic plasticity. Annu. Rev. Ecol. Syst. 24:35-68.

Schimper, A.F.W. and W.R. Fisher. 1903. Plant-geography upon a physiological basis. Vol. 2. Clarendon Press,.

Schlapfer, F. 1999. Expert estimates about effects of biodiversity on ecosystem processes and services. Oikos 84:346-352.

Schnorr, C. P. 1971. A unified approach to the definition of a random sequence. Mathematical Systems Theory 5(3):246–258.

Schnorr, C. P. 1973. Process complexity and effective random tests". Journal of Computer and System Sciences 7(4):376–388.

Schonfeld, M., T. Yaacoby, O. Michael and B. Rubin. 1987. Triazine resistance without reduced vigor in *Phalaris paradoxa*. Plant Physiology 83:329-333.

Schopenhauer, A. 1969 (1844). The world as will and representation; vol. 1-2. Dover Publications, NY.

Schreiber, M. M. 1965a. Development of giant foxtail under several temperatures and photoperiods. Weeds 13:40-43.

Schreiber, M.M. 1965b. Effect of date of planting and stage of cutting on seed production of giant foxtail. Weeds 13:60-62.

Schreiber, M. M. 1977. Longevity for foxtail taxa in undisturbed sites. Weed Science. 25:66-72.

Schreiber, M.M. and L.R. Oliver. 1971. Two new varieties of *Setaria viridis*. Weed Sci. 19:424-427.

Schroeder, M. 1991. Fractals, chaos, power laws – Minutes from and infinite paradise. Dover, Mineola, NY.

Schulz B., Boyle C. 2005. The endophytic continuum. Mycological Research 109(Pt 6):661–86.

Schulze, E.D. and H.A. Mooney. 1993. Ecosystem function of biodiversity: a summary. In: Schulze, E.D. and Mooney, H.A. (Editors), Biodiversity and ecosystem function. Springer-Verlag, Berlin.

Schutte, B.J., B.J. Tomasek, A.S. Davis, L. Andersson, D.L. Benoit, A. Cirujeda, J. Dekker, F. Forcella, J.L. Gonzalez-Andujar, F. Graziana, A.J. Murdock, P. Neve, I.A. Rasmussen, B. Sera, J. Salonen, F. Tei, K.S. Torresen, J.M. Urbano. 2014. An investigation to enhance understanding of the stimulation of weed seedling emergence by soil disturbance. Weed Research 54(16):1-12.

Seger, J. and H.J. Brockman. 1987. What is bet hedging? Oxford Surveys in Evolutionary Biology 4:182-211.

Sells, G.D. 1965. $CO_2$-$O_2$ ratios in relation to weed seed germination. Ph.D. Thesis, Iowa State University, Ames.

Selosse M.A., Richard F., He X., Simard S.W. 2006. Mycorrhizal networks: des liaisons dangereuses?. Trends in Ecology and Evolution 21(11):621–628.

Sexton, R., and J.A. Roberts. 1982. Cell biology of abscission. Annual Review of Plant Physiology 33: 133-162.

Shakespeare, W. 1623. As you like it. Act II, Scene VII; Mr. William Shakespeares Comedies, Histories, & Tragedies. Stationers Company, E. Blount, W. and I. Jaggard, Publishers.

Shannon, C.E. and W. Weaver. 1949. The mathematical theory of communication. P.31. University of Illinois Press, Urbana.

Shannon, C.E. and W. Weaver. 1963. The mathematical theory of communication. University of Illinois Press, Chicago, IL.

Sharkey, T.D. 1990. Feedback limitation of photosynthesis and the physiological role of ribulose bisphosphate carboxylase carbamylation. Botanical Magazine-Tokyo Special Issue 2:87-105.

Shinomura, T. 1997. Phytochrome regulation of seed germination. Journal of Plant Research 110(1):151-161.

Shinozaki, K. and T. Kira. 1956. Intraspecific competition among higher plants. VII. Logistic theory of the C-D effect. J. Inst. Polytech. Osaka City Univ. D7:35-72.

Shinozaki, K. and T. Kira. 1961. The C-D rule, its theory and practical uses. J. Biol. Osaka City Univ. 12:69-82.

Silvertown, J. 1984. Phenotypic variety in seed germination behaviour: the ontogeny and evolution of somatic polymorphism in seeds. American Naturalist 124:1-16.

Silvertown, J. and D. Charlesworth. 2001. Introduction to plant population biology; 4th edition. Blackwell Science, Iowa State Press, Ames.

Silvertown, J. and M. Dodd. 1997. Comparing plants and connecting traits. In: Plant life histories: ecology, phylogeny and evolution; J. Silvertown, M. Franco and J.L. Harper, editors. Cambridge University Press, UK; pp. 3-16.

Silvertown, J.W. and J.L. Doust. 1993. Introduction to plant population biology. Blackwell Science, Oxford, UK.

Simberloff, D. and B. Von Holle. 1999. Positive interactions of nonindigenous species: invasion meltdown? Biological Invasions 1:21-32.

Simpson, G. M. 1990. Seed dormancy in grasses. Cambridge University Press, Cambridge.

Slife, F.W. 1954. A new *Setaria* species in Illinois. Proc. North Cent. Weed Cont. Conf. 11:6-7.

Smith, C.C. and S.D. Fretwell. 1974. The optimal balance between the size and number of offspring. American Naturalist 108:499-506.

Smith, C.E. Jr. 1968. The New World centers of origin of cultivated plants and the archaeological evidence. Econ. Bot. 22:253-266.

Smith, Harry. 1986. The perception of light quality. In: Photomorphogenesis in plants; R.E. Kendrick and G.H.M. Kronenberg (eds.); pp. 187-217; Martinus Nijhoff Publishers, Dordrecht, NL

Snyder, Laura J. 2011. The philosophical breakfast club: Four remarkable friends who transformed science and changed the world. Broadway Paperbacks, NY.

Solbrig, O.T., R. Sarandon, and W. Bossert. 1988. A density-dependent growth model of a perennial herb, *Viola fimbriatula*. American Naturalist 131:385-400.

Solomonoff, R. 1964a. A Formal Theory of Inductive Inference, Part I". Information and Control 7(1):1-22.

Solomonoff, R. 1964b. A Formal Theory of Inductive Inference, Part II. Information and Control 7(2):224-254.

Solomonoff, R. 1997. The Discovery of Algorithmic Probability. Journal of Computer and System Sciences 55(1):73-88.

Stace, C.A. (Ed.) 1975. Hybridization and the flora of the British Isles. Pp. 626, Academic Press, London.

Stachowicz, J.J. 2001. Mutualism, facilitation, and the structure of ecological communities. BioScience 51:235-246.

Stapf, O. and C.K. Hubbard. 1930. Setaria. In: Flora of tropical Africa 9:768-866, Prain (Ed.), London.

Stanway, V. 1971. Laboratory germination of giant foxtail, *Setaria faberii* Herrm., at different stages of germination. Proceedings Assoc. Official Seed Analysts; 61:85-90.

Stebbins, G.L. 1958. Longevity, Habitat, and Release of Genetic Variability in the Higher Plants. Cold Spring Harbor Symp. Quant. Biol. 23:365-378.

Steel, M.G., P.B. Cavers, S.M. Lee, S.M. 1983. The biology of Canadian weeds. 59. *Setaria glauca* (L.) Beauv. and *S. verticillata* (L.) Beauv. Canadian Journal of Plant Science 63:711-725.

Steel, R.G.D. and J.H. Torrie. 1960. Principles and procedures of statistics. McGraw-Hill, NY.

Steel, R.G.D. and J.H. Torrie. 1980. Principles and procedures of statistics, 2$^{nd}$ edition. McGraw-Hill, NY.

Steinback, K. K.L. McIntosh, L. Bogorad, and C. Arntzen. 1981. Identification of the triazine receptor protein as a chloroplast gene product. Proceedings of the National Academy of Science USA 78: 7463-7463.

Stevens, O.A. 1960. Weed development notes. Res. Rep. 1, Pp.22, North Dakota Agric. Exp. Sta.

Stoller, E. W. and L. M. Wax. 1974. Dormancy changes fate of some annual weed seeds in the soil. Weed Science. 22:154155.

Stoltenberg, D.E., and R.J. Wiederholt. 1995. Giant foxtail (*Setaria faberi*) resistance to arlyoxyphenoxypropionate and cyclohexanedione herbicides. Weed Sci. 43:527-535.

Stowe, A.E., J.S. Holt. 1988. Comparison of triazine-resistant and - susceptible biotypes of *Senecio vulgaris* and their F1 hybrids. Plant Physiology 87: 183-189.

Stoyan, D., and H. Stoyan. 1994. Fractals, random shapes and point fields. John Wiley & Sons, UK.

Stuart, A. and K. Ord. 2009. Kendall's Advanced Theory of Statistics, Volume 1: Distribution Theory, 6th Ed. Edward Arnold, UK.

Stubbendieck, J., G.Y. Frissoe, and M.R. Bolick. 1995. Weeds of Nebraska and the Great Plains, 2$^{nd}$ edition. Nebraska Department of Agriculture, Lincoln, NE.

Stucky, J.M., T.J. Monaco, and A.D. Worsham. 1980. Identifying seedling and mature common weeds of the southeastern United States. University of North Carolina, Raleigh.

Symons, S.J., R.E. Angold, M. Black, and J.M. Chapman. 1983. Changes in the growth capacity of the developing wheat embryo. I. The influences of the enveloping tissues and premature drying. Journal of Experimental Botany 34: 1541-1550.

Taiz, L. and E. Zeiger. 2002. Plant Physiology. 3rd edition; Sinauer Associates.

Takahashi, N. and T. Hoshino. 1934. Natural crossing in *Setaria italica* (Beauv.). Proc. Crop Sci. Soc., Japan 6:3-19.

Taleb, N.N. 2010. The bed of Procrustes: Philosophical and practical aphorisms. Random House, USA.

Taleb, N.N. 2010. The Black Swan, The impact of the highly improbable, 2$^{nd}$ edition. Random House Trade Paperbacks, NY.

Taleb, N.N. 2012. Antifragile: things that gain from disorder. Random House, NY.

Tansley, A.G. 1935. The use and abuse of vegetational concepts and terms. Ecology 16:284-307.

Tegmark, M. 2009. Parallel universes. Scientific American special report SCA45026.

Taylorson, R. B. 1966. Control of seed production in three annual grasses by dimethylarsinic acid. Weeds 14: 207-210.

Taylorson, R. B. 1986. Water stress induced germination of giant foxtail (*Setaria faberi*) seeds. Weed Science. 34:871-875.

Thomas, A. G., J. D. Banting and G. Bowes. 1986. Longevity of green foxtail seeds in a Canadian prairie soil. Can. J. Plant Sci. 66:189-192.

Thomas, W.L. Jr. (Ed.). 1956. Man's role in changing the face of the earth. University of Chicago Press, Chicago, IL.

Thompson, D.W. 1917. On growth and form. Cambridge University Press, Cambridge, MA.

Thompson, K., S.R. Band, and J.G. Hodgson. 1993. Seed size and shape predict persistence in soil. Functional Ecology 7:236-241.

Thornhill, R. and J. Dekker. 1993. Mutant weeds of Iowa: V. S-triazine resistant giant foxtail (*Setaria faberii* Hermm.). Journal of the Iowa Academy of Science 100:13-14.

Till-Bottraud, I., X. Reboud, P. Brabant, M. Lefranc, B. Rherissi, F. Vedel, and H. Darmency. 1992. Outcrossing and hybridization in wild and cultivated foxtail millets: consequences for the release of transgenic crops. Theor. Appl. Genet. 83:940-946.

Tirado, R. and F. I. Pugnaire. 2005. Community structure and positive interactions in constraining environments. OIKOS 111: 437-444.

Toole, E.H. and E. Brown. 1946. Final results of the Duvel buried seed experiment. J. Agric. Res. 72:201-210.

Tranel, D.M., and J.H. Dekker. 1996. Differential seed germinability in triazine resistant and susceptible giant foxtail (*Setaria faberi*). NCWSS Proceedings 51:165.

Tranel, D. and J. Dekker. 2002. Differential seed germinability in triazine-resistant and -susceptible giant foxtail (*Setaria faberii*). Asian Journal of Plant Sciences 1(4):334-336.

Trewavas, A.J. 1986. Timing and memory processes in seed embryo dormancy-A conceptual paradigm for plant development questions. BioEssays 6(2), 87-92.

Trewavas, A.J. 2000. Signal perception and transduction. In: Buchanan, B.B.B., W. Gruissem and R.L. Jones (Eds.) Biochemistry and molecular biology of plants. Maryland: American Society of Plant Physiologists, 930-988.

Trivers, R.L. 1971. The evolution of reciprocal altruism. Quarterly Review of Biology 46:35–57.

Trewavas, A.J. 2000. Signal perception and transduction. In: Biochemistry and molecular biology of plants; Buchanan, B.B.B., W. Gruissem, and R.L. Jones, editors. Maryland: American Society of Plant Physiologists, pp. 930-988.

Uglow, Jenny. 2002. The lunar men: The inventors of the modern world 1730-1810. Faber and Faber, London, UK.

Uetz, G.W., J. Halaj, A.B. Cady, 1999. Guild structure of spiders in major crops. The Journal of Arachnology 27: 270-280. Proceedings of the XIV International Congress of Arachnology and a Symposium on Spiders in Agroecosystems, pp. 270-280

Uvarov, E.B., D.R. Chapman, and A. Issacs (Eds.). 1982. The Penguin Dictionary of Science, 5th edition. Penguin Books, UK.

Vanden Born, W.H. 1971. Green foxtail: seed dormancy, germination and growth. Can. J. Plant Sci 51:53-59.

van der Heijden M.G., Streitwolf-Engel R., Riedl R., Siegrist S., Neudecker A., Ineichen K., Boller T., Wiemken A., Sanders I.R. 2006. The mycorrhizal contribution to plant productivity, plant nutrition and soil structure in experimental grassland. New Phytologist 172(4):739–752.

van Oorshot, J.L.P., P.H. van Leeuwen. 1984. Comparison of the photosynthetic capacity between intact leaves of triazine-resistant and -susceptible biotypes of six weed species. Zeitschrift fur Naturforschung 39c:440- 442.

van Valen, L. 1973. A New Evolutionary Law. Evolutionary Theory 1:1-30.

Vaughn, K.C. 1986. Characterization of triazine-resistant and susceptible isolines of canola (*Brassica napus* L.). Plant Physiology 82: 859-863.

Vaughn, K.C. and S.O. Duke. 1984. Ultrastructural alterations to chloroplasts in triazine-resistant weed biotypes. Physiologia Plantarum 62:510-520.

Vavilov, N.I. 1992. Origin and geography of cultivated plants; D. Love, translator; Cambridge University Press, UK. Original papers from 1920 to 1949 first published in Russian as: Proiskhozhdenie I geografia kul'turnykh rastenii; 1987; Nauka, Leningrad Branch.

Venable, D.L. and J.S. Brown. 1988. The selective interactions of dispersal, dormancy, and seed size as adaptations for reducing risk in variable environments. American Naturalist 131:360-384.

Verhulst, P.F. 1839. Notice sur la Loi que la Population Suit dans son Accroissement Corr. Math. Phys., A. Quetelet, Paris, 10 (pp. 1–113).

Violle, C., M.-L. Navas, D. Vile, E. Kazakou, C. Fortunel, I. Hummel, E. Garnier. 2007. Let the concept of trait be functional! Oikos 116(3):882-892.

Vitousek, P.M., Mooney, H. A., and Lubchenco, J. et al. 1997. Human domination of Earth's ecosystems. Science 277:494-499.

Volenberg, D.S., D.E. Stoltenberg, and C.M. Boerboom. 2001. Biochemical mechanism and inheritance of cross-resistance to acetolactate synthase inhibitors in giant foxtail. Weed Sci. 49:635-641.

Waddington C.H. 1942. Canalization of development and the inheritance of acquired characters. Nature 150(3811):563–565

Waismann, F. 1930. Logische analyse des Wahrscheinlichkeitsbegriffs. Erkenntnis 1(1):228-248.

Waldron, L.R. 1904. Weed studies. Vitality and growth of buried weed seed. N. Dak. Agric. Exp. Sta. Bull. 62:439-446.

Waldrop, M.M. 1994. Complexity: the emerging science at the edge of order and chaos. Penguin Books, Harmondsworth, UK.

Walker, B., Holling, C. S., Carpenter, S. R., Kinzig, A. 2004. Resilience, adaptability and transformability in social–ecological systems. Ecology and Society 9(2):5.

Walker, B.H. 1992. Biodiversity and ecological redundancy. Conservation Biology 6:18-23.

Walker, P.M.B. (Ed.). 1988. Chambers Science and Technology Dictionary. W & R Chambers Ltd., Edinburgh and Cambridge University Press, UK.

Wallace, D.F. 2011. The Pale King. Hamish Hamilton, Penguin, UK.

Wang, R.L. and J. Dekker. 1995. Weedy adaptation in *Setaria* spp.: III. Variation in herbicide resistance in *Setaria* spp. Pesticide Biochemistry and Physiology 51:99-116.

Wang, R.L, J. Wendell and J. Dekker. 1995a. Weedy adaptation in *Setaria* spp.: I. Isozyme analysis of the genetic diversity and population genetic structure in *S. viridis.*. American Journal of Botany 82(3):308-317.

Wang, R.L, J. Wendell and J. Dekker. 1995b. Weedy adaptation in *Setaria* spp.: II. Genetic diversity and population genetic structure in *S. glauca*, *S. geniculata* and *S. faberii*. American Journal of Botany 82(8):1031-1039.

Wang T., Chen H.B., Reboud X., Darmency H. (1997) Pollen-mediated gene flow in an autogamous crop : Foxtail millet (*Setaria italica*). Plant Breed 116:579-583

Warwick, S.I. 1990. Allozyme and life history variation in five northwardly colonizing North American weed species. Pl. Syst. Evol. 169:41-54.

Watters, E. 2006. DNA is not destiny. Discover Magazine, November; p.33-37; 75.

Weaver, S.E. and M.J. Lechowicz. 1983. Biology of Canadian weeds. 56. *Xanthium strumarium* L. Cocklebur. Canadian Journal of Plant Science 63, 211-225.

Weis E, Berry JA (1987) Quantum efficiency of photosystem II in relation to "energy" dependent quenching of chlorophyll fluorescence. Biochimica et Biophysica Acta 894:198-208.

Wiener, N. 1948. Cybernetics: or, control and communication in the animal and the machine. MIT Press, Cambridge, MA.

Willey, R.W. and S.B. Heath, 1969. The quantitative relationships between plant population and crop yield. Advances in Agronomy 21:281-321.

Williams, J.T. and J.L. Harper. 1965. Seed polymorphism and germination. 1. The influence of nitrates and low temperatures on the germination of *Chenopodium album*. Weed Research 5, 141-150.

Williams R.D., Schreiber M.M. 1976. Numerical and chemotaxonomy of the green foxtail complex. Weed Sci 24:331-335

Willweber-Kishimoto, E. 1962. Interspecific relationships in the genus *Setaria*. Contrib. Bio. Kyoto Univ.

Wilson, R.G. Jr. 1980. Dissemination of weed seeds by surface irrigation water in western Nebraska. Weed Sci. 28:87-92.

Whitelam, G.C., and P.F. Devlin. 1997. Roles of different phytochromes in *Arabidopsis* photomorphogenesis. Plant, Cell & Environment 20:752-758.

Whitson, T.D., L.C. Burrill, S.A. Dewey, D.W. Cudney, B.E. Nelson, R.D. Lee, and R. Parker. 1991. Weeds of the West. Western Society of Weed Science.

Wikipedia: The Free Encyclopedia. Wikimedia Foundation, Inc., Web: numerous citations by topic and month.day (e.g. 3.12, March, 2012)

Wilson, E.O. 1965. The challenge from related species. In: Genetics of colonizing species, H.G. Baker and G.L. Stebbins, editors. Academic Press, New York, NY.

Wilson, R.A. and F.C. Keil. 1999. The MIT encyclopedia of the cognitive sciences. MIT Press, Cambridge, MA.

Wolfram, S. 2002. A new kind of science. Wolfram Media, Champaign, IL.

Wolschke-Bulmahn, J. 1995. Political landscapes and technology: Nazi Germany and the landscape design of the Reichsautobahnen (Reich Motor Highways); Selected CELA Annual Conference Papers: Nature and Technology, Iowa State University, Sept. 9-12, Vol. 7.

Wootton, D. 2006. Bad Medicine. Oxford University Press, UK.

Wright, S. 1932. The roles of mutation, inbreeding, crossbreeding, and selection in evolution. Proceedings of the Sixth International Congress on Genetics; pp. 355–366.

Yenish, J.P., J.D. Doll, and D.D. Buhler. 1992. Effects of tillage on vertical distribution and viability of weed seed in soil. Weed Science 40:429-433.

Yoda, K., T. Kira, and K. Hozumi. 1957. Intraspecific competition among higher plants. IX. Further analysis of the competitive interaction between adjacent individuals. J. Inst.Polytech. Osaka Cilty Univ. 8:161-178.

Yoda, K., T. Kira, H. Ogawa, and K. Hozumi. 1963. Self-thinning in overcrowded pure stands under cultivation and natural conditions. Journal of Biology Osaka City University 14:107-129.

Yule, G. U. 1925. A mathematical theory of evolution, based on the conclusions of Dr. J. C. Willis, F.R.S. Philosophical Transactions of the Royal Society B 213 (402–410): 21–87.

Zohary 1965 (Polyploid species clusters) In: Genetics of colonizing species; H.G. Baker and G.L Stebbins (Eds.). Academic Press, NY.

Zee, S.-Y. 1972. Vascular tissue and transfer cell distribution in the rice spikelet. Aust. J. Biol. Sci. 25:411-414.

Zee, S.-Y. and T.P. O'Brien. 1971. Aleurone transfer cells and other structureal features of the spikelet of millet. Aust. J. Biol. Sci. 24:391-395.

# INDEXED GLOSSARY

**adaptation:**
1: the process of adjustment of an individual organism to environmental stress; adaptability;
2: process of evolutionary modification which results in improved survival and reproductive efficiency;
3: any morphological, physiological, developmental or behavioral character that enhances survival and reproductive success of an organism
4: a positive characteristic of an organism that has been favored by natural selection and increases the fitness of its possessor (Wikipedia, 5.08)
5: any property of an organism believed to add to its fitness (Mayr, 2001)
(see ch. 2.5)

**after-ripening:**
1: changes in seeds, spores and buds that cannot germinate (even under favorable conditions) in the period after dispersal which allow it to germinate
2: the poorly understood chemical and/or physical changes which must occur inside the dry seeds of some plants after shedding or harvesting, if germination is to take place after the seeds are moistened (Chambers Scientific and Technology dictionary)
3: the period after a seed has been dispersed when it cannot germinate, even if conditions are favorable, and during which physiological changes occur so that it can germinate (Henderson's Dictionary of Biological Terms).
(see ch. 7.3.1)

**age structure:**
1: the number or percentage of individuals in each age class of a population; age distribution; age composition
2: structure in populations where recruitment is a frequent event compared to the lifespan of adults (e.g. forest trees) (Silvertown and Charlesworth, 2001)
(see ch. 5.2.3)

**aggregrate species:**
1: a group of species that are so closely related that they are regarded as a single species.
2: a named species that represents a range of very closely related organisms
(see ch. 5.4.4.3)

**agroecosystem**: an agricultural ecosystem: row crop (i.e. corn), solid planted crop (i.e. wheat), perennial forage, managed forest, rangeland, etc.; crop rotation

**agroecotype**: an edaphic ecotype adapted to cultivated soils

**agrestal**: growing on arable land (Lincoln et al., 1998)
(see ch. 1.2)

**allele**: any of the different forms of a gene occupying the same locus (q.v. on homologous chromosomes), and which undergo meiotic pairing (q.v. and can mutate one to another)

**amensalism:**

1: an interspecific interaction in which one organism, population or species is inhibited, typically by toxin produced by another (amensal), which is unaffected (Lincoln et al., 1998)
2: an interaction between two species in which one impedes or restricts the success of the other while the other species is unaffected; a type of symbiosis (Wikipedia, 12.10).
(see ch. 9.2.2.2)

**annual**:
1: having a yearly periodicity; living for one year (Lincoln et al., 1998)
2: plants which germinate, bloom and produce seeds, and die in one growing season (Alex and Switzer, 1976)
(see ch. 6.2.1)

> **winter annual**:
> 1: usually germinate in the autumn and overwinter as seedlings before renewing growth in the spring (Stubbendieck et al., 1995)
> 2: plants which germinate and produce a leafy rosette in the autumn and then bloom, seed, and die the following summer (Alex and Switzer, 1976)
> (see ch. 6.2.1)

> **summer annual**: plants which germinate and grow in the spring or summer, bloom, produce seed and die in the autumn
> (see ch. 6.2.1)

**apomixis**:
1: seeds have the same genotype as their mother; in angiosperms pollination and fertilization of the endosperm (psuedogamy) is common
2: apomixis (also called apogamy) is asexual reproduction, without fertilization. In plants with independent gametophytes (notably ferns), apomixis refers to the formation of sporophytes by parthenogenesis of gametophyte cells. Apomixis also occurs in flowering plants, where it is also called agamospermy. Apomixis in flowering plants mainly occurs in two forms: agamogenesis, adventitious embryony
(see ch. 5.4.4.3)

> **agamogenesis**: (also called gametophytic apomixis), the embryo arises from an unfertilized egg that was produced without meiosis.
> (see ch. 5.4.4.3)

> **adventitious embryony**, a nucellar embryo is formed from the surrounding nucellus tissue.
> (see ch. 5.4.4.3)

**archetype**:
1: a perfect or typical specimen
2: an original model or pattern; prototype

**biennial**: lasting for two years; occurring every two years; requiring two years to complete the life cycle (Lincoln et al., 1998)
(see ch. 6.2.1)

**biodiversity**
1: the variety of organisms considered at all levels, from genetic variants of a single species through arrays of species to arrays of genera, families and still higher taxonomic levels
2: the variety of ecosystems, which comprise both the communities of organisms within particular habitats and the physical conditions under which they live;
3: the totality of biological diversity
(see ch. 10.5)

**biotic resistance hypothesis**: the biological diversity (number and variety of species) in a community or an ecosystem, conveys a degree of "biotic resistance" that preserves the integrity of an ecosystem over time, a protection against invasions; more biologically diverse communities are more stable in the face of perturbations
(see ch. 10.5.3.2)

**Black Swan event**:
1. unpredictability; it is an outlier, it lies outside the realm of regular expectations because nothing in the past can convincingly point to its possibility
2. consequences; it carries and extreme impact
3. retrospective explainability; in spite of its outlier status, human nature makes us concoct explanations for its occurrence after the fact, making it explainable and predictable
(see ch. 14.3.1)

**bottleneck (genetic)**: a sudden decrease in the size of a population with corresponding reduction of total genetic variability
(see ch. 4.2.1.2)

**bottleneck effect**:
1: population becomes very small and then recovers in size; can occur at the foundation of the deme or later (e.g. disturbance)
2: fluctuations in gene frequencies occurring when a large population passes through a contracted stage and then expands again with an altered gene pool (usually one with reduced variability) as a consequence of genetic drift
3: a conceptual occurence of genetic drift in populations reduced in size through fluctuations in abundance.
(see ch. 4.2.1.2)

**canalization**: a measure of the ability of a population to produce the same phenotype regardless of variability of its environment or genotype
(see ch. 4.3)

**catastrophes**:
1: an event subverting the order or system of things; significant population decrease, possible local extinction
2: disaster, a horrible event (Wikipedia, 5.08)
(see ch. 3.3.4.4)

**catastrophe theory**: a field of mathematics that studies how the behaviour of dynamic systems can change drastically with small variations in specific parameters (Wikipedia, 5.08)
(see ch. 3.3.4.4)

**chaperone species**: a species that may facilitate the invasion of another species either directly assisting it or by inhibiting a third species (Kelly, 1994)
(see ch. 10.4.4)

**climate**: the statistics of temperature, humidity, atmospheric pressure, wind, rainfall, atmospheric particle count and other meteorological elements in a given region over long periods of time (Wikipedia, 2010)
(see ch. 3.4.3.2.1)

> **macroclimate**: the regional climate of a broad area, it can include an area on the scale of tens to hundreds of kilometers (Wikipedia, 2010)
> (see ch. 3.4.3.2.1)

> **mesoclimate**: the climate of a particular site, generally restricted to a space of a tens or hundreds of meters (Wikipedia, 2010)
> (see ch. 3.4.3.2.1)

> **microclimate**: the specific environment in a small restricted spaces, such as a crop row or the environment around an individual plant (Wikipedia, 2010)
> (see ch. 3.4.3.2.1)

**coexistence**: the continuing occurrence of two or more species in the same area or habitat, usually used of potential competitors (Lincoln et al., 1998)

**colonization**:
1: the successful invasion of a new habitat by a species (Lincoln et al., 1998)
2: the occupation of bare soil by seedlings or sporelings (Lincoln et al., 1998)
3: (of plants and animals) to become established in (a new environment) (Anonymous, 1979)
(see ch. 3.5.2)

**colonizing species**: a plant, typically 'r'-selected, which invades and colonizes a new habitat or territory (Lincoln et al., 1998)
(see ch. 1.2)

**commensalism**: symbiosis in which one species derives benefit from a common food supply whilst the other species is not adversely affected (Lincoln et al., 1998)
(see ch. 9.2.3.2)

**communication**: the activity of conveying information through the exchange of messages, or information, as by signals or behavior (Wikipedia, 10.12)
(see ch. 13.1.1)

**community**
1: any group of organisms belonging to a number of different species that co-occur in the same habitat or area and interact through trophic and spatial relationships; typically characterized by reference to one or more dominant species (Lincoln)

2: in ecology, an assemblage of populations of different species, interacting with one another; sometimes limited to specific places, times, or subsets of organisms; at other times based on evolutionary taxonomy and biogeography; other times based on function and behavior regardless of genetic relationships (Wikipedia, 6.08)
3: in ecology, assemblages or associations of populations of two or more different species occupying the same geographical area and in a particular time; in its simplest form, groups of organisms in a specific place or time (Wikipedia, 6.14)
(see ch. 9.1.2; 10.2)

>**biological community**
>1: biocoenosis, biocoenose, biocenose
>2: all the interacting organisms living together in a specific habitat (or biotope); biotic community, ecological community; the extent or geographical area of a biocenose is limited only by the requirement of a more or less uniform species composition. (Wikipedia, 6.08)
>(see ch. 10.2)

>**local community**: all the interacting demes in a locality
>(see ch. 2.2.2; 10.2)

**competition**:
1: the tendency of neighbouring plants to utilize the same quantum of light, ion of a mineral nutrient, molecule of water, or volume of space (Grime, 1979)
2: an interaction between species in a mixture in which each lowers the net reproductive rate of the other (Silvertown and Charlesworth, 2001)
3: the simultaneous demand by two or more organisms or species for an essential common resource that is actually or potentially in limited supply (exploitation competition), or the detrimental interaction between two or more organisms or species seeking a common resource that is not limiting (interference competition) (Lincoln et al., 1998)
4: a mutually detrimental interaction between individuals, populations or species (Wikipedia, 12.10)
(see ch. 9.2.2.1)

**competitive exclusion**: the exclusion of one species by another when they compete for a common resource that is in limited supply
(see ch. 3.2.2)

**competitive exclusion principle**:
1: two species competing for the same resources cannot coexist if other ecological factors are constant (Hardin, 1960); complete competitors cannot coexist
2: a theory which states that two species competing for the same resources cannot stably coexist, if the ecological factors are constant; complete competitors cannot coexist. Either of the two competitors will always take over the other which leads to either the extinction of one of the competitors or its evolutionary or behavioural shift towards a different ecological niche (Wikipedia, 5.08)
(see ch. 9.3.2.2.2)

**complex**:
1: made up of various interconnected parts; composite; intricate or involved; a whole made up of interconnected or related parts

2: a whole that comprehends a number of intricate parts, especially one with interconnected or mutually related parts
(see ch. 12.1)

**complexity**:
1: the state or quality of being intricate; something intricate; complication
2: something with many parts where those parts interact with each other in multiple ways phenomena which emerge from a collection of interacting objects
(see ch. 12.1)

> **Kolmogorov complexity**:
> 1: the complexity of a number, message, set of data is the inverse of simplicity and order, and it corresponds to information
> 2: calculating complexity in terms of algorithms; the complexity of an object is the size of the smallest computer program needed to generate it;
> 3: complexity as the amount of information, the degree of randomness, in a message; information = randomness = complexity
> (see ch. 13.1.3.2)

**conjecture**:
1: the formation of conclusions from incomplete evidence; guess; the inference or conclusions so formed
2: In mathematics, a conjecture is a conclusion or proposition based on incomplete information, for which no proof has been found (Wikipedia, 6.16)
(see U6.1)

**constructal law**: for a finite-size flow system to persist in time (to live), its configuration must evolve in such a way that provides easier access to the currents that flow through it (given freedom from constraints)
(see ch. 11.1.2)

**contagious distribution**: a distribution pattern in which values, observations or individuals are more aggregated or clustered than in a random distribution, indicating that the presence of one individual or value increases the probability of another occurring nearby (Lincoln et al., 1998)
(see ch. 12.3.3)

**cooperation**: the process of working or acting together, intentional and non-intentional; acting in harmony, side by side, or in more complicated ways; the alternative to working separately in competition
(see ch. 9.2.3)

**crash**:
1: an event of collapse or sudden failure
2: a decrease in a population, within the time scale [life history] of an organism
(see ch. 3.3.4.4)

**creativity**:
1: the capacity to create new explanations

2: a phenomenon whereby something new and somehow valuable is formed; the created item may be intangible (such as an idea, a scientific theory, a musical compostion or a joke) or a physical object (such as an invention, a literary work or a painting) (Wikipedia, 6.16)
3: a process of becoming sensitive to problems, deficiencies, gaps in knowledge, missing elements, disharmonies, and so on; identifying the difficulty; searching for solutions, making guesses, or formulating hypotheses about the deficiencies: testing and retesting these hypotheses and possibly modifying and retesting them; and finally communicating the results (Wikipedia, 6.16, Torrance, P.)
(see U6.1)

**cycle**:
1: happening at regular intervals
2: an interval of space or time in which one set of events or phenomena is completed
3: a complete rotation of anything
4: a process that returns to its beginning and then repeats itself in the same sequence
(Wikipedia, 5.08)
(see ch. 3.3.4.4)

**deme**: a local population of potentially interbreeding individuals of a species at a given locality
(see ch. 2.2.2)

**demographic space**: demographic space exists in two forms:
1: a particular locality, the specific area over which a plant interacts with neighbors, the ecological and genetic neighborhood
2: plant numerical density, the number of plants per unit area; density is used as an average measure of space occupancy
(see ch. 9.4.1)

**demography**: the study of populations, especially of growth rates and age structure

**density dependence**: a change in the influence of an environmental factor (a density dependent factor) that affects population growth as a population density changes, tending to retard population growth (by increasing mortality or decreasing fecundity) as density increases or to enhance population growth (by decreasing mortality or increasing fecundity) as density increases
(see ch. 5.2.6)

**density independent factor**: any factor affecting population density, the influence of which is independent of population density
(see ch. 5.2.6)

**development**: the biological study of the process by which organs grow and develop; closely related to ontogeny; genetic control of cell growth, differentiation and morphogenesis, which is the process that gives rise to tissues, organs and anatomy; developmental genetics studies the temporal and spatial control of gene expression, the effect that genes have in a phenotype, given normal or abnormal epigenetic parameters (Wikipedia, 5.14)
(see ch. 6.1)

**dioecious**:

1: separate male and female sex individuals
2: having unisexual reproductive units with male and female plants (flowers, conifer cones, or functionally equivalent structures) occurring on different individuals; from Greek for "two households". Individual plants are not called dioecious: they are either gynoecious (female plants) or androecious (male plants)
(see ch. 5.4.4.2)

> **androdioecious**: individuals either male or hermaphrodite
> (see ch. 5.4.4.2)
>
> **gynodioecious**: individuals either female or hermaphrodite
> (see ch. 5.4.4.2)
>
> **subdioecious**: a tendency in some dioecious species to produce monoecious plants. The population produces normally male or female plants but some are hermaphroditic, with female plants producing some male or hermaphroditic flowers or vice-versa. The condition is thought to represent a transition between hermaphroditism and dioecy
> (see ch. 5.4.4.2)
>
> **androecious**: plants producing male flowers only, produce pollen but no seeds, the male plants of a dioecious species
> (see ch. 5.4.4.2)
>
> **subandroecious**: plant has mostly male flowers, with a few female or hermaphrodite flowers
> (see ch. 5.4.4.2)
>
> **gynoecious**: plants producing female flowers only, produces seeds but no pollen, the female of a dioecious species. In some plant species or populations all individuals are gynoecious with non sexual reproduction used to produce the next generation
> (see ch. 5.4.4.2)
>
> **subgynoecious**: plant has mostly female flowers, with a few male or hermaphrodite flowers
> (see ch. 5.4.4.2)

**dispersal**
1: outward spreading of organisms or propagules from their point of origin or release; one-way movement of organisms from one home site to another (Lincoln et al., 1998)
2: the act of scattering, spreading, separating in different directions (Anonymous, 2001)
3: the spread of animals, plants, or seeds to new areas (Anonymous, 1979)
4: the outward extension of a species' range, typically by a chance event; accidental migration (Lincoln et al., 1998)
5: the search by plant propagules (e.g. seeds, buds) for opportunity in space and time
(see ch. 3.5.2; 8.1)

**disturbance**
1: the act of disturbing or the state of being disturbed (Anonymous, 1979, 2001)
2: an interruption or intrusion (Anonymous, 1979, 2001)
3: destruction of biomass by any natural or human agency (Silvertown and Charlesworth, 2001)

4: an interruption or intrusion with direct and indirect spatial, temporal, biological or abiological effects that alters or destroys a biological individual or community
(see ch. 3.3.4)

> **perturb**:
> 1: to disturb the composure of (Hanks et al., 1979)
> 2: to throw into disorder (Hanks et al., 1979)
> (see ch. 3.3.4)

> **perturbation**:
> 1: the act of perturbing or the state of being perturbed (Hanks et al., 1979)
> 2: a cause of disturbance or upset (Hanks et al., 1979)
> (see ch. 3.3.4)

**ecological guild**
1: a group of species having similar ecological resource requirements and foraging strategies, and therefore having similar roles (niches) in the community
2: a group of species that exploit the same class of environmental resources in a similar way (Root, 1967)
3: groups of species that exploit resources in a particular way (Silvertown, 2001)
4: groups of functionally similar species in a community
(see ch. 10.3.3.1)

**ecological succession**
1: the chronological distribution of organisms within an area
2: the geological, ecological or seasonal sequence of species within a habitat or community
3: the development of plant communities leading to a climax
4: a process of continuous colonization of and extinction on a site by species populations (Bazzaz, 1979)
(see ch. 10.4.4)

> **succession**: a directional, progressive, orderly change in vegetation that would ultimately converge on a stable, predictable climax community (Clements, 1916, 1936)
> (see ch. 10.4.4)

> **successional trajectory**: the likely sequence of plant and animal communities predicted to occur in a habitat following a particular type of disturbance
> (see ch. 10.4.4)

**ecological stability**:
1: connoting a continuum, ranging from resilience (returning quickly to a previous state) to constancy (lack of change) to persistence (simply not going extinct); the precise definition depends on the ecosystem in question, the variable or variables of interest, and the overall context
2: in conservation ecology, populations that do not go extinct; in mathematical models of systems, Lyapunov stability (dynamical system that start out near an equilibrium point stay there forever) (Wikipedia, 6.08)
(see ch. 10.5.3.2)

> **community stability**: how communities resist change in response to disturbance or stress (Booth and Swanton, 2002)

(see ch. 10.5.3.2)

**ecological stability-constancy and persistence**: living systems that can remain unchanged in observational studies of ecosystems (Wikipedia, 6.08)
(see ch. 10.5.3.2)

**ecological stability-resistance and inertia (or persistence)**:
1: a system's response to some perturbation (disturbance; any externally imposed change in conditions, usually happening in a short time period)
2: resistance is a measure of how little the parameter of interest changes in response to external pressures
3: inertia (or persistence) implies that the living system is is able to resist external fluctuations (Wikipedia, 6.08)
(see ch. 10.5.3.2)

**ecological stability-resilience, elasticity and amplitude**:
1: the tendency of a system to return to a previous state after a perturbation
2: 'engineering resilience' is the time required for an ecosystem to return to an equilibrium or steady-state following a perturbation; 'engineering resilience' (Hollings, 1973)
3: 'ecological resilience' is the capacity of a system to absorb disturbance and reorganize while undergoing change so as to still retain essentially the same function, structure, identity, and feedbacks; it presumes the existence of multiple stable states or domains (Walker et al., 2004)
4: elasticity and amplitude are measures of resilience; elasticity is the speed with which a system returns; amplitude is a measure of how far a system can be moved from the previous state and still return (Wikipedia, 6.08)
(see ch. 10.5.3.2)

**ecology**: the study of the interrelationships between living organisms and their environment

**ecosystem**
1: a community of organisms and their physical environment interacting as an ecological unit; the entire biological and physical content of a biotope
2: an ecosystem is a natural unit consisting of all plants, animals and micro-organisms (biotic factors) in an area functioning together with all of the non-living physical (abiotic) factors of the environment (Wikipedia, 6.08)
3: a biotic community along with its physical environment (Tansley, 1935)
(see ch. 10.2)

**ecotype**:
1: a locally adapted population; a race or infraspecific group having distinctive characters which result from the selective pressures of the local environment; ecological race;
2: a subunit capable of interbreeding with members of that and other ecotypes q.v. comprising individuals capable of interbreeding with members of that and other ecotypes within the ecospecies but remaining distict through selection and isolation;
3: biotype

**emergence:**
1: class of high-level phenomena that can be well explained in terms of each other alone, with no direct reference to anything at (the atomic) (lower) level or below: the behavior of that whole class of high-level phenomena is quasi-autonomous (almost self-contained)
2: resolution into explicability of phenomena at a higher (quasi-autonomous) level
(see U6.1)

**enforced dormancy**
1: an inability to germinate due to an environmental restraint (e.g. shortage of water, low temperature, poor aeration)
2: unfavorable conditions prevent germination of a seed that would otherwise be able to germinate
3: conditional dormancy; germination only in a narrow range of conditions (both in the loss of dormancy and reacquisition of dormancy in dormancy cycling through the year)
(see ch. 7.3.1)

**entropy:**
1: lack of pattern or organization; disorder (Hanks et al., 1979)
2: the macroscopic variable representing the degree of disorder with a system (Borowski and Borwein, 1991)
3: (Statistical mechanics) the increase in entropy in a closed system to a maximum at equilibrium is the consequence of the trend from a less probable to a more probable state (Walker, 1988)
4: the entropy of a system is a measure of its degree of disorder. The total entropy of any isolated system can never decrease in any change; it must either increase (irreversible process) or remain constant (reversible process). The total entropy of the *Universe* therefore is increasing, tending towards a maximum, corresponding to complete disorder of the particles in it (assuming that it may be regarded as an isolated system). (Uvarov et al., 1982)
5: a measure of the efficiency of a system, such as a code or language, in transmitting information (Hanks et al., 1979)
6: (information theory) a measure of the uncertainty associated with a random variable; Shannon entropy (Wikipedia, 3.12)

> **Shannon entropy** (Shannon and Weaver, 1963):
> 1: quantity of the expected value of the information contained in a message (a specific realization of the random variable), usually in units such as bits
> 2: a measure of the average information content one is missing when one does not know the value of the random variable
> 3: an absolute limit on the best possible lossless compression of any communication

**epistasis:**
1: a class of interactions between pairs of genes in their phenotypic effects; technically the interactions are non-additive which means, roughly, that the combined effect of the two genes is not the same as the sum of their separate effects; for instance, one gene might mask the effects of the other. The word is mostly used of genes at different loci, but some authors use it to include interactions between genes at the same locus, in which case dominance/recessiveness is a special case (Dawkins,1999)
2: the interaction of non-allelic genes in which one gene (epistatic gene) masks the expression of another at a different locus (Lincoln, et al.)
3: the nonreciprocal interaction of nonallelic genes; the situation in which one gene masks the expression of another

4: interactions between two or more genes (Mayr, 2001)
5: the interaction between genes; the effects of one gene are modified by one or several other genes (modifier genes) (Wikepedia, 5.09)
(see ch. 4.1.2)

**epigenesis**:
1: 'in addition to' genetic information encoded in DNA sequence
2: heritable changes in gene function without DNA change
3: a theory of the development of an organism by progressive differentiation of an initially undifferentiated whole
4: the unfolding development in an organism, and in particular the development of a plant, fungus or animal from a seed, spore or egg through a sequence of steps in which cells differentiate and organs form (Wikipedia, 4.14)
5: each embryo or organism is gradually produced from an undifferentiated mass by a series of steps and stages during which new parts are added
(see ch. 4.1.2)

**epigenetics**
1: the study of the mechanism that produces phenotypic effects from gene activity, processes involved in the unfolding development of an organism, during differentiation and development, or heritable changes in gene expression that do not involve changes in gene sequence.
2: the study of how environmental factors affecting a parent can result in changes in the way genes are expressed in the offspring, heritable changes in gene function without DNA change.
3: the study of reversible heritable changes in gene function that occur without a change in the sequence of nuclear DNA: how gene-regulatory information that is not expressed in DNA sequences is transmitted from one generation (of cells or organisms) to the next
4: changes in gene expression due to mechanisms other than changes in DNA sequence
5: the study of the mechanisms of temporal and spatial control of gene activity during the development of complex organisms; thus epigenetic can be used to describe anything other than DNA sequence that influences the development of an organism (Holliday, 1990)
(see ch. 4.1.2)

**equilibrium**: the condition of a system in which competing influences are balanced
(see ch. 10.5.3.3)

**error**:
1: the uncertainty in a measurement or estimate of a quantity (Daintith and Clark, 1999)
2: the difference between some quantity and an approximation to or estimate of it, often expressed as an absolute or relative range (Borowski and Borwein, 1991)
3: variation produced by disturbing factors, both known and unknown; important effects man be wholly or partially obscured by experimental error; conversely, the experimenter may be misled into believing in effects that do not exist (Box, et al., 1978)
(see ch. 14.2.4)

> **random error**:
> 1: any deviation, the magnitude and direction of which cannot be predicted (Lincoln et al., 1998)

2: an error that occurs in any direction, cannot be predicted, and cannot be compensated for; it includes the limitations in the accuracy of the measuring instrument and the limitations in reading it (Daintith and Clark, 1999)
(see ch. 14.2.4)

**systematic error**: error arising from the faults or changes in conditions that can be corrected for (Daintith and Clark, 1999)
(see ch. 14.2.4)

### type I error:
1: error of rejecting a true hypothesis (false positive) [but not explicitly stating it is false?]
2: false positive
3: a true null hypothesis was incorrectly rejected
(see ch. 14.2.4)

### type II error:
1: error of not rejecting a false hypothesis [ but not explicitly stating it is true?]
2: false negative
3: one fails to reject a false null hypothesis
(see ch. 14.2.4)

**establishment**: growing and reproducing successfully in a given area (Lincoln et al., 1998)
(see ch. 3.5.2)

**evolution**
1: any gradual directional change
2: change in the properties of populations of organisms over time (Mayr, 2001)
3: any cumulative change in characteristics of organisms or populations from generation to generation; descent or development with modification
4: the opportunistic process of change in the characteristics (traits) of individual organisms/phenotypes and their local populations (demes) from generation to generation by the process of natural selection
5: change in the frequency of genes in a population
6: the genetic turnover of the individuals of every population from generation to generation (Mayr, 2001)
7: the gradual process by which the living world has been developing following the origin of life (Mayr, 2001)
(see ch. 2.2)

### co-evolution:
1: reciprocal evolution as a consequence of two (or more) kinds of organisms interacting with each other such that each exerts a selection pressure on the other; much of the process of evolution occurs through coevolution (Mayr, 2001)
2: parallel evolution of two kinds of organisms that are interdependent, like flowers and their pollinators, or where at least one depends on the other, like predators on prey or parasites and their hosts, and where any change in one will result in an adaptive response in the other (Mayr, 2001)
3: change of a biological object triggered by the change of a related object (Wikepedia, 5.09)
(see ch. 2.2)

**microevolution:**
1: minor evolutionary events usually viewed over a short period of time, consisting of changes in gene frequencies, chromosome structure or number within a population over a few generations (Lincoln)
2: the occurrence of small-scale changes in allele frequencies in a population, over a few generations, also known as change at or below the species level (Wikipedia, 5.08)
3: evolution at or below the species level (Mayr, 2001)
(see ch. 2.2.1)

**macroevolution:**
1: major evolutionary events or trends usually viewed through the perspective of geological time; the origin of higher taxonomic categories; transspecific evolution; macrophylogenesis; megaevolution (Lincoln et al., 1998)
2: a scale of analysis of evolution in separated gene pools; change that occurs at or above the level of species; the occurrence of large-scale changes in gene frequencies in a population over a geological time period (Wikipedia, 5.08)
3: evolution above the species level; the evolution of higher taxa and the production of evolutionary novelties, such as new structures
(see ch. 2.2.1)

**variational evolution:** a population or species changes through continuous production of new genetic variation and through elimination of most members of each generation because they are less successful either in the process of nonrandom elimination of individuals or in the process of sexual selection (i.e. they have less reproductive success) (Mayr, 2001)
(see ch. 2.2)

**exaptation:** shifts in the function of a trait during evolution; for example, a trait can evolve because it served one particular function, but subsequently it may come to serve another (Wikipedia, 11.10)
(see ch. 3.1.2)

**explanation:**
1: statement about what is there, what it does, and how and why
2: a set of statements constructed to describe a set of facts which clarifies the causes, context, and consequences of those facts (Wikipedia, 6.16)
(see U6.1)

**extinction**
1: the process of elimination, as of less fit genotypes
2: the disappearance of a species or taxon from a given habitat or biota, not precluding later recolonization from elsewhere
(see ch. 3.5.4)

**fecundity**
1: the potential reproductive capacity of an organism or population, measured by the number of gametes or asexual propagules (Lincoln et al., 1998)
2: potential fertility or the capability of repeated fertilization. Specifically the term refers to the quantity of gametes, generally eggs, produced per individual over a defined period of time
(see ch. 2.4.3)

**feral plants**: a plant that has reverted to the wild from a state of cultivation or domestication; wild, not cultivated or domesticated (Lincoln et al., 1998)
(see ch. 1.2)

**fitness**:
1: the average number of offspring produced by individuals with a certain genotype, relative to the numbers produced by individuals with other genotypes.
2: the relative competitive ability of a given genotype conferred by adaptive morphological, physiological or behavioral characters, expressed and usually quantified as the average number of surviving progeny of one genotype compared with the average number of surviving progeny of competing genotypes; a measure of the contribution of a given genotype to the subsequent generation relative to that of other genotypes (Lincoln, et al., 1998)
3: the relative ability of an organism to survive and transmit its genes to the next generation (or some defined number of future generations)
(see ch. 2.4.3)

> **Darwinian fitness**:
> 1: the relative probability of survival and reproduction for a genotype
> 2: a measure of the relative contribution of an individual to the gene pool of the next generation
> 3: the relative reproductive success of a genotype as measured by survival; fecundity or other life history parameters
> (see ch. 2.4.3)

**flow (flow systems)**:
1: flow represents the movement of one entity (in the channel) relative to another (the background)
2: flow is described by what the flow carries (fluid, heat, mass, information), how much it carries (mass flow rate, heat current, traffic, etc.) and where the stream is located
(see ch. 11.1.3)

**foraging**:
1: a manifestation of phenotypic plasticity (or behavior) that increases resource capture in an environment that is spatially heterogeneous for the resource (Silvertown and Charlesworth, 2001)
2: behavior associated with seeking, capturing and consuming food (Lincoln et al., 1998)
(see ch. 9.3.1)

**forbs**: herbaceous plants other than grasses and grasslike plants; herbaceous plants die after one season of growth (annuals) or die back near the soil surface (perennials or biennials); generally forbs have solid or pithy stems and broad leaves that are usually net-veined; flowers may be small or large, colored, and showy
(see ch. 6.2.2.1)

**founder effect**:
1: changes in allele frequencies caused by sampling at the foundation of new demes (Silvertown and Charlesworth, 2001)
2: that only a small fraction of the genetic variation of a parent population or species is present in a small number of founder members of a new colony or population (Lincoln et al., 1998)

3: the principle that when a small sample of a larger population establishes itself in a newly isolated entity, its gene pool carries only a fraction of the genetic diversity represented in the parental population. The evolutionary fates of the parental and derived populations are thus likely to be set along different pathways because the different evolutionary pressures in the different areas occupied by the two populations will be operating on different gene pools (King and Stansfield, 1985)
(see ch. 4.2.1.2)

**fractal:**
1: geometry of the rough and broken; from Latin *fractus* (Taleb, 2009)
2: (Math) a set with non-integral Hausdorff dimension (Borowski and Borwein, 1991)
3: a curve or surface that has a fractal dimension and is formed by the limit of a series of successive operations; generated by an iterative process and that a small part of the figure contains the information that could produce the whole figure; self-similar (Daintith and Clark, 1999)
4: mathematical models for very irregular, very detailed sets, their degree of 'wildness' or 'roughness' characterized by fractal dimension (higher is rougher) (Stoyan and Stoyan, 1994)
5: (Math) a set such that two standard definitions of its dimension give different answers; a curve is one-dimensional according to one definition but a curve with arbitrarily small irregularies may have a larger dimension by another definition, and it is then called a fractal curve (Walker et al., 1998)
(see ch. 14.2.6.3)

**fragile:**
1: vulnerability of something to the disorder, volatility, uncertainty of things that affect it with time
2: easily broken, damaged, or destroyed; physically weak; frail; delicate; tenuous; flimsy
3: a concave sensitivity to stressors, leading to a negative sensitivity to increases in volatility (or variability, stress, dispersion of outcomes, or uncertainty: the "disorder cluster") (Taleb, 2012)
(see ch. 11.2.1)

> **antifragile**: a convex response to a stressor or source of harm (for some range of variation), leading to a positive sensitivity (positive convexity effects) to increases in volatility (or variability, stress, dispersion of outcomes, or uncertainty: the "disorder cluster"); shocks bring more benefits (less harm) as their intensity increases (up to a point) (Taleb, 2012)
> (see ch. 11.2.1)

**frequency-dependent selection**: selection occurring in the situation in which the relative fitness of alternative genotypes is related to their frequency of occurrence within a population

**frequency distribution**: (statistics) the function of the distribution of a sample that corresponds to the probability density function of the given underlying population and tends to it as the sample size increases; the set of relative frequencies of the sample points falling within given intervals of the range of the random variable (Borowski and Borwein, 1991)
(see ch. 14.2.4)

**gene flow:**
1: the exchange of genetic factors within and between populations by interbreeding or migration; incorporation of characteristics into a population from another population
2: in population genetics, gene flow (also known as gene migration) is the transfer of alleles of genes from one population to another (Wikipedia, 5.08).

(see ch. 4.2.1.1)

**genet:**
1: a unit or group derived by asexual reproduction from a single original zygote, such as a seedling or a clone
2: a clonal colony, a group of genetically identical individuals that have grown in a given location, all originating vegetatively (not sexually) from a single ancestor
(see ch. 5.4.4.4)

**genetic drift:**
1: the occurrence of random changes in the gene frequencies of small isolated populations, not due to selection, mutation or immigration; drift; Sewall Wright effect; equivalent to static noise in system; adaptive alleles can be lost in process, especially in small populations
2: in population genetics, genetic drift (or more precisely allelic drift) is the evolutionary process of change in the allele frequencies (or gene frequencies) of a population from one generation to the next due to the phenomena of probability in which purely chance events determine which alleles (variants of a gene) within a reproductive population will be carried forward while others disappear (Wikipedia, 5.08)
(see ch. 4.2.1.2)

**genotype:**
1: the hereditary or genetic constitution of an individual; all the genetic material of a cell, usually referring only to the nuclear material
2: all the individuals sharing the same genetic constitution; biotype
3: the set of genes of an individual (Mayr, 2001)
4: the specimen on which a genus-group taxon is based; the primary type of the type species
(see ch. 4.1)

**grasses:** hollow or occasionally solid culms or stems with nodes; leaves are two-ranked and have parallel veins; flowers are small and inconspicuous
(see ch. 6.2.2.1)

**grasslike plants:** resembling grasses, but generally have solid stems without nodes; the leaves are 2- or 3-ranked and have parallel veins; flowers small and inconspicuous
(see ch. 6.2.2.1)

**guild:**
1: a group of species having similar ecological resource requirements and foraging strategies, and therefore similar roles (niches) in the community
2: groups of species that exploit resources in a particular way (Silvertown, 2001)

**habitat:**
1: the locality, site and particular type of local environment occupied by an organism
2: local environment
3: the physical conditions that surround a species, or species population, or assemblage of species, or community (Clements and Shelford, 1939).
4: an ecological or environmental area that is inhabited by a particular species; the natural environment in which an organism lives, or the physical environment that surrounds (influences and is utilized by) a population (Wikipedia, 5.08)

(see ch. 3.2.1)

>**microhabitat:**
>1: a physical location that is home to very small organisms (e.g. a seed in the soil); microenvironment is the immediate surroundings and other physical factors of an individual plant or animal within its habitat
>2: a very localized habitat (e.g. on the size scale of an individual seed in the soil seed bank)
>(see ch. 3.2.1)

>**microsite:** analogous to a microhabitat; e.g. the local spatial environment immediately perceived by a seed in the seed bank, or a seedling in a field
>(see ch. 3.2.1)

**Hardy-Weinberg equilibrium:** the maintenance of more or less constant allele frequencies in a population through successive generations; genetic equilibrium
(see ch. 4.2.1)

**Hardy-Weinberg law:** that allele frequencies will tend to remain constant from generation to generation and that genotypes will reach an equilibrium frequency in one generation of random mating and will remain at that frequency thereafter; demonstrating that meiosis and recombination do not alter gene frequencies
(see ch. 4.2.1)

**hedge-betting:** strategy of spreading risks to reduce the variance in fitness, even though this reduces intrinsic mean fitness; favored in unpredictable environments where the risk of death is high because it allows a species to survive despite recurring, fatal, disturbances; risks can be spread in time or space by either behavior or physiology; risk spreading can be conservative (risk avoidance by a single phenotype) or diversified (phenotypic variation within a single genotype) (Jovaag et al., 2008C)

**heredity:** the mechanism of transmission of specific characters or traits from parent to offspring
(see ch. 2.4.4)

**heritability:** an attribute of a quantitative trait in a population that expresses how much of the total phenotypic variation is due to genetic variation.
>a: in the **broad sense**, heritability is the degree to which a trait is genetically determined, and it is expressed as the ration of the total genetic variance to the phenotypic variance.
>b: in the **narrow sense**, heribability is the degree to which a trait is transmitted from parents to offspring, and is expressed as the ration of the additive genetic variance to the total phenotypic variance. The concept of additive genetic variance makes no assumption concerning the mode of gene action (expression) involved (King and Stansfield, 1985); narrow-sense heritability can be estimated from the resemblance between relatives (Silvertown and Charlesworth, 2001)
>(see ch. 5.4)

**hermaphrodite:**
1: flowers have both male and female functions
2: plant that has only bisexual reproductive units (flowers, conifer cones, or functionally equivalent structures); i.e. perfect flowers. In angiosperm terminology a synonym is monoclinous from the Greek "one bed"
(see ch. 5.4.4.1)

**holism**: the misconception that all significant explanations are of components in terms of wholes rather than vice versa
(see U6.1)

**homeostasis**:
1: the property of a system (typically a living organism), either open or closed, that regulates its internal environment and tends to maintain a stable, constant condition; multiple dynamic equilibrium adjustment and regulation mechanisms make homeostasis possible (Wikipedia, 12.10)
2: the maintenance of metabolic equilibrium within an animal by a tendency to compensate for disrupting changes (Collins English Dictionary, 1979; Collins, Glasgow, UK)
3: the tendency for the internal environment of an organism to be maintained constant (Chambers Sci Dictionary)
(see ch. 4.1.2; 10.5.3.3; 12.4.4)

**hybridization**: any crossing of individuals of different genetic compostion, typically belonging to separate species, resulting in hybrid offspring

**induced dormancy**
1: an acquired condition of inability to germinate caused by some experience after abscission
2: acquired dormancy after abscission due to some environmental condition(s)
(see ch. 7.3.1)

**inductivism**: the misconception that scientific theories are obtained by generalizing or extrapolating repeated experiences, and that the more often a theory is confirmed by observation the more likely it becomes
(see U6.1)

**induction**: the non-existent process of 'obtaining' referred to in inductivism
(see U6.1)

> **principle of induction**: the idea that 'the future will resemble the past' combined with the misconception that this asserts anything about the future
> (see U6.1)

**information**:
1: message being conveyed; a sequence of symbols that can be interpreted as a message, recorded as signs, or transmitted as signals (Wikipedia, 10.12)
2: any kind of event that affects the state of a dynamic system (Wikipedia, 10.12)
3: negative entropy, choice and uncertainty, surprise, difficulty; anything that changes probabilities or reduces uncertainty (Gleik, 2011)
4: a mathematical abstraction of the content of any meaningful statement or data, enabling the study of the most efficient way of recording or transmitting them (Borowski and Borwein, 1991)
5: also called **uncertainty**; formally, a real-valued function of events in a probability space that depends only on the probability of events, and is such that events of probability one have zero uncertainty, that uncertainty increases as probability drops, and that the uncertainty of the simultaneous occurrence of two independent events is the sum of their individual uncertainties (Borowski and Borwein, 1991)

6: knowledge acquired through experience or study; knowledge of specific and timely events or situations; news (Hanks et al., 1979)
(see ch. 13.1.1)

**information theory**:
1: information theory is a branch of probability theory that deals with uncertainty, accuracy, and information content in the transmission of messages; random signals (noise) are often added, altering the message during transmission; information theory determines the probability that a sent signal is the same as that received; redundancy, repeating the signal, is needed to overcome limitations in the transmission system; the statistics of choosing a message out of all possibilities determines the amount of information it contains (Daintith and Clark, 1999)
2: a collection of mathematical theories, based on probability theory, that are concerned with methods of coding, decoding, storing and retrieving information, and with the likelihood of a given degree of accuracy in the transmission of a message through a channel that is subject to probabilities of failure that are described by a probability law (Borowski and Borwein, 1991)
(see ch. 13.1.1)

**inheritance**: the transmission of genetic information from ancestors or parents to descendants or offspring
(see ch. 2.4.4)

**innate dormancy**
1: unable to germinate under any normal set of environmental conditions
2: the condition of seeds (born with) as they leave the parent plant, and is a viable state but prevented from germinating when exposed to warm, moist aerated conditions by some property of the embryo or the associated endosperm or parental structures (e.g. envelopes) (implies a time lag) (after-ripening)
2: characteristic of genotype, species; dormancy when leaving the parent
3: innate dormancy is caused by: incomplete development; biochemical 'trigger'; removal of an inhibitor; physical restriction of water and/or gas access
(see ch. 7.3.1)

**introgression, introgessive hybridization**:
1: the spread of genes of one species into the gene pool of another by hybridization and backcrossing (Lincoln et al., 1998)
2: the incorporation of genes of one species into the gene pool of another. If the ranges of two species overlap and fertile hybrids are produced, they tend to backcross with the more abundant species. This process results in a population of individuals most of whom resemble the more abundant parents but which possess also some characters of the other parent species (King and Stansfield, 1985)
(see ch. 4.2.1.1)

**invasive species**:
1: organism undergoing a mass movement or encroachment from one area to another (Lincoln et al., 1998)
2: an alien species whose introduction does or is likely to cause economic or environmental harm or harm to human health (Anonymous, 1999)
3: a species that is non-native (or alien) to the ecosystem under consideration and whose introduction causes or is likely to cause economic or environmental harm or harm human health (Anonymous, 2004)
(see ch. 1.2)

**iteroparity**:
1: organisms that have repeated reproductive cycles (Lincoln et al., 1998)
2: repeated periods of reproduction during the life of an individual (King and Stansfield, 1985)
(see ch. 5.2.4)

**keystone species**:
1: species whose importance is not obvious from its size or abundance; a weed that alters nutrient cycles, soil properties or provides food for invasive animals (Booth and Swanton, 2002)
2: a keystone species has a disproportionate effect on local community function relative to it biomass (Paine, 1966, 1969)
3: some species and functions are extremely important to the total ecosystem function (Walker, 1992)
(see ch. 10.3.2.2)

**keystone trait**:
a functional phenotypic trait possessed by an organism disproportionately large effect on its life history events and individual fitness
(see ch. 10.3.2.2)

**knowledge**:
1: the facts, feelings, or experiences known by a person or group of people
2: awareness, consciousness, or familiarity gained by experiene or learning
3: specific information about a subject
4: familiarity, awareness or understanding of someone or something, such as facts, information, descriptions, or skills, which is acquired through experience or education by perceiving, discovering, or learning; a theoretical or practical understanding of a subject. (Wikipedia, 6.16)
(see U6.1)

**kurtosis**:
1: (statistical) a measure of the concentration of a distribution about its mean (Hanks et al., 1979)
2: departed from a normal distribution; 'peakedness' of the distribution; *leptokurtic*, more observations near the mean and the tails and less in the intermediate regions relative to a normal distribution; *platykurtic*, fewer observations at the mean and the tails but with more in the intermediate regions relative to a normal distribution (Sokal and Rohlf, 1981)
(see ch. 14.2.6.2)

**law of constant final yield**: at high density the total plant yield (biomass, seed weight) of a specified area tends to approach a constant weight
(see ch. 9.4.2)

**life cycle**
1: the sequence of events from the origin as a zygote, to the death of an individual
2: those stages through which an organism passes between the production of gametes by one
(see ch. 6.1)

**life history**:
1: the significant features of the life cycle through which an organism passes, with particular reference to strategies influencing survival and reproduction

2: how long it typically lives, how long it usually takes to reach reproductive size, how often it reproduces and a number of other attributes that have demographic and fitness consequences (Silvertown & Charlesworth, 2001)
(see ch. 6.1)

**life history strategy**: a specialized phenotype of correlated traits that has evolved independently in different populations or species exposed to similar selection pressures (Silvertown and Charlesworth, 2001)
(see ch. 6.2.3)

**life span**: the life span of a weed is its longevity, the maximum or mean duration of life of an individual or group (Lincoln et al., 1998)
(see ch. 6.2.1)

**locality**: the geographic position of an individual population or collection
(see ch. 3.2.1)

**Markov chain or process**: a sequence of events, usually called states, the probability of which is dependent only on the event immediately preceeding it (Borowski and Borwein, 1991)
(see ch. 6.1.1)

**metapopulation**:
1: a group of partially isolated populations belonging to the same species; subpopulations of natural populations that are partially isolated one from another and are connected by pathways of immigration and emigration; exchange of individuals occurs between such populations and individuals are able to recolonize sites from which the species has recently become extinct (Lincoln et al., 1998)
2: a network of local populations connected by dispersal, whose persistence depends on reciprocal but unsynchronized migration between populations (Silvertown and Charlesworth, 2001)
(see ch. 10.3.1)

**microenvironment**: the immediate surroundings and other physical factors of an individual plant or animal within its habitat; analogous to a microhabitat
(see ch. 3.2.1)

**microhabitat**:
1: a physical location that is home to very small organisms (e.g. a seed in the soil); microenvironment is the immediate surroundings and other physical factors of an individual plant or animal within its habitat
2: a very localized habitat (e.g. on the size scale of an individual seed in the soil seed bank)
(see ch. 3.2.1)

**microsite**: analogous to a microhabitat; e.g. the site perceived by a seed in the seed bank, or a seedling in a field
(see ch. 3.2.1)

**microspecies**:
1: genotype that is perpetuated by apomixis; small population with limited genetic variability (Wiktionary, 5.14)
2: apomictic plants genetically identical over generations in which each generic lineage has some characters of a true species; distinctly different apomictic lineages having intra-generic differences much smaller than normal (Wikipedia, 5.14)
(see ch. 5.4.4.3)

**monoecious**:
1: separate sex flowers on the same individual plants
2: having separate male and female reproductive units (flowers, conifer cones, or functionally equivalent structures) on the same plant; from Greek for "one household". Individuals bearing separate flowers of both sexes at the same time are called simultaneously or synchronously monoecious. Individuals that bear flowers of one sex at one time are called consecutively monoecious; plants may first have single sexed flowers and then later have flowers of the other sex
(see ch. 5.4.4.2)

> **andromonoecious**: both hermaphrodite and male (female sterile) flower structures occur on the same individuals
> (see ch. 5.4.4.2)

> **gynomonoecious**: both hermaphrodite and female (male-sterile) flower structures occur on the same individuals
> (see ch. 5.4.4.2)

> **trimonoecious (polygamous)**: male, female, and hermaphrodite structures all appear on the same plant
> (see ch. 5.4.4.2)

**mortality**: death rate as a proportion of the population expressed as a percentage or as a fraction; mortality rate; often used in a general sense as equivalent to death; often divided into these partially overlapping concepts:
(see ch. 5.2.6)

> **density-dependent mortality**:
> 1: the increasing risk of death associated with increasing population density
> 2: a decrease in population density (numbers per unit area) due to the effects of population density (self-thinning)
> (see ch. 5.2.6)

> **density-independent mortality**:
> 1: the increasing risk of death not associated with population density change
> 2: a decrease in population density due to any factor which is independent of population density
> (see ch. 5.2.6)

**mutation**
1: a sudden heritable change in the genetic material, most often an alteration of a single gene by duplication, replacement or deletion of a number of DNA base pairs;
2: an individual that has undergone such a mutational change; mutant
(see ch. 4.2.1.1)

**mutualism**: a symbiosis in which both organisms benefit, frequently a relationship of complete dependence; interdependent association; synergism (Lincoln et al., 1998)
(see ch. 9.2.3.1)

**natural selection**:
1: process by which forms of organisms in a population that are best adapted to the environment increase in frequency relative to less well-adapted forms over a number of generations
2: the process by which phenotypes of individual organisms in a local population that are best adapted to the local opportunity spacetime increase in frequency relative to less well-adapted phenotype neighbors over a number of generations
3: the non-random and differential reproduction of different genotypes acting to preserve favorable variants and to eliminate less favorable variants; viewed as the creative force that directs the course of evolution by preserving those variants or traits best adapted in the face of natural competition
4: essence of theory of evolution by natural selection is that genotypes with higher fitness leave a proportionately greater number of offspring, and consequently their genes will be present in a higher frequency in the next generation
(see ch. 2.3.2)

>    **artificial selection**: natural selection by humans; domestication; selective breeding
>    (see ch. 2.3.2)
>
>    **directional selection**: selection for an optimum phenotype resulting in a directional shift in gene frequencies of the character concerned and leading to a state of adaptation in a progressively changing environment; dynamic selection; progressive selection
>    (see ch. 5.2.6)
>
>    **disruptive selection**: selection for phenotypic extremes in a polymorphic population, which preserves and accentuates discontinuity; centrifugal selection; diversifying selection
>    (see ch. 5.2.6)
>
>    **stabilizing selection**: selecting for the mean, mode or intermediate phenotype with the consequent elimination of peripheral variants, maintaining an existing state of adaptation in a stable environment; centripetal selection; normalizing selection
>    (see ch. 5.2.6)

**neighborhood**:
1: used in population genetics for the area around an individual within which the gametes that produced the individual can be considered to have been drawn at random (Lincoln et al., 1998)
2: the plants in an area adjacent to an individual target plant that a exert competitive influence (Silvertown and Charlesworth, 2001
(see ch. 9.1.2)

**niche**:
1: the ecological role of a species in a community; conceptualized as the multidimensional space, of which the coordinates are the various parameters representing the condition of existence of the species, to which it is restricted by the presence of competitor species;
2: loosely as an equivalent of microhabitat
3: the relational position of a species or population in its ecosystem; how an organism makes a living; how an organism or population responds to the distribution of resources and neighbors and how it in turn alters those same factors (Wikipedia, 5.08)
(see ch. 3.2.2)

**fundamental niche:** the entire multidimensional space that represents the total range of conditions within which an organism can function and which it could occupy in the absence of competitors or other interacting (neighbor) species
(see ch. 3.2.2)

**realized niche:** that part of the fundamental niche q.v. actually occupied by a species in the presence of competitive or interactive (neighbor) species
(see ch. 3.2.2)

**n-dimensional niche hypervolume:**
1: the multi-dimensional space of resources and conditions available to, and specifically used by, organisms in a locality
2: the phenotype is described by the niche hypervolume; phenotype = G x E = realized niche; the selection pressure consequence of the G x E interaction
3: the limits or borders within which which a species has adapted,
4: experimentally defined by the testable parameters (dimensions) one can evaluate; the parameters determining the form of existence of a plant
(see ch. 3.2.3)

**normal distribution:**
1: Gaussian distribution; the type of statistical distribution where the variation of a quantity $x$ about its mean value of the same measurement taken several times is entirely random; the graph of the normal distribution is bell-shaped and symmetrical about the mean. (Daintith and Clark, 1999)
2: Gaussian distribution; a distribution that is continuous and symmetrical with mean, median and mode coincident; many quantitative measurements appear to be approximately normally distributed; this is in part a consequence of the *central limit theorem* (Borowski and Borwein, 1991)
3: an exceptionally' thin tail distribution'
(see ch. 14.2.5)

> **central limit theorem:** if sufficiently many samples are successively drawn from any population, the sum or mean of the sample values can be thought of, approximately, as an outcome from a normally distributed random variable (Borowski and Borwein, 1991)
> (see ch. 14.2.5)

**ontogeny:** the origination and development of an organism, usually from the time of fertilization of the egg to the organism's mature form; also the study of the entirety of an organism's lifespan (ontogenesis or morphogenesis) (Wikipedia, 5.14)
(see ch. 6.1)

**opportunity spacetime:** locally habitable space for an organism at a particular time which includes its available resources (e.g. light, water, nutrients, gases), pervasive conditions (e.g. heat, terroir), disturbance history (e.g. tillage, herbicides), and neighboring organisms (e.g. crops, other weed species) with which it interacts
(see ch. 3.1.2)

**demographic space**: demographic space exists in two forms:
1: a particular locality, the specific area over which a plant interacts with neighbors, the ecological and genetic neighborhood
2: plant numerical density, the number of plants per unit area; density is used as an average measure of space occupancy (Silvertown and Charlesworth, 2001)

**space (physical)**: location in geographical space and time of agents and artifacts (Wikipedia, 12.10)

**space (conceptual)**: "location" in a set of categories structured so that "nearby" agents will tend to interact (Wikipedia, 12.10)

**ortet**: the original organism from which a clone was derived
(see ch. 5.4.4.4)

**oskar**:
1: stunted juveniles that linger around the photosynthetic compensation point waiting for favorable growing conditions (Silvertown and Charlesworth, 2001)
2: juveniles that age but do not grow (Grass, 2010)
(see ch. 5.2.3)

**outcrossing mating system**: breeding system with self-incompatibility, spatial separation of male and female reproductive organs, or temporal separation of male and female flower activity
(see ch. 5.4.4.2)

**parasitism**: an obligatory symbiosis between individuals of two different species, in which the parasite is metabolically dependent on the host, and in which the host is typically adversely affected by rarely killed (Lincoln et al., 1998)
(see ch. 9.2.2.3.1)

**pathogen**: parasite which causes disease
(see ch. 9.2.2.3.3)

**pathogenesis** : the course or development of a disease disease: any impairment of normal physiological function affecting all or part of an organism, esp. a specific pathological change caused by infection, stress, etc. producing characteristic symptoms
(see ch. 9.2.2.3.3)

**perennial**:
1: used of plants that persist for several years with a period of growth each year;
2: occurring throughout the year (Lincoln et al., 1998)
(see ch. 6.2.1)

**herbaceous perennial**: plants which live a number of years, develop each year from underground stems, or roots, or crowns, usually flowering each year but dying back to the ground each year (Alex and Switzer, 1976)
(see ch. 6.2.1)

**woody perennial**: trees, shrubs and woody vines which live for many years, producing new growth each year from their aboveground stems, branches and twigs, and in some cases from underground stems, or roots, or crowns (Alex and Switzer, 1976)
(see ch. 6.2.1)

**phenology**: the study of periodic plant and animal life cycle events and how these are influenced by seasonal and interannual variations in climate and habitat factors; principally concerned with the dates of first occurrence of biological events in their annual cycle (e.g. date of emergence of leaves and flowers); in ecology, the term is used more generally to indicate the time frame for any seasonal biological phenomena (Wikipedia, 5.14)
(see ch. 6.1)

**phenotype:**
1: the sum total of observable structural and functional properties of an organism; the product of the interaction between the genotype and the environment.
2: the total of all observable features of a developing or developed individual (including its anatomical, physiological, biochemical, and behavioral characteristics). The phenotype is the result of interaction between genotype and the environment (Mayr, 2001)
3: the characters of an organism, whether due to the genotype or environment
4: the manifested attributes of an organism, the joint product of its genes and their environment during ontogeny; the conventional phenotype is the special case in which the effects are regarded as being confined to the individual body in which the gene sits (Dawkins,1999)
5: phenotypic function consists always of two universes, the physical (quantitative, formal structure; physiological) and the phenomenal (quantities that constitute a 'world') (Sacks, 1998, p.129)
(see ch. 2.2.2; 4.1)

> **extended phenotype**: all effects of a gene upon the world; 'effect' of a gene is understood as meaning in comparison with its alleles; the concept of phenotype is extended to include functionally important consequences of gene differences, outside the bodies in which the genes sit; in practice it is convenient to limit 'extended phenotype' to cases where the effects influence the survival chances of the gene, positively or negatively (Dawkins, 1999)
> (see ch. 4.1.1)

> **central theorem of the extended phenotype**: an animal's behavior tends to maximize the survival of the genes "for" that behavior, whether or not those genes happen to be in the body of the particular animal performing it
> (see ch. 4.1.1)

**phenotypic plasticity:**
1: The capacity of an organism to vary morphologically, physiologically or behaviorally as a result of environmental flucuations; reaction type
2: the capacity for marked variation in the phenotype as a result of environmental influences on the genotype during development [during the plants life history]
(see ch. 4.3.1)

**pleiotropy:**
1: pertaining to how a gene may affect several aspects of the phenotype (Mayr, 2001)

2: a single gene influences multiple phenotypic traits; a mutation in a gene may affect some all or all the traits simultaneously; selection on one trait may favors one allele while selection on another trait favors another allele (Wikepedia, 5.09)
(see ch. 4.1.2)

**ploidy:** the number of sets of chromosomes present (e.g. haploid, diploid, tetraploid)
(see ch. 4.2.2.3.6)

**polygenic inheritance (polygeny):**
1: inheritance of a trait govered by several genes (polygenes or multiple factors); their effect is cumulative (Mayr, 2001)
2: quantitative inheritance; multifactorial inheritance; inheritance of a phenotypic characteristic (trait) that is attributable to two or more genes and their interaction with the environment; polygenic traits do not follow patterns of Mendelian inheritance (qualitative traits). Instead their phenotypes typically vary along a continuous gradient depicted by a bell curve (Gaussian frequency distribution) (Wikepedia, 5.09)
(see ch. 4.1.2)

**polyploid complex, polyploid series:**
1: a genus, section, or species group containing (e.g.) diploids and tetraploids (Grant, 1971)
2: a group of interrelated and interbreeding plants that also have differing levels of ploidy that can allow genetic exchanges between unrelated species
(see ch. 4.2.2.3.6)

**polyploidy:** multiple sets of homologous chromosomes in an organism (e.g. tetraploid, octaploid)

**population:**
1: all individuals of one or more species within a prescribed area
2: a group of organisms of one species, occupying a defined area and usually isolated to some degree from other similar groups
(see ch. 2.2.2)

**population biology:** Study of the spatial and temporal distributions of organisms

**population genetics:** Study of gene frequencies and selection pressures in populations

**population genetic structure:**
1: the genetic composition and gene frequencies of individuals in a population
2: any pattern in the genetic makeup of individuals within a population which can be characterised by their genotype and allele frequencies (Wikipedia, 12.10)
(see ch. 10.3.2.2)

**population dynamics:** the study of changes within populations and of the factors that cause or influence those changes; the study of populations as functioning systems.

**preadaptation:**
1: the possession by an organism of characters or traits that would favor its survival in a new or changed environment (Lincoln et al., 1998)

2: a situation where a species evolves to use a preexisting structure inherited from an ancestor for a potentially unrelated function (Wikipedia, 11.10)
3: a new character or function arising from a mutation in an organism which becomes a useful adaptation after an environmental change, often occurring many generations later (King and Stansfield, 1985)
(see ch. 3.1.2)

**predation:**
1: a biological interaction where a predator (an organism that is hunting) feeds on its prey (the organism that is attacked)–predators may or may not kill their prey prior to feeding on them, but the act of predation always results in the death of its prey and the eventual absorption of the prey's tissue through consumption (Begon et al., 1996)
2: an interaction between organisms in which one organism captures biomass from another; widely defined it includes all forms of one organism eating another, regardless of trophic level (e.g. herbivory), closeness of association (e.g. parasitism, parasitoidism) and harm done to prey (e.g. grazing) (Wikipedia, 12.10)
(see ch. 9.2.2.3.2)

**predator**
1: the consumption of one animal (the prey) by another animal (the predator) (carnivore)
2: also used to include the consumption of plants by animal (herbivore), and the partial consumption of a large prey organism by a smaller predator (micropredation) (parasites; parasitoids)
(see ch. 9.2.2.3.2)

**principle:**
1: a fundamental or general truth or law
2: the essence of something; a source of fundamental cause, origin
3: a rule or law concerning a natural phenomenon or behavior of a system
(Hanks et al., 1979)
(see ch. 8.4.6.2)

**principle of strategic allocation:** organisms under natural selection optimize partitioning of limited resources and time in a way that maximizes fitness (by means of tradeoffs between partitions over time)
(see ch. 7.4.2)

**priority effect:** in ecology, the impact that a particular species can have on community development due to prior arrival at a site:
>**inhibitory priority effects** occur when a species that arrives first at a site negatively impacts a species that arrives later by reducing the availability of space or resources
>**facilitative priority effects** occur when a species that arrives first at a site alters abiotic or biotic conditions in ways that positively impact a species arriving later
>(see ch. 9.1.2)

**probability:**
1: (statistics) a measure of the relative frequency or likelihood of occurrence of an event; values lie between zero (impossibility) and one (certainty) and are derived from a theoretical distribution or from observations (Hanks et al., 1979)
2: the likelihood of a given event occurring (Daintith and Clark, 1999)

3: a measure or estimate of the degree of confidence one may have in the occurrence of an event, measured on a scale from 0 (impossibility) to 1 (certainty) (Borowski and Borwein, 1991)
4: (mathematics) the proportion of favourable outcomes to the total number of possible outcomes if these are indifferent (Borowski and Borwein, 1991)
5: the likelihood that events will occur in random, stochastic phenomena (Gullberg, 1997)
(see ch. 14.2.1)

> **frequency probability**: the relative frequency of a certain class of occurrences within a certain other class (axiomatically, the probability of $x_1$ with regard to $x_2$) (Popper, 2009)
> (see ch. 14.2.2.1)

> **Baysian probability**: an abstract concept, a quantity that we assign theoretically, for the purpose of representing a state of knowledge, or that we calculate from previously assigned probabilities (Jaynes, 1986; Wikipedia, 4.12)
> (see ch. 14.2.2.2)

> **axiomatic probability**:
> 1: a study of *probability* in terms of *probability measure*
> 2: the formal study of random or chance events, usually in terms of probability measure, independent random variables and their generalizations (Borowski and Borwein, 1991)
> 3: a primitive statement of the rules of a *deductive formal system* for the symbolic calculation of probability
> (see ch. 14.2.3.2)

> **probability measure**:
> 1: a mapping, P, of a sample space, X, into the interval [0, 1] subject to the conditions that P maps the whole space to 1; P is then the probability density function (a function representing the relative distribution of the frequency of a continuous random variable) (Borowski and Borwein, 1991)
> 2: a mapping of the entire sample space of the relative frequency distribution of a continuous random variable
> (see ch. 14.2.3.2)

**productivity**
1: the potential rate of incorporation or generation of energy or organic matter by an individual, population or trophic unit per unit time per unit area or volume; rate of carbon fixation;
2: often used loosely for the organic fertility or capacity of a given area or habitat
(see ch. 2.4.3)

**protoandrous**: describes individuals that function first as males and then change to females
(see ch. 5.4.4.2)

**protogynous**: describes individuals that function first as females and then change to males
(see ch. 5.4.4.2)

**ramet**:
1: an individual in a plant genet

2: a member or modular unit of a clone, that may follow an independent existence if separated from the parent organism
(see ch. 5.4.4.4)

**random (non-order):**
1: haphazard, no recognizable pattern (Lincoln et al., 1998)
2: lacking any definite plan or prearranged order; haphazard (Hanks et al., 1979)
3: having no definite aim or purpose; not sent or guided in a particular direction; made, done, occurring, etc., without method or conscious choice; haphazard (ref OED)
4: a non-order or non-coherence in a sequence of symbols or steps, such that there is no intelligible pattern or combination (Wikipedia, 3.12)
(see ch. 14.1.1)

**random (unpredictable):**
1: (Statistics) having a value that cannot be determined before the value is taken, but only described probabilistically (Borowski and Borwein, 1991)
2: (Statistics) with an *a priori* probability of occurrence other than zero or unity (Lincoln et al., 1998)
3: a lack of predictability, but admitting regularities in the occurrences of events whose outcomes are not certain; situations in which the certainty of the outcome is at issue and notions of haphazardness are irrelevant (Wikipedia, 3.12
(see ch. 14.1.1)

>  **epistemic randomness (unknown):**
>  1: random because my knowledge about causation is incomplete, not necessarily because the process has truly unpredictable properties; uncertainty due to incomplete information; what I cannot guess (Taleb, 2010)
>  2: randomness is unknowledge, the world is opaque, appearances fool us (Taleb, 2010)
>  3: something in which none of our standard methods of perception and analysis can detect regularities; something in which no simple program can detect regularities (Wolfram, 2002)
>  (see ch. 14.1.1)
>
>  **ontological randomness (ahistorical):** the type of randomness where the future is not implied by the past (or not even implied by anything); with ontic uncertainty the future is created continuously by the complexity of ongoing actions; ontic uncertainty is much more fundamental than epistemic which comes from imperfections in knowledge (Taleb, 2010)
>  (see ch. 14.1.1)
>
>  **algorithmic randomness (Kolmogorov complexity)** (Schnorr 1973, Levin 1973):
>  1: a random sequence is incompressible: no prefix can be produced by a program much shorter than the prefix; a measure of the computational resources needed to specify the object
>  2: a string (usually of bits) is Kolmogorov random if and only if it is shorter than any computer program that can produce that string; a random string is also an "incompressible" string, in the sense that it is impossible to give a representation of the string using a program whose length is shorter than the length of the string itself
>  (see ch. 14.1.1)

**algorithmic randomness (constructive null covers)** (Martin-Löf 1966):
1: a random real number should not have any property that is "uncommon"
2: a random sequence is not contained in any constructive null cover set
(see ch. 14.1.1)

**algorithmic randomness (constructive martingales)** (Schnorr 1971):
1: no effective procedure should be able to make money betting against a random sequence.
2: a martingale is a betting strategy; the martingale characterization says that no betting strategy implementable by any computer can make money betting on a random sequence.
3: in probability theory, a martingale is a model of a fair game where no knowledge of past events can help to predict future winnings; a martingale is a sequence of random variables (a stochastic process) for which, at a particular time in the realized sequence, the expectation of the next value in the sequence is equal to the present observed value even given knowledge of all prior observed values at a current time; martingales exclude the possibility of winning strategies based on game history, and thus they are a model of fair games
(see ch. 14.1.1)

**random variable**:
1: formally, a measurable function defined on a probability space and having range in the interval [0,1] (Borowski and Borwein, 1991)
2: a quantity that may take any of a range of values (either continuous or discrete) that cannot be predicted with certainty but only described probabilistically (Borowski and Borwein, 1991)
3: (Statistics) a variable that can assume any of a given set of values with assigned probability (Lincoln, et al., 1998)
4: assignment of a numerical value to each possible outcome of an event space (Wikipedia, 3.12)
(see ch. 13.1.1)

**reach**:
1: the ability of some explanations to solve problems beyond those that they were created to solve
2: an intrinsic attribute of an explanation, not an empirical or inductive assumption
3: the reach of ideas into the world of abstractions is a property of the knowledge that they contain (not of the brain in which they may happen to be instantiated)
4: reach always has an explanation, the explanation is not yet known
(see U6.1)

**jump to universality**: the tendency of gradually improving systems to undergo a sudden large increase in functionality (reach), becoming universal in some domain
(see U6.1)

**reaction norm**:
1: set of phenotypes expressed by a singe genotype, when a trait changes continuously under different environmental and developmental conditions
2: phenotype space; opportunity space; hedge-bet structure
3: a norm of reaction describes the pattern of phenotypic expression of a single genotype across a range of environments (Wikipedia, 5.08)
(see ch. 4.3.1)

**recombination**
1: any process that gives rise to a new combination of hereditary determinants, such as the reassortment of parental genes during meiosis through crossing over; mixing in the offspring of the genes and chromosomes of their parents.
2: event, occurring by crossing over of chromosomes during meiosis, in which DNA is exchanged between a pair of chromosomes of a pair. Thus, two genes that were previously unlinked, being on different chromosomes, can become linked because of recombination, and linked genes may become unlinked
(see ch. 4.2.1.1)

**recruitment**
1: seedling and bud shoot emergence
2: the influx of new members into a population by reproduction or immigration (Lincoln et al., 1998)
(see ch. 3.5.2)

**Red Queen principle**: for an evolutionary system, continuing development is needed just in order to maintain its fitness relative to the systems it is co-evolving with (Van Valen, 1973)
(see ch. 9.2.2.3.4)

**reductionism**: the misconception that science must or should always explain things by analyzing them into components (and hence that higher-level explanations cannot be fundamental)
(see U6.1)

**relay floristics**:
1: earlier successional species prepare the environment for later species (Egler, 1954)
2: vegetation in each successional pathway alters the environment and ameliorates it for later invading species, early species facilitate the invasion of later species (Clements, 1916, 1936)
(see ch. 10.4.4)

**reproduction**: the act or process of producing offspring
(see ch. 2.4.4)

**reproductive isolating mechanism:** a cytological, anatomical, physiological, behavioral, or ecological difference,or a geographic barrier which prevents successful mating between two or more related groups of organisms
(see ch. 4.2.2.2)

**reproductive isolation**
1: the absence of interbreeding between members of different species
2: the condition in which interbreeding between two or more populations is prevented by intrinsic factors
(see ch. 4.2.2.2)

**robust**:
1: strong in constitution; hardy; vigorous; sturdily built; requiring or suited to physical strength
2: immunity from one's external circumstances, good or bad, and an absence of fragility; resilience, resists shocks but remains the same

3: the ability of a system to resist change without adapting its initial stable configuration; the persistence of a system's characteristic behavior, or a characteristic or trait (canalization), under perturbations or unusual or conditions of uncertainty (Wikipedia, 6.14)
(see ch. 11.2.1)

**ruderal:** a plant inhabiting a disturbed site (Lincoln et al., 1998)
(see ch. 1.2)

**scenescence:** a decline in physiological state with age, manifested by an increase in mortality with age (Silvertown and Charlesworth, 2001)
(see ch. 5.2.4)

**seed dormancy** (and bud dormancy)
1: a state in which viable seeds (or buds; spores) fail to germinate under conditions of moisture, temperature and oxygen favorable for vegetative growth (Amen, 1968);
2: a state of relative metabolic quiescence; interruption in growth sequences, life cycle (usually in embryonic stages)
3: dispersal in time
(see ch. 7.3.1)

**seed dormancy capacity:** the dormancy-germinability state of an individual seed at abscission, as well as the amount of environmental signals (water, heat, oxygen, etc.) required to stimulate a change in state (after-ripening)
(see ch. 7.3.3)

**seed germinability:** the capacity of an seed, bud or spore to germinate under some set of conditions
(see ch. 7.3.1)

**seed heteroblasty:** variability in dormancy-germinability capacity among individual seeds at abscission shed by a single parent plant; germination heteromorphism
(see ch. 7.3.3)

> **somatic polymorphism of seed dormancy:** the occurrence of several different forms (amounts) of seed dormancy produced by an individual plant; distinctively different dormancy forms (amounts) adapted to different conditions
> (see ch. 7.3.3)

**segregation distortion:** the unequal segregation of genes in a heterozygote due to:
1: an aberrant meiotic mechanism; e.g. meiotic drive: any mechanism operating differentially during meiosis in a heterozygote to produce the two kinds of gametes with unequal frequencies;
2: other phenomena that result in altered gametic transmission ratios; e.g. in pollen competition where one allele results in a more slowly growing pollen tube than an alternate allele. Gametes bearing this allele will therefore show up in zygotes at a frequency less than 50%, as will all genes linked to the slow growing pollen tube allele (Wendel, pers. comm., 1998)
(see ch. 4.2.1.1)

**selection:**
1: gametic and zygotic differential mortality; non-random differential reproduction of different genotypes in a population
2: certain traits or alleles of a species may be subject to selection in the context of evolution. Under selection, individuals with advantageous or "adaptive" traits tend to be more successful than their peers reproductively: they contribute more offspring to the succeeding generation than others do. When these traits have a genetic basis, selection can increase the prevalence of those traits, because offspring will inherit those traits from their parents. When selection is intense and persistent, adaptive traits become universal to the population or species, which may then be said to have evolved (Wikipedia, 5.08)

> **natural selection:** the non-random and differential reproduction of different genotypes acting to preserve favorable variants and to eliminate less favorable variants; viewed as the creative force that directs the course of evolution by preserving those variants or traits best adapted in the face of natural competition
> (see ch. 2.3.2)
>
> **artificial selection:** selection by humans; domestication; selective breeding
> (see ch. 2.3.2)
>
> **directional selection:** selection for an optimum phenotype resulting in a directional shift in gene frequencies of the character concerned and leading to a state of adaptation in a progressively changing environment; dynamic selection; progressive selection
> (see ch. 5.2.6)
>
> **disruptive selection:** selection for phenotypic extremes in a polymorphic population, which preserves and accentuates discontinuity; centrifugal selection; diversifying selection
> (see ch. 5.2.6)
>
> **stabilizing selection:** selecting for the mean, mode or intermediate phenotype with the consequent elimination of peripheral variants, maintaining an existing state of adaptation in a stable environment; centripetal selection; normalizing selection
> (see ch. 5.2.6)

**self-similarity:** an invariance with respect to scaling; invariance not with additive translations, but invariance with multiplicative changes of scale
(see ch. 12.3.2)

**semelparity:**
1: organisms that have only one brood during the lifetime; big-bang reproduction (Lincoln et al., 1998)
2: reproduction that occurs only once in the life of and individual (King and Stansfield, 1985)
(see ch. 5.2.4)

**signal transduction:** a mechanism that converts a mechanical/chemical stimulus to a cell into a specific cellular response. Signal transduction starts with a signal to a receptor, and ends with a change in cell function (Wikipedia, 11.10)
(see ch. 3.4.4)

**skewness:**
1: (statistical) a measure of the symmetry of a distribution about its mean (Hanks et al., 1979)
2: observed asymmetric frequency distribution departed from a normal distribution in which one tail of the distribution is more drawn out more than the other; mean and median do not coincide (Sokal and Rohlf, 1981)
(see ch. 14.2.6.2)

**somatic polymorphism**
1: production of different plant parts, or different plant behaviors, within the same individual plant; the expression of somatic polymorphism traits is not much altered by the environmental conditions it encounters (as opposed to phenotypic plasticity)
2: the occurrence of several different forms of a structure-organ of a plant body; distinctively different forms adapted to different conditions
(see ch. 4.3.2)

**speciation:**
1: The formation of new species;
2: the splitting of a phylogenetic lineage;
3: acquistion of reproductive isolating mechanisms producing discontinuities between populations;
4: process by which a species splits into 2 or more species
5: the evolutionary process by which new biological species arise (Wikipedia, 5.08)
(see ch. 4.2.2)

**species:**
1: a group of organisms, minerals or other entities formally recognized as distinct from other groups;
2: a taxon of the rank of species; in the hieracrchy of biological classification the category below genus; the basic unit of biological classification; the lowest principal category of zoological classification
3: a group of morphologically similar organisms of common ancestry that under natural conditions are potentially capable of interbreeding
4: a species is a group of interbreeding natural populations that are reproductively isolated from other such groups (Lincoln)
5: the basic units of biological classification and a taxonomic rank; a group of organisms capable of interbreeding and producing fertile offspring (Wikipedia, 5.08)
(see ch. 4.2.2)

> **aggregrate species:**
> 1: a group of species that are so closely related that they are regarded as a single species.
> 2: a named species that represents a range of very closely related organisms (Wiktionary, 5.14)
> 3: a named species consisting of a large group of identified and named microspecies (Wikipedia, 5.14)
> (see ch. 5.4.4.3; 10.3.2.3.3)

> **microspecies:**
> 1: genotype that is perpetuated by apomixis; small population with limited genetic variability (Wiktionary, 5.14)
> 2: apomictic plants genetically identical over generations in which each generic lineage has some characters of a true species; distinctly different apomictic lineages having intra-generic differences much smaller than normal (Wikipedia, 5.14)
> (see ch. 5.4.4.3)

**species diversity-stability hypothesis**:
1: species diversity allows more ecosystem stability; complex trophic structures result in more stable communities (MacArthur, 1955)
2: human-disturbed communities wherein many species have gone extinct are prone to pest outbreaks and unpredictable fluctuations in population; there exists a negative relationship between native species richness, biodiversity, and community invasibility; communities with many species would be invasion resistant, while species poor communities will be highly invasible (Elton, 1958)
(see ch. 10.5.3.2)

**species-group**:
1: a group of closely related species, usually with partially overlapping ranges; sometimes used as an equivalent of superspecies (Lincoln et al., 1998).
2: an informal taxonomic rank into which an assemblage of closely-related species within a genus are grouped because of their morphological similarities and their identity as a biological unit with a single monophyletic origin (Wikipedia, 12.10)
(see ch. 10.3.2.3.1)

> **superspecies**: a group of at least two more or less distinctive species with approximately parapatric distributions; a superspecies consisting of two sister [sic] species is called a "species pair" (Wikipedia, 12.10)
> (see ch. 10.3.2.3.1)

**species richness**: the number of different species in a given area
(see ch. 10.5.3.3.1)

> **biodiversity (species richness)**: the number and relative abundance of species in an area or community (Booth and Swanton, 2002)
> (see ch. 10.5.3.3.1)

> **biodiversity (species evenness)**: abundance of each species in a community (Booth and Swanton, 2002)
> (see ch. 10.5.3.3.1)

**stability**: [see ecological stability]

**stage structure**: individuals of the same age but with different local environments may be at different stages of their life life cycles as a consequence of highly plastic plant rates of growth and development (Silvertown and Charlesworth, 2001)
(see ch. 5.2.3)

**succulents**: fleshy plants that have thick, water-retaining stems that may resemble pads; large amounts of water may be stored in the stems or pads and utilized by the plants during periods of insufficient soil moisture; flowers are often showy, and pads are frequently armed with sharp spines; examples include cacti
(see ch. 6.2.2.1)

**sustainability**:

1: humanity's investment in a system of living, projected to be viable on an ongoing basis that provides quality of life for all individuals of sentient species and preserves natural ecosystems
2: a characteristic of a process or state that can be maintained at a certain level indefinitely. 3: environmental, the potential longevity of vital human ecological support systems, such as the planet's climatic system, systems of agriculture, industry, forestry, fisheries, and the systems on which they depend
4: how long human ecological systems can be expected to be usefully productive; emphasis on human systems and anthropogenic problems (Wikipedia, 6.08)
(see ch. 10.5.3.2)

**symbiosis**: a prolonged, close association between organisms
(see ch. 9.3.2.3.1)

**system**: the region in space or the quantity of mass selected by an observer for analysis and discussion defined by a sharp and precise boundary around the entity
(see ch. 11.1.3)

**terroir**: the special characteristics that geography and geology bestow upon particular plants; "a sense of place," which is embodied in certain characteristic qualities; the sum of the effects that the local environment has had on the plant (Wikipedia, 2010)
(see ch. 3.4.3.2)

**trait:**
1: a character: any detectable phenotypic property of an organism
2: any character or property of an organism
3: a characteristic feature or quality distinguishing a particular person or thing
4: predictors (proxies) of organismal performance (Darwin, 1859)
(see ch. 2.4.3)

> **functional trait**: morpho-physio-phenological traits which impact fitness indirectly via their effects on growth, reproduction or survival, the three components of individual performance (Violle et al., 2007)
> (see ch. 2.4.3)

**transduction**:
1: (biophysics) conveyance of energy from a donor electron to a receptor electron, during which the class of energy changes; a committed process, no return path
2: (physiology) the transportation of stimuli to a biological receptor; conversion of a stimulus from one form to another.
3: (biology) a mechanical/physical stimulus alerting events is converted into an action potential which is transmitted along a biological channel where it is integrated
(see ch. 13.1.2)

> **signal transduction**:
> 1: signal transduction is any biological process by which a transducer converts one kind of signal/energy/stimulus into another (Wikipedia, 10.12)
> 2: the original form of a signal is converted to another form of energy using a transducer, implying the use of a sensor/detector; a transducer is a device that converts one form of energy to another (electrical,

mechanical, electromagnetic (including light), chemical, acoustic or thermal energy) (Wikipedia, 10.12)
(see ch. 13.1.2)

**variable:**
1: characteristics which show variability or variation; chance variables, random variables; quantitative or qualitative; continuous or discrete (Steel and Torrie, 1980)
2: a property with respect to which individuals within a sample differ in some discernible way (Lincoln, et al., 1998)
3: an expression that can be assigned any of a set of values (Borowski and Borwein, 1991)
4: the actual property, character, measured by individual observations in an experiment (Sokal and Rohlf, 1981)
5: a changing quantity that might have any one of a range of possible values (Daintith and Clark, 1999)
(see ch. 14.2.4)

### random variable:
1: (Statistics) a variable that can assume any of a given set of values with assigned probability (Lincoln et al., 1998)
2: chance variable; *stochastic* variable; a quantity that can take any one of a number of unpredicted values. A *discrete random* variable has a definite set of possible values with corresponding probabilities. A *continuous random* variable can take any value over a continuous range, the probabilities of a particular value occurring is given by a probability density function. (Daintith and Clark, 1999)
3: a quantity that may take any of a range of values (either continuous or discrete) that cannot be predicted with certainty but only described probabilistically. Formally, a measurable function defined on a probability space and having range in the interval [0, 1] (Borowski and Borwein, 1991)
4: the mathematical representation of a variate associated with a stochastic phenomenon (Walker, 1988)
(see ch. 14.2.4)

### independent variable:
1: a variable that has no dependent relationship on another variable (Daintith and Clark, 1999)
2: a variable in a mathematical equation or statement whose value determines that of the dependent variable; the argument (Borowski and Borwein)
3: (Statistics) the variable that defines a set of distinct experimental conditions, or that an experimenter deliberately manipulates in order to observe its relationship with some other quantity; the predictor (Borowski and Borwein)
(see ch. 14.2.4)

### dependent variable:
1: a variable whose value depends on another chosen variable's value (Daintith and Clark, 1999)
2: a variable whose value is determined by that taken by the independent variables (Borowski and Borwein
3: (Statistics) a variable whose values are observed for different values of the independent variable; response variable; predicted variable (Borowski and Borwein)
(see ch. 14.2.4)

**variance:**
1: a measure of the dispersion of a statistical sample (Daintith and Clark, 1999)
2: a measure of the dispersion of the distribution of a random variable, obtained by taking the expected value of the square of the difference between the random variable and its mean; the square of the standard deviation (Borowski and Borwein, 1991)
(see ch. 14.2.4)

> **standard deviation:**
> 1: a measure of the dispersion of a statistical sample, equal to the square root of the variance (Daintith and Clark, 1999)
> 2: a measure of the dispersion of a distribution, given by the square root of the variance; the unit whose multiples are used to describe the divergence of a random variable from its mean; the interval within one $sd$ of the mean contains approximately 68% of the population (Borowski and Borwein, 1991)
> (see ch. 14.2.4)
>
> **analysis of variance:**
> 1: any of a number of techniques for resolving the observed variance between sets of data into components, especially to determine whether the difference between two or more samples is explicable as random sampling variation with the same underlying population (Borowski and Borwein, 1991)
> 2: a statistical test whether two or more sample means could have been obtained from populations with the same parametric mean (Sokal and Rohlf, 1981)
> (see ch. 14.2.4)

**viviparous:** producing offspring from within the body of the parent
(see ch. 7.3.1)

**vivipary:** germinating while still attached to the parent plant; in crops its called pre-harvest sprouting, or precocious germination
(see ch. 7.3.1)

**weather:** the state of the atmosphere, to the degree that it is hot or cold, wet or dry, calm or stormy, clear or cloudy; the present condition of climate and their variations over periods up to two weeks (Wikipedia, 2010)
(see ch. 3.4.3.2.1)

**weed:**
1: any plant that is objectionable or interferes with the activities or welfare of man (Anonymous, 1994)
2: a plant out of place, or growing where it is not wanted (Blatchley, 1912)
3: a plant growing where it is not desired (Buchholtz, 1967)
4: a very unsightly plant of wild growth, often found in land that has been cultivated (Thomas, 1956)
5: useless, unwanted, undesirable (Bailey and Bailey, 1941)
6: a herbaceous plant, not valued for use or beauty, growing wild and rank, and regarded as cumbering the ground or hindering growth of superior vegetation (Murray et al., 1961)
7: a plant whose virtues have not yet been discovered (Emerson, 1878)
8: a generally unwanted organism that thrives in habitats disturbed by man (Harlan and deWet, 1965)
9: opportunistic species that follow human disturbance of the habitat (Pritchard, 1960)
10: a plant that grows spontaneously in a habitat greatly modified by human action (Harper, 1944)

11: a plant is a weed if, in any specified geographical area, its populations grow entirely or predominantly in situations markedly disturbed my man (without, of course, being deliberately cultivated plants) (Baker, 1965, p. 147)

12: pioneers of secondary succession of which the weedy arable field is a special case (Bunting, 1960)

13: competitive and aggressive behavior (Brenchley, 1920)

14: appearing without being sown or cultivated (Brenchley, 1920)

15: persistence and resistance to control (Gray, 1879)

(see ch. 1.2)

**woody plants**: trees and shrubs; solid stems (with growth rings) and secondary growth from aerial stems which live throughout the year; the aerial stems may be dormant part of the year; leaves of broadleaf trees are generally net-veined, while evergreen trees usually have needle-like leaves; flowers may be either inconspicuous or showy (see ch. 6.2.2.1)

"For the snark was a Boojum, you see"
Dodgeson, C.L (Lewis Carroll), 1876

# Weeds-Я-Us

www.ingramcontent.com/pod-product-compliance
Lightning Source LLC
Chambersburg PA
CBHW080648190526
45169CB00006B/2025